Geotechnical Engineering

C. VENKATRAMAIAH
Professor, Dept. of Civil Engineering
S.V. University Colleg of Engineering
TIRUPATI-INDIA

JOHN WILEY & SONS
NEW YORK CHICHESTER BRISBANE TORONTO SINGAPORE

First Published in 1993 by
WILEY EASTERN LIMITED
4835/24 Ansari Road, Daryaganj
New Delhi 110 002, India

Distributors:

Australia and New Zealand :
JACARANDA WILEY LIMITED
PO Box 1226, Milton Old 4046, Australia

Canada :
JOHN WILEY & SONS CANADA LIMITED
22 Worcester Road, Rexdale, Ontario, Canada

Europe and Africa :
JOHN WILEY & SONS LIMITED
Baffins Lane, Chichester, West Sussex, England

South East Asia :
JOHN WILEY & SONS (PTE) LIMITED
05-04, Block B, Union Industrial Building
37 Jalan Pemimpin, Singapore 2057

Africa and South Asia :
WILEY EASTERN LIMITED
4835/24 Ansari Road, Daryaganj
New Delhi 110 002, India

North and South America and rest of the world :
JOHN WILEY & SONS INC.
605, Third Avenue, New York, NY 10158, USA

Copyright © 1993 WILEY EASTERN LIMITED
New Delhi, India

Library of Congress Cataloging-in-Publication Data

ISBN 0-470-21730-8 John Wiley & Sons Inc.
ISBN 81-224-0376-X Wiley Eastern Limited

Printed in India by S.P. Printers, Noida-201301

Geotechnical Engineering

PREFACE

The author does not intend to be apologetic for adding yet another book to the existing list in the field of Geotechical engineeing. For one thing, the number of books available cannot be considered too large, although certain excellent reference books by stalwarts in the field are available. For another, the number of books by Indian Authors is only a few. Specifically speaking, the number of books in this field in the S.I. System of Units is small, and books from Indian Authors are virtually negligible. This fact, coupled with the author's observation that not many books are available designed specifically to meet the requirements of undergraduate curriculum in Civil Engineering and Technology, has been the motivation to undertake this venture.

The special features of this book are as follows :

1. The S.I. System of Units is adopted along with the equivalents in the MKS Units in some instances. (A note on the S.I. Units commonly used in Geotechnical Engineering is included).
2. Reference is made to the relevant Indian Standards[+], wherever applicable, and extracts from these are quoted for the benefit of the student as well as the practising engineer.
3. A few illustrative problems and problems for practice are given in the M.K.S. Units to facilitate those who continue to use these Units during the transition period.
4. The number of illustrative problems is fairly large compared to that in other books. This aspect would be helpful to the student to appreciate the various types of problems likely to be encountered.
5. The number of problems for practice at the end of each chapter is also fairly large. The answers to the numerical problems are given at the end of the book.
6. The illustrative examples and problems are graded carefully with regard to the toughness.
7. A few objective questions are also included at the end. This feature would be useful to students even during their preparation for Competitive and other Examinations such as GATE.
8. "Summary of Main Points", given at the end of each Chapter, would be very helpful to a student trying to brush up his preparation on the eve of the examination.
9. Chapter-wise references are given; this is considered a better way to encourage further reading than a big Bibliography at the end.

[+]*Note* : References are invited to the latest editions of these specifications for further details. These standards are available from Indian Standards Institution, New Delhi and its Regional Branch and Inspection Offices at Ahmedabad, Bangalore, Bhopal, Bhubaneshwar, Bombay, Calcutta, Chandigarh, Hyderabad, Jaipur, Kanpur, Madras, Patna, Pune and Trivandrum.

10. The sequence of topics and subtopics is sought to be made as logical as possible. Symbols and Nomenclature adopted are such that they are consistent (without significant variation from Chapter to Chapter), while being in close agreement with the internationally standardized ones. This would go a long way in minimising the possible confusion in the mind of the student.
11. The various theories, formulae, and schools of thought are given in the most logical sequence, laying greater emphasis on those that are most commonly used, or are more sound from a scientific point of view.
12. The author does not pretend to claim any originality for the material; however, he does claim some degree of special effort in the style of presentation, in the degree of lucidity sought to be imparted, and in his efforts to combine the good features of previous books in the field. All sources are properly acknowledged.

The book has been designed as a Text-book to meet the needs of undergraduate curricula of Indian Universities in the two conventional courses -- "Soil Mechanics" and "Foundation Engineering". Since a text always includes a little more than what is required, a few topics marked by asterisks may be omitted on first reading or by undergraduates depending on the needs of a specific syllabus.

The author wishes to express his grateful thanks and acknowledgement to :
(i) The Indian Standards Institution, for according permission to include extracts from a number of relevant Indian Standard Codes of Practice in the field of Geotechnical Engineering ;
(ii) The authors and publishers of various Technical papers and books, referred to in the appropriate places;
(iii) The Sri Venkateswara University, for permission to include questions and problems from their University Question Papers in the subject (some cases, in a modified system of Units) :

The author specially acknowledges his colleague, Prof. K. Venkata Ramana, for critically going through most of the Manuscript and offering valuable suggestions for improvement.

Efforts will be made to rectify errors, if any, pointed out by readers, to whom the author would be grateful. Suggestions for improvement are also welcome.

The author thanks the publishers for bringing out the book nicely

The author places on record the invaluable support and unstinted encouragement received from his wife, Mrs. Lakshmi Suseela and his daughter, Ms. Sarada and Ms. Usha Padmini, during the period of preparation of the manuscript.

C. VENKATRAMAIAH

TIRUPATI,
INDIA

Purpose and Scope of the Book
'GEOTECHNICAL ENGINEERING'

There are not many books which cover both soil mechanics and foundation engineering and which a student can use for his paper on Geotechnical Engineering. This paper is studied compulsorily and available books, whatever few are there, have not been found satisfactory. Students are compelled to refer to three or four books to meet their requirements. The author has been prompted by the lack of a good comprehensive textbook to write this present work. He has made a sincere effort to sum up his experience of thirty-three years of teaching in the present book. The notable features of the book are as follows :

1. The S. I. (Standard International) System of Units, which is a modification of the Metric System of units, is adopted mostly. A note on the S.I. Units is included by way of elucidation.
   However, to facilitate even those who continue to use the Metric System during the transitory period, some of the numerical problems are given in this system of units.
2. Reference is made to the relevant Indian Standards, wherever applicable.
3. The number of illustrative problems as well as the number of practice problems is made as large as possible so as to cover the various types of problems likely to be encountered. The problems are carefully graded with regard to their toughness.
4. A few "objective questions" are also included.
5. "Summary of Main Points" is given at the end of each Chapter.
6. References are given at the end of each Chapter.
7. Symbols and nomenclature adopted are mostly consistent, while being in close agreement with the internationally standardised ones.
8. The sequence of topics and subtopics is made as logical as possible.
9. The author does not pretend to claim any originality for the material, the sources being appropriately acknowledged; however, he does claim some degree of it in the presentation, in the degree of lucidity sought to be imparted, and in his efforts to combine the good features of previous works in this field.

In view of the meagre number of books in this field in S.I. Units, this can be expected to be a valuable contribution to the existing literature.

CONTENTS

Preface

PART I

CHAPTER 1   SOIL AND SOIL MECHANICS
    1.1 Introduction 1
    1.2 Development of Soil Mechanics 2
    1.3 Fields of Application of Soil Mechanics 3
    1.4 Soil Formation 4
    1.5 Residual and Transported Soils 6
    1.6 Some Commonly Used Soil Designations 7
    1.7 Structure of Soils 8
    1.8 Texture of Soils 11
    1.9 Major Soil Deposits of India 11
    Summary of Main points 11
    References 12
    Questions 12

CHAPTER 2   COMPOSITION OF SOIL-TERMINOLOGY AND DEFINITIONS 14
    2.1 Composition of Soil 14
    2.2 Basic Terminology 15
    2.3 Certain Important Relationships 20
    2.4 Illustrative Examples 24
    Summary of Main Points 32
    References 32
    Questions and problems 32

CHAPTER 3   INDEX PROPERTIES AND CLASSIFICATION TESTS 35
    3.1 Introduction 35
    3.2 Soil Colour 35
    3.3 Particle Shape 36
    3.4 Specific Gravity of Soil Solids 37
    3.5 Water Content 40
    3.6 Density Index 43
    3.7 In-Situ Unit Weight 47
    3.8 Particle Size Distribution (Mechanical Analysis) 51
    3.9 Consistency of Clay Soils 66
    3.10 Activity of Clays 81
    3.11 Unconfined Compression Strength and Sensitivity of Clays 82
    3.12 Thixotropy of Clays 83
    3.13 Illustrative Examples 83
    Summary of Main Points 100
    References 101
    Questions and Problems 102

x   Contents

CHAPTER 4   IDENTIFICATION AND CLASSIFICATION OF SOILS
                                                                                 106
    4.1 Introduction 106
    4.2 Field Identification of Soils 106
    4.3 Soil Classification-the need 108
    4.4 Engineering Soil Classification-Desirable Features 108
    4.5 Classification Systems-More Common Ones 109
    4.6 Illustrative Examples 120
    Summary of Main Points 124
    References 125
    Questions and Problems 125

CHAPTER 5   SOIL MOITURE-PERMEABILITY AND CAPILIARITY
                                                                                 128
    5.1 Introduction 128
    5.2 Soil Moisture-Modes of Occurrence 128
    5.3 Neutral and Effective Pressures 130
    5.4 Flow of Water Through Soil-permeability 133
    5.5 Determination of Permcability 138
    5.6 Factors Affecting Permeability 149
    5.7 Values of Permeability 153
    5.8 Permeability of Layered Soils 153
    *5.9 Capillarity 156
    5.10 Illustrative Examples 168
    Summary of Main points 185
    References 186
    Questions and Problems 188

CHAPTER 6   SEEPAGE AND FLOWNETS   192
    6.1 Introduction 192
    6.2 Flownet for One-Dimensional Flow 192
    6.3 Flownet for Two-Dimensional Flow 195
    6.4 Basic Equation for Seepage 200
    *6.5 Seepage Through Nonhomogeneous and Anisotropic Soil 205
    6.6 Top Flow Line in an Earth Dam 208
    *6.7 Radial Flow Nets 219
    6.8 Methods of Obtaining Flow Nets 222
    6.9 Quick Sand 224
    6.10 Seepage Forces 226
    6.11 Effective Stress in a Soil Mass under Seepage 226
    6.12 Illustrative Problems :
    Summary of Main Points 233
    References 233
    Questions and Problems 234

CHAPTER 7   COMPRESSIBILITY AND CONSOLIDATION OF SOILS
                                                                                 237
    7.1 Introduction 237
    7.2 Compressibility of Soils 237

Contents xi

    7.3 A Mechanistic Model for Consolidation 257
    7.4 Terzaghi's Theory of One-Dimensional Consolidation 262
    7.5 Solution of Terzaghi's Equation for One-Dimentional Consolidation 267
    7.6 Graphical Presentation of Consolidation Relationships 270
    7.7 Evaluation of Coefficient of Consolidation from Ocdometer Test Data 274
   *7.8 Secondary Consolidation 279
    7.9 Illustrative Examples 280
        Summary of Main Points 288
        References 289
        Questions and Problems 290

CHAPTER 8 SHEARING STRENGTH OF SOILS   295

    8.1 Introduction 295
    8.2 Friction 295
    8.3 Principal Planes and Principal Stresses-Mohr's Circle 297
    8.4 Strength Theories for Soils 303
    8.5 Shearing Strength-a Function of Effective Stress 307
   *8.6 Hvorslev's True Shear Parameters 308
    8.7 Types of Shear Tests Based on Drainage Conditions 308
    8.8 Shearing Strength Tests 310
   *8.9 Pore Pressure Parameters 327
   8.10 Shearing Characteristics of Sands 329
   8.11 Shearing Characteristics of Clays 335
   8.12 Illustrative Examples 344
        Summary of Main Points 364
        References 365
        Questions and Problems 366

CHAPTER 9 STABILITY OF EARTH SLOPES   370

    9.1 Introduction 370
    9.2 Infinite Slopes 370
    9.3 Finite Slopes 378
    9.4 Illustrative Examples 398
        Summary of Main Points 407
        References 408
        Questions and Problems 408

PART II

CHAPTER 10 STRESS DISTRIBUTION IN SOIL   410

    10.1 Introduction 410
    10.2 Point Load 411
    10.3 Line Load 421
    10.4 Strip Load 423
    10.5 Uniform Load on Circular Area 427

xii *Contents*

    10.6 Uniform Load on Rectangular Area 431
    10.7 Uniform Load on Irregular Areas-Newmark's Chart 435
    10.8 Approximate Methods 438
    10.9 Illustrative Examples 440
        Summary of Main Points 449
        References 449
        Questions and Problems 450

CHAPTER 11 SETTLEMENT ANALYSIS   **453**
    11.1 Introduction 453
    11.2 Data for Settlement Analysis 453
    11.3 Settlement 456
    11.4 Corrections to Computed Settlement 463
    11.5 Further Factors Affecting Settlement 466
    11.6 Other Factors Pertinent to Settlement 468
    11.7 Settlement Records 471
    11.8 Contact Pressure and Active Zone from Pressure bulb Concept 472
    11.9 Illustrative Examples 476
        Summary of Main Points 486
        References 487
        Questions and Problems 488

CHAPTER 12 COMPACTION OF SOILS   **490**
    12.1 Introduction 490
    12.2 Compaction Phenomenon 490
    12.3 Compaction Test 491
    12.4 Saturation (Zero Air-voids) Live 493
    12.5 Laboratory Compaction Tests 494
    12.6 In-Situ or Field Compaction 500
    12.7 Compaction of Sand 506
    12.8 Compaction versus Consolidation 507
    12.9 Illustrative Examples 507
        Summary of Main Points 514
        References 514
        Questions and Problems 515

CHAPTER 13 LATERAL EARTH PRESSURE AND STABILITY OF RETAINING WALLS   **518**
    13.1 Introduction 518
    13.2 Types of Earth-Retaining Structures 518
    13.3 Lateral Earth Pressures 520
    13.4 Earth Pressure at Rest 523
    13.5 Earth Pressure Theories 525
    13.6 Rankine's Theory 526
    13.7 Coulomb's Wedge Theory 543
    13.8 Stability Considerations for Retaining Walls 578
    13.9 Illustrative Examples 591

Summary of Main Points 617
References 619
Questions and Problems 620

CHAPTER 14 BEARING CAPACITY   623
14.1 Introduction and Definitions 623
14.2 Bearing Capacity 624
14.3 Methods of Determining Bearing Capacity 625
14.4 Bearing Capacity from Building Codes 625
14.5 Analytical Methods of Determining Bearing Capacity 627
14.6 Effects of Water Table on Bearing Capacity 655
14.7 Safe Bearing Capacity 657
14.8 Foundation Settlements 659
14.9 Plate Load Tests 660
14.10 Bearing Capacity from Penetration Tests 666
*14.11 Bearing Capacity from Model Tests-Housel's Approach 667
14.12 Bearing Capacity from Laboratory Tests 668
14.13 Bearing Capacity of Sands 668
14.14 Bearing Capacity of Clays 673
14.15 Recommended Practice 674
14.16 Illustrative Examples 674
Summary of Main Points 696
References 697
Questions and Problems 698

CHAPTER 15 SHALLOW FOUNDATIONS   702
15.1 Introduction Concepts on Foundations 702
15.2 General Types of Foundations 702
15.3 Choice of Foundation Type of Preliminary Selection 711
15.4 Spread Footings 715
15.5 Strap Footings 730
15.6 Combined Footings 732
15.7 Raft Foundations 735
*15.8 Foundations on Non-Uniform Soils 471
15.9 Illustrative Examples 743
Summary of Main Points 750
References 751
Questions and Problems 752

CHAPTER 16 PILE FOUNDATIONS   754
16.1 Introduction 754
16.2 Classification of Piles 754
16.3 Use of Piles 756
16.4 Pile Driving 757
16.5 Pile Capacity 760
16.6 Pile Groups 782
16.7 Settlement of Piles and Pile Groups 788
*16.8 Laterally Loaded Piles 791
16.9 Batter Piles 793

16.10 Design of Pile Foundations 794
16.11 Construction of Pile Foundations 795
16.12 Illustrative Examples 796
    Summary of Main Points 801
    References 801
    Questions and Problems 803

CHAPTER 17 SOIL STABILIZATION   **805**
17.1 Introduction 805
17.2 Classification of the Methods of Stabilization 805
17.3 Stabilization of Soil without Additives 806
17.4 Stabilization of Soil with additives 810
17.5 California Bearing Ratio 818
17.6 Illustrative Examples 824
    Summary of Main Points 827
    References 828
    Questions and Problsms 829

CHAPTER 18 SOIL EXPLORATION   **831**
18.1 Introduction 831
18.2 Site Investigation 831
18.3 Soil Exploration 833
18.4 Soil Sampling 840
18.5 Sounding and Penetration Tests 846
18.6 Indirect Methods-Geophysical Methods 855
18.7 Illustrative Examples 860
    Summary of Main Points 862
    References 863
    Questions and Problems 863
    Answers to Numerical Problems 866
    Appendix A: A Note on SI Units   870
    Appendix B: Notation   874
    Author Index   887
    Subject Index   889

1
SOIL AND SOIL MECHANICS

1.1 INTRODUCTION

The term 'Soil' has different meanings in different scientific fields. It has originated from the Latin word *Solum*. To an agricultural scientist, it means "the loose material on the earth's crust consisting of disintegrated rock with an admixture of organic matter, which supports plant life". To a geologist, it means the disintegrated rock material which has not been transported from the place of origin. But, to a civil engineer, the term 'soil' means, the loose unconsolidated inorganic material on the earth's crust produced by the disintegration of rocks, overlying hard rock with or without organic matter. Foundations of all structures have to be placed on or in such soil, which is the primary reason for our interest as Civil Engineers in its engineering behaviour.

Soil may remain at the place of its origin or it may be transported by various natural agencies. It is said to be 'residual' in the earlier situation and 'transported' in the latter.

"Soil Mechanics" is the study of the engineering behaviour of soil when it is used either as a construction material or as a foundation material. This is a relatively young discipline of civil engineering, systematised in its modern form by Karl Von Terzaghi (1925), who is rightly regarded as the "Father of Modern Soil Mechanics".[*]

An understanding of the principles of mechanics is essential to the study of soil mechanics. A knowledge and application of the principles of other basic sciences such as physics and chemistry would also be helpful in the understanding of soil behaviour. Further, laboratory and field research have contributed in no small measure to the development of soil mechanics as a discipline.

The application of the principles of soil mechanics to the design and construction of foundations for various structures is known as "Foundation Engineering". "Geotechnical Engineering" may be considered to include both soil mechanics and foundation engineering. In fact, according to Terzaghi, it is difficult to draw a distinct line of demarcation between soil mechanics and foundation engineering; the latter starts where the former ends.

---

[*]According to him, "Soil Mechanics is the application of the laws of mechanics and hydraulics to engineering problems dealing with sediments and other unconsolidated accumulations of soil particles produced by the mechanical and chemical disintegration of rocks regardless of whether or not they contain an admixture of organic constituents".

Until recently, a civil engineer has been using the term 'soil' in its broadest sense to include even the underlying bedrock in dealing with foundations. However, of late, it is well-recognised that the study of the engineering behaviour of rock material distinctly falls in the realm of 'rock mechanics', research into which is gaining impetus the world over.

1.2 DEVELOPMENT OF SOIL MECHANICS

The use of soil for engineering purposes dates back to prehistoric times. Soil was used not only for foundations but also as construction material for embankments. The knowledge was empirical in nature and was based on trial and error, and experience.

The hanging gardens of Babylon were supported by huge retaining walls, the construction of which should have required some knowledge, though empirical, of earth pressures. The large public buildings, harbours, aqueducts, bridges, roads and sanitary works of Romans certainly indicate some knowledge of the engineering behaviour of soil. This has been evident from the writings of Vitruvius, the Roman Engineer in the first century, B.C. Mansar and Viswakarma, in India, wrote books on 'construction science' during the medieval period. The Leaning Tower of Pisa, Italy, built between 1174 and 1350 A.D., is a glaring example of a lack of sufficient knowledge of the behaviour of compressible soil, in those days.

Coulomb, a French Engineer, published his wedge theory of earth pressure in 1776, which is the first major contribution to the scientific study of soil behaviour. He was the first to introduce the concept of shearing resistance of the soil as composed of the two components—cohesion and internal friction. Poncelet, Culmann and Rebhann were the other Frenchmen who extended the work of Coulomb. D'Arcy and Stokes were notable for their laws for the flow of water through soil and settlement of a solid particle in liquid medium, respectively. These laws are still valid and play an important role in soil mechanics. Rankine gave his theory of earth pressure in 1857; he did not consider cohesion, although he knew of its existence.

Boussinesq, in 1885, gave his theory of stress distribution in an elastic medium under a point load on the surface.

Mohr, in 1871, gave a graphical representation of the state of stress at a point, called 'Mohr's Circle of Stress'. This has an extensive application in the strength theories applicable to soil.

Atterberg, a Swedish soil scientist, gave in 1911 the concept of 'consistency limits' for a soil. This made possible the understanding of the physical properties of soil. The Swedish method of slices for slope stability analysis was developed by Fellenius in 1926. He was the chairman of the Swedish Geotechnical Commission.

Prandtl gave his theory of plastic equilibrium in 1920 which became the basis for the development of various theories of bearing capacity.

Terzaghi gave his theory of consolidation in 1923 which became an important development in soil mechanics. He also published, in 1925, the first treatise on

Soil Mechanics, a term coined by him. (*Erd bau mechanik*, in German). Thus, he is regarded as the 'Father of modern soil mechanics'. Later on R.R. Proctor and A. Casagrande and a host of others were responsible for the development of the subject as a full-fledged discipline.

Twelve International Conferences have been held till now under the auspices of the International Society of Soil Mechanics and Foundation Engineering at Harvard (Massachusetts, U.S.A.) 1936, Rotterdam (The Netherlands) 1948, Zurich (Switzerland) 1953, London (U.K.) 1957, Paris (France) 1961, Montreal (Canada) 1965, Mexico city (Mexico) 1969, Moscow (U.S.S.R.) 1973, Tokyo (Japan) 1977, Stockholm (Sweden) 1981, San Francisco (U.S.A.) 1985, and Rio de Janeiro (Brazil) 1989. The thirteenth is proposed to be held in New Delhi in 1994.

These conferences have given a big boost to research in the field of Soil Mechanics and Foundation Engineering.

1.3 FIELDS OF APPLICATION OF SOIL MECHANICS

The knowledge of soil mechanics has application in many fields of Civil Engineering.

1.3.1 Foundations

The loads from any structure have to be ultimately transmitted to a soil through the foundation for the structure. Thus, the foundation is an important part of a structure, the type and details of which can be decided upon only with the knowledge and application of the principles of soil mechanics.

1.3.2 Underground and Earth-retaining Structures

Underground structures such as drainage structures, pipe lines, and tunnels and earth-retaining structures such as retaining walls and bulkheads can be designed and constructed only by using the principles of soil mechanics and the concept of 'soil-structure interaction'.

1.3.3 Pavement Design

Pavement Design may consist of the design of flexible or rigid pavements. Flexible pavements depend more on the subgrade soil for transmitting the traffic loads. Problems peculiar to the design of pavements are the effect of repetitive loading, swelling and shrinkage of sub-soil and frost action. Consideration of these and other factors in the efficient design of a pavement is a must and one cannot do without the knowledge of soil mechanics.

1.3.4 Excavations, Embankments and Dams

Excavations require the knowledge of slope stability analysis; deep excavations may need temporary supports—'timbering' or 'bracing', the design of which requires knowledge of soil mechanics. Likewise the construction of embankments and earth dams where soil itself is used as the construction material, requires a thorough knowledge of the engineering behaviour of soil especially in the presence of water. Knowledge of slope stability, effects of seepage, consolidation and consequent settlement as well as compaction characteristics for achieving maximum unit weight of the soil *in-situ*, is absolutely essential for efficient design and construction of embankments and earth dams.

The knowledge of soil mechanics, assuming the soil to be an ideal material —elastic, isotropic, and homogeneous material—coupled with the experimental determination of soil properties, is helpful in predicting the behaviour of soil in the field.

Soil being a particulate and heterogeneous material, does not lend itself to simple analysis. Further, the difficulty is enhanced by the fact that soil strata vary in extent as well as in depth even in a small area.

A thorough knowledge of soil mechanics is a prerequisite to be a successful foundation engineer. It is difficult to draw a distinguishing line between Soil Mechanics and Foundation Engineering; the latter starts where the former ends.

1.4 SOIL FORMATION

Soil is formed by the process of 'Weathering' of rocks, that is, disintegration and decomposition of rocks and minerals at or near the earth's surface through the actions of natural or mechanical and chemical agents into smaller and smaller grains.

The factors of weathering may be atmospheric, such as changes in temperature and pressure; erosion and transportation by wind, water and glaciers; chemical action such as crystal growth, oxidation, hydration, carbonation and leaching by water, especially rainwater with time.

Obviously, soils formed by mechanical weathering (that is, disintegration of rocks by the action of wind, water and glaciers) bear a similarity in certain properties to the minerals in the parent rock, since chemical changes which could destroy their identity do not take place.

It is to be noted that 95% of the earth's crust consists of igneous rocks, and only the remaining 5% consists of sedimentary and metamorphic rocks. However, sedimentary rocks are present on 80% of the Earth's surface area. Feldspars are the minerals abundantly present (60%) in igneous rocks. Amphiboles and pyroxenes, quartz and micas come next in that order.

Rocks are altered more by the process of chemical weathering than by mechanical weathering. In chemical weathering some minerals disappear partially or fully, and new compounds are formed. The intensity of weathering depends upon the presence of water and temperature and the dissolved materials in water. Carbonic acid and oxygen are the most effective dissolved materials found in water which cause the weathering of rocks. Chemical weathering has the maximum intensity in humid and tropical climates.

'Leaching' is the process whereby water-soluble parts in the soil such as Calcium Carbonate, are dissolved and washed out from the soil by rainfall or percolating subsurface water. 'Laterite' soil, in which certain areas of Kerala abound, is formed by leaching.

Harder minerals will be more resistant to weathering action, for example, Quartz present in igneous rocks. But, prolonged chemical action may affect even such relatively stable minerals, resulting in the formation of secondary products of weathering, such as clay minerals—illite, kaolinite and montmorillonite. 'Clay Mineralogy' has grown into a very complicated and broad subject (Ref: 'Clay Mineralogy' by R.E. Grim).

Soil Profile

A deposit of soil material, resulting from one or more of the geological processes described earlier, is subjected to further physical and chemical changes which are brought about by the climate and other factors prevalent subsequently. Vegetation starts to develop and rainfall begins the processes of leaching and eluviation of the surface of the soil material. Gradually, with the passage of geological time profound changes take place in the character of the soil. These changes bring about the development of 'soil profile'.

Thus, the soil profile is a natural succession of zones or strata below the ground surface and represents the alterations in the original soil material which have been brought about by weathering processes. It may extend to different depths at different places and each stratum may have varying thickness.

Generally, three distinct strata or horizons occur in a natural soil profile; this number may increase to five or more in soils which are very old or in which the weathering processes have been unusually intense.

From top to bottom these horizons are designated as the A-horizon, the B-horizon and the C-horizon. The A-horizon is rich in humus and organic plant residue. This is usually eluviated and leached; that is, the ultrafine colloidal material and the soluble mineral salts are washed out of this horizon by percolating water. It is dark in colour and its thickness may range from a few centimetres to half a metre. This horizon often exhibits many undesirable engineering characteristics and is of value only to agricultural soil scientists.

The B-horizon is sometimes referred to as the zone of accumulation. The material which has migrated from the A-horizon by leaching and eluviation gets deposited in this zone. There is a distinct difference of colour between this zone and the dark top soil of the A-horizon. This soil is very much chemically active at the surface and contains unstable fine-grained material. Thus, this is important in highway and airfield construction work and light structures such as single storey residential buildings, in which the foundations are located near the ground surface. The thickness of B-horizon may range from 0.50 to 0.75 m.

The material in the C-horizon is in the same physical and chemical state as it was first deposited by water, wind or ice in the geological cycle. The thickness of this horizon may range from a few centimetres to more than 30 m. The upper region of this horizon is often oxidised to a considerable extent. It is from this horizon that the bulk of the material is often borrowed for the construction of large soil structures such as earth dams.

Each of these horizons may consist of sub-horizons with distinctive physical and chemical characteristics and may be designated as A_1, A_2, B_1, B_2, etc. The transition between horizons and sub-horizons may not be sharp but gradual. At a certain place, one or more horizons may be missing in the soil profile for special reasons. A typical soil profile is shown in Fig. 1.1.

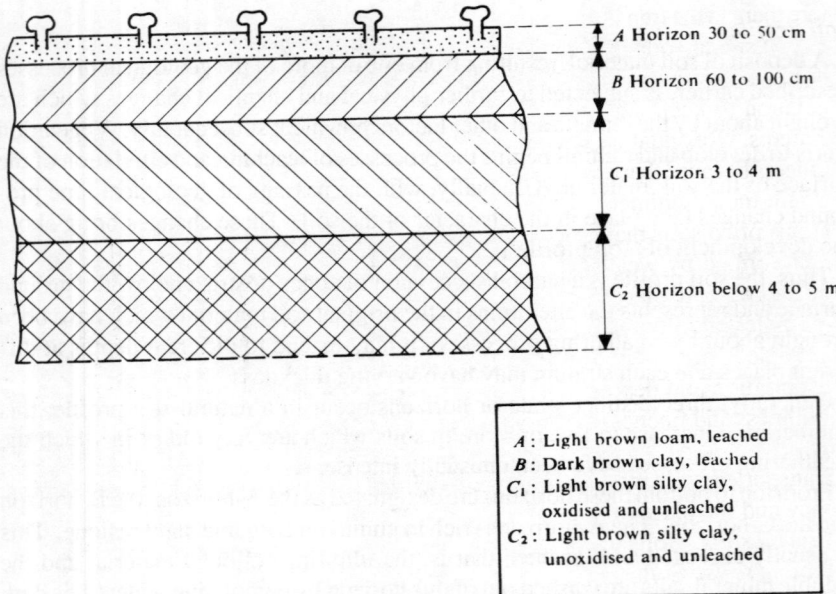

Fig. 1.1 A typical soil profile.

The morphology or form of a soil is expressed by a complete description of the texture, structure, colour and other characteristics of the various horizons, and by their thicknesses and depths in the soil profile. For these and other details the reader may refer "Soil Engineering" by M.G. Spangler.

1.5 RESIDUAL AND TRANSPORTED SOILS

Soils which are formed by weathering of rocks may remain in position at the place of origin. In that case these are 'Residual Soils'. These may get transported from the place of origin by various agencies such as wind, water, ice, gravity, etc. In this case these are termed "Transported soils". Residual soils differ very much from transported soils in their characteristics and engineering behaviour. The degree of disintegration may vary greatly throughout a residual soil mass and hence, only a gradual transition into rock is to be expected. An important characteristic of these soils is that the sizes of grains are not definite because of the partially disintegrated condition. The grains may break into smaller grains with the application of a little pressure.

Soil and Soil Mechanics

The residual soil profile may be divided into three zones: (*i*) The upper zone in which there is a high degree of weathering and removal of material; (*ii*) the intermediate zone in which there is some degree of weathering in the top portion and some deposition in the bottom portion; and (*iii*) the partially weathered zone where there is the transition from the weathered material to the unweathered parent rock. Residual soils tend to be more abundant in humid and warm zones where conditions are favourable to chemical weathering of rocks and have sufficient vegetation to keep the products of weathering from being easily transported as sediments. Residual soils have not received much attention from geotechnical engineers because these are located primarily in undeveloped areas. In some zones in South India, sedimentary soil deposits range from 8 to 15 m in thickness.

Transported soils may also be referred to as 'Sedimentary' soils since the sediments, formed by weathering of rocks, will be transported by agencies such as wind and water to places far away from the place of origin and get deposited when favourable conditions like a decrease of velocity occur. A high degree of alteration of particle shape, size, and texture as also sorting of the grains occurs during transportation and deposition. A large range of grain sizes and a high degree of smoothness and fineness of individual grains are the typical characteristics of such soils.

Transported soils may be further subdivided, depending upon the transporting agency and the place of deposition, as under:

Alluvial soils. Soils transported by rivers and streams: Sedimentary clays.
Aeoline soils. Soils transported by wind: loess.
Glacial soils. Soils transported by glaciers: Glacial till.
Lacustrine soils. Soils deposited in lake beds: Lacustrine silts and lacustrine clays.
Marine soils. Soils deposited in sea beds: Marine silts and marine clays.

Broad classification of soils may be:
1. Coarse-grained soils, with average grain-size greater than 0.075 mm, e.g., gravels and sands.
2. Fine-grained soils, with average grain-size less than 0.075 mm, e.g., silts and clays.

These exhibit different properties and behaviour but certain general conclusions are possible even with this categorisation. For example, fine-grained soils exhibit the property of 'cohesion'—bonding caused by inter-molecular attraction while coarse-grained soils do not; thus, the former may be said to be cohesive and the latter non-cohesive or cohesionless.

Further classification according to grain-size and other properties is given in later chapters.

1.6 SOME COMMONLY USED SOIL DESIGNATIONS

The following are some commonly used soil designations, their definitions and basic properties:

Bentonite. Decomposed volcanic ash containing a high percentage of clay mineral—montmorillonite. It exhibits high degree of shrinkage and swelling.

Black cotton soil. Black soil containing a high percentage of montmorillonite and colloidal material; exhibits high degree of shrinkage and swelling. The name is derived from the fact that cotton grows well in the black soil.

Boulder clay. Glacial clay containing all sizes of rock fragments from boulders down to finely pulverised clay materials. It is also known as 'Glacial till'.

Caliche. Soil conglomerate of gravel, sand and clay cemented by calcium carbonate.

Hard pan. Densely cemented soil which remains hard when wet. Boulder clays or glacial tills may also be called hard-pan—very difficult to penetrate or excavate.

Laterite. Deep brown soil of cellular structure, easy to excavate but gets hardened on exposure to air owing to the formation of hydrated iron oxides.

Loam. Mixture of sand, silt and clay size particles approximately in equal proportions; sometimes contains organic matter.

Loess. Uniform wind-blown yellowish brown silt or silty clay; exhibits cohesion in the dug condition, which is lost on wetting. Near vertical cuts can be made in the dry condition.

Marl. Mixtures of calcareous sands or clays or loam; clay content not more than 75% and lime content not less than 15%.

Moorum. Gravel mixed with red clay.

Top-soil. Surface material which supports plant life.

Varved clay. Clay and silt of glacial origin, essentially a lacustrine deposit; *varve* is a term of Swedish origin meaning thin layer. Thicker silt varves of summer alternate with thinner clay varves of winter.

1.7 STRUCTURE OF SOILS

The 'structure' of a soil may be defined as the manner of arrangement and state of aggregation of soil grains. In a broader sense, consideration of mineralogical composition, electrical properties, orientation and shape of soil grains, nature and properties of soil water, and the interaction of soil water and soil grains, also may be included in the study of soil structure, which is typical for transported or sedimented soils. Structural composition of sedimental soils influences, many of their important engineering properties such as permeability, compressibility and shear strength. Hence, a study of the structure of soils is important.

The following types of structure are commonly studied:
(a) Single-grained structure
(b) Honey-comb structure
(c) Flocculent structure

1.7.1 Single-grained Structure

Single-grained structure is characteristic of coarse-grained soils, with a particle size greater than 0.02 mm. Gravitational forces predominate the surface forces

Soil and Soil Mechanics 9

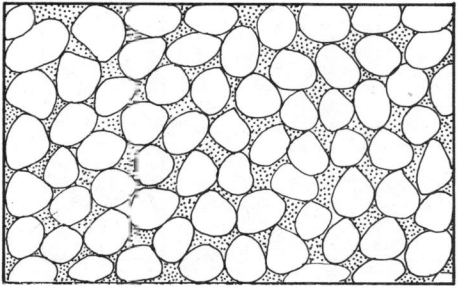

Fig. 1.2 Single-grained structure.

and hence grain to grain contact results. The deposition may occur in a loose state, with large voids or in a dense state, with less of voids.

1.7.2 Honey-comb Structure

This structure can occur only in fine-grained soils, especially in silt and rock flour. Due to the relatively smaller size of grains, besides gravitational forces, inter-particle surface forces also play an important role in the process of settling down. Miniature arches are formed, which bridge over relatively large void spaces. This results in the formation of a honey-comb structure, each cell of a honey-comb being made up of numerous individual soil grains. The structure has a large void space and may carry high loads without a significant volume change. The structure can be broken down by external disturbances.

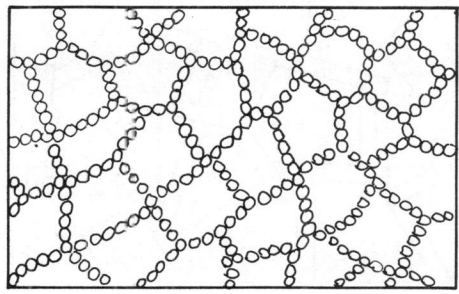

Fig. 1.3 Honey-comb structure.

1.7.3 Flocculent Structure

This structure is characteristic of fine-grained soils such as clays. Inter-particle forces play a predominant role in the deposition. Mutual repulsion of the particles

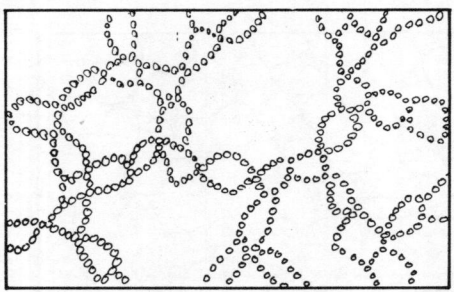

Fig. 1.4 Flocculent structure.

may be eliminated by means of an appropriate chemical; this will result in grains coming closer together to form a 'floc'. Formation of flocs is 'flocculation'. But the flocs tend to settle in a honey-comb structure, in which in place of each grain, a floc occurs.

Thus, grains grouping around void spaces larger than the grain-size are flocs and flocs grouping around void spaces larger than even the flocs result in the formation of a 'flocculent' structure.

Very fine particles or particles of colloidal size (< 0.001 mm) may be in a flocculated or dispersed state. The flaky particles are oriented edge-to-edge or edge-to-face with respect to one another in the case of a flocculated structure. Flaky particles of clay minerals tend to form a card house structure (Lambe, 1953), when flocculated. This is shown in Fig. 1.5.

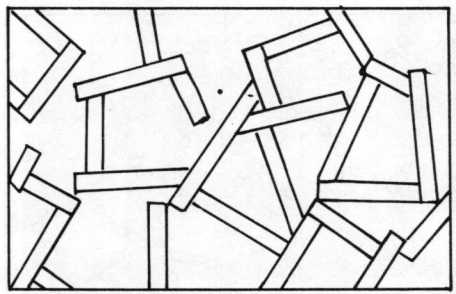

Fig. 1.5 Card-house structure of flaky particles.

When inter-particle repulsive forces are brought back into play either by remoulding or by the transportation process, a more parallel arrangement or reorientation of the particles occurs, as shown in Fig. 1.6. This means more face-to-face contacts occur for the flaky particles when these are in a dispersed state. In practice, mixed structures occur, especially in typical marine soils.

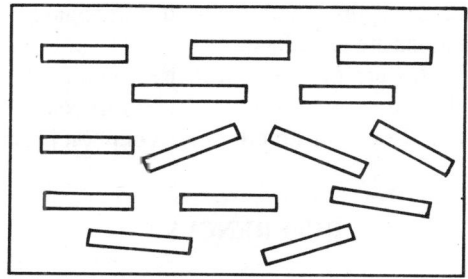

Fig. 1.6 Dispersed structure.

1.8 TEXTURE OF SOILS

The term 'Texture' refers to the appearance of the surface of a material, such as a fabric. It is used in a similar sense with regard to soils. Texture of a soil is reflected largely by the particle size, shape, and gradation. The concept of texture of a soil has found some use in the classification of soils to be dealt with later.

1.9 MAJOR SOIL DEPOSITS OF INDIA

The soil deposits of India can be broadly classified into the following five types:
1. Black cotton soils, occurring in Maharashtra, Gujarat, Madhya Pradesh, Karnataka, parts of Andhra Pradesh and Tamil Nadu. These are expansive in nature. On account of high swelling and shrinkage potential these are difficult soils to deal with in foundation design.
2. Marine soils, occurring in a narrow belt all along the coast, especially in the Rann of Kutch. These are very soft and sometimes contain organic matter, possess low strength and high compressibility.
3. Desert soils, occurring in Rajasthan. These are deposited by wind and are uniformly graded.
4. Alluvial soils, occurring in the Indo-Gangetic plain, north of the Vindhyachal ranges.
5. Lateritic soils, occurring in Kerala, South Maharashtra, Karnataka, Orissa and West Bengal.

SUMMARY OF MAIN POINTS

1. The term 'Soil' is defined and the development of soil mechanics or geotechnical engineering as a discipline in its own right is traced.

2. Foundations, underground and earth-retaining structures, pavements, excavations, embankments and dams are the fields in which the knowledge of soil mechanics is essential.
3. The formation of soils by the action of various agencies in nature is discussed, residual soils and transported soils being differentiated. Some commonly used soil designations are explained.
4. The structure and texture of soils affect their nature and engineering performance. Single-grained structure is common in coarse-grained soils and honey-combed and flocculent structures are common in fine-grained soils.

REFERENCES

1. A. Atterberg: *Über die physikalische Boden untersuchung, und über die plastizität der Tone*, Internationale Mitteilungen für Bodenkunde, Verlag für Fachliteratur, G.m.b.H. Berlin, 1911.
2. J.V. Boussinesq: *Application des potentiels à 1 etude de l' équilibre et du mouvement des solides élastiques"*, Paris, Gauthier Villars, 1885.
3. C.A. Couloumb: *Essai sur une application des régles de maximis et minimis à quelques problèmes de statique relatifs á l'architecture*. Mémoires de la mathématique et de physique, présentés à 1' Academie Royale des sciences, par divers Savans, et lûs dans sés Assemblées, Paris, De L' Imprimerie Royale, 1776.
4. W. Fellenius: *Calculation of the Stability of Earth Dams*, Trans. 2nd Congress on large Dams, Washington, 1979.
5. T.W. Lambe: *The Structure of Inorganic Soil*, Proc. ASCE, Vol. 79, Separate No. 315, Oct., 1953.
6. O. Mohr: *Techniche Mechanik*, Berlin, William Ernst und Sohn, 1906.
7. L. Prandtl: *Über die Härte plastischer Körper*, Nachrichten von der Königlichen Gesellschaft der Wissenschaften zu Göttingen (*Mathematisch —physikalische Klasse aus dem Jahre* 1920, Berlin, 1920).
8. W.J.M. Rankine: *On the Stability of Loose Earth*, Philosophical Transactions, Royal Society, London, 1857,
9. M.G. Spangler: *Soil Engineering*, International Textbook Company, Scranton, USA, 1951.
10. K. Terzaghi: *Erdbaumechanik auf bodenphysikalischer Grundlage*, Leipzig und Wien, Franz Deuticke Vienna, 1925.

QUESTIONS

1.1 (a) Differentiate between 'residual' and 'transported' soils. In what way does this knowledge help in soil engineering practice?
(b) Write brief but critical notes on 'texture' and 'structure' of soils.
(c) Explain the following materials:

(*i*) Peat, (*ii*) Hard pan, (*iii*) Loess, (*iv*) Shale, (*v*) Fill, (*vi*) Bentonite, (*vii*) Kaolinite, (*viii*) Marl, (*ix*) Caliche.
(S.V.U.—B.Tech. (Part-time)—June, 1981)

1.2 Distinguish between 'Black Cotton Soil' and 'Laterite' from an engineering point of view.
(S.V.U.—B.E., (R.R.)—Nov., 1974)

1.3 Briefly describe the processes of soil formation.
(S.V.U.—B.E., (R.R.)—Nov., 1973)

1.4 (a) Explain the meanings of 'texture' and 'structure' of a soil.
   (b) What is meant by 'black cotton soil'? Indicate the geological and climatic conditions that tend to produce this type of soil.
(S.V.U.—B.E., (R.R.)—May, 1969)

1.5 (a) Relate different formations of soils to the geological aspects.
   (b) Describe different types of texture and structure of soils.
   (c) Bring out the typical characteristics of the following materials:
      (*i*) Peat, (*ii*) Organic soil, (*iii*) Loess, (*iv*) Kaolinite, (*v*) Bentonite, (*vi*) Shale, (*vii*) Black cotton soil.
(S.V.U.—B.Tech., (Part-time)—April, 1982)

1.6 Distinguish between
   (*i*) Texture and Structure of soil.
   (*ii*) Silt and Clay.
   (*iii*) Aeoline and Sedimentary deposit.
(S.V.U.—B.Tech., (Part-time)—May, 1983)

2
COMPOSITION OF SOIL— TERMINOLOGY AND DEFINITIONS

2.1 COMPOSITION OF SOIL

Soil is a complex physical system. A mass of soil includes accumulated solid particles or soil grains and the void spaces that exist between the particles. The void spaces may be partially or completely filled with water or some other liquid. Void spaces not occupied by water or any other liquid are filled with air or some other gas.

'Phase' means any homogeneous part of the system different from other parts of the system and separated from them by abrupt transition. In other words, each physically or chemically different, homogeneous, and mechanically separable part of a system constitutes a distinct phase. Literally speaking, phase simply means appearance and is derived from Greek. A system consisting of more than one phase is said to be heterogeneous.

Since the volume occupied by a soil mass may generally be expected to include material in all the three states of matter—solid, liquid and gas, soil is, in general, referred to as a "three-phase system".

A soil mass as it exists in nature is a more or less random accumulation of soil particles, water and air-filled spaces as shown in Fig. 2.1(a). For purposes of analysis it is convenient to represent this soil mass by a block diagram, called 'Phase-diagram', as shown in Fig. 2.1(b). It may be noted that the separation of

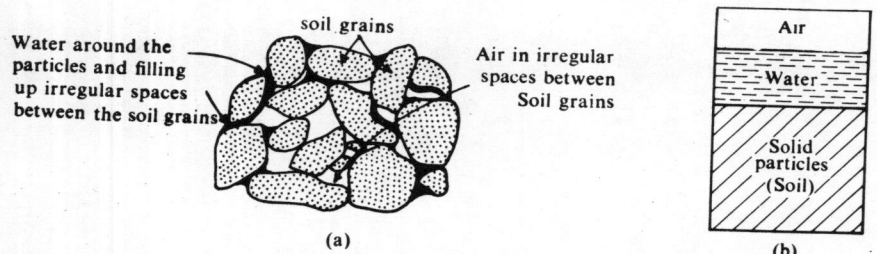

Fig. 2.1 (a) Actual soil mass, (b) Representation of soil mass by phase diagram.

solids from voids can only be imagined. The phase-diagram provides a convenient means of developing the weight-volume relationship for a soil.

When the soil voids are completely filled with water, the gaseous phase being absent, it is said to be 'fully saturated' or merely 'saturated'. When there is no water at all in the voids, the voids will be full of air, the liquid phase being absent; the soil is said to be dry. (It may be noted that the dry condition is rare in nature and may be achieved in the laboratory through oven-drying). In both these cases, the soil system reduces to a 'Two-phase' one as shown in Fig. 2.2(a) and (b). These are merely special cases of the three-phase system.

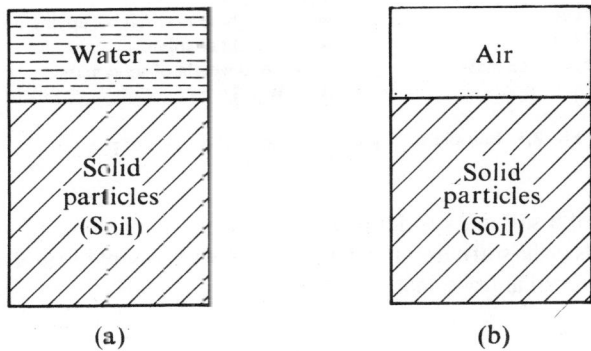

Fig. 2.2 (a) Saturated soil, (b) Dry soil represented as two-phase systems.

2.2 BASIC TERMINOLOGY

A number of quantities or ratios are defined below, which constitute the basic terminology in soil mechanics. The use of these quantities in predicting the engineering behaviour of soil will be demonstrated in later chapters.

The general three-phase diagram for soil will help in understanding the terminology and also in the development of more useful relationships between the various quantities. Conventionally, the volumes of the phases are represented on the left-side of the phase-diagram, while weights are represented on the right-side as shown in Fig. 2.3.

Porosity

'Porosity' of a soil mass is the ratio of the volume of voids to the total volume of the soil mass. It is denoted by the letter symbol n and is commonly expressed as a percentage:

$$n = \frac{V_v}{V} \times 100 \qquad \text{...(Eq. 2.1)}$$

Here
$$V_v = V_a + V_w; \quad V = V_a + V_w + V_s$$

16 Geotechnical Engineering

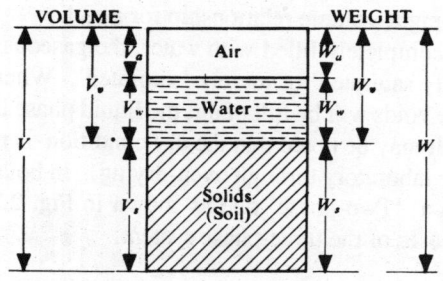

V_a = Volume of air
V_w = Volume of water
V_v = Volume of voids
V_s = Volume of solids
V = Total volume of soil mass

W_a = Weight of air (negligible or zero)
W_w = Weight of water
W_v = Weight of material occupying void space
W_s = Weight of solids
W = Total weight of solid mass
$W_v \approx W_w$

Fig. 2.3. Soil-phase diagram (volumes and weights of phases).

Void Ratio

'Void ratio' of a soil mass is defined as the ratio of the volume of voids to the volume of solids in the soil mass. It is denoted by the letter symbol *e* and is generally expressed as a decimal fraction:

$$e = \frac{V_v}{V_s} \qquad \ldots(\text{Eq. 2.2})$$

Here $\qquad V_v = V_a + V_w$

'Void ratio' is used more than 'Porosity' in soil mechanics to characterise the natural state of soil. This is for the reason that, in void ratio, the denominator, V_s, or volume of solids, is supposed to be relatively constant under the application of pressure, while the numerator, V_v, the volume of voids alone changes; however, in the case of porosity, both the numerator V_v and the denominator V change upon application of pressure.

Degree of Saturation

'Degree of saturation' of a soil mass is defined as the ratio of the volume of water in the voids to the volume of voids. It is designated by the letter symbol S and is commonly expressed as a percentage:

$$S = \frac{V_w}{V_v} \times 100 \qquad \ldots(\text{Eq. 2.3})$$

Here $\qquad V_v = V_a + V_w$

For a fully saturated soil mass, $V_w = V_v$.
Therefore, for a saturated soil mass $S = 100\%$.
For a dry soil mass, V_w is zero.
Therefore, for a perfectly dry soil sample S is zero.

Composition of Soil — Terminology and Definitions

In both these conditions, the soil is considered to be a two-phase system.

The degree of saturation is between zero and 100%, the soil mass being said to be 'partially' saturated—the most common condition in nature.

Percent Air Voids

'Percent air voids' of a soil mass is defined as the ratio of the volume of air voids to the total volume of the soil mass. It is denoted by the letter symbol n_a and is commonly expressed as a percentage:

$$n_a = \frac{V_a}{V} \times 100 \qquad \text{...(Eq. 2.4)}$$

Air Content

'Air content' of a soil mass is defined as the ratio of the volume of air voids to the total volume of voids. It is designated by the letter symbol a_c and is commonly expressed as a percentage:

$$a_c = \frac{V_a}{V_v} \times 100 \qquad \text{...(Eq. 2.5)}$$

Water (Moisture) Content

'Water content' or 'Moisture content' of a soil mass is defined as the ratio of the weight of water to the weight of solids (dry weight) of the soil mass. It is denoted by the letter symbol w and is commonly expressed as a percentage:

$$w = \frac{W_w}{W_s (\text{or } W_d)} \times 100 \qquad \text{...(Eq. 2.6)}$$

$$= \frac{(W - W_d)}{W_d} \times 100 \qquad \text{...[Eq. 2.6(a)]}$$

In the field of Geology, water content is defined as the ratio of weight of water to the total weight of soil mass; this difference has to be borne in mind.

For the purpose of the above definitions, only the free water in the pore spaces or voids is considered. The significance of this statement will be understood as the reader goes through the later chapters.

Bulk (Mass) Unit Weight

'Bulk unit weight' or 'Mass unit weight' of a soil mass is defined as the weight per unit volume of the soil mass. It is denoted by the letter symbol γ.

Hence, $\qquad\qquad \gamma = W/V \qquad\qquad$ ...(Eq. 2.7)

Here $\qquad\qquad W = W_w + W_s$

and $\qquad\qquad V = V_a + V_w + V_s$

The term 'density' is loosely used for 'unit weight' in soil mechanics, although, strictly speaking, density means the mass per unit volume and not weight.

Unit Weight of Solids

'Unit weight of solids' is the weight of soil solids per unit volume of solids alone. It is also sometimes called the 'absolute unit weight' of a soil. It is denoted by the letter symbol γ_s:

$$\gamma_s = \frac{W_s}{V_s} \qquad \ldots\text{(Eq. 2.8)}$$

Unit Weight of Water

'Unit weight of water' is the weight per unit volume of water. It is denoted by the letter symbol γ_w:

$$\gamma_w = \frac{W_w}{V_w} \qquad \ldots\text{(Eq. 2.9)}$$

It should be noted that the unit weight of water varies in a small range with temperature. It has a convenient value at 4°C, which is the standard temperature for this purpose. γ_o is the symbol used to denote the unit weight of water at 4°C.

The value of γ_o is 1 g/cm³ or 1000 kg/m³ or 9.81 kN/m³.

Saturated Unit Weight

The 'Saturated unit weight' is defined as the bulk unit weight of the soil mass in the saturated condition. This is denoted by the letter symbol γ_{sat}.

Submerged (Buoyant) Unit Weight

The 'Submerged unit weight' or 'Buoyant unit weight' of a soil is its unit weight in the submerged condition. In other words, it is the submerged weight of soil solids $(W_s)_{sub}$ per unit of total volume, V of the soil. It is denoted by the letter symbol γ':

$$\gamma' = \frac{(W_s)_{sub}}{V} \qquad \ldots\text{(Eq. 2.10)}$$

$(W_s)_{sub}$ is equal to the weight of solids in air minus the weight of water displaced by the solids. This leads to:

$$(W_s)_{sub} = W_s - V_s \cdot \gamma_w \qquad \ldots\text{(Eq. 2.11)}$$

Since the soil is submerged, the voids must be full of water; the total volume V, then, must be equal to $(V_s + V_w)$. $(W_s)_{sub}$ may now be written as:

$$(W_s)_{sub} = W - W_w - V_s \cdot \gamma_w$$

$$= W - V_w \cdot \gamma_w - V_s \gamma_w$$

$$= W - \gamma_w (V_w + V_s)$$

$$= W - V \cdot \gamma_w$$

Dividing throughout by V, the total volume,

$$\frac{(W_s)_{sub}}{V} = (W/V) - \gamma_w$$

or
$$\gamma' = \gamma_{sat} - \gamma_w \qquad \ldots(\text{Eq. 2.12})$$

It may be noted that a submerged soil is invariably saturated, while a saturated soil need not be submerged.

Equation 2.12 may be written as a direct consequence of Archimedes' Principle which states that the apparent loss of weight of a substance when weighed in water is equal to the weight of water displaced by it.

Thus, $\gamma' = \gamma_{sat} - \gamma_w$

since these are weights of unit volumes.

Dry Unit Weight

The 'Dry unit weight' is defined as the weight of soil solids per unit of total volume; the former is obtained by drying the soil, while the latter should be got prior to drying. The dry unit weight is denoted by the letter symbol γ_d and is given by:

$$\gamma_d = \frac{W_s (\text{or } W_d)}{V} \qquad \ldots(\text{Eq. 2.13})$$

Since the total volume is a variable with respect to packing of the grains as well as with the water content, γ_d is a relatively variable quantity, unlike γ_s, the unit weight of solids.*

Mass Specific Gravity

The 'Mass specific gravity' of a soil may be defined as the ratio of mass or bulk unit weight of soil to the unit weight of water at the standard temperature (4°C). This is denoted by the letter symbol G_m and is given by:

$$G_m = \frac{\gamma}{\gamma_o} \qquad \ldots(\text{Eq. 2.14})$$

This is also referred to as 'bulk specific gravity' or 'apparent specific gravity'.

Specific Gravity of Solids

The 'specific gravity of soil solids' is defined as the ratio of the unit weight of solids (absolute unit weight of soil) to the unit weight of water at the standard temperature (4°C). This is denoted by the letter symbol G and is given by:

$$G = \frac{\gamma_s}{\gamma_o} \qquad \ldots(\text{Eq. 2.15})$$

This is also known as 'Absolute specific gravity' and, in fact, more popularly as 'Grain Specific Gravity'. Since this is relatively constant value for a given soil, it enters into many computations in the field of soil mechanics.

---

*The term 'density' is loosely used for 'unit weight' in soil mechanics, although the former really means mass per unit volume and not weight per unit volume.

Specific Gravity of Water

'Specific gravity of water' is defined as the unit weight of water to the unit weight of water at the standard temperature (4°C). It is denoted by the letter symbol, G_w and is given by:

$$G_w = \frac{\gamma_w}{\gamma_o} \qquad \ldots(\text{Eq. 2.16})$$

Since the variation of the unit weight of water with temperature is small, this value is very nearly unity, and in practice is taken as such.

In view of this observation, γ_o in Eqs. 2.14 and 2.15 is generally substituted by γ_w, without affecting the results in any significant manner.

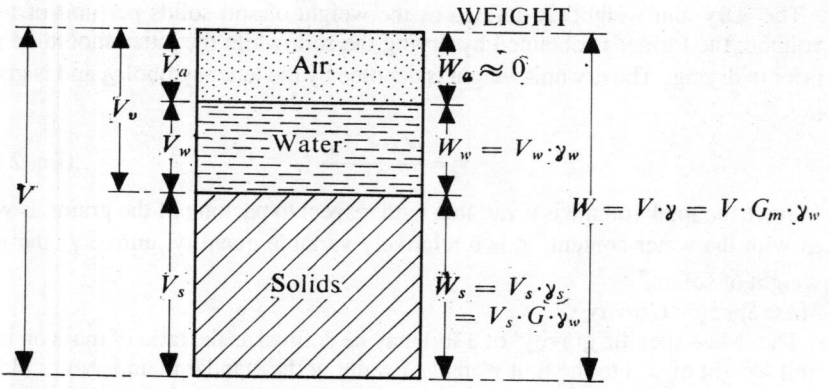

Fig. 2.4 Soil phase diagram showing additional equivalents on the weight side.

2.3 CERTAIN IMPORTANT RELATIONSHIPS

In view of foregoing definitions, the soil phase diagram may be shown as follows, with additional equivalents on the weight side.

A number of useful relationships may be derived based on the foregoing definitions and the soil-phase diagram.

2.3.1 Relationships Involving Porosity, Void Ratio, Degree of Saturation, Water Content, Percent Air Voids and Air Content

$$n = \frac{V_v}{V}, \text{ as a fraction}$$

$$= \frac{V - V_s}{V} = 1 - \frac{V_s}{V} = 1 - \frac{W_s}{G\gamma_w V}$$

$$\therefore \quad n = 1 - \frac{W_d}{G\gamma_w V} \quad \ldots(Eq.\ 2.17)$$

This may provide a practical approach to the determination of n.

$$e = \frac{V_v}{V_s}$$

$$= \frac{(V - V_s)}{V_s} = \frac{V}{V_s} - 1 = \frac{VG\gamma_w}{W_s} - 1$$

$$\therefore \quad e = \frac{V.G.\gamma_w}{W_d} - 1 \quad \ldots(Eq.\ 2.18)$$

This may provide a practical approach to the determination of e.

$$n = \frac{V_v}{V} \qquad e = \frac{V_v}{V_s}$$

$$1/n = V/V_v = \frac{V_s + V_v}{V_v} = \frac{V_s}{V_v} + \frac{V_v}{V_v} = 1/e + 1 = \frac{(1+e)}{e}$$

$$\therefore \quad n = \frac{e}{(1+e)} \quad \ldots(Eq.\ 2.19)$$

$$e = n/(1-n),\ \text{by algebraic manipulation} \quad \ldots(Eq.\ 2.20)$$

These interrelationships between n and e facilitate computation of one if the other is known.

$$\therefore \quad a_c = \frac{V_a}{V_v} \quad \text{and} \quad n = \frac{V_v}{V}$$

$$\therefore \quad na_c = \frac{V_a}{V} = n_a$$

or

$$n_a = n.a_c \quad \ldots(Eq.\ 2.20)$$

By definition,

$w = W_w/W_s$, as fraction; $S = V_w/V_v$, as fraction; $e = V_v/V_s$

$$S.e. = V_w/V_s$$

$$w = W_w/W_s = \frac{V_w.\gamma_s}{V_s.\gamma_s} = \frac{V_w.\gamma_s}{V_s.G.\gamma_w} = V_w/V_s G = S.e/G$$

$$\therefore \quad w.G = S.e \quad \ldots(Eq.\ 2.21)$$

(**Note:** This is valid even if both w and S are expressed as percentages). For saturated condition, $S = 1$

$$\therefore \quad w_{sat} = e/G \text{ or } e = w_{sat}.G \quad \ldots(Eq.\ 2.22)$$

$$n_a = \frac{V_a}{V} = \frac{V_v - V_w}{V_s + V_v} = \frac{\frac{V_v}{V_s} - \frac{V_w}{V_s}}{1 + \frac{V_v}{V_s}} = \frac{e - \frac{V_w}{V_s}}{1+e}$$

But $\qquad S.e = V_w/V_s$

$$\therefore \quad n_a = \frac{e - S.e}{1+e} = \frac{e(1-S)}{1+e} \qquad \ldots\text{(Eq. 2.23)}$$

Also $\qquad n_a = (e/\overline{1+e})\ (1-S) = n(1-S) \qquad \ldots\text{(Eq. 2.24)}$

$$a_c = V_a/V_v$$

$$S = V_w/V_v$$

$$a_c + S = \frac{(V_a + V_w)}{V_v} = V_v/V_v = 1$$

$$\therefore \quad a_c = (1-S) \qquad \ldots\text{(Eq. 2.25)}$$

In view of Eq. 2.25, Eq. 2.24 becomes $n_a = n.a_c$, which is Eq. 2.20.

2.3.2 Relationships Involving Unit Weights, Grain Specific Gravity, Void Ratio, and Degree of Saturation

$$\gamma = W/V = \frac{W_s + W_w}{V_s + V_v} = \frac{W_s(1 + W_w/W_s)}{V_s(1 + V_v/V_s)}$$

But $W_w/W_s = w$, as a fraction; $\dfrac{V_v}{V_s} = e$; and

$$\frac{W_s}{V_s} = \gamma_s = G.\gamma_w$$

$$\therefore \quad \gamma = G\gamma_w \frac{(1+w)}{(1+e)} \quad (w \text{ as a fraction}) \qquad \ldots\text{(Eq. 2.26)}$$

Further,

$$\gamma = \frac{(G + wG)}{(1+e)} \gamma_w$$

But $\quad w.G = S.e$

$$\therefore \quad \gamma = \frac{(G + S.e)}{1+e}.\gamma_w \quad (S \text{ as a fraction}) \qquad \ldots\text{(Eq. 2.27)}$$

This is a general equation from which the unit weights corresponding to the saturated and dry states of soil may be got by substituting $S = 1$ and $S = 0$ respectively (as a fraction).

$$\therefore \quad \gamma_{sat} = \left(\frac{G+e}{1+e}\right)\cdot\gamma_w \quad \ldots\text{(Eq. 2.28)}$$

and
$$\gamma_d = \frac{G\cdot\gamma_w}{(1+e)} \quad \ldots\text{(Eq. 2.29)}$$

(Note. γ_{sat} and γ_d may be derived from first principles also in just the same way as γ.)

The submerged unit weight γ' may be written as:

$$\gamma' = \gamma_{sat} - \gamma_w \quad \ldots\text{(Eq. 2.12)}$$

$$= \frac{(G+e)}{(1+e)}\cdot\gamma_w - \gamma_w$$

$$= \gamma_w\left[\frac{(G+e)}{(1+e)} - 1\right]$$

$$\therefore \quad \gamma' = \frac{(G-1)}{(1+e)}\cdot\gamma_w \quad \ldots\text{(Eq. 2.30)}$$

$$\gamma_d = \frac{W_s}{V}$$

But,
$$w = \frac{W_w}{W_s}, \text{ as a fraction}$$

$$(1+w) = \frac{(W_w + W_s)}{W_s} = W/W_s$$

Whence
$$W_s = W/(1+w)$$

$$\therefore \quad \gamma_d = \frac{W}{V(1+w)} = \frac{\gamma}{(1+w)} \quad (w \text{ as a fraction})$$

$$\ldots\text{(Eq. 2.31)}$$

$$G_m = \frac{\gamma}{\gamma_w} = \frac{(G+S.e)}{(1+e)} \quad \ldots\text{(Eq. 2.32)}$$

Solving for e,
$$e = \frac{(G-G_m)}{(G_m - S)} \quad \ldots\text{(Eq. 2.33)}$$

2.3.3 Unit-phase Diagram

The soil-phase diagram may also be shown with the volume of solids as unity; in such a case, it is referred to as the 'Unit-phase Diagram'.

It is interesting to note that all the interrelationships of the various quantities enumerated and derived earlier may conveniently be obtained by using the unit-phase diagram also.

24 Geotechnical Engineering

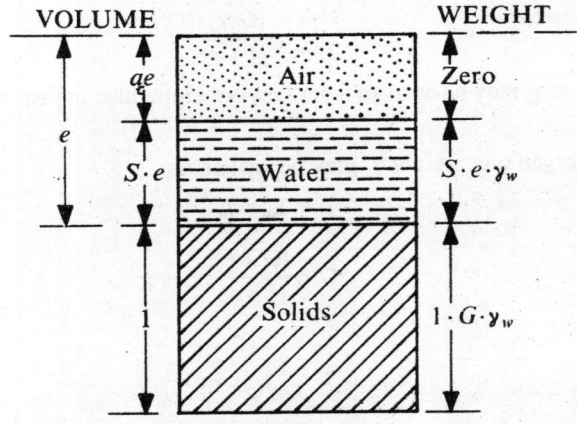

Fig. 2.5 Unit-phase diagram.

For example: Porosity,

$$n = \frac{\text{volume of voids}}{\text{total volume}} = e/(1+e)$$

Water content, $w = \dfrac{\text{weight of water}}{\text{weight of solids}} = \dfrac{S.e.\gamma_w}{G.\gamma_w} = S.e/G$

or
$$w.G = S.e$$

$$\gamma = \frac{\text{total weight}}{\text{total volume}} = \frac{(G+S.e)}{(1+e)}.\gamma_w = \frac{G\gamma_w(1+w)}{(1+e)}$$

$$\gamma_d = \frac{\text{weight of solids}}{\text{total volume}} = \frac{G.\gamma_w}{(1+e)}$$

and so on.

The reader may in a similar manner prove the other relationships also.

2.4 ILLUSTRATIVE EXAMPLES

Example 2.1: One cubic metre of wet soil weighs 19.80 kN. If the specific gravity of soil particles is 2.70 and water content is 11%, find the void ratio, dry density and degree of saturation.   (S.V.U.—B.E.(R.R.)—Nov. 1975)

Bulk unit weight, = 19.80 kN/m³
Water content, $w = 11\% = 0.11$

Dry unit weight, $\gamma_d = \dfrac{\gamma}{(1+w)} = \dfrac{19.80}{(1+0.11)}$ kN/m³ = **17.84 kN/m³**

Specific gravity of soil particles $G = 2.70$

Composition of Soil — Terminology and Definitions

$$\gamma_d = \frac{G \cdot \gamma_w}{1+e}$$

Unit weight of water, $\gamma_w = 9.81$ kN/m³

$$\therefore \quad 17.84 = \frac{2.70 \times 9.81}{(1+e)}$$

$$(1+e) = \frac{2.70 \times 9.81}{17.84} = 1.485$$

Void ratio, $e = 0.485$
Degree of Saturation, $S = wG/e$

$$\therefore \quad S = \frac{0.11 \times 2.70}{0.485} = 0.6124$$

$\therefore$ Degree of Saturation = **61.24%**

Example 2.2: Determine the (*i*) Water content, (*ii*) Dry density, (*iii*) Bulk density, (*iv*) Void ratio and (*v*) Degree of saturation from the following data:
Sample size 3.81 cm dia. × 7.62 cm ht.
Wet weight = 166.8 g
Oven-dry weight = 140 g
Specific gravity = 2.7   (S.V.U.—B. Tech. (Part-time)—June, 1981)
Wet weight, $W = 166.8$ g
Oven-dry weight, $W_d = 140.0$ g

Water content, $w = \dfrac{(166.8 - 140.0)}{140} \times 100\% = \mathbf{19.14\%}$

Total volume of soil sample, $V = \dfrac{\pi}{4} \times (3.81)^2 \times 7.62$ cm³

$$= 86.87 \text{ cm}^3$$

Bulk unit weight, $\quad \gamma = W/V = \dfrac{166.80}{86.87} = 1.92$ g/cm³

$$= 18.84 \text{ kN/m}^3$$

Dry unit weight, $\gamma_d = \dfrac{\gamma}{(1+w)} = \dfrac{18.84}{(1+0.1914)}$ kN/m³ = **15.81 kN/m³**

Specific gravity of solids, $G = 2.70$

$$\gamma_d = \frac{G \cdot \gamma_w}{(1+e)} \qquad \gamma_w = 9.81 \text{ kN/m}^3$$

$$15.81 = \frac{2.7 \times 9.81}{(1+e)} \qquad (1+e) = \frac{2.7 \times 9.81}{15.81} = 1.675$$

$\therefore$ Void ratio, $e = \mathbf{0.675}$

Degree of saturation, $S = \dfrac{wG}{e} = \dfrac{0.1914 \times 2.70}{0.675} = 0.7656 = \mathbf{76.56\%}$

Example 2.3: A soil has a bulk density of 20.1 kN/m³ and water content of 15%. Calculate the water content if the soil partially dries to a density of 19.4 kN/m³ and the void ratio remains unchanged. (S.V.U.—B.E. (R.R.)—Dec., 1971)

Bulk unit weight, $\gamma = 20.1$ kN/m³

Water content, $w = 15\%$

Dry unit weight, $\gamma_d = \dfrac{\gamma}{(1+w)} = \dfrac{20.1}{(1+0.15)}$ kN/m³ $= 17.5$ kN/m³

But $\gamma_d = \dfrac{G \cdot \gamma_w}{(1+e)}$;

if the void ratio remains unchanged while drying takes place, the dry unit weight also remains unchanged since G and γ_w do not change.

New value of $\gamma = 19.4$ kN/m³

$$\gamma_d = \dfrac{\gamma}{(1+w)}$$

∴ $\gamma = \gamma_d(1+w)$

or $19.4 = 17.5 \ (1+w)$

$$(1+w) = \dfrac{19.4}{17.5} = 1.1086$$

$$w = 0.1086$$

Hence the water content after partial drying = **10.86%**

Example 2.4: The porosity of a soil sample is 35% and the specific gravity of its particles is 2.7. Calculate its void ratio, dry density, saturated density and submerged density. (S.V.U.—B.E. (R.R.)—May, 1971)

Porosity, $n = 35\%$

Void ratio, $e = n/(1-n) = 0.35/0.65 = \mathbf{0.54}$

Specific gravity of soil particles = 2.7

Dry unit weight, $\gamma_d = \dfrac{G \cdot \gamma_w}{(1+e)}$

$= \dfrac{2.7 \times 9.81}{1.54}$ kN/m³ $= \mathbf{17.20}$ **kN/m³**

Saturated unit weight, $\gamma_{sat} = \dfrac{(G+e)}{(1+e)} \cdot \gamma_w$

$$= \frac{(2.70+0.54)}{1.54} \times 9.81 \text{ kN/m}^3$$

$$= 20.64 \text{ kN/m}^3$$

Submerged unit weight, $\gamma' = \gamma_{sat} - \gamma_w$

$$= (20.64 - 9.81) \text{ kN/m}^3$$

$$= 10.83 \text{ kN/m}^3.$$

Example 2.5: (*i*) A dry soil has a void ratio of 0.65 and its grain specific gravity is = 2.80. What is its unit weight?

(*ii*) Water is added to the sample so that its degree of saturation is 60% without any change in void ratio. Determine the water content and unit weight.

(*iii*) The sample is next placed below water. Determine the true unit weight (not considering buoyancy) if the degree of saturation is 95% and 100% respectively.

(S.V.U.—B.E.(R.R.)—Feb, 1976)

(*i*) **Dry Soil**
Void ratio, $e = 0.65$
Grain specific gravity, $G = 2.80$
Unit weight, $\gamma_d = \dfrac{G \cdot \gamma_w}{(1+e)} = \dfrac{2.80 \times 9.8}{1.65}$ kN/m^3 = **16.65 kN/m^3**.

(*ii*) **Partial saturation of the soil**
Degree of saturation, $S = 60\%$
Since the void ratio remained unchanged, $e = 0.65$
Water content, $w = \dfrac{S \cdot e}{G} = \dfrac{0.60 \times 0.65}{2.80} = 0.1393$

$$= 13.93\%$$

Unit weight $= \dfrac{(G + Se)}{(1+e)} \cdot \gamma_w = \dfrac{(2.80 + 0.60 \times 0.65)}{1.65} \, 9.81$ kN/m^3

$$= 18.97 \text{ kN/m}^3$$

(*iii*) **Sample below water**
High degree of saturation $S = 95\%$
Unit weight $= \dfrac{(G + S \cdot e)}{(1+e)} \cdot \gamma_w = \dfrac{(2.80 + 0.95 \times 0.65)}{1.65} \, 9.81$ kN/m^3

$$= 20.32 \text{ kN/m}^3$$

Full saturation, $S = 100\%$
Unit weight $= \dfrac{(G+e)}{(1+e)} \cdot \gamma_w = \dfrac{(2.80 + 0.65)}{1.65} \, 9.81$ kN/m^3

$$= 20.51 \text{ kN/m}^3$$

28   Geotechnical Engineering

Example 2.6: A sample of saturated soil has a water content of 35%. The specific gravity of solids is 2.65. Determine its void ratio, porosity, saturated unit weight and dry unit weight.   (S.V.U.—B.E.(R.R.)—Dec., 1970)

Saturated soil
Water content, $w = 35\%$
Specific gravity of solids, $G = 2.65$
Void ratio, $e = wG$, in this case.

$$\therefore \quad e = 0.35 \times 2.65 = \mathbf{0.93}$$

Porosity,
$$n = \frac{e}{1+e} = \frac{0.93}{1.93} = 0.482 = \mathbf{48.20\%}$$

Saturated unit weight,
$$\gamma_{sat} = \frac{(G+e)}{(1+e)} \cdot \gamma_w$$

$$= \frac{(2.65 + 0.93)}{(1+0.93)}$$

$$= \mathbf{1.85 \text{ g/cm}^3}$$

Dry unit weight,
$$\gamma_d = \frac{G \cdot \gamma_w}{(1+e)}$$

$$= \frac{2.65 \times 1}{1.93}$$

$$= \mathbf{1.37 \text{ g/cm}^3}$$

Example 2.7: A saturated clay has a water content of 39.3% and a bulk specific gravity of 1.84. Determine the void ratio and specific gravity of particles.
             (S.V.U.—B.E.(R.R.)—May, 1969)

Saturated clay
Water content, $w = 39.3\%$
Bulk specific gravity, $G_m = 1.84$
Bulk unit weight, $\gamma = G_m \cdot \gamma_w$

$$= 1.84 \times 1 = \mathbf{1.84 \text{ g/cm}^3}$$

In this case, $\gamma_{sat} = 1.84 \text{ g/cm}^3$

$$\gamma_{sat} = \frac{(G+e)}{(1+e)} \cdot \gamma_w$$

For a saturated soil,
$$e = wG$$

or
$$e = 0.393 \, G$$

$$\therefore \quad 1.84 = \frac{(G + 0.393\, G)}{(1 + 0.393\, G)} \cdot 1$$

whence
$$G = 2.74$$
Specific gravity of soil particles = **2.74**
Void ratio = $0.393 \times 2.74 = \mathbf{1.08}$

Example 2.8: The mass specific gravity of a fully saturated specimen of clay having a water content of 30.5% is 1.96. On oven drying, the mass specific gravity drops to 1.60. Calculate the specific gravity of clay.
(S.V.U.—B.E.(R.R.)—Nov. 1972)

Saturated clay
 Water content, $w = 30.5\%$
 Mass specific gravity, $G_m = 1.96$
$$\therefore \quad \gamma_{sat} = G_m \cdot \gamma_w = 1.96\,\gamma_w$$
On oven-drying, $\quad G_m = 1.60$
$$\therefore \quad \gamma_d = G_m \cdot \gamma_w = 1.60\,\gamma_w$$

$$\gamma_{sat} = 1.96 \cdot \gamma_w = \frac{(G+e)\gamma_w}{(1+e)} \qquad \ldots(i)$$

$$\gamma_d = 1.60 \cdot \gamma_w = \frac{G \cdot \gamma_w}{(1+e)} \qquad \ldots(ii)$$

For a saturated soil, $e = wG$
$$\therefore \quad e = 0.305\,G$$

From (i),
$$1.96 = \frac{(G + 0.305\,G)}{(1 + 0.305\,G)} = \frac{1.305\,G}{(1 + 0.305\,G)}$$
$$\Rightarrow \quad 1.96 + 0.598\,G = 1.305\,G$$
$$\Rightarrow \quad G = \frac{1.960}{0.707} = \mathbf{2.77}$$

From (ii),
$$1.60 = G/(1+e)$$
$$\Rightarrow \quad G = (1 + 0.305\,G)\,1.6$$
$$\Rightarrow \quad G = 1.6 + 0.485\,G$$
$$\Rightarrow \quad 0.512\,G = 1.6$$
$$\Rightarrow \quad G = 1.6/0.512 = \mathbf{3.123}$$

The latter part should not have been given (additional and inconsistent data).

30 Geotechnical Engineering

Example 2.9: A sample of clay taken from a natural stratum was found to be partially saturated and when tested in the laboratory gave the following results. Compute the degree of saturation. Specific gravity of soil particles = 2.6; wet weight of sample = 250 g; dry weight of sample = 210 g; and volume of sample = 150 cm³.

(S.V.U.—B.E.(R.R.)—Nov., 1974)

Specific gravity of soil particles, $G = 2.60$
Wet weight, $W = 250$ g; Volume, $V = 150$ cm³
Dry weight, $W_d = 210$ g

$$\text{Water content, } w = \frac{(W - W_d)}{W_d} \times 100 = \frac{(250 - 210)}{210} \times 100\%$$

$$= \frac{40}{210} \times 100\% = 19.05\%$$

Bulk unit weight, $\gamma = W/V = 250/150 = 1.67$ g/cm³

$$= 1.67 \times 9.81 \text{ kN/m}^3$$

$$= 16.38 \text{ kN/m}^3$$

Dry unit weight, $\gamma_d = \dfrac{\gamma}{(1+w)} = \dfrac{16.38}{(1+0.1905)}$ kN/m³

$$= 13.76 \text{ kN/m}^3$$

[Also, $\gamma_d = \dfrac{W_d}{V} = 210/150$ g/cm³ $= 1.4$ g/cm³ $= 13.734$ kN/m³]

But
$$\gamma_d = \frac{G \cdot \gamma_w}{(1+e)}$$

$$13.76 = \frac{2.6 \times 9.81}{(1+e)}$$

$$(1+e) = \frac{2.6 \times 9.81}{13.76} = 1.854$$

$$e = 0.854$$

Degree of saturation, $S = \dfrac{wG}{e} = \dfrac{0.1905 \times 2.6}{0.854} = 0.58$

$$= \mathbf{58\%}$$

Aliter. From the phase-diagram (Fig. 2.6)

$$V = 150 \text{ cc}$$

$$W = 250 \text{ g}$$

$$W_d = W_s = 210 \text{ g}$$

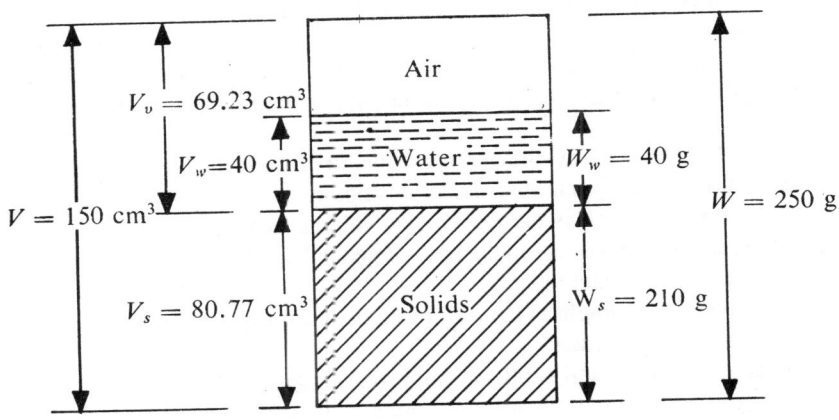

Fig. 2.6 Phase diagram (Example 2.9).

$$W_w = (250 - 210) \text{ g}$$

$$= 40 \text{ g}$$

$$V_w = \frac{W_w}{\gamma_w} = 40/1 = 40 \text{ cm}^3,$$

since $\gamma_w = 1 \text{ g/cm}^3$

$$V_s = \frac{W_s}{\gamma_s} = \frac{W_s}{G.\gamma_w} = \frac{210}{2.6 \times 1} = 80.77 \text{ cm}^3$$

$$V_v = (V - V_s) = (150 - 80.77) = 69.23 \text{ cm}^3$$

Degree of saturation, $S = \dfrac{V_w}{V_v}$

∴ $\quad S = 40/69.23 = 0.578$

∴ Degree of saturation = **57.8%**

Thus, it may be observed that it may sometimes be simpler to solve numerical problems by the use of the soil-phase diagram.

Note. All the illustrative examples may be solved with the aid of the soil-phase diagram or the unit-phase diagram also; however, this may not always be simple.

SUMMARY OF MAIN POINTS

1. Soil is a complex physical system; generally speaking, it is a three-phase system, mineral grains of soil, pore water and pore air, constituting the three phases. If one of the phases such as pore water or pore air is absent, it is said to be dry or saturated in that order; the system then reduces to a two-phase one.
2. Phase-diagram is a convenient representation of the soil which facilitates the derivation of useful quantitative relationships involving volumes and weights.

   Void ratio, which is the ratio of the volume of voids to that of the soil solids, is a useful concept in the field of geotechnical engineering in view of its relatively invariant nature.
3. Submerged unit weight is the difference between saturated unit weight and the unit weight of water.
4. Specific gravity of soil solids or grain specific gravity occurs in many relationships and is one of the most important values for a soil.

REFERENCES

1. Alam Singh & B.C. Punmia: *Soil Mechanics and Foundations*, Standard Book House, Delhi-6, 1970.
2. A.R. Jumikis: *Soil Mechanics*, D. Van Nostrand Co., Princeton, NJ, USA, 1962.
3. T.W. Lambe and R.V. Whitman: *Soil Mechanics*, John Wiley & Sons, Inc., NY, 1969.
4. D.F. McCarthy: *Essentials of Soil Mechanics and Foundations*, Reston Publishing Company, Reston, Va, USA, 1977.
5. V.N.S. Murthy: *Soil Mechanics and Foundation Engineering*, Dhanpat Rai and Sons, Delhi-6, 2nd ed., 1977.
6. S.B. Sehgal: *A text book of Soil Mechanics*, Metropolitan Book Co., Ltd., Delhi, 1967.
7. G.N. Smith: *Essentials of soil mechanics for civil and mining engineers*, Third Edition, Metric, Crosby Lockwood Staple, London, 1974.
8. M.G. Spangler: *Soil Engineering*, International Textbook Company, Scranton, USA, 1951.
9. D.W. Taylor: *Fundamentals of Soil Mechanics*, John Wiley & Sons, Inc., New York, 1948.

QUESTIONS AND PROBLEMS

2.1 (*a*) Define:
  (*i*) Void ratio, (*ii*) Porosity, (*iii*) Degree of saturation, (*iv*) Water content, (*v*) Dry density, (*vi*) Bulk density, (*vii*) Submerged density.

Composition of Soil — Terminology and Definitions 33

(b) Derive from fundamentals:
   (i) $S.e = w.G$,
   where
      S represents degree of saturation,
      e represents void ratio,
      w represents water content, and
      G represents grain specific gravity
   (ii) Derive the relationship between dry density and bulk density in terms of water content. (S.V.U.—B. Tech., (Part-time)—June, 1981)

2.2 Sketch the phase diagram for a soil and indicate the volumes and weights of the phases on it. Define 'Void ratio', 'Degree of saturation', and 'Water content'. What is a unit phase diagram?
(S.V.U.—B.E., (R.R.)—Feb., 1976)

2.3 Establish the relationship between degree of saturation, soil moisture content, specific gravity of soil particles, and void ratio.

The volume of an undisturbed clay sample having a natural water content of 40% is 25.6 cm^3 and its wet weight is 43.5 g. Calculate the degree of saturation of the sample if the grain specific gravity is 2.75.
(S.V.U.—B.E., (R.R.)—May, 1975)

2.4 (a) Distinguish between Black cotton soil and Laterite from an engineering point of view.
   (b) Defining the terms 'Void ratio', 'Degree of saturation' and 'Water content', explain the engineering significance of determining these properties. (S.V.U.—B.E., (R.R.)—Nov., 1974)

2.5 A piece of clay taken from a sampling tube has a wet weight of 155.3 g and volume of 95.3 cm^3. After drying in an oven for 24 hours at 105°C, its weight is 108.7 g. Assuming the specific gravity of the soil particles as 2.75, determine the void ratio and degree of saturation of the clay sample.
(S.V.U.—B.E., (R.R.)—Nov., 1973)

2.6 Derive the formula between soil moisture content (w), degree of saturation (S), specific gravity (G), and void ratio (e).

A saturated clay has a water content of 40% and bulk specific gravity of 1.90. Determine the void ratio and specific gravity of particles.
(S.V.U.—B.E., (R.R.)—May, 1970)

2.7 Derive the relation between void ratio (e), specific gravity of particles (G) and moisture content at full saturation (w).

A certain sample of saturated soil in a container weighs 65 g. On drying in an oven in the container it weighs 60 g. The weight of container is 35 g. The grain specific gravity is 2.65. Determine the void ratio, water content, and bulk unit weight.
(S.V.U.—B.E., (R.R.)—Nov., 1969)

2.8 (a) Define 'Soil Texture' and 'Soil structure'. What are the various terms used to describe the above properties of the soil?

(b) A clay sample containing natural moisture content weighs 34.62 g. The specific gravity of soil particles is 2.70. After oven drying, the soil weighs 20.36 g. If the displaced volume of the wet soil sample is 24.26 cm³ calculate: (i) the moisture content of the sample, (ii) its void ratio, and (iii) degree of saturation.

(S.V.U.—B.E., (N.R.)—Sep., 1968)

2.9 (a) The porosity and the specific gravity of solids of 100% saturated soil are known. In terms of these quantities and with the aid of a properly drawn sketch, derive a formula for the moisture content of the soil.

(b) A highly sensitive volcanic clay was investigated in the laboratory and found to have the following properties:

(i) γ_{wet} = 1.28 g/cm³ (12.56 kN/m³)
(ii) G = 2.75
(iii) e = 9.0
(iv) w = 311%

In rechecking the above values, one was found to be inconsistent with the rest. Find the inconsistent value and report it correctly.

(S.V.U.—B.E., (N.R.)—April, 1966)

2.10 A partially saturated soil from an earth fill has a natural water content of 19% and a bulk unit weight of 19.33 kN/m³. Assuming the specific gravity of soil solids as 2.7, compute the degree of saturation and void ratio. If subsequently the soil gets saturated, determine the dry density, buoyant unit weight and saturated unit weight.

(S.V.U.— B. Tech., (Part-time)—April, 1982)

2.11 In a field density test, the volume and wet weight of soil obtained are 785 cm³ and 1580 g respectively. If the water content is found to be 36%, determine the wet and dry unit weights of the soil. If the specific gravity of the soil grains is 2.6, compute the void ratio.

(S.V.U.—B. Tech. (Part-time)—May, 1983)

2.12 A clay sample, containing its natural moisture content, weighs 33.30 g. The specific gravity of solids of this soil is 2.70. After oven-drying, the soil sample weighs 20.25 g. The volume of the moist sample, before oven-drying, found by displacement of mercury is 24.30 cm³. Determine the moisture content, void ratio and degree of saturation of the soil.

3
INDEX PROPERTIES AND CLASSIFICATION TESTS

3.1 INTRODUCTION

As an aid for the soil and foundation engineer, soils have been divided into basic categories based upon certain physical characteristics and properties. The categories have been relatively broad in scope because of the wide range of characteristics of the various soils that exist in nature. For a proper evaluation of the suitability of soil for use as foundation or construction material, information about its properties, in addition to classification, is frequently necessary. Those properties which help to assess the engineering behaviour of a soil and which assist in determining its classification accurately are termed 'Index Properties'. The tests required to determine index properties are in fact 'classification tests'. Index properties include indices that can be determined relatively quickly and easily, and which will have a bearing on important aspects of engineering behaviour such as strength or load-bearing capacity, swelling and shrinkage, and settlement. These properties may be relating to individual soil grains or to the aggregate soil mass. The former are usually studied from disturbed or remoulded soil samples and the latter from relatively undisturbed samples, i.e., from soil *in-situ*.

Some of the important physical properties, which may relate to the state of the soil or the type of the soil include soil colour, soil structure, texture, particle shape, grain specific gravity, water content, *in-situ* unit weight, density index, particle size distribution, and consistency limits and related indices. The last two are classification tests, strictly speaking. These, and a few properties peculiar to clay soils, will be studied in the following sections, except soil structure and texture, which have already been dealt with in Chapter 1.

3.2 SOIL COLOUR

Colour of a soil is one of the most obvious of its features. Soil colour may vary widely, ranging from white through red to black; it mainly depends upon the mineral matter, quantity and nature of organic matter and the amount of colouring oxides of iron and manganese, besides the degree of oxidation.

Iron compounds of some minerals get oxidised and hydrated, imparting red, brown or yellow colour of different shades to the soil. Manganese compounds and decayed organic matter impart black colour to the soil. Green and blue colours may be imparted by famous compounds such as pyrite. Absence of coloured compounds will lead to grey and white colours of the soil. Quartz, kaolinite and a few other clay minerals may induce these colours. Light grey colour may be imparted by small amounts of organic matter as well. Soil colour gets darkened by an increase in organic content.

Change in moisture content leads to lightening of soil colour. A dark coloured soil turns lighter on oven-drying. For identification and descriptive purposes, the colour should be that of moist state and, preferably, of the undisturbed state. In general, clays are darker in colour than sands and silts because of the capacity of the former for retention of water.

3.3 PARTICLE SHAPE

Shape of individual soil grains is an important qualitative property. In the case of coarse-grained soils, including silts, the grains are bulky in nature, indicating that the three principal dimensions are approximately of the same order.

Individual particles are frequently very irregular in shape, depending on the parent rock, the stage of weathering and the agents of weathering. The particle shape of bulky grains may be described by terms such as 'angular', 'sub-angular', 'sub-rounded', 'rounded' and 'well-rounded' (Fig. 3.1). Silt particles rarely break down to less than 2μ size (one μ = one micron = 0.001 mm), because of their mineralogical composition.

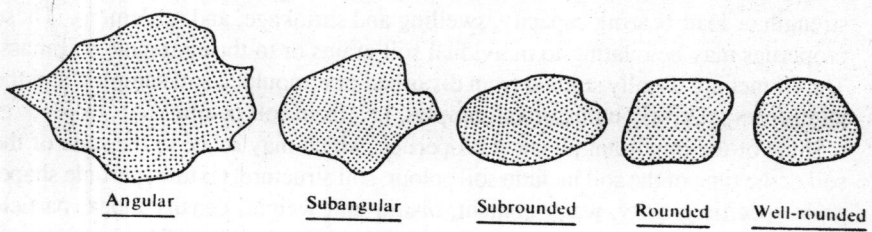

Fig. 3.1 Shapes of granular soil particles

The mineralogical composition of true clay is distinctly different from the mineral components of other soil types, thus necessitating the distinction between clay minerals and non-clay minerals. Clay particles are invariably less than 2μ size. Microscopic studies of such soils reveal that the particle shape is flake-like or needle-like; clay minerals are invariably crystalline in nature, having an orderly, sheet-like molecular structure. Clay particles, in fact, may consist of several such sheets on top of one another. The clay minerals, kaolinite, illite, and montmorillonite, show such sheet structure and flaky particle shape.

3.4 SPECIFIC GRAVITY OF SOIL SOLIDS

Specific gravity of the soil solids is useful in the determination of void-ratio, degree of saturation, etc., besides the 'Critical Hydraulic Gradient', and 'Zero-air-voids' in compaction. It is useful in computing the unit weight of the soil under different conditions and also in the determination of particle size by wet analysis. Hence, the specific gravity of soil solids should be determined with great precision.

The grain specific gravities of some common soils are listed in Table 3.1, which should serve as a guideline to the engineer :

Table 3.1 Grain specific gravities of some soils

S.No.	Soil Type	Grain specific gravity
1.	Quartz sand	2.64 – 2.65
2.	Silt	2.68 – 2.72
3.	Silt with organic matter	2.40 – 2.50
4.	Clay	2.44 – 2.92
5.	Bentonite	2.34
6.	Loess	2.65 – 2.75
7.	Lime	2.70
8.	Peat	1.26 – 1.80
9.	Humus	1.37

The standardised detailed procedure for the determination of the specific gravity of soil solids is contained in the Indian Standard Specification – "IS:2720 (Part-III)-1980, First Revision–Methods of Test for Soils, Part III, Determination of specific gravity". (Section 1 for fine-grained soils and section 2 for fine, medium and coarse grained soils).

However, the general procedure is set out below:

A 50-cc density bottle or a 500-cc pycnometer may be used. While the density bottle is the more accurate and suitable for all types of soils, the pycnometer (Fig. 3.2) is used only for coarse-grained soils. The sequence of observations and the procedure are similar in both cases.

First, the weight of the empty pycnometer is determined (W_1), in the dry condition. Then the sample of oven-dried soil, cooled in the desiccator, is placed in the pycnometer and its weight with the soil is determined (W_2). The remaining volume of the pycnometer is then gradually filled with distilled water or kerosene. The entrapped air should be removed either by gentle heating and vigorous shaking or by applying vacuum. The weight of the pycnometer, soil and water is obtained (W_3) carefully. Lastly, the bottle is emptied, thoroughly cleaned and filled with distilled water or kerosene, and its weight taken (W_4).

With the aid of these four observations, the grain specific gravity may be determined as follows:

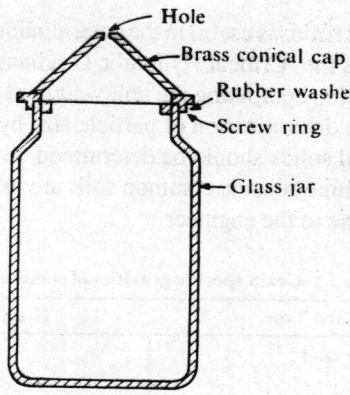

Fig. 3.2 Pycnometer

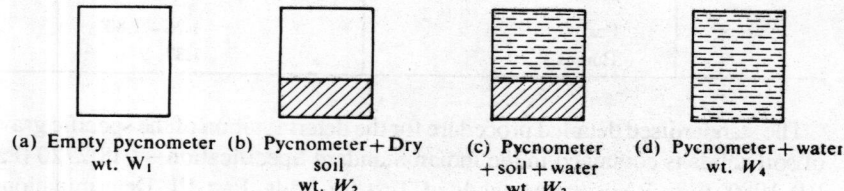

Fig. 3.3 Determination of grain specific gravity

From the readings, the wt of solids $W_s = W_2 - W_1$, from (a) and (b)

Wt of water = $W_3 - W_2$, from (b) and (c)

Wt of distilled water = $W_4 - W_1$, from (a) and (d)

∴ Weight of water having the same volume as that of soil solids = $(W_4 - W_1) - (W_3 - W_2)$

By definition, and by Archimedes' principle,

$$G = \frac{\text{Weight of soil solids}}{\text{Weight of water of volume equal to that of solids}}$$

$$= \frac{(W_2 - W_1)}{(W_4 - W_1) - (W_3 - W_2)}$$

$$= \frac{(W_2 - W_1)}{(W_2 - W_1) - (W_3 - W_4)}$$

$$\therefore \quad G = \frac{W_s}{W_s - (W_3 - W_4)} \qquad \ldots \text{(Eq. 3.1)}$$

W_s is nothing but the dry weight of the soil.

Aliter. If the soil solids are removed from W_3 and replaced by water of equal volume, W_4 is obtained.

Volume of solids $= \dfrac{W_s}{G}$

$$\therefore \quad W_4 = W_3 - W_s + \frac{W_s}{G}$$

Hence, $\quad G = \dfrac{W_s}{(W_s) - (W_3 - W_4)}$, same as Eq. 3.1.

If kerosene is used,

$$G = \frac{W_s \cdot G_k}{W_s - (W_3 - W_4)} \qquad \ldots \text{(Eq. 3.2)}$$

Where G_k = Specific gravity of kerosene at the temperature of the test.

Kerosene is used in preference to distilled water, if a density bottle is used; kerosene has better wetting capacity, which may be needed if the soil sample is of clay. In the case of clay, de-airing should be done much more carefully by placing the bottle in a vacuum desiccator for about 24 hours. This procedure should be resorted to for obtaining the weights W_3 and W_4.

Conventionally, the specific gravity is reported at a temperature of 27°C. If the room temperature *at the time of testing is different from this, then temperature* correction becomes necessary. Alternatively, the weights W_3 and W_4 should be taken after keeping the bottle in a constant temperature bath at the desired temperature of 27°C.

If the specific gravity, determined at a temperature of T_1°C, is G_{T_1}, and it is desired to obtain the specific gravity G_{T_2} at a temperature of T_2°C, the following equation may be used:

$$G_{T_2} = G_{T_1} \cdot \frac{(G_w)_{T_2}}{(G_w)_{T_1}} \qquad \ldots \text{(Eq. 3.3)}$$

where $(G_w)_{T_1}$ and $(G_w)_{T_2}$ are the specific gravities of water at temperatures T_1°C and T_2°C respectively. (These should be the values for kerosene if that liquid has been used in place of water).

In other words, the grain specific gravity is directly proportional to the specific gravity of the water at the test temperature. In the light of this observation, Eq. 3.1 is sometimes modified to read as follows:

$$G = \frac{W_s \cdot (G_w)_T}{W_s - (W_3 - W_4)} \qquad \ldots \text{(Eq. 3.4)}$$

where $(G_w)_T$ is the specific gravity of water at the test temperature.

If $(G_w)_T$ is taken as unity, which is true only at 4°C, Eq. 3.4 reduces to Eq. 3.1; that is to say, Eq. 3.1 may be used if one desires to report the value of G at 4°C and if one would like to ignore the effect of temperature. (The proof of the equations 3.3 and 3.4 is not difficult and is left to the reader).

Since the specific gravity of water varies only in a small range (1.0000 at 4°C and 0.9922 at 40°C), the temperature correction in the determination of grain specific gravity is quite often ignored. However, errors due to the presence of entrapped air can be significant.

3.5 WATER CONTENT

'Water content' or 'moisture content' of a soil has a direct bearing on its strength and stability. The water content of a soil in its natural state is termed its 'Natural moisture content', which characterises its performance under the action of load and temperature. The water content may range from a trace quantity to that sufficient to saturate the soil or fill all the voids in it. If the trace moisture has been acquired by the soil by absorption from the atmosphere, then it is said to be 'hygroscopic moisture'.

The knowledge of water content is necessary in soil compaction control, in determining consistency limits of soil, and for the calculation of stability of all kinds of earth works and foundations.

The method for the determination of water content, recommended by the Indian Standards Institution (I.S.I.), is set out in "IS: 2720 (Part-II) – 1964, Methods of Test for soils – Part II Determination of Moisture content", and is based on oven-drying of the soil sample.

The following methods will be given here:
 (*i*) Oven-drying method
 (*ii*) Pycnometer method
 (*iii*) Rapid moisture Tester method

3.5.1 Oven-drying Method

The most accurate approach is that of oven-drying the soil sample and is adopted in the laboratory.

A clean container of non-corrodible material is taken and its empty weight along-with the lid is taken. A small quantity of moist soil is placed in the container, the lid is replaced, and the weight is taken.

The lid is then removed and the container with the soil is placed in a thermostatically-controlled oven for 24 hours, the temperature being maintained between 105-110°C. After drying, the container is cooled in a desiccator, the lid is replaced and the weight is taken. For weighing a balance with an accuracy of 0.01 g is used.

Thus, the observations are:

Weight of an empty container with lid = W_1

Weight of container with lid + wet soil = W_2

Weight of container with lid + dry soil = W_3

The calculations are as follows:

Weight of dry soil = $W_3 - W_1$

Weight of water in the soil = $W_2 - W_3$

$$\text{Water content, } w = \frac{\text{Wt of water}}{\text{Wt of dry soil}} \times 100\%$$

$$\therefore \quad w = \frac{(W_2 - W_3)}{(W_3 - W_1)} \times 100\% \qquad \ldots(\text{Eq. 3.5})$$

Sandy soils need only about four hours of drying, while clays need at least 15 hours. To ensure complete drying, 24 hours of oven drying is recommended. A temperature of more than 110°C may result in the loss of chemically bound water around clay particles and hence should not be used. A low value such as 50°C is preferred in the case of organic soils such as peat to prevent oxidation of the organic matter. If gypsum is suspected to be present in the soil, drying at 80°C for longer time is preferred to prevent the loss of water of crystallisation of gypsum.

To obtain quick results in the field, sometimes heating on a sand-bath for about one hour is resorted to instead of oven-drying. This is considered to be a crude method since there is no temperature control.

3.5.2 Pycnometer Method

This method may be used when the specific gravity of solids is known. This is a relatively quick method and is considered suitable for coarse-grained soils only.

The following are the steps involved:

(i) The weight of the empty pycnometer (Fig. 3.2) with its cap and washer is found (W_1).

(ii) The wet soil sample is placed in the pycnometer (upto about 1/4 to 1/3 of the volume) and its weight is obtained (W_2).

(iii) The pycnometer is gradually filled with water, stirring and mixing thoroughly with a glass rod, such that water comes flush with the hole in the conical cap. The pycnometer is dried on the outside with a cloth and its weight is obtained (W_3).

(iv) The pycnometer is emptied and cleaned thoroughly; it is filled with water upto the hole in the conical cap, and its weight is obtained (W_4).

The water content of the soil sample may be calculated as follows:

$$W = \left[\frac{(W_2 - W_1)}{(W_3 - W_4)} \left(\frac{G-1}{G} \right) - 1 \right] \times 100\% \qquad \ldots(\text{Eq. 3.6})$$

This can be easily derived from the schematic phase diagrams shown in Fig. 3.4: If the solids from (iii) are replaced with water, we get W_4 of (iv).

Volume of solids $= \dfrac{W_s}{G}$

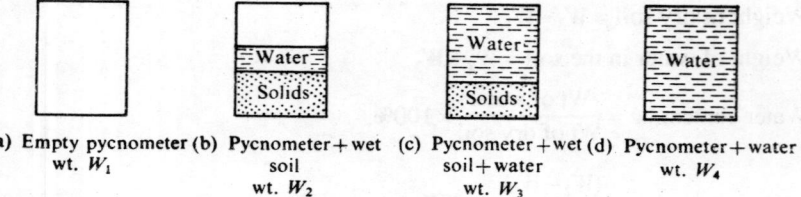

(a) Empty pycnometer wt. W_1 (b) Pycnometer + wet soil wt. W_2 (c) Pycnometer + wet soil + water wt. W_3 (d) Pycnometer + water wt. W_4

Fig. 3.4 Determination of Water Content

$$W_4 = W_3 - W_s + \frac{W_s}{G}$$

$$W_s\left(1 - \frac{1}{G}\right) = W_3 - W_4$$

$$\therefore \quad W_s = (W_3 - W_4)\ [G/(G-1)]$$

Weight of water W_w in the soil sample is given by:

$$W_w = (W_2 - W_1) - W_s$$

Water content, $w = \dfrac{W_w}{W_s}$

$$\therefore \quad w = \frac{W_2 - W_1 - W_s}{W_s}$$

$$= \frac{(W_2 - W_1)}{W_s} - 1$$

$$= \frac{(W_2 - W_1)}{(W_3 - W_4)} \cdot \frac{(G-1)}{G} - 1$$

$$\therefore \quad w = \left[\frac{W_2 - W_1}{W_3 - W_4}\left(\frac{G-1}{G}\right) - 1\right] \times 100\% \qquad \ldots(\text{Eq. 3.6})$$

It may be noted that this method is suitable for coarse-grained soils only, since W_3 cannot be determined accurately for fine-grained soils.

3.5.3 Rapid Moisture Tester Method

A device known as 'Rapid Moisture Tester' has been developed for rapid determination of the water content of a soil sample. The principle of operation is based on the reaction that occurs between a carbide reagent and soil moisture. The wet soil sample is placed in a sealed container with calcium carbide, and the acetylene gas produced exerts pressure on a sensitive diaphragm placed at the end of the container. This pressure is correlated to the moisture content and is calibrated on a dial gauge on the other side of the diaphragm.

However, the reading gives the moisture expressed as a percentage of the wet weight of the soil. It may be converted to the moisture content expressed as a percentage of the dry weight by the following relationship:

$$w = \frac{w_r}{(1-w_r)} \times 100\% \qquad \ldots(\text{Eq. 3.7})$$

where w_r = moisture content obtained by the rapid moisture tester, expressed as a decimal fraction.

The method is rapid and results may be got in about ten minutes.

The field kit consists of the moisture tester, a single small pan weighing balance, a bottle of calcium carbide and a brush.

This method is becoming popular in the field control of compaction (Chapter 12) where quick results are imperative.

Even nuclear approaches have been developed for the determination of moisture content. Sometimes, penetration resistance is calibrated against water content and is determined by a penetrometer needle. (Chapter 12.)

3.6 DENSITY INDEX

Density Index (or relative density according to older terminology) of a soil, I_D, indicates the relative compactness of the soil mass. This is used in relation to coarse-grained soils or sands.

In a dense condition, the void ratio is low whereas in a loose condition, the void ratio is high. Thus, the in-place void ratio may be determined and compared, with the void ratio in the loosest state or condition and that in the densest state or condition (Fig. 3.5).

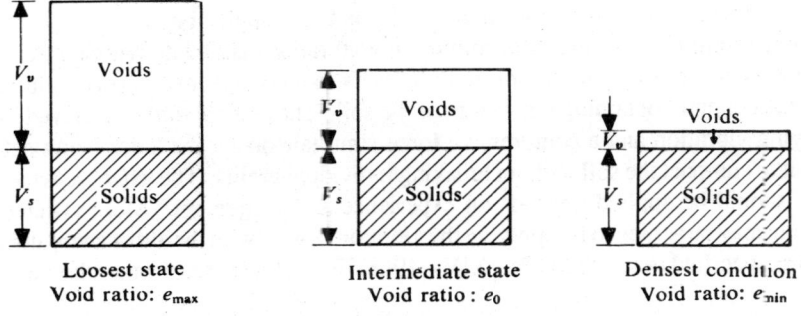

Fig. 3.5 Relative states of packing of a coarse-grained soil

44 *Geotechnical Engineering*

The density index may be considered zero if the soil is in its loosest state and unity if it is in the densest state. Consistent with this idea, the density index may be defined as follows:

$$I_D = \frac{(e_{max} - e_0)}{(e_{max} - e_{min})} \qquad \text{...(Eq. 3.8)}$$

where,

e_{max} = maximum void ratio or void ratio in the loosest state.

e_{min} = minimum void ratio or void ratio in the densest state.

e_0 = void ratio of the soil mass in the natural state or the condition under question.

e_{max} and e_{min} are referred to as the limiting void ratios of the soil.

Sometimes I_D is expressed as a percentage also. Equation 3.8 may be recast in terms of the dry unit weights as follows:

$$I_D = \left(\frac{1}{\gamma_{min}} - \frac{1}{\gamma_0}\right) \bigg/ \left(\frac{1}{\gamma_{min}} - \frac{1}{\gamma_{max}}\right) \qquad \text{...(Eq. 3.9)}$$

$$= \left(\frac{\gamma_{max}}{\gamma_0}\right)\left(\frac{\gamma_0 - \gamma_{min}}{\gamma_{max} - \gamma_{min}}\right) \qquad \text{...(Eq. 3.10)}$$

These forms are more convenient since the dry unit weights may be determined directly.

However, if it is desired to determine the void ratio in any state, the following relationships may be used:

$$e = \frac{G \cdot \gamma_w}{\gamma_d} - 1 \qquad \text{...(Eq. 3.11)}$$

$$e = \frac{V \cdot G \cdot \gamma_w}{W_s} - 1 \qquad \text{...(Eq. 3.12)}$$

A knowledge of the specific gravity of soil solids is necessary for this purpose. The determination of the volume of the soil sample may be a source of error in the case of clay soils; however, this is not so in the case of granular soils, such as sands, for which alone the concept of density index is applicable.

The maximum unit weight (or minimum void ratio) may be determined in the laboratory by compacting the soil in thin layers in a container of known volume and subsequently obtaining the weight of the soil. The compaction is achieved by applying vibration and a compressive force simultaneously, the latter being sufficient to compact the soil without breaking individual grains. The extent to which these should be applied depends on experience and judgement. More efficient packing may be achieved by applying the vibratory force (with the aid of a vibratory table as specified in IS-2720 (Part XIV)-1968)[*] in the presence of water; however

---

[*]"I.S.-2720 (Part XIV)-1968 Methods of Test for Soils - Part XIV Determination of Density Index (Relative density) for soils" gives two approaches - the dry method and the wet method for the determination of the maximum density.

this needs a proper drainage arrangement at the base of the cylinder used for the purpose, and also the application of vacuum to remove both air and water. It should be noted, however, that it is not possible to obtain a zero volume of void spaces, because of the irregular size and shape of the soil particles. Practically speaking, there will always be some voids in a soil mass, irrespective of the efforts (natural or external) at densification.

In the dry method, the mould with the dry soil in it is placed on a vibratory table and vibrated for 8 minutes at a frequency of 60 vibrations per second, after having placed a standard surcharge weight on top.

In the wet method, the mould should be filled with wet soil and a sufficient quantity of water added to allow a small quantity of water to accumulate on the surface. During and just after filling, it should be vibrated for a total of 6 minutes. Amplitude of vibration may be reduced during this period to avoid excessive boiling. The mould should be again vibrated for 8 minutes after adding the surcharge weight. Dial gauge readings are recorded on the surcharge base plate to facilitate the determination of the final volume.

The wet method should be preferred if it is found to give higher maximum densities than the dry method; otherwise, the latter may be employed as quicker results are secured by this approach.

Other details are contained in the relevant Indian Standard, and its revised versions.

The minimum unit-weight (or maximum void ratio) can be determined in the laboratory by carefully letting the soil flow slowly into the test cylinder through a funnel. Once this task has been carefully performed, the top surface is struck level with the top of the cylinder by a straight edge and the weight of the soil of known volume may be found in this state, which is considered to be the loosest. Oven-dried soil is to be used. Even the slightest disturbance may cause slight densification, thus affecting the result.

If proper means are available for the determination of the final volume of vibrated sand, the known weight of sand in the loosest state may itself be used for the determination of the void ratio in the densest state. In that case the sequence of operations will change.

Thus, it may be understood, that there is some degree of arbitrariness involved in the determination of the void ratio or unit weight in the densest as well as in the loosest state.

The concept of Density Index is developed somewhat as follows:
Assuming that the sand is in the loosest state:

$$e_{max} = \frac{V_{v\,max}}{V_{s\,min}},$$

for which the corresponding value of density index is taken as zero.

Assuming that the sand is in the densest state:

$$e_{min} = \frac{V_{v\,min}}{V_{s\,max}},$$

for which the corresponding value of density index is taken as unity.

It can be understood that the density index is a function of the void ratio:
$$I_D = f(e) \qquad ...(Eq.\ 3.13)$$
This relation between e and I_D may be expressed graphically as follows.

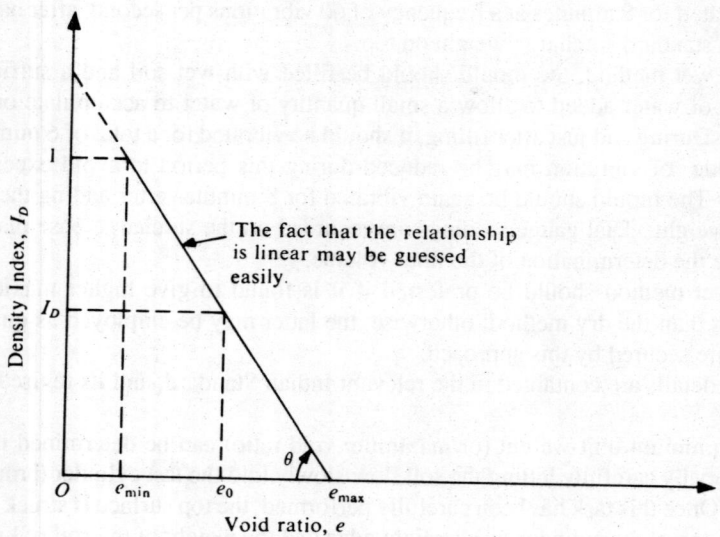

Fig. 3.6 Void ratio-Density Index relationship

It may be seen that:
$$\tan \theta = \frac{1}{(e_{max} - e_{min})}$$
$$\therefore \quad \cot \theta = (e_{max} - e_{min}) \qquad ...(Eq.\ 3.14)$$
For any intermediate value e_0,
$$(e_{max} - e_0) = I_D \cdot \cot \theta \qquad ...(Eq.\ 3.15)$$
$$\therefore \quad I_D = \frac{(e_{max} - e_0)}{\cot \theta}$$
Substituting for $\cot \theta$ from Eq. 3.14
$$I_D = \frac{(e_{max} - e_0)}{e_{max} - e_{min}} \qquad ...(Eq.\ 3.8)$$

Obviously, if $e_0 = e_{max}$, $I_D = 0$,
and if $e_0 = e_{min}$, $I_D = 1$.

For very dense gravelly sand I_D sometimes comes out to be greater than unity. This would only indicate that the natural packing does not permit itself to be repeated or simulated in the laboratory.

Representative values of density index and typical range of unit weights are given in Table 3.2.

Table 3.2 Representative values of Density Index and typical unit weights (Mc Carthy, 1977)

Descriptive Condition	Density Index, %	Typical range of unit weight, kN/m^3
Loose	Less than 35	Less than 14
Medium dense	35 to 65	14 to 17
Dense	65 to 85	17 to 20
Very dense	Greater than 85	Above 20

Depending upon the texture, two sands with the same void ratio may display different abilities for densification; hence the density index gives a better idea of the unit weight than the value of the void ratio itself.

The density index concept finds application in compaction of granular material, in various soil vibration problems associated with earth works, pile driving, foundations of machinery, vibrations transmitted to sandy soils by automobiles and trains, etc. Density index value gives us an idea, in such cases, whether or not such undesirable consequences can be expected from engineering operations which might affect structures or foundations due to vibration settlement.

3.7 *IN-SITU* UNIT WEIGHT

The *in-situ* unit weight refers to the unit weight of a soil in the undisturbed condition or of a compacted soil in-place.

Determination of *in-situ* unit weight is made on borrow-pit soils so as to estimate the quantity of soil required for placing and compacting a certain fill or embankment. During the construction of compacted fills, it is standard practice to make *in-situ* determination of a unit weight of the soil after it is placed to ensure that the compaction effort has been adequate.

Two important methods for the determination of the *in-situ* unit weight are being given:
  (*i*) Sand-replacement method.
  (*ii*) Core-cutter method.

3.7.1 Sand-replacement Method

The principle of the sand replacement method consists in obtaining the volume of the soil excavated by filling in the hole *in-situ* from which it is excavated, with sand, previously calibrated for its unit weight, and thereafter determining the weight of the sand required to fill the hole.

The apparatus* consists of the sand pouring cylinder (Fig. 3.7), tray with a central circular hole, container for calibration, balance, scoop, etc.

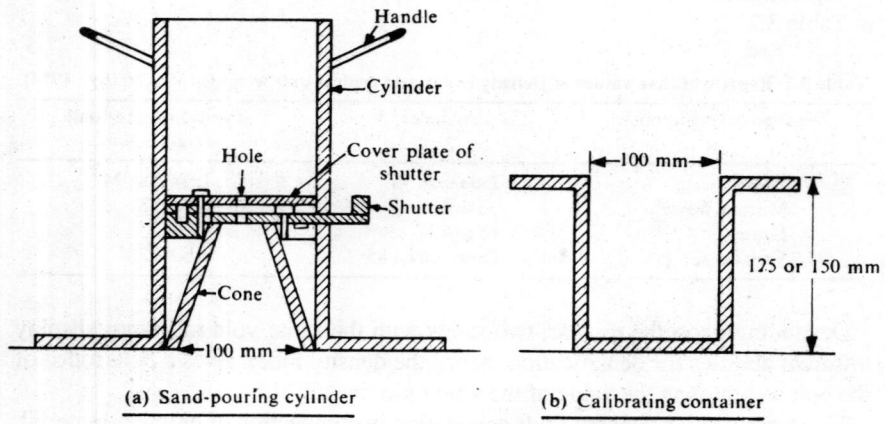

Fig. 3.7 Sand pouring cylinder

The procedure consists of calibration of the cylinder and later, the measurement of the unit weight of the soil.

(a) *Calibration of the cylinder and sand:* This consists in obtaining the weight of sand required to fill the pouring cone of the cylinder and the bulk unit weight of the sand. Uniformly graded, dry, clean sand is used. The cylinder is filled with sand almost to the top and the weight of the cylinder with the sand is taken (W_1). The sand is run out of the cylinder into the conical portion by pulling out the shutter. When no further sand runs out, the shutter is closed. The weight of the cylinder with the remaining sand is found (W_2). The weight of the sand collected in the conical portion may also be found separately for a check (W_c), which should be equal to ($W_1 - W_2$).

The cylinder is placed centrally above the calibrating container such that the bottom of the conical portion coincides with the top of the container. The sand is allowed to run into the container as well as the conical portion until both are filled, as indicated by the fact that no further sand runs out; then the shutter is closed. The weight of the cylinder with the remaining sand is found (W_3). The weight of the sand filling the calibrating container (W_{cc}) may be found by deducting the weight of sand filling the conical portion (W_c) from the weight of sand filling this and the container ($W_2 - W_3$). Since the volume of the cylindrical calibrating container (V_{cc}) is known precisely from its dimensions, the unit weight of the sand

---

*"IS–2720 (Part XXVIII)–1974 (First revision) – Methods of Test for Soils – Determination of in-place density by sand-replacement method" contains the complete details of the apparatus and the recommended procedure in this regard.

Index Properties and Classification Tests   49

may be obtained by dividing the weight W_{cc}, by the volume V_{cc}. (W_{cc} may also be found directly by striking-off the sand level with the top of the container and weighing it).

The observations and calculations relating to this calibration part of the work will be as follows:

Initial weight of cylinder + sand = W_1

Weight of cylinder + sand, after running sand into the conical portion = W_2

∴ Weight of sand occupying conical portion, W_c = $(W_1 - W_2)$

Weight of cylinder + sand, after running sand into the conical portion and calibrating container = W_3

∴ Weight of sand occupying conical portion and calibrating container = $(W_2 - W_3)$

∴ Weight of sand filling the calibrating container,

$$W_{cc} = (W_2 - W_3) - W_c$$
$$= (W_2 - W_3) - (W_1 - W_2)$$
$$= (2W_2 - W_1 - W_3)$$

Volume of the calibrating container = V_{cc}

∴ Unit weight of the sand: $\gamma_s = \dfrac{W_{cc}}{V_{cc}}$

(b) *Measurement of Unit Weight of the Soil:* The site at which the *in-situ* unit weight is to be determined is cleaned and levelled. A test hole, about 10 cm diameter and for about the depth of the calibrating container (15 cm), is made at the site, the excavated soil is collected and its weight is found (W). The sand pouring cylinder is filled with sand to about 3/4 capacity and is placed over the hole, after having determined its initial weight with sand (W_4), and the sand is allowed to run into it. The shutter is closed when no further movement of sand takes place. The weight of the cylinder and remaining sand is found (W_5). The weight of the sand occupying the test hole and the conical portion will be equal to $(W_4 - W_5)$. The weight of the sand occupying the test hole, W_s, will be obtained by deducting the weight of the sand occupying the conical portion, W_c, from this value. The volume of the test hole, V, is then got by dividing the weight, W_s by the unit weight of the sand.

The *in-situ* unit weight of the soil, γ, is then obtained by dividing the weight of the soil, W, by its volume, V. If the moisture content, w, is also determined, the dry unit weight of the soil, γ_d, is obtained as $\dfrac{\gamma}{(1+w)}$. Thus, the observations and calculations for this part may be set out as follows:

Initial weight of cylinder + sand = W_4

Weight of cylinder + sand, after running sand
into the test hole and the conical portion = W_5

∴ Weight sand occupying the test hole and the
conical portion = $(W_4 - W_5)$

∴ Weight of sand occupying the test hole, W_s = $(W_4 - W_5) - W_c$
= $(W_4 - W_5) - (W_1 - W_2)$

Volume of test hole, $V = \dfrac{W_s}{\gamma_s}$

In-situ unit weight of the soil, = W/V
Dry unit weight, $\gamma_d = \gamma/(1+w)$,
where, w = water content (fraction).

In an alternative approach, the volume of the test hole may be determined more directly by inflating a rubber balloon into the hole, making it fit the hole snugly, and reading off the fall in water level in a graduated Lucite cylinder which is properly connected to the balloon.

3.7.2 Core-cutter Method

The apparatus consists of a mild steel-cutting ring with a dolly to fit its top and a metal rammer.

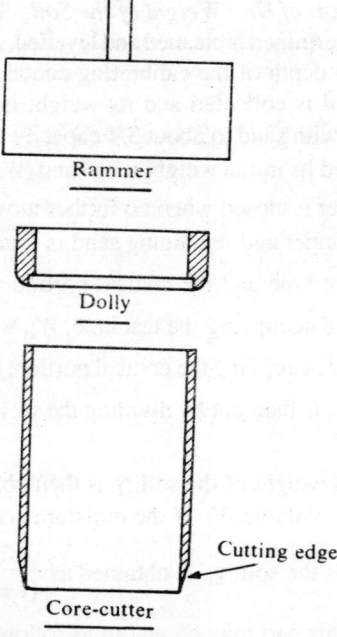

Fig. 3.8 Core-cutter apparatus

Index Properties and Classification Tests

The core-cutter is 10 cm in diameter and 12.5 cm in length. The dolly is 2.5 cm long. The bottom 1 cm of the ring is sharpened into a cutting edge. The empty weight (W_1) of the core-cutter is found. The core-cutter with the dolly is rammed into the soil with the aid of a 14-cm diameter metal rammer. The ramming is stopped when the top of the dolly reaches almost the surface of the soil. The soil around the cutter is excavated to remove the cutter and dolly full of soil, from the ground. The dolly is also removed later, and the soil is carefully trimmed level with the top and bottom of the core-cutter. The weight of the core-cutter and the soil is found (W_2). The weight of the soil in the core-cutter, W, is then got as ($W_2 - W_1$). The volume of this soil is the same as that of the internal volume of the cutter, V, which is known.

The *in-situ* unit weight of the soil, γ, is given by W/V. If the moisture content, w, is also found, the dry-unit weight, γ_d, may be found as $\gamma_d = \gamma/(1+w)$.

This method* is suitable for soft cohesive soils. It cannot be used for stiff clays, sandy soils and soils containing gravely particles, which could damage the cutting edge.

In an alternative approach, the volume, V, of a clay soil sample which can be trimmed into a more or less regular-shaped piece, can be obtained by coating it with paraffin and then immersing it in a graduated jar filled with water. The rise in water level in the jar gives the volume of the sample together with the paraffin. The volume of the paraffin can be got by dividing the weight of paraffin by its known unit weight. It can then be subtracted from this to obtain the volume of the soil sample. The weight of the soil sample, W, would have been obtained earlier before coating it with paraffin. The weight of the paraffin can also be got as the increase in weight of the sample on coating it with the paraffin. The *in-situ* unit weight of the soil may now be got as $\gamma = W/V$.

Paraffin, being water-proof, prevents the entry of water into the soil sample, thus affording a simple means to determine the volume of the sample.

3.8 PARTICLE SIZE DISTRIBUTION (MECHANICAL ANALYSIS)

This classification test determines the range of sizes of particles in the soil and the percentage of particles in each of these size ranges. This is also called 'grain-size distribution'; 'mechanical analysis' means the separation of a soil into its different size fractions.**

The particle-size distribution is found in two stages:
  (i) Sieve analysis, for the coarse fraction.
  (ii) Sedimentation analysis or wet analysis, for the fine fraction.

'Sieving' is the most direct method for determining particle sizes, but there are practical lower limits to sieve openings that can be used for soils. This lower limit

---

*"IS: 2720 (Part XXIX)–1966 – Methods of Test for Soils – Determination of in-place density by the core-cutter method" contains the complete details of the apparatus and the recommended procedure in this regard.

**Determination of the textural composition of the soil is also known as 'granulometry'.

is approximately at the smallest size attributed to sand particles (75 μ or 0.075 mm).

Sieving is a screening process in which coarser fractions of soil are separated by means of a series of graded mesh. Mechanical analysis is one of the oldest test methods for soils.

3.8.1 Nomenclature of Grain Sizes

Natural soils are mixtures of particles of various sizes and it is necessary to have a nomenclature for the various fractions comprising particles lying between certain specified size limits. Particle size is customarily expressed in terms of a single diameter. This is taken as the size of the smallest square hole in a sieve, through which the particle will pass.

The Indian standard nomenclature is as follows:

Gravel	...	80 mm to 4.75 mm
Sand	...	4.75 mm to 0.075 mm
Silt	...	0.075 mm to 0.002 mm
Clay	...	Less than 0.002 mm

3.8.2 Sieve Analysis

Certain sieve sizes have been standardised by certain Standard Organisations such as the British Standards Organisation (B.S.), American Society for Testing Materials (A.S.T.M.), and Indian Standards Institution (I.S.I.); the first two, in F.P.S. units and the third, in M.K.S. units. Sieve designation is specified by the number of openings per inch in the B.S. and A.S.T.M. standards, while it is specified by the size of the aperture in mm or microns in the I.S. standard. (IS: 460–1962 Revised).

Information with regard to important I.S. Sieves in common use is given in Table 3.3.

Table 3.3 Certain I.S. Sieves and their aperture sizes

Designation	Aperture mm	Designation		Aperture mm
50 mm	50.0	600–μ	(60)**	0.600
40 mm	40.0	*500–μ		0.500
		425–μ		0.425
20 mm	20.0	*355–μ		0.355
		300–μ	(30)**	0.300
10 mm	10.0	250–μ		0.250
*5.6 mm	5.6	*180–μ		0.180
		150–μ	(15)**	0.150
*4.0 mm	4.0	*125–μ		0.125
*2.8 mm	2.8	*90–μ		0.090
240**	2.36			
*2.0 mm	2.0	75–μ	(8)**	0.075
*1.4 mm	1.4	*63–μ		0.063
120**	1.18			
*1.0 mm	1.0	*45–μ		0.045

* Proposed as an International Standard (ISO). μ = micron = 0.001 mm.
** Old I.S. Designations were based on nearest one-hundredths of a mm.

The test procedure for sieve analysis has been standardised by ISI as given in IS:2720 (Part IV) – 1965.

The general procedure may be summarised as follows:

A series of sieves* having different-size openings are stacked with the larger sizes over the smaller. A receiver is kept at the bottom and a cover is kept at the top of the assembly. The soil sample to be tested is dried, clumps are broken if necessary, and the sample is passed through the series of sieves by shaking. The fractions retained on and passing 2 mm IS Sieve are tested separately. An automatic sieve-shaker, run by an electric motor, may be used; about 10 to 15 minutes of shaking is considered adequate. Larger particles are caught on the upper sieves, while the smaller ones filter through to be caught on one of the smaller underlying sieves.

The material retained on any particular sieve should naturally include that retained on the sieves on top of it, since the sieves are arranged with the aperture size decreasing from top to bottom. The weight of material retained on each sieve is converted to a percentage of the total sample. The percentage material finer than a sieve size may be got by subtracting this from 100. The material passing the bottom-most sieve, which is usually the 75–μ sieve, is used for conducting sedimentation analysis for the fine fraction.

If the soil is clayey in nature the fine fraction cannot be easily passed through the 75–μ sieve in the dry condition. In such a case, the material is to be washed through it with water (preferably mixed with 2 gm of sodium hexametaphosphate per litre), until the wash water is fairly clean. The material which passes through the sieve is obtained by evaporation. This is called 'wet sieve analysis, and may be required in the case of cohesive granular soils.

Soil grains are not of an equal dimension in all directions. Hence, the size of a sieve opening will not represent the largest or the smallest dimension of a particle, but some intermediate dimension, if the particle is aligned so that the greatest dimension is perpendicular to the sieve opening.

The resulting data are conventionally presented as a "Particle-size distribution curve" (or "Grain-size distribution curve") – the two terms being used synonymously hereafter) plotted on semi-log co-ordinates, where the sieve size is on a horizontal 'logarithmic' scale, and the percentage by weight of the size smaller than a particular sieve-size is on a vertical 'arithmetic' scale. The "reversed" logarithmic scale is only for convenience in presenting coarser to finer particles from left to right. A typical presentation is shown in Fig. 3.9. (Results may be presented in tabular form also).

Logarithmic scales for the particle diameter gives a very convenient representation of the sizes because a wide range of particle diameter can be shown in a single plot; also a different scale need not be chosen for representing the fine fraction with the same degree of precision as the coarse fraction.

---

*The sieves may be 600-micron, 212-micron and 75-micron I.S. Sieves. These correspond to the limits of coarse, medium and fine sand. Other sieves may be introduced depending upon the additional information desired to be obtained from the analysis.

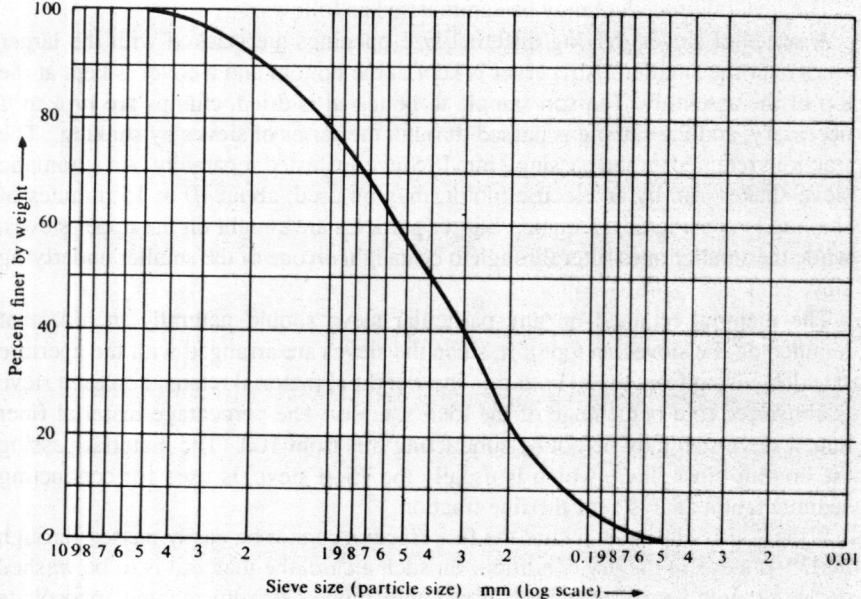

Fig. 3.9 Particle-size distribution curve

The characteristics of grain-size distribution curves will be studied in a later sub-section.

3.8.3 Sedimentation Analysis (Wet Analysis)

The soil particles less than 75-μ size can be further analysed for the distribution of the various grain-sizes of the order of silt and clay by 'sedimentation analysis' or 'wet analysis'. The soil fraction is kept in suspension in a liquid medium, usually water. The particles descend at velocities, related to their sizes, among other things.

The analysis is based on 'Stokes' Law' for what is known as the 'terminal velocity' of a sphere falling through an infinite liquid medium. If a single sphere is allowed to fall in an infinite liquid medium without interference, its velocity first increases under the influence of gravity, but soon attains a constant value. This constant velocity, which is maintained indefinitely unless the boundary conditions change, is known as the 'terminal velocity'. The principle is obvious; coarser particles tend to settle faster than finer ones.

By Stokes' law, the terminal velocity of the spherical particle is given by

$$v = (1/18).[(\gamma_s - \gamma_1)/\mu_1].D^2 \qquad \ldots\text{(Eq. 3.16)}$$

which is dimensionally consistent.

Thus, if

γ_s = unit weight of the material of falling sphere in g/cm³,
γ_1 = unit weight of the liquid medium in g/cm³,

Index Properties and Classification Tests 55

μ_1 = viscosity of the liquid medium in g sec/cm^2,

and D = diameter of the spherical particle in cm,
v, the terminal velocity, is obtained in cm/s.
In S.I. units,
if γ_s and γ_1 are expressed in kN/m^3,
  μ_1 in kN sec/m^2,
  D in metres,
v will be obtained in m/sec.
Since, usually D is to be expressed in mm, while v is to be expressed in cm/sec, and μ_1 in N-sec/m^2, Eq. 3.15 may be rewritten as follows:

$$v = \frac{1}{180} \frac{(\gamma_s - \gamma_1)}{\mu_1} \cdot D^2 \qquad \ldots(\text{Eq. 3.17})$$

Here γ_s and γ_1 are in kN/m^3, μ_1 in N-sec/m^2, and D in mm; v will then be in cm/sec. Usually, the liquid medium is water; then γ_1 and μ_1 will be substituted by γ_w and μ_w. Then Eq. 3.16 will become:

$$v = \frac{1}{180} \frac{(\gamma_s - \gamma_w)}{\mu_w} \cdot D^2 \qquad \ldots(\text{Eq. 3.18})$$

It should be noted that γ_w and μ_w vary with temperature, the latter varying more significantly than the former.
Noting that $\gamma_s = G \cdot \gamma_w$,

$$v = \frac{1}{180} \cdot \frac{\gamma_w(G - 1)}{\mu_w} \cdot D^2 \qquad \ldots(\text{Eq. 3.19})$$

At 20°C, $\gamma_w = 0.9982$ g/cm^3 = 0.9982×9.810 kN/m^3

$$= 9.792 \text{ kN/m}^3$$

$$\mu_w = 0.001 \text{ N-sec/m}^2$$

Assuming $G = 2.67$, on an average,

$$v = \frac{1}{180} \times \frac{(9.792)(2.67 - 1)}{0.001} \cdot D^2 = 90.85 \, D^2$$

$$\therefore \quad v \approx 91 \, D^2 \qquad \ldots(\text{Eq. 3.20})$$

where D is in mm and v is in cm/sec.

Using this approximate version of Stokes' law, one can determine the time required for a particle of a specified diameter to settle through a particular depth; e.g., a particle of 0.06 mm diameter settles through 10 cm in about 1/2 minute, while one of 0.002 mm diameter settles in about 7 hours 38 minutes.
From Eq. 3.19,

$$v = \frac{1}{180} \cdot \frac{\gamma_w}{\mu_w} (G-1) \cdot D^2$$

$$\therefore D = \sqrt{\frac{180 \cdot \mu_w \cdot v}{\gamma_w (G-1)}}$$

If the particle falls through H cm in t minutes

$$v = H/60t \text{ cm/sec.}$$

$$\therefore D = \sqrt{\frac{180 \mu_w \cdot H}{\gamma_w (G-1) \cdot 60 t}}$$

$$= \sqrt{\frac{3\mu_w}{\gamma_w(G-1)}} \times \sqrt{\frac{H}{t}}$$

$$\therefore D = K\sqrt{H/t} \qquad \ldots \text{(Eq. 3.21)}$$

where
$$K = \sqrt{\frac{3\mu_w}{\gamma_w(G-1)}}$$

Here,

G = grain specific gravity of the soil particles,
γ_w = unit weight of water in kN/m^3 ⎱ at the particular
μ_w = viscosity of water in N-sec/m^2 ⎰ temperature.

H = fall in cm, and t = time in min.

The factor K can be tabulated or graphically represented for different values of temperature and grain specific gravity.

Stokes' Law is considered valid for particle diameters ranging from 0.2 to 0.0002 mm.

For particle sizes greater than 0.2 mm, turbulent motion is setup and for particle sizes smaller than 0.002 mm, Brownian motion is setup. In both these cases Stokes' law is not valid.

The general procedure for sedimentation analysis, which may be performed either with the aid of a pipette or a hydrometer is as follows:

An appropriate quantity of an oven-dried soil sample, finer than 75–μ size, is mixed with a known volume (V) of distilled water in a jar. The sample is pretreated with an oxidising agent and an acid to remove organic matter and calcium compounds. Addition of hydrogen peroxide and heating would remove organic matter. Treatment with 0.2 N hydrochloric acid would remove calcium compounds. Later, a deflocculating or a dispersing agent, such as sodium hexametaphosphate is added to the solution. (Further details regarding the preparation of the sample may be obtained from IS: 2790 (Part IV)–1965 and its revised versions). The mixture is shaken thoroughly by means of a mechanical stirrer and the test is started, keeping the jar vertical. The soil particles are assumed to be uniformly distributed throughout the suspension, at the instant of commencement of the test. After the lapse of time t, only those particles which have settled less than depth H would remain in suspension. The size of the particles, finer than those which have settled to depth H or more at this instant, can be found from Eqs. 3.21 and 3.22. Hence, sampling at different time intervals (by pipette), or determining the specific gravity of the suspension (by hydrometer), at this sampling depth, would provide the means of determining the content of particles of different sizes. (The logic would become much clearer if all particles are considered to be of the same size.) Since, the soil particles are dispersed uniformly throughout the suspension, and according to Stokes' law, particles of the same size settle at the same rate, particles of a given size, wherever they exist, have the same degree of concentration as at the commencement of the test. As such, particles smaller than a given size will be present in the same degree of concentration as at the start, and particles larger than this size would have settled already below the sampling depth, and hence are not present at that depth. The percentage of particles finer than a specified size may be got by determining their concentration at that depth at different times either with the aid of a pipette or of a hydrometer.

The limitations of sedimentation analysis, based on Stokes' law, or the assumptions are as follows:

(*i*) The finer soil particles are never perfectly spherical. Their shape is flake-like or needle-like. However, the particles are assumed to be spheres, with equivalent diameters, the basis of equivalence being the attainment of the same terminal velocity as that in the case of a perfect sphere.

(*ii*) Stokes' law is applicable to a sphere falling freely without any interference, in an infinite liquid medium. The sedimentation analysis is conducted in a one-litre jar, the depth being finite; the walls of the jar could provide a source of interference to the free fall of particles near it. The fall of any particle may be affected by the presence of adjacent particles; thus, the fall may not be really free.

However, it is assumed that the effect of these sources of interference is insignificant if suspension is prepared with about 50 g of soil per litre of water.

(*iii*) All the soil grains may not have the same specific gravity. However, an average value is considered all right, since the variation may be insignificant in the case of particles constituting the fine fraction.

(*iv*) Particles constituting the fine soil fraction may carry surface electric charges, which have a tendency to create 'flocs'. Unless these flocs are broken, the sizes calculated may be those of the flocs. Flocs can be a source of erroneous results.

A deflocculating agent, such as sodium silicate, sodium oxalate, or sodium hexa-metaphosphate, is used to get over this difficulty.

Pipette Analysis

The sedimentation analysis may be conducted with the aid of a pipette in the laboratory. A pipette, sedimentation jar, and a number of sampling bottles are necessary for the test. A boiling tube of 500 ml capacity kept in a constant temperature bath may also be used in place of a sedimentation jar. The capacity of the sampling pipette is usually 10 ml.

The method consists in drawing off 10 ml samples of soil suspension by means of the sampling pipette from a standard depth of 10 cm at various time intervals after the start of sedimentation. The soil-water suspension should have been prepared as has been mentioned earlier. The usual total time intervals at which the samples are drawn are 30 s, 1 min., 2 min., 4 min., 8 min., 15 min., 30 min., 1 h, 2 h, and 4 h from the start of sedimentation. The pipette should be inserted about 20 seconds prior to the chosen instant and the process of sucking should not take more than 20 seconds. Each of the samples taken is transferred to a sampling bottle and dried in an oven. The weight solids, W_D in the suspension, finer than a certain size D, related to the time of sampling, may be found by careful weighing, from the concentration of these solids in the pipette sample. Let W_s be the weight of soil (fine fraction) used in the suspension of volume V, and W_D be the weight of soil particles finer than size D in the entire suspension. Also, let W_p be the weight of solids in the pipette sample of volume V_p.

Then, by the argument presented in the general procedure for sedimentation analysis,

$$\frac{W_D}{V} = \frac{W_p}{V_p} \qquad \ldots(\text{Eq. 3.23})$$

or

$$W_D = \frac{W_p}{V_p} \cdot V = W_p \left(\frac{V}{V_p}\right) \qquad \ldots(\text{Eq. 3.24})$$

The calculation will be somewhat as follows:
From Equations 3.21 and 3.22, for the known values of H and t, we obtain the size D.

Let the weight of solids per ml in the pipette sample be multiplied by the total volume of the suspension; this would give W_D as defined in Eq. 3.24.

Percentage of particles finer than the size D, in the fine-fraction, N_f, is given by:

$$N_f = \frac{W_D}{W_s} \times 100 \qquad \ldots(\text{Eq. 3.25})$$

Substituting for W_D from Eq. 3.24,

$$N_f = \left(\frac{W_p}{W_s}\right)\left(\frac{V}{V_p}\right) \times 100 \qquad \ldots(\text{Eq. 3.26})$$

This is to be corrected if a dispersing agent is added. If w is the weight of the dispersing agent added,

$$N_f = \frac{(W_D - w)}{W_s} \times 100 \qquad \ldots(\text{Eq. 3.27})$$

For a combined sieve and sedimentation analysis, if W is the total dry weight of the soil originally taken, the over-all percentage, N, of particles, finer than D, is given by:

$$N = N_f \times \frac{W_f}{W} \qquad \ldots(\text{Eq. 3.28})$$

where W_f = Weight of fine soil fraction out of the total weight of a soil sample, W, taken for the combined sieve and sedimentation analysis.

The calculations completed for all the samples would provide information for the grain-size distribution curve.

The pipette analysis, although very simple and direct in principle, is tedious and requires very sensitive weighing apparatus. Accurate results are rather difficult to obtain. For this reason, the hydrometer analysis is preferred in the laboratory.

Hydrometer Analysis

The hydrometer method differs from the pipette analysis in that the weights of solids per ml in the suspension at the chosen depth at chosen instants of time are

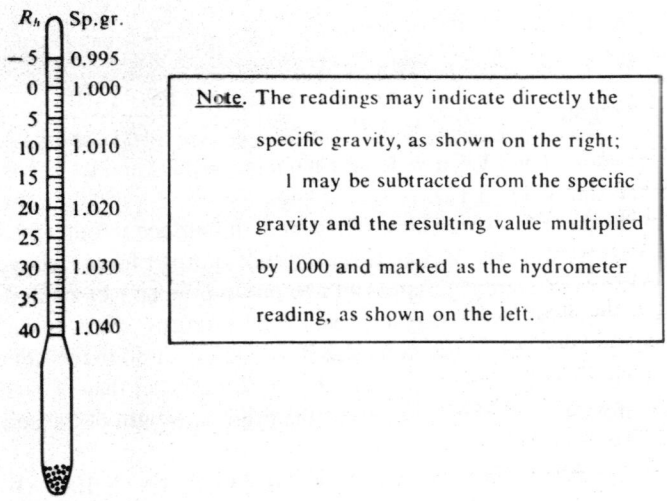

Note. The readings may indicate directly the specific gravity, as shown on the right; 1 may be subtracted from the specific gravity and the resulting value multiplied by 1000 and marked as the hydrometer reading, as shown on the left.

Fig. 3.10 Hydrometer

obtained indirectly by reading the specific gravity of the soil suspension with the aid of a hydrometer.

Hydrometer is a device which is used to measure the specific gravity of liquids (Fig. 3.10). However, for a soil suspension, the particles start settling down right from the start, and hence the unit weight of the suspension varies from top to bottom.

It can be established that measurement of unit weight of the suspension at a known depth at a particular time provides a point on the grain-size distribution curve.

Let W be weight of fine soil fraction mixed in water
V be the volume of suspension

Initially, the weight of solids per unit volume of suspension
$$= W/V$$

Volume of solids per unit volume of suspension $= \dfrac{W}{V.G.\gamma_w}$

Volume of water per unit volume of suspension $= 1 - \dfrac{W}{V.G.\gamma_w}$

Weight of water per unit volume of suspension $= \gamma_w\left(1 - \dfrac{W}{V.G.\gamma_w}\right)$

$$= \gamma_w - \dfrac{W}{V.G}$$

Initial weight of a unit volume of suspension $= \dfrac{W}{V} + \left(\gamma_w - \dfrac{W}{VG}\right)$

Initial unit weight, $\quad \gamma_i = \gamma_w + \dfrac{(G-1)}{G} \cdot \dfrac{W}{V}$ ...(Eq. 3.29)

Let us consider a level XX, at a depth z from the surface and let t be the elapsed time from the start (Fig. 3.11).

Size D of the particles which have fallen from the surface through the depth z in time t may be got from Eqs. 3.21 and 3.22, by substituting z for H. Above the level XX, no particle of size greater than D will be present. In an elemental depth dz at this depth z, the suspension may be considered uniform, since, initially it was uniform, and all particles of the same size have settled through the same depth in the given time. In this element, particles of the size smaller than D exist. Let the percentage of weight of these particles to the original weight of the soil particles in the suspension be N.

Wt of solids per unit volume of the suspension at depth $z = (N/100) \cdot (W/V)$

By similar reasoning as for Eq. 3.29, the unit weight of the suspension, γ_z, at depth z and at time t, is given by:

$$\gamma_z = \gamma_w + \dfrac{(G-1)}{G} \cdot \dfrac{N}{100} \cdot \dfrac{W}{V} \qquad \text{...(Eq. 3.30)}$$

$$\therefore \quad N = \dfrac{100\,G}{(G-1)} \cdot (\gamma_z - \gamma_w) \cdot \dfrac{V}{W} \qquad \text{...(Eq. 3.31)}$$

Hence, if γ_z is obtained by a hydrometer, N can be got, thus getting a point on the grain-size distribution curve.

In pipette analysis, the sampling depth is kept fixed; however, in the hydrometer analysis, the sampling depth (known as the effective depth) goes on increasing with time since the particles are allowed to settle. It is, therefore, necessary to calibrate the hydrometer with respect to the sedimentation jar, with a view to determining the value of the effective depth for any particular reading of the hydrometer. If the same hydrometer and the same sedimentation jar are used for a number of tests, one calibration chart will serve the purpose for all the tests.

Index Properties and Classification Tests 61

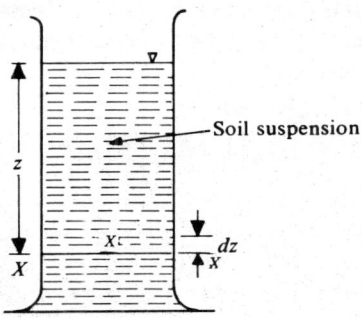

Fig. 3.11 Suspension Jar

Calibration

The method of calibration can be easily understood from Fig. 3.12

Let h be the height of the bulb and H be the height of any reading R_h from the top of the bulb or neck. The jar with a soil suspension is shown in Fig. 3.12 (b); the surface is xx and the level at which the specific gravity of the suspension is being measured is designated yy, the depth being H_e, the effective depth.

As shown in Fig. 3.12 (c), on immersion of the hydrometer into the suspension in the jar, the levels xx and yy will rise to $x'x'$ and $y'y'$ respectively.

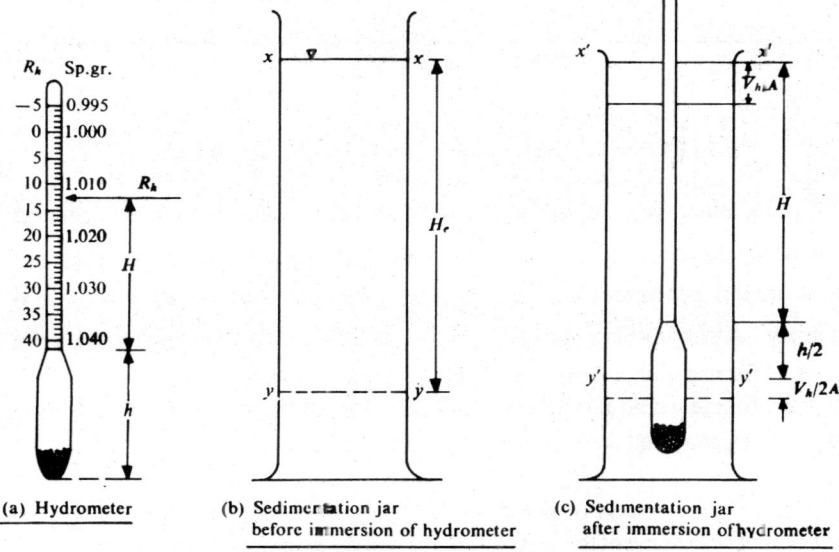

Fig. 3.12 Calibration of hydrometer with respect to sedimentation jar

If V_h is the volume of the hydrometer and A is the area of cross-section of the jar containing the suspension, the rise in the level xx is given by V_h/A. The rise in the level yy will be approximately $v_h/2A$ since the effective depth is reckoned to the middle level of the hydrometer bulb. The level $y'y'$ correspond to this mid-level, but the soil particles at this level are in the same concentration as they were at yy, as the level yy has merely risen to $y'y'$ consequent to the immersion of the hydrometer in the suspension.

Therefore, on correlating (b) and (c),

$$H_e = \left(H + \frac{h}{2} + \frac{V_h}{2A}\right) - \frac{V_h}{A}$$

or
$$H_e = H + \frac{1}{2}\left(h - \frac{V_h}{A}\right) \qquad \text{... (Eq. 3.32)}$$

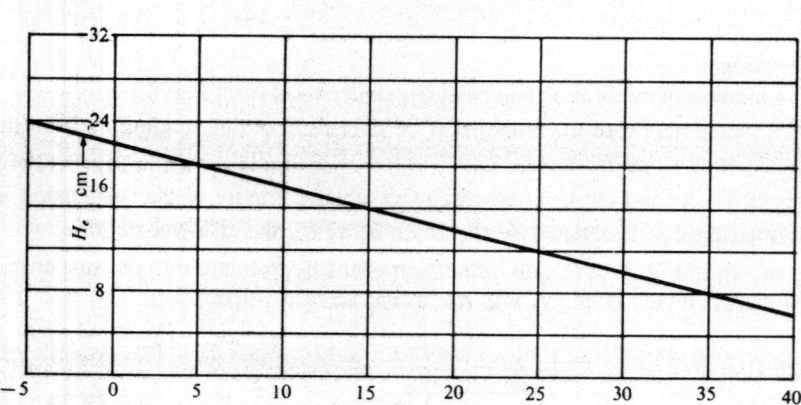

Fig. 3.13 Calibration graph between hydrometer reading and effective depth

Thus, the effective depth H_e at which the specific gravity is measured depends upon H and, hence, upon the observed hydrometer reading, R_h'. However, h, V_h, and A are independent of R_h', and may be easily obtained with a fair degree of accuracy. A calibration graph between R_h' and H_e can be prepared as shown in Fig. 3.13 for use with a particular hydrometer and a particular suspension jar.

For the first few readings for about 2 minutes, the hydrometer may not go down much and H_e may be taken approximately to be equal $(H+h/2)$.

Procedure: The method of preparation of the soil suspension has already been indicated. The volume of the suspension is 1000 ml in this case. The sedimentation jar is shaken vigorously and is then kept vertical over a firm base and stopwatch is started simultaneously. The hydrometer is slowly inserted in the jar and readings taken at elapsed times 30 s, 1 min and 2 min. The hydrometer is then taken out. Further readings are taken at elapsed times of 4 min, 8 min, 15 min, 30 min, 1 h, 2 h, 4 h, etc., by inserting the hydrometer about 20 seconds prior to the desired instant. The hydrometer should be stable without oscillations at the time each

Index Properties and Classification Tests 63

reading is taken. Since the soil suspension is opaque, the reading is taken corresponding to the upper level of the meniscus. The temperature is recorded. For good results, the variation should not be more than 2° celsius.

Certain corrections are required to be applied to the recorded hydrometer readings before these could be used for the computation of the unit weight of the suspension.

Corrections to Hydrometer Readings

The following three corrections are necessary:
(1) Meniscus correction
(2) Temperature correction
(3) Deflocculating agent correction

Meniscus Correction

The reading should be taken at the lower level of the meniscus. However, since the soil suspension is opaque, the reading is taken at the upper meniscus. Therefore, a correction is required to be applied to the observed reading. Since the hydrometer readings increase downward on the stem, the meniscus correction (C_m) is obviously positive.

The magnitude of the correction can be got by placing the hydrometer in distilled water in the same jar and noting the difference in reading at the top and bottom levels of the meniscus.

Temperature Correction

Hydrometers are usually calibrated at a temperature of 27°C. If the temperature at the time of conducting the test is different, a correction will be required to be applied to the hydrometer reading on this account. For this purpose, the hydrometer is placed in clean distilled water at different temperatures and a calibration chart prepared for the corrections required. If the temperature at the time of test is more than that of calibration of the hydrometer, the observed reading will be less and the correction (C_t) would be positive and vice versa.

Deflocculating Agent Correction

The addition of the deflocculating agent increases the density of the suspension and thus necessitates a correction (C_d) which is always negative. This is obtained by immersing the hydrometer, alternately in clean distilled water and a solution of the deflocculating agent in water (with the same concentration as is to be used in the test), and noting the difference in the reading.

A composite correction for all the above may be obtained by noting the hydrometer readings in a solution of the deflocculating agent at different temperatures. These with reversed sign give the composite correction.

The corrected hydrometer reading R_h may be got from the observed reading R'_h by applying the composite correction 'C':

64   *Geotechnical Engineering*

$$R_h = R'_h \pm C \quad \text{...(Eq. 3.33)}$$

where
$$C = C_m - C_d \pm C_t$$

The next step is to determine the percentage finer of the particles of a specified size related to any hydrometer reading.

Calculations: After obtaining the corrected hydrometer readings R'_h at various elapsed times t, and the corresponding effective depths H_e, Equations 3.20 and 3.21 may be used (H_e being used for H), to obtain the corresponding particle size. Now Eq. 3.27 may be used as follows for determining the percent finer than the particle size D:

$$N = \frac{100\,G}{(G-1)} \cdot (\gamma_z - \gamma_w) \cdot V/W$$

But $\gamma_z = G_{ss}\,\gamma_w$, where G_{ss} = specific gravity of soil suspension

$$= \left(1 + \frac{R_h}{1000}\right) \cdot \gamma_w, \text{ since } R_h = 1000(G_{ss} - 1).$$

$$\therefore \quad N = \frac{100\,G \cdot \gamma_w}{(G-1)} \cdot \frac{R_h}{1000} \times \frac{V}{W} = \frac{G\gamma_w}{(G-1)} \cdot \frac{V}{W} \cdot \frac{R_h}{10}$$

$$\therefore \quad N = \frac{G\gamma_w}{(G-1)} \cdot \frac{V}{W} \cdot \frac{R_h}{10} \quad \text{...(Eq. 3.34)}$$

Thus for each hydrometer reading, R_h, we obtain a set of values for D and N, fixing one point on the grain-size distribution curve.

The grain-size distribution may thus be completed by the sedimentation analysis in conjunction with sieve analysis for the coarse fraction.

3.8.4 Characteristics of Grain-size Distribution Curves

Grain-size distribution curves of soils primarily indicate the type of the soil, the history and stage of its deposition, and the gradation of the soil. If the soil happens to be predominantly coarse-grained or predominantly fine-grained, this will be very clearly reflected in the curve.

Typical grain-size distribution curves are shown in Fig. 3.14. A soil is said to be "well-graded", if it contains a good representation of various grain-sizes. Curve marked (a) indicates a well-graded soil. If the soil contains grains of mostly one size, it is said to be "uniform" or "poorly graded". Curve marked (b) indicates a uniform soil. A soil is said to be "gap-graded", if it is deficient in a particular range of particle sizes. Curve marked (c) indicates a gap-graded soil. A "young residual" soil is indicated by curve marked (d); in course of time, as the particles get broken, the soil may show a curve nearing type (a). Curve (e) indicates a soil which is predominantly coarse-grained, while curve (f) indicates a soil which is predominantly fine-grained. The more uniform a soil is, the steeper is its grain-size distribution curve.

Index Properties and Classification Tests 65

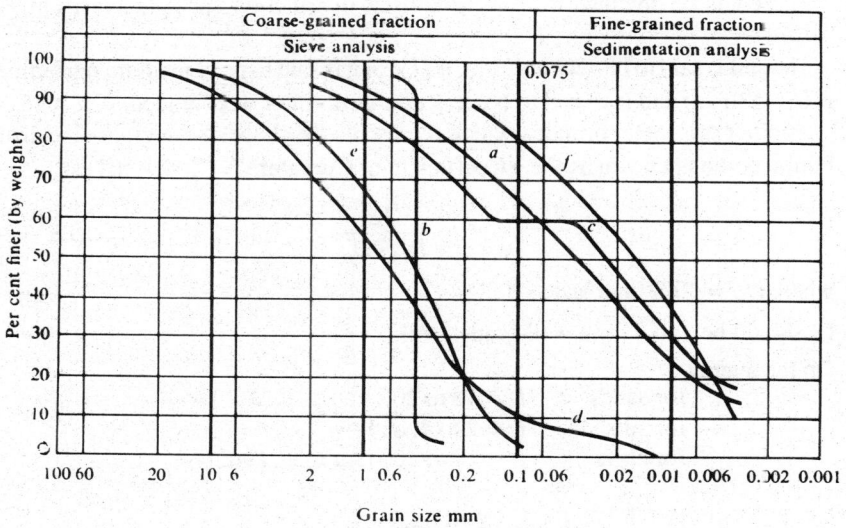

Fig. 3.14 Typical grain-size distribution curves

River deposits may be well-graded, uniform, or gap-graded, depending upon the velocity of the water, the volume of suspended solids, and the zone of the river where the deposition occurred.

Certain properties of granular or coarse-grained soils have been related to particle diameters. Allen Hazen (1892) tried to establish the particular diameter in actual spheres that would cause the same effect as a given soil, and opined that the diameter for which 10% was finer would give this equivalence. It may be recalled that the effective diameter of a soil particle is the diameter of a hypothetical sphere that is assumed to act in the same way as the particle of an irregular shape, and that data obtained from sedimentation analysis using Stokes' law lead to effective diameters, D_e, of the soil particles. Thus, Allen Hazen's $D_{10} = D_e$. The effective diameter is also termed the "Effective Size" of the soil. It is this size that is related to permeability and capillarity. D_{10} may be easily determined by reading-off from the grain-size distribution curve for the soil.

An important property of a granular or coarse-grained soil is its "degree of uniformity". The grain-size distribution curve of the soil itself indicates, by its shape, the degree of soil uniformity. A steeper curve indicates more uniform soil. Because of this, the grain-size distribution curve is also called the 'uniformity curve'.

Quantitatively speaking, the uniformity of a soil is defined by its "Coefficient of Uniformity" U:

$$U = \frac{D_{60}}{D_{10}} \quad \quad \text{...(Eq. 3.35)}$$

where D_{60} = 60% finer size.

and D_{10} = 10% finer size, or effective size.

These can be obtained from the grain-size distribution curve as shown in Fig. 3.15.

The soil is said to be very uniform, if $U < 5$; it is said to be of medium uniformity, if $U = 5$ to 15; and it is said to be very non-uniform or well-graded, if $U > 15$.

Another parameter or index which represents the shape of the grain-size distribution curve is known as the "Coefficient of Curvature", C_c, defined as:

$$C_c = \frac{(D_{30})^2}{D_{10} \cdot D_{60}} \qquad \text{...(Eq. 3.36)}$$

where D_{30} = 30% finer size.

C_c should be 1 to 3 for a well-graded soil.

On the average,
- for sands $U = 10$ to 20,
- for silts $U = 2$ to 4, and
- for clays $U = 10$ to 100 (Jumikis, 1962)

3.9 CONSISTENCY OF CLAY SOILS

'Consistency' is that property of a material which is manifested by its resistance to flow. In this sense, consistency of a soil refers to the resistance offered by it against forces that tend to deform or rupture the soil aggregate; in other words, it represents the relative ease with which the soil may be deformed. Consistency may also be looked upon as the degree of firmness of a soil and is often directly related to strength. This is applicable specifically to clay soils and is generally related to the water content.

Consistency is conventionally described as soft, medium stiff (or medium firm), stiff (or firm), or hard. These terms are unfortunately relative and may convey different meaning to different persons. In the case of *in-situ* or undisturbed clays, it is reasonable and practical to relate consistency to strength, for purposes of standardisation. (A little more of this aspect will be studied in one of the later sections.)

In the remoulded state, the consistency, of a clay soil varies with the water content, which tends to destroy the cohesion exhibited by the particles of such a soil. As the water content is reduced from a soil from the stage of almost a suspension, the soil passes through various states of consistency, as shown in Fig. 3.16. A. Atterberg, a Swedish Soil Scientist, in 1911, formally distinguished the following states of consistency – liquid, plastic, semi-solid, and solid. The water contents at which the soil passes from one of these states to the next have been arbitrarily designated as 'consistency limits' – Liquid limit, Plastic limit and Shrinkage limit, in that order. These are called 'Atterberg limits' in honour of the originator of the concept.

Index Properties and Classification Tests 67

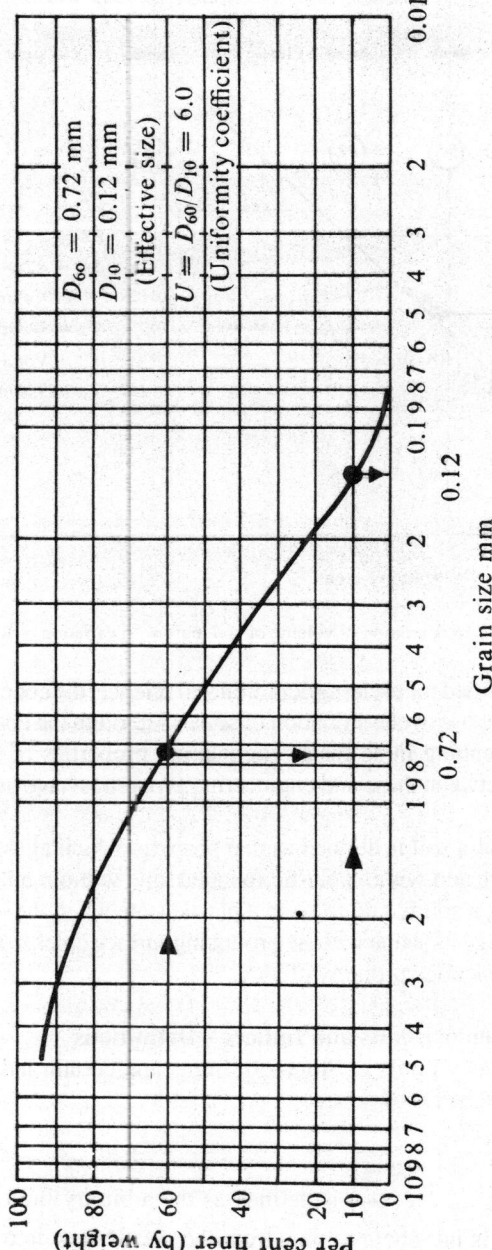

Fig. 3.15 Effective size and uniformity coefficient from grain-size distribution curve

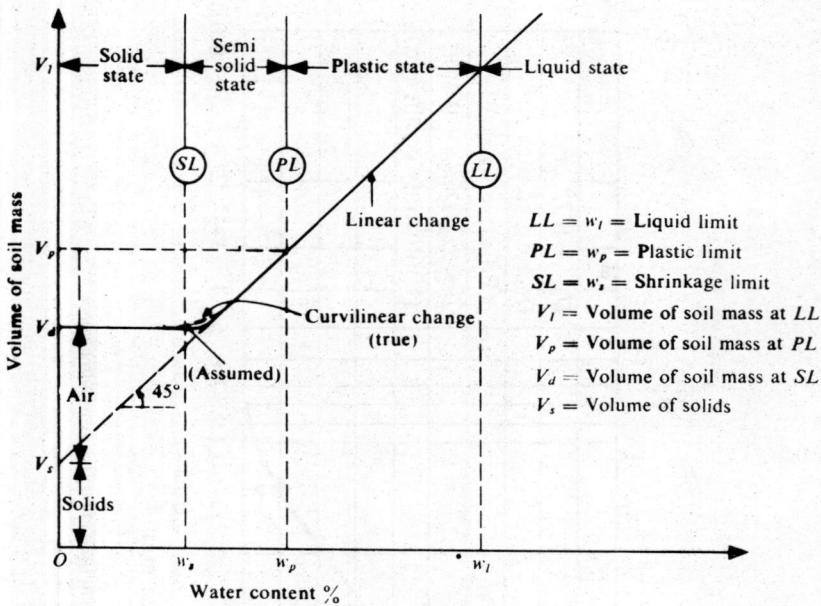

Fig. 3.16 Variation of volume of soil mass with variation of water content

Initially intended for use in agricultural soil science, the concept was later adapted for engineering use in classification of soils. Although the consistency limits have little direct meaning in so far as engineering properties of soils are concerned, correlations between these and engineering properties have been established since then.

'Plasticity' of a soil is defined as that property which allows it to be deformed, without rupture and without elastic rebound, and without a noticeable change in volume. Also, a soil is said to be in a plastic state when the water content is such that it can change its shape without producing surface cracks. Plasticity is probably the most conspicuous property of clay.

3.9.1 Consistency Limits and Indices—Definitions

The consistency limits or Atterberg limits and certain indices related to these may be defined as follows:

Liquid Limit

'Liquid limit' (LL or w_L) is defined as the arbitrary limit of water content at which the soil is just about to pass from the plastic state into the liquid state. At this limit, the soil possesses a small value of shear strength, losing its ability to flow as a liquid. In other words, the liquid limit is the minimum moisture content at which the soil tends to flow as a liquid.

Plastic Limit

'Plastic limit' (PL or w_p) is the arbitrary limit of water content at which the soil tends to pass from the plastic state to the semi-solid state of consistency. Thus, this is the minimum water content at which the change in shape of the soil is accompanied by visible cracks, i.e., when worked upon, the soil crumbles.

Shrinkage Limit

'Shrinkage limit' (SL or w_s) is the arbitrary limit of water content at which the soil tends to pass from the semi-solid to the solid state. It is that water content at which a soil, regardless of further drying, remains constant in volume. In other words, it is the maximum water content at which further reduction in water content will not cause a decrease in volume of the soil mass, the loss in moisture being mostly compensated by entry of air into the void space. In fact, it is the lowest water content at which the soil can still be completely saturated. The change in colour upon drying of the soil, from dark to light also indicates the reaching of shrinkage limit.

Upon further drying, the soil will be in a partially saturated solid state; and, ultimately, the soil will reach a perfectly dry state.

Plasticity Index

'Plasticity index' (PI or I_p) is the range of water content within which the soil exhibits plastic properties; that is, it is the difference between liquid and plastic limits.

$$\text{PI (or } I_p) = (\text{LL} - \text{PL}) = (w_L - w_p) \qquad \ldots(\text{Eq. 3.37})$$

When the plastic limit cannot be determined, the material is said to be non-plastic (NP). Plasticity index for sands is zero.

Burmister (1949) classified plastic properties of soils according to their plasticity indices as follows:

Table 3.4 Plasticity characteristics

Plasticity index	Plasticity
0	Non-plastic
1 to 5	Slight
5 to 10	Low
10 to 20	Medium
20 to 40	High
> 40	Very high

At the liquid limit the soil grains are separated by water just enough to deprive the soil mass of shear strength. At the plastic limit the soil moisture does not separate the soil grains, and has enough surface tension to effect contact between the soil grains, causing the soil mass to behave as a semi-solid.

70 Geotechnical Engineering

For proper evaluation of the plasticity properties of a soil, it has been found desirable to use both the liquid limit and the plasticity index values. As will be seen in the next chapter, engineering soil classification systems use these values as a basis for classifying the fine-grained soils.

Shrinkage Index

'Shrinkage Index' (SI or I_s) is defined as the difference between the plastic and shrinkage limits of a soil; in other words, it is the range of water content within which a soil is in a semi-solid state of consistency.

$$\text{SI}(\text{or } I_s) = (\text{PL} - \text{SL}) = (w_p - w_s) \quad \ldots(\text{Eq. 3.38})$$

Consistency Index

'Consistency Index' or 'Relative consistency' (CI or I_c) is defined as the ratio of the difference between liquid limit and the natural water content to the plasticity index of a soil:

$$\text{CI}(\text{or } I_c) = \frac{(\text{LL} - w)}{\text{PI}} = \frac{(w_L - w)}{I_p} \quad \ldots(\text{Eq. 3.39})$$

where w = natural water content of the soil (water content of a soil in the undisturbed condition in the ground).

If $I_c = 0, w = \text{LL}$

$I_c = 1, w = \text{PL}$

$I_c > 1$, the soil is in semi-solid state and is stiff.

$I_c < 0$, the natural water content is greater than LL, and the soil behaves like a liquid.

Liquidity Index

'Liquidity Index (LI or I_L)' or 'Water-plasticity ratio' is the ratio of the difference between the natural water content and the plastic limit to the plasticity index:

$$\text{LI}(\text{or } I_L) = \frac{(w - \text{PL})}{\text{PI (or } I_p)} = \frac{w - w_p}{I_p} \quad \ldots(\text{Eq. 3.40})$$

If

$I_L = 0, w = \text{PL}$

$I_L = 1, w = \text{LL}$

$I_L > 1$, the soil is in liquid state.

$I_L < 0$, the soil is in semi-solid state and is stiff.

Obviously,

$$\text{CI} + \text{LI} = 1 \quad \ldots(\text{Eq. 3.41})$$

For the purpose of convenience, the consistency of a soil in the field may be stated or classified as follows on the basis of CI or LI values:

Index Properties and Classification Tests 71

Table 3.5 Consistency classification

C.I.	LI	Consistency
1.00 to 0.75	0.00 to 0.25	Stiff
0.75 to 0.50	0.25 to 0.50	Medium-soft
0.50 to 0.25	0.50 to 0.75	Soft
0.25 to 0.00	0.75 to 1.00	Very soft

3.9.2 Laboratory Methods for the Determination of Consistency Limits and Related Indices

The definitions of the consistency limits proposed by Atterberg are not, by themselves, adequate for the determination of their numerical values in the laboratory, especially in view of the arbitrary nature of these definitions. In view of this, Arthur Casagrande and others suggested more practical definitions with special reference to the laboratory devices and methods developed for the purpose of the determination of the consistency limits.

In this sub-section, the laboratory methods for determination of the liquid limit, plastic limit, shrinkage limit, and other related concepts and indices will be studied, as standardized and accepted by the Indian Standard Institution and incorporated in the codes of practice.

Determination of Liquid Limit

The liquid limit is determined in the laboratory with the aid of the standard mechanical liquid limit device, designed by Arthur Casagrande and adopted by the ISI, as given in IS:2720 (Part V)–1970. The apparatus required are the mechanical liquid limit device, grooving tool, porcelain evaporating dish, flat glass plate, spatula, palette knives, balance, oven, wash bottle with distilled water and containers. The soil sample should pass 425–μ IS Sieve. A sample of about 120 g should be taken. Two types of grooving tools—Type A (Casagrande type) and Type B (ASTM type)—are used depending upon the nature of the soil. (Fig. 3.17).

The cam raises the brass cup to a specified height of 1 cm from where the cup drops upon the block exerting a blow on the latter. The cranking is to be performed at a specified rate of two rotations per second. The grooving tool is meant to cut a standard groove in the soil sample just prior to giving blows.

Air-dried soil sample of 120 g passing 425–μ I.S. Sieve is taken and is mixed with water and kneaded for achieving uniformity. The mixing time is specified as 5 to 10 min. by some authorities. The soil paste is placed in the liquid limit cup, and levelled off with the help of the spatula. A clean and sharp groove is cut in the middle by means of a grooving tool. The crank is rotated at about 2 revolutions per second and the number of blows required to make the halves of the soil pat separated by the groove meet for a length of about 12 mm is counted. The soil cake before and after the test are shown in Fig. 3.17. The water content is determined from a small quantity of the soil paste.

This operation is repeated a few more times at different consistencies or moisture contents. The soil samples should be prepared at such consistencies that the number of blows or shocks required to close the groove will be less and more than 25. The

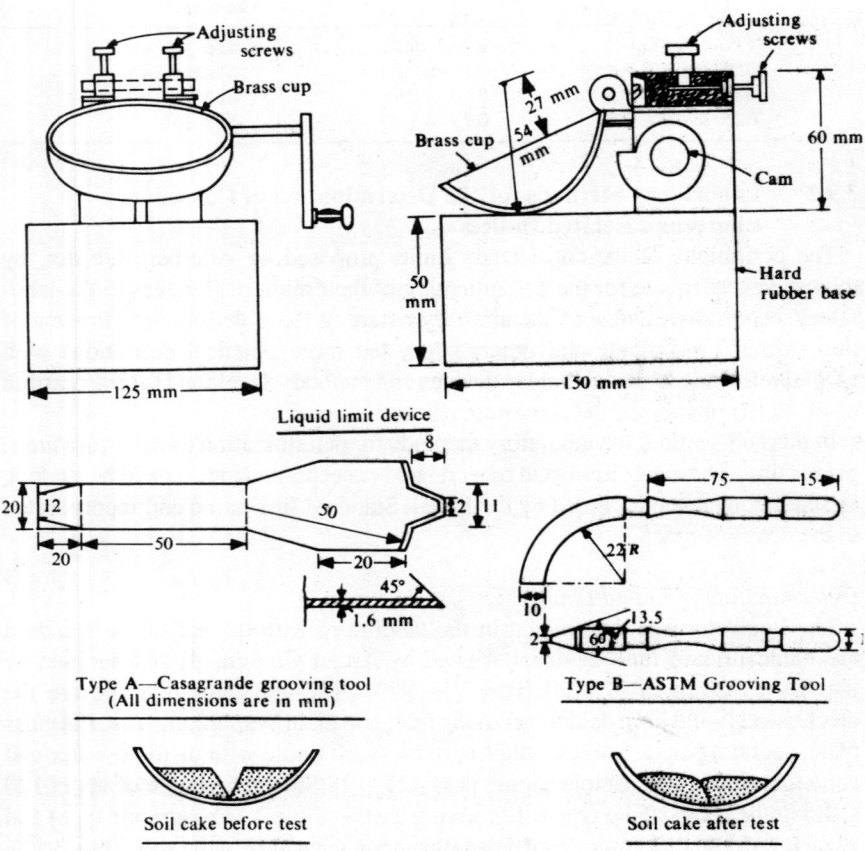

Fig. 3.17 Liquid limit apparatus

relationship between the number of blows and corresponding moisture contents thus obtained are plotted on semi-logarithmic graph paper, with the logarithm of the number of blows on the x-axis, and the moisture content on the y-axis. The graph thus obtained, i.e., the best fit straight line, is referred to as the 'Flow-graph' or 'Flow curve'. (Fig. 3.18).

The moisture content corresponding to 25 blows from the flow curve is taken as the liquid limit of the soil. This is the practical definition of this limit with specific reference to the liquid limit apparatus and the standard procedure recommended. Experience indicates that such a curve is actually a straight line.
The equation to this straight line will be

$$(w_2 - w_1) = I_f \log_{10} \frac{N_1}{N_2} \qquad \ldots \text{(Eq. 3.42)}$$

where w_1 and w_2 are the water contents corresponding to the number of blows N_1 and N_2 and I_f is the slope of the flow curve, called the 'flow index'.

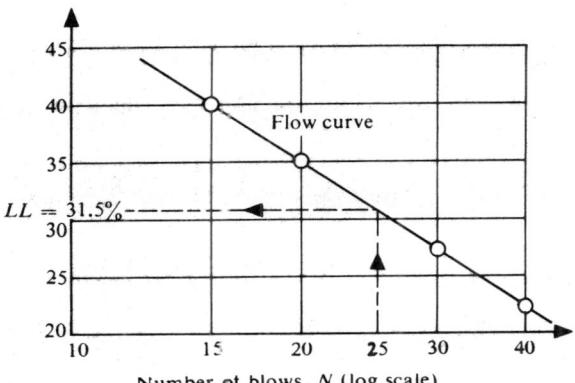

Fig. 3.18 Flow graph

$$I_f = (w_2 - w_1)/\log_{10}(N_1/N_2) \qquad \ldots \text{(Eq. 3.43)}$$

If the flow curve is extended such that N_1 and N_2 correspond to one log-cycle difference, I_f will be merely the difference of the corresponding water contents.

One-point Method

Attempts have been made to simplify the trial and error procedure of the determination of liquid limit described above. One such is the 'One-point method' which aims at determining the liquid limit with just one reading of the number of the blows and the corresponding moisture content.

The trial moisture content should be as near the liquid limit as possible. This can be done with a bit of experience with the concerned soils. For soils with liquid limit between 50 and 120%, the accepted range shall require 20 to 30 drops to close the groove. For soils with liquid limit less than 50%, a range of 15 to 35 drops is acceptable. At least two consistent consecutive closures shall be observed before taking the moisture content sample for calculation of the liquid limit. The test shall always proceed from the drier to the wetter condition of the soil. (IS: 2720, Part V-1970.)

The water content w_N of the soil of the accepted trial shall be calculated. The liquid limit w_L of the soil shall be calculated by the following relationship.

$$w_L = w_N (N/25)^x \qquad \ldots \text{(Eq. 3.44)}$$

where

N = number of drops required to close the groove at the moisture content w_N. Preliminary work indicates that $x = 0.092$ for soils with liquid limit less than 50% and $x = 0.120$ for soils with liquid limit more than 50%.

Note: The liquid limit should be reported to the nearest whole number. The history of the soil sample, that is, natural state, air-dried, oven-dried, the method used, the period of soaking should also be reported.

Cone Penetration Method

This method is based on the principle of static penetration. The apparatus is a 'Cone Penetrometer', consisting of a metallic cone with half angle of $15°30' \pm 15'$ and 30.5 mm coned depth (IS: 2720, Part V–1970). It shall be fixed at the end of

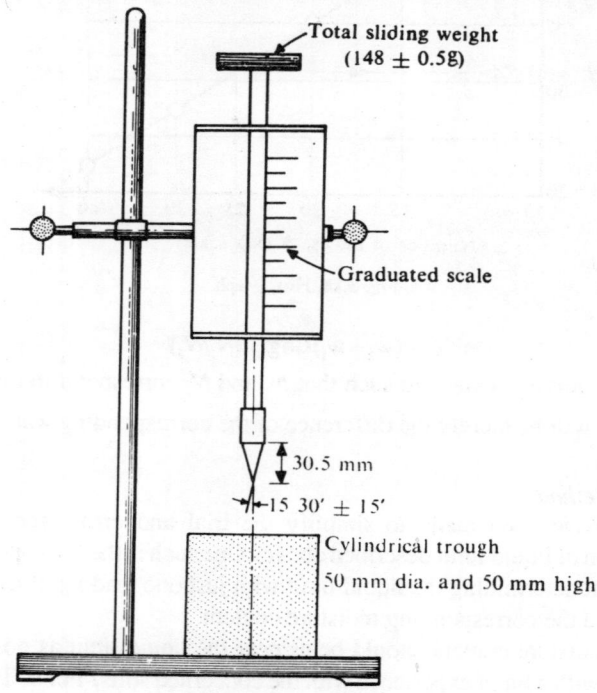

Fig. 3.19 Cone penetrometer

a metallic rod with a disc at the top of the rod so as to have a total sliding weight of 148 ± 0.5 g. The rod shall pass through two guides (to ensure vertical movement), fixed to a stand as shown in Fig. 3.19.

Suitable provision shall be made for clamping the vertical rod at any desired height above the soil paste in the trough. A trough 50 mm in diameter and 50 mm high internally shall be provided. The soil sample shall be prepared as in the case of other methods.

In the case of clay soils, it is recommended that the soil shall be kept wet and allowed to stand for a sufficient time (24 hrs) to ensure uniform distribution of moisture.

The wet soil paste shall then be transferred to the cylindrical trough of the cone penetrometer and levelled to the top of the trough. The penetrometer shall be so adjusted that the cone point just touches the surface of the soil paste in the trough. The penetrometer scale shall then be adjusted to zero and the vertical rod released so that the cone is allowed to penetrate into the soil paste under its weight. The penetration shall be noted after 30 seconds from the release of the cone. If the penetration is less than 20 mm, the wet soil from the trough shall be taken out and more water added and thoroughly mixed. The test shall then be repeated till a penetration between 20 mm and 30 mm is obtained. The exact depth of penetration between these two values obtained during the test shall be noted. The moisture content of the corresponding soil paste shall be determined in accordance with IS procedure, referred to earlier.

The liquid limit of the soil which corresponds to the moisture content of a paste which would give 25 mm penetration of the cone shall be determined from the following formula:

$$w_L = w_\alpha + 0.01\,(25 - \alpha)(w_\alpha + 15) \qquad \ldots(\text{Eq. 3.45})$$

where
w_L = liquid limit of the soil,

w_α = moisture content corresponding to penetration α,

α = depth of penetration of cone in mm.

The liquid limit may also be read out directly from a ready-made nomographic chart.

The expression is based on the assumption that, at the liquid limit, the shear strength of the soil is about 17.6 g/cm^2, which the penetrometer gives for a depth of 25 mm under a total load of 148 g.

Determination of Plastic Limit

The apparatus consists of a porcelain evaporating dish, about 12 cm in diameter (or a flat glass plate, 10 mm thick and about 45 cm square), spatula, about 8 cm long and 2 cm wide (or palette knives, with the blade about 20 cm long and 3 cm wide, for use with flat glass plate for mixing soil and water), a ground-glass plate, about 20 × 15 cm, for a surface for rolling, balance, oven, containers, and a rod, 3 mm in diameter and about 10 cm long. (IS: 2720, Part V–1970).

A sample weighing about 20 g from the thoroughly mixed portion of the material passing 425–µ IS sieve is to be taken. The soil shall be mixed with water so that the mass becomes plastic enough to be easily shaped into a ball. The mixing shall be done in an evaporating dish or on the flat glass plate. In the case of clayey soils, sufficient time (24 hrs) should be given to ensure uniform distribution of moisture throughout the soil mass. A ball shall be formed and rolled between the fingers and the glass plate with just enough pressure to roll the mass into a thread of uniform diameter throughout its length. The rate of rolling shall be between 80 and 90 strokes per minute, counting a stroke as one complete motion of the hand forward and back to the starting position again. The rolling shall be done till the threads are of 3 mm diameter. The soil shall then be kneaded together to a uniform mass and rolled again. This process of alternate rolling and kneading shall be continued until the thread crumbles under the pressure required for rolling and the soil can no longer be rolled into a thread. At no time shall attempt be made to produce failure at exactly 3 mm diameter. The crumbling may occur at a diameter greater than 3 mm; this shall be considered a satisfactory end point, provided the soil has been rolled into a thread of 3 mm in diameter immediately before. The pieces of crumbled soil thread shall be collected and the moisture content determined, which is the 'plastic limit'. The history of the soil sample shall also be reported.

'Toughness Index' (I_T) is defined as the ratio of the plasticity index to the flow index:

$$I_T = \frac{I_p}{I_f} \qquad \text{...(Eq. 3.46)}$$

where I_p = Plasticity index

I_f = Flow index.

Determination of Shrinkage Limit

If a saturated soil sample is taken (with water content, a little over the liquid limit) and allowed to dry up gradually, its volume will go on decreasing till a stage will come after which the reduction in the water content will not result in further reduction in the total volume of the sample; the water content corresponding to this stage is known as the shrinkage limit.

By analysing the conditions of the soil pat at the initial stage, at the stage of shrinkage limit, and at the completely dry state, one can arrive at an expression for the shrinkage limit as follows (Fig. 3.20):

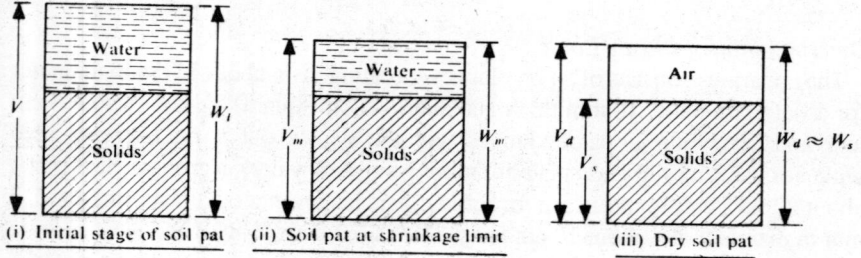

Fig. 3.20 Determination of shrinkage limit

Weight of water initially = $(W_i - W_d)$
Loss of water from the initial stage to the
state of shrinkage limit = $(V_i - V_m)\gamma_w$
∴ Weight of water at shrinkage limit = $(W_i - W_d) - (V_i - V_m)\gamma_w$

∴ Shrinkage limit, $w_s = \left[\dfrac{(W_i - W_d) - (V_i - V_m)\gamma_w}{W_d} \right] \times 100\%$

...(Eq. 3.47)

or
$$w_s = \left[w_i - \frac{(V_i - V_d)\gamma_w}{W_d} \right] \times 100\% \qquad ...(Eq.\ 3.48)$$

where w_i = initial water content

V_i = initial volume of the soil pat

$V_d = V_m$ = dry volume of the soil pat

W_d = dry weight of the soil sample

This equation suggests a laboratory method for the determination of the shrinkage limit.

The equipment or apparatus consists of a porcelain evaporating dish of about 12 cm diameter with flat bottom, a shrinkage dish of stainless steel with flat bottom,

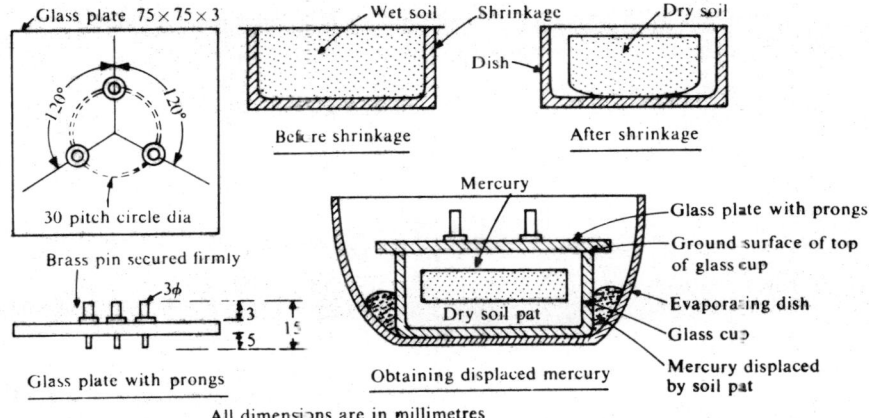

Fig. 3.21 Apparatus for determining volume-change in the shrinkage limit test

45 mm in diameter and 15 mm high, two glass plates, each 75 mm × 75 mm, 3 mm thick—one plain glass, and the other with three metal prongs, glass cup 50 mm in diameter and 25 mm high, with its top rim ground smooth and level, straight edge, spatula, oven, mercury desiccator, balances and sieves. (Fig. 3.21) (IS: 2720, Part VI—1972).

If the test is to be made on an undisturbed sample, it is to be trimmed to a pat about 45 mm in diameter and 15 mm in height. If it is to be conducted on a remoulded sample, about 100g of a thoroughly mixed portion of the material passing 425—μ – IS sieve is to be used.

The procedure is somewhat as follows:

The volume of the shrinkage dish is first determined by filling it with mercury, removing the excess by pressing a flat glass plate over the top and then weighing the dish filled with mercury. The weight of the mercury divided by its unit weight (13.6 g/cm³) gives the volume of the dish which is also the initial volume of the wet soil pat (V_i). The inside of the dish is coated with a thin layer of vaseline. The dish is then filled with the prepared soil paste in instalments. Gentle tapping is given to the hard surface to eliminate entrapped air. The excess soil is removed with the aid of a straight edge and any soil adhering to the outside of the dish is wiped off. The weight of the wet soil pat of known volume is found (W_i). The dish is then placed in an oven and the soil pat is allowed to dry up. The weight of the dry soil pat can be found by weighing (W_d).

The glass cup is filled with mercury and excess is removed by pressing the glass plate with three prongs firmly over the top. The dry soil pat is placed on the surface of the mercury in the cup and carefully pressed by means of the glass plate with prongs. The weight of the displaced mercury is found and divided by its unit weight to get the volume of the dry soil pat (V_d). The shrinkage limit may then be obtained by Eq. 3.47 or 3.48.

Alternative approach: Shrinkage limit may also be determined by an alternative approach if the specific gravity of the soil solids has already been determined. From Fig. 3.20 (*iii*),

$$w_s = \frac{(V_d - V_S)\gamma_w}{W_d} \times 100$$

$$= \frac{\left(V_d - \frac{W_d}{\gamma_s}\right)\gamma_w}{W_d} \times 100$$

$$\therefore \quad w_s = \left(\frac{V_d \gamma_w}{W_d} - 1/G\right) \times 100 \qquad \ldots\text{(Eq. 3.49)}$$

This may also be written as

$$w_s = [(\gamma_w / \gamma_d) - 1/G] \times 100 \qquad \ldots\text{(Eq. 3.50)}$$

where γ_d = dry unit weight based on the minimum or dry volume.

Substituting $\gamma_d = \frac{G \cdot \gamma_w}{(1+e)}$ in Eq. 3.46,

$$w_s = \frac{(1+e)}{G} - \frac{1}{G} = \frac{e}{G} \qquad \ldots\text{(Eq. 3.51)}$$

where e is the void ratio of the soil at its minimum volume. (This also indicates that the soil is still saturated at its minimum volume). Initial wet weight and initial wet volume are not required in this approach.

Another alternative approach is to determine the weight and volume of the soil pat at a series of decreasing moisture contents using air-drying (and ultimately oven-drying to obtain the values in the dry state) and plotting the volume observations against the moisture contents as follows (Fig. 3.22):

The straight line portion of the curve is produced to meet the horizontal through the point representing the minimum or dry volume. The water content corresponding to this meeting point is the shrinkage limit, w_s. However, this approach is too laborious.

Index Properties and Classification Tests 79

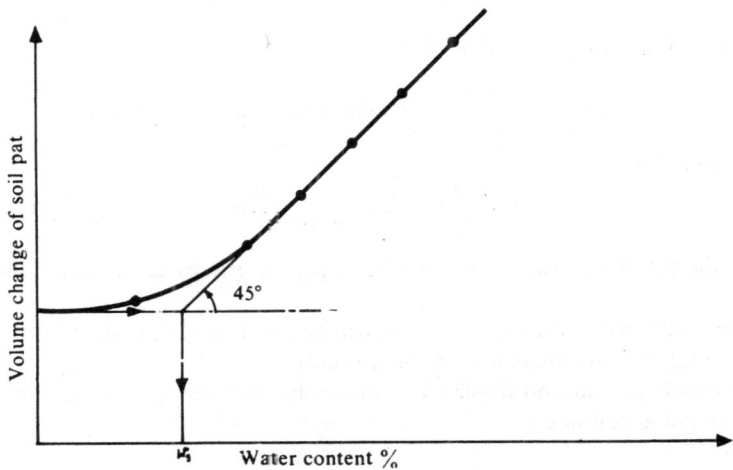

Fig. 3.22 Water content vs volume of soil pat

Approximate value of G from Shrinkage limit

The approximate value of G may be got from this test as follows:

$$\gamma_s = G \cdot \gamma_w = \frac{W_s}{V_s} = \frac{W_d}{V_s}$$

or

$$G = \frac{W_d}{V_s \gamma_w}.$$

From Fig. 3.20 (i),

$$V_s = V_i - \frac{(W_i - W_d)}{\gamma_w}$$

$$\therefore \quad G = \frac{W_d}{V_i \gamma_w - (W_i - W_d)} \qquad \ldots \text{(Eq. 3.52)}$$

Also from Eq. 3.47, if the shrinkage limit is already determined by the first approach,

$$G = \frac{1}{\left(\frac{\gamma_w}{\gamma_d} - \frac{w_s}{100}\right)} \qquad \ldots \text{(Eq. 3.53)}$$

w_s being in per cent.

Shrinkage Ratio (R)

'Shrinkage ratio' (R) is defined as the ratio of the volume change expressed as per cent of the dry volume to the corresponding change in moisture content from the initial value to the shrinkage limit:

$$R = \frac{(V_i - V_d)}{V_d} \times 100/(w_i - w_s) \qquad \ldots(\text{Eq. 3.54})$$

w_i and w_s being expressed as percentage.

$$(w_i - w_s) = \frac{(V_i - V_d)\gamma_w}{W_d} \times 100 \quad \text{from Fig. 3.20 (ii) and (iii).}$$

Substituting in Eq. 3.54,

$$R = \frac{W_d}{V_d \gamma_w} = \frac{\gamma_d}{\gamma_w} = \gamma_{m\,(dry)} = \frac{W_d}{V_d} \qquad \ldots(\text{Eq. 3.55})$$

Thus, the shrinkage ratio is also the mass specific gravity of the soil in the dry state.

The test data from shrinkage limit test can be substituted directly either in Eq. 3.54 or in Eq. 3.55 to obtain the shrinkage ratio.

If the shrinkage limit and shrinkage ratio are obtained, the approximate value of G may be got as follows:

$$G = \frac{1}{[(1/R) - w_s/100]} \qquad \ldots(\text{Eq. 3.56})$$

w_s being in per cent.

(**Note:** This relationship is easily derived by recasting Eq. 3.46 and recognising that $R = \gamma_d/\gamma_w$).

Volumetric Shrinkage (V_s)

The 'Volumetric Shrinkage' (or Volumetric change V_s) is defined as the decrease in the volume of a soil mass, expressed as a percentage of the dry volume of the soil mass, when the water content is reduced from an initial value to the shrinkage limit:

$$V_s = \frac{(V_i - V_d)}{V_d} \times 100 \qquad \ldots(\text{Eq. 3.57})$$

Obviously, the numerator of Eq. 3.54 is V_s.

$$\therefore \quad R = \frac{V_s}{(w_i - w_s)}$$

or
$$V_s = R(w_i - w_s) \qquad \ldots(\text{Eq. 3.58})$$

Degree of Shrinkage (S_r)

'Degree of Shrinkage' (S_r) is expressed as the ratio of the difference between initial volume and final volume of the soil sample to its initial volume.

$$S_r = \frac{(V_i - V_d)}{V_i} \times 100 \qquad \ldots(\text{Eq. 3.59})$$

The only difference between this and the volumetric shrinkage is in the denominator, which is the initial volume in this case.

Schedig classifies the soil qualitatively based on its degree of shrinkage as follows:

Good soil	...	$S_r < 5\%$
Medium soil	...	S_r — 5 to 10%
Poor soil	...	S_r — 10 to 15%
Very poor soil	...	$S_r > 15\%$

Linear Shrinkage (L_s)

'Linear Shrinkage (L_s)' is defined as the decrease in one dimension of the soil mass expressed as a percentage of the initial dimension, when the water content is reduced from a given value to the shrinkage limit. This is obtained as follows:

$$L_s = \left(1 - \sqrt[3]{\frac{100}{V_s + 100}}\right) \times 100 \qquad \ldots(\text{Eq. 3.60})$$

Shrinkage Limit of Undisturbed Soil (w_{su})

The shrinkage limit of undisturbed soil specimen is obtained as follows:

$$w_{su} = \left(\frac{V_{du}}{W_{du}} - \frac{1}{G}\right) \times 100 \qquad \ldots(\text{Eq. 3.61})$$

where V_{du} and W_{du} are the volume in ml and weight in g, respectively, of the oven-dry soil specimen.

3.10 ACTIVITY OF CLAYS

The presence of even small amounts of certain clay minerals can have significant effect on the properties of the soil. The identification of clay minerals requires special techniques and equipment. The techniques include microscopic examination, X-ray diffraction, differential thermal analysis, optical property determination and electron micrography. Even qualitative identification of the various clay minerals is adequate for many engineering purposes. Detailed treatment of clay minerals is considered out of scope of the present text.

An indirect method of obtaining information on the type and effect of clay mineral in a soil is to relate plasticity to the quantity of clay-size particles. It is known that for a given amount of clay mineral the plasticity resulting in a soil will vary for the different types of clays.

'Activity (A)' is defined as the ratio of plasticity index to the percentage of clay-sizes:

$$A = \frac{I_p}{c} \qquad \ldots(\text{Eq. 3.62})$$

where c is the percentage of clay sizes, i.e., of particles of size less than 0.002 mm.

82 *Geotechnical Engineering*

Activity can be determined from the results of the standard laboratory tests such as the wet analysis, liquid limit and plastic limit. Clays containing kaolinite will have relatively low activity and those containing montmorillonite will have high activity.

A qualitative classification based on activity is given in Table 3.6:

Table 3.6 Activity classification

Activity	Classification
Less than 0.75	Inactive
0.75 to 1.25	Normal
Greater than 1.25	Active

3.11 UNCONFINED COMPRESSION STRENGTH AND SENSITIVITY OF CLAYS

The unconfined compression strength of a clay soil is obtained by subjecting an unsupported cylindrical clay sample to axial compressive load, and conducting the test until the sample fails in shear. The compressive stress at failure, giving due allowance to the reduction in area of cross-section, is termed the 'unconfined compression strength' (q_u). In the field, a vane-shear device or a pocket penetrometer may be used for quick and easy determination of strength values, which may be related to qualitative terms indicating consistency. (These and other aspects of strength will be studied in greater detail in Chapter 8).

It has been established that the strength of a clay soil is related to its structure. If the original structure is altered by reworking or remoulding or chemical changes, resulting in changes in the orientation and arrangement of the particles, the strength of the clay gets decreased, even without alteration in the water content. (It is known that the strength of a remoulded clay soil is affected by water content.)

'Sensitivity (S_t)' of a clay is defined as the ratio of its unconfined compression strength in the natural or undisturbed state to that in the remoulded state, without any change in the water content:

$$S_t = \frac{q_u \text{ (undisturbed)}}{q_u \text{ (remoulded)}} \qquad \text{...(Eq. 3.63)}$$

The classification of clays based on sensitivity, in a qualitative manner, is given in Table 3.7:

Table 3.7 Sensitivity classification

Sensitivity	Classification	Remarks
2 to 4	Normal or less sensitive	Honeycomb structure
4 to 8	Sensitive	Honeycomb or flocculent structure
8 to 16	Extra-sensitive	Flocculent structure
> 16	Quick	Unstable

* 3.12 THIXOTROPY OF CLAYS

When clays with a flocculent structure are used in construction, these may lose some strength as a result of remoulding. With passage of time, however, the strength increases, though not back to the original value. This phenomenon of strength loss-strength gain, with no change in volume or water content, is called 'Thixotropy'. This may also be said to be "a process of softening caused by remoulding, followed by a time-dependent return to the original harder state".

The loss of strength on remoulding is partly due to the permanent destruction of the structure in the *in-situ* condition, and partly due to the reorientation of the molecules in the adsorbed layers. The gain in strength is due to the rehabilitation of the molecular structure of the soil. The strength loss due to destruction of structure cannot be recouped with time.

'Thixis' means the tough, the shaking, and *'tropo'* means to turn, to change. Thus, thixotropy means "to change by touch"; it may also be defined, basically, as a reversible gel-sol-gel transformation in certain colloidal systems brought about by a mechanical disturbance followed by a period of rest.

The loss in strength on remoulding and the extent of strength gain over a period of time are dependent on the type of clay minerals involved; generally, the clay minerals that absorb large quantities of water into their lattice structures, such as montmorillonites, experience greater thixotropic effects than other more stable clay minerals.

For certain construction situations, thixotropy is considered a beneficial phenomenon, since with passage of time, the earth structure gets harder and presumably safer. However, it has its problems—handling of materials and equipments may pose difficulties. Thixotropic influences have affected piles, a type of foundation construction, driven in soils. The disturbance may cause temporary loss in strength of the surrounding soil. Driving must be fully done before thixotropic recovery becomes pronounced. Thixotropic fluids used in drilling operations are called 'drilling muds'.

3.13 ILLUSTRATIVE EXAMPLES

Example 3.1: In a specific gravity test with pyknometer, the following observed readings are available:
 Weight of the empty pyknometer = 750 g
 Weight of pyknometer + dry soil = 1,730 g
 Weight of pyknometer + dry soil + water filling the remaining volume = 2,245 g
 Weight of pyknometer + water = 1,630 g
Determine the specific gravity of the soil solids, ignoring the effect of temperature.

The given weights are designated W_1 to W_4 respectively.
Then,
 the weight of dry soil solids, $W_s = W_2 - W_1$
$$= (1730 - 750) \text{ g} = 980 \text{ g}$$

84 *Geotechnical Engineering*

The specific gravity of soil solids is given by Eq. 3.1:

$$G = \frac{W_s}{W_s - (W_3 - W_4)} \text{ (ignoring the effect of temperature)}$$

$$= \frac{980}{980 - (2245 - 1630)}$$

$$= \frac{980}{(980 - 615)}$$

$$= 980/365 = 2.685$$

∴ Specific gravity of the soil solids = **2.685**.

Example 3.2: In a specific gravity test, the weight of the dry soil taken is 66 g. The weight of the pyknometer filled with this soil and water is 675.60 g. The weight of the pyknometer full of water is 633.95 g. The temperature of the test is 30°C. Determine the grain specific gravity, taking the specific gravity of water at 30°C as 0.99568.

Applying the necessary temperature correction, report the value of G which would be obtained if the test were conducted at 4°C and also at 27°C. The specific gravity values of water at 4°C and 27°C are respectively 1 and 0.99654.

Weight of dry soil taken, W_s = 66 g

Weight of pyknometer + soil + water (W_3) = 675.60 g

Weight of pyknometer + Water (W_4) = 633.95 g

Temperature of the Test (T) = 30°C

Specific gravity of water at 30°C (G_{w_T}) = 0.99568

By Eq. 3.4,

$$G = \frac{W_s \cdot G_{w_T}}{W_s - (W_3 - W_4)}$$

$$= \frac{66 \times 0.99568}{66 - (675.60 - 633.95)} = 2.69876$$

$$\approx 2.70$$

If the tests were conducted at 4°C, G_{w_T} = 1

$$\therefore \quad G = \frac{W_s \cdot 1}{W_s - (W_3 - W_4)} = \frac{66 \times 1}{66 - (675.60 - 633.95)} = \mathbf{2.71}$$

If the tests were conducted at 27°C, G_{w_T} = 0.99654

$$\therefore \quad G = \frac{W_s \times 0.99654}{W_s - (W_3 - W_4)} = \frac{66 \times 0.99654}{66 - (675.60 - 633.95)} = 2.7011$$

$$\approx 2.70$$

Example 3.3: In a specific gravity test in which a density bottle and kerosene were used, the following observations were made:
 Weight of empty density bottle = 60.25 g
 Weight of bottle + clay sample = 81.60 g
 Weight of bottle + clay + kerosene filling the remaining volume = 257.34 g
 Weight of bottle + kerosene = 242.17 g
 Temperature of the test = 27°C
 Specific gravity of kerosene at 27°C = 0.773
Determine the specific gravity of the soil solids.
What will be the value if it has to be reported at 4°C?
Assume the specific gravity of water at 27°C as 0.99654.

Let the weights be designated as W_1 through W_4 in that order.

Wt of dry clay sample, $W_s = (W_2 - W_1) = (81.60 - 60.25)$ g $= 21.35$ g

By Eq. 3.2,

$$G = \frac{W_s \cdot G_k}{W_s - (W_3 - W_4)}$$

G_k here is given as 0.773.

$$\therefore \quad G = \frac{21.35 \times 0.773}{21.35 - (257.34 - 242.17)} \approx 2.67$$

If the value has to be reported at 4°C, by Eq. 3.3,

$$G_{T_2} = G_{T_1} \cdot \frac{G_{wT_2}}{G_{wT_1}}$$

$$\therefore \quad G_{4°} = G_{27°} \cdot \frac{1}{0.99654} = \frac{2.67 \times 1}{0.99654} = 2.68$$

Example 3.4: In a specific gravity test, the following observations were made:
 Weight of dry soil : 104.0 g
 Weight of bottle + soil + water : 538.0 g
 Weight of bottle + water : 475.6 g
What is the specific gravity of soil solids. If, while obtaining the weight 538.0 g, 3 ml of air remained entrapped in the suspension, will the computed value of G be higher or lower than the correct value? Determine also the percentage error. Neglect temperature effects.

Neglecting temperature effects, by Eq. 3.1,

$$G = \frac{W_s}{W_s - (W_3 - W_4)}.$$

In this case, $W_s = 104.0$ g; $W_3 = 538.0$ g; $W_4 = 475.6$ g

$$\therefore G = \frac{104}{104.0 - (538.0 - 475.6)} = 104.0/41.6 = \mathbf{2.50}$$

If some air is entrapped while the weight W_3 is taken, the observed value of W_3 will be lower than if water occupied this air space. Since W_3 occurs with a negative sign in Eq. 3.1 in the denominator, the computed value of G would be *lower* than the correct value.

Since the air entrapped is given as 3 ml, this space, if occupied by water, would have enhanced the weight W_3 by 3 g.

$$\therefore \text{Correct value of } G = \frac{104}{104.0 - (541.0 - 475.6)} = 104.0/38.6$$

$$= \mathbf{2.694}$$

Percentage error $= \dfrac{(2.694 - 2.500)}{2.694} \times 100 = \mathbf{7.2}$

Example 3.5: A pyknometer was used to determine the water content of a sandy soil. The following observations were obtained:
  Weight of empty pyknometer = 800 g
  Weight of pyknometer + wet soil sample = 1160 g
  Weight of pyknometer + wet soil + water filling the remaining volume: 2000 g
  Weight of pyknometer + water = 1800 g
  Specific gravity of soil solids (determined by a separate test) = 2.66
Compute the water content of the soil sample.

The weights may be designated W_1 through W_4 in that order.
By Eq. 3.6,

$$w = \left[\frac{(W_2 - W_1)}{(W_3 - W_4)} \left(\frac{G-1}{G}\right) - 1\right] \times 100$$

Substituting the values,

$$w = \left[\frac{(1160 - 800)}{(2000 - 1800)} \times \frac{(2.66 - 1)}{2.66} - 1\right] \times 100$$

$$= \left[\frac{360}{200} \times \frac{1.66}{2.66} - 1\right] \times 100$$

$$= (1.1233 - 1) \times 100$$

$$= 12.33$$

$\therefore$ Water content of the soil sample = **12.33%**

Example 3.6: A soil sample with a grain specific gravity of 2.67 was filled in a 1000 ml container in the loosest possible state and the dry weight of the sample was found to be 1475 g. It was then filled at the densest state obtainable and the weight was found to be 1770 g. The void ratio of the soil in the natural state was 0.63. Determine the density index in the natural state.

$G = 2.67$
Loosest state:
 Weight of soil = 1475 g
 Volume of solids = $1475/2.67$ cm^3 = 552.4 cm^3
 Volume of voids = $(1000 - 552.4)$ cm^3 = 447.6 cm^3
 Void ratio, $e_{max} = 447.6/552.4 = 0.810$

Densest state:
 Weight of soil = 1770 g
 Volume of solids = $1770/2.67$ cm^3 = 662.9 cm^3
 Volume of voids = $(1000 - 662.9)$ cm^3 = 337.1 cm^3
 Void ratio, $e_{min} = \dfrac{337.1}{662.9} = 0.508$

Void ratio in the natural state, $e = 0.63$

Density Index, $I_D = \dfrac{(e_{max} - e)}{(e_{max} - e_{min})}$, by (Eq. 3.8)

$$= \dfrac{0.810 - 0.630}{0.810 - 0.508} = \dfrac{0.180}{0.302} = 0.596$$

$$\therefore I_D = 59.6\%$$

Example 3.7: The dry unit weight of a sand sample in the loosest state is 13.34 kN/m^3 and in the densest state, it is 21.19 kN/m^3. Determine the density index of this sand when it has a porosity of 33%. Assume the grain specific gravity as 2.68.

γ_{min} (loosest state) = 13.34 kN/m^3
γ_{max} (densest state) = 21.19 kN/m^3
Porosity, $n = 33\%$

Void ratio, $e_0 = \dfrac{n}{(1-n)} = 33/67 = 0.49$

$$\gamma_0 = \dfrac{G \cdot \gamma_w}{(1+e_0)} = \dfrac{2.68 \times 9.81}{(1+0.49)} \text{ kN/m}^3 = 17.64 \text{ kN/m}^3$$

Density Index, I_D (by Eq. 3.10)

$$= \dfrac{(\gamma_{max})}{(\gamma_0)} \left(\dfrac{\gamma_0 - \gamma_{min}}{\gamma_{max} - \gamma_{min}} \right)$$

88 Geotechnical Engineering

$$= \frac{21.19}{17.64} \times \frac{(17.64-13.34)}{(21.19-13.34)} = \frac{21.19}{17.64} \times \frac{4.30}{7.85} = 0.658$$

$$= 65.8\%$$

Alternatively:

$$\gamma_{min} = \frac{G \cdot \gamma_w}{(1+e_{max})} \text{ or } 13.34 = \frac{2.68 \times 9.81}{(1+e_{max})}$$

$$\therefore \quad e_{max} = 0.971$$

$$\gamma_{max} = \frac{G \cdot \gamma_w}{(1+e_{min})} \text{ or } 21.19 = \frac{2.68 \times 9.81}{(1+e_{min})}$$

$$\therefore \quad e_{min} = 0.241$$

$$\therefore \quad I_D = \frac{(e_{max}-e_0)}{(e_{max}-e_{min})}, \text{ (by Eq. 3.8)}$$

$$= \frac{(0.971-0.49)}{(0.971-0.241)} = \frac{0.48}{0.73} = 65.8\%$$

Example 3.8: The following data were obtained during an *in-situ* unit wt determination of an embankment by the sand-replacement method:

Volume of calibrating can	= 1000 ml
Weight of empty can	= 900 g
Weight of can + sand	= 2500 g
Weight of sand filling the conical portion of the sand-pouring cylinder	= 450 g
Initial weight of sand-pouring cylinder + sand	= 5400 g
Weight of cylinder + sand, after filling the excavated hole	= 4140 g
Wet weight of excavated soil	= 936 g
In-situ water content	= 9%

Determine the *in-situ* unit weight and *in-situ* dry unit weight.

Sand-replacement method of *in-situ* unit weight determination:

Weight of sand filling the calibrating can of volume 1000 ml	= (2500 – 900) g = 1600 g
Unit weight of sand	= 1600/1000 g/cm³ = 1.60 g/cm³
Weight of sand filling the excavated hole and conical portion of the sand pouring cylinder	= (5400 – 4140) g = 1260 g
Weight of sand filling the excavated hole	= (1260 – 450) g = 810 g
Volume of the excavated hole	= 810/1.60 cm³ = 506.25 cm³
Weight of excavated soil	= 936 g

Index Properties and Classification Tests 89

In-situ unit weight, γ = 936.00/506.25 g/cm³
= 1.85 gm/cm³
= 18.15 kN/m³
Water content, w = 9%

In-situ dry unit weight, $\gamma_d = \dfrac{\gamma}{(1+w)}$ = $\dfrac{1.85}{(1+0.09)}$ g/cm³
= 16.67 kN/m³

Example 3.9: A field density test was conducted by core-cutter method and the following data was obtained:
Weight of empty core-cutter = 2280 g
Weight of soil and core-cutter = 5005 g
Inside diameter of the core-cutter = 90.0 mm
Height of core-cutter = 180.0 mm
Weight of wet sample for moisture determination = 54.05 g
Weight of oven-dry sample = 51.12 g
Specific gravity of soil grains = 2.72
Determine (a) dry density, (b) void-ratio, and (c) degree of saturation.
(S.V.U.—B.E.(Part-time)—FE—April, 1982)

Weight of soil in the core-cutter (W) = (5005 – 2280) g = 2725 g
Volume of core-cutter (V) = $(\pi/4) \times 9^2 \times 18$ cm³ = 1145.11 cm³

Wet unit weight of soil (γ) = $W/V = \dfrac{2725.00}{1145.11}$ g/cm³ = 23.34 kN/m³

Weight of oven-dry sample = 51.12 g
Weight of moisture = (54.05 – 51.12) g = 2.93 g

Moisture content, w = $\dfrac{2.93}{51.12} \times 100\% = 5.73\%$

Dry unit weight, γ_d = $\dfrac{\gamma}{(1+w)} = \dfrac{23.34}{(1+0.0573)}$ kN/m³
= 22.075 kN/m³

Grain specific gravity, G = 2.72

$\gamma_d = \dfrac{G \cdot \gamma_w}{(1+e)}$ or $22.075 = \dfrac{2.72 \times 9.81}{(1+e)}$

whence, the void-ratio, e ≈ 0.21

Degree of saturation, $S = wG/e$ = $\dfrac{0.0573 \times 2.72}{0.21} = 74.6\%$

Example 3.10: A soil sample consists of particles ranging in size from 0.6 mm to 0.02 mm. The average specific gravity of the particles is 2.66. Determine the time of settlement of the coarsest and finest of these particles through a depth of 1 metre. Assume the viscosity of water as 0.001 N-sec/m² and the unit weight as 9.8 kN/m³.

By Stokes' law (Eq. 3.19),
$$v = (1/180) \cdot (\gamma_w/\mu_w)(G-1)D^2$$

where v = terminal velocity in cm/sec,
γ_w = unit weight of water in kN/m³,
μ_w = viscosity of water in N-sec/m²,
G = grain specific gravity, and
D = size of particle in mm.

$$\therefore \quad v = \frac{1}{180} \times \frac{9.80}{0.001} (2.66-1) D^2 \approx 90 D^2$$

For the coarsest particle, $D = 0.6$ mm

$$v = 90 \times (0.6)^2 \text{ cm/sec} = 32.4 \text{ cm/sec.}$$

$$t = h/v = 100.0/32.4 \text{ sec} = 3.086 \text{ sec.}$$

For the finest particle, $D = 0.02$ min.

$$v = 90 (0.02)^2 \text{ cm/sec} = 0.036 \text{ cm/sec.}$$

$$t = h/v = \frac{100.000}{0.036} \text{ sec} = 2777.78 \text{ sec} = 46 \text{ min. } 17.78 \text{ sec.}$$

This time of settlement of the coarsest and finest particles through one metre are nearly **4 sec.** and **46 min. 18 sec.** respectively.

Example 3.11: In a pipette analysis, 50 g of dry soil (fine fraction) of specific gravity 2.72 were mixed in water to form half a litre of uniform suspension. A pipette of 10 ml capacity was used to obtain a sample from a depth of 10 cm after 10 minutes from the start of sedimentation. The weight of solids in the pipette sample was 0.32 g. Assuming the unit weight of water and viscosity of water at the temperature of the test as 9.8 kN/m³ and 0.001 N-sec/m² respectively, determine the largest size of the particles remaining at the sampling depth and percentage of particles finer than this size in the fine soil fraction taken. If the percentage of fine fraction in the original soil was 50, what is the percentage of particles finer computed above in the entire soil sample?

By Eqs. 3.21 and 3.22,
$$D = K\sqrt{H/t}$$

Index Properties and Classification Tests 91

where
$$K = \sqrt{\frac{3\mu_w}{\gamma_w(G-1)}}$$

$$\therefore \quad K = \sqrt{\frac{3 \times 0.001}{9.8 \times (2.72-1)}} = 0.0133 \text{ mm } \frac{\sqrt{\min}}{\sqrt{\text{cm}}}$$

$$D = 0.0133\sqrt{\frac{10}{10}} \text{ mm} = 0.0133 \text{ mm}$$

This is the largest size remaining at the sampling depth.

By Eq. 3.26,

$$N_f = \left(\frac{W_p}{W_s}\right)\left(\frac{V}{V_p}\right) \times 100$$

Here $W_p = 0.32$ g; $W_s = 50$ g; $V = 500$ ml; $V_p = 10$ ml.

$$\therefore \quad N_f = \frac{0.32}{50} \times \frac{500}{10} \times 100 = 32\%$$

Thus, the percentage of particles finer than 0.0133 mm is 32.

By Eq. 3.28, $N = N_f(W_f/W)$

Here W_f/W is given as 0.50.

$$\therefore \quad N = 32 \times 0.50 = 16$$

Hence, this percentage is 16, based on the entire sample of soil.

Example 3.12: In a hydrometer analysis, the corrected hydrometer reading in a 1000 ml uniform soil suspension at the start of sedimentation was 28. After a lapse of 30 minutes, the corrected hydrometer reading was 12 and the corresponding effective depth 10.5 cm. The specific gravity of the solids was 2.70. Assuming the viscosity and unit weight of water at the temperature of the test as 0.001 N-s/m² and 9.8 kN/m³ respectively, determine the weight of solids mixed in the suspension, the effective diameter corresponding to the 30-min. reading and the percentage of particles finer than this size.

The corrected hydrometer reading initially, $R_{h_i} = 28$

$$\therefore \quad \gamma_i = 1.028 \text{ g/cm}^3$$

But, by Eq. 3.29,

$$\gamma_i = \gamma_w + [(G-1)/G] \cdot W/V$$

92 Geotechnical Engineering

Substituting,

$$1.028 = 1 + \frac{(2.7-1)}{2.7} \times \frac{W}{1000}$$

Whence $W = \dfrac{28 \times 2.7}{1.7}$ g $= 44.5$ g

∴ The weight of solid mixed in the suspension = **44.5 g**.

By Eqs 3.21 and 3.22,

$$D = K\sqrt{H/t}$$

where $K = \sqrt{\dfrac{3 \mu_w}{\gamma_w (G-1)}}$

∴ $K = \sqrt{\dfrac{3 \times 0.001}{9.8 \times (2.7-1)}} = 0.01342$ mm $\dfrac{\sqrt{\min}}{\sqrt{\text{cm}}}$.

$D = 0.01342 \sqrt{\dfrac{10.5}{30}}$ mm $= 0.00794$ mm ≈ 0.008 mm

∴ The effective diameter corresponding to the 30-min. reading = **0.008 mm**

By Eq. 3.34,

$$N = \frac{G \cdot \gamma_w}{(G-1)} \cdot \frac{V}{W} \cdot \frac{R_h}{10}$$

∴ $N = \dfrac{2.7}{1.7} \times 1 \times \dfrac{1000}{44.5} \times \dfrac{12}{10} = 42.83$

∴ The percentage of particles finer than 0.008 mm is **43**.

Example 3.13: The liquid limit of a clay soil is 56% and its plasticity index is 15%. (*a*) In what state of consistency is this material at a water content of 45%? (*b*) What is the plastic limit of the soil? (*c*) The void ratio of this soil at the minimum volume reached on shrinkage, is 0.88. What is the shrinkage limit, if its grain specific gravity is 2.71?

Liquid limit, $W_L = 56\%$
Plasticity index, $I_p = 15\%$
$I_p = w_L - w_p$, by Eq. 3.37.
∴ $15 = 56 - w_p$
Whence the plastic limit, $w_p = (56 - 15) = \mathbf{41\%}$

∴ At a water content of 45%, the soil is in the *plastic state* of consistency.
Void ratio at minimum volume, $e = 0.88$
Grain specific gravity, $G = 2.71$
Since at shrinkage limit, the volume is minimum and the soil is still saturated,

Index Properties and Classification Tests 93

or
$$e = w_s G$$
$$w_s = e/G = 0.88/2.71 = 32.5\%$$

∴ Shrinkage limit of the soil = 32.5%

Example 3.14: A soil has a plastic limit of 25% and a plasticity index of 30. If the natural water content of the soil is 34%, what is the liquidity index and what is the consistency index? How do you describe the consistency?

Plastic limit, $w_p = 25\%$

Plasticity index, $I_p = 30$

By Eq. 3.37, $I_p = w_L - w_p$

∴ Liquid limit, $w_L = I_p + w_p = 30 + 25 = 55\%$.

By Eq. 3.40,
$$\text{Liquidity index, } I_L = \frac{(w - w_p)}{I_p}$$

where w is the natural moisture content.

∴ Liquidity index, $I_L = \frac{(34 - 25)}{30} = 0.30$

By Eq. 3.39,
$$\text{Consistency index, } I_c = \frac{(w_L - w)}{I_p}$$

∴ Consistency index, $I_c = \frac{(55 - 34)}{30} = 0.70$

The consistency of the soil may be described as 'medium soft' or 'medium stiff'.

Example 3.15: A fine grained soil is found to have a liquid limit of 90% and a plasticity index of 50. The natural water content is 28%. Determine the liquidity index and indicate the probable consistency of the natural soil.

Liquid limit, $w_L = 90\%$

Plasticity index, $I_p = 50$

By Eq. 3.37, $I_p = w_L - w_p$

∴ Plastic limit, $w_p = w_L - I_p = 90 - 50 = 40\%$

The natural water content, $w = 28\%$.

Liquidity index, I_L, by Eq. 3.40, is given by
$$I_L = \frac{w - w_p}{I_p}$$
$$= \frac{28 - 40}{50} = -\frac{12}{50} = -0.24 \text{ (negative)}$$

Since the liquidity index is negative, the soil is in the semi-solid state of consistency and is stiff; this fact can be inferred directly from the observation that the natural moisture content is less than the plastic limit of the soil.

Example 3.16: A clay sample has void ratio of 0.50 in the dry condition. The grain specific gravity has been determined as 2.72. What will be the shrinkage limit of this clay?

The void ratio in the dry condition also will be the void ratio of the soil even at the shrinkage limit: but the soil has to be saturated at this limit.
For a saturated soil,
$$e = wG$$
or
$$w = e/G$$
$$\therefore \quad w_s = e/G = 0.50/2.72 = 18.4\%,$$
Hence the shrinkage limit for this soil is **18.4%**.

Example 3.17: The following are the data obtained in a shrinkage limit test:
- Initial weight of saturated soil = 95.6 g
- Initial volume of the saturated soil = 68.5 cm³
- Final dry volume = 24.1 cm³
- Final dry weight = 43.5 g

Determine the shrinkage limit, the specific gravity of grains, the initial and final dry unit weight, bulk unit weight, and void ratio.
(S.V.U.—B.Tech. (Part-time)—Sept., 1982)

From the data,

Initial water content, $w_i = \dfrac{(95.6 - 43.5)}{43.5} \times 100 = 119.77\%$

By Eq. 3.48, the shrinkage limit is given by
$$w_s = \left[w_i - \dfrac{(V_i - V_d)}{W_d} \cdot \gamma_w\right] \times 100$$

$$= \left[1.1977 - \dfrac{(68.5 - 24.1)}{43.5} \times 1\right] \times 100 = \mathbf{17.70\%}$$

Final dry unit weight = $\dfrac{43.5}{24.1}$ g/cm³ = 1.805 g/cm³ = **17.71 kN/m³**

Initial bulk unit weight = $\dfrac{95.6}{68.5}$ g/cm³ = 1.396 g/cm³ = **13.70 kN/m³**

By Eq. 3.53,

Grain specific gravity = $\dfrac{1}{\left(\dfrac{\gamma_w}{\gamma_d} - \dfrac{w_s}{100}\right)} = \dfrac{1}{(1/1.805) - \left(\dfrac{17.70}{100}\right)}$

$$= \mathbf{2.65}$$

Initial dry unit weight $= \dfrac{\gamma_{bi}}{(1+w_i)} = \dfrac{1.396}{(1+1.1977)} = 0.635$ g/cm^3

$\hspace{7cm} = 6.23$ kN/m^3

Initial void ratio $= w_i \cdot G = 1.1977 \times 2.64 = \mathbf{3.16}$
Final void ratio $= w_s \cdot G = 0.1770 \times 2.64 = \mathbf{0.47}$

Example 3.18: The Atterberg limits of a clay soil are: Liquid limit = 75%; Plastic limit = 45%; and Shrinkage limit = 25%. If a sample of this soil has a volume of 30 cm^3 at the liquid limit and a volume 16.6 cm^3 at the shrinkage limit, determine the specific gravity of solids, shrinkage ratio, and volumetric shrinkage.

The phase diagrams at liquid limit, shrinkage limit, and in the dry state are shown in Fig. 3.23:

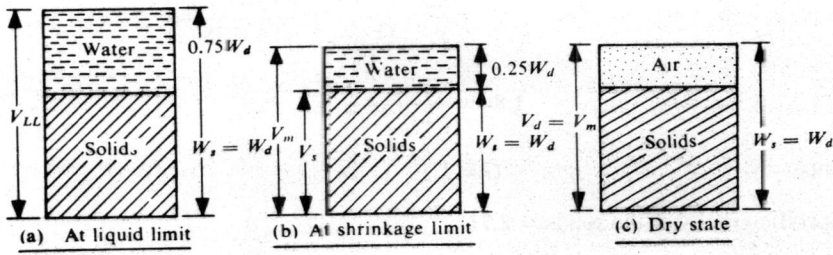

Fig. 3.23 Phase diagrams of the clay soil (Example 3.18)

Difference in the volume of water at LL and SL = $(30 - 16.6)$ cm^3 = 13.4 cm^3
Corresponding difference in weight of water = 13.4 g
But this is $(0.75 - 0.25) W_d$ or $0.50 W_d$ from Fig. 3.23.
$\hspace{4cm} \therefore \hspace{1cm} 0.50 W_d = 13.4$

$\hspace{7cm} W_d = 13.4/0.5 = 26.8$ g

Weight of water at SL = $0.25 W_d = 0.25 \times 26.8 = 6.7$ g
$\therefore$ Volume of water at SL = 6.7 cm^3
Volume of solids, V_s = Total volume at SL − Volume of water at SL.
$\hspace{4cm} = (16.6 - 6.7)$ cm^3 = 9.9 cm^3.
Weight of solids, $W_d = 26.8$ g

$\therefore$ Specific gravity of solids $= \dfrac{\gamma_s}{\gamma_w} = \dfrac{26.8}{9.9} = \mathbf{2.71}$

96 *Geotechnical Engineering*

Shrinkage ratio, $\quad R = \dfrac{W_d}{V_d} = \dfrac{26.8}{16.6} = 1.61$

Volumetric shrinkage, $V_s = R(w_i - w_s) = R(w_L - w_s)$, here

$$= 1.61\,(75 - 25) = 80.5\%$$

Example 3.19: The mass specific gravity of a saturated specimen of clay is 1.84 when the water content is 38%. On oven drying the mass specific gravity falls to 1.70. Determine the specific gravity of solids and shrinkage limit of the clay.

For a saturated soil,
$$e = w \cdot G$$
$$\therefore \quad e = 0.38\,G$$

Mass specific gravity in the saturated condition

$$= \dfrac{\gamma_{sat}}{\gamma_w} = \dfrac{(G+e)}{(1+e)} = \dfrac{(G+0.38\,G)}{(1+0.38\,G)}$$

$$\therefore \quad 1.84 = \dfrac{1.38\,G}{1+0.38\,G}$$

whence $\quad G = 2.71$

∴ Specific gravity of the solids $= 2.71$

By Eq. 3.50, the shrinkage limit is given by

$$w_s = \left(\dfrac{\gamma_w}{\gamma_d} - \dfrac{1}{G}\right) \times 100$$

where $\gamma_d =$ dry unit weight in dry state.

But, $\gamma_d =$ (mass specific gravity in the dry state) γ_w

$$= 1.70\,\gamma_w$$

$$\therefore \quad w_s = \left(\dfrac{\gamma_w}{1.70\,w} - \dfrac{1}{G}\right)100 = \left(\dfrac{1}{1.70} - \dfrac{1}{2.71}\right)100$$

$$= 21.9\%$$

∴ Shrinkage limit of the clay $= 21.9\%$

Example 3.20: A saturated soil sample has a volume of 23 cm^3 at liquid limit. The shrinkage limit and liquid limit are 18% and 45%, respectively. The specific gravity of solids is 2.73. Determine the minimum volume which can be attained by the soil.

The minimum volume which can be attained by the soil occurs at the shrinkage limit. The phase diagrams of the soil at shrinkage limit and at liquid limit are shown in Fig. 3.24:

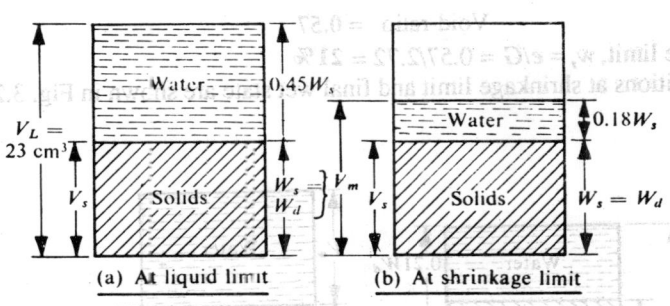

Fig. 3.24 Phase diagrams (Example 3.20)

At liquid limit,
 Volume of water = 0.45 W_s cm³, if W_s is the weight of solids in g.

$$\text{Volume of solids} = \frac{W_s}{G \gamma_w} = \frac{W_s}{2.73} \text{ cm}^3$$

$$\text{Total volume} = \frac{W_s}{2.73} + 0.45 W_s = 23$$

whence $W_s = 28.18$ g

At shrinkage limit,
 the volume, $V_m = V_s + 0.18 W_s$

$$= \frac{28.18}{2.73} + 0.18 \times 28.18 \text{ cm}^3$$

$$= 15.4 \text{ cm}^3$$

Example 3.21: An oven-dry soil sample of volume 225 cm³ weighs 390 g. If the grain specific gravity is 2.72, determine the void-ratio and shrinkage limit. What will be the water content which will fully saturate the sample and also cause an increase in volume equal to 8% of the original dry volume?

Dry unit weight of the oven-dry sample $= \frac{390}{225}$ g/cm³

$$= 1.733 \text{ g/cm}^3$$

But $\qquad \gamma_d = \dfrac{G \cdot \gamma_w}{(1+e)}$

98 *Geotechnical Engineering*

$$\therefore \quad 1.733 = \frac{2.72 \times 1}{(1+e)}$$

$$(1+e) = 2.720/1.733 = 1.569$$

$$\therefore \quad e \approx 0.57$$

$$\therefore \quad \text{Void-ratio} = 0.57$$

Shrinkage limit, $w_s = e/G = 0.57/2.72 = 21\%$

The conditions at shrinkage limit and final wet state are shown in Fig. 3.25:

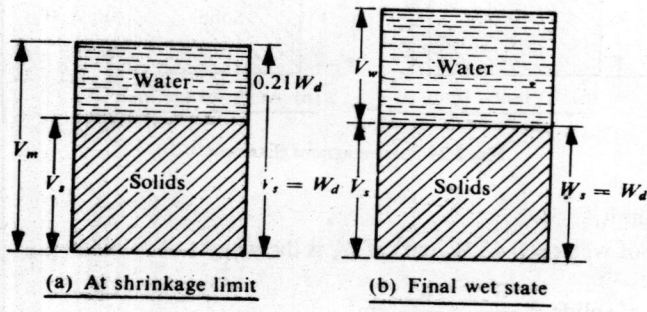

Fig. 3.25 Phase-diagrams of soil (Example 3.21)

$$V_s = \frac{W_d}{G} = 390/2.72 \text{ cm}^2 = 143.38 \text{ cm}^3$$

Volume in the final wet state, $V = (225 + 0.08 \times 225) = 243 \text{ cm}^3$

Volume of water in the final wet state, $V_w = (243 - 143.38) \text{ cm}^3 = 99.62 \text{ cm}^3$

Weight of water in the final wet state = 99.62 g

Water content in the final wet state = $\dfrac{99.62}{390} = 25.5\%$

Example 3.22: The plastic limit and liquid limit of a soil are 33% and 45% respectively. The percentage volume change from the liquid limit to the dry state is 36% of the dry volume. Similarly, the percentage volume change from the plastic limit to the dry state is 24% of the dry volume. Determine the shrinkage limit and shrinkage ratio.

The data are incorporated in Fig. 3.26 for making things clear:

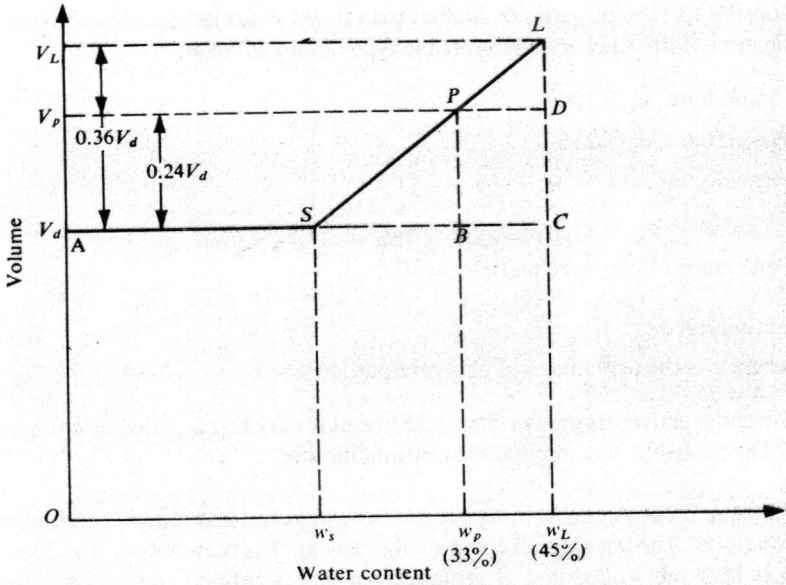

Fig. 3.26 Water content *vs* volume of soil (Example 3.22)

Say, V_d is the dry volume.
PB = $0.24\ V_d$
LC = $0.36\ V_d$
LD = $0.12\ V_d$
PD = 12%
From the triangles LPD and LSC, which are similar,
PD/SC = LD/LC

$$\frac{12}{SC} = \frac{0.12\ V_d}{0.36\ V_d} = 1/3$$

∴ SC = 36%
∴ Shrinkage limit, $w_s = w_L - (SC) = (45 - 36) = 9\%$

Shrinkage ratio $= \dfrac{(V_L - V_d)}{(V_d)}\ 100/(w_L - w_s)$

$$= \left(\frac{1.36\ V_d - V_d}{V_d}\right) 100/(45 - 9) = 36\%/36\% = \mathbf{1.00}$$

Example 3.23: The liquid limit and plastic limit of a clay are 100% and 25%, respectively. From a hydrometer analysis it has been found that the clay soil consists of 50% of particles smaller than 0.002 mm. Indicate the activity classification of this clay and the probable type of clay mineral.

Liquid limit, $w_L = 100\%$
Plastic limit, $w_p = 25\%$
Plasticity Index, $I_p = (w_L - w_p)$

$$= (100 - 25)\% = 75\%$$

Percentage of clay-size particles = 50

Activity, $A = \dfrac{I_p}{c}$ (Eq. 3.62)

where c is the percentage of clay-size particles.
∴ $A = 75/50 = 1.50$
Since the activity is greater than 1.25, the clay may be classified as being **active**.
The probable clay mineral is **montmorillonite**.

Example 3.24: A clay soil sample has been obtained and tested in its undisturbed condition. The unconfined compression strength has been obtained as 200 kN/m². It is later remoulded and again tested for its unconfined compression strength, which has been obtained as 40 kN/m². Classify the soil with regard to its sensitivity and indicate the possible structure of the soil.

Unconfined compression strength in the undisturbed state, $q_{uu} = 200$ kN/m²
Unconfined compression strength in the remoulded state, $q_{ur} = 40$ kN/m²

Sensitivity, $S_t = \dfrac{q_{uu}}{q_{ur}}$

$$= 200/40 = 5$$

Since the sensitivity falls between 4 and 8, the soil may be soil to be **"sensitive"**. The possible structure of the soil may be **'honeycombed'** or **'flocculent'**.

SUMMARY OF MAIN POINTS

1. Certain physical properties such as colour, structure, texture, particle shape, grain specific gravity, water content, *in-situ* unit weight, density index, particle size distribution, consistency limits and related indices are termed index properties, some of which are useful as classification tests.
2. Grain specific gravity or the specific gravity of soil solids is useful in the determination of many other quantitative characteristics of soil and is, as such, considered basic to the study of geotechnical engineering. Water content assumes importance because the presence of water can significantly alter the engineering behaviour of soil.

3. Density index, which indicates the relative compactness, is an important characteristic of a coarse grained soil; it has bearing on its engineering behaviour.
4. Grain size analysis is a useful index for textural classification; it consists of sieve analysis applicable to coarse fraction and wet analysis applicable to fine fraction. Stokes' law is the basis for wet analysis. Effective size and uniformity coefficient indicate the average grain-size and degree of gradation.
5. Consistency limits or Atterberg limits provide the main basis for the classification of cohesive soils; plasticity index indicating the range of water content over which the soil exhibits plasticity, is the most important index.
6. Activity, sensitivity and thixotropy are properties which are typical of cohesive soils.

REFERENCES

1. Allen Hazen: *Some Physical Properties of Sands and Gravels, with special reference to their use in Filtration*, 24th Annual Report of the State Board of Health of Massachusetts, U.S.A., 1892.
2. A. Atterberg: *Die plastizität der Tone, Internationale Mitteilungen für Bodenkunde*, Vol.I, no.1, Verlag für Fachliteratur G.m.b.H., Berlin, 1911.
3. D.M. Burmister: *Principles and Techniques of Soil Identification*, Proceedings, Annual Highway Research Board Meeting, Washington, D.C., 1949.
4. IS: 460-1978: (Second Revision) *Specifications for Test Sieves*.
5. IS: 2720 (Part II)—1964: *Methods of Test for Soils—Part II Determination of Moisture Content*.
6. IS: 2720 (Part-III)—1980 First Revision: *Methods of Test for Soil—Part III Determination of Specific Gravity* (Sections 1 and 2)
7. IS: 2720 (Part IV)—1965: *Methods of Test for Soils—Part IV Grain size Analysis*.
8. IS: 2720 (Part V)—1970: *Methods of Test for Soils—Part V Determination of Liquid and Plastic Limits*.
9. IS: 2720 (Part VI)—1972: *Methods of Test for Soils—Part VI Determination of Shrinkage Factors*.
10. IS : 2720 (Part-XIV)—1968: *Methods of Test for Soils—Part XIV Determination of Density Index* (Relative Density) for soils.
11. IS : 2720 (Part XXVIII)—1974: First Revision: *Methods of Test for Soils – Determination of In-place Density by Sand-replacement Methods*.
12. IS : 2720 (Part XXIX)—1966: *Methods of Test for Soils—Determination of In-place density by Core-cutter Method*.
13. A.R. Jumikis: *Soil Mechanics*, D. Van Nostrasnd Co., Princeton, NJ, 1962.
14. D.F. McCarthy: *Essentials of Soil Mechanics and Foundations*. Reston Publishing Co., Inc., Reston, Virginia, USA.
15. D.W. Taylor: *Fundamentals of Soil Mechanics*, John Wiley & Sons., Inc., New York, 1948.

QUESTIONS AND PROBLEMS

3.1 Distinguish between:
   (i) Silt size and clay size
   (ii) Degree of sensitivity and degree of saturation.
   (S.V.U.—B.Tech., (Part-time)—May, 1983)

3.2 Sketch typical complete grain-size distribution curves for (i) well graded soil and (ii) uniform silty sand. From the curves, determine the uniformity coefficient and effective size in each case. What qualitative inferences can you draw regarding the engineering properties of each soil?
   (S.V.U.—B.Tech., (Part-Time)—May, 1983)

3.3 Define the following:
   (i) Flow index, (ii) Toughness index, (iii) Liquidity index, (iv) Shrinkage index, (v) Plasticity index, (vi) Uniformity coefficient (vii) Relative density (Density index), (viii) Sensitivity, (ix) Activity.
   (S.V.U.—B.Tech., (Part-Time)—Sep., 1982)

3.4 Write short notes on the "Methods of determination of Atterberg limits".
   (S.V.U.—B.Tech., (Part-Time)—Apr., 1982)

3.5 Write a short note on 'Relative density'.
   (S.V.U.—B.Tech., (Part-Time)—June, 1982)

3.6 Define and explain the following:
   (i) Uniformity coefficient, (ii) Relative density, (iii) Stokes' law, (iv) Flow index.
   (S.V.U.—B.Tech., (Part-Time)—Apr., 1982)

3.7 Write short note on 'Consistency of clayey soils'.
   (S.V.U.—B.E., (R.R.)—June, 1972)

3.8 Define and explain: Liquid limit; Plastic limit; Shrinkage limit; and Plasticity index.
   Briefly describe the procedure to determine the Liquid Limit of a soil.
   (S.V.U.—B.E., (R.R.)—Dec., 1977)

3.9 Distinguish between:
   (i) Liquid limit and liquidity index,
   (ii) Density and relative density.
   (S.V.U.—B.E., (R.R.)—Sept., 1967)

3.10 An oven-dried soil weighing 189 g is placed in a pyknometer which is then filled with water. The total weight of the pyknometer with water and soil is 1581 g. The pyknometer filled with water alone weighs 1462 g. What is the specific gravity of the soil, if the pyknometer is calibrated at the temperature of the test?

3.11 In a specific gravity test, 117 g of oven-dried soil was taken. The weight of pyknometer, soil and water was obtained as 651 g. The weight of pyknometer full of water alone was 580 g. What is the value of the specific gravity of solids at the temperature of the test? If, while determining the weight of pyknometer, soil, and water, 2 cm^3 of air got entrapped, what is the correct value of the specific gravity and what is the percentage of error?

Index Properties and Classification Tests 103

3.12 In order to determine the water content of a wet sand, a sample weighing 400 g was put in a pyknometer. Water was then poured to fill it and the weight of the pyknometer and its contents was found to be 2250 g. The weight of pyknometer with water alone was 2030 g. The grain specific gravity of the sand was known to be 2.67. Determine the water content of the sand sample.

3.13 An undisturbed sample of sand has a dry weight of 18.9 N and a volume of 1143 cm^3. The solids have a specific gravity of 2.72. Laboratory tests indicate void ratios of 0.40 and 0.90 at the maximum and minimum unit weights, respectively. Determine the density index of the sand sample.

3.14 A sand at a borrow pit is determined to have an *in-situ* dry unit weight of 18.4 kN/m^3. Laboratory tests indicate the maximum and minimum unit weight values of 19.6 kN/m^3 and 16.32 kN/m^3, respectively. What is the density index of the natural soil?

3.15 The following observations were recorded in a Field density determination by sand-replacement method:
Volume of Calibrating can = 1000 cm^3
Weight of empty can = 1000 g
Weight of can + sand = 2660 g
Weight of sand required to fill the excavated hole = 828 g
Weight of excavated soil = 990 g
In-situ water content = 10%
Determine the *in-situ* bulk unit weight and the *in-situ* dry unit weight.

3.16 A core-cutter 12.6 cm in height and 10.2 cm in diameter weighs 1071 g when empty. It is used to determine the *in-situ* unit weight of an embankment. The weight of core-cutter full of soil is 2970 g. If the water content is 6%, what are *in-situ* dry unit weight and porosity? If the embankment gets fully saturated due to heavy rains, what will be the increase in water content and bulk unit weight, if no volume change occurs? The specific gravity of the soil solids is 2.69.

3.17 Using Stokes' law, determine the time of settlement of a sand particle of 0.2 mm size (specific gravity 2.67) through a depth of water of 30 cm. The viscosity of water is 0.001 N-sec/m^2 and unit weight is 9.80 kN/m^3.

3.18 In a pipette analysis, 50 g of dry soil of the fine fraction was mixed in water to form one litre of uniform suspension. A pipette of 10 ml capacity was used to obtain a sample from a depth of 10 cm, 40 min. from the start of sedimentation. The weight of solids in the pipette sample was 0.20 g. Determine the co-ordinates of the corresponding point on the grain-size distribution curve. Assume the grain-specific gravity as 2.70, the viscosity of water as 0.001 N-sec/m^2, and the unit weight of water as 9.8 kN/m^3.

3.19 A litre of suspension containing 50 g of soil with a specific gravity of 2.70 is prepared for a hydrometer test. When no temperature correction is considered necessary, what should be the hydrometer reading if the hydrometer could be immersed and read at the instant sedimentation begins?

3.20 In a hydrometer analysis, 50 g. of soil was mixed in water to form one litre of uniform suspension. The corrected hydrometer reading after a lapse of 60 min. from the start of sedimentation was 1.010, and the corresponding effective depth was 10.8 cm. The grain specific gravity was 2.72. Assuming the viscosity of water as 0.001 N-sec/m^2 and the unit weight of water as 9.8 kN/m^3, determine the co-ordinates of the corresponding point on the grain-size distribution curve.

3.21 The liquid limit and plastic limit of a soil are 75% and 33% respectively. What is the plasticity index? The void ratio of the soil on oven-drying was found to be 0.63. What is the shrinkage limit? Assume grain specific gravity as 2.7.

3.22 A piece of clay taken from a sampling tube has a wet weight of 155.3 g and volume of 95.3 cm^3. After drying in an oven for 24 hours at 105°C, its weight is 108.7 g. The liquid and plastic limits of the clay are respectively 56.3% and 22.5%. Determine the liquidity index and void ratio of the sample. Assume the specific gravity of soil particles as 2.75.

(S.V.U.—B.E., (R.R.)—Nov., 1973)

3.23 A completely saturated sample of clay has a volume of 31.25 cm^3 and a weight of 58.66 g. The same sample after drying has a volume of 23.92 cm^3 and a weight of 42.81 g. Compute the porosity of the initial soil sample, specific gravity of the soil grains and shrinkage limit of the sample.

(S.V.U.—Four-year B.Tech.—June, 1982)

3.24 In a shrinkage limit test, the following data were obtained:
Initial weight of the saturated soil = 192.8 g
Initial volume of saturated soil = 106.0 cm^3
Weight after complete drying = 146.2 g
Volume after complete drying = 77.4 cm^3

Determine the shrinkage limit, the specific gravity of soil grains, initial void ratio, bulk unit weight, and dry unit weight and final void ratio and unit weight.

(S.V.U.—B.Tech., (Part-Time)—April, 1982)

3.25 The Atterberg limits of a clay soil are: Liquid limit = 63%, Plastic limit = 40%, and shrinkage limit = 27%. If a sample of this soil has a volume of 10 cm^3 at the liquid limit and a volume of 6.4 cm^3 at the shrinkage limit, determine the specific gravity of solids, shrinkage ratio, and volumetric shrinkage.

3.26 A 100 cm^3-clay sample has a natural water content of 30%. Its shrinkage limit is 18%. If the specific gravity of solids is 2.72, what will be the volume of the sample at a water content of 15%?

3.27 The oven-dry weight of a pat of clay is 117 g and the weight of mercury displaced by it is 85.5 g. Assuming the specific gravity of solids as 2.71, determine the shrinkage limit and shrinkage ratio. What will be the water content, at which the volume increase is 10% of the dry volume?

3.28 The liquid limit of a soil is 40% and its plasticity index is 15%. When the soil is fully dried from the plastic limit, the decrease in volume is 20% of the volume at the plastic limit. Similarly, when the soil is fully dried from the liquid limit, the decrease in volume is 40% of the volume at liquid limit. Determine the shrinkage limit and shrinkage ratio.

3.29 The liquid limit and plastic limit of a clay are 50% and 20% respectively. The soil consists of 30% of particles smaller than clay size. Indicate the activity classification of this clay.

3.30 A clay soil has an unconfined compression strength of 180 kN/m^2 in the undisturbed state and 18 kN/m^2 after remoulding. Classify the soil with regard to its sensitivity and indicate the possible structure of the soil.

4
IDENTIFICATION AND CLASSIFICATION OF SOILS

4.1 INTRODUCTION

It has already been stated that certain terms such as 'Gravel', 'Sand', 'Silt' and 'Clay' are used to designate a soil and are based on the average grain-size or particle-size. Most natural soils are mixtures of two or more of these types, with or without organic matter. The minor component of a soil mixture is prefixed as an adjective to the major one—for example, 'silty sand', 'sandy clay', etc. A soil consisting of approximately equal percentages of sand, silt, and clay is referred to as 'Loam'. The differentiation between 'coarse-grained soils' and 'fine-grained soils' has already been brought out in Chapters 1 and 3.

In this chapter, certain procedures for field identification of the nature of a soil, as well as certain generalised procedures for classification of a soil with the help of one of the systems to be dealt with, will be studied in some detail.

4.2 FIELD IDENTIFICATION OF SOILS

Basically, coarse-grained and fine-grained soils are distinguished based on whether the individual soil grains can be seen with naked eye or not. Thus, grain-size itself may be adequate to distinguish between gravel and sand; but silt and clay cannot be distinguished by this technique.

Field identification of soils becomes easier if one understands how to distinguish gravel from sand, sand from silt, and silt from clay. The procedures are given briefly hereunder:

Gravel from Sand

Individual soil particles larger than 4.75 mm and smaller than 80 mm are called 'Gravel'; soil particles ranging in size from 4.75 mm down to 0.075 mm are called 'Sand'. (Refer IS: 1498-1970 "Classification and Identification of Soils for General Engineering Purposes"—First Revision). These limits, although arbitrary in nature, have been accepted widely. The shape of these particles is also important and may be described as angular, sub-angular, rounded, etc., as given in Sec. 3.3. Field identification of sand and gravel should also include identification of mineralogical composition, if possible.

Sand from Silt

Fine sand cannot be easily distinguished from silt by simple visual examination. Silt may look a little darker in colour. However, it is possible to differentiate between the two by the 'Dispersion Test'. This test consists of pouring a spoonful of sample in a jar of water. If the material is sand, it will settle down in a minute or two, but, if it is silt, it may take 15 minutes to one hour. In both these cases, nothing may be left in the suspension ultimately.

Silt from Clay

Microscopic examination of the particles is possible only in the laboratory. In the absence of one, a few simple tests are performed.

(i) Shaking Test

A part of the material is shaken after placing it in the palm. If it is silt, water comes to the surface and gives it a shining appearance. If it is kneaded, the moisture will re-enter the soil and the shine disappears.

If it is clay, the water cannot move easily and hence it continues to look dark. If it is a mixture of silt and clay, the relative speed with which the shine appears may give a rough indication of the amount of silt present. This test is also known as 'dilatancy test'.

(ii) Strength Test

A small briquette of material is prepared and dried. Then one has to try to break it. If it can be broken easily, the material is silt. If it is clay, it will require effort to break. Also, one can dust off loose material from the surface of the briquette, if it is silt. When moist soil is pressed between fingers, clay gives a soapy touch; it also sticks, dries slowly, and cannot be dusted off easily.

(iii) Rolling Test

A thread is attempted to be made out of a moist soil sample with a diameter of about 3 mm. If the material is silt, it is not possible to make such a thread without disintegration and crumbling. If it is clay, such a thread can be made even to a length of about 30 cm and supported by its own weight when held at the ends. This is also called the 'Toughness test'.

(iv) Dispersion Test

A spoonful of soil is poured in a jar of water. If it is silt, the particles will settle in about 15 minutes to one hour. If it is clay, it will form a suspension which will remain as such for hours, and even for days, provided flocculation does not take place.

A few other miscellaneous identification tests are as follows:

Organic Content and Colour

Fresh, wet organic soils usually have a distinctive odour of decomposed organic matter, which can be easily detected on heating. Another distinctive feature of such soils is the dark colour.

Acid Test

This test, using dilute hydrochloric acid, is primarily for checking the presence of calcium carbonate. For soils with a high value of dry strength, a strong reaction may indicate the presence of calcium carbonate as cementing material rather than colloidal clay.

Shine Test

When a lump of dry or slightly moist soil is cut with a knife, a shiny surface is imparted to the soil if it is a highly plastic clay, while a dull surface may indicate silt or a clay of low plasticity.

Considerable experience is required for field identification of soils. A knowledge of the geology of the area will be invaluable in this process.

The identification procedures given above may have bearing on preliminary classification of the soil, dealt with in later sections.

4.3 SOIL CLASSIFICATION – THE NEED

Natural soil deposits are never homogeneous in character; wide variations in properties and behaviour are commonly observed. Deposits that exhibit similar average properties, in general, may be grouped together, as a class. Classification of soils is necessary since through it an approximate, but fairly accurate, idea of the average properties of a soil group or a soil type is obtainable, which is of great convenience in any routine type of soil engineering project. From engineering point of view, classification may be made based on the suitability of a soil for use as a foundation material or as a construction material.

There is, however, some difference of opinion among soil engineers as to the importance of soil classification and the broad generalisation of the properties of various groups. This is largely because of different points of view and emanates primarily from the difficulty in forming soil groups in view of very wide variations in engineering properties which are too large in number. Thus, it is inevitable that in any classification there will be border cases which may fall into two or more groups. Similarly, the same soil may be placed into groups that appear radically different under different systems of classification.

In view of this, classification is to be taken merely as a preliminary guide to the engineering behaviour of the soil, which cannot be fully or solely predicted from the classification alone; certain important soil engineering tests should necessarily be conducted in connection with the use of soil in any important project, since different properties govern the soil behaviour in different situations. The understanding of the engineering behaviour of a soil should be the more important issue.

4.4 ENGINEERING SOIL CLASSIFICATION—DESIRABLE FEATURES

The general requirements of an ideal and effective system of soil classification are as follows:

(*a*) The system should have scientific approach.

Identification and Classification of Soils 109

(b) It should be simple and subjective element in rating the soil should be eliminated as far as possible.
(c) A limited number of different groupings should be used, which should be on the basis of only a few similar properties.
(d) The properties considered should have meaning for the engineering profession.
(e) It should have fair accuracy in indicating the probable performance of a soil under certain field conditions.
(f) It should be based on a generally accepted uniform soil terminology so that the classification is done in commonly well-understood and well-conversant terms.
(g) It should be such as to permit classification of a soil by simple visual and manual tests, or at least only by a few simple tests.
(h) The soil group boundaries should be drawn as closely as possible where significant changes in soil properties are known to occur.
(i) It should be acceptable to all engineers.

These are rather very ambitious requirements for a soil classification system and cannot be expected to be met cent per cent by any system, primarily because soil is a complex material in nature and does not lend itself to a simple classification. Therefore, an engineering soil classification system is probably satisfactory only for that specific kind of geotechnical engineering project for which it is hopefully developed.

4.5 CLASSIFICATION SYSTEMS—MORE COMMON ONES

A number of systems of classification of soils have been evolved for engineering purposes. Certain of these have been developed specifically in connection with ascertaining the suitability of soil for use in particular soil engineering projects. Some are rather preliminary in character while a few are relatively more exhaustive, although some degree of arbitrariness is necessarily inherent in each of the systems.

The more common classification systems, some of which will be dealt with in greater detail in later sections, are enumerated below:
1. Preliminary classification by soil types or Descriptive classification.
2. Geological classification or Classification by origin.
3. Classification by structure.
4. Grain-size classification or Textural classification.
5. Unified Soil Classification System.
6. Indian Standard Soil Classification System.

4.5.1 Preliminary Classification by Soil Types or Descriptive Classification

Familiarity with common soil types is necessary for an understanding of the fundamentals of soil behaviour. In this approach, soils are described by designations such as 'Boulders', 'Gravel', 'Sand', 'Silt', 'Clay', 'Rockflour', 'Peat', 'China clay', 'Fill', 'Bentonite', 'Black Cotton Soils', 'Boulder clay', 'Caliche', 'Hardpan', 'Laterite', 'Loam', 'Loess', 'Marl', 'Moorum', 'Topsoil', and 'Varved clay'. All of these except the first nine have already been described in Chapter 1.

Boulders, gravel and sand belong to the category of coarse-grained soils, distinguished primarily, by the particle-size; these do not exhibit the property of cohesion, and so may be said to be 'cohesionless' or 'non-cohesive' soils. The sizes are in the decreasing order.

'Silt' refers to a soil with particle-sizes finer than sand. If it is inorganic in nature, it is called 'Rock flour' and it is non-plastic, generally. It may exhibit slight plasticity when wet and slight compressibility if the particle shape is plate-like. Organic silts contain certain amounts of fine decomposed organic matter, are dark in colour, have peculiar odour, and exhibit some degree of plasticity and compressibility.

'Clay' consists of soil particles smaller than 0.002 mm in size and exhibit plasticity and cohesion over a fairly large range of moisture contents. They may be called 'lean clays' or 'fat clays' depending on the degree of plasticity. These are basically secondary products of weathering, produced by prolonged action of water on silicate minerals; three of the major clay minerals are 'Kaolinite', 'Illite' and 'Montmorillonite'. Organic variety of clay, called 'Peat', containing partially carbonised organic matter, is recognised by its dark colour, odour of decay, fibrous nature, very low specific gravity (0.5 to 0.8), and very high compressibility and ranks lowest as a foundation material.

'China clay', also called 'Kaolin', is a pure white clay, used in the ceramic industry.

All man-made deposits ranging from rock dumps to sand and gravel fills are termed 'Fill'; thus, fills may consist of every imaginable material.

4.5.2 Geological Classification or Classification by Origin

Soils may be classified on the basis of their geological origin. The origin of a soil may refer either to its constituents or to the agencies responsible for its present state.

Based on constituents, soils may be classified as:
1. Inorganic soils
2. Organic soils $\begin{cases} \text{Plant life} \\ \text{Animal life} \end{cases}$

Based on the agencies responsible for their present state, soils may be classified as:
1. Residual soils
2. Transported soils
   a. Alluvial or sedimentary soils   (transported by water)
   b. Aeoline soils   (transported by wind)
   c. Glacial soils   (transported by glaciers)
   d. Lacustrine soils   (deposited in lakes)
   e. Marine soils   (deposited in seas)

These have been dealt with in Chapter 1.

Over the geologic cycle, soils are formed by disintegration and weathering of rocks. These are again reformed by compaction and cementation by heat and pressure.

4.5.3 Classification by Structure

Depending upon the average grain-size and the conditions under which soils are formed and deposited in their natural state, they may be categorised on the basis of their structure, as:
1. Soils of single-grained structure
2. Soils of honey-comb structure
3. Soils of flocculent structure

These have also been treated in detail in Chapter 1.

4.5.4 Grain-size or Textural Classifications

In the grain-size classifications, soils are designated according to the grain-size or particle-size. Terms such as gravel, sand, silt and clay are used to indicate certain ranges of grain-sizes. Since natural soils are mixtures of all particle-sizes, it is preferable to call these fractions as 'Sand size', 'Silt size', etc. A number of grain-size classifications have been evolved, but the commonly used ones are:
1. U.S. Bureau of Soils and Public Roads Administration (PRA) System of U.S.A.
2. International classification, proposed at the International Soil Congress at Washington, D.C., in 1927.
3. Massachusetts Institute of Technology (MIT) system of classification of U.S.A.
4. Indian Standard Classification (IS:1498-1970).

These are shown diagrammatically on page 112 (Fig. 4.1)

A soil classification system purely based on grain-size is called a 'Textural classification'. One such is the U.S. Bureau of Soils and P.R.A. Classification depicted by a 'Triangular chart' (Fig. 4.2), which ignores the fraction coarser than sand:

Any soil with the three constituents—sand, silt, and clay—can be represented by one point on the Triangular chart. For example, a soil with 25% sand, 25% silt and 50% clay will be represented by the point 'S', obtained by the dotted lines, as shown by the arrows. Certain zones on the chart are marked to represent certain soils such as sand, silt, clay, sandy clay, silty clay, loam, sandy loam, etc. These have been marked rather arbitrarily. ('Loam' is primarily an agricultural term).

Textural or grain-size classifications are inadequate primarily because plasticity characteristics—consistency limits and indices—do not find any place in these classifications.

4.5.5 Unified Soil Classification System

The Unified soil classification system was originally developed by A. Casagrande and adopted by the U.S. Corps of Engineers in 1942 as 'Airfield classification'. It was later revised for universal use and redesignated as the "Unified Soil Classification" in 1958.

In this system (Table 4.1), soils are classified into three broad categories:

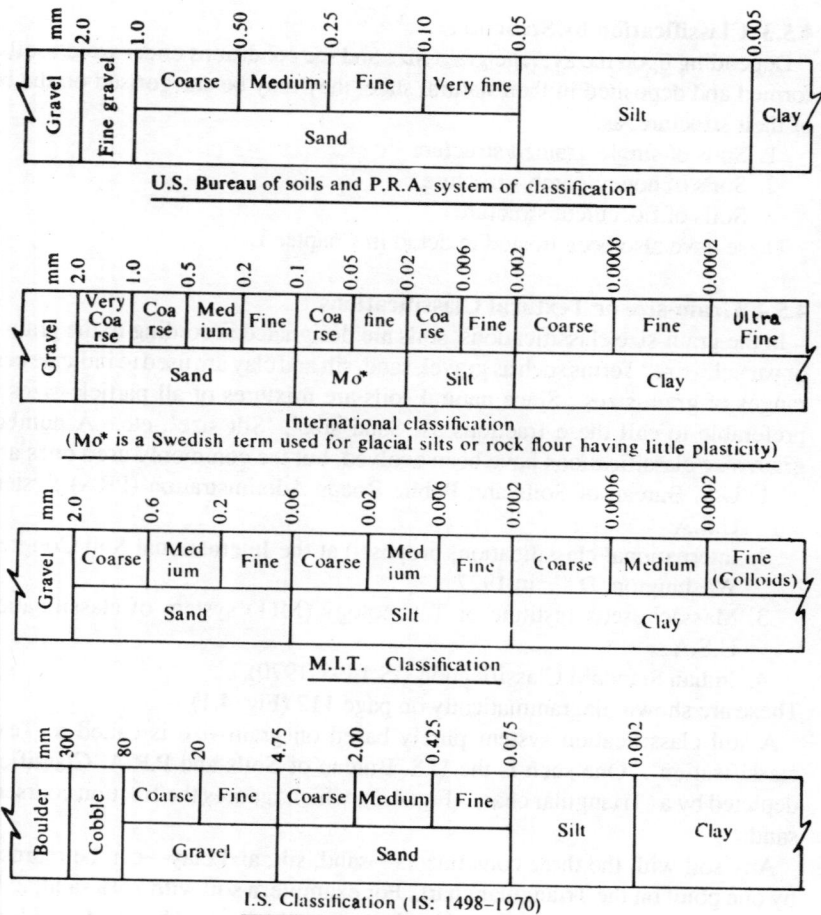

Fig. 4.1 Grain-size classifications

1. Coarse-grained soils with up to 50% passing No. 200 ASTM Sieve
(No. 75-µ IS Sieve)
2. Fine-grained soils with more than 50% passing No. 200 ASTM Sieve
(No. 75-µ IS Sieve)
3. Organic soils

The first two categories can be distinguished by their plasticity characteristics. The third can be easily identified by its colour, odour and fibrous nature.

Each soil component is assigned a symbol as follows:

Gravels: G   Silt: M (from the Swedish   Organic: O
                         word 'Mo', for silt)
Sand: S   Clay: C   Peat: *Pt*

Identification and Classification of Soils 113

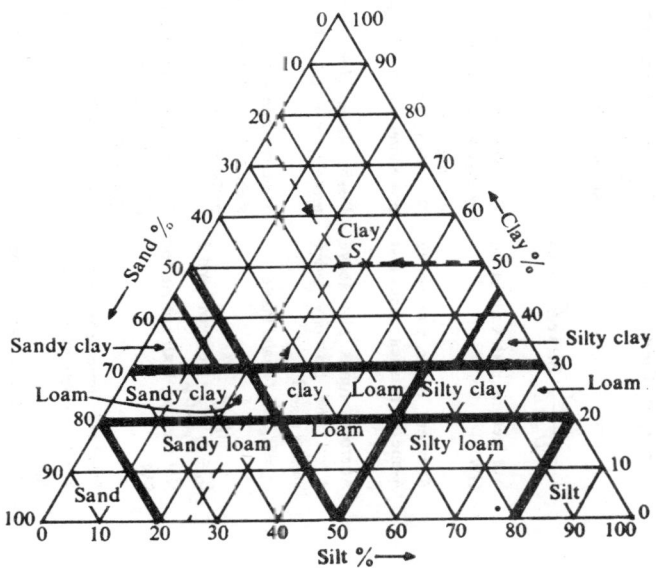

Fig. 4.2 Bureau of soils triangular chart for textural classification

Coarse-grained soils are further subdivided into well-graded (W) and poorly graded (P) varieties, depending upon the Uniformity coefficient (C_u) and coefficient of Curvature (C_c):

Well-graded gravel,	$C_u > 4$
Well-graded sand,	$C_u > 6$
Well-graded soil, C_c	$= 1$ to 3

Note that C_u is the same as U defined in Eq. 3.35.

Fine-grained soils are subdivided into those with low plasticity (L), with liquid limit less than 50%, and those with high plasticity (H), with liquid limits more than 50%. Dilatancy, dry strength and toughness tests are to be used for filed identification.

The plasticity chart devised by Casagrande is used for identification of fine-grained soils (Fig. 4.3).

4.5.6 Indian Standard Soil Classification System

IS:1498-1970 describes the Indian Standard on Classification and Identification of Soils for general engineering purposes (first revision). Significant provisions of this system are given below :

Soils shall be broadly divided into three divisions :

1. *Coarse-grained soils:* More than 50% of the total material by weight is larger than 75-μ IS Sieve size.

114 Geotechnical Engineering

Table 4.1 Unified Soil Classification System

Major Division			Group symbol	Typical Name	Classification Criteria	Field Identification Procedures (Excluding particles larger than 8 cm)
Coarse grained soils More than 50% retain on No. 200 ASTM Sieve	More than 50% retain on No.4 ASTM Sieve	Clean Gravels	GW*	Well-graded gravels and gravel-sand mixtures, little or no fines.	$C_u = D_{60}/D_{10}$ Greater than 4 $C_c = \dfrac{(D_{30})^2}{D_{10} \times D_{60}}$ Between 1 and 3	Wide range in grain-size and substantial amounts of all intermediate particle sizes
			GP	Poorly graded gravels and gravel-sand mixtures, little or no fines	Not meeting both Criteria for GW	Predominantly one size or a range of sizes with some intermediate size missing
		Gravels with fines	GM	Silty gravels, gravel-sand-silt mixtures	Atterberg limits plot below A-line and plasticity index less than 4	Non-plastic fines (for identification procedures see ML below)
			GC	Clayey gravels, gravel-sand-clay mixtures	Atterberg limits plot above A-line and plasticity index greater than 7	Plastic fines (for identification procedures see CL below)
	More than 50% of coarse fraction passes No.4 ASTM sieve	Clean sands	SW	Well-graded sands and gravelly sands, little or no fines	$C_u = D_{60}/D_{10}$ Greater than 6 $C_c = \dfrac{(D_{30})^2}{D_{10} \times D_{60}}$ Between 1 and 3	Wide range in grain-sizes and substantial amounts of all intermediate particle sizes
			SP	Poorly graded sands and gravelly sands, little or no fines	Not meeting both criteria for SW	Predominantly one size or a range of sizes with some intermediate sizes missing
		Sands with fines	SM	Silty sands, sand-silt mixtures	Atterberg limits plot below A-line or plasticity index less than 4	Non-plastic fines (for identification procedures see ML below)
			SC	Clayey sands, sand-clay mixtures	Atterberg limits plot above A-line and plasticity index greater than 7	Plastic fines (for identification procedures see CL below)

Identification and Classification of Soils 115

				Identification procedures on Fraction smaller than No.40 ASTM Sieve		
				Dry Strength	Dilatancy	Toughness
Fine grained soils 50% or more passes No. 200 ASTM Sieve	Silt and clays Liquid limit 50% or less	Inorganic Silts, very fine sands, rock flour, silty or clayey fine sands	ML	None to slight	Quick to slow	None
		Inorganic clays of low to medium plasticity, gravelly clays, sandy clays, silty clays, lean clays	CL	Medium to high	None to very slow	Medium
		Organic silts and organic silty clays of low plasticity	OL	Slight to Medium	Slow	Slight
	Silt and clays Liquid limit greater than 50%	Inorganic silts, micaceous or diatomaceous fine sands or silts, elastic silts	MH	Slight to Medium	Slow to None	Slight to Medium
		Inorganic clays of high plasticity, fat clays	CH	High to very high	None	High
		Organic clays of medium to high plasticity	OH	Medium to high	None to very slow	Slight to Medium
Highly organic clays		Peat, muck and other highly organic soils	Pt	Readily identified by colour, odour, spongy feel, and frequently by fibrous texture. Fibrous organic matter, will char, burn, or glow		

Note: "Boundary classification": soils possessing characteristics of two groups are designated by combinations of group symbols, for example, GW-GC, Well-graded, gravel-sand mixture with clay binder.

* Classification on the basis of percentage of fines
  Less than 5% passing No. 200 ASTM Sieve..GW, GP, SW, SP
  More than 12% passing No. 200 ASTM Sieve..GM, GC, SM, SC
  5% to 12% passing No. 200 ASTM Sieve..Border line classification requiring use of dual symbols.

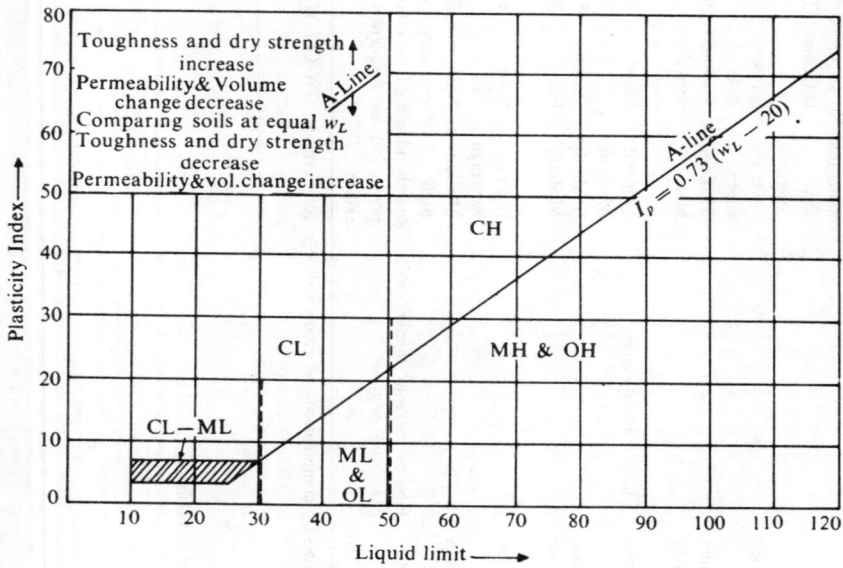

Fig. 4.3 Plasticity chart (unified soil classification)

2. *Fine-grained soils:* More than 50% of the total material by weight is smaller than 75-μ IS Sieve size.
3. *Highly Organic Soils and other Miscellaneous Soil Materials:* These soils contain large percentages of fibrous organic matter, such as peat, and particles of decomposed vegetation. In addition, certain soils containing shells, concretions, cinders and other non-soil materials in sufficient quantities are also grouped in this division.

Coarse-grained soils shall be divided into two sub-divisions:
(a) *Gravels*: More than 50% of coarse fraction (+75 μ) is larger than 4.75 mm IS Sieve size.
(b) *Sands*: More than 50% of Coarse fraction (+75 μ) is smaller than 4.75 mm IS Sieve size.

Fine-grained soils shall be divided into three sub-divisions:
(a) Silts and clays of low compressibility: Liquid limit less than 35% (L).
(b) Silts and clays of medium compressibility: Liquid limit greater than 35% and less than 50% (I).
(c) Silts and clays of high compressibility: Liquid limit greater than 50 (H).

The coarse-grained soils shall be further sub-divided into eight basic soil groups, and the fine-grained soils into nine basic soil groups; highly organic soils and other miscellaneous soil materials shall be placed in one group. The various sub-divisions, groups and group symbols are set out in Table 4.2.

Boundary Classification for Coarse-grained Soils

Coarse-grained soils with 5% to 12% fines are considered as border-line cases between clean and dirty gravels or sands as, for example, GW-GC, or SP-SM. Similarly, border-line cases might occur in dirty gravels and dirty sands, where I_p is between 4 and 7, as for example, GM-GC or SM-SC. It is, therefore, possible to have a border line case of a border line case. The rule for correct classification in such cases is to favour the non-plastic classification. For example, a gravel with 10% fines, a C_u of 20, a C_c of 2.0, and I_p of 6 would be classified GW—GM rather than GW—GC.

Classification Criteria for Fine-grained Soils

The plasticity chart (Fig. 4.4) forms the basis for the classification of fine-grained soils, based on the laboratory tests. Organic silts and clays are distinguished from inorganic soils which have the same position on the plasticity chart, by odour and colour. In case of any doubt, the material may be oven-dried, remixed with water and retested for liquid limit. The plasticity of fine-grained organic soils is considerably reduced on oven-drying. Oven-drying also affects the liquid limit of inorganic soils, but only to a small extent. A reduction in liquid limit after oven-drying to a value less than three-fourth of the liquid limit before oven-drying is positive identification of organic soils.

Boundary Classification for Fine-grained Soils

The fine-grained soils whose plot on the plasticity chart falls on, or practically on A-line, $w_L = 35$-lines, $w_L = 50$ and lines shall be assigned the proper boundary classification. Soils which plot above the A-line, or practically on it, and which have a plasticity index between 4 and 7 are classified ML—CL.

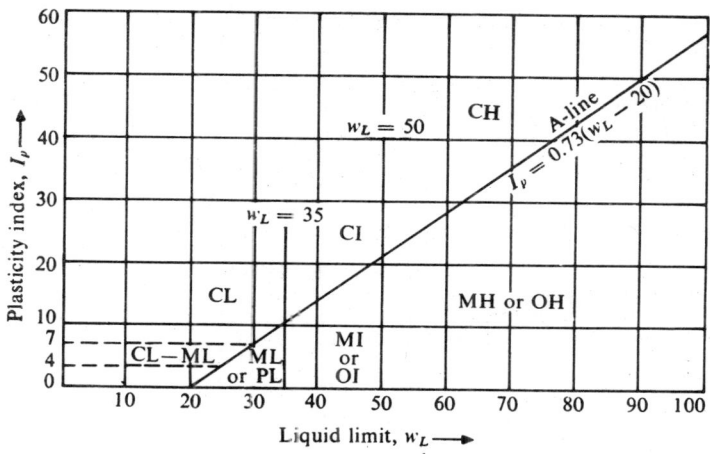

Fig. 4.4 Plasticity chart (I.S. soil classification).

Table 4.2 Soil Classification Including Field Identification and Description (IS:1498-1970)

Division	Sub-division		Group Letter Symbol	Typical Names	Field Identification procedures (Excluding particles larger than 80 mm)	Information required for describing soils
Coarse grained soils (More than 50% of material larger than 75-IS Sieve. This sieve is about the smallest particle visible to the naked eye.)	Gravels (More than 50% of the coarse fraction is larger than 4.75 mm IS Sieve size)	Clean Gravels (Little or no fines)	GW	Well-graded gravels, gravel-sand mixtures; little or no fines.	Wide range in grain sizes and substantial amounts of all intermediate particle sizes	Give typical name; indicate approximate percentages of sand and gravel; maximum size; angularity, surface condition, and hardness of the coarse grains; local or geologic name and other pertinent descriptive information; and symbol in parenthesis.
			GP	Poorly graded gravels or gravel-sand mixtures; little or no fines.	Predominantly one size or a range of sizes with some intermediate sizes missing.	
		Gravels with fines	GM	Silty gravels, poorly graded gravel-sand-silt mixtures	Non-plastic fines or fines with low plasticity (for identification procedures see ML and MI below).	For undisturbed soils and information on stratification; degree of compactness, cementation, moisture conditions and drainage characteristics.
			GC	Clayey gravels, poorly graded gravel-sand-clay mixtures.	Plastic fines (for identification procedures, see CL and CI below).	
	Sands (More than 50% of the coarse fraction is smaller than 4.75 mm IS Sieve size)	Clean sands (Little or no fines)	SW	Well-graded sands, gravelly sands; little or no fines.	Wide range in grain-size and substantial amounts of all intermediate particle sizes.	Example: Silty sand, gravelly; about 20% hard angular gravel particles, 10 mm max. size; rounded and sub-angular sand grains; about 15% non-plastic fines with low dry strength; well compacted and moist; in place; alluvial sand (SM)
			SP	Poorly graded sands or gravelly sands; little or no fines.	Predominantly one size or a range of sizes with some intermediate sizes missing.	
		Sands with fines	SM	Silty sands, poorly graded sand-silt-mixtures.	Non-plastic fines or fines with low plasticity (for identification procedures, see ML and MI below).	
			SC	Clayey sands, poorly graded sand-clay mixtures.	Plastic fines (for identification procedures, see CL and CI below).	

Identification and Classification of Soils

	Group Symbol	Typical Names	Dry Strength	Dilatancy	Toughness	Information Required for Describing Soils
				Identification procedures (on fraction smaller than 425 μ IS Sieve sizes)		
Fine-grained Soils — More than 50% of the material is smaller than 75-μ IS Sieve Size						Give typical name; indicate degree and character of plasticity, amount and max. size of coarse grains; colour in wet condition, odour, if any; local or geologic name and other pertinent descriptive information and symbol in parenthesis. For undisturbed soils add information on structure, stratification, consistency in undisturbed and remoulded states, moisture and drainage conditions. Example: Clayey silt, brown; slightly plastic; small percentage of fine sand; numerous vertical root holes; firm and dry in place; loess (ML).
Silts and clays with low compressibility and liquid limit less than 35%	ML	Inorganic silts and very fine sands, rock flour, silty or clayey fine sands, or clayey silts with none to low plasticity.	None to low	Quick	None	
	CL	Inorganic silts, gravelly clays, sandy clays, silty clays, lean clays of low plasticity.	Medium	None to very slow	Medium	
	OL	Organic silts and organic silty clays of low plasticity.	Low	Slow	Low	
Silts and clays with medium compressibility and liquid limit greater than 35% and less than 50%	MI	Inorganic silts, silty or clayey fine sands, or clayey silts of medium plasticity.	Low	Quick to slow	None	
	CI	Inorganic clays, gravelly clays, sandy clays, silty clays, lean clays of medium plasticity.	Medium to high	None	Medium	
	OI	Organic silts and organic silty clays of medium plasticity.	Low to medium	Slow	Low	
Silts and clays with high compressibility and liquid limit greater than 50%	MH	Inorganic silts of high compressibility, micaceous or diatomaceous fine sandy or silty soils, elastic silts.	Low to medium	Slow to none	Low to medium	
	CH	Inorganic clays of high plasticity, fat clays.	High to very high	None	High	
	OH	Organic clays of medium to high plasticity.	Medium to high	None to very slow	Low to medium	
Highly organic soils	Pt	Peat and other highly organic soils with very high compressibility.	Readily identified by colour, odour, spongy feel and frequently by fibrous texture.			

NOTE: "Boundary classification": Soil possessing characteristics of two groups are designated by combinations of group symbols, for example, GW-GC, well-graded, gravel-sand mixture with clay binder.

120 *Geotechnical Engineering*

Black Cotton Soils

These are inorganic clays of medium to high compressibility. These are characterised by high shrinkage and swelling properties. When plotted on the plasticity chart, these lie mostly along a band above A-line. Some may lie below the A-line also. 'Kaolin' behaves as inorganic silt and usually lies below A-line; thus, it shall be classified as such (ML, MI, MH), although it is clay from the mineralogical standpoint.

The classification procedures for coarse-grained soils and for fine-grained soils, using this system, may be set out in the form of flow diagrams as shown in Figs. 4.5 and 4.6.

Relative Suitability for General Engineering Purposes

The characteristics of the various soil groups pertinent to roads and airfields – value as subgrade, sub-base and base material, compressibility and expansion, drainage characteristics, and compaction equipment (in qualitative terms), ranges of unit dry weight, CBR value percent, and sub-grade modulus—are also tabulated.

Characteristics pertinent to embankments and foundations—value as embankment material, compaction characteristics, value as foundation material, requirements for seepage control (in qualitative terms), ranges of permeability and unit dry weight—are also tabulated.

Characteristics pertinent to suitability for canal sections—compressibility, workability as a construction material and shearing strength when compacted and saturated are also given in relative or qualitative terms.

This information is supposed to serve the purpose of a guideline or an indication of the suitability of a soil based on the I.S. Classification System. Important and large projects will need detailed investigation of the soil behaviour—first-hand. A comparison between IS Classification and Unified Classification shows many points of similarity but only a few points of difference, especially in classifying fine-grained soils.

4.6 ILLUSTRATIVE EXAMPLES

Example 4.1: Two soils S_1 and S_2 are tested in the laboratory for the consistency limits. The data available is as follows:

	Soil S_1	*Soil S_2*
Plastic Limit, w_p	18%	20%
Liquid Limit, w_L	38%	60%
Flow index, I_f	10	5
Natural moisture content, w	40%	50%

(a) Which soil is more plastic?
(b) Which soil is better foundation material when remoulded?
(c) Which soil has better strength as a function of water content?
(d) Which soil has better strength at the plastic limit?
(e) Could organic material be present in these soils?

Identification and Classification of Soils 121

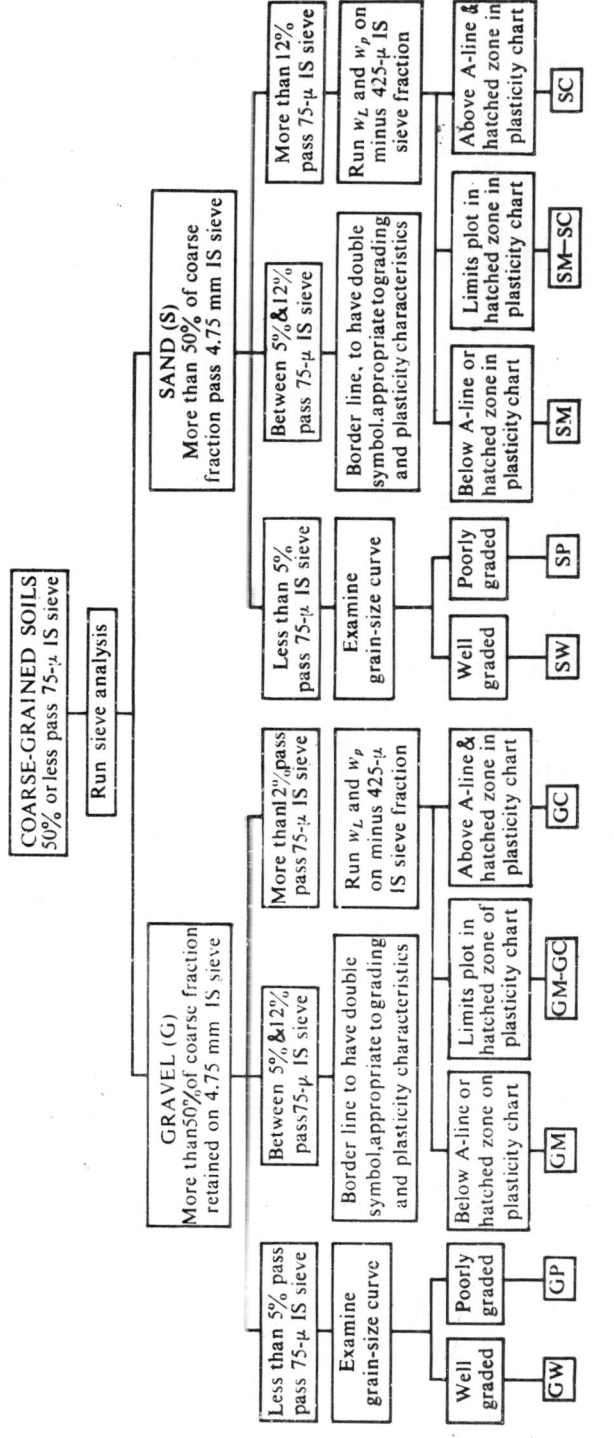

Fig. 4.5 Flow chart for classification of coarse-grained soils

122 Geotechnical Engineering

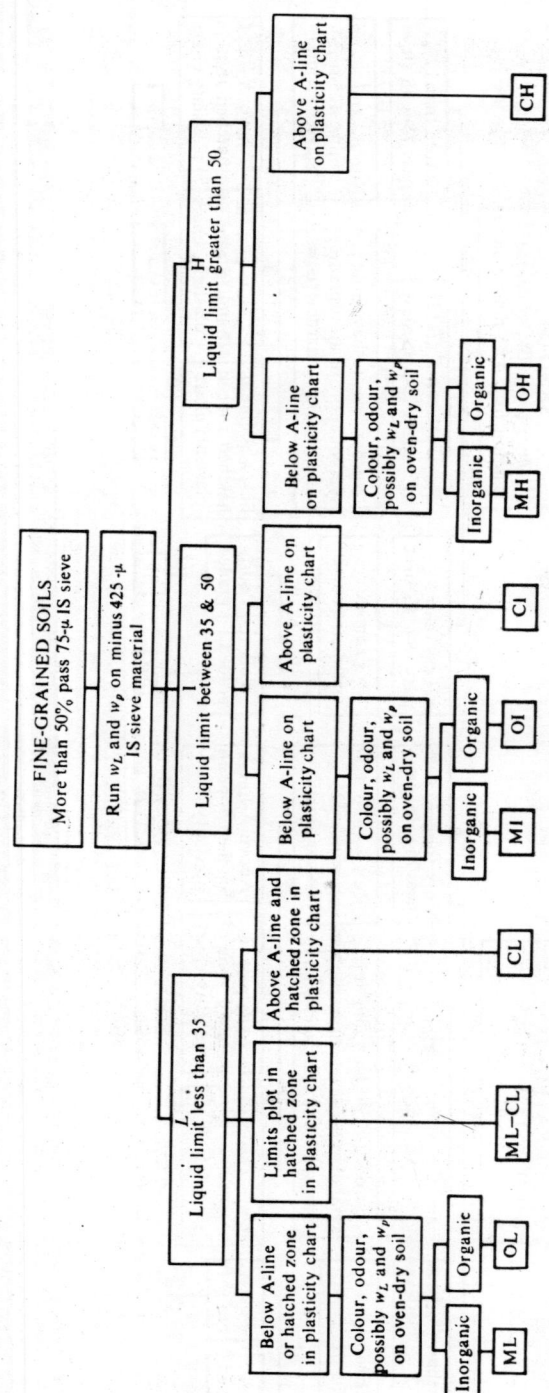

Fig. 4.6 Flow chart for classification of fine-grained soils

Plot the positions of these soils on the Casagrande's plasticity chart and try to classify them as per IS Classification.

(a) Plasticity index, I_p for soil $S_1 = w_L - w_p = (38 - 18) = 20$
I_p for soil $S_2 = w_L - w_p = (60 - 20) = 40$

Obviously, Soil S_2 is the more plastic.

As per Burmister's classification of the degree of plasticity, S_1 borders between low-to-medium plasticity and S_2 between medium-to-high plasticity.

(b) Consistency index, I_c for soil $S_1 = \dfrac{(w_L - w)}{I_p} = \dfrac{(38 - 40)}{20} = -0.1$

I_c for soil $S_2 = \dfrac{(60 - 50)}{40} = 0.25$

Since the consistency index for soil S_1 is negative it will become a slurry on remoulding; therefore, soil S_2 is likely to be a better foundation material on remoulding.

(c) Flow index, I_f for soil $S_1 = 10$
I_f for soil $S_2 = 5$

Since the flow index for soil S_2, is smaller than that for S_1, soil S_2 has better strength as a function of water content.

(d) Toughness index, I_T for soil $S_1 = I_p/I_f = 20/10 = 2$
I_T for soil $S_2 = 40/5 = 8$

Since toughness index is greater for soil S_2, it has a better strength at plastic limit.

(e) Since the plasticity indices are low for both the soils, the probability of the presence of organic material is small.

These conclusions may be mostly confirmed from the following:
The soils are marked on Casagrande's plasticity chart as shown in Fig. 4.7.

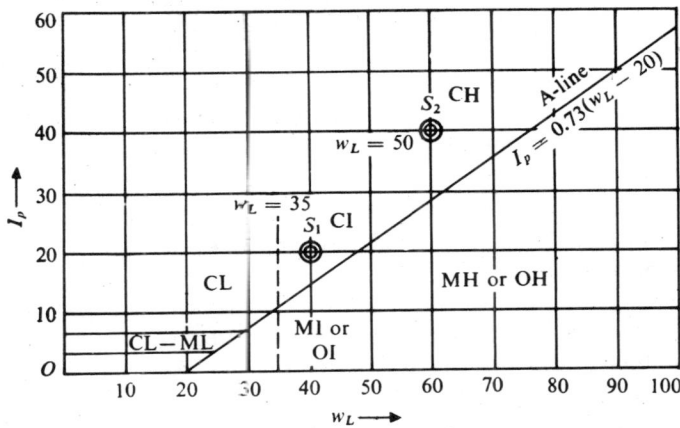

Fig. 4.7 Plasticity chart, soils S_1 and S_2 plotted (Example 4.1)

124 *Geotechnical Engineering*

S_1 and S_2 are respectively in the zone of CI and CH (inorganic clays of medium and high plasticity).

Example 4.2: A soil sample has a liquid limit of 20% and plastic limit of 12%. The following data are also available from sieve analysis:

Sieve size	% passing
2.032 mm	100
0.422 mm	85
0.075 mm	38

Classify the soil approximately according to Unified Classification or IS Classification.

(S.V.U.—Four year B.Tech.—June, 1982).

Since more than 50% of the material is larger than 75-μ size, the soil is a coarse-grained one.

100% material passes 2.032 mm sieve; the material passing 0.075 mm sieve is also included in this. Since this latter fraction any way passes this sieve, a 100% of coarse fraction also passes this sieve.

Since more than 50% of coarse fraction is passing this sieve, it is classified as a sand. (This will be the same as the per cent passing 4.75 mm sieve.)

Since more than 12% of the material passes the 75-μ sieve, it must be SM or SC.

Now it can be seen that the plasticity index, I_p, is $(20 - 12) = 8$, which is greater than 7.

Also, if the values of w_L and I_p are plotted on the plasticity chart, the point falls above *A*-line.

Hence the soil is to be classified as SC, as per IS classification.

Even according to Unified Classification System, this will be classified as SC, which may be checked easily.

SUMMARY OF MAIN POINTS

1. Certain generalised procedures have been evolved for identification of soils in the laboratory and in the field, and for classification of soils. The need for classification arises from the fact that natural soil deposits vary widely in their properties and engineering behaviour.
2. The requirements or desirable features of an engineering soil classification system are so ambitious that it is almost impossible to evolve an ideal system satisfying all of these.
3. Preliminary classification procedures include descriptive and geological classifications, and also classification by structure.
4. Textural classifications are used as part of the more systematic and exhaustive systems such as the Unified Soil Classification.

Identification and Classification of Soils 125

5. The Indian Standard Soil Classification bears many similarities to the Unified Soil Classification although there are a few points of difference, especially with regard to the classification of fine-grained soils.
6. Grain-size is the primary criterion for the classification of coarse-grained soils, while plasticity characteristics, incorporated in the plasticity chart, are the primary criterion for the classification of fine-grained soils.

REFERENCES

1. A. Casagrande: *Classification and Identification of Soils*, Transactions of ASCE, Vol. 113, 1948.
2. IS:1498-1970 (First Revision): *Classification of Soils for General Engineering Purposes.*
3. A.R. Jumikis: *Soil Mechanics*, D. Van Nostrand Co., Princeton, NJ, 1962.
4. D.W. Taylor: *Fundamentals of Soil Mechanics*, John Wiley & Sons Inc., NY, U.S.A, 1948.
5. *The Unified Soil Classification System*, Appendix B, Technical Memorandum No.3-357, March, 1953, revised June, 1957. U.S. Army Engineer Waterways Experiment Station, Corps of Engineers, Vicksburg, Mississippi, U.S.A.
6. M. Whitney: *Methods of the Mechanical Analysis of Soils*, U.S. Department of Agriculture, Division of Agricultural Soils, Bulletin No. 4, Washington, D.C., Government Printing Office, 1896.

QUESTIONS AND PROBLEMS

4.1 (a) Four soil samples collected from a borrow area, to form a low earth dam, are classified as GW, CL, SC and SM. What is your inference?
 (b) The following data relate to five soil samples.
    LL (%) ...   25   45   50   60   80
    PL (%) ...   15   23   25   35   36
  Plot these on Casagrande's A-line chart and classify the soils.
                    (S.V.U.—B.Tech. (Part-time)— May, 1983)
4.2 The following data refer to a sample of soil:
  Percent passing 4.75 mm IS Sieve   = 64
  Percent passing 75-μ IS Sieve = 6
  Uniformity Coefficient   = 7.5
  Coefficient of Curvature   = 2.7
  Plasticity index   = 2.5
  Classify the soil.
4.3 A certain soil has 99% by weight finer than 1 mm, 80% finer than 0.1 mm, 25% finer than 0.01 mm, 8% finer than 0.001 mm. Sketch the grain-size distribution curve and determine the percentage of sand, silt and clay fractions as per IS nomenclature. Determine Hazen's effective size and uniformity coefficient.

4.4 (a) Write a brief note on Textural classification.
   (b) Sketch neatly the Casagrande's plasticity chart indicating various aspects. How would you use it in classifying the fine grained soils? Give a couple of examples. How would you differentiate between organic and inorganic soils?
   (S.V.U.—B.Tech., (Part-Time)—Sept., 1982)
4.5 (a) Bring out the salient aspects of Indian Standard Classification System.
   (b) Write a brief note on the Textural classification.
   (c) How would you distinguish if a material is:
      (i) GW or GP or GM or GC
      (ii) SW or SP or SM or SC
   (S.V.U.—B.Tech., (Part-time)—April, 1982)
4.6 Describe in detail the Indian System of soil classification. When would you use dual symbols for soils?
   (S.V.U.—Four year B.Tech.—June, 1982)
4.7 (a) Draw neatly the IS plasticity chart and label the symbol of various soils.
   (b) What are the limitations of any soil classification system?
   (c) Explain the following tests with their significance.
      (i) Dilatancy, (ii) Thread Test, (iii) Dry Strength Test.
   (S.V.U.—B.Tech. (Part-Time)—April, 1982)
4.8 (a) Why is classification of soils required?
   (b) What are common classification tests?
   (c) How do you classify a soil by the I.S. Classification system?
   (d) How would you differentiate between SC and SF soils
   (S.V.U.—B.Tech. (Part-Time)—June, 1982)
   (*Hint:* The symbol 'F' was used in the older versions of the Unified classification, i.e., in the Airfield classification, to denote 'Fines'. Thus, SF and GF were used in place of SM and GM).
4.9 What physical properties of soil distinguish between cohesive and cohesionless soils? Also explain the principle of sub-dividing cohesive and cohesionless deposits for the purpose of soil classification.
   (S.V.U.—B.E., (R.R.)—May, 1975)
4.10 (a) Describe the U.S. Bureau of Soils Textural classification.
   (b) Describe field identification tests to distinguish between clay and silt.
   (S.V.U.—B.E., (R.R.)—November, 1974)
4.11 (a) Explain why soils are classified and outline the salient features of Casagrande's airfield classification.
   (S.V.U.—B.E., (R.R.)—November, 1973)
   (*Hint:* Casagrande's airfield classification was developed earlier and formed the basis for the Unified Classification. Symbols SF and GF were used in place of SM and GM, which were introduced later.)
4.12 (a) State the various classification systems of soils for general engineering purposes.
   (b) Briefly describe the "Unified Soil Classification".
   (S.V.U.—B.E., (R.R.)—Dec., 1971)

4.13 (*a*) Describe the method of field identification of soils.
 (*b*) How do you use the A-line to distinguish between various types of clays?
（S.V.U.—B.E., (N.R.)—May, 1969)
4.14 How do you distinguish between clay and silt in the field? State the purpose of identification and classification of soils. List any three important engineering classification systems and describe one in detail, clearly bringing out its limitations.
（S.V.U.—B.E., (N.R.)—Sept., 1967)

5
SOIL MOISTURE–PERMEABILITY AND CAPILLARITY

5.1 INTRODUCTION

Natural soil deposits invariably include water. Under certain conditions soil moisture or water in the soil is not stationary but is capable of moving through the soil. Movement of water through soil affects the properties and behaviour of the soil, rather in a significant way. Construction operations and the performance of completed construction could be influenced by soil water. Ground water is frequently encountered during construction operations; the manner in which movement of water through soil can occur and its effects are, therefore, of considerable interest in the practice of geotechnical engineering.

5.2 SOIL MOISTURE AND MODES OF OCCURRENCE

Water present in the void spaces of a soil mass is called 'Soil water'. Specifically, the term 'soil moisture' is used to denote that part of the sub-surface water which occupies the voids in the soil above the ground water table.

Soil water may be in the forms of 'free water' or 'gravitational water' and 'held water', broadly speaking. The first type is free to move through the pore space of the soil mass under the influence of gravity; the second type is that which is held in the proximity of the surface of the soil grains by certain forces of attraction.

5.2.1 Gravitational Water

'Gravitational water' is the water in excess of the moisture that can be retained by the soil. It translocates as a liquid and can be drained by the gravitational force. It is capable of transmitting hydraulic pressure.

Gravitational water can be subdivided into (*a*) free water (bulk water) and (*b*) Capillary water. Free water may be further distinguished as (*i*) Free surface water and (*ii*) Ground water.

(*a*) **Free water (bulk water):** It has the usual properties of liquid water. It moves at all times under the influence of gravity, or because of a difference in hydrostatic pressure head.

Soil Moisture—Permeability and Capillarity 129

(*i*) *Free surface water:* Free surface water may be from precipitation, run-off, flood-water, melting snow, water from certain hydraulic operations. It is of interest when it comes into contact with a structure or when it influences the ground water in any manner.

Rainfall and run-off are erosive agents which are capable of washing away soil and causing certain problems of strength and stability in the field of geotechnical engineering. The properties of free surface water correspond to those of ordinary water.

(*ii*) *Ground water:* Ground water is that water which fills up the voids in the soil up to the ground water table and translocates through them. It fills coherently and completely all voids. In such a case, the soil is said to be saturated. Ground water obeys the laws of hydraulics. The upper surface of the zone of full saturation of the soil, at which the ground water is subjected to atmospheric pressure, is called the 'Ground water table'. The elevation of the ground water table at a given point is called the 'Ground water level'.

(*b*) **Capillary water:** Water which is in a suspended condition, held by the forces of surface tension within the interstices and pores of capillary size in the soil, is called 'capillary water'. The phenomenon of 'Capillarity' will be studied in some detail in a later section.

5.2.2 Held Water

'Held water' is that water which is held in soil pores or void spaces because of certain forces of attraction. It can be further classified as (*a*) Structural water and (*b*) Absorbed water. Sometimes, even 'capillary water' may be said to belong to this category of held water since the action of capillary forces will be required to come into play in this case.

(*a*) **Structural water:** Water that is chemically combined as a part of the crystal structure of the mineral of the soil grains is called 'Structural water'. Under the loading encountered in geotechnical engineering, this water cannot be separated by any means. Even drying at $105° - 110°C$ does not affect it. Hence structural water is considered as part and parcel of the soil grains.

(*b*) **Adsorbed water:** This comprises, (*i*) hygroscopic moisture and (*ii*) film moisture.

(*i*) *Hygroscopic moisture:* Soils which appear quite dry contain, nevertheless, very thin films of moisture around the mineral grains, called 'hygroscopic moisture', which is also termed 'contact moisture' or 'surface bound moisture'. This form of moisture is in a dense state, and surrounds the surfaces of the individual soil grains as a very thin film. The soil particles derive their hygroscopic moisture not only from water but also from the atmospheric air by the physical force of attraction of unsatisfied ionic bonds on their surfaces. The weight of an oven-dried sample, when exposed to atmosphere, will increase up to a limit, depending upon its maximum hygroscopicity, which, in turn, depends upon the temperature and relative humidity of air, and the characteristics of the soil grains. Coarse-grained soils have relatively low hygroscopic moisture due to their low 'specific surface', or surface area per unit volume. The average hygroscopicity of sands, silts and clays is 1%, 7% and 17% respectively; the high value for clays is because of the very small grain-size and consequent high specific surface. The thickness of the absorbed layer may vary from 200 Å for silts to 30 Å for clays (1 Å = 10^{-7} mm). The hygroscopic moisture film is known to be bound rigidly to the soil grains with an immense force—up to about 10,000 Atmospheres. The nearer the hygroscopic soil moisture is attracted to the surface of the soil grain, the more it is densified. These physical forces are now established to be electro-chemical in nature.

Hygroscopic moisture is affected neither by gravity nor by capillary forces and would not move in the liquid form. It cannot be evaporated ordinarily. However, hygroscopic moisture can be removed by oven-drying at 105° – 110°C. Moisture in this form has properties which differ considerably from those of liquid water —Hygroscopic moisture has greater density, higher boiling point, greater viscosity, greater surface tension, and a much lower freezing point than ordinary water.

Hygroscopic moisture has a pronounced effect on the cohesion and plasticity characteristics of a clayey soil; it also affects the test results of grain specific gravity of the soil. This is because the volume of the displaced water is too low by the amount of hygroscopic moisture, thus leading to higher values of specific gravity than the correct value. (The error could range from 4% to 8% depending upon the hygroscopicity).

(ii) *Film moisture:* Film moisture forms on the soil grains because of the condensation of aqueous vapour; this is attached to the surface of the soil particle as a film upon the layer of the hygroscopic moisture film. This film moisture is also held by molecular forces of high intensity but not as high as in the case of the hygroscopic moisture film. Migration of film moisture can be induced by the application of an external energy potential such as thermal or electric potential; the migration will then be from points of higher temperature/higher potential to points of lower temperature/lower potential. Film moisture does not transmit external hydrostatic pressure. It migrates rather slowly. The greater the specific surface of the soil, the more is the film moisture that can be contained. When the film moisture corresponds to the maximum molecular moisture capacity of the soil, the soil possesses its maximum cohesion and stability.

5.3 NEUTRAL AND EFFECTIVE PRESSURES

As a prerequisite, let us see something about "Geostatic Stresses".

5.3.1 Geostatic Stresses

Stresses within a soil mass are caused by external loads applied to the soil and also by the self-weight of the soil. The pattern of stresses caused by external loads is usually very complicated; the pattern of stresses caused by the self-weight of the soil also can be complicated. But, there is one common situation in which the self-weight of the soil gives rise to a very simple pattern of stresses—that is, when the ground surface is horizontal and the nature of the soil does not vary significantly

Soil Moisture—Permeability and Capillarity 131

in the horizontal directions. This situation exists frequently in the case of sedimentary deposits. The stresses in such a situation are referred to as 'Geostatic Stresses'.

Further, in this situation, there can be no shear stresses upon vertical and horizontal planes within the soil mass. Therefore, the vertical geostatic stress may be computed simply by considering the weight of the soil above that depth.

If the unit weight of the soil is constant with depth,

$$\sigma_v = \gamma \cdot z \qquad \text{...(Eq. 5.1)}$$

where $\sigma_v = $ vertical geostatic stress

$\gamma = $ unit weight of soil

$z = $ depth under consideration

The vertical geostatic stress, thus, varies linearly with depth in this case (Fig. 5.1).

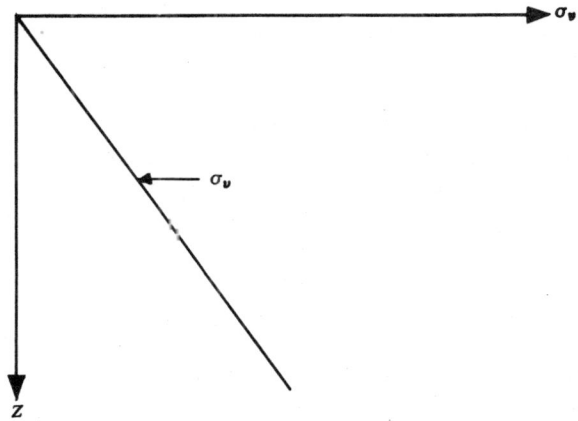

Fig. 5.1 Vertical geostatic stress in soil with horizontal surface.

However, it is known that the unit weight of soil is seldom constant with depth. Usually a soil becomes denser with depth owing to the compression caused by the geostatic stresses. If the unit weight of soil varies continuously with depth, σ_v can be evaluated by means of the integral:

$$\sigma_v = \int_0^z \gamma \cdot dz \qquad \text{...(Eq. 5.2)}$$

If the soil is stratified, with different unit weights for each stratum, σ_v may be computed conveniently by summation:

$$\sigma_v = \Sigma \gamma \cdot \Delta z \qquad \text{...(Eq. 5.3)}$$

5.3.2 Effective and Neutral Pressures

The total stress, either due to self-weight of the soil or due to external applied forces or due to both, at any point inside a soil mass is resisted by the soil grains as also by water present in the pores or void spaces in the case of a saturated soil. (By 'stress' here, we mean the macroscopic stress, i.e., force/total area; the 'contact stresses' at the grain-to-grain contacts will be very high owing to a very small area of contact in relation to the area of cross-section and these are not relevant to this context).

'Neutral stress' is defined as the stress carried by the pore water and it is the same in all directions when, there is static equilibrium since water cannot take static shear stress. This is also called 'pore water pressure' and is designated by u. This will be equal to $\gamma_w \cdot z$ at a depth z below the water table:

$$u = \gamma_w \cdot z \qquad \ldots(\text{Eq. 5.4})$$

'Effective stress' is defined as the difference between the total stress and the neutral stress; this is also referred to as the intergranular pressure and is denoted by:

$$\bar{\sigma} = \sigma - u \qquad \ldots(\text{Eq. 5.5})$$

Equation 5.5 is the 'Effective Stress Equation'.

The effective stress has influence in decreasing the void ratio of the soil and in mobilising the shear strength, while the neutral stress does not have any influence on the void ratio and is ineffective in mobilising the shearing strength.

Thus the 'Effective Stress Principle' may be stated as follows:

(*i*) The effective stress is equal to the total stress minus the pore pressure.

(*ii*) The effective stress controls certain aspects of soil behaviour, notably compressibility and shear strength.

[*Note:* Latest research on the effective stress concept indicates that the effective stress equation has to be modified in the case of saturated clays and highly plastic, dispersed systems such as montmorillonite by introducing the term $(R - A)$, where R is related to the repulsive forces between adjacent clay particles due to electrical charges and A is related to Van der Waals' attractive forces between these particles. Similarly, Bishop *et al.* (1960) proposed a different effective stress equation for partially saturated soils. However, these concepts are of an advanced nature and are outside the scope of the present work.]

For a situation where the water table is at the ground surface, the conditions of stress at a depth from the surface will be as follows:

$$\sigma = \gamma_{sat} \cdot z \qquad \ldots(\text{Eq. 5.6})$$

$$u = \gamma_w \cdot z. \qquad \ldots(\text{Eq. 5.4})$$

By Eq. 5.5,

$$\bar{\sigma} = (\sigma - u) = \gamma_{sat} \cdot z - \gamma_w \cdot z = z(\gamma_{sat} - \gamma_w).$$

Since $(\gamma_{sat} - \gamma_w) = \gamma'$, the submerged unit weight,

$$\bar{\sigma} = \gamma' \cdot z \qquad \ldots(\text{Eq. 5.7})$$

Therefore, the effective stress is computed with the value of the buoyant or effective unit weight.

5.4 FLOW OF WATER THROUGH SOIL-PERMEABILITY

It is necessary for a Civil Engineer to study the principles of fluid flow and the flow of water through soil in order to solve problems involving, – (a) The rate at which water flows through soil (for example, the determination of rate of leakage through an earth dam); (b) Compression (for example, the determination of the rate of settlement of a foundation; and (c) Strength (for example, the evaluation of factors of safety of an embankment). The emphasis in this discussion is on the influence of the fluid on the soil through which it is flowing; in particular on the effective stress.

Soil, being a particulate material, has many void spaces between the grains because of the irregular shape of the individual particles; thus, soil deposits are porous media. In general, all voids in soils are connected to neighbouring voids. Isolated voids are impossible in an assemblage of spheres, regardless of the type of packing; thus, it is hard to imagine isolated voids in coarse soils such as gravels, sands, and even silts. As clays consist of plate-shaped particles, a small percentage of isolated voids would seem possible. Modern methods of identification such as electron micrography suggest that even in clays all voids are interconnected.

Water can flow through the pore spaces in the soil and the soil is considered to be 'permeable'; thus, the property of a porous medium such as soil by virtue of which water (or other fluids) can flow through it is called its 'permeability'. While all soils are permeable to a greater or a smaller degree, certain clays are more or less 'impermeable' for all practical purposes. Permeability is one of the most important of soil properties. The path of flow from one point to another is considered to be a straight one, on a macroscopic scale and the velocity of flow is considered uniform at an effective value; this path, in a microscopic scale, is invariably a tortuous and erratic one because of the random arrangement of soil particles, and the velocity of flow may vary considerably from point to point depending upon the size of the pore and other factors.

According to fundamental hydraulics flowing water may assume either of two characteristic states of motion—the 'laminar flow' and the 'turbulent flow'. In laminar flow each particle travels along a definite path which never crosses the path of other particles; while, in turbulent flow the paths are irregular and twisting, crossing and recrossing at random. Osborne Reynolds, from his classic experiments on flow through pipes, established a lower limit of velocity at which the flow changes from laminar to a turbulent one; it is called the 'lower critical velocity'. In laminar flow, the resistance to flow is primarily due to the viscosity of water and the boundary conditions are not of much significance; in turbulent flow, however, the boundary conditions have a major influence and the effect of viscosity is insignificant.

The lower critical velocity v_c is governed by a dimensionless number, known as Reynold's number:

$$R = \frac{v \cdot D}{\nu} \qquad \ldots(Eq.\ 5.8)$$

or

$$R = \frac{v \cdot D \cdot \gamma_w}{\mu \cdot g} \qquad \ldots(Eq.\ 5.9)$$

Where R = Reynold's number
v = Velocity of flow
D = Diameter of pipe/pore
ν = Kinemnatic viscosity of water
γ_w' = Unit weight of water
μ = Viscosity of water, and
g = Acceleration due to gravity

Reynolds found that v_c is governed by:

$$R = \frac{v_c \cdot D}{\nu} = 2000 \qquad \ldots(Eq.\ 5.10)$$

It is difficult to study the conditions of flow in an individual soil pore; only average conditions existing at any cross-section in a soil mass can be studied. Since pores of most soils are small, flow through them is invariably 'laminar'; however, in the case of soils coarser than coarse sand, the flow may be turbulent. (Assuming uniform particle size, laminar flow may be considered to occur up to an equivalent particle diameter of 0.5 mm).

5.4.1 Darcy's Law

H. Darcy of France performed a classical experiment in 1856, using a set-up similar to that shown in Fig. 5.2, in order to study the properties of the flow of water through a sand filter bed.

By measuring the value of the rate of flow or discharge, q for various values of the length of the sample, L, and pressure of water at top and bottom the sample, h_1 and h_2, Darcy found that q was proportional to $(h_1 - h_2)/L$ or the hydraulic gradient, i:

$$q = k[(h_1 - h_2)/L] \times A = k \cdot i \cdot A \qquad \ldots(Eq.\ 5.11)$$

where
q = the rate of flow or discharge
k = a constant, now known as Darcy's coefficient of permeability
h_1 = the height above datum which the water rose in a standpipe inserted at the entrance of the sand bed,
h_2 = the height above datum which the water rose in a stand pipe inserted at the exit end of the sand bed.
L = the length of the sample.

A = the area of cross-section of the sand bed normal to the general direction of flow.

$i = (h_1 - h_2)/L$, the hydraulic gradient.

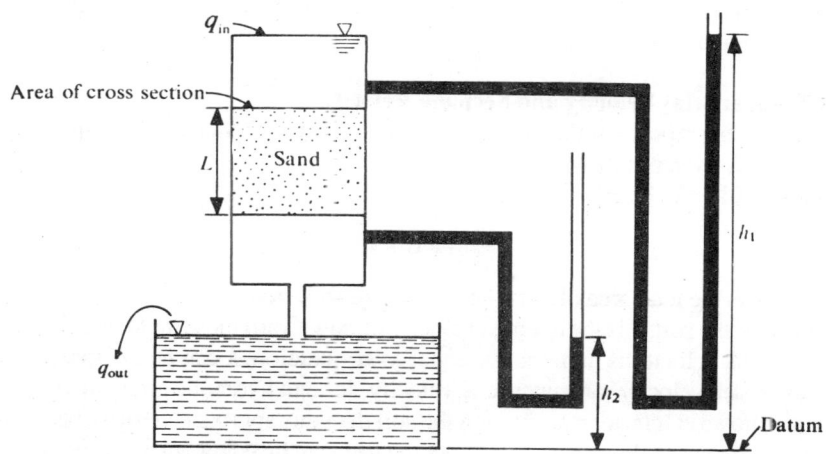

Fig. 5.2 Darcy's Experiment

Equation 5.11 is known as Darcy's law and is valid for laminar flow. It is of utmost importance in geotechnical engineering in view of its wide range of applicability.

Later researchers have established the validity of Darcy's law for most types of fluid flow in soils; Darcy's law becomes invalid only for liquid flow at high velocity or gas flow at very low or at very high velocity.

Darcy's coefficient of permeability provides a quantitative means of comparison for estimating the facility with which water flows through different soils.

It can be seen that k has the dimensions of velocity; it can also be looked upon as the velocity of flow for a unit hydraulic gradient. k is also referred to as the 'coefficient of permeability' or simply 'permeability'.

5.4.2 Validity of Darcy's Law

Reynolds found a lower limit of critical velocity for transition of the flow from laminar to a turbulent one, as already given by Eq. 5.10.

Many researchers have attempted to use Reynolds' concept to determine the upper limit of the validity of Darcy's law. (Muskat, 1946; Scheidegger, 1957). The values of R for which the flow in porous media become turbulent have been measured as low as 0.1 and as high as 75. According to Scheidegger, the probable reason that porous media do not exhibit a definite critical Reynold's number is because soil can by no means be accurately represented as a bundle of straight tubes. He further discussed several reasons why flow through very small openings may not follow Darcy's law.

There is overwhelming evidence which shows that Darcy's law holds in silts as well as medium sands and also for a steady state flow through clays. For soils more pervious than medium sand, the actual relationship between the hydraulic gradient and velocity should be obtained only through experiments for the particular soil and void ratio under study.

5.4.3 Superficial Velocity and Seepage Velocity

Darcy's law represents the macroscopic equivalent of Navier-Stokes' equations of motion for viscous flow.

Equation 5.11 can be rewritten as:

$$\frac{q}{A} = k.i. = v \qquad \ldots(\text{Eq. 5.12})$$

Since A is the total area of cross-section of the soil, same as the open area of the tube above the soil, v is the average velocity of downward movement of a drop of water. This velocity is numerically equal to ki; therefore k can be interpreted as the 'approach velocity' or 'superficial velocity' for unit hydraulic gradient. A drop of water flows at a faster rate through the soil than this approach velocity because the average area of flow channel through the soil is reduced owing to the presence of soil grains. This reduced flow channel may be schematically represented as shown in Fig. 5.3:

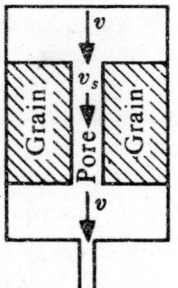

Fig. 5.3 Flow channel

By the principle of continuity, the velocity of approach, v, may be related to the seepage velocity or average effective velocity of flow, v_s, as follow:

$$q = A \cdot v = A_v \cdot v_s$$

where A_v = area of cross-section of voids

$$\therefore \quad v_s = v \cdot \frac{A}{A_v} = v \cdot \frac{AL}{A_v L} = v \cdot \frac{V}{V_v} = \frac{v}{n}$$

$$v_s = v/n = ki/n \qquad \ldots(\text{Eq. 5.13})$$

where n = porosity (expressed as a fraction).

Thus, seepage velocity is the superficial velocity divided by the porosity. This gives the average velocity of a drop of water as it passes through the soil in the direction of flow; *this is the straight dimension of the soil in the direction of flow divided by the time required for the drop to flow through this distance.* As pointed out earlier, a drop of water flowing through the soil takes a winding path with varying velocities; therefore, v_s is a fictitious velocity obtained by assuming that the drop of water moves in a straight line at a constant velocity through the soil.

Even though the superficial velocity and the seepage velocity are both fictitious quantities, they can be used to compute the time required for water to move through a given distance in soil.

Equation 5.13 indicates that the seepage velocity is also proportional to the hydraulic gradient. It may be rewritten as follows:

$$v_s = ki/n = (k/n) i = k_p \cdot i \qquad \text{...(Eq. 5.14)}$$

where k_p, the constant of proportionality, is called the 'Coefficient of Percolation', and is given by:

$$k_p = k/n \qquad \text{...(Eq. 5.15)}$$

5.4.4 Energy Heads

In the study of fluid flow it is convenient to express energy in any form in terms of 'head', which is energy per unit mass:

1. Pressure head, h_p = the pressure divided by the unit weight of fluid = p/ρ.

$$\left(\text{Pressure energy} = \frac{p \cdot M}{\rho} = \frac{M}{L^2} \cdot \frac{M \cdot L^3}{M} = ML, \text{ Head} = \frac{\text{energy}}{\text{mass}} = \frac{ML}{M} = L \right)$$

2. Elevation or datum head, h_e = the height from the datum (Elevation or potential energy = ML)

3. Velocity Head, h_v = square of velocity divided by twice acceleration due to gravity = $\dfrac{v^2}{2g}$

$$\left(\text{Kinetic energy} = \frac{Mv^2}{2g} = \frac{ML^2T^2}{T^2L} = ML \right).$$

Here,

M = Mass, v = Velocity, g = Acceleration due to Gravity
L = Length, T = Time, p = Pressure.

In dealing with problems involving fluid flow in soil, the velocity head is taken to be negligible, and as such, the total head will be the sum of pressure head and elevation head. In dealing with problems involving pipe and channel flow, total head is defined as the sum of pressure head, elevation head and velocity head; the sum of pressured head and elevation head is usually called the 'Piezometric' head. In the case of flow through soils. the total head and the piezometric head are equal.

Since both pressure head and elevation head can contribute to the movement of fluid through soils, it is the total head that determines flow, and the hydraulic gradient to be used in Darcy's law is computed from the difference in total head. Unless there is a gradient of total head, no flow can occur.

Pressure head or water pressure at a point in a soil mass can be determined by a piezometer; the height to which water rises in the piezometer above the point is the pressure head at that point. The manometer or standpipe and the Bourdon pressure gauge are two simple piezometers, which require a flow of water from the soil into the measuring system to actuate each device. This flow may require a significant time-lag if the soil is a relatively impermeable one such as silt or clay. To measure pore pressures under 'no-flow' conditions various types of piezometers have been developed. (Lambe, 1948; Bishop 1961; Whitman *et al.* 1961). Piezometers such as 'Casagrande Piezometer' have been developed for use in field installations.

When instantaneous stress release occurs, even a negative value of pore pressure can develop, with values which might go below even absolute zero (minus 1 Atmosphere).

In general, it is more convenient to determine first the elevation and total heads and then compute the pressure head by subtracting the elevation head from the total head.

Thus, the following are the interesting points in energy heads:
(*a*) The velocity head in soils is negligible.
(*b*) Negative pore pressure can exist.
(*c*) Direction of flow is determined by the difference in total head.
(*d*) Elevation and total heads are determined first, and then the pressure head by difference.
(*e*) Absolute magnitude of elevation head, which depends upon the location of the datum, is not important.

5.5 THE DETERMINATION OF PERMEABILITY

The permeability of a soil can be measured in either the laboratory or the field; laboratory methods are much easier than field methods. Field determinations of permeability are often required because permeability depends very much both on the microstructure—the arrangement of soil-grains—and on the macrostructure —such as stratification, and also because of the difficulty of getting representative soil samples. Laboratory methods permit the relationship of permeability to the void ratio to be studied and are thus usually run whether or not field determinations are made.

The following are some of the methods used in the laboratory to determine permeability.
1. Constant head permeameter
2. Falling or variable head permeameter
3. Direct or indirect measurement during an Oedometer test
4. Horizontal capillarity test

The following are the methods used in the field to determine permeability.
1. Pumping out of wells
2. Pumping into wells

In both these cases, the aquifer or the water-bearing stratum, can be 'confined' or 'unconfined'.

Permeability may also be computed from the grain-size or specific surface of the soil, which constitutes an indirect approach.

The various methods will be studied in the following sub-sections.

5.5.1 Constant-Head Permeameter

A simple set-up of the constant-head permeameter is shown in Fig. 5.4:

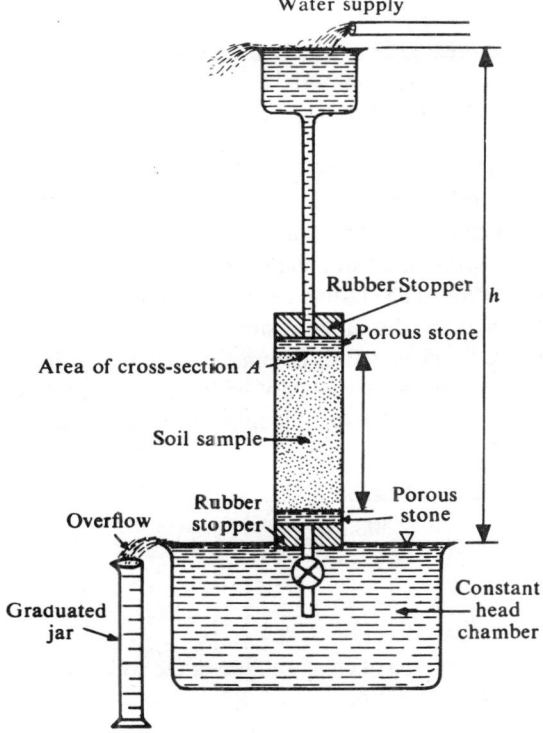

Fig. 5.4 Set-up of the constant-head permeameter

The principle in this set-up is that the hydraulic head causing flow is maintained constant; the quantity of water flowing through a soil specimen of known cross-sectional area and length in a given time is measured. In highly impervious soils the quantity of water that can be collected will be small and, accurate measurements are difficult to make. Therefore, the constant head permeameter is mainly applicable to relatively pervious soils, although, theoretically speaking, it can be used for any type of soil.

If the length of the specimen is large, the head lost over a chosen convenient length of the specimen may be obtained by inserting piezometers at the ends of the specified length.

If Q is the total quantity of water collected in the measuring jar after flowing through the soil in an elapsed time t, from Darcy's law,

$$q = Q/t = k.i.A$$

$$\therefore \quad k = (Q/t).(1/iA) = (Q/t).(L/Ah) = QL/thA \quad \ldots\text{(Eq. 5.16)}$$

where
    $k =$ Darcy's coefficient of permeability
    L and $A =$ length and area of cross-section of soil specimen
    $h =$ hydraulic head causing flow.

The water should be collected only after a steady state of flow has been established.

The constant head permeameter is widely used owing to its simplicity in principle. However, certain modifications will be required in the set-up in order to get reasonable precision in the case of soils of low permeability.

5.5.2 Falling or Variable Head Permeameter

A simple set-up of the falling, or variable head permeameter is shown in Fig. 5.5

A better set-up in which the top of the standpipe is closed, with manometers and vacuum supply, may also be used to enhance the accuracy of the observations (Lambe and Whitman, 1969). The falling head permeameter is used for relatively less permeable soils where the discharge is small.

The water level in the stand-pipe falls continuously as water flows through the soil specimen. Observations should be taken after a steady state of flow has reached. If the head or height of water level in the standpipe above that in the constant head chamber falls from h_0 to h_1, corresponding to elapsed times t_0 and t_1, the coefficient of permeability, k, can be shown to be:

$$k = \frac{2.303 \, aL}{A(t_1 - t_0)} \cdot \log_{10}(h_0/h_1) \quad \ldots\text{(Eq. 5.17)}$$

where
    $a =$ area of cross-section of standpipe
    L and $A =$ length and area of cross-section of the soil sample and the other quantities as defined.

This can be derived as follows:

Let $- dh$ be the change in head in a small interval of time dt. (Negative sign indicates that the head decreases with increase in elapsed time).
From Darcy's law,

$$Q = (-a.dh)/dt = k.i.a$$

$$-adh/dt = K.A.h/L$$

Soil Moisture—Permeability and Capillarity 141

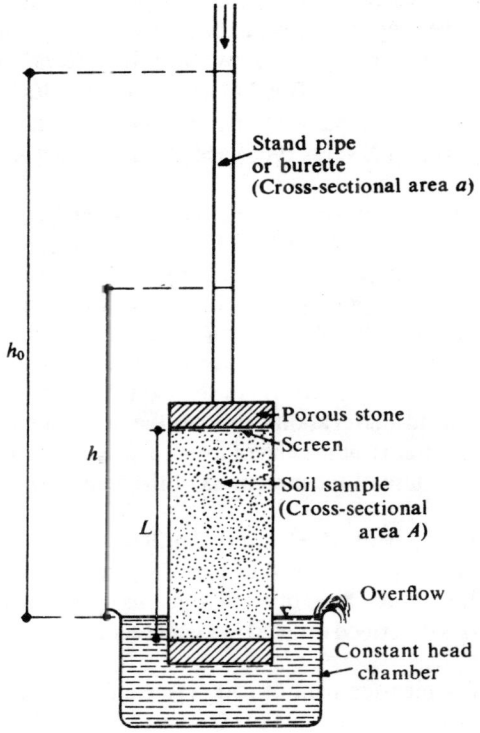

Fig. 5.5 Falling, or variable, head permeameter

$$\therefore \quad (kh/L) \cdot A = -a \cdot \frac{dh}{dt}$$

or
$$(kA/aL) \cdot dt = -dh/h$$

Integrating both sides and applying the limits t_0 and t_1 for t, and h_0 and h_1 for h,

$$\frac{kA}{aL} \int_{t_0}^{t_1} dt = -\int_{h_0}^{h_1} \frac{dh}{h} = \int_{h_1}^{h_0} \frac{dh}{h}$$

$$\therefore \quad (kA/aL)(t_1 - t_0) = \log_e (h_0/h_1)$$

$$= 2.3 \log_{10} (h_0/h_1).$$

Transposing the terms,

$$k = \frac{2.303 \, aL}{A \, (t_1 - t_0)} \cdot \log_{10} (h_0/h_1)$$

which is Eq. 5.17.

The 'Jodhpur permeameter' developed at the M.B.M. Engineering College, Jodhpur, may be conveniently used for conducting the falling head as well as constant head tests on remoulded as well as undisturbed specimens. Remoulded specimens may be prepared by static or dynamic compaction. The apparatus has been patented and manufactured by 'AIMIL'. (M/s. Associated Instrument Manufacturers India Limited, Bombay). The detailed description of the apparatus and the procedure for the permeability tests are given in the relevant Indian standards. [(IS: 2720 Part XVII—1966) and (IS: 2720 Part XXXVI—1975)].

5.5.3 Direct or Indirect Measurement During an Oeodometer Test

As discussed in Chapter 7, the rate of consolidation of a soil depends directly on the permeability. The permeability can be computed from the measured rate of consolidation by using appropriate relationships. Since there are several quantities in addition to permeability that enter into the rate of consolidation-permeability relationship, this method is far from precise since these quantities cannot easily be determined with precision. Instead of the indirect approach, it would be better to run a constant-head permeability test on the soil sample in the oedometer or consolidation apparatus, at the end of a compression increment. This would yield precise results because of the directness of the approach.

5.5.4 Permeability from Horizontal Capillarity Test

The 'Horizontal capillarity test' or the 'Capillarity-Permeability test', used for determining the capillary head of a soil, can also be used to obtain the permeability of the soil. This is described in detail in a later section dealing with the phenomenon of 'Capillarity'.

The laboratory measurement of soil permeability, although basically straightforward, requires good technique to obtain reliable results. The reader is referred to Lambe (1951) for an exhaustive treatment of measurement of permeability.

5.5.5 Determination of Permeability—Field Approach

The average permeability of a soil deposit or stratum in the field may be somewhat different from the values obtained from tests on laboratory samples; the former may be determined by pumping tests in the field. But these are time-consuming and costlier.

A few terms must be understood in this connection. 'Aquifer' is a permeable formation which allows a significant quantity of water to move through it under field conditions. Aquifers may be 'Unconfined aquifers' or 'Confined aquifers'. Unconfined aquifer is one in which the ground water table is the upper surface of the zone of saturation and it lies within the test stratum. It is also called 'free', 'phreatic' or 'non-artesian' aquifer. Confined aquifer is one in which ground water remains entrapped under pressure greater than atmospheric, by overlying relatively impermeable strata. It is also called 'artesian aquifer'. 'Coefficient of Transmissibility' is defined as the rate of flow of water through a vertical strip of aquifer of unit width and extending the full height of saturation under unit hydraulic gradient. This coefficient is obtained by multiplying the field coefficient of permeability by the thickness of the aquifer.

Soil Moisture—Permeability and Capillarity 143

When a well is penetrated into a homogeneous aquifer, the water table in the well initially remains horizontal. When water is pumped out from the well, the aquifer gets depleted of water, and the water table is lowered resulting in a circular depression in the phreatic surface. This is referred to as the 'Drawdown curve' or 'Cone of depression'. The analysis of flow towards such a well was given by Dupuit (1863) and modified by Thiem (1870).

In pumping-out tests, drawdowns corresponding to a steady discharge are observed at a number of observation wells. Pumping must continue at a uniform rate for an adequate time to establish a steady-state condition, in which the drawdown changes negligibly with time.

The following assumptions are relevant to the discussion that would follow:
(i) The aquifer is homogeneous with uniform permeability and is of infinite areal extent.
(ii) The flow is laminar and Darcy's law is valid.
(iii) The flow is horizontal and uniform at all points in the vertical section.
(iv) The well penetrates the entire thickness of the aquifer.
(v) Natural groundwater regime affecting the aquifer remains constant with time.
(vi) The velocity of flow is proportional to the tangent of the hydraulic gradient (Dupuit's assumption).

Unconfined Aquifer

A well penetrating an unconfined aquifer to its full depth is shown in Fig. 5.6.

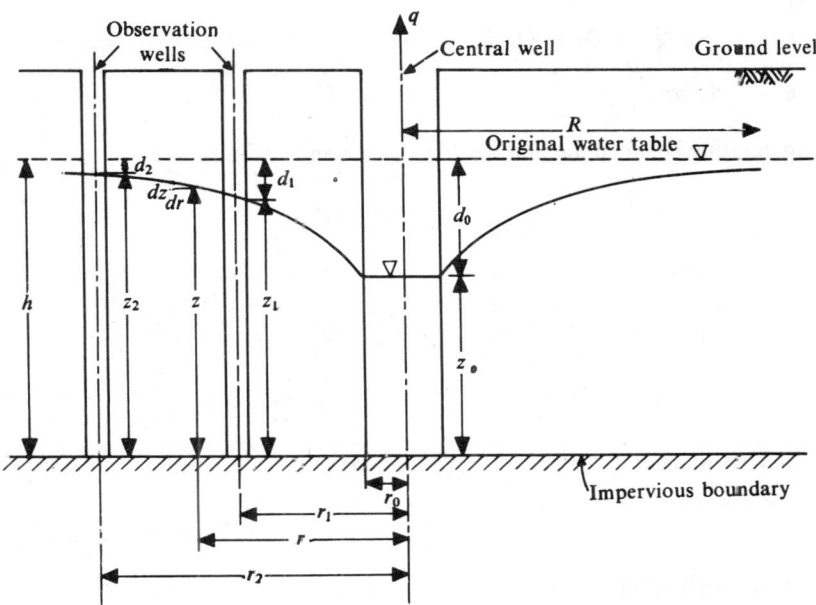

Fig. 5.6 Flow toward a well in an unconfined aquifer

Let r_0 be radius of central well,

r_1 and r_2 be the radial distances from the central well to two of the observation wells,

z_1 and z_2 be the corresponding heights of a drawdown curve above the impervious boundary,

z_0 be the height of water level after pumping in the central well above the impervious boundary,

d_0, d_1 and d_2 be the depths of water level after pumping from the initial level of water table, or the respective drawdowns at the central well and the two observation wells respectively,

h be the initial height of the water table above the impervious layer ($h = z_0 + d_0$, obviously) and,

R be the radius of influence or the radial distance from the central well of the point where the drawdown curve meets the original water table.

Let r and z be the radial distance and height above the impervious boundary at any point on the drawdown curve.

By Darcy's law, the discharge q is given by:

$q = k.A.dz/dr$, since the hydraulic gradient, i, is given by dz/dr by Dupuit's assumption.

Here,

k is the coefficient of permeability.

But $A = 2\pi rz$.

$\therefore \quad q = k.2\pi rz.dz/dr$

or $\quad k.z\,dz = \left(\dfrac{q}{2\pi}\right) \cdot \dfrac{dr}{r}$

Integrating between the limits r_1 and r_2 for r and z_1 and z_2 for z,

$$k\left\{\dfrac{z^2}{2}\right\}_{z_1}^{z_2} = (q/2\pi)\{\log_e r\}_{r_1}^{r_2}$$

$\therefore \quad k \cdot \dfrac{(z_2^2 - z_1^2)}{2} = (q/2\pi)\left(\log_e \dfrac{r_2}{r_1}\right)$

$\therefore \quad k = \dfrac{q}{\pi(z_2^2 - z_1^2)} \log_e \dfrac{r_2}{r_1} = \dfrac{q}{1.36\,(z_2^2 - z_1^2)} \log_{10} \dfrac{r_2}{r_1}$ ...(Eq. 5.18)

k can be evaluated if z_1, z_2, r_1 and r_2 are obtained from observations in the field. It can be noted that $z_1 = (h - d_1)$ and $z_2 = (h - d_2)$.

If the extreme limits z_0 and h at r_0 and R are applied,

Equation 5.18 reduces to

$$k = \dfrac{q}{1.36\,(h^2 - z_0^2)} \cdot \log_{10} \dfrac{R}{r_0} \quad \ldots\text{(Eq. 5.19)}$$

This may also be put in the form

$$k = \frac{q}{1.36\, d_0(d_0 + 2z_0)} \cdot \log_{10} \frac{R}{r_0} \qquad \ldots(\text{Eq. 5.20})$$

For one to be in a position to use (Eq. 5.19) or (Eq. 5.20), one must have an idea of the radius of influence R. The selection of a value for R is approximate and arbitrary in practice. Sichart gives the following approximate relationship between R, d_0 and k;

$$R = 3000\, d_0 \sqrt{k} \qquad \ldots(\text{Eq. 5.21})$$

where,

d_0 is in metres,
k is in metres/sec,
and R is in metres.

One must apply an approximate value for the coefficient of permeability here, which itself is the quantity sought to be determined.

Two observation wells may not be adequate for obtaining reliable results. It is recommended that a few symmetrical pairs of observation wells be used and the average values of the drawdowns which, strictly speaking, should be equal for observation wells, located symmetrically with respect to the central well, be employed in the computations. Several values may be obtained for the coefficient of permeability by varying the combination of the wells chosen for the purpose. Hence, the average of all these is treated to be a more precise value than when just two wells are observed.

Alternatively, when a series of wells is used, a semi-logarithmic graph may be drawn between r to the logarithmic scale and z^2 to the natural scale, which will be a straight line. From this graph, the difference of ordinates y, corresponding to the limiting abscissae of one cycle is substituted in the following equation to obtain the best fit value of k for all the observations:

$$k = q/y \qquad \ldots(\text{Eq. 5.22})$$

This is a direct consequence of Eq. 5.18, observing that $\log_{10}(r_2/r_1) = 1$ and denoting $(z_2^2 - z_1^2)$ by y.

Confined Aquifer

A well penetrating a confined aquifer to its full depth is shown in Fig. 5.7:

The notation in this case is precisely the same as that in the case of the unconfined aquifer; in addition, H denotes the thickness of the confined aquifer, bounded by impervious strata.

By Darcy's law, the discharge q is given by:

$q = k \cdot A \cdot dz/dr$, as before.

But the cylindrical surface area of flow is given by $A = 2\pi r H$, in view of the confined nature of the aquifer.

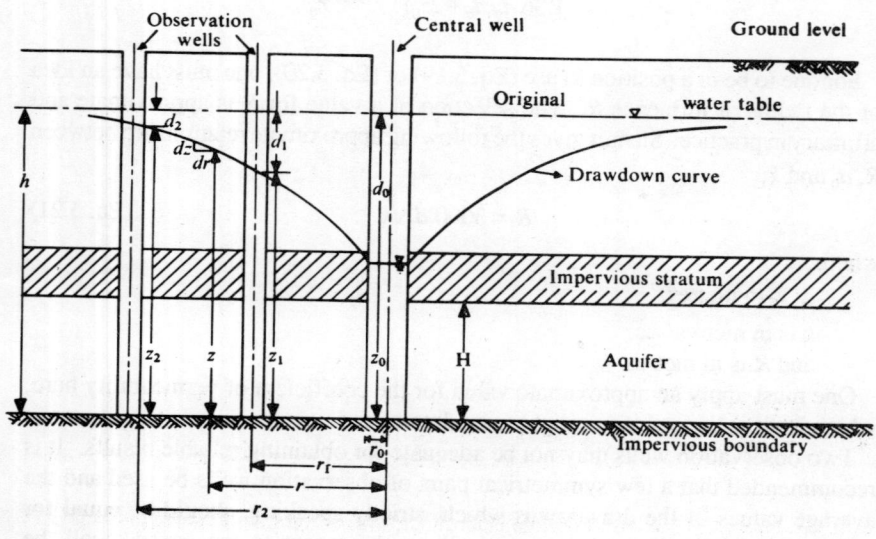

Fig. 5.7 Flow toward a well in a confined aquifer

$$\therefore \quad q = k \cdot 2\pi r H \cdot dz/dr$$

or
$$k \cdot dz = \frac{q}{2\pi H} \cdot \frac{dr}{r}.$$

Integrating both sides within the limits z_1 and z_2 for z, and r_1 and r_2 for r,

$$k[z]_{z_1}^{z_2} = \frac{q}{2\pi H} [\log_e r]_{r_1}^{r_2}$$

or
$$k(z_2 - z_1) = \frac{q}{2\pi H} \cdot \log_e \frac{r_2}{r_1}$$

or
$$k = \frac{q}{2\pi H (z_2 - z_1)} \cdot \log_e \frac{r_2}{r_1}$$

Since $z_1 = (h - d_1)$ and $z_2 = (h - d_2)$, $(z_2 - z) = (d_1 - d_2)$

Substituting, we have :

$$k = \frac{q}{2\pi H (d_1 - d_2)} \cdot \log_e \frac{r_2}{r_1} = \frac{q}{2.72 \, H (d_1 - d_2)} \cdot \log_{10} \frac{r_2}{r_1} \quad \ldots \text{(Eq. 5.23)}$$

Since the coefficient of transmissibility, T, by definition, is given by kH,

Soil Moisture—Permeability and Capillarity 147

$$T = \frac{q}{2.72\,(d_1 - d_2)} \cdot \log_{10}(r_2/r_1) \qquad \ldots(\text{Eq. 5.24})$$

If the extreme limits z_0 and h at r_0 and R are applied, we get:

$$k = \frac{q}{2\pi H\,(h - z_0)} \cdot \log_e (R/r_0)$$

But $(h - z_0) = d_0$

$$\therefore \quad k = \frac{q}{2\pi H \cdot d_0} \cdot \log_e (R/r_c) = \frac{q}{2.72\,H d_0} \cdot \log_{10}(R/r_0) \qquad \ldots(\text{Eq. 5.25})$$

Since $T = kH$,

$$T = \frac{q}{2.72\,d_0} \cdot \log_{10} \frac{R}{r_0} \qquad \ldots(\text{Eq. 5.26})$$

The field practice is to determine the average value of the coefficient of transmissibility from the observation of drawdown values from a number of wells. A convenient procedure for this is as follows:

A semi-logarithmic graph is plotted with r to the logarithmic scale as abscissa and d to the natural scale as ordinate, as shown in Fig. 5.8:

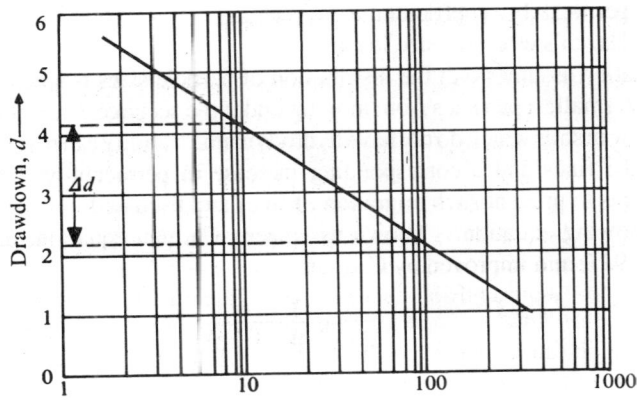

Fig. 5.8 Determination of T

From Equation 5.24,

$$T = \frac{q}{2.72 \cdot \Delta d} \qquad \ldots(\text{Eq. 5.27})$$

if r_2 and r_1 are chosen such that $r_2/r_1 = 10$ and Δd is the corresponding value of the difference in drawdowns, $(d_2 - d_1)$.

Thus, from the graph, d may be got for one logarithmic cycle of abscissa and substituted in Eq. 5.27 to obtain the coefficient of transmissibility, T.

The coefficient of permeability may then be computed by using the relation $k = T/H$, where H is the thickness of the confined aquifer.

Pumping-in tests have been devised by the U.S. Bureau of Reclamation (U.S.B.R.) for a similar purpose.

Field testing, though affording the advantage of obtaining the *in-situ* behaviour of soil deposits, is laborious and costly.

5.5.6 Indirect Methods of Determination of Permeability

These constitute the methods which relate the permeability with grain-size and with specific surface of the soil. The influence of grain-size on permeability is also referred to in Sec. 5.6 and Sub-sec. 5.6.2.

It is logical that the smaller the grain-size, smaller are the voids, which constitute the flow channels, and hence the lower is the permeability. A relationship between permeability and particle-size is much more reasonable in silts and sands than in clays, since the particles are nearly equidimensional in the case of silts and sands. From his experimental work on sands, Allen Hazen (1892, 1911) proposed the following equation:

$$k = 100 D_{10}^2 \quad \ldots\text{(Eq. 5.28)}$$

where

k = permeability coefficient in cm/sec, and

D_{10} = Hazen's effective size in cm.

This relation assumes that the distribution of particle sizes is spread enough to prevent the smallest particles from moving under the seepage force of the flowing water. Flow in soils which do not have hydrodynamic stability can result in washing away of the fines and a corresponding increase in permeablity. Particle-size requirements to prevent such migration of fines are discussed in Chapter 6.

The following equation is known as Kozeny-Carman equation, proposed by Kozeny (1927) and improved by Carman:

$$k = \frac{1}{k_0 \cdot S^2} \cdot \frac{\gamma}{\mu} \cdot \frac{e^3}{(1+e)} \quad \ldots\text{(Eq. 5.29)}$$

where,

e = void ratio,

γ = unit weight of fluid,

μ = viscosity of fluid,

S = specific surface area, and

k_0 = factor depending on pore shape and ratio of length of actual flow path to the thickness of soil bed.

Loudon (1952) developed the following empirical relationship:

$$\log_{10}(kS^2) = a + bn \quad \ldots\text{(Eq. 5.30)}$$

Here a and b are constants, their values being 1.365 and 5.15 respectively at 10°C, and n is the porosity.

5.6 FACTORS AFFECTING PERMEABILITY

Since permeability is the property governing the ease with which a fluid flows through the soil, it depends on the characteristics of the fluid, or permeant, as well as those of the soil.

An equation reflecting the influence of the characteristics of the permeant fluid and the soil on permeability was developed by Taylor (1948) based on Poiseuille's law for laminar flow through a circular capillary tube. The flow through a porous medium is considered similar to a flow through a bundle of straight capillary tubes. The equation is:

$$k = D_s^2 \cdot \frac{\gamma}{\mu} \cdot \frac{e^3}{(1+e)} \cdot C \qquad \ldots(\text{Eq. 5.31})$$

in which,
k = Darcy's coefficient of permeability
D_s = effective particle-size
γ = unit weight of permeant
μ = viscosity of permeant
e = void ratio
C = shape factor

This equation helps one in analysing the variables affecting permeability. The characteristics of the permeant are considered first and those of the soil next.

5.6.1 Permeant Fluid Properties

Equation 5.31 indicates that the permeability is influenced by both the viscosity and the unit weight of the permeant fluid. In the field of soil mechanics, the engineer will have occasion to deal with only water as the common permeant fluid. The unit weight of water does not significantly vary, but its viscosity does vary significantly with temperature. It is easy to understand that the permeability is directly proportional to the unit weight and inversely proportional to the viscosity of the permeant fluid.

It is common practice to determine the permeability at a convenient temperature in the laboratory and reduce the results to a standard temperature; this standard temperature is 27°C as per I.S. Code of practice. (IS: 2720 Part XVII-1966 and its revised versions). This is done by using the following equation:

$$k_{27} = k_T \cdot \frac{\mu_T}{\mu_{27}} \qquad \ldots(\text{Eq. 5.32})$$

where k_T and μ_T are the permeability of soil and the viscosity of water at the test temperature of t°C and, k_{27} and μ_{27} are the permeability and viscosity at the standard temperature, i.e., 27°C.

According to Muskat (1937), these two permeant characteristics, that is, viscosity and unit weight, can be eliminated as variables by defining a more general permeability, K, as follows:

$$K = \frac{k \cdot \mu}{\gamma} \qquad \ldots\text{(Eq. 5.33)}$$

where

K = specific, absolute, or physical permeability
μ = viscosity of the permeant
γ = unit weight of the permeant
k = Darcy's coefficient of permeability

Since k has the units L/T, K has the units L^2. K is also expressed in Darcy's; 1 darcy being equal to 0.987×10^{-8} cm^2. K has the same value for a particular soil, for all fluids and at all temperatures as long as the void ratio and structure of the soil remain unaltered.

Viscosity and unit weight are considered to be the only variables of the permeant fluid that influence the permeability of pervious soils; however, other permanent characteristics can have a major influence on the permeability of relatively impervious soils. The effects of viscosity and unit weight may be eliminated by expressing the permeability in terms of the absolute permeability. It has been found by Michaels and Lin (1954) that the values of absolute permeability of Kaolinite varies significantly with the nature of the permeant fluid, when the comparisons are made at the same void ratio. Further, they found that the variation was large when the kaolinite was moulded in the fluid which was to be used as the permeant than when water was used as the moulding fluid and initial permeant, each succeeding permeant displacing the preceding one. These differences in permeability at the same void ratio have been attributed to the changes in the soil fabric resulting from a sample preparation in the different fluids. The effect of the soil fabric will be discussed in next sub-section.

5.6.2 Soil Characteristics

The following soil characteristics have influence on permeability:
1. Grain-size
2. Void ratio
3. Composition
4. Fabric or structural arrangement of particles
5. Degree of saturation
6. Presence of entrapped air and other foreign matter

Equation 5.31 indicates directly only grain-size and void ratio as having influence on permeability. The other characteristics are considered indirectly or just ignored. Unfortunately, the effects of one of these are difficult to isolate in view of the fact that these are closely interrelated; for example, fabric usually depends on grain-size, void ratio and composition.

Grain-size

Equation 5.31 suggests that the permeability varies with the square of particle diameter. It is logical that the smaller the grain-size the smaller the voids and thus the lower the permeability. A relationship between permeability and grain-size is more appropriate in case of sands and silts than that of other soils since the grains are more nearly equidimensional and fabric changes are not significant.

As already stated in sub-section 5.5.6, Allen Hazen proposed,

$k = 100 \, D_{10}^2$ where D_{10} is in cm and k is in cm/s.

Void Ratio

Equation 5.31 indicates that a plot of k versus $e^3/(1+e)$ should be a straight line. This is more true of coarse grained soils since the shape factor C does not change appreciably with the void ratio for these soils.

Other theoretical equations have suggested that k versus $e^2/(1+e)$ or k versus e^2 should be a straight line. It is interesting to note that, as indicated in Fig. 5.9, a plot of log k versus e approximates a straight line for many soils within a wide range of permeability values.

This suggests a simple method for the permeability of a soil at any void ratio when values of permeability are known at two or more void ratios. Once the line is drawn, the permeability at any void ratio may be read directly.

Increase in the porosity leads to an increase in the permeability of a soil for two distinct reasons. Firstly, it causes an increase in the percentage of cross-sectional area available for flow. Secondly, it causes an increase in the dimension of the pores, which increases the average velocity, through an increase in the hydraulic mean radius, which enters the derivation of Eq. 5.31, and which, in turn, is dependent on the void ratio.

Composition

The influence of soil composition on permeability is generally of little significance in the case of gravels, sands, and silts, unless mica and organic matter are present. However, this is of major importance in the case of clays. Montmorillonite has the least permeability; in fact, with sodium as the exchangeable ion, it has the lowest permeability (less than 10^{-7} cm/s, even at a very high void ratio of 15). Therefore, sodium montmorillonite is used by the engineer as an additive to other soils to make them impermeable. Kaolinite is a hundred times more permeable than montmorillonite.

Fabric or Structural Arrangement of Particles

The fabric or structural arrangement of particles is an important soil characteristic influencing permeability, especially of fine-grained soils. At the same void ratio, it is logical to expect a soil in the most flocculated state will have the highest permeability, and the one in the most dispersed state will have the lowest permeability. Remoulding of a natural soil invariably reduces the permeability.

152 *Geotechnical Engineering*

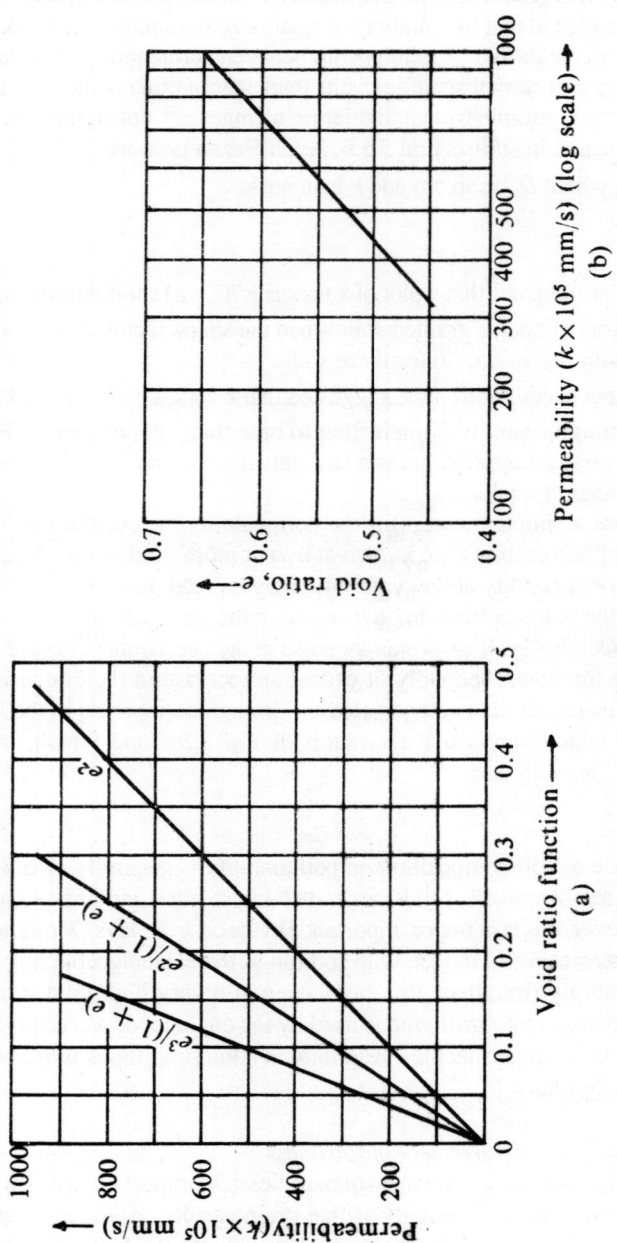

Fig. 5.9 Permeability-void ratio relationships

Stratification or macrostructure also has great influence; the permeability parallel to stratification is much more than that perpendicular to stratification, as will be shown in a later section.

Degree of Saturation

The higher the degree of saturation, the higher the permeability. In the case of certain sands the permeability may increase three-fold when the degree of saturation increases from 80% to 100%.

Presence of Entrapped Air and Other Foreign Matter

Entrapped air has pronounced effect on permeability. It reduces the permeability of a soil. Organic foreign matter also has the tendency to move towards flow channels and choke them, thus decreasing the permeability. Natural soil deposits in the field may have some entrapped air or gas for several reasons. In the laboratory, air-free distilled water may be used a vacuum applied to achieve a high degree of saturation. However, this may not lead to a realistic estimate of the permeability of a natural soil deposit.

The importance of duplicating or simulating field conditions is emphasised by the preceding discussion on the factors affecting permeability, when the aim is to determine field permeability in the laboratory.

5.7 VALUES OF PERMEABILITY

Table 5.1 Typical Values of Permeability (S.B. Sehgal, 1967)

Soil description	Coefficient of permeability mm/s	Degree of permeability (After Terzaghi & Peck, 1948)
Coarse gravel	Greater than 1	High
Fine gravel—fine sand	1 to 10^{-2}	Medium
Silt-sand admixtures, loose silt, rock flour, and loess	10^{-2} to 10^{-4}	Low
Dense silt, clay-silt admixtures, non-homogeneous clays	10^{-4} to 10^{-6}	Very low
Homogeneous clays	Less than 10^{-6}	Almost impervious

5.8 PERMEABILITY OF LAYERED SOILS

Natural soil deposits may exhibit stratification. Each layer may have its own coefficient of permeability, assuming it to be homogeneous. The 'average permeability' of the entire deposit will depend upon the direction of flow in relation to the orientation of the bedding planes.

Two cases will be considered—the first one with flow perpendicular to the bedding planes and the next with flow parallel to the bedding planes.

Flow Perpendicular to the Bedding Planes

Let the flow be perpendicular to the bedding planes as shown in Fig. 5.10.

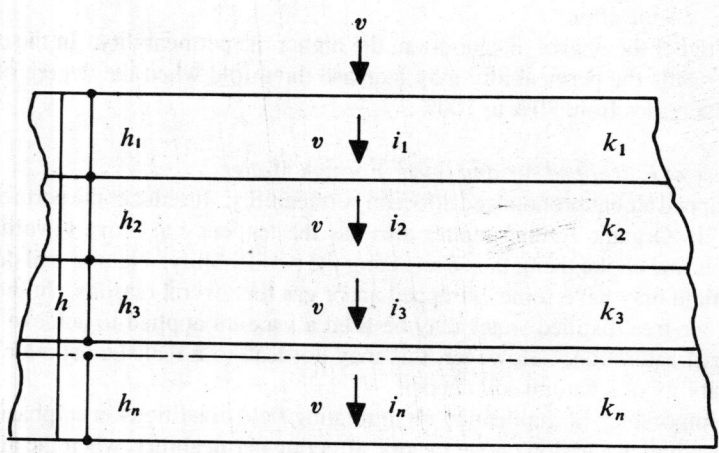

Fig. 5.10 Flow perpendicular to bedding planes

Let h_1, h_2, h_3 ... h_n be the thicknesses of each of the n layers which constitute the deposit, of total thickness h. Let k_1, k_2, k_3 ... k_n be the Darcy coeffieients of permeability of these layers respectively.

In this case, the velocity of flow v, and hence the discharge q, is the same through all the layers, for the continuity of flow.

Let the total head lost be Δh and the head lost in each of the layers be Δh_1, Δh_2, Δh_3, ... Δh_n.

$$\Delta h = \Delta h_1 + \Delta h_2 + \Delta h_3 + ... \Delta h_n.$$

The hydraulic gradients are:

$$i_1 = \Delta h_1/h_1$$
$$i_2 = \Delta h_2/h_2$$
$$i_n = \Delta h_n/h_n$$

If i is the gradient for the deposit, $i = \Delta h/h$

Since q is the same in all the layers, and area of cross-section of flow is the same, the velocity is the same in all layers.

Let k_z be the average permeability perpendicular to the bedding planes.

Now
$$k_z \cdot i = k_1 i_1 = k_2 i_2 = k_3 i_3 = \; ... \; k_n i_n = v$$

$$\therefore \quad k_z \Delta h/h = k_1 \Delta h_1/h_1 = k_2 \Delta h_2/h_2 = \; ... \; \frac{k_n \Delta h_n}{h_n} = v$$

Substituting the expressions for Δh_1, Δh_2, ... in terms of v in the equation for Δh, we get:

$$v h / k_z = v h_1 / k_1 + v h_2 / k_2 + \ldots + v h_n / k_n$$

or

$$k_z = \frac{h}{(h_1/k_1 + h_2/k_2 + \ldots + h_n/k_n)} \qquad \ldots(\text{Eq. 5.34})$$

This is the equation for average permeability for flow perpendicular to the bedding planes.

Flow Parallel to the Bedding Planes

Let the flow be parallel to the bedding planes as shown in Fig. 5.11:

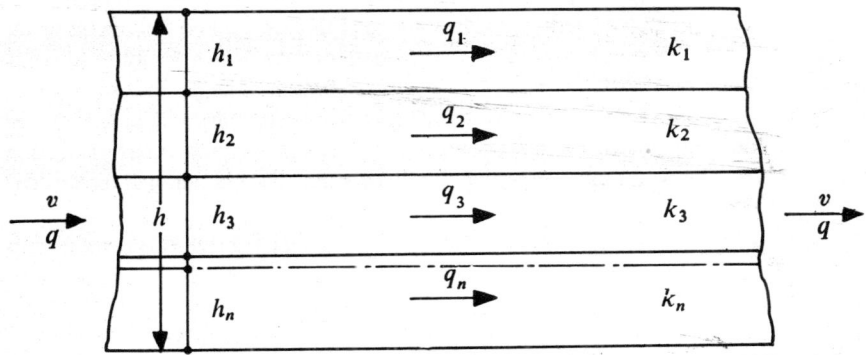

Fig. 5.11 Flow parallel to the bedding planes

With the same notation as in the first case, the hydraulic gradient i will be the same for all the layers as for the entire deposit. Since $v = ki$, and k is different for different layers, v will be different for the layers, say, $v_1, v_2, \ldots v_n$.

Also, $v_1 = k_1 i$; $v_2 = k_2 i$ ... and so on.

Considering unit dimension perpendicular to the plane of the paper, the areas of flow for each layer will be plane of the paper, the areas of flow for each layer will be $h_1, h_2, \ldots h_n$ respectively, and it is h for the entire deposit.

The discharge through the entire deposit is equal to the sum of the discharge through the individual layers. Assuming k_x to be the average permeability of the entire deposit parallel to the bedding planes, and applying the equation:

$$q = q_1 + q_2 + \ldots + q_n,$$

we have, $k_x \cdot ih = k_1 i \cdot h_1 + k_2 i \cdot h_2 + \ldots k_n i \cdot h_n$.

156   *Geotechnical Engineering*

$$\therefore \quad k_x = \left(\frac{k_1 h_1 + k_2 h_2 + \ldots k_n h_n}{h} \right) \qquad \ldots\text{(Eq. 5.35)}$$

where $h = h_1 + h_2 + \ldots + h_n$.

In other words, k_x is the weighted mean value, the weights being the thickness for each layer.

It can be shown that k_x is always greater than k_z for a given situation.

5.9 CAPILLARITY

The phenomenon in which water rises above the ground water table against the pull of gravity, but is in contact with the water table as its source, is referred to as 'Capillary rise' with reference to soils. The water associated with capillary rise is called 'capillary moisture'. The phenomenon by virtue of which a liquid rises in capillary tubes is, in general, called 'capillarity'.

All voids in soil located below the ground water table would be filled with water (except possibly for small pockets of entrapped air or gases). In addition, soil voids for a certain height above the water table will also be completely filled with water. This zone of saturation above the water table is due to capillary rise in soil. Even above this zone of full saturation, a condition of partial saturation exists. The zone of soil above the water table in which capillary water rises is denoted as the 'capillary fringe'.

Capillus literally means hair in Greek, indicating that the size of opening with which the phenomenon of capillarity is connected or related, is of this order of magnitude.

5.9.1 Rise of Water in Capillary Tubes

The principle of capillary rise in soils can be related to the rise of water in glass capillary tubes in the laboratory. When the end of a vertical capillary tube is inserted into a source of water, the water rises in the tube and remains there. This rise is attributed to the attraction between the water and the glass and to surface tension which develops at the air-water interface at the top of the water column in the capillary tube.

The surface tension is analogous to a stretched membrane, or a very thin but tough film. The water is "pulled up" in the capillary tube to a height, dependent upon the diameter of the tube, the magnitude of surface tension, and the unit weight of water.

The attraction between the water and capillary tube, or the tendency of water to wet the walls of the tube affects the shape of the air-water interface at the top of the column of water. For water and glass, the shape is concave as seen from top, that is, the water surface is lower at the centre of the column than at the walls of the tube. The resulting curved liquid surface is called the 'meniscus'. The surface of the liquid meets that of the tube at a definite angle, known as the 'contact angle'. This angle, incidentally, is zero for water and glass (Fig. 5.12).

Soil Moisture—Permeability and Capillarity

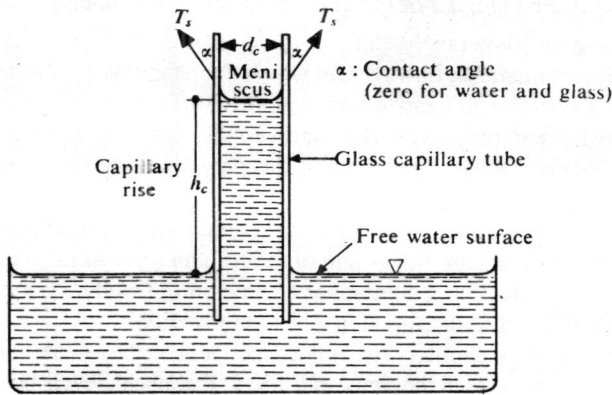

Fig. 5.12 Capillary rise of water in a glass-tube

The column of water in the capillary tube rises, against the pull of gravity, above the surface of the water source. For equilibrium, the effect of the downward pull of gravity on the capillary column of water has to be resisted by surface tension of the water film adhering to the wall of the tube to hold the water column.

If T_s is the surface tension, in force units per unit length, the vertical component of the force is given by $\pi d_c \cdot T_s \cdot \cos \alpha$ where α is the contact angle and d_c is the diameter of the capillary tube. With water and glass, the meniscus is tangent to the wall surface, so that the contact angle, α, is zero.

Therefore, the weight of a column of water, that is capable of being supported by the surface tension, is $\pi d_c \cdot T_s$. But the weight of water column in the capillary tube is $\dfrac{\pi d_c^2}{4} \cdot h_c \cdot \gamma_w$,

where γ_w is the unit weight of water and h_c is the capillary rise.

$$\therefore \quad \pi d_c \cdot T_s = \frac{\pi d_c^2}{4} \cdot h_s \cdot \gamma_w$$

or
$$h_c = \frac{4 T_s}{\gamma_w \cdot h_c} \qquad \ldots(\text{Eq. 5.36})$$

This equation helps one in computing the capillary rise of water in a glass capillary tube.

The value of T_s for water varies with temperature. At ordinary or room temperature, T_s is nearly 7.3 dynes/mm or 73×10^{-6} N/mm and γ_w may be taken as 9.81×10^{-6} N/mm³.

$$\therefore \quad h_c = \frac{4 \times 73 \times 10^{-6}}{9.81 \times 10^{-6} d_c} \approx \frac{30}{d_c} \quad \ldots \text{(Eq. 5.37)}$$

where d_c is the diameter of the glass capillary in mm, and h_c is the capillary rise of water in the glass tube in mm.

There *are* situations, however, in which the temperature effects should be considered. Generally, as temperature increases, surface tension decreases, indicating a decrease in capillary rise under warm conditions or an increase in capillary rise under cold conditions. The effect of this on soil will be discussed in a later subsection.

As the column of water stands in the capillary tube, supported by the surface tension at the meniscus, the weight of the column is transmitted to the walls of the capillary tube creating a compressive force on the walls. The effect of such an action on soil is also discussed in a later sub-section.

The height of the capillary rise is not dependent on the orientation of the capillary tube, or on variations in the shape and size of the tube at levels below the meniscus as shown in Fig. 5.13.

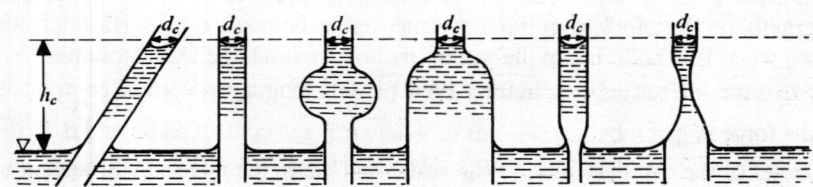

Fig. 5.13 Capillary heights of capillary tubes of various shapes
(These are equal if the diameters of their menisci are the same)

However, for water migrating up a capillary tube, a large opening can prevent further movement up an otherwise smaller diameter tube. The determining factor is the relation between the size of the opening and the particular height of its occurrence above the source of water.

In case the capillary rise computed on the basis of a larger opening is more than the height of this section of the tube, the water would rise further, and the final level will depend upon the capillary rise, computed and based upon the smaller opening above. In other words, the capillary rise would be dependent upon the diameter of the meniscus in such cases. However, in case the capillary rise computed on the basis of a larger opening is less than the height of this section, the water would rise no further, even if the section above is of a smaller size.

The hydrostatic pressure in the capillary tube at the level of the free surface of the water supply is zero, that is, it is equal to the atmospheric pressure. It is known that below the free water surface hydrostatic pressure u increases linearly with depth:

$$u = \gamma_w \cdot z$$

where γ_w = Unit weight of water.

Conversely, hydrostatic pressure measured in the capillary column *above* the free water surface is considered to be negative, that is, below atmospheric. This negative pressure, called capillary tension, is given by:

$$u = -\gamma_w \cdot h$$

where h is the height measured from the free water surface. The maximum value of the capillary tension u_c is:

$$u_c = \gamma_w \cdot h_c = \frac{4 T_s}{d_c} \qquad \ldots(\text{Eq. 5.38})$$

Capillary rise is not limited to tubes. If two vertical glass plates are placed so that they touch along one end and form a 'V' in plan, a wedge of water will rise in the V because of the phenomenon of capillarity (Fig. 5.14).

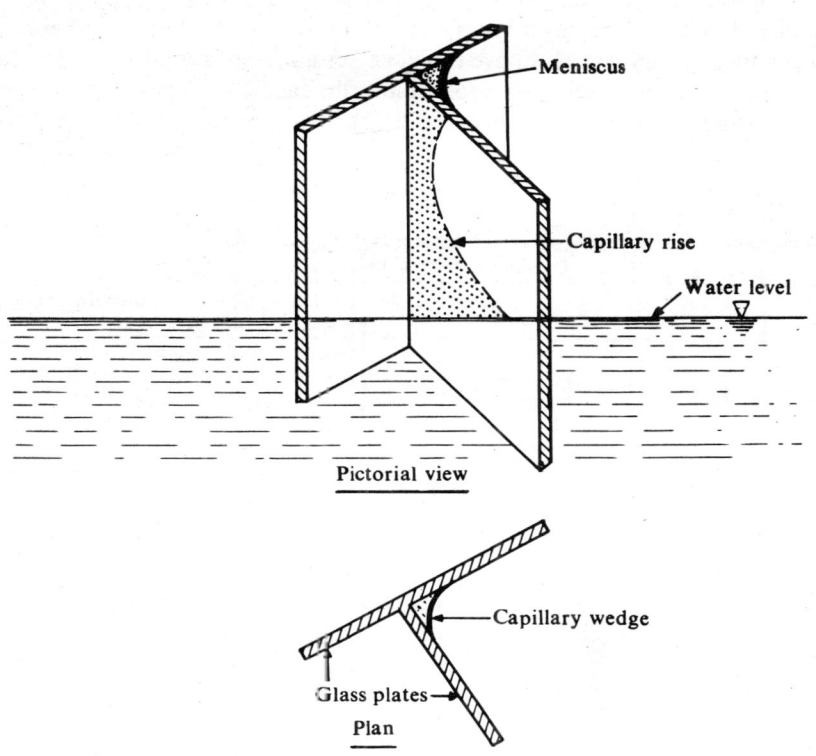

Fig. 5.14 Capillary rise in the corner formed by glass plates

160 *Geotechnical Engineering*

The height of such rise is related not only to the attraction between the water and plates and the physical properties of water, as in tubes, but also to the angle formed by the 'V'.

5.9.2 Capillary Rise in Soil

The rise of water in soils above the ground water table is analogous to the rise of water into capillary tubes placed in a source of water. However, the void spaces in a soil are irregular in shape and size, as they interconnect in all directions. Thus, the equations derived for regular shaped capillary tubes cannot be, strictly speaking, directly applicable to the capillary phenomenon associated with soil water. However, the features of capillary rise in tubes facilitate an understanding of factors affecting capillarity and help determine the order of magnitude for a capillary rise in the various types of soils.

Equation 5.36 indicates that even relatively large voids will be filled with capillary water if soil is close to the ground water table. As the height above the water table increases, only the smaller voids would be expected to be filled with capillary water. The larger voids represent interference to an upward capillary flow and would not be filled. The soil just above the water table may become fully saturated with capillary water, but even this is questionable since it is dependent upon a number of factors. The larger pores may entrap air to some extent while getting filled with capillary water. Above this zone lies a zone of partial saturation due to capillarity. In both these zones constituting the capillary fringe, even adsorbed water contributes to the pore water (Fig. 5.15).

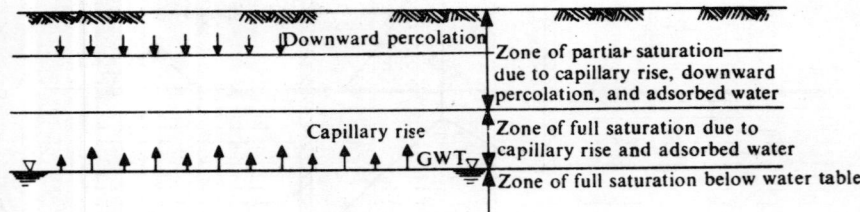

Fig. 5.15 Capillary fringe with zones of full and partial saturation

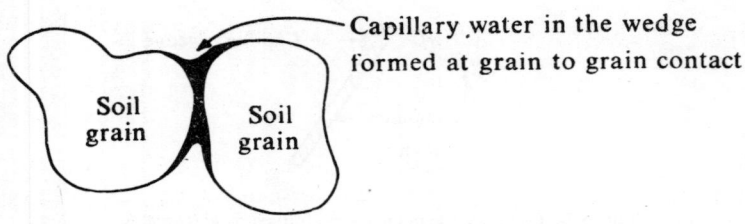

Fig. 5.16 Wedge of capillary water at the contact of soil grains

Soil Moisture—Permeability and Capillarity

In the zone of partial saturation due to the capillary phenomenon, capillary movement of water may occur even in the wedges of the capillary V formed wherever soil grains come into contact (similar to the V formed by vertical plates discussed in the preceding sub-section) Fig. 5.16. This is referred to as "Contact Moisture".

Since void spaces in soil are of the same order of magnitude as the particle sizes, it follows that the capillary rise would be greater in fine-grained soils than in coarse-grained soils. Relative values of capillary rise in various soils are given in Table 5.2 for an idea of orders of magnitude.

Table 5.2 Typical ranges of capillary rise in soils (Mc Carthy, 1977)

Soil Designation	Approximate Capillary height in mm
Fine Gravel	20 – 100
Coarse sand	150
Fine sand	300 – 1000
Silt	1000 – 10000
Clay	10000 – 30000

Temperature plays an important role in the capillary rise in soil. At lower temperature capillary rise is more and vice versa. Capillary flow may also be induced from a warm zone towards a cold zone.

It is to be noted that the negative pressures in the pore water in the capillary zone transfers a compressive stress of equal magnitude on to the mineral skeleton of the soil. Thus, the maximum increase in intergranular pressure in the capillary zone is given by:

$$\overline{\sigma}_c = h_c \cdot \gamma_w \qquad \text{...(Eq. 5.39)}$$

This is also loosely referred to as the 'capillary pressure' in the soils. This leads to shrinkage effects in fine-grained soils such as a clay. Representative values of capillary pressures are given in Table 5.3:

Table 5.3 Representative values of capillary pressures (Mc Carthy, 1977)

Soil	Capillary pressure – kN/m^2
Silt	10 to 100
Clay	100 to 300

5.9.3 Time Rate of Capillary Rise

In cases where a fill or an embankment is placed for highways, buildings or other purposes, the time necessary for the capillary rise to gain maximum height requires consideration. On the basis of typical sizes of voids, clay and fine silt will have a significant capillary rise. However, the time required for the rise to occur may be so great that other influences, such as evaporation and change in ground water level, may also play their part.

162 Geotechnical Engineering

The "Capillary Conductivity" or "Capillary permeability" is the property which indicates the rate of capillary rise. The factors known to effect the capillary permeability of a soil are size of voids, water content and temperature of the soil. This property is quantitatively greater for higher water contents and lower temperatures. The relative rates of capillary conductivity are similar to the comparative values for Darcy's permeability —that is, more for coarse soils and low for silts and clays. Absolute values of capillary conductivity are not available.

5.9.4 Suspended Capillaries

Percolating surface water due to rain or pore water resulting from a formerly higher water table can be held in a suspended state in the soil voids because of the surface tension phenomenon responsible for capillary rise. There would be menisci at both ends of the suspended column, each meniscus being in tension. The length of such a column would be controlled by the same factors that affect capillary rise.

5.9.5 Removal of Capillary Water in Soil

The existence of an air-water interface is a prerequisite for the occurrence of capillary rise. Since capillary water can exist only above the water table, it follows that capillarity will cease to exist where submergence of a soil zone exists.

Evaporation is another means of removing capillary water. This capillary water is very mobile as evaporation is continually replaced by capillary water.

5.9.6 Effects of Surface Tension and Capillarity

At the level of the meniscus the surface tension imposes a compressive force on to the soil grains in contact with the meniscus of magnitude equal to the weight of water in the capillary column, as indicated in an earlier sub-section. This effect applies to both a meniscus resulting from capillary rise and for pore water suspended above a capillary zone. The compressive force imposed on the soil in contact with the held column of water causes compression or shrinkage of the soil.

When the ground water drops subsequent to the time of formation of a clay deposit, internal compressive stresses in the clay mass due to the surface tension and capillary forces make it firm and strong. This is referred to as drying by desiccation and the clays are therefore called "desiccated clays". Sometimes, such desiccated clays may overlie soft and weak deeper clays. However, the strength and thickness of the desiccated zone may be such that roads and light buildings could be satisfactorily supported by it.

Since the intergranular pressure in the capillary zone is increased by capillary pressures, the procedure for determination of the effective stress when such a zone overlies a saturated soil mass, gets modified as illustrated below (Fig. 5.17):

Let a saturated soil mass of depth h_s be overlain by a capillary zone of height h_c assumed saturated by capillarity.

At level PP:

Total stress $\sigma = (h_c + h_s) \gamma_{sat}$

Neutral stress $u = h_s \gamma_w$

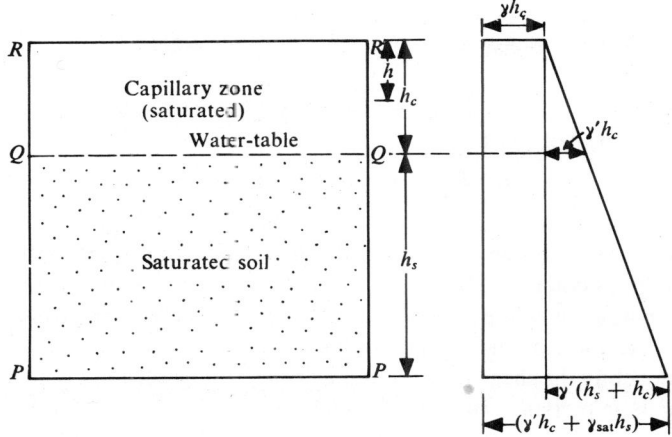

Fig. 5.17 Capillary zone-computation of effective stress

Effective stress $\sigma' = \sigma - u = h_c \gamma_{sat} + h_s \gamma_{sat} - h_s \gamma_w = h_c \gamma_{sat} + h_s \gamma'$

At level QQ:

Total stress $\sigma = h_c \cdot \gamma_{sat}$

Neutral stress $u =$ zero

Effective stress $\sigma' = \sigma - u = h_c \cdot \gamma_{sat}$.

This is because the capillary phenomenon increases the effective or intergranular stress by a magnitude equal to the negative pore pressure $h_c \cdot \gamma_w$ at the top of the capillary fringe, the pore pressure being zero at the bottom of the capillary fringe.

This is interesting because the effective stress increases from $h_c \cdot \gamma'$ to $h_c \cdot \gamma_{sat}$ at the bottom of the capillary zone, when the saturation is by capillarity and not by submergence.

At Level RR:

Effective stress $\sigma' =$ capillary pressure $= h_c \gamma_w$

The effective stress diagram is shown in Fig. 5.17.

The effect of a capillary fringe of height h_c is analogous to that of a surcharge $h_c \cdot \gamma_w$ placed on the saturated soil mass.

At depth h below the surface ($n < h_c$):

Effective stress $\sigma' = h \gamma' + h_c \gamma_w$

This may be shown as follows:

Total stress $\sigma = h \cdot \gamma_{sat}$

Neutral stress, $u = -$ (Pressure due to weight of water hanging below that level)

$$= -(h_c - h)\gamma_w$$

164   *Geotechnical Engineering*

$$\therefore \text{Effective stress } \sigma' = \sigma - u$$
$$= h\gamma_{sat} + h_c\gamma_w - h\gamma_w$$
$$= h\gamma' + h_c\gamma_w$$

When $h = h_c$, this becomes:

$$\sigma' = h_c\gamma' + h_c\gamma_w = h_c\gamma_{sat}, \text{ as earlier.}$$

The surface tension phenomenon also contributes to the strength of the soil mass in partially saturated coarse-grained soils. The moisture will be in the shape of wedges at grain contacts while the central portion of the void is filled with air. Thus, an air-water interface is formed. The surface tension in this meniscus imposes a compressive force on the soil grains, increasing the friction between the grains and consequently the shear strength (more of this will be seen in Chapter 8). This strength gain in partially saturated granular soils due to surface tension is termed 'Apparent Cohesion' (Terzaghi). This gain can be significant in some situations. This apparent cohesion disappears on full saturation and hence cannot always be relied upon.

5.9.7 Horizontal Capillarity Test

The 'Horizontal capillary test', also known as the 'Permeability – capillary test', is based on the determination of the rate of horizontal capillary saturation of a dry soil sample subjected to a hydraulic head from one end. The set-up for this test is shown in Fig. 5.18.

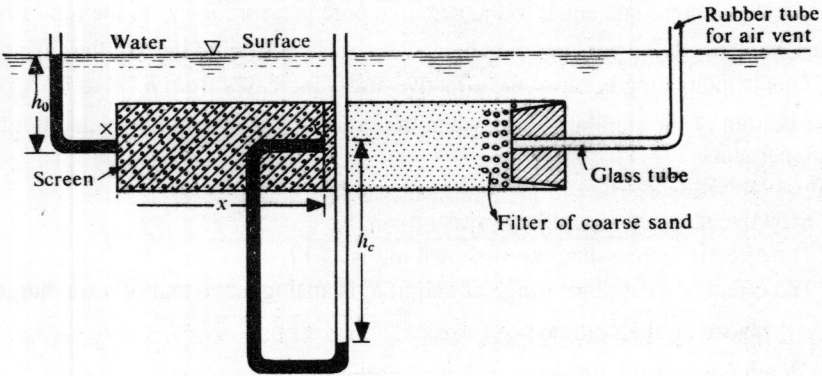

Fig. 5.18 Horizontal capillarity test (After Taylor, 1948)

If a dry and powdered soil sample is thoroughly mixed and packed into a glass tube with a screen at one end and a vented stopper at the other and then the tube is immersed in a shallow depth of water in a horizontal position, the water is sucked into the soil by capillary action. The distance to which the sample gets saturated may be expressed as a function of time.

In this situation, the menisci are developed to the maximum curvature possible for the void sizes in the sample; the corresponding capillary head constant for the soil at a given void ratio. The pressure difference between either end of the line of saturation or the menisci in the pore water is the capillary tension at all times.

Let the area of cross-section of the tube be A, and the porosity of the sample be n. If the line of saturation has proceeded a distance x, the hydraulic head expended may be formed as follows:

At point X:

Elevation head: $-h_0$ (water surface is the datum assumed)

Pressure head: $+h_0$

Total head $= -h_0 + h_0 = 0$

At point Y:

Elevation head $= -h_0$

Pressure head $= -h_c$

Total head $= -(h_0 + h_c)$

Therefore, the head expended from X to $Y = 0 + (h_0 + h_c) = (h_0 + h_c)$

If we imagine that standpipes could be inserted at X and Y, water would rise to the elevations shown in the figure, and the difference between these elevations would be $(h_0 + h_c)$.

If a fine-grained sample is placed just below water surface, h_0 may be negligible compared to h_c. When the size of the tube is of appreciable magnitude, h_0 varies for different points and the head lost is greater at the bottom than it is at the top of the sample; however, it may be made almost constant for all points in the cross-section if the tube is revolved about its axis as saturation proceeds:

If the degree of saturation is S, Darcy's Law for a particular value of x, gives:

$$S.v = k.i$$

In terms of seepage velocity v_s, this reduces to:

$$S.n.v_s = k.i, \quad n \text{ being the porosity.}$$

Here, v_s = the seepage velocity parallel to X-direction $= dx/dt$

$$\therefore \quad S.n.\, dx/dt = k.\frac{(h_0 + h_c)}{x}$$

or

$$xdx = \frac{k}{S.n}(h_0 + h_c)\, dt$$

Integrating between the limits x_1 and x_2 for x, and t_1 and t_2 for t,

$$\int_{x_1}^{x_2} x.dx = \int_{t_1}^{t_2} \frac{k}{S.n}(h_s + h_c)dt$$

$$\therefore \quad \left(\frac{x_2^2 - x_1^2}{t_2 - t_1}\right) = \frac{2k}{S \cdot n}(h_0 + h_c) \qquad \ldots\text{(Eq. 5.40)}$$

The degree of saturation, may be found from the dry weight, volume, grain specific gravity, and the wet weight at the end of the test. The porosity may also be computed from these.

In case the degree of saturation is assumed to be 100%, we may write:

$$\frac{(x_2^2 - x_1^2)}{(t_2 - t_1)} = \frac{2k}{n}(h_0 + h_c) \qquad \ldots\text{(Eq. 5.41)}$$

There are two unknowns k and h_c in this equation. The usual procedure recommended for their solution is as follows:

The first stage of the test is done with a certain value of the head, h_{0_1} when the sample is saturated for about one-half of its length, the values of x being recorded for different time lapses t. The second stage of the test is conducted with a much larger value of the head, h_{0_2}; this large value of the head is best imposed by clamping a head water tube to the left end of the glass tube containing the soil sample.

A plot of t versus x^2 gives a straight line which has different slopes for the two stages as shown in Fig. 5.19.

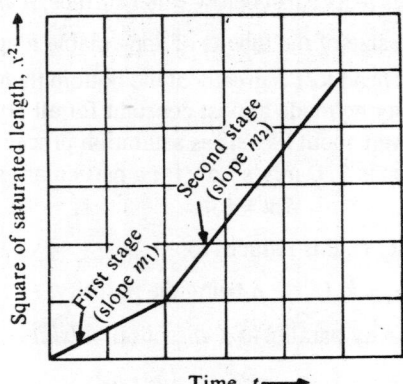

Fig. 5.19 Plot of t vs. x^2 in horizontal capillarity test

The left-hand side of Eq. 5.41 is nothing but the slope m of this plot; thus, we have two values m_1 and m_2 for the l.h.s. of this equation for the two values of head h_{0_1} and h_{0_2} for the two stages.

Hence, we have:

$$m_1 = 2k/n \left(h_{0_1} + h_c\right)$$

$$m_2 = 2k/n \left(h_{0_2} + h_c\right)$$

Soil Moisture—Permeability and Capillarity 167

The solution of these two simultaneous equations would yield the values for h_0 and k.

This two-stage test is also termed 'Capillarity-permeability test' as it affords a procedure for the determination of the permeability in addition to the capillary head.

5.9.8 Vertical Capillarity Test

There is little choice between the horizontal capillarity test and the vertical capillarity test, which will be described in this sub-section. However, the horizontal capillarity test is generally preferred.

The set-up for the vertical capillarity test is shown in Fig. 5.20:

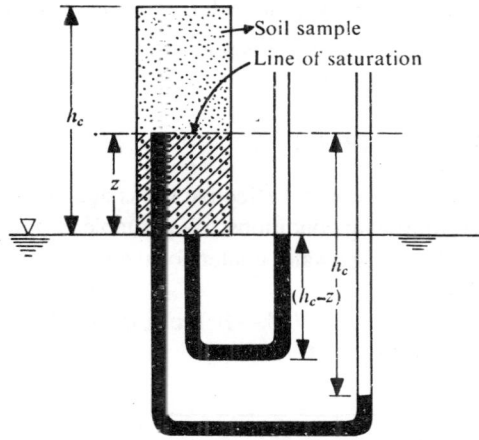

Fig. 5.20 Vertical capillarity test (After Taylor, 1948)

The saturation proceeds by capillary action vertically upward into the tube:
At point X:
Both elevation head and pressure head are zero.
At point Y:
Elevation head = z
Pressure head = $-h_c$

Total head = $(z - h_c)$

∴ Hydraulic gradient $= \dfrac{(h_c - z)}{z}$

The corresponding differential equation for the speed of saturation is obtained by using Darcy's law:

$$n \cdot \frac{dz}{dt} = k \cdot \frac{(h_c - z)}{z}$$

168 Geotechnical Engineering

If z is zero when t is zero, the solution of this differential equation is:

$$t = \frac{nh_c}{k}\left[-\log_e\left(1-\frac{z}{h_c}\right)-z/h_c\right] \qquad \ldots \text{(Eq. 5.42)}$$

Here h_c and k are assumed constants. However, as soon as z becomes larger than the height of the bottom zone of partial capillary saturation the increasing air content leads to variations in both h_c and k. The permeability decreases and eventually becomes so small that the tests required to determine the height of the zone of a partial capillary saturation needs a very long period of time. The height to the top of the zone of partial saturation has no relation to the constant capillary head h_c, since the former depends on the size of the smaller pores, and the latter acts only when there is maximum saturation and depends upon the size of the larger pores.

It has been found from laboratory experiments that for values z less than about 20% of h_c, the saturation is relatively high, and h_c and k are essentially constant.

For different sets of values of z and t, it is possible to solve for h_c and k simultaneously.

5.10 ILLUSTRATIVE EXAMPLES

Example 5.1: Determine the neutral and effective stress at a depth of 16m below the ground level for the following conditions: Water table is 3 m below ground level; $G = 2.68$; $e = 0.72$; average water content of the soil above water table is 8%.

(S.V.U.—B.Tech., (Part-time)—April, 1982)

The conditions are shown in Fig. 5.21:

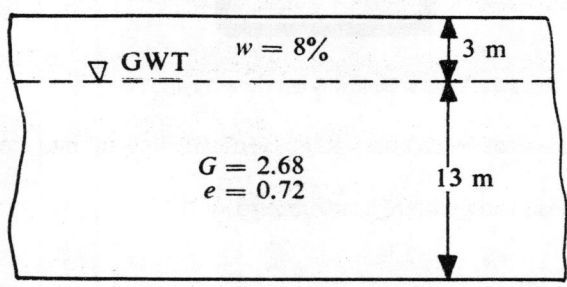

Fig. 5.21 Soil Profile (Example 5.1)

$G = 2.68$

$e = 0.72$

$w = 8\%$ for soil above water table.

$$\gamma = \frac{G(1+w)}{(1+e)} \cdot \gamma_w$$

$$= 2.68 \times \frac{1.08}{1.72} \times 9.81 \text{ kN/m}^3$$

$$= 16.51 \text{ kN/m}^3.$$

$$\gamma_{sat} = \left(\frac{G+e}{1+e}\right) \cdot \gamma_w$$

$$= \frac{(2.68 + 0.72)}{1.72} \times 9.81 \text{ kN/m}^3 = 19.39 \text{ kN/m}^3$$

Total pressure at a depth of 16 m —

$$\sigma = (3 \times 16.51 + 13 \times 19.39) = 301.6 \text{ kN/m}^2.$$

Neutral pressure at this depth —

$$u = 13 \times 9.81 = \mathbf{127.5 \text{ kN/m}^2}$$

∴ Effective stress at 16 m below the ground level —

$$\sigma' = (\sigma - u) = (301.6 - 127.5) = \mathbf{174.1 \text{ kN/m}^2}.$$

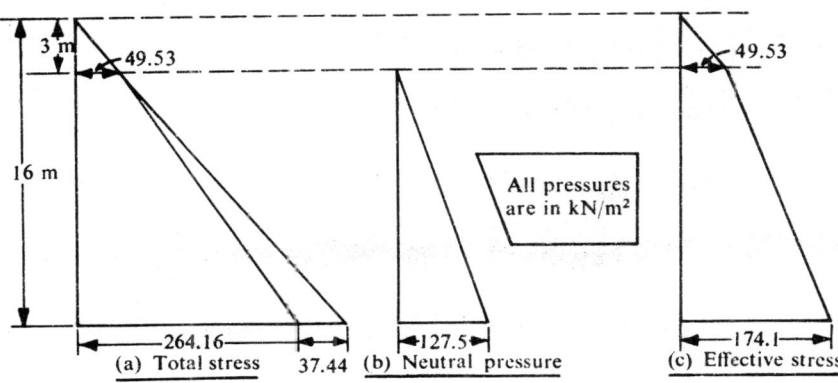

Fig. 5.22 Pressure Diagrams (Example 5.1)

Example 5.2: A saturated sand layer over a clay stratum is 5 m in depth. The water is 1.5 m below ground level. If the bulk density of saturated sand is 1.8, calculate the effective and neutral pressure on the top of the clay layer.

(S.V.U.—B.E., (R.R.)—Nov., 1969)

170 Geotechnical Engineering

The conditions given are shown in Fig. 5.23.

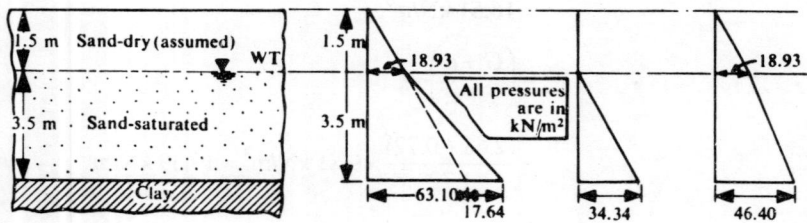

Fig. 5.23 Soil Profile (Example 5.2)

(a) Total pressure (b) Neutral pressure
(c) Effective pressure
Fig. 5.24 Pressure Diagrams

Let us, in the absence of data, assume that the sand above the water table is dry.
Bulk density of saturated sand,

$$\gamma_{sat} = 1.8 \text{ g/cm}^3 = 1.8 \times 9.81 \text{ kN/m}^3$$

$$= 17.66 \text{ kN/m}^3.$$

$$\gamma_{sat} = \left(\frac{G+e}{1+e}\right)\gamma_w$$

Let us assume : $G = 2.65$

or $$1.8 = \frac{(2.65+e)}{(1+e)} \cdot 1$$

or $$1.8 + 1.8\,e = 2.65 + e$$

or $$0.8\,e = 0.85$$

$$\therefore \quad e = \frac{0.85}{0.80} = 1.06$$

$$\gamma_d = \frac{G\gamma_w}{(1+e)} = \frac{2.65}{1+1.06} \times 9.81 \text{ kN/m}^3 = 12.62 \text{ kN/m}^3$$

Total stress at the top of clay layer:

$$\sigma = 1.5 \times 12.62 + 3.5 \times 17.66 = 80.74 \text{ kN/m}^2$$

Neutral stress at the top of clay layer :

$$u = 3.5 \times 9.81 = 34.34 \text{ kN/m}^2$$

Effective stress at the top of clay layer :

$$\sigma' = (\sigma - u) = 80.74 - 34.34 = 46.40 \text{ KN/m}^2.$$

Example 5.3: Compute the total, effective and pore pressure at a depth of 15 m below the bottom of a lake 6 m deep. The bottom of the lake consists of soft clay with a thickness of more than 15 m. The average water content of the clay is 40% and the specific gravity of soils may be assumed to be 2.65.

(S.V.U.—B.E., (R.R.)—April, 1966)

The conditions are shown in Fig. 5.25:

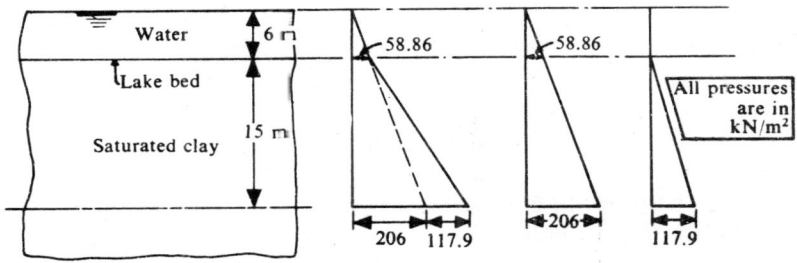

(a) Total pressure (b) Neutral pressure
(c) Effective pressure

Fig. 5.25 Clay layer below lake bed (Example 5.3)

Fig. 5.26 Pressure Diagrams (Example 5.3)

Water content $w_{sat} = 40\%$

Specific gravity of solids, $G = 2.65$

Void ratio,
$$e = w_{sat} \cdot G$$
$$= 0.4 \times 2.65$$
$$= 1.06$$

$$\gamma_{sat} = \left(\frac{G+e}{1+e}\right)\gamma_w$$

$$= \frac{(2.65 + 1.06)}{(1 + 1.06)} \times 9.81 \text{ kN/m}^3$$

$$= 17.67 \text{ kN/m}^3.$$

Total stress at 15 m below the bottom of the lake :

$$\sigma = 6 \times 9.81 + 15 \times 17.67 = 323.9 \text{ kN/m}^2$$

Neutral stress at 15 m below the bottom of the lake:

$$u = 21 \times 9.81 \text{ KN/m}^3 = \textbf{206.0 kN/m}^2$$

Effective stress at 15 m below the bottom of the lake:

$$\sigma' = 323.9 - 206.0 = \textbf{117.9 kN/m}^2$$

Also, $\quad \sigma' = 15 \times v' = 15 \times (\gamma_{sat} - \gamma_w)$

$$= 15 \,(17.67 - 9.81)$$

$$= 117.9 \text{ kN/m}^2$$

The pressure diagrams are shown in Fig. 5.26.

Example 5.4: A uniform soil deposit has a void ratio 0.6 and specific gravity of 2.65. The natural ground water is at 2.5 m below natural ground level. Due to capillary moisture, the average degree of saturation above ground water table is 50%. Determine the neutral pressure, total pressure and effective pressure at a depth of 6 m. Draw a neat sketch.

(S.V.U.—B.Tech., (Part-time)—April, 1982)

The conditions are shown in Fig. 5.27:

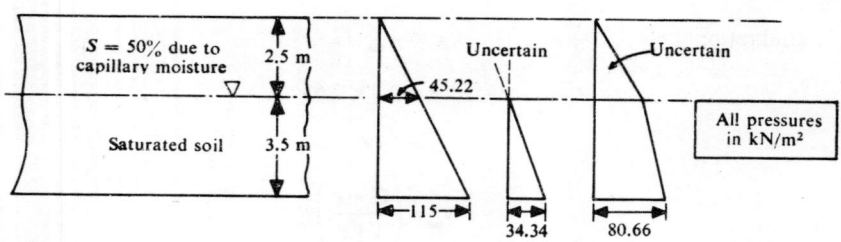

Fig. 5.27 Soil Profile (Example 5.4)   (a) Total pressure (b) Neutral pressure
(c) Effective pressure
Fig. 5.28 Pressure Diagrams (Example 5.4)

Void ratio, $e = 0.6$

Specific gravity $G = 2.65$

Soil Moisture—Permeability and Capillarity 173

$$\gamma_{sat} = \left(\frac{G+e}{1+e}\right)\cdot \gamma_w = \frac{(2.65+0.60)}{(1+0.60)} \times 9.81 \text{ kN/m}^3$$

$$= 19.93 \text{ kN/m}^3$$

γ at 50% saturation

$$= \frac{(G+Se)}{(1+e)}\cdot \gamma_w = \frac{(2.65\times 0.5\times 0.60)}{(1+0.60)} \times 9.81 \text{ kN/m}^3 = 18.09 \text{ kN/m}^3.$$

Total pressure, σ at 6 m depth $= 2.5 \times 18.09 + 3.5 \times 19.93$

$$= 115 \text{ kN/m}^2$$

Neutral pressure, u at 6 m depth $= 3.5 \times 9.81 = \mathbf{34.34 \text{ kN/m}^2}$

Effective pressure, σ' at 6 m depth $= (\sigma - u)$

$$= 115.00 - 34.34 = \mathbf{80.66 \text{ kN/m}^2}$$

The pressure diagrams are shown in Fig. 5.28.

It may be pointed out that the pore pressure in the zone of partial capillary saturation is difficult to predict and hence the effective pressure in this zone is also uncertain. It may be a little more than what is given here.

Example 5.5: Estimate the coefficient of permeability for a uniform sand where a sieve analysis indicates that the D_{10} size is 0.12 mm.

$D_{10} = 0.12$ mm $= 0.012$ cm.
According to Allen Hazen's relationship,
$$k = 100 \, D_{10}^2$$
where k is permeability in cm/s and D_{10} is effective size in cm.

$\therefore \quad k = 100 \times (0.012)^2 = 100 \times 0.000144 = 0.0144$ cm/s

$\therefore \quad$ Permeability coefficient $= \mathbf{1.44 \times 10^{-1} \text{ mm/s}}$

Example 5.6: Determine the coefficient of permeability from the following data:
Length of sand sample = 25 cm
Area of cross section of the sample = 30 cm^2
Head of water = 40 cm
Discharge = 200 ml in 110 s.

(S.V.U.—B.Tech., (Part time)—June, 1981)

$L = 25$ cm

$A = 30$ cm^2

$h = 40$ cm (assumed constant)

$Q = 200$ ml. $t = 110$ s

$q = Q/t = 200/110$ ml/s $= 20/11 = 1.82$ cm^3/s

$i = h/L = 40/25 = 8/5 = 1.60$

$q = k \cdot i \cdot A$

$$k = q/iA = \frac{20}{11 \times 1.6 \times 30} \text{ cm/s}$$

$= 0.03788$ cm/s

$= \mathbf{3.788 \times 10^{-1}}$ **mm/s**

Example 5.7: The discharge of water collected from a constant head permeameter in a period of 15 minutes is 500 ml. The internal diameter of the permeameter is 5 cm and the measured difference in head between two gauging points 15 cm vertically apart is 40 cm. Calculate the coefficient of permeability.

If the dry weight of the 15 cm long sample is 486 g and the specific gravity of the solids is 2.65, calculate the seepage velocity.

(S.V.U.—B.E., (N.R.)—May, 1969)

$Q = 500$ ml; $t = 15 \times 60 = 900$ s.

$A = (\pi/4) \times 5^2 = 6.25\pi$ cm^2; $L = 15$ cm; $h = 40$ cm;

$$k = \frac{QL}{Ath} = \frac{500 \times 15}{6.25 \times \pi \times 900 \times 40} \text{ cm/s} = \mathbf{0.016 \text{ mm/s}}$$

Superficial velocity $v = Q/At = \dfrac{500}{900 \times 6.25\pi}$ cm/s

$= 0.0283$ cm/s

$= 0.283$ mm/s

Dry weight of sample = 486 g.

Volume of sample $= A \cdot L = 6.25 \times \pi \times 15$ cm^3 = 294.52 cm^3

Dry density, $\gamma_d = \dfrac{486}{294.52}$ g/cm^3 = 1.65 g/cm^3

$$\gamma_d = \frac{G\gamma_w}{(1+e)}$$

$$(1+e) = \frac{2.65 \times 1}{1.65} = 1.606$$

$$e = 0.606$$

$$n = \frac{e}{(1+e)} = 0.3773 = 37.73\%$$

∴ Seepage velocity, $v_s = v/n = \dfrac{0.283}{0.3773} = \mathbf{0.750\ mm/s}$

Example 5.8: A glass cylinder 5 cm internal diameter and with a screen at the bottom was used as a falling head permeameter. The thickness of the sample was 10 cm. With the water level in the tube at the start of the test as 50 cm above the tail water, it dropped by 10 cm in one minute, the tail water level remaining unchanged. Calculate the value of k for the sample of the soil. Comment on the nature of the soil.

(S.V.U.—B.E., (R.R.)—May, 1969)

Falling head permeability test:

$$h_1 = 50\ cm;\quad h_2 = 40\ cm$$

$$t_1 = 0;\quad t_2 = 60\ s\ ...\ t = t_2 - t_1 = 60\ s$$

$$A = (\pi/4) \times 5^2 = 6.25\pi\ cm^2;\ L = 10\ cm$$

Since a is not given, let us assume $a = A$.

$$k = 2.303 \frac{aL}{At} \cdot \log_{10}(h_1/h_2)$$

$$= 2.303 \times (10/60)\ \log_{10}(50/40)\ cm/s$$

$$= 0.0372\ cm/s$$

$$= 3.72 \times 10^{-1}\ mm/s$$

The soil may be coarse sand or fine graved.

Example 5.9: In a falling head permeability test, head causing flow was initially 50 cm and it drops 2 cm in 5 minutes. How much time required for the head to fall to 25 cm?

(S.V.U.—B.E., (R.R.)—Feb., 1976)

Falling head permeability test:

We know: $k = 2.303 \dfrac{aL}{At} \log_{10}(h_1/h_2)$

176 *Geotechnical Engineering*

Designating $2.303 \dfrac{aL}{A}$ as a constant C

$$k = C \cdot \dfrac{1}{t} \cdot \log_{10}(h_1/h_2)$$

When $h_1 = 50$, $h_2 = 48$, $t = 300$ s

$$\therefore \quad \dfrac{k}{C} = \dfrac{1}{300}\log_{10}(50/48)$$

When $h_1 = 50$; $h_2 = 25$; substituting:

$$\dfrac{1}{300}\log_{10}(50/48) = (1/t)\log_{10}(50/25)$$

$$\therefore \quad t = 300\dfrac{\log_{10} 2}{\log_{10}(25/24)} = 5093.4\text{ s} = \textbf{84.9 min.}$$

Example 5.10: A sample in a variable head permeameter is 8 cm in diameter and 10 cm high. The permeability of the sample is estimated to be 10×10^{-4} cm/s. If it is desired that the head in the stand pipe should fall from 24 cm to 12 cm in 3 min., determine the size of the standpipe which should be used.

(S.V.U.—B.E., (R.R.)—Dec., 1970)

Variable head permeameter:
Soil sample diameter = 8 cm
height (length) = 10 cm
Permeability (approx.) = 10×10^{-4} cm/s
$h_1 = 24$ cm, $h_2 = 12$ cm, $t = 180$ s
Substituting in the equation

$$k = 2.303\dfrac{aL}{At}\log_{10}(h_1/h_2),$$

$$10^{-5} = \dfrac{2.303 \times a \times 10}{\pi \times 16 \times 180}\log_{10}(24/12)$$

$$\therefore \quad a = \dfrac{\pi \times 16 \times 180}{2.303 \times 10^6(\log_{10} 2)}\text{ cm}^2 = 0.01305\text{ cm}^2$$

If the diameter of the standpipe is d cm

$$a = (\pi/4)\, d^2$$

$$\therefore \quad d = \sqrt{\dfrac{4 \times 0.01305}{\pi}}\text{ cm} = 0.129\text{ cm}$$

∴ The standpipe should be **1.30 mm** in diameter.

Example 5.11: A horizontal stratified soil deposit consists of three layers each uniform in itself. The permeabilities of these layers are 8×10^{-4} cm/s, 52×10^{-4} cm/s, and 6×10^{-4} cm/s, and their thicknesses are 7, 3 and 10 m respectively. Find the effective average permeability of the deposit in the horizontal and vertical directions.

(S.V.U.—B.Tech., (Part-time)—April, 1982)

The deposit is shown in Fig. 5.29:

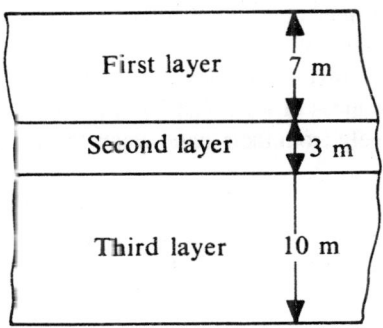

Fig. 5.29 Soil Profile (Example 5.11)

$k_1 = 8 \times 10^{-4}$ cm/s $h_1 = 7$ m

$k_2 = 52 \times 10^{-4}$ cm/s $h_2 = 3$ m

$k_3 = 6 \times 10^{-4}$ cm/s $h_3 = 10$ m

$$k_h \text{ (or } k_x) = \frac{(k_1 h_1 + k_2 h_2 + k_3 h_3)}{(h_1 + h_2 + h_3)}$$

$$= \frac{(8 \times 7 + 52 \times 3 + 6 \times 10)}{20} \times 10^{-4}$$

$$= 13.6 \times 10^{-4} \text{ cm/s}$$

∴ Effective average permeability in the horizontal direction

$$= 13.6 \times 10^{-3} \text{ mm/s}$$

$$k_v \text{ (or } k_z) = \frac{h}{\left(\frac{h_1}{k_1} + \frac{h_2}{k_2} + \frac{h_3}{k_3}\right)}$$

178 Geotechnical Engineering

$$= \frac{20}{\frac{1}{10^{-4}}[7/8 + 3/52 + 10/6]}$$

$$= 7.7 \times 10^{-4} \text{ cm/s}$$

∴ Effective average permeability in the vertical direction

$$= 7.7 \times 10^{-3} \text{ mm/s}$$

Example 5.12: An unconfined aquifer is known to be 32 m thick below the water table. A constant discharge of 2 cubic metres per minute is pumped out of the aquifer through a tubewell till the water level in the tubewell becomes steady. Two observation wells at distances of 15 m and 70 m from the tubewell show falls of 3 m and 0.7 m respectively from their static water levels. Find the permeability of the aquifer.

(S.V.U.—B.Tech., (Part-time)—Apr., 1982)

The conditions given are shown in Fig. 5.30:

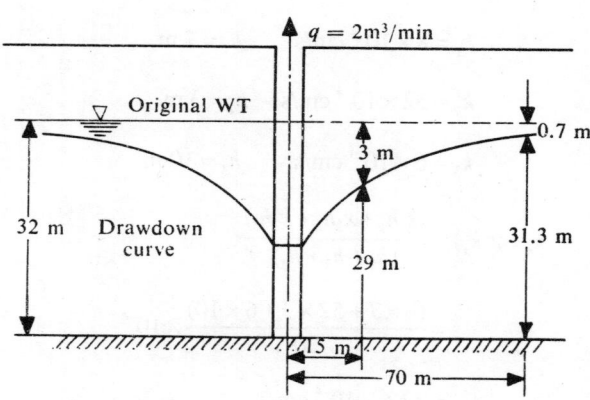

Fig. 5.30 Unconfined aquifer (Example 5.12)

We know,

$$k = \frac{q \log_e (r_2/r_1)}{\pi(z_2^2 - z_1^2)}$$

$$= \frac{2.303 \times 2 \log_{10} (70/15) \times 100}{60 \times \pi (31.3^2 - 29^2)} \text{ cm/s}$$

$$= 1.18 \times 10^{-1} \text{ mm/s}$$

Example 5.13: Determine the order of magnitude of the composite shape factor in the Poiseulle's equation adapted for flow of water through uniform sands that have spherical grains and a void ratio of 0.9, basing this determination on Hazen's approximate expression for permeability.

Poiseulle's equation adapted for the flow of water through soil is:

$$k = D_s^2 \cdot \frac{\gamma}{\mu} \cdot \frac{e^3}{(1+e)} \cdot C$$

with the usual notation, C being the composite shape factor.

By Allen Hazen's relationship,

$$k = 100 \, D_{10}^2$$

D_{10} is the same as the diameter of grains, D_s, for uniform sands.

$\therefore k = 100 \, D_s^2$. Here D_s is in cm while k is in cm/s.

Substituting –

$$100 \, D_s^2 = D_s^2 \cdot \frac{\gamma}{\mu} \cdot \frac{e^3}{(1+e)} \cdot C$$

$$C = 100 \cdot \frac{\mu}{\gamma} \cdot \frac{(1+e)}{e^3} ;$$

here μ is in N-Sec/cm² and γ is in N/cm³.

$$C = \frac{100 \times 10^{-7} \times 10^6 \times 1.9}{9.81 \times 10^3 \times (0.9)^3}$$

since $\mu = 10^{-3}$ N-sec/cm² (at 20°C) and $\gamma = 9.81$ kN/m³

$$= 0.002657.$$

Example 5.14: A cohesionless soil has a permeability of 0.036 cm per second at a void ratio of 0.36. Make predictions of the permeability of this soil when at a void ratio of 0.45 according to the two functions of void ratio that are proposed.

$$k_1 : k_2 = \frac{e_1^3}{(1+e_1)} : \frac{e_2^3}{(1+e_2)}$$

$$0.036 : k_2 = \frac{(0.36)^3}{1.36} : \frac{(0.45)^3}{1.45} = 0.546 : 1$$

$$\therefore \quad k_2 = \frac{1}{0.546} \times 0.36 \text{ mm/s} = 6.60 \times 10^{-1} \text{ mm/s}$$

Also,

$$k_1 : k_2 = e_1^2 : e_2^2$$

$$0.036 : k_2 = (0.36)^2 : (0.45)^2$$

$$= 0.1296 : 0.2025$$

$$\therefore \quad k_2 = \frac{0.2025}{0.1296} \times 0.36 = \mathbf{5.625 \times 10^{-1}} \ \mathbf{mm/s}.$$

Example 5.15: Permeability tests on a soil sample gave the following data:

Test No.	Void ratio	Temperature °C	Permeability in 10^{-3} mm/s
1	0.63	20	0.36
2	1.08	36	1.80

Estimate the coefficient of permeability at 27°C for a void ratio of 0.90.

Viscosity at 20°C = 1.009×10^{-9} N.sec/mm^2
Viscosity at 36°C = 0.706×10^{-9} N.sec/mm^2 } Unit weight $\gamma = 9.81 \times 10^{-6}$ N/mm^3
Viscosity at 27°C = 0.855×10^{-9} N.sec/mm^2

According to Poisuelle's equation adapted to flow through soil,

$$k = D_s^2 \cdot \frac{\gamma}{\mu} \cdot \frac{e^3}{(1+e)} \cdot C$$

From test 1,

$$\frac{0.36 \times 10^{-3}}{D_s^2 \cdot C} = \frac{9.81 \times 10^{-6}}{1.009 \times 10^{-9}} \cdot \frac{(0.63)^3}{(1.63)} = 1491.46$$

$$\therefore \quad D_s^2 \cdot C = 2.414 \times 10^{-8}$$

From test 2,

$$\frac{1.80 \times 10^{-3}}{D_s^2 \cdot C} = \frac{9.81 \times 10^{-6}}{0.706 \times 10^{-9}} \cdot \frac{(1.08)^3}{2.08} = 8415.35$$

$$\therefore \quad D_s^2 \cdot C = 2.139 \times 10^{-8}$$

Average value of $D_s^2 \cdot C = 2.2765 \times 10^{-8}$

$$\therefore \quad \text{At 27°C and a void ratio of 0.90,}$$

$$k = 2.2765 \times 10^{-8} \times \frac{9.81 \times 10^{-6}}{0.855 \times 10^{-9}} \times \frac{(0.90)^3}{1.9} = \mathbf{1.002 \times 10^{-3}} \ \mathbf{mm/s}$$

Example 5.16: To what height would water rise in a glass capillary tube of 0.01 mm diameter? What is the water pressure just under the meniscus in the capillary tube?

Capillary rise, $h_c = \dfrac{4T_s}{\gamma_w d_c}$, where T_s is surface tension of water and γ_w is its unit weight and d_c is the diameter of the capillary tube.

$\therefore h_c = \dfrac{4 \times 73 \times 10^{-6}}{9.81 \times 10^{-6} \times 0.01}$ mm.

assuming $T_s = 73 \times 10^{-6}$ N/mm and $\gamma_w = 9.81 \times 10^{-6}$ N/mm^3.

$= 30/0.01$ mm

$= 3000$ mm or **3 m**.

Water pressure just under the meniscus in the tube

$= 3000 \times 9.81 \times 10^{-6}$ N/mm^2

$= 3 \times 9.81$ kN/m^2

$= \mathbf{29.43}$ **kN/m^2**

Example 5.17: The D_{10} size of a soil is 0.01mm. Assuming (1/5) D_{10} as the pore size, estimate the height of capillary rise assuming surface tension of water as 75 dynes/cm.

(S.V.U.—B.Tech., (Part-time)—April, 1982)

The effective size D_{10} of the soil $= 0.01$ mm

Pore size, $d_c = (1/5)D_{10} = (1/5) \times 0.01$ mm $= 0.002$ mm.

Surface tension of water $= 75$ dynes/cm $= 75 \times 10^{-6}$ N/mm.

Height of Capillary rise in the soil:

$$h_c = \dfrac{4T_s}{\gamma_w d_c}$$

$= \dfrac{4 \times 75 \times 10^{-6}}{9.81 \times 10^{-6} \times 0.002}$ mm, since $\gamma_w = 9.81 \times 10^{-6}$ N/mm^3

$= \dfrac{4 \times 75}{9.81 \times 0.002 \times 1000}$ m

$= \mathbf{15.3}$ **m**

Example 5.18: What is the height of capillary rise in a soil with an effective size of 0.06 mm and void ratio of 0.63?

Effective size $= 0.06$ mm

Solid volume $\propto (0.06)^3$

$\therefore$ Void volume per unit of solid volume $\propto 0.63 (0.06)^3$

182 *Geotechnical Engineering*

Average void size, $d_c = (0.63)^{1/3} \times 0.06$ mm $= 0.857 \times 0.06 = 0.0514$ mm

Capillary rise, $h_c = \dfrac{4 T_s}{\gamma_w d_c}$

$$= \dfrac{4 \times 73 \times 10^{-6}}{9.81 \times 10^{-6} \times 0.0514} \text{ mm}$$

$\approx \mathbf{0.58 \ m}$

Example 5.19: The effective sizes of two soils are 0.05 mm and 0.10 mm, the void ratio being the same for both. If the capillary rise in the first soil is 72 cm, what would be the capillary rise in the second soil?

Effective size of first soil = 0.05 mm
Solid volume $\propto (0.05)^3$
$\therefore$ Void volume $\propto e\,(0.05)^3$
Average pore size, $d_c = e^{1/3} \times 0.05$ mm

Capillary rise of $h_c = \dfrac{4 T_s}{\gamma_w d_c}$

$\therefore \quad h_c \propto 1/d_c$

Since the void ratio is the same for the soils, average pore size for the second soil $= e^{1/3} \times 0.10$ mm

Substituting, $h_c = \dfrac{4 T_s}{\gamma_w d_c} = \mathbf{36 \ cm,}$

since d_c for the second soil is double that of the first soil and since $h_c \propto \dfrac{1}{d_c}$.

Example 5.20: The figure (Fig. 5.31) shows a tube of different diameter at different sections. What is the height to which water will rise in this tube? If this tube is dipped in water and inverted, what is the height to which water will stand? What are the water pressures in the tube at points *X, Y* and *Z* ?

Let us denote the top level of each section above the water level as h and the height of capillary rise based on the size of the tube in that section, as h_c.

Using $h_c \text{ (mm)} = \dfrac{30}{d_c \text{ (mm)}}$,

we can obtain the following as if that section is independently immersed in water:

d_c (mm)	h_c (mm)	h (mm)	Remarks
2	15	10	Water enters the next section
1	30	20	Water enters the next section
0.75	40	35	Water enters the next section
0.50	60	70	Water does not enter the next section

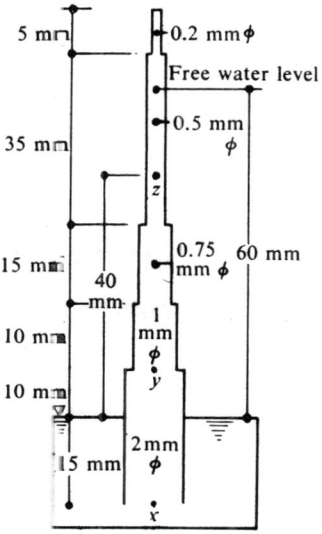

Fig. 5.31 Tube with varying section (Example 5.20)

Therefore water enters and stands at **60 mm** above free water level.
If the tube is dipped in water and inverted,
$$h_c = 30/d = 30/0.2 = 150 \text{ mm}.$$
Since this is greater than the height of the tube, it will be completely filled.

Pressure at $X = + h_c \gamma_w = \dfrac{1.5 \times 9.81}{100} = 0.147$ kN/m² = **147 N/m²**

Pressure at $Y = - h_c \gamma_w = \dfrac{-1 \times 9.81}{100} = -0.098$ kN/m² = **– 98 N/m²**

Pressure at $Z = - h_c \gamma_w = \dfrac{-4 \times 9.81}{100} = -0.392$ kN/m² = **– 392 N/m²**

Example 5.21: Sketch the variation in total stress, effective stress, and pore water pressure up to a depth of 6 m below ground level, given the following data. The water table is 2 m below ground level. The dry density of the soil is 1.8 g/cm³, water content is 12%; specific gravity is 2.65. What would be the change in these stresses, if water-table drops by 1.0 m?

(S.V.U.—B.Tech., (Part-time)—May, 1983)

$$\gamma_d = 1.8 \text{ g/cm}^3 = 1.8 \times 9.81 \text{ kN/m}^3 = 17.66 \text{ kN/m}^3$$

$$G = 2.65$$

$w = 12\%$ (assumed that it is at saturation)

$$\gamma_d = \frac{G\,\gamma_w}{(1+e)}$$

$$1.8 = \frac{2.65}{(1+e)} \quad \therefore \quad (1+e) = \frac{2.65}{1.80} = 1.472$$

$\therefore \quad e = 0.472$

$$\gamma_{sat} = \frac{2.65 + 0.472}{1.472} \times 9.81 \text{ kN/m}^3 = 20.81 \text{ kN/m}^3$$

Total stress at 2 m below GL = 2×17.66 kN/m² = 35.32 kN/m²
Total stress at 6 m below GL = $(2 \times 17.66 + 4 \times 20.81)$ = 118.56 kN/m²
Neutral stress at 2 m below GL = zero.
Neutral stress at 6 m below GL = $4 \times \gamma_w = 4 \times 9.81$ kN/m² = 39.24 kN/m²

Effective stress at 2 m below GL, $\bar{\sigma} = \sigma - u = 35.32 - 0 = 35.32$ kN/m²
Effective stress at 6 m below GL, $\bar{\sigma} = \sigma - u = 118.56 - 39.24 = 79.32$ kN/m²

The variation in the total, neutral, and effective stresses with depth is shown in Fig. 5.32.

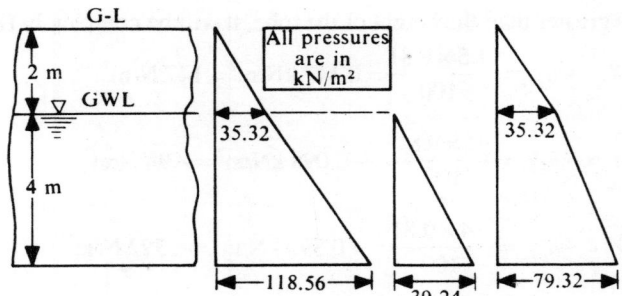

(a) Initial conditions (b) Total stress (c) Neutral stress (d) Effective stress

Fig. 5.32 Soil profile and pressure diagrams (Example 5.21)

Immediately after the water table is lowered by 1 m, the conditions are shown in Fig. 5.33.
The top 2 m is assumed to be dry.
The next 1 m is under capillary conditions.
With suspended water it may be assumed to be 100% saturated.
The next 3 m is submerged

Total stress:

σ at 2 m below GL = 2×17.66 kN/m^2 = 35.32 kN/m^2

σ at 3 m below GL = $(2 \times 17.66 + 1 \times 20.81)$ = 56.13 kN/m^2

σ at 6 m below GL = $(2 \times 17.66 + 4 \times 20.81)$ = 118.56 kN/m^2

Variation is linear.

Neutral stress:

u up to 2 m below GL is uncertain.

u at 2 m below GL is due to capillary meniscus.

It is given by -1×9.81 kN/m^2.

u at 3 m below GL is zero.

u at 6 m below GL = $+3 \times 9.81$ kN/m^3 = 29.43 kN/m^2

Effective stress:

$\bar{\sigma}$ at 2 m below GL = $35.32 - (-9.81)$ = 45.13 kN/m^2

$\bar{\sigma}$ upto 2 m below GL is uncertain.

$\bar{\sigma}$ at 3 m below GL = 56.13 kN/m^2.

$\bar{\sigma}$ at 6 m below GL = $118.56 - 29.43$ = 89.13 kN/m^2

The variation of total, neutral, and effective stresses is shown in Fig. 5.33.

The variation of the latter two from the surface up to 2 m depth is uncertain because the capillary conditions in this zone cannot easily be assessed.

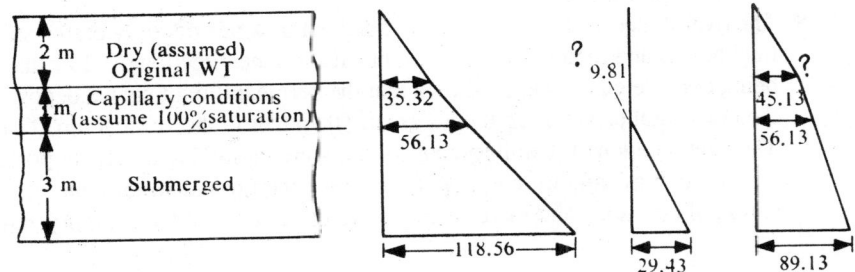

(a) Conditions after lowering the water table by 1 m (b) Total stress (c) Neutral stress (d) Effective stress

Fig. 5.33 Conditions and pressure diagrams (Example 5.21)

SUMMARY OF MAIN POINTS

1. Soil moisture or water in soil occurs in several forms, the free water being the most important.

2. The total stress applied to a saturated soil mass will be shared by the pore water and the solid grains; that which is borne by pore water is called the 'neutral stress' (computed as $\gamma_w \cdot h$), and that which is borne through grain-to-grain contact is called the 'effective stress'; the effective stress, obtained indirectly by subtracting the neutral stress from the total stress, is significant in the mobilisation of shear strength.
3. Permeability is the property of a porous medium such as soil, by virtue of which water or any other fluid can flow through the medium.
4. Darcy's law ($Q = k \cdot i \cdot A$) is valid for the flow of water through most soils except in the case of very coarse gravelly ones. The macroscopic velocity obtained by dividing that total discharge by the total area of cross-section is called 'superficial velocity'. In contrast to this, the microscopic velocity obtained by considering the actual pore space available for flow is referred to as the 'seepage velocity'.
5. Energy may be expressed in the form of three distinct energy heads, i.e., the pressure head, the elevation head, and the velocity head. The direction of flow is determined by the difference in total head between two points.
6. The constant head permeameter and the variable head permeameter are used in the laboratory for the determination of the coefficient of permeability of a soil.

   Pumping tests are used in the field for the same purpose, using the principles of well hydraulics.
7. Permeant properties, such as viscosity and unit weight, and soil properties, such as grain-size, void ratio, degree of saturation, and presence of entrapped air, affect permeability.
8. The overall permeability of a layered deposit depends not only on the strata thicknesses and their permeabilities, but also on the direction of flow that is being considered. It can be shown that the permeability of such a deposit in the horizontal direction is always greater than that in the vertical direction.
9. The phenomenon of 'capillary rise' of moisture in soil has certain important effects such as saturation of soil even above the ground water table, desiccation of clay soils and increase in the effective stress in the capillary zone.

REFERENCES

1. Alam Singh and B.C. Punmia: *Soil Mechanics and Foundations*, Standard Book House, Delhi-6.
2. A.W. Bishop: *The Measurement of Pore pressure in the Triaxial Test*, Pore pressure and Suction in soils, Butterworths, London, 1961.
3. A.W. Bishop, I. Alpan, E.E. Blight and I.B. Donald: *Factors controlling the strength of Partly Saturated Cohesive Soils*, Proc. ASCE Research conference on shear strength of cohesive soils, Boulder, Colorado, USA, 1960.
4. H.Darcy: *Les fontaines pulaliques de la ville de Dijon*, Paris: Dijon, 1856.

5. J. Dupuit: *Etudes théoretiques et pratiques sur la mouvement des eaux dans les canaux découvert et a travers les terrains perméables*, 2nd edition, Paris, Dunod, 1863.
6. A. Hazen: *Some Physical Properties of Sand and Gravels with Special Reference to Their Use in Filtration*, Massachusetts State Board of Health, 24th Annual Report, 1892.
7. A. Hazen: *Discussion of 'Dams on Sand Foundations'*, by A.C. Koenig, Transactions, ASCE, 1911.
8. IS: 2720(Part XVII)-1966: *Methods of test for soils – Laboratory Determination of Permeability*.
9. IS: 2720 (Part XXXVI)—1975: *Methods of test for soils—Laboratory Determination of Permeability of Granular Soils (constant head)*.
10. A.R. Jumikis: *Soil Mechanics*, D. Van Nostrand Co., Princeton, NJ, USA, 1962.
11. J.S. Kozeny: *Über Kapillare Leitung des wassers in Boden*, Berlin Wein Akademie, 1927.
12. T.W. Lambe: *The Measurement of Pore Water Pressures in Cohesionless Soils*, Proc 2nd Internal Conference SMFE, Roterdam, 1948.
13. T.W. Lambe: *Soil Testing for Engineers*, John Wiley and Sons, Inc., NY, USA, 1951.
14. T.W. Lambe and R.V. Whitman: *Soil Mechanics*, John Wiley and Sons, Inc., NY, USA, 1969.
15. A.G. Loudon: *The Computation of Permeability from Simple Soil Tests*, Geotechnique, 1952.
16. D.F. McCarthy: *Essentials of Soil Mechanics and Foundations*, Reston Publishing Co., Reston, VA, USA, 1977.
17. A.S. Michaels and C.S. Lin: *The Permeability of Kaolinite*—Industrial and Engineering Chemistry, 1952.
18. M. Muskat: *The Flow of Homogeneous Fluids through Porous Media*, McGraw-Hill Book Co., New York, USA, 1937.
19. M. Muskat: *The Flow of Homogeneous Fluids Through Porous Media*, J.W. Edwards, 1946.
20. A.E. Scheidegger: *The Physics of Flow Through Porous Media*, The MacMillan Co., New York, USA, 1957.
21. S.B. Sehgal: *A Textbook of Soil Mechanics*, Metropolitan Book Co., Pvt., Ltd., Delhi-6, 1967.
22. G.N. Smith: *Elements of Soil Mechanics for Civil and Mining Engineers*, 3rd edition, Metric, Crosby Lockwood Staples, London, 1974.
23. M.G. Spangler: *Soil Engineering*, International Text Book Company, Scranton, USA, 1951.
24. D.W. Taylor: *Fundamentals of Soil Mechanics*, John Wiley and Sons, Inc., New York, USA, 1948.
25. K. Terzaghi and R.B. Peck: *Soil Mechanics in Engineering Practice*, John Wiley and Sons, Inc., 1943.

26. A. Thiem: *Über die Ergiebig Keit artesicher Bohrlocher, Schachtbrunnen und Filtergalerien*, Journal für Gasbeleuchtung und Wasseracersorgung, 1870.
27. R.V. Whitman, A.M. Richardson, and K.A. Healy: *Time-lags in Pore pressure Measurements*, 5th International Conference SMFE, Paris, 1961.

QUESTIONS AND PROBLEMS

5.1 Define 'neutral' and 'effective' pressure in soils.
(S.V.U.—B.E., (R.R.)—Nov., 1969)
5.2 Write short notes on 'neutral' and 'effective' pressure. What is the role of effective stress in soil mechanics ?. (S.V.U.—B.E., (R.R.)—Dec., 1970)
(S.V.U.—B.E., (R.R.) – Nov., 1975)
5.3 A uniform homogeneous sand deposit of specific gravity 2.60 and void ratio 0.65 extends to a large depth. The ground water table is 2 m from G.L. Determine the effective, neutral, and total stress at depths of 2 m and 6 m. Assume that the soil from 1 m to 2 m has capillary moisture leading to degree of saturation of 60%.
(S.V.U.— B.Tech., (Part-time) —Sep., 1962)
5.4 Describe clearly with a neat sketch how you will determine the coefficient of permeability of a clay sample in the laboratory and derive the expression used to compute the permeability coefficient. Mention the various precautions, you suggest, to improve the reliability of the test results.
(S.V.U.—B.Tech., (Part-time)—May, 1983)
5.5 Define 'permeability' and explain how would you determine it in the field.
(S.V.U.—B.Tech., (Part-time)—May, 1983)
5.6 What are the various parameters that affect the permeability of soil in the field? Critically discuss.
(S.V.U—B.Tech., (Part-time)—April, 1982)
5.7 What are the various factors that affect the permeability of a soil stratum? If k_1, k_2, k_3 are the permeabilities of layers h_1, h_2, h_3 thick, what is its equivalent permeability in the horizontal and vertical directions? Derive the formulae used.
(S.V.U—Four-year B.Tech.,—June, 1982)
5.8 Differentiate between 'Constant Head type' and 'Variable Head type' permeameters. Are both of them required in the laboratory? If so, why? Derive the expression for the coefficient of permeability as obtained from the variable head permeameter.
(S.V.U.—B. Tech. (Part-Time)—June, 1981)
5.9 Estimate the coefficient of permeability for a uniform sand with $D_{10} = 0.18$ mm.
5.10 Explain the significance of permeability of soils. What is Darcy's law? Explain how the permeability of a soil is affected by various factors.
(S.V.U.—B.R., (R.R)—Feb., 1976)

5.11 Distinguish between superficial velocity and seepage velocity. Describe briefly how they are determined for sand and clay in the laboratory.
(S.V.U.—B.E., (R.R)—Nov., 1974)

5.12 List the factors that affect the permeability of a given soil. State the precautions that should be taken so that satisfactory permeability data may be obtained.

Explaining the test details of a falling head permeameter, derive the formula used in the computation. Also evaluate the types of soil material for which falling head and variable head permeameters are used in the laboratory. Compare the relative merits and demerits of laboratory and field methods of determining the coefficient of permeability.
(S.V.U.—B.E., (R.R)—Nov., 1973)

5.13 Calculate the coefficient of permeability of a soil sample, 6 cm in height and 50 cm^2 in cross-sectional area, if a quantity of water equal to 430 ml passed down in 10 minutes, under an effective constant head of 40 cm.
(S.V.U.—B.E., (R.R)—Dec. 1971)

5.14 Define coefficient of permeability and list four factors on which the permeability depends.

A falling head permeability test is to be performed on a soil sample whose permeability is estimated to be about 3×10^{-5} cm/s. What diameter of the standpipe should be used if the head is to drop from 27.5 cm to 20.0 cm in 5 minutes and if the cross-sectional area and length of the sample are respectively 15 cm^2 and 8.5 cm ? Will it take the same time for the head to drop from 37.5 cm to 30.0 cm?
(S.V.U.—B.E., (R.R.)—Nov., 1973)

5.15 (a) What are the conditions necessary for Darcy's law to be applicable for flow of water through soil ?

(b) Why is the permeability of a clay soil with flocculated structure greater than that for it in the remoulded state?

5.16 (a) State the principle of Darcy's law for laminar flow of water through saturated soil.

(b) Demonstrate that the coefficient of permeability has the dimension of velocity.

(c) The discharge of water collected from a constant head permeameter in a period of 15 minutes is 400 ml. The internal diameter of the permeameter is 6 cm and the measured difference in heads between the two gauging points 15 cm apart is 40 cm. Calculate the coefficient of permeability and comment on the type of soil.
(S.V.U.—B.E., (N.R.)—Sept., 1967)

5.17 A glass cylinder 5 cm internal diameter with a screen at the bottom is used as a falling head permeameter. The thickness of the sample is 10 cm. The water level in the tube at the start of the test was 40 cm above tail water level and it dropped by 10 cm in one minute while the level of tail water remained unchanged. Determine the value of the coefficient of permeability.
(S.V.U.—B.E., (N.R.)—Sep., 1968)

5.18 (a) State the important factors that affect the permeability of a soil.

(b) A permeameter of 8.2 cm diameter contains a sample of soil of length 35 cm. It can be used either for constant head or falling head tests. The standpipe used for the latter has a diameter of 2.5 cm. In the constant head test the loss of head was 116 cm measured on a length of 25 cm when the rate of flow was 2.73 ml/s. Find the coefficient of permeability of the soil.

If a falling head test were then made on the same soil, how much time would be taken for the head to fall from 150 to 100 cm ?

(S.V.U.—B.E., (N.R.)—March, 1966)

5.19 The initial head is 300 mm in a falling head permeability test. It drops by 10 mm in 3 minutes. How much longer should the test continue, if the head is to drop to 180 mm?

5.20 Determine the average horizontal and vertical permeabilities of a soil mass made up of three horizontal strata, each 1 m thick, if the coefficients of permeability are 1×10^{-1} mm/s, 3×10^{-1} mm/s, and 8×10^{-2} mm/s for the three layers.

5.21 The coefficient of permeability of a soil sample is found to be 9×10^{-2} mm/s at a void ratio of 0.45. Estimate its permeability at a void ratio of 0.63.

5.22 In a falling head permeability test the time intervals noted for the head to fall from h_1 to h_2 and from h_2 to h_3 have been found to be equal. Show the h_2 is the geometric mean of h_1 and h_3.

5.23 A sand deposit of 12 m thick overlies a clay layer. The water table is 3 m below the ground surface. In a field permeability pump-out test, the water is pumped out at a rate of 540 litres per minute when steady state conditions are reached. Two observation wells are located at 18 m and 36 m from the centre of the test well. The depths of the drawdown curve are 1.8 m and 1.5 m respectively for these two wells. Determine the coefficient of permeability.

5.24 The following data relate to a pump-out test:
Diameter of well = 24 cm
Thickness of confined aquifer = 27 m
Radius of circle of influence = 333 m
Draw down during the test = 4.5 m
Discharge = 0.9 m^3/s.
What is the permeability of the aquifer?

5.25 (a) Why is the capillary rise greater for fine grained soils than for coarse-grained soils?
(b) What is the effect of temperature on the capillary rise of water in soil?
(c) How is capillarity related to the firm condition of fine-grained soils' near surface?
(d) How can the effects of capillarity be removed from a soil?

5.26 A glass tube of 0.02 mm diameter. What is the height to which water will rise in this tube by capillarity action? What is the pressure just under the meniscus?

5.27 The effective size of a soil is 0.05 mm. Assuming the average void size to be (1/5) D_{10}, determine the capillary rise of pore water in this soil.

5.28 The effective size of a silt soil is 0.01 mm. The void ratio is 0.72. What is the height of capillary rise of water in this soil?

5.29 The effective sizes of two sands are 0.09 mm and 0.54 mm. The capillary rise of water in the first sand is 480 mm. What is the capillary rise in the second sand, if the void ratio is the same for both sands?

5.30 The water table is lowered from a depth of 3 m to a depth of 6 m in a deposit of silt. The silt remains saturated even after the water table is lowered. What would be the increase in the effective stress at a depth of 3 m and at 10 m on account of lowering of the water table? Assume the water content as 27% and grain specific gravity 2.67.

6
SEEPAGE AND FLOW NETS

6.1 INTRODUCTION

'Seepage' is defined as the flow of a fluid, usually water, through a soil under a hydraulic gradient. A hydraulic gradient is supposed to exist between two points if there exists a difference in the 'hydraulic head' at the two points. By hydraulic head is meant the sum of the position or datum head and pressure head of water. The discussion on flow nets and seepage relates to the practical aspect of controlling groundwater during and after construction of foundations below the groundwater table, earth dam and weirs on permeable foundations.

In Chapter 5, the discussion was confined to one-dimensional flow. This chapter considers two-dimensional flow, including the cases of non-homogeneous and anisotropic soil. The following approach is adopted: (a) the 'flow net' is introduced in an intuitive manner with the aid of a simple one-dimensional flow situation; (b) the flow nets for several two-dimensional situations are given; (c) the theoretical basis for the flow net is derived; (d) seepage through non-homogeneous and anisotropic soil is treated; and (e) seepage forces and their practical consequences are dealt with.

6.2 FLOW NET FOR ONE-DIMENSIONAL FLOW

Figure 6.1 (a) shows a tube of square cross-section (400 mm × 400 mm) through which steady-state vertical flow is occurring. The total head, elevation head and pressure head are plotted in Fig. 6.1 (b). The rate of seepage through the tube may be computed by Darcy's law:

$$q = k.i.A = 0.5 \times \frac{1600}{1000} \times 400 \times 400 = 1.28 \times 10^5 \text{ mm}^3/\text{s}$$

as the situation is one of simple one-dimensional flow.

If a dye is placed at the top of the soil and its movement through the soil is traced on a macroscopic scale, a vertical 'flow line', 'flow path', or 'stream line' would be obtained; that is to say, each drop of water that goes through the soil follows a flow line. An infinite number of flow lines can be imagined in the tube. The vertical edges of the tube are flow lines automatically; in addition to these, three more flow lines are shown at equal distances apart, for the sake of convenience. These five flow lines divide the vertical cross-section of the tube into four 'flow channels' of equal size. Since the flow is purely vertical, there cannot be flow

Seepage and Flow Nets 193

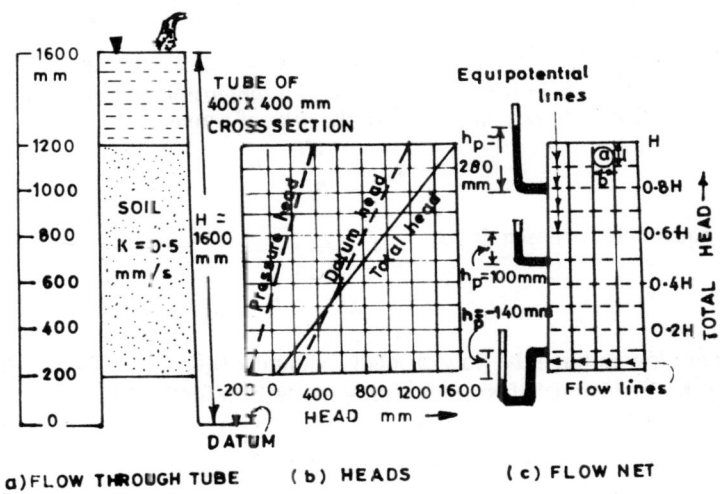

Fig. 6.1 One-dimensional flow

from one channel into another. Since there are four flow channels in, say, the x-direction, and since the tube is a square one, there are also four flow channels in the y-direction, i.e., perpendicular to the page. Thus there will be a total of 16 flow channels. If the flow in one channel is found the total flow is obtained by multiplying it by 16.

In the figure, dashed lines indicate the lines along which the total head is a constant. These lines through points of equal total head are known as 'equipotential lines'. Just as the number of flow lines is infinite, the number of equipotential lines is also infinite.

If equipotential lines are drawn at equal intervals, it means that the head loss between any two consecutive equipotential lines is the same.

A system of flow lines and equipotential lines, as shown in Fig. 6.1 (c), constitutes a 'flow net'. In isotropic soil, the flow lines and equipotential lines intersect at right angles, indicating that the direction of flow is perpendicular to the equipotential lines. An orthogonal net is formed by the intersecting flow lines and equipotential lines. The simplest of such patterns is one of the squares. From a flow net three very useful items of information may be obtained: rate of flow or discharge; head; and hydraulic gradient.

First, let us see how to determine the rate of flow or discharge from the flow net. Consider square a in the flow net – Fig. 6.1 (c). The discharge q_a through this square is

$$q_a = k \cdot i_a \cdot A_a$$

The head lost in square a is given by H/n_d, where H is the total head lost and n_d is the number of head drops in the flow net. i_a is then equal to $\dfrac{H}{n_d \cdot l}$, where l is the vertical dimension of square a. The cross-sectional area A_a of square a, as seen in plan, is b as shown in the figure, since a unit dimension perpendicular to the plane of the paper is to be considered for the sake of convenience.

$$\therefore \quad q_a = k \cdot \frac{H}{n_d \cdot l} \cdot b$$

Since a square net is chosen, $b = l$.

$$\therefore \quad q_a = kH \cdot \frac{l}{n_d}.$$

Since all flow through the flow channel containing square a must pass through square a, the flow through this square represents the flow for the entire flow channel.

In order to obtain the flow per unit of length L perpendicular to the paper, q_a should be multiplied by the number of flow channels, say, n_f:

$$\therefore \quad q/L = q_a \cdot n_f = kH \cdot \frac{n_f}{n_d}$$

$$\therefore \quad q/L = k \cdot H \cdot \frac{n_f}{n_d} = kH \cdot s \qquad \ldots(\text{Eq. 6.1})$$

the ratio $s = \dfrac{n_f}{n_d}$ is a characteristic of the flow net and is independent of the permeability k and the total loss of head H. It is called the 'shape factor' of the flow net. It should be noted that n_f and n_d need not necessarily be integers; these may be fractional, in which case the net may involve a few rectangles instead of squares.

The value of s in this case is

$$s = \frac{n_f}{n_d} = 4/10 = 0.4$$

and,

$$q/L = k \cdot H \cdot s = 0.5 \times 1600 \times 0.4 \text{ mm}^3/\text{s/mm}$$

$$= 320 \text{ mm}^3/\text{s/mm}$$

Then,

$$q = (q/L) \times (400) = 320 \times 400 = 128000$$

$$= 1.28 \times 10^5 \text{ mm}^3/\text{s}.$$

The value of total seepage is, of course, the same as that obtained by the initial computation using Darcy's law directly.

Next, let us see how to use the flow net to determine the head at any point. Since there are ten equal head drops, $\frac{1}{10}H$ is lost from one equipotential to the next. Since it is the total head that controls the flow, it should be noted that equipotentials are drawn through points of equal *total head*. The pressure head may be readily determined since the total head and elevation head are known.

For example, at elevation 1000 mm,

The total head, $h = (8/10) \times H = (8/10) \times 1600$ mm $= 1280$ mm

Elevation head, $h_e = 1000$ mm

∴ Pressure head, $h_p = (1280 - 1000)$ mm $= 280$ mm.

Since this is also the piezometric head, water will rise to a height of 280 mm in a piezometer installed at this elevation, as shown by the side of the flow net. The pore pressure at this elevation is $280 \times 9.81 \times 10^{-6}$ N/mm^2 or 2.75×10^{-3} N/mm^2. Similarly, the pressure heads at elevations 700 mm and 300 mm are 100 mm and −140 mm, respectively, as shown at the left of the flow net.

Finally, let us see how to use the flow net to determine the hydraulic gradient at any point in the flow net. The gradient for any square is given by h/l, where h is the head lost for the square and l is the length in which it is lost. Thus for all the squares in the flow net which are of the same size, the hydraulic gradient is given by $(H/10) \times (1/l)$ or

$$\frac{1600}{10} \times \frac{1}{100} = 1.6.$$

The example selected is so simple that these quantities could have been obtained easily even without the flow net. For complex two-dimensional flow situations, the techniques just described may be applied even for complex flow net patterns.

The values of flow, head, and gradient are exactly correct when obtained from an exactly correct flow net; thus, the results can be only as accurate as the flow net itself is. The flow net is a valuable tool in that it gives insight into the flow problem.

6.3 FLOW NET FOR TWO-DIMENSIONAL FLOW

It may be necessary to use flow nets to evaluate flow, where the directions of flow are irregular, or where the flow boundaries are not well-defined. Flow nets are a pictorial method of studying the path of the moving water.

In moving between two points, water tends to travel by the shortest path. If changes in direction occur, the changes take place along smooth curved paths. Equipotential lines must cross flow lines at right-angles since they represent pressure normal to the direction of flow. The flow lines and equipotential lines together form the flow net and are used to determine the quantities and other effects of flow through soils.

During seepage analysis, a flow net can be drawn with as many flow lines as desired. The number of equipotential lines will be determined by the number of flow lines selected. Generally speaking, it is preferable to use the fewest flow lines that still permit reasonable depiction of the path along the boundaries and within the soil mass. For many problems, three or four flow channels (a channel being the space between adjacent flow lines) are sufficient.

In this section the flow nets for three situations involving two-dimensional fluid flow are discussed. The first and second – flow under a sheet pile wall and flow under a concrete dam – are cases of confined flow since the boundary conditions are completely defined. The third – flow through an earth dam – is unconfined flow since the top flow line is not defined in advance of constructing the flow net. The top flow line or the phreatic line has to be determined first. Thereafter, the flow net may be completed as usual.

6.3.1 Flow under Sheet Pile Wall

Figure 6.2 shows a sheet pile wall driven into a silty soil. The wall runs for a considerable length in a direction perpendicular to the paper; thus, the flow underneath the sheet pile wall may be taken to be two-dimensional.

The boundary conditions for the flow under the sheet pile wall are; **m b**, upstream equipotential; **jn**, downstream equipotential; **bej**, flow line and **pq**, flow line. The flow net shown has been drawn within these boundaries. With the aid of the flow net, we can compute the seepage under the wall, the pore pressure at any point and the hydraulic gradient at any point. A water pressure plot, such as that shown in Fig. 6.2 is useful in the structural design of the wall.

6.3.2 Flow under Concrete Dam

Figures 6.3 to 6.7 show a concrete dam resting on an isotropic soil. The sections shown are actually those of the spillway portion. The upstream and tail water elevations are shown. The first one is with no cut-off walls, the second with cut-off wall at the heel as well as at the toe, the third with cut off-wall at the heel only, the fourth with cut-off wall at the toe only and the fifth is with upstream impervious blanket. The boundary flow lines and equipotentials are known in each case and the flow nets are drawn as shown within these boundaries. The effect of the cut off walls is to reduce the under seepage, the uplift pressure on the underside of the dam and also the hydraulic gradient at the exit, called the 'exit gradient'. A flow net can be understood to be a very powerful tool in developing a design and evaluating various schemes.

6.3.3 Flow through Earth Dam

The flow through an earth dam differs from the other cases in that the top flow line is not known in advance of sketching the flow net. Thus, it is a case of unconfined flow. The determination of the top flow line will be dealt with in a later section.

The top flow line as well as the flow net will be dependent upon the nature of internal drainage for the earth dam. Typical cases are shown in Fig. 6.8; the top flow line only is shown here.

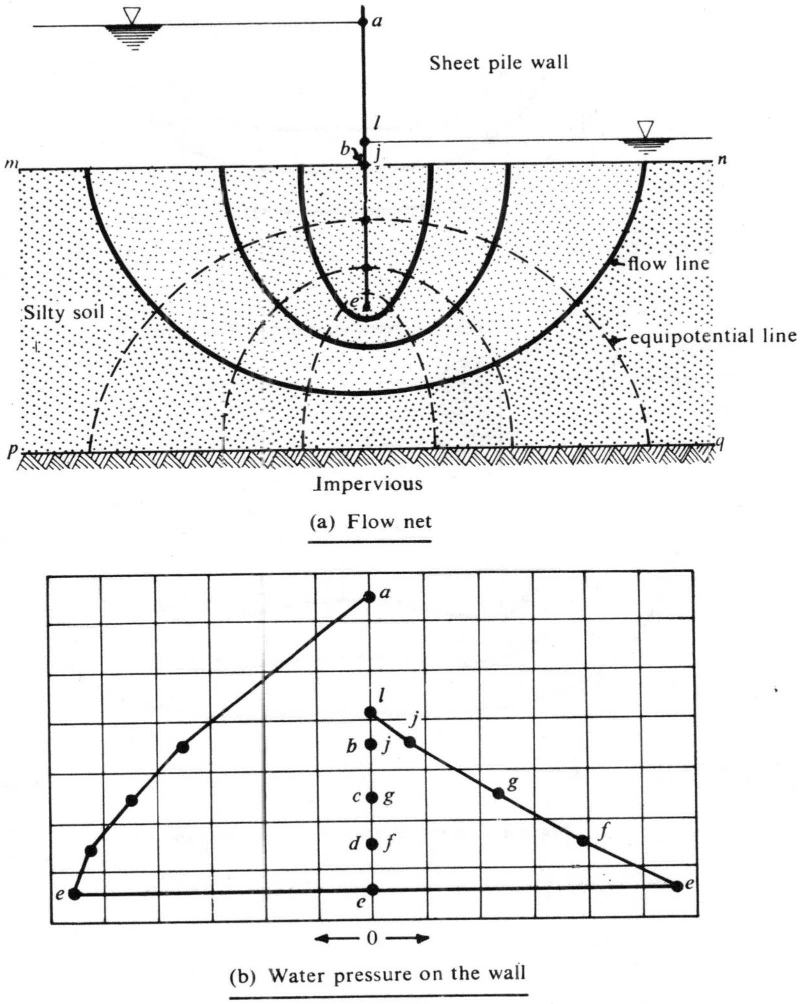

(a) Flow net

(b) Water pressure on the wall

Fig. 6.2 Sheet pile wall

Assuming that the top flow line is determined, a typical flow net for an earth dam with a rock toe, resting on an impervious foundation is shown in Fig. 6.9:

AB is known to be an equipotential and **AD** a flow line. **BC** is the top flow line; at all points of this line the pressure head is zero. Thus **BC** is also the 'phreatic line'; or, on this line, the total head is equal to the elevation head. Line **CD** is neither an equipotential nor a flow line, but the total head equals the elevation head at all points of **CD**.

198 *Geotechnical Engineering*

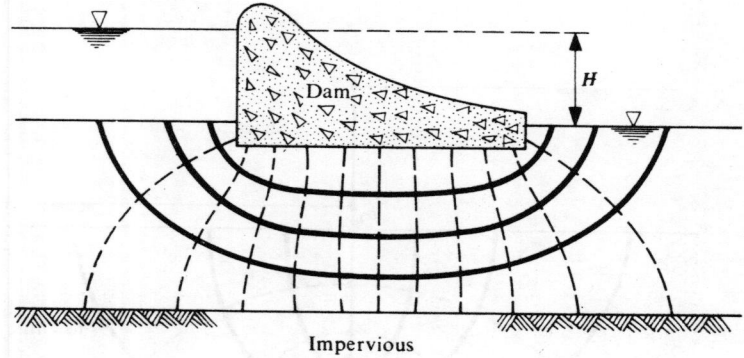

Fig. 6.3 Concrete dam with no cut-off walls

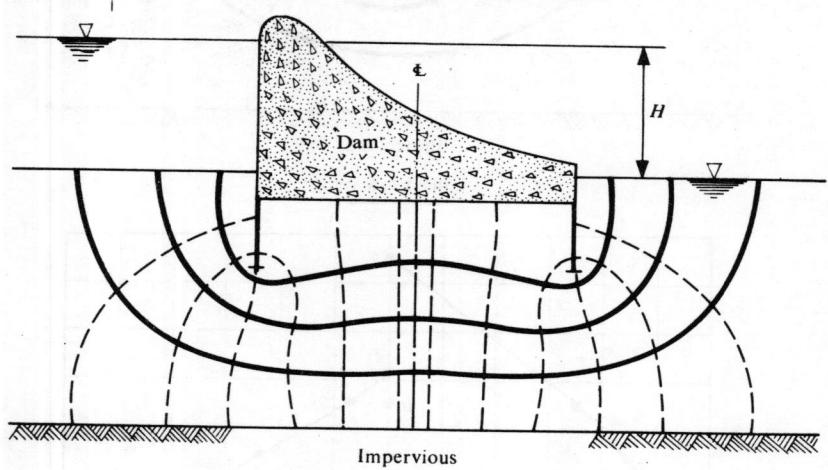

Fig. 6.4 Concrete dam with cut-offs at heel and at toe

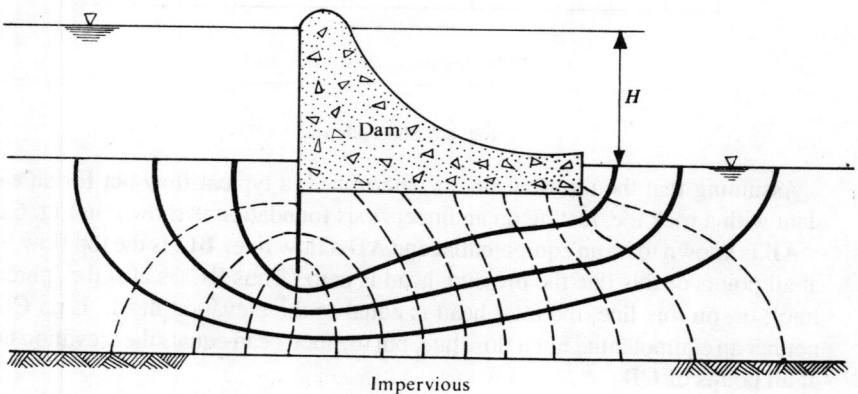

Fig. 6.5 Concrete dam with cut-off wall at heel

Seepage and Flow Nets 199

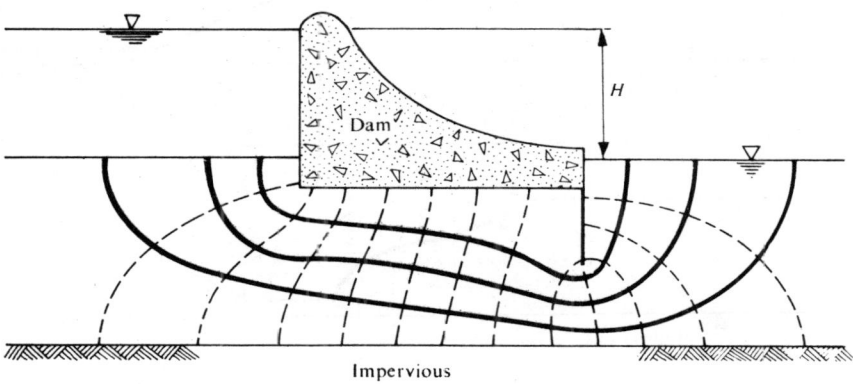

Fig. 6.6 Concrete dam with cut-off wall at toe

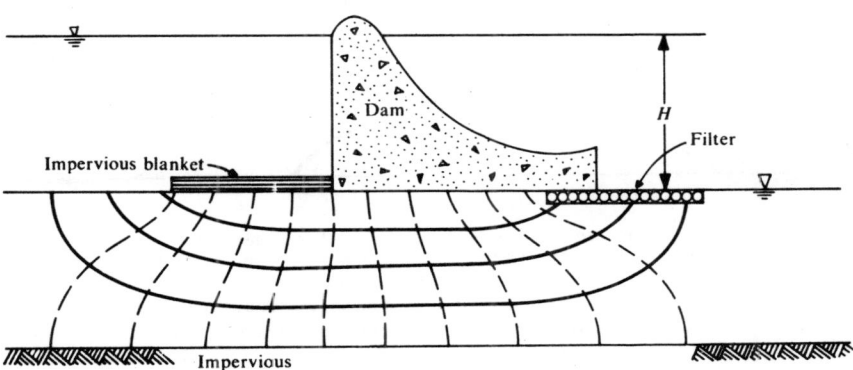

Fig. 6.7 Concrete dam with impervious blanket on the upstream side and filter on the downstream side

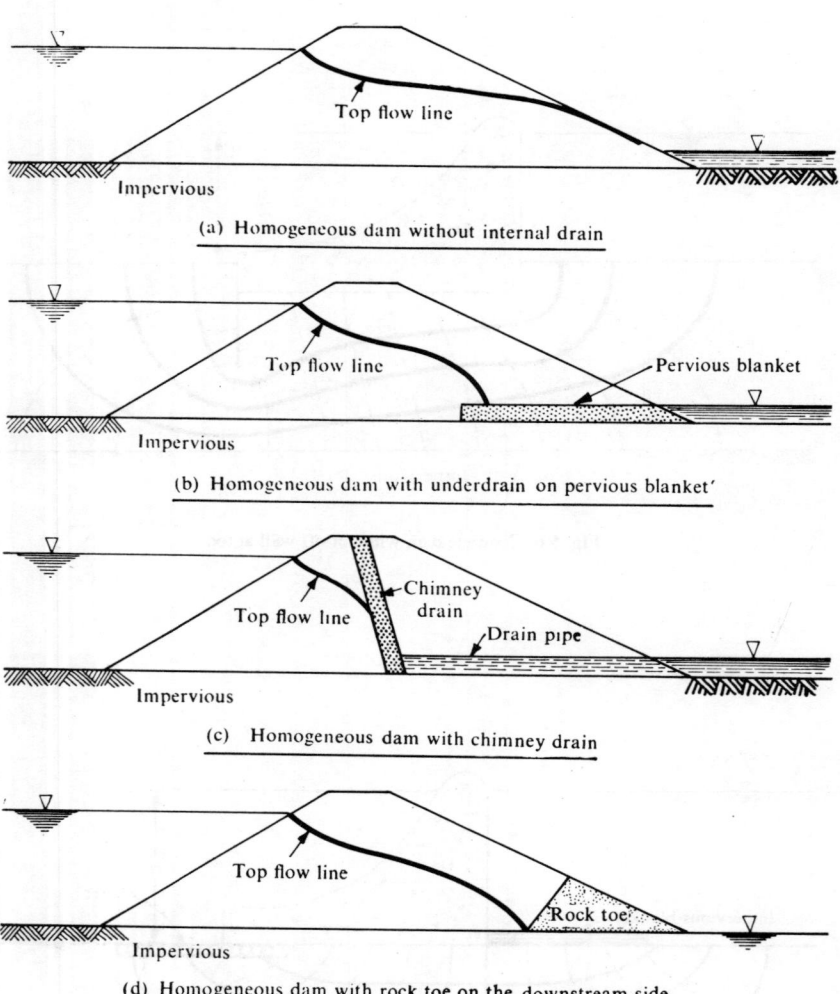

Fig. 6.8 Top flow line for typical cases of homogeneous earth dams on impervious foundation with different internal drainage arrangements

6.4 BASIC EQUATION FOR SEEPAGE

The flow net was introduced in an intuitive manner in the preceding sections. The equation for seepage through soil which forms the theoretical basis for the flow net as well as other methods of solving flow problems will be derived in this section.

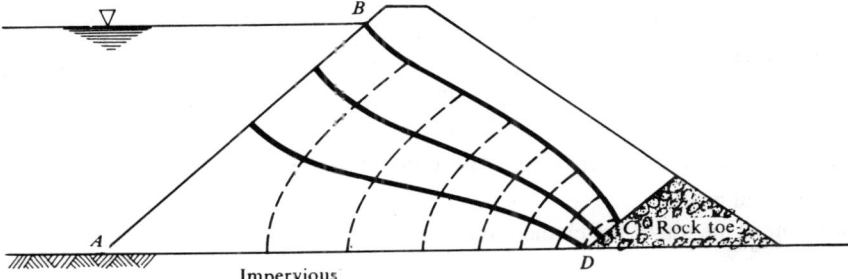

Fig. 6.9 Flow net for an earth dam with rock toe (for steady state seepage)

The following assumptions are made:
1. Darcy's law is valid for flow through soil.
2. The hydraulic boundary conditions are known at entry and exit of the fluid (water) into the porous medium (soil).
3. Water is incompressible.
4. The porous medium is incompressible.

These assumptions have been known to be very nearly or precisely valid.

Let us consider an element of soil as shown in Fig. 6.10, through which laminar flow of water is occurring:

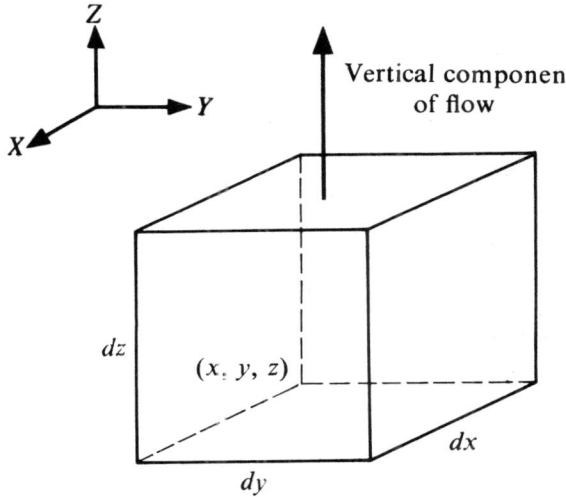

Fig. 6.10 Flow through an element of soil

Let q be the discharge with components, q_x, q_y and q_z in the X-, Y- and Z-directions respectively.

$q = q_x + q_y + q_z$, obviously.

202 Geotechnical Engineering

By Darcy's law,

$q_z = k \cdot i \cdot A$, where A is the area of the bottom face and q_z is the flow into the bottom face.

$$= k_z \left(-\frac{\partial h}{\partial z} \right) dx \cdot dy,$$

where k_z is the permeability of the soil in the Z-direction at the point (x, y, z) and h is the total head.

Flow out of the top of the element is given by:

$$q_z + \Delta q_z = \left(k_z + \frac{\partial k_z}{\partial z} \cdot dz \right) \left(-\frac{\partial h}{\partial z} - \frac{\partial^2 h}{\partial z^2} \cdot dz \right) \cdot dx \, dy$$

Net flow into the element from vertical flow:

$$\Delta q_z = \text{in flow} - \text{out flow}$$

$$= k_z \left(-\frac{\partial h}{\partial z} \right) dx \, dy - \left(k_z + \frac{\partial k_z}{\partial z} \cdot dz \right) \left(\frac{\partial h}{-\partial z} - \frac{\partial^2 h}{\partial z^2} \cdot dz \right) dx \, dy$$

$$\therefore \quad \Delta q_z = \left(k_z \cdot \frac{\partial^2 h}{\partial z^2} + \frac{\partial k_z}{\partial z^2} \frac{\partial h}{} + \frac{\partial k_z}{\partial z} \cdot dz \cdot \frac{\partial^2 h}{\partial z^2} \right) dx \, dy \, dz$$

Assuming the permeability to be constant at all points in a given direction, (that is, the soil is homogeneous),

$$\frac{\partial k_z}{\partial z} = 0$$

$$\therefore \quad \Delta q_z = \left(k_z \frac{\partial^2 h}{\partial z^2} \right) dx \, dy \, dz$$

Similarly, the net in flow in the X-direction is:

$$\Delta q_x = \left(k_x \cdot \frac{\partial^2 h}{\partial x^2} \right) dx \, dy \, dz$$

For two-dimensional flow, $q_y = 0$

$$\therefore \quad \Delta q = q_x + q_z = \left(k_x \cdot \frac{\partial^2 h}{\partial x^2} + k_z \cdot \frac{\partial^2 h}{\partial z^2} \right) dx \, dy \, dz$$

Δq may be obtained in a different manner as follows:

The volume of water in the element is:

$$V_w = \frac{S \cdot e}{(1+e)} \cdot dx \, dy \, dz$$

Δq = rate of change of water in the element with time:

$$= \frac{\partial V_w}{\partial t} = (\partial/\partial t)\left[\frac{S \cdot e}{(1+e)} \cdot dx\, dy\, dz\right]$$

$\dfrac{dx\, dy\, dz}{(1+e)}$ is the volume of solids, which is constant.

$$\therefore \quad \Delta q = \frac{dx\, dy\, dz}{(1+e)}(\partial/\partial t)(S \cdot e)$$

Equating the two expressions for Δq, we have:

$$\left(k_x \cdot \frac{\partial^2 h}{\partial x^2} + k_z \cdot \frac{\partial^2 h}{\partial z^2}\right) dx\, dy\, dx = \frac{dx\, dy\, dz}{(1+e)} \cdot (\partial/\partial t)(S \cdot e)$$

or

$$k_x \frac{\partial^2 h}{\partial x^2} + k_z \cdot \frac{\partial^2 h}{\partial z^2} = \frac{1}{(1+e)}\left(e \cdot \frac{\partial S}{\partial t} + S \frac{\partial e}{\partial t}\right) \quad \ldots(\text{Eq. 6.2})$$

This is the basic equation for two-dimensional laminar flow through soil.

The following are the possible situations:
(*i*) Both e and S are constant.
(*ii*) e varies, S remaining constant.
(*iii*) S varies, e remaining constant.
(*iv*) Both e and S vary.

Situation (*i*) represents steady flow which has been treated in Chapter 5 and this chapter. Situation (*ii*) represents 'Consolidation' or 'Expansion', depending upon whether e decreases or increases, and is treated in Chapter 7. Situation (*iii*) represents 'drainage' at constant volume or 'imbibition', depending upon whether S decreases or increases. Situation (*iv*) includes problems of compression and expansion. Situations (*iii*) and (*iv*) are complex flow conditions for which satisfactory solutions have yet to be found. (Strictly speaking, Eq. 6.2 is applicable only for small strains).

For situation (*i*), Eq. 6.2 reduces to:

$$k_x \cdot \frac{\partial^2 h}{\partial x^2} + k_z \cdot \frac{\partial^2 h}{\partial z^2} = 0 \quad \ldots(\text{Eq. 6.3})$$

If the permeability is the same in all directions, (that is, the soil is isotropic),

$$\frac{\partial^2 h}{\partial x^2} + \frac{\partial^2 h}{\partial z^2} = 0 \quad \ldots(\text{Eq. 6.4})$$

This is nothing but the Laplace's equation in two-dimensions. In words, this equation means that the change of gradient in the X-direction plus that in the Z-direction is zero.

From Eq. 6.3,

$$(\partial/\partial x)\left(k_x \cdot \frac{\partial h}{\partial z}\right) + (\partial/\partial z)\left(k_z \cdot \frac{\partial h}{\partial z}\right) = 0$$

But $k_x \cdot \partial h/\partial x = v_x$ and $k_z \cdot \partial h/\partial z = v_z$, by Darcy's law.

$$\therefore \quad \frac{\partial v_x}{\partial x} + \frac{\partial v_z}{\partial z} = 0 \quad \ldots\text{(Eq. 6.5)}$$

This is called the 'Equation of Continuity' in two-dimensions and can be got by setting $\Delta q = 0$ (or net inflow is zero) during the derivation of Eq. 6.2.

The flow net which consists of two sets of curves – a series of flow lines and of equipotential lines – is obtained merely as a solution to the Laplace's equation – Eq. 6.4. The fact that the basic equation of steady flow in isotropic soil satisfies Laplace's equation, suggests that, the flow lines and equipotential lines intersect at right-angles to form an orthogonal net – the 'flow net'. In other words, the flow net as drawn in the preceding sections is a theoretically sound solution to the flow problems.

The 'velocity potential' is defined as a scalar function of space and time such that its derivative with respect to any direction gives the velocity in that direction.

Thus, if the velocity potential, ϕ is defined as kh, ϕ being a function of x and z,

$$\left. \begin{array}{l} \dfrac{\partial \phi}{\partial x} = k \cdot \dfrac{\partial h}{\partial x} = v_x \\[2mm] \dfrac{\partial \phi}{\partial z} = k \cdot \dfrac{\partial h}{\partial z} = v_y \end{array} \right\} \quad \ldots\text{(Eq. 6.6)}$$

Similarly,

In view of Eq. 6.4 for an isotropic soil and in view of the definition of the velocity potential, we have:

$$\frac{\partial^2 \phi}{\partial x^2} + \frac{\partial^2 \phi}{\partial z^2} = 0 \quad \ldots\text{(Eq. 6.7)}$$

This is to say the head as well as the velocity potential satisfy the Laplace's equation in two-dimensions.

The equipotential lines are contours of equal head or potential. The direction of seepage is always at right angles to the equipotential lines.

The 'stream function' is defined as a scalar function of space and time such that the partial derivative of this function with respect to any direction gives the component of velocity in a direction inclined at $+90°$ (clockwise) to the original direction.

If the stream function is designated as $\psi(x, z)$,

$$\left. \begin{array}{l} \dfrac{\partial \psi}{\partial z} = v_x \\[2mm] \dfrac{\partial \psi}{\partial z} = -v_z \end{array} \right\} \quad \ldots\text{(Eq. 6.8)}$$

and

by definition.

By Eqs. 6.6 and 6.8, we have:

$$\left. \begin{array}{l} \partial \phi/\partial x = \partial \psi/\partial z \\[2mm] \partial \phi/\partial z = -\partial \psi/\partial x \end{array} \right\} \quad \ldots\text{(Eq. 6.9)}$$

and

These equations are known as Cauchy-Riemann equations. Substituting the relevant values in terms of ψ in the continuity equation (Eq. 6.5) and Laplace's equation (Eq. 6.7), we can show easily that the stream function $\psi(x, z)$ satisfies both these equations just as $\phi(x, z)$ does.

Functions ϕ and ψ are termed 'Conjugate harmonic functions'. In such a case, the curves "$\phi(x, z) =$ a constant" will be orthogonal trajectories of the curves "$\psi(x, z) =$ a constant".

Flow nets will be useful for the determination of rate of seepage, hydrostatic pressure, seepage pressure and exit gradient. These aspects have already been discussed in Sec. 6.2.

The following important properties of the flow nets are useful to remember:
  (i) The flow lines and equipotential lines intersect at right angles to each other.
  (ii) The spaces between consecutive flow and equipotential lines form elementary squares (a circle can be inscribed touching all four lines).
  (iii) The head drop will be the same between successive equipotentials; also, the flow in each flow channel will be the same.
  (iv) The transitions are smooth, being elliptical or parabolic in shape.
  (v) The smaller the size of the elementary square, the greater will be the velocity and the hydraulic gradient.

These are correct for homogeneous and isotropic soils.

6.5 SEEPAGE THROUGH NON-HOMOGENEOUS AND ANISOTROPIC SOIL

Although Eq. 6.2 was derived for general conditions, the preceding examples considered only soil that does not vary in properties from point to point horizontally or vertically–homogeneous soil–and one that has similar properties at a given location on planes at all inclinations–isotropic soil. Unfortunately, soils are invariably non-homogeneous and anisotropic.

The process of formation of sedimentary soils is such that the vertical compression is larger than the horizontal compression. Because of the higher vertical effective stress in a sedimentary soil, the clay platelets tend to have a horizontal alignment resulting in lower permeability for vertical flow than for horizontal flow.

In man-made as well as natural soil, the horizontal permeability tends to be larger than the vertical. The method of placement and compaction of earth fills is such that stratifications tend to be built into the embankments leading to anisotropy.

Nonhomogeneous Soil
In case of flow perpendicular to soil strata, the loss of head and rate of flow are influenced primarily by the less pervious soil whereas in the case of flow parallel to the strata, the rate of flow is essentially controlled by comparatively more pervious soil.

Figure. 6.11 shows a flow channel and part of a flow net, from soil **A** to soil **B**. The permeability of soil **A** is greater than that of soil **B**. By the principle of continuity, the same rate of flow exists in the flow channel in soil **A** as in soil **B**. By means of this, relationship between the angles of incidence of the flow paths with the boundary of the two flow channels can be determined. Not only does the direction of flow change at the boundary between soils with different permeabilities, but also the geometry of the figures in the flow net changes. As can be seen from Fig. 6.11, the figures in soil **B** are not squares as in soil **A**, but are rectangles.

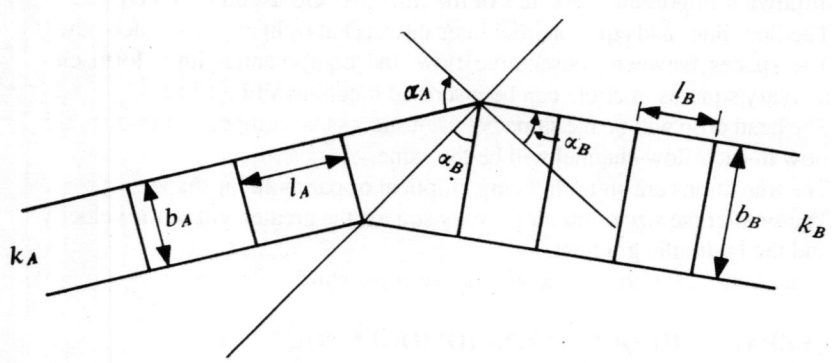

Fig. 6.11 Flow at the boundary between two soils

$$q_A = q_B$$

But
$$q_A = k_A \cdot \frac{\Delta h}{l_A} \cdot b_A$$

$$q_B = k_B \cdot \frac{\Delta h}{l_B} \cdot b_B$$

$$\therefore \quad k_A \cdot \frac{\Delta h}{l_A} \cdot b_A = k_B \cdot \frac{\Delta h}{l_B} \cdot b_B$$

$$\frac{l_A}{b_A} = \tan \alpha_A \qquad \frac{l_B}{b_B} = \tan \alpha_B$$

$$\frac{k_A}{\tan \alpha_A} = \frac{k_B}{\tan \alpha_B}$$

$$\frac{\tan \alpha_A}{\tan \alpha_B} = \frac{k_A}{k_B}$$

Anisotropic Soil

Laplace's equation for flow through soil, Eq. 6.4, was derived under the assumption that permeability is the same in all directions. Before stipulating this condition in the derivation, the equation was:

$$k_x \cdot \frac{\partial^2 h}{\partial x^2} + k_z \cdot \frac{\partial^2 h}{\partial z^2} = 0 \qquad \ldots\text{(Eq. 6.3)}$$

This may be reduced to the form:

$$\frac{\partial^2 h}{\partial z^2} + \frac{\partial^2 h}{\left(\frac{k_z}{k_x}\right) \cdot \partial x^2} = 0 \qquad \ldots\text{(Eq. 6.10)}$$

By changing the co-ordinate x to x_T such that $x_T = \sqrt{\frac{k_z}{k_x}} \cdot x$, we get

$$\frac{\partial^2 h}{\partial z^2} + \frac{\partial^2 h}{\partial x_T^2} = 0 \qquad \ldots\text{(Eq. 6.11)}$$

which is once again the Laplace's equation in x_T and z.

In other words, the profile is to be transformed according to the relationship between x and x_T and the flow net sketched on the transformed section.

From the transformed section, the rate of seepage can be determined using Eq. 6.1 with the exception that k_e is to be substituted for k (see Fig. 6.12):

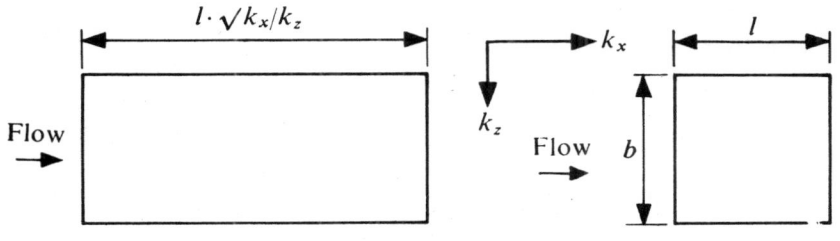

Natural scale   Transformed scale

Fig. 6.12 Flow in anisotropic soil

Transformed Section:

$$q_T = k_e \cdot iA = k_e \cdot \frac{\Delta h}{l} \cdot b = k_e \cdot \Delta h$$

Natural Section:

$$q_N = k_x \cdot i_N \cdot A = k_x \cdot \frac{\Delta h}{l\sqrt{k_x/k_z}} \cdot b = k_x \cdot \frac{\Delta h}{\sqrt{k_x/k_z}}$$

Since $q_T = q_N$,

$$k_e \Delta h = k_x \cdot \frac{\Delta h}{\sqrt{k_x/k_z}}$$

$$\therefore \quad k_e = \sqrt{k_x k_z} \qquad \qquad \ldots \text{(Eq. 6.12)}$$

k_e is said to be the effective permeability.

The transformed section can also be used to determine the head at any point. However, when determining a gradient, it is important to remember that the dimensions on the transformed section must be corrected while taking the distance over which the head is lost. To compute the gradient, the head loss between equipotentials is divided by the distance l_N, the perpendicular distance between equipotentials on the natural scale, and not by l_T, the distance between equipotentials on the transformed scale (Fig. 6.13).

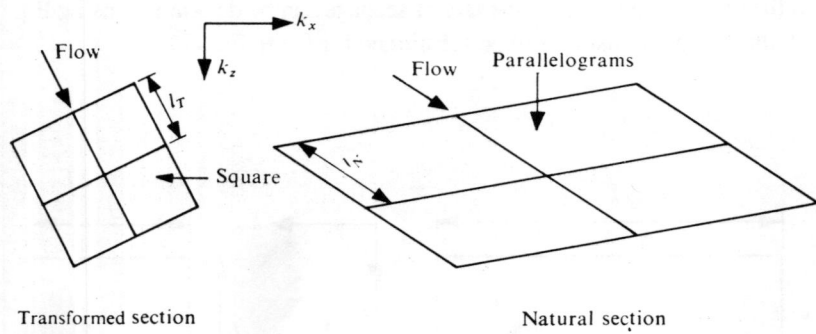

Fig. 6.13 Portion of flow net in anisotropic soil

Also, note that flow is perpendicular to equipotentials in only isotropic soils.

6.6 TOP FLOW LINE IN AN EARTH DAM

The flow net for steady seepage through an earth dam can be obtained by any one of the methods available, including the graphical approach. However, since this is the case of an unconfined flow, the top flow line is not known and hence should be determined first. The top flow line is also known as the 'phreatic line', as the pressure is atmospheric on this line. Thus, the pressures in the dam section below the phreatic line are positive hydrostatic pressures.

The top flow line may be determined either by the graphical method or by the analytical method. Although the typical earth dam will not have a simple homogeneous section, such sections furnish a good illustration of the conditions that must be fulfilled by any top flow line. Furthermore, the location of a top flow line in a simple case can often be used for the first trial in the sketching of a flow net for a more complicated case.

The top flow line must obey the conditions illustrated in Fig. 6.14.

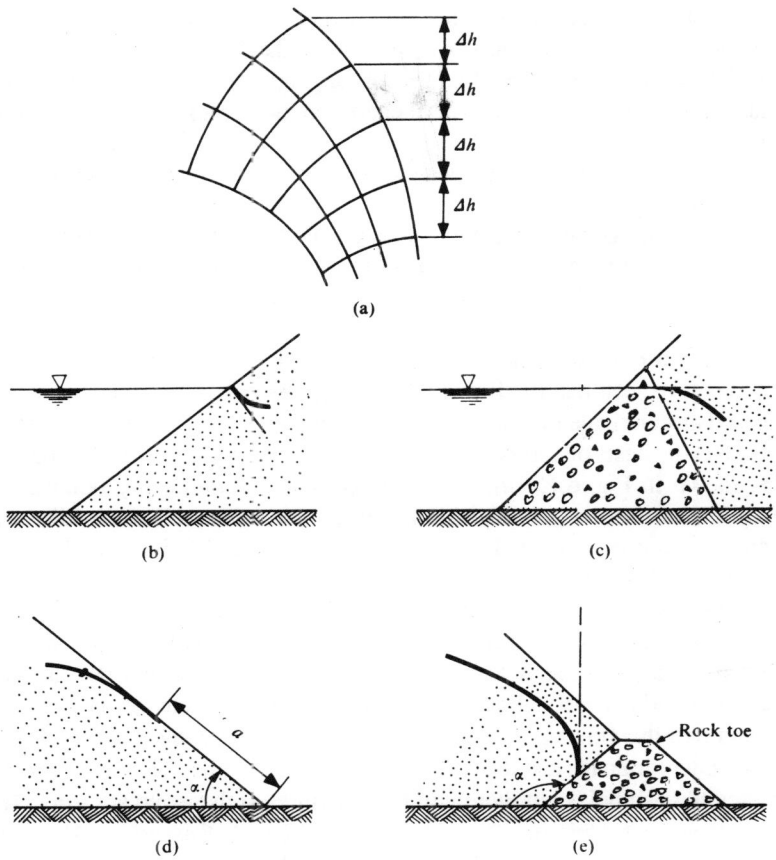

Fig. 6.14 Characteristics of top flow lines (After Taylor, 1948)

Since the top flow line is at atmospheric pressure, the only head that can exist along it is the elevation head. Therefore, there must be equal drops in elevation between the points at which successive equipotentials meet the top flow line, as in Fig. 6.14 (*a*).

At the starting point, the top flow line must be normal to the upstream slope, which is an equipotential line, as shown in Fig. 6.14 (b). However, an exception occurs when the coarse material at the upstream face is so pervious that it does not offer appreciable resistance to flow, as shown in Fig. 6.14 (c). Here, the upstream equipotential is the downstream boundary of the coarse material. The top flow line cannot be normal to this equipotential since it cannot rise without violating the condition illustrated in Fig. 6.14 (a). Therefore, this line starts horizontally and zero initial gradient and zero velocity occur along it. This zero condition relieves the apparent inconsistency of deviation from a 90-degree intersection.

At the downstream end of the top flow line the particles of water tend to follow paths which conform as nearly as possible to the direction of gravity, as shown in Fig. 6.14 (d); the top flow line here is tangential to the slope at the exit. This is also illustrated by the vertical exit condition into a rock-toe as shown in Fig. 6.14 (e).

Two simple cases will be dealt with in regard to the determination of the top flow line:
1. Discharge takes place into a horizontal filter inside the downstream toe.
2. Downstream slope of the dam forms in itself a medium for discharge and a horizontal filter is outside the downstream toe.

6.6.1 Top Flow Line for an Earth Dam with a Horizontal Filter

We have seen that the flow lines and equipotentials, analytically speaking, are based on conjugate functions; one of the simplest examples of conjugate functions is given by nests of confocal parabolas, shown in Fig. 6.15 (b). A simple parabola is shown in Fig. 6.15 (a). It is defined as the curve, every point on which is equidistant from a point called the 'focus' and a line called the 'directrix'. If a cross-section of an earth dam can be conceived to satisfy the boundary conditions so that the flow lines and equipotentials conform to this parabolic shape, Fig. 6.15 (b) gives a flow net for this dam (Kozeny, 1931).

Figure 6.15(c) shows an earth dam cross-section for which a flow net consisting of confocal parabolas holds rigorously. In this case, **BC** and **DF** are flow lines, and **BD** and **FC** are equipotentials. The upstream equipotential is the only unusual feature of this flow net. Figure 6.15 (d) shows the common case of an earth dam with underdrainage and the corresponding top flow line. The flow net for this will resemble parabolas but there will be departures at the upstream side. There will be reverse curvature near **B** for a short distance. A. Casagrande (1937) suggests that **BA** is approximately equal to 0.3 times **BE** where **B** is the starting point of the Kozeny parabola at the upstream water level and **E** is on the upstream water level vertically above the heel **D** of the dam.

Thus, the top flow line may be obtained by constructing the parabola with focus at **F**, the starting point of the filter, and passing through **A**, as per Casagrande's suggestion. The short section of reversed curvature can be easily sketched by visual judgement.

The following are the steps in the graphical determination of the top flow line:

(i) Locate the point **A**, using **BA** = 0.3 (**BE**). A will be the starting point of the Kozeny parabola.

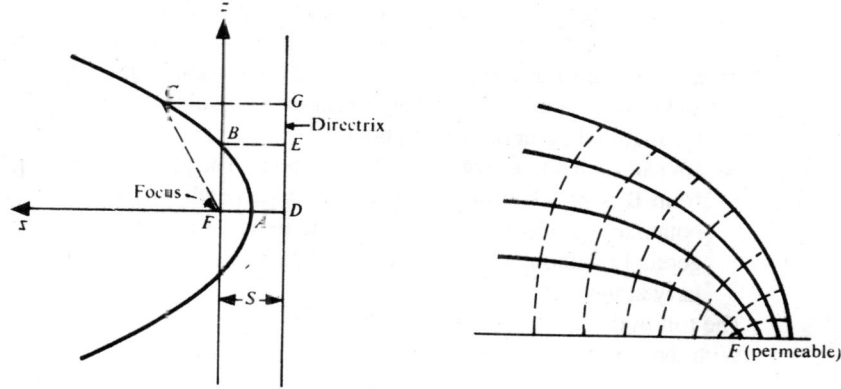

(a) Parabola

(b) Conjugate confocal parabolas (Kozeny, 1931)

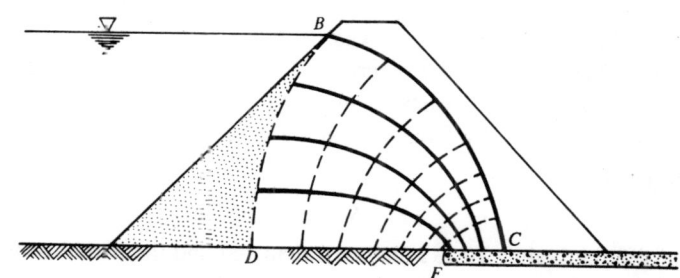

(c) Earth dam with flow net consisting of confocal parabolas

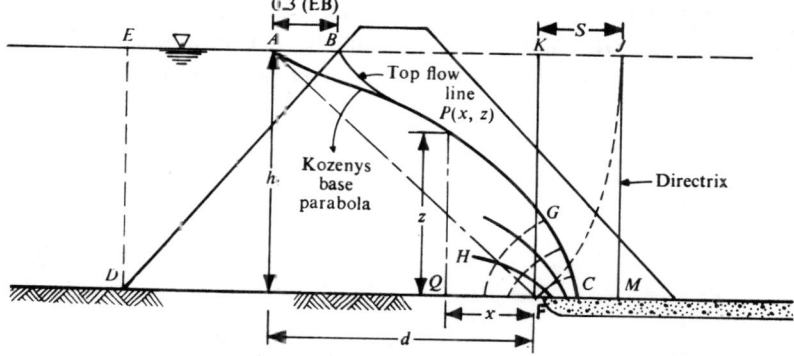

(d) Common case of earth dam and the top flow line (A. Casagrande, 1940)

Fig. 6.15 Flow net consisting of confocal parabolas (After Taylor, 1948)

(ii) With **A** as centre and **AF** as radius, draw an arc to cut the water surface (extended) in **J**. The vertical through **J** is the directrix. Let this meet the bottom surface of the dam in **M**.

(iii) The vertex **C** of the parabola is located midway between **F** and **M**.

(iv) For locating the intermediate points on the parabola the principle that it must be equidistant from the focus and the directrix will be used. For example, at any distance x from **F**, draw a vertical and measure **QM**. With **F** as center and **QM** as radius, draw an arc to cut the vertical through **Q** in **P**, which is the required point on the parabola.

(v) Join all such points to get the base parabola. The portion of the top flow line from **B** is sketched in such that it starts perpendicular to **BD**, which is the boundary equipotential and meets the remaining part of the parabola tangentially without any kink. The base parabola meets the filter perpendicularly at the vertex **C**.

The following analytical approach also may be used:

With the origin of co-ordinates at the focus [Fig. 6.15(d)], **PF = QM**

$$\sqrt{x^2+z^2}=x+S \qquad \ldots\text{(Eq. 6.13)}$$

$$\therefore \quad x=\frac{(z^2-S^2)}{2S} \qquad \ldots\text{(Eq. 6.14)}$$

This is the equation to the parabola.

The value of S may be determined either graphically or analytically. The graphical approach consists in measuring **FM** after the determination of the directrix. Analytically S may be got by substituting the coordinates of **A** (d, h_i) in Eq. 6.13:

$$S=\sqrt{d^2+h_i^2}-d \qquad \ldots\text{(Eq. 6.15)}$$

For different values of x, z may be calculated and the parabola drawn. The corrections at the entry may then be incorporated.

A simple expression may be got for the rate of seepage. In Fig. 6.15 (d), the head at the point G equals S, that along **FC** is zero; hence the head lost between equipotentials **GH** and **FC** is S. Equation 6.1 may now be applied to the part of the flow net **GCFH**; n_f and n_d are each equal to 3 for this net.

$$\therefore \quad q=k.H.3/3=k.S. \qquad \ldots\text{(Eq. 6.16)}$$

Alternatively, the expression for q may be got analytically as follows:

$$q=k.i.A$$

$$=k\cdot\frac{d_z}{d_x}\cdot z \text{ for unit length of the dam.}$$

But $z=(2xS+S^2)^{\frac{1}{2}}$ from equation 6.13.

$$\therefore \quad \frac{d_z}{d_x}=\frac{1}{2}\cdot\frac{2S}{(2xS+S^2)^{\frac{1}{2}}}=\frac{S}{(2xS+S^2)^{\frac{1}{2}}}$$

Substituting,

$$q = k \cdot \frac{S}{(2x\,S + S^2)^{\frac{1}{2}}} \cdot (2x\,S + S^2)^{\frac{1}{2}} = k \cdot S,$$

as obtained earlier (Eq. 6.16).

6.6.2 Top Flow Line for a Homogeneous Earth Dam Resting on an Impervious Foundation

In the case of a homogeneous earth dam resting on an impervious foundation with no drainage filter directly underneath the dam, the top flow line ends at some point on the downstream face of the dam; the focus of the base parabola in this case happens to be the downstream toe of the dam itself as shown in Fig. 6.16.

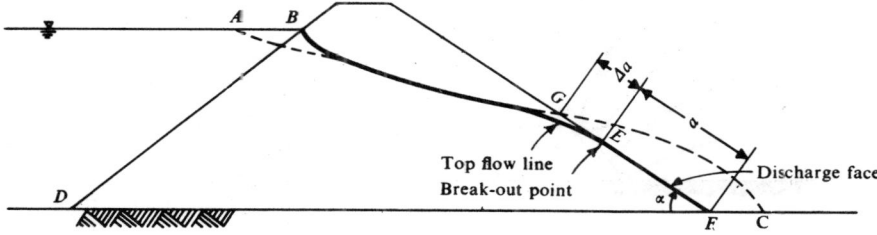

Fig. 6.16 Homogeneous earth dam with no drainage filter

The slope of the 'discharge face', **EF**, with the base of the dam is designated α, measured clockwise. This can have values of 90° or more also depending upon the provision of rock-toe or a drainage face. If the points at which the base parabola and the actual top flow line meet the downstream slope are designated **G** and **E** respectively, **EF** and **GE** are designated a and Δa respectively. The underdrainage case is defined by value of 180° for α. The top flow line meets the downstream face tangentially at the breakout point, **E**.

The top flow lines for typical inclinations of the discharge faces are shown in Fig. 6.17.

The values of $\dfrac{\Delta a}{(a + \Delta a)}$ for different values of α, as given by A. Casagrande, are plotted in Fig. 6.18.

The top flow line in each case is in close agreement with the parabola, except for a short portion of its path in the end. It may be noted that the line of length a, which is a boundary of the flow net, is neither a flow line not an equipotential. Since it is at atmospheric pressure, it is a boundary along which the head at any point is equal to its elevation.

214 Geotechnical Engineering

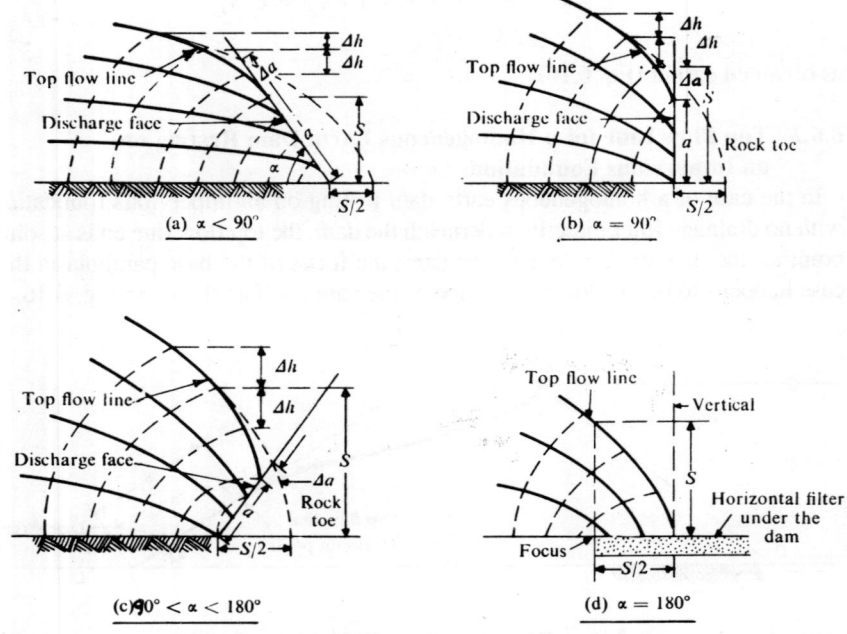

Fig. 6.17 Exit conditions for different slopes of discharge faces

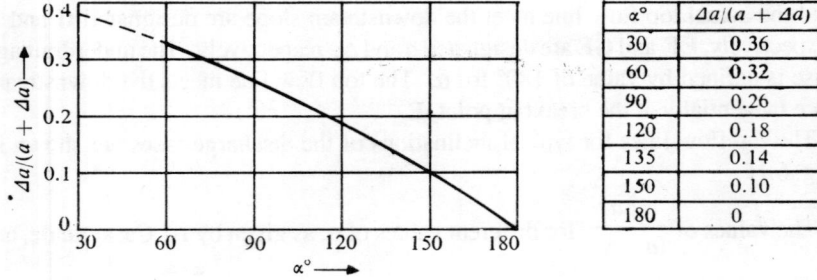

$\alpha°$	$\Delta a/(a + \Delta a)$
30	0.36
60	0.32
90	0.26
120	0.18
135	0.14
150	0.10
180	0

Fig. 6.18 Relation between α and $\Delta a/(a + \Delta a)$ (after A. Casagrande)

The following are the steps in the graphical determination of the top flowline for a homogeneous dam resting on an impervious foundation:

(i) Sketch the base parabola with its focus at the downstream toe of the dam, as described earlier.

(ii) For dams with flat slopes, this parabola will be correct for the central portion of the top flow line. Necessary corrections at the entry on the upstream side and at the exit on the downstream side are to be effected. The portion of the top flow line at entry is sketched visually to meet the boundary condition there i.e., the perpendicularity with the upstream face, which is a boundary equipotential and the tangentiality with the base parabola.

(iii) The intercept $(a + \Delta a)$ is now known. The breakout point on the downstream discharge face may be determined by measuring out Δa from the top along the face, Δa may be obtained from Fig. 6.18.

(iv) The necessary correction at the downstream end may be made making use of one of the boundary conditions at the exit, as shown in Fig. 6.17.

The seepage through all dams with flat slopes may be determined with good accuracy from the simple equation (Eq. 6.16) which holds true for parabolic nets.

L. Casagrande's Solution for a Triangular Dam

For triangular dams on impervious foundations with discharge faces at 90° or less to the horizontal, L. Casagrande gives a simple and reasonably accurate solution for the top flow line (Fig. 6.19):

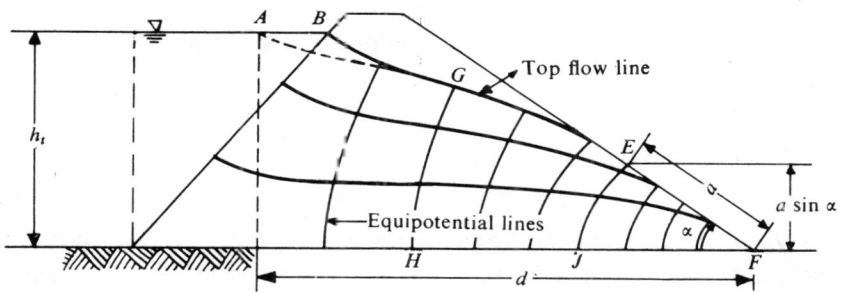

Fig. 6.19 L. Casagrande's method for the determination of top flow line

The top flow line starts at B instead of the theoretical starting point of the parabola, A; the necessary correction at the entry is made as usual. The top flow line ends at E, the location of which is desired, and is defined by the distance a.

Let z be the vertical co-ordinate measured from the tail water or foundation level. The general equation for flow across any equipotential such as GH is given by:

$$q = k \cdot i_{av} (GH)$$

One of the assumptions in the method is that length GH is equal to its projection on the vertical, that is, to the z-co-ordinate of point G. The inaccuracy introduced is insignificant for dams with flat slopes. The other assumption is with regard to the expression used for the gradient. Along the top flow line the gradient is $(-dz/ds)$, since the only head is the elevation head. Since the variation in the size of the square in the vicinity of the equipotential line **GH** is small, the gradient must be approximately constant. It is assumed that the gradient at the top flow line is the average gradient for all points of the equipotential line.

With these assumptions, we have:

$$q = k\left(-\frac{dz}{ds}\right) \cdot z \qquad \ldots(\text{Eq. 6.17})$$

At point **E** the gradient equals $\sin \alpha$, and z equals $a \sin \alpha$. Thus, for the equipotential **EJ**, we have:

$$q = k \cdot a \cdot \sin^2 \alpha \qquad \ldots(\text{Eq. 6.18})$$

Equating the expressions for q given by Eqs. 6.17 and 6.18, and rearranging and integrating between appropriate limits,

$$a \sin^2 \alpha \int_0^{(S-a)} ds = -\int_{h_t}^{a \sin \alpha} z \cdot dz$$

s is the distance along the flow line.

Starting at **A**, the value of α is zero at **A** and $(S - a)$ at **E**. The value of z at **A** is h_t and that at **E** is $a \sin \alpha$.

Solving, we get

$$a = S - \sqrt{S - (h_t \csc \alpha)^2} \qquad \ldots(\text{Eq. 6.19})$$

The value of S differs only slightly from the straight distance **AF**. Using this approximation,

$$S = \sqrt{h_t^2 + d^2} \qquad \ldots(\text{Eq. 6.20})$$

Substituting this in Eq. 6.19, we have:

$$a = \sqrt{h_t^2 + d^2} - \sqrt{d^2 - h_t^2 \cot^2 \alpha} \qquad \ldots(\text{Eq. 6.21})$$

A graphical solution for the distance a, based on Eq. 6.21, was developed by L. Casagrande and is given in Fig. 6.20, which is self-explanatory.

G. Gilboy developed a solution for the distance $a \sin \alpha$, which avoids the approximation of Eq. 6.20; but the equation developed is too complicated for practical use. Instead, use of a chart developed by him is made.

A graphical method of sketching of the top flow line after the determination of the breakout point, which is based on L. Casagrande's differential equation, is given in Fig. 6.21.

Combining Eqs. 6.17 and 6.18, we have:

$$\Delta S = z\left(\frac{\Delta z}{a \sin^2 \alpha}\right) \qquad \ldots(\text{Eq. 6.22})$$

If **C** is plotted such that **CF** $= a \sin \alpha$, the height of **C** above **F** is obviously $a \sin^2 \alpha$.

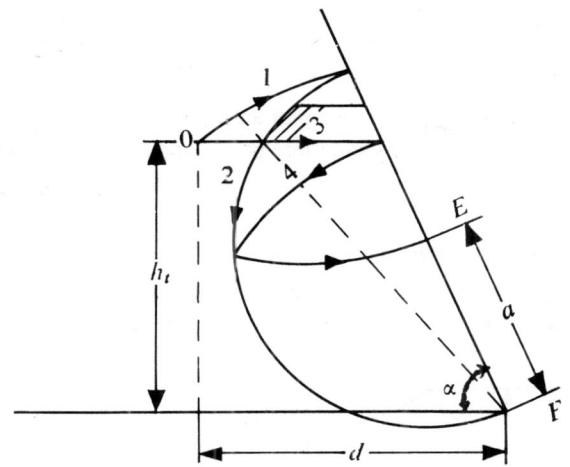

Fig. 6.20 Graphical solution based on L. Casagrande's method

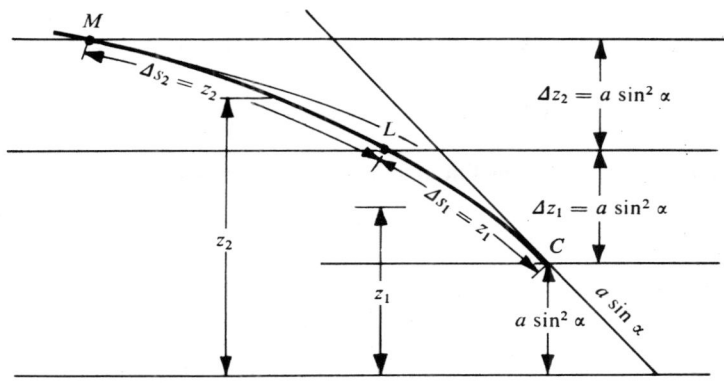

Fig. 6.21 Graphical method for sketching the top flow line, based on L. Casagrande's method (After Taylor, 1948)

Let us assume that the top flow line is divided into sections of constant head drop, say, Δz (a convenient choice is a given fraction of $a \sin^2 \alpha$).
From Eq. 6.22,

if $\Delta z = C_1 \, a \sin^2 \alpha$, where C_1 is a constant,

$$s = C_1 \cdot Z \qquad \ldots \text{(Eq. 6.23)}$$

C_1 has been conveniently chosen as unity, for the illustration given in Fig. 6.21.

218 Geotechnical Engineering

The head drop Δz_1 is laid-off equal to $a \sin^2 \alpha$, z_1 being the average ordinate $\left(a \sin^2 \alpha + \frac{1}{2} a \sin^2 \alpha\right)$. By setting Δs_1 equal to z_1, point **L** on the top flow line is obtained. By repeating this process, we plot a number of points on the top flow line. A smaller value of C_1 would yield more points and a better determination of the top flow line.

Schaffernak and Iterson's Solution for $\alpha < 30°$

Shcaffernak and Iterson (1917) assumed the energy gradient as $\tan \alpha$ or dz/dx. This is approximately true as long as the slope is gentle–say $\alpha < 30°$.

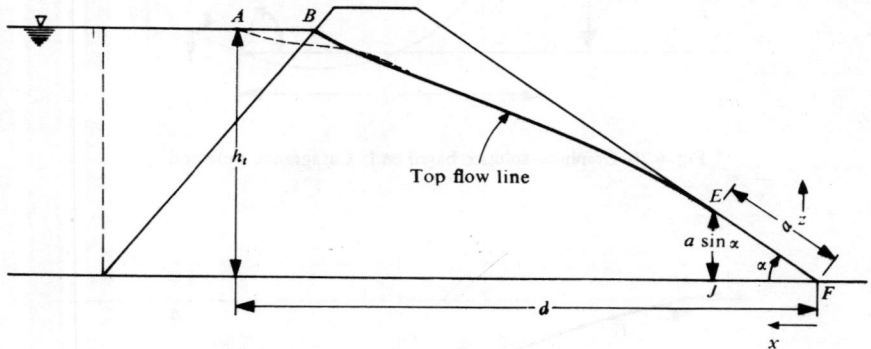

Fig. 6.22 Schaffernak and Iterson's solution for $\alpha < 30°$

Referring to Fig. 6.22, flow through the vertical section **EJ** is given by

$$q = k \cdot \frac{dz}{dx} \cdot z$$

But $dz/dx = \tan \alpha$ and $z = a \sin \alpha$

$$\therefore \quad q = k \cdot a \sin \alpha \tan \alpha \quad \ldots(Eq.\ 6.24)$$

Also,

$$k \cdot \frac{dz}{dx} \cdot z \cdot dx = k a \sin \alpha \tan \alpha \cdot dx$$

or
$$a \sin \alpha \tan \alpha \, dx = z \, dz$$

Integrating between the limits $x = d$ to $x = a \cos \alpha$

and $z = h_t$ to $z = a \sin \alpha$

We have

$$a \sin \alpha \tan \alpha \int_{d}^{a \cos \alpha} dx = \int_{h_t}^{a \sin \alpha} z \cdot dz$$

$$a \sin \alpha \, \tan \alpha \, (a \cos \alpha - d) = \frac{(a^2 \sin^2 \alpha - h_t^2)}{2}$$

From this, we have:

$$a = \frac{d}{\cos \alpha} - \sqrt{\frac{d^2}{\cos^2 \alpha} - \frac{h_t^2}{\sin^2 \alpha}} \qquad \ldots \text{(Eq. 6.25)}$$

6.7 RADIAL FLOW NETS

A case of two-dimensional flow, called 'radial flow', occurs when the flow net is the same for all radial cross-sections through a given axis. This axis may be the centre line of a well, or any other opening that acts as a boundary of cylindrical shape to saturated soil. The direction of flow at every point is towards some point on the axis of symmetry. The flow net for such a section is called 'radial flow net'.

A well with an impervious wall that extends partially through a pervious stratum constitutes an example of radial flow and is shown in Fig. 6.23.

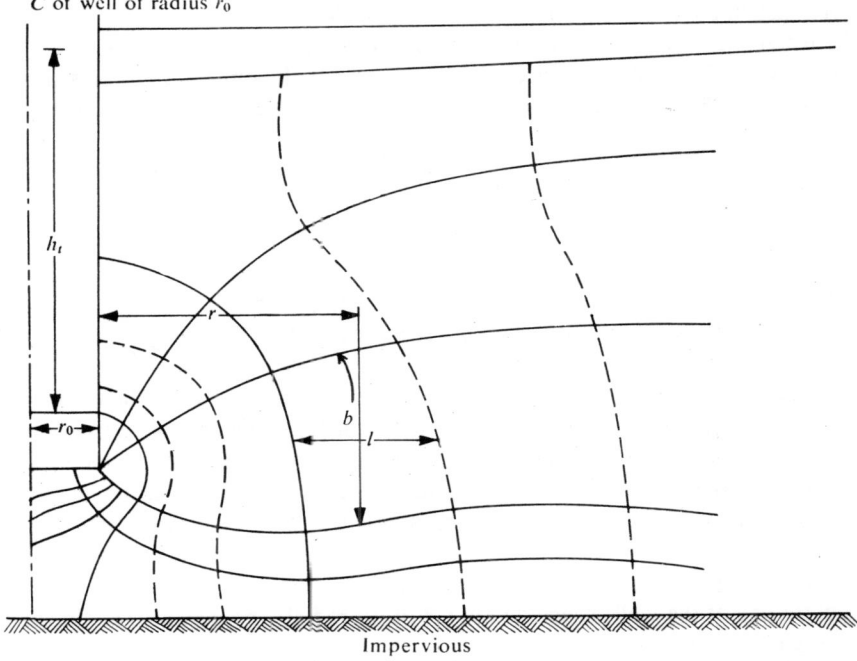

Fig. 6.23 Radial flow net for seepage into a well (After Taylor, 1948)

The width of flow path, b, multiplied by the distance normal to the section, $2\pi r$, is the area of flow $2\pi r b$. The flow through any figure of the flow net is

$$\Delta Q = k \cdot \frac{\Delta h}{l} \cdot 2\pi r b$$

or
$$\Delta Q = 2\pi k (\Delta h)(rb/l) \qquad \ldots\text{(Eq. 6.26)}$$

If ΔQ and Δh are to have the same values for every figure of the flow net, rb/l must be the same for all the figures. Thus the requirement for a radial flow net is that the b/l ratio for each figure must be inversely proportional to the radius, whereas in the other type of the ordinary flow net this ratio must be a constant.

Also $\qquad Q = n_f \cdot \Delta Q$ and $h_t = n_d \cdot \Delta h$

Substituting in Eq. 6.26, we have

$$Q = 2\pi k h_t \cdot \frac{n_f}{n_d} \cdot \frac{r_b}{l} \qquad \ldots\text{(Eq. 6.27)}$$

Here, Q is the total time-rate of seepage for the well.

Two simple cases of radial flow lend themselves to easy mathematical manipulation.

The first one—the simplest case of radial flow—is that into a well at the centre of a round island, penetrating through a pervious, homogeneous, horizontal stratum of constant thickness. It is illustrated in Fig. 6.24.

When the water level is above the level of the pervious stratum, the flow everywhere is radial and horizontal; the gradient at all points is dh/dr for such a flow. The flow across any vertical cylindrical surface at radius r is given by:

$$Q = k \cdot \frac{dh}{dr} \cdot 2\pi r \cdot Z$$

whence
$$h = \frac{Q}{2\pi k Z} \log_e \frac{r}{r_0} \qquad \ldots\text{(Eq. 6.28)}$$

Here h is the head loss between radius r and radius of the well rim.

Here Q is the seepage through the entire thickness Z of the pervious stratum. Eq. 6.1 for q/L may be used with the flow net drawn. The value of Q obtained from the flow net would agree reasonably well with that obtained from the theoretical Eq. 6.28, depending upon the accuracy with which the flow net is sketched.

It is interesting to note that this is a one-dimensional case, since the only space variable is the radius. The seepage pattern is the same for any horizontal plane through the pervious stratum; therefore, a flow net of the type shown in Fig. 6.24 (b) may be drawn.

Further, the seepage pattern is the same on all radial vertical planes through the centre line of the well; hence a flow net of the radial type may also be drawn as shown in Fig. 6.24 (c).

This simple case of horizontal, radial flow is the only one for which it is possible to draw both types of flow nets.

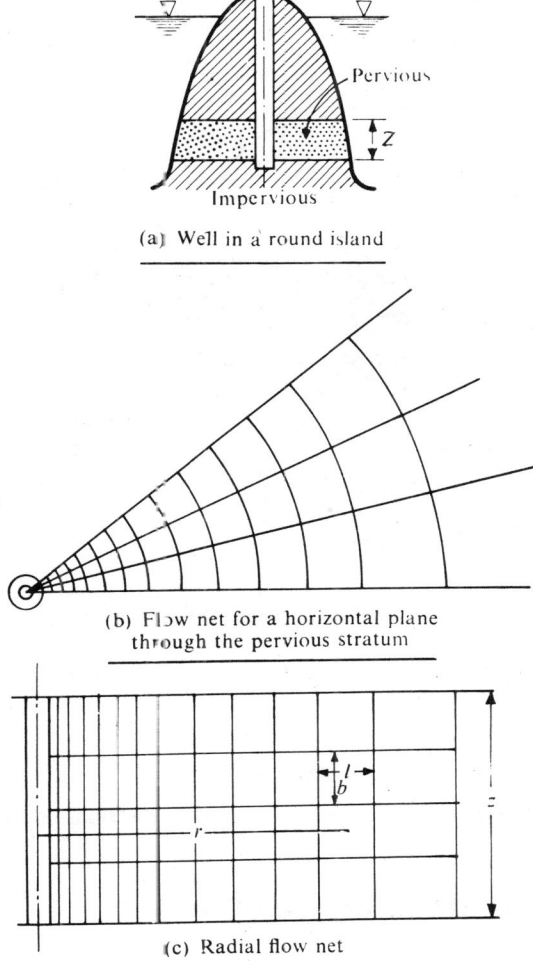

(a) Well in a round island

(b) Flow net for a horizontal plane through the pervious stratum

(c) Radial flow net

Fig 6.24 Radial horizontal flow

The special feature of radial flow is that a relatively large proportion of the head loss occurs in the near vicinity of the well. In view of this, the radial extent of the cross section, the depth below the well and the depth of the soil have little effect on the results.

The second one is the case of spherically radial flow, when the flow everywhere is directed toward a single point. In this case the expression for flow is best written in spherical coordinates; the area across which flow occurs at any given radius is

222 Geotechnical Engineering

the spherical surface of area $4\pi(r')^2$, where r' is the spherical radius. Referring to Fig. 6.25, the discharge may be written

$$Q = k \cdot \frac{dh}{dr'} \cdot 4\pi(r')^2$$

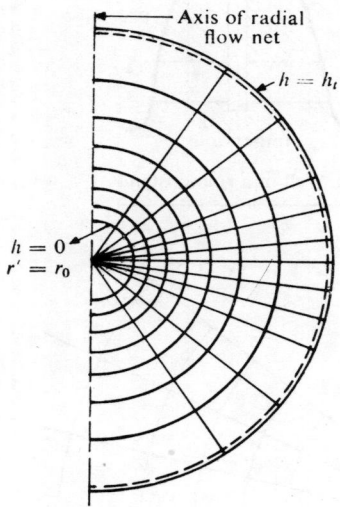

Fig. 6.25 Radial flow net for spherically radial flow (After Taylor, 1948)

If the head is h at radius r' and is zero at radius r_0' the solution of the differential equation is

$$h = \frac{Q}{4\pi k r_0'}\left(1 - \frac{r_0'}{r'}\right) \qquad \ldots\text{(Eq. 6.29)}$$

In this case also, practically all the head is lost in the final portions of the flow paths.

6.8 METHODS OF OBTAINING FLOW NETS

The following methods are available for the determination of flow nets:
1. Graphical solution by sketching
2. Mathematical or analytical methods
3. Numerical analysis
4. Models
5. Analogy methods

All the methods are based on Laplace's equation.

Seepage and Flow Nets 223

6.8.1 Graphical Solution by Sketching

A flow net for a given cross-section is obtained by first transforming the cross-section (if the subsoil is anisotropic), and then sketching by trial and error, taking note of the boundary conditions. The properties of flow nets such as the orthogonality of the flow lines and equipotential lines, and the spaces being elementary squares, and the various rules concerning boundary conditions and smooth transitions must be observed.

Sketching by trial and error was first suggested by Forchheimer (1930) and further developed by A. Casagrande (1937). The following suggestions are made by Casagrande for the benefit of the sketcher:

(a) Every opportunity to study well-constructed flow nets should be utilised to get the feel of the problem.
(b) Four to five flow channels are usually sufficient for the first attempt.
(c) The entire flow net should be sketched roughly before details are adjusted.
(d) The fact that all transitions are smooth and are of elliptical or parabolic shape should be borne in mind.
(e) The boundary flow lines and boundary equipotentials should first be recognised and sketched.

This method has the advantage that it helps the sketcher to get a feel of the problem. The undesirable feature is that the technique is difficult. Many people are not inherently talented in sketching. This difficulty is partially offset by the happy fact that the solution of a two-dimensional flow problem is relatively insensitive to the quality of the flow net. Even a crudely drawn flow net generally permits an accurate determination of seepage, pore pressure and gradient. In addition, the available literature in geotechnical engineering contains good flow nets for a number of common situations.

6.8.2 Mathematical or Analytical Methods

In a few relatively simple cases the boundary conditions may be expressed by equations and solutions of Laplace's equation may be obtained by mathematical procedure. This approach is largely of academic interest because of the complexity of mathematics even for relatively simple problems.

Perhaps the best known theoretical solution was given by Kozeny (1933) and later extended further by A. Casagrande, for flow through an earth dam with a filter drain at the base towards the downstream side. This flownet consists of confocal parabolas.

Another problem for which a theoretical solution is available is a sheet pile wall (Harr, 1962).

6.8.3 Numerical Analysis

Laplace's equation for two-dimensional flow can be solved by numerical techniques in case the mathematical solution is difficult. Relaxation methods involving successive approximation of the total heads at various points in a mesh or net work are used. The Laplace's differential equation is put in its finite difference form and a digital computer is used for rapid solution.

224 Geotechnical Engineering

6.8.4 Models

A flow problem may be studied by constructing a scaled model and analysing the flow in the model. Earth dam models have been used quite frequently for the determination of flow lines. Such models are commonly constructed between two parallel glass or lucite sheets. By the injection of spots of dye at various points, the flow lines may be traced. This approach facilitates the direct determination of the top flow line. Piezometer tubes may be used for the determination of the heads at various points.

Models are specially suited to illustrate the fundamentals of fluid flow. Models are of limited use in the general solution of flow problems because of the time and effort required to construct such models and also because of the difficulties caused by capillarity. The capillary flow in the zone above the top flow line may be significant in a model, although it is of little significance in the prototypes.

6.8.5 Analogy Methods

Laplace's equation for fluid flow also holds for electrical and heat flows. The use of electrical models for solving complex fluid flow problems is more common. In an electrical model, voltage corresponds to total head, current to velocity and conductivity to permeability. Ohm's Law is analogous to Darcy's Law. Measuring voltage, one can locate the equipotentials. The flow pattern can be sketched later.

The versatility of electrical analogy in taking into account boundary conditions too difficult to deal with by other methods, makes the method suitable for solving complex flow situations. Electrical models are considered convenient for instructional purposes, especially in connection with the determination of the top flow line and flow nets for earth dams.

6.9 QUICKSAND

Let us consider the upward flow of water through a soil sample as shown in Fig. 6.26.

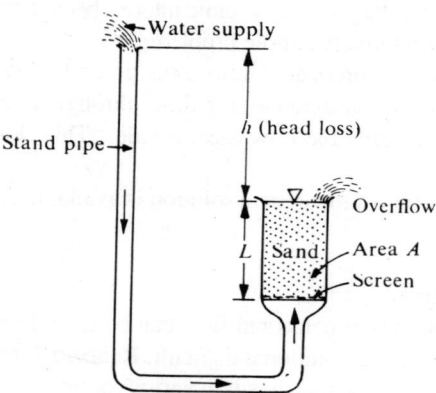

Fig. 6.26 Upward flow of water through soil

Total upward water force on the soil mass at the bottom surface
$$= (h + L) \gamma_w \cdot A$$

Total downward force at the bottom surface = Weight of the soil in the saturated condition
$$= \gamma_{sat} \cdot L \cdot A$$
$$= \frac{(G+e)}{(1+e)} \cdot \gamma_w \cdot L \cdot A$$

Assuming that there is no friction at the sides, it is evident that the soil will be washed out if a sufficiently large value of h is applied. Such a boiling condition will become imminent if the upward water force just equals the weight of the material acting downward; that is,

$$(h+L)\gamma_w \cdot A = \frac{(G+e)}{(1+e)} \cdot \gamma_w L \cdot A \qquad \ldots \text{(Eq. 6.30)}$$

whence $i = h/L = (G-1)/(1+e)$ ...(Eq. 6.31)

This means that an upward hydraulic gradient of magnitude $(G-1)/(1+e)$ will be just sufficient to start the phenomenon of "boiling" in sand. This gradient is commonly referred to as the "Critical hydraulic gradient", i_c. Its value is approximately equal to unity. A saturated sand becomes "Quick" or "Alive" at this gradient; this is only a condition and not a type of sand.

According to Darcy's law, the velocity at which water flows varies as the permeability, in order to maintain a specified hydraulic gradient such as unity. This explains the fact that quicksand conditions occur more commonly in fine sands with low permeability. In the case of gravels with high permeability, much higher velocity of flow will be required to cause the "quicksand condition".

Quicksand conditions are likely to occur in nature in a number of instances; however, the widespread belief that animals and man could be sucked into the quicksand is a myth, since the unit weight of the saturated sand is nearly double that of water. However, quicksand conditions present constructional difficulties. When the exit gradient for a hydraulic structure like a dam assumes the critical value, boiling occurs. This may lead to the phenomenon of progressive backward erosion in the form of a pipe or closed channel underneath the structure and ultimately failure of the structure. This is called "piping". The ratio of the critical gradient to the actual exit gradient is called the "factor of safety against piping".

In summary, we may note:
1. "Quick" refers to a condition and not to a material.
2. Two factors are required for a soil to become quick: Strength must be proportional to effective stress and the effective stress must be equal to zero.
3. The upward gradient needed to cause a quick condition in a cohesionless soil is equal to γ'/γ_w, and is approximately equal to unity. This leads to boiling, piping and ultimate failure of the structure.
4. The amount of flow required to maintain quick condition increases as the permeability of the soil increases.

6.10 SEEPAGE FORCES

Quicksand conditions are caused by seepage forces. These forces have importance in many situations, even when there is no quick condition. Seepage forces are present in clays through which flow occurs, but cohesion prevents the occurrence of boiling.

Referring to Fig. 6.26, the head h is expanded in forcing water through the pores of the soil. This head is dissipated in viscous friction, a drag being exerted in the direction of motion.

The effective weight of the submerged mass is the submerged weight $\gamma' \cdot LA$ or $\dfrac{(G-1)}{(1+e)} \cdot \gamma_w \cdot LA$. An upward force $h \cdot \gamma_w \cdot A$ is dissipated, or transferred by viscous friction into an upward frictional drag *on the particles*. When quick condition is incipient, these forces are equal:

$$h\gamma_w \cdot A = \frac{(G-1)}{(1+e)} \cdot \gamma_w \cdot L.A. \qquad \ldots\text{(Eq. 6.32)}$$

which again leads to equation 6.31. The only difference between Eqs. 6.30 and 6.32 is that both sides of Eq. 6.30 include a force $\gamma_w LA$.

The upward force of seepage is $h \cdot \gamma_w A$, as is in the left hand side of Eq. 6.32. In uniform flow it is distributed uniformly throughout the volume of soil $L.A$, and hence the seepage force j per unit volume is $h\gamma_w \cdot A/LA$, which equals $i \cdot \gamma_w$.

$$j = i \cdot \gamma_w \qquad \ldots\text{(Eq. 6.33)}$$

Thus, the seepage force in an isotropic soil acts *in the direction of flow* and is given by $i.\gamma_w$ per unit volume. The vector sum of seepage forces and gravitational forces is called the resultant body force. This combination may be accomplished either by a combination of the total weight and the boundary neutral force or by a combination of the submerged weight and the seepage force. The two approaches give identical results.

6.11 EFFECTIVE STRESS IN A SOIL MASS UNDER SEEPAGE

The effective stress in the soil at any point is decreased by an amount equal to the seepage force at that point for upward flow; correspondingly, the neutral pressure is increased by the same amount, the total stress remaining unaltered.

Similarly, the effective stress is increased by an amount equal to the seepage force for downward flow; correspondingly, the neutral pressure is decreased by the same amount, the total stress remaining unaltered.

This is due to the fact that the seepage force is the viscous drag transmitted to the particles while the seeping water is being pushed through the pores, the surfaces of the particles serving as the walls surrounding the pores. This, in addition to the fact that seepage force acts in the direction of flow, will enable one to determine the effective stress in a soil mass under steady state seepage.

6.12 ILLUSTRATIVE EXAMPLES

Example 6.1: What is the critical gradient of a sand deposit of specific gravity = 2.65 and void ratio = 0.5?

(S.V.U.—B.Tech., (Part-Time)—Sep., 1982)

$G = 2.65 \quad e = 0.50$

Critical hydraulic gradient, $i_c = (G - 1)/(1 + e)$.

$$= \frac{(2.65 - 1)}{(1 + 0.50)} = \frac{1.65}{1.50} = 1.1$$

Example 6.2: A 1.25 m layer of the soil ($G = 2.65$ and porosity = 35%) is subject to an upward seepage head of 1.85 m. What depth of coarse sand would be required above the soil to provide a factor of safety of 2.0 against piping assuming that the coarse sand has the same porosity and specific gravity as the soil and that there is negligible head loss in the sand.

(S.V.U.—B.E.(R.R.)—Sep., 1978)

$G = 2.65; \; n = 35\% = 0.35; \; e = \dfrac{n}{(1-n)} = \dfrac{0.35}{0.65} = 7/13$

Critical hydraulic gradient, $i_c = \dfrac{(G-1)}{(1+e)} = \dfrac{(2.65-1)}{(1+7/13)}$

$$= \frac{1.65 \times 13}{20} = 1.0725$$

With a factor of safety of 2.0 against piping,

Gradient, $i = \dfrac{i_c}{2} = \dfrac{1.0725}{2} = 0.53625$

But $\quad i = h/L$

$L = h/i = 1.050/0.53625 \text{ m} = 3.45 \text{ m}$

Available flow path = thickness of soil = 1.25 m.

∴ Depth of coarse sand required = **2.20 m**.

Example 6.3: A glass container with pervious bottom containing fine sand in loose state (void ratio = 0.8) is subjected to hydrostatic pressure from underneath until quick condition occurs in the sand. If the specific gravity of sand particles = 2.65, area of cross-section of sand sample = 10 cm² and height of sample = 10 cm, compute the head of water required to cause quicksand condition and also the seepage force acting from below.

(S.V.U.—B.E.(R.R.)—Nov. 1974)

$e = 0.8$, $G = 2.65$

$$i_c = \frac{(G-1)}{(1+e)} = \frac{(2.65-1)}{(1+0.8)} = \frac{1.65}{1.80} = 11/12 = 0.92$$

$L = 10$ cm. $h = L \cdot i_c = \dfrac{10 \times 11}{12} = 55/6 = $ **9.17 cm.**

Seepage force per unit volume $= i \cdot \gamma_w$

$$= \frac{11}{12} \times 9.81 \text{ kN/m}^3$$

Total seepage force $= \dfrac{11}{12} \times 9.81 \times \dfrac{10 \times 10}{100 \times 100 \times 100}$ kN

$$\approx 0.0009 \text{ kN} = \mathbf{0.9 \text{ N}}$$

Example 6.4: A large excavation was made in a stratum of stiff clay with a saturated unit weight of 18.64 kN/m³. When the depth of excavation reached 8 m, the excavation failed as a mixture of sand and water rushed in. Subsequent borings indicated that the clay was underlain by a bed of sand with its top surface at a depth of 12.5 m. To what height would the water have risen above the stratum of sand into a drill hole before the excavation was started?

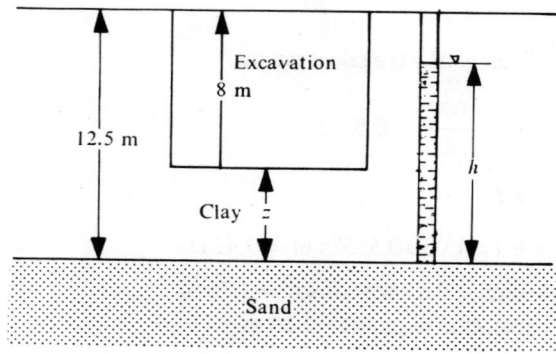

Fig. 6.27 Excavation of clay underlain by sand (Example 6.4)

Referring to Fig. 6.27, the effective stress at the top of sand stratum goes on getting reduced as the excavation proceeds due to relief of stress, the neutral pressure in sand remaining constant.

The excavation would fail when the effective stress reached zero value at the top of sand.

Effective stress at the top of sand stratum,
$$\bar{\sigma} = z \cdot \gamma_{sat} - h \cdot \gamma_w$$

If this is zero, $h\gamma_w = z \cdot \gamma_{sat}$

or
$$h = \frac{z \cdot \gamma_{sat}}{\gamma_w} = \frac{(12.5 - 8) \times 18.64}{9.81} = 8.55 \text{ m}$$

Therefore, the water would have risen to a height of **8.55 m** above the stratum of sand into the drill hole before excavation under the influence of neutral pressure.

Example 6.5: Water flows at the rate of 0.09 ml/s in an upward direction through a sand sample with a coefficient of permeability of 2.7×10^{-2} mm/s. The thickness of the sample is 120 mm and the area of cross-section is 5400 mm^2. Taking the saturated unit weight of the sand as 18.9 kN/m^3, determine the effective pressure at the middle and bottom of the sample.

Here, $q = 0.09$ ml/s $= 90$ mm^3/s, $k = 2.7 \times 10^{-2}$ mm/s

$A = 5400$ mm^2

$$i = q/kA = \frac{90}{2.7 \times 10^{-2} \times 5400} = 0.6173$$

$\gamma' = \gamma_{sat} - \gamma_w = (18.90 - 9.81)$ kN/m^3 $= 9.09$ kN/m^3

$$= 9.09 \times 10^{-6} \text{ N/mm}^3$$

For the bottom of the sample, $z = 120$ mm

$$\bar{\sigma} = \gamma'_z - iz\gamma_w,$$

for downward flow, considering the effect of seepage pressure.

$\therefore \bar{\sigma} = (9.09 \times 10^{-6} \times 120 - 0.6173 \times 120 \times 9.81 \times 10^{-6})$ N/mm^2

$= 120 \times 10^{-6} (9.09 - 0.6173 \times 9.81) = 0.364 \times 10^{-3}$ N/mm^2

$$= 364 \text{ N/m}^2$$

For the middle of the sample, $z = 60$ mm

$$\bar{\sigma} = \gamma'_z - iz \gamma_w$$

$= (9.09 \times 10^{-6} \times 60 - 0.6173 \times 60 \times 9.81 \times 10^{-6})$ N/mm^2

$= 0.182 \times 10^{-3}$ N/mm^2 $= \mathbf{182}$ **N/m^2**

Example 6.6: A deposit of cohesionless soil with a permeability of 3×10^{-2} cm/s has a depth of 10 m with an impervious ledge below. A sheet pile wall is driven into this deposit to a depth of 7.5 m. The wall extends above the surface of the soil and a 2.5 m depth of water acts on one side. Sketch the flow net and determine the seepage quantity per metre length of the wall.

(S.V.U.—B.E.(R.R.)—Nov., 1973)

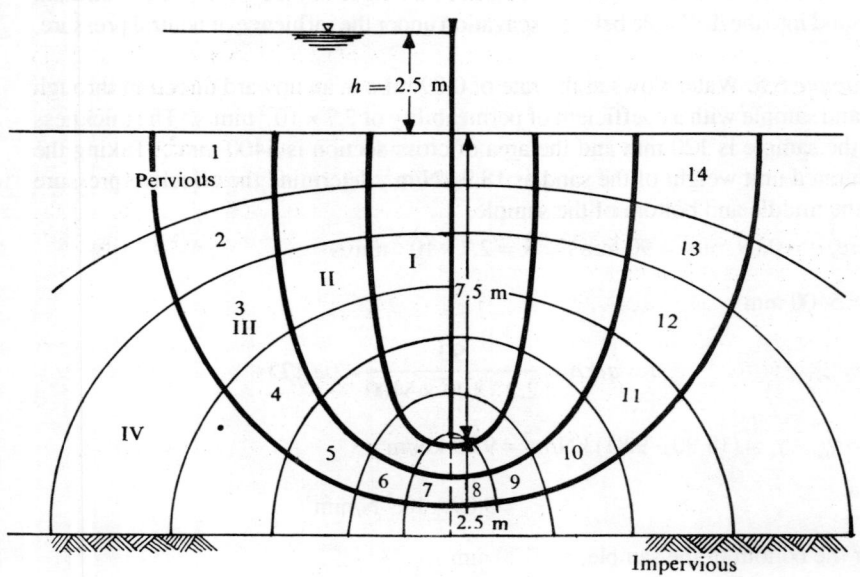

Fig. 6.28 Sheet pile wall (Example 6.6)

The flow net is shown.

Number of flow channels, $n_f = 4$

Number of equipotential drops, $n_d = 14$

Quantity of seepage per meter length of wall $\quad \Big\} \quad q = k . H . \dfrac{n_f}{n_d}$

$$= 3 \times 10^{-4} \times 2.5 \times \frac{4}{14} \text{ m}^3/\text{sec/metre run}$$

$$= 2.143 \times 10^{-4} \text{ m}^3/\text{sec/meter run}$$

$$= \mathbf{214.3 \text{ ml/sec/metre run}}$$

Example 6.7: An earth dam of homogeneous section with a horizontal filter is shown in Fig. 6.29. If the coefficient of permeability of the soil is 3×10^{-3} mm/s, find the quantity of seepage per unit length of the dam.

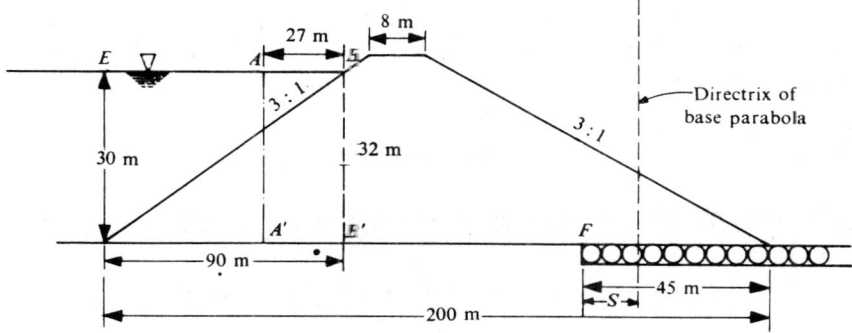

Fig. 6.29 Homogeneous earth dam with horizontal filter at toe (Example 6.7)

$AB = 0.3(EB) = 3.0 \times 90 = 27$ m

With respect to the focus, **F** (the end of the filter), as origin, the co-ordinates of **A**, the starting point of the base parabola, are : $x = FA' = 200 - 90 - 45 + 27 = 92$ m

$$z = A'A = 30 \text{ m}$$

The equation to the parabola is

$$\sqrt{x^2 + z^2} = x + S,$$

where S is the distance to the directrix from the focus, **F**.

$$\therefore \quad \sqrt{92^2 + 30^2} = 92 + S$$

or $\quad S = (\sqrt{92^2 + 30^2} - 92) \text{ m} = 4.77 \text{ m}$

The quantity of seepage per metre unit length of the dam

$$q = k \cdot S$$

$$= 3 \times 10^{-2} \times 10^{-3} \times 4.77 \text{ m}^3/\text{s}$$

$$= 14.31 \times 10^{-5} \text{ m}^3/\text{s}$$

$$= \mathbf{143.1 \text{ ml/s}}$$

Example 6.8: For a homogeneous earth dam 32 m high and 2 m free board, a flow net was constructed with four flow channels. The number of potential drops was 20. The dam has a horizontal filter at the base near the toe. The coefficient of permeability of the soil was 9×10^{-2} mm/s. Determine the anticipated seepage, if the length of the dam is 100 metres.

232 Geotechnical Engineering

Head of water, $H = (32 - 2)$ m $= 30$ m
Permeability of the soil $= 9 \times 10^{-2}$ mm/s
Number of flow channels $= 4$
Number of head drops $= 20$

Discharge per metre length $= k \cdot H \cdot \dfrac{n_f}{n_d}$

$$= \dfrac{9 \times 10^{-2}}{10^3} \times 30 \times \dfrac{4}{20} \text{ m}^3/\text{s}$$

Seepage anticipated for the entire length of the dam

$$= \dfrac{100 \times 9 \times 10^{-2} \times 30 \times 4}{1000 \times 20} \text{ m}^3/\text{s}$$

$$= 0.054 \text{ m}^3/\text{s} = \mathbf{541 \text{ ml/s}}$$

Example 6.9: An earth dam is built on an impervious foundation with a horizontal filter at the base near the toe. The permeability of the soil in the horizontal and vertical directions are 3×10^{-2} mm/s and 1×10^{-2} mm/s respectively. The full reservoir level is 30 m above the filter. A flow net constructed for the transformed section of the dam, consists of 4 flow channels and 16 head drops. Estimate the seepage loss per metre length of the dam.

$k_h = 3 \times 10^{-2}$ mm/s $k_v = 1 \times 10^{-2}$ mm/s

$k_h = 3 \times 10^{-5}$ m/s $k_v = 10^{-5}$ m/s

$H = 30$ m

Equivalent permeability $k_e = \sqrt{k_h k_v}$

$$= \sqrt{3 \times 10^{-5} \times 1 \times 10^{-5}} \text{ m/s}$$

$$= 1.732 \times 10^{-5} \text{ m/s}$$

Seepage loss per metre length of the dam

$$= k_e \cdot H \cdot \dfrac{n_f}{n_d}$$

$$= 0.433 \times 10^{-5} \times 30 \times \dfrac{4}{16} \text{ m}^3/\text{s}$$

$$= 1.299 \times 10^{-4} \text{ m}^3/\text{s}$$

$$= \mathbf{129.9 \text{ ml/s}}$$

SUMMARY OF MAIN POINTS

1. Flow of water through soil under a hydraulic gradient is called "Seepage".
2. A flow net is a system of squares or rectangles formed by flow lines intersecting equipotential lines.
3. The rate of flow is given by $q = k \cdot H \cdot s$, where s, the shape factor ($= n_f/n_d$), is provided by the flow net.
4. The basic equation for seepage is Laplace's equation: $\dfrac{\partial^2 h}{\partial x^2} + \dfrac{\partial^2 h}{\partial z^2} = 0$ (for isotropic soil). A flow net is simply a solution of this equation for a given set of boundary conditions.
5. Functions which satisfy Laplace's equation are called 'Conjugate harmonic functions'.
6. From a flow net one can obtain (a) rate of flow, (b) pore pressure, and (c) gradient.
7. In anisotropic soil, the section must be transformed, using $X_T = \sqrt{k_z/k_x} \cdot x$; the effective permeability, k_e, is then given by $\sqrt{k_x k_z}$.
8. The flow through an earth dam is bounded by a top flow line or phreatic line, which is determined first. The location depends upon the drainage conditions at the downstream toe and the inclination of the discharge face. The phreatic line mostly follows the base parabola of Kozeny, with slight modifications at the beginning and the end, as given by A. Casagrande.
9. The flow around a circular well constitutes a radial flow net.
10. Graphical method of sketching by trial and error and analogy methods are important among the methods of obtaining flow nets.
11. "Quick" refers to a condition wherein a cohesionless soil loses its strength because the upward flow of water makes the effective stress zero; the hydraulic gradient at which this condition is imminent is called the critical one, and is given by $i_c = \dfrac{(G-1)}{(1+e)}$.
12. The seepage force per unit volume of the soil is $i \cdot \gamma_w$ and (for isotropic soil) acts in the direction of flow.

REFERENCES

1. A. Casagrande: *Seepage through Dams*, Jl. New England Water Works Association, June, 1937. Reprinted by Boston Society of Civil Engineers, Contributions to Soil Mechanics, 1925 to 1940.
2. P. Forchheimer: *Hydraulik*, 3rd ed., Leipzig, Teubner, 1930.
3. G. Gilboy: *Hydraulic fill dams*, Proc. Int. Comm. on Large Dams, World Power Conference, Stockholm, 1933.

4. M.E. Harr: *Ground Water and Seepage*, McGraw Hill, 1962.
5. A.R. Jumikis: *Soil Mechanics*, D. Van Nostrand Co., Princeton, NJ, 1962.
6. K.S. Kozeny: *Grundwasserbewegung bei freiem Spiegel, Fluss und Kanalversicherung*, Wasserkraft und Wasserwirtschaft, No. 3, 1931.
7. T.W. Lambe and R.V. Whitman: *Soil Mechanics*, John Wiley & Sons, Inc., NY, 1969.
8. D.F. McCarthy: *Essentials of Soil Mechanics and Foundations*, Reston Publishing Company, Reston, Va, USA, 1977.
9. Schaffernak: *Uberdie Stansicherheit durchlaessiger ges chuetterter Damme*, Allgem, Bauzeitung, 1917.
10. G.N. Smith: *Essentials of Soil Mechanics for Civil and Mining Engineers*, third edition Metric, Crosby Lockwood Staples, London, 1974.
11. M.G. Spangler: *Soil Engineering*, International Textbook Company, Scranton, USA, 1951.
12. D.W. Taylor: *Fundamentals of Soil Mechanics*, John Wiley & Sons, Inc., NY, 1948.

QUESTIONS AND PROBLEMS

6.1 Write short notes on 'Flow nets'.

(S.V.U.—Four year B.Tech.—Apr., 1983)

6.2 (*a*) What are the principles of a flow net? What are its uses?
  (*b*) Explain the phenomenon of "Piping".

(S.V.U.—B.Tech., (Part-Time)—May, 1983)

6.3 What are the salient characteristics of a flow net?
Describe a suitable procedure of drawing flow net.

(S.V.U.—B.Tech., (Part-Time)—Sep., 1982)

6.4 Write short notes on: Critical hydraulic gradient, Phreatic line.

(S.V.U.—B.Tech., (Part-Time)—Apr., 1982)

6.5 Write short notes on: Quicksand conditions, Laplace systems.

(S.V.U.—Four year B.Tech.,—June, 1982)

6.6 Briefly discuss:
  (*i*) Properties and utility of the flow net.
  (*ii*) Seepage force
  (*iii*) Electrical analogy method.

(S.V.U.—B.Tech., (Part-Time)—Apr., 1982)

6.7 (*a*) Explain the meaning of the term "Seepage pressure".
  (*b*) Show how the effective pressure is altered when water is flowing through the soil vertically downwards and vertically upwards.

(S.V.U.—B.E., (R.R.)—Sep., 1978)

6.8 Stating the basic principles of flow nets describe the trial sketching method of obtaining a flow net with particular reference to a homogeneous earth dam.

(S.V.U.—B.E., (R.R.)—Nov., 1975)

6.9 (*a*) Define "Critical hydraulic gradient" and explain how "piping" is produced.

(b) Explain the principle of drawing flow nets and derive the expression to calculate the amount of flow of water in the case of a sheet pile wall with a head of water h on one side.

(S.V.U.—B.E., (R.R.)—May, 1970)

6.10 The hydraulic gradient for an upward flow of water through a sand mass is 0.90. If the specific gravity of soil particles is 2.65 and the void ratio is 0.50, will quicksand conditions develop?

6.11 The foundation soil at the toe of a masonry dam has a porosity of 40% and the specific gravity of grains is 2.70. To assure safety against piping, the specifications state that the upward gradient must not exceed 25% of the gradient at which a quick condition occurs. What is the maximum permissible upward gradient?

6.12 In a container filled with each of the following materials, at a porosity of 40%, determine the upward gradient required to cause quick condition:
   (a) lead shot with a specific gravity of 11.35;
   (b) fibre beads with a specific gravity of 1.55;
   (c) sand with a specific gravity 2.65.

6.13 An excavation is to be performed in a stratum of clay, 9 m thick, underlain by a bed of sand. In a trial bore hole, the ground-water is observed to rise up to an elevation 3 m below ground surface. Find the depth to which the excavation can be safely carried out without the bottom becoming unstable under uplift pressure of ground-water. The specific gravity of clay particles is 2.70 and the void ratio is 0.70. If the excavation is to be safely carried to a depth of 7 m, how much should the water table be lowered in the vicinity of the trench?

6.14 There is an upward flow of 0.06 ml/s through a sand sample with a coefficient of permeability 3×10^{-2} mm/s. The thickness of the sample is 150 mm and the cross-sectional area is 4500 mm². Determine the effective stress at the bottom and middle of the sample, if the saturated unit weight of the sample is 18.9 kN/m³.

6.15 A deposit of cohesionless soil with a permeability of 0.3 mm/s has a depth of 12 m with impervious ledge below. A sheet pile wall is driven into this deposit to a depth of 8 m. The wall extends above the surface of the soil, and a 3 m depth of water acts on one side. Sketch the flow net and determine the seepage quantity.

6.16 A homogeneous earth dam has a top width of 6 m and a height of 42 metres with side slopes of 3 to 1 and 4 to 1 on the upstream side and downstream side respectively. The free board is 2 m. There is a horizontal filter at the base on the downstream side extending for a length of 60 m from the toe. If the coefficient of permeability of the soil is 9×10^{-2} mm/s, find the quantity of seepage per day for 100 metre length of the dam.

6.17 A double wall sheet pile coffer dam retains a height of water 9 m on one side. A flow net constructed for this structure, driven into a pervious deposit overlain by an impervious ledge, consists of five flow channels and fifteen potential drops. The length of the coffer dam is 90 metres. If the coefficient of permeability of the pervious deposit is 10^{-2} mm/s determine the seepage in cubic metres per day.

6.18 A concrete dam retains water to a height of 9 m. It has rows of sheet piling at both heel and toe which extend half way down to an impervious stratum. From a flow net sketched on a transformed section, it is found that there are four flow channels and sixteen head drops. The average horizontal and vertical permeabilities of the soil are 6×10^{-3} mm/s and 2×10^{-3} mm/s, respectively. What is the seepage per day, if the length of the dam is 150 metres?

7
COMPRESSIBILITY AND CONSOLIDATION OF SOILS

7.1 INTRODUCTION

When a structure is placed on a foundation consisting of soil, the loads from the structure cause the soil to be stressed. The two most important requirements for the stability and safety of the structure are: (1) The deformation, especially the vertical deformation, called 'settlement' of the soil, should not be excessive and must be within tolerable or permissible limits; and, (2) The shear strength of the foundation soil should be adequate to withstand the stresses induced.

The first of these requirements needs consideration and study of the aspect of the "Compressibility and Consolidation of soils" and forms the subject matter of this chapter. The second needs consideration of the aspects of shear strength and bearing capacity of soil, which are dealt with in subsequent chapters.

The nature of the deformation of soil under compressible loads may be elastic, plastic or compressive, or a combination of these. Elastic deformation causes lateral bulging with little change of porosity and the material recovers fully upon removal of the load. Plastic deformation is due to the lateral flow of the soil under pressure with negligible rebound after removal of load. 'Plasticity' is the property by which the material can undergo considerable deformation before failure. Clays exhibit this property to a greater or smaller degree at moisture contents greater than the plastic limit. Compressive deformation occurs when the particles are brought closer together by pressure causing volume changes in the soil. The property of a soil by virtue of which volume decrease occurs under applied pressure is termed its 'Compressibility'.

Since natural soil deposits are laterally confined on all sides, deformation under stress is primarily associated with volume changes, specifically, volume decrease.

7.2 COMPRESSIBILITY OF SOILS

A soil is a particulate material, consisting of solid grains and void spaces enclosed by the grains. The voids may be filled with air or other gas, with water or other liquid, or with a combination of these.

The volume decrease of a soil under stress might be conceivably attributed to:
(1) Compression of the solid grains;
(2) Compression of pore water or pore air;

(3) Expulsion of pore water or pore air from the voids, thus decreasing the void ratio or porosity.

Under the loads usually encountered in geotechnical engineering practice, the solid grains as well as pore water may be considered to be incompressible. Thus, compression of pore air and expulsion of pore water are the primary sources of volume decrease of a soil mass subjected to stresses. A partially saturated soil may experience appreciable volume decrease through the compression of pore air before any expulsion of pore water takes place; the situation is thus more complex for such a soil. However, it is reasonable to assume that volume decrease of a saturated soil mass is, for all practical purposes, only due to expulsion of pore water by the application of load. Sedimentary deposits and submerged clay strata are invariably found in nature in the fully saturated condition and problems involving volume decrease and the consequent ill-effects are associated with these.

Specifically, the compressibility of a soil depends on the structural arrangement of the soil particles, and in fine-grained soils, the degree to which adjacent particles are bonded together. A structure which is more porous, such as a honey-combed structure, is more compressible than a dense structure. A soil which is composed predominantly of flat grains is more compressible than one with mostly spherical grains. A soil in a remoulded state may be more compressible than the same soil in its natural state.

When the pressure is increased, volume decrease occurs for a soil. If the pressure is later decreased some expansion will take place, but the rebound or recovery will not occur to the full extent. This indicates that soils show some elastic tendency, but only to a small degree. It is rather difficult to separate the elastic and inelastic compression in soils.

There is another kind of volume rebound shown by fine-grained soils. Water held between the flaky particles by certain forces gets squeezed out under compression. When the stress is removed, these forces cause the water to be sucked in again, resulting in the phenomenon of 'swelling'. Expulsion and sucking of water may take a very long time.

The process of gradual compression due to the expulsion of pore water under steady pressure is referred to as 'Consolidation', which is dealt with in later sections. This is a time-dependent phenomenon, especially in clays. Thus, the volume change behaviour has two distinct aspects: first, the magnitude of volume change leading to a certain total compression or settlement, and secondly, the time required for the volume change to occur under a particular stress.

The process of mechanical compression resulting in reduction or compression of pore air and consequent densification of soil is referred to as 'Compaction', and it is dealt with in a later chapter.

In sands, consolidation may be generally considered to keep pace with construction; while, in clays, the process of consolidation proceeds long after the construction has been completed and thus needs greater attention.

7.2.1 One-dimensional Compression and Consolidation

The previous discussion refers to compression in general. The general case is complex, but an analysis of the case in which the compression takes place in one direction only is relatively simple. The simple type of one-dimensional compression, to be described in a later sub-section, holds in the laboratory except for minor variations caused by side friction. The compression at shallow elevations underneath a loaded structure is definitely three-dimensional, but the compression in deep strata is essentially one-dimensional. Besides, there are other practical situations in which the compressions approach a truly one-dimensional case. In view of this, one-dimensional analysis of compression and consolidation has significant practical applications.

Escape of pore water must occur during the compression or one-dimensional consolidation of a saturated soil; this escape takes place according to Darcy's law. The time required for the compression or consolidation is dependent upon the coefficient of permeability of the soil and may be quite long if the permeability is low. The applied pressure which is initially borne by the pore water goes on getting transferred to the soil grains during the transient stage and gets fully transferred to the grains as effective stress, reducing the excess pore water pressure to zero at the end of the compression under the applied stress. Thus, 'Consolidation', may be defined as the gradual and time-dependent process involving expulsion of pore water from a saturated soil mass, compression and stress transfer. This definition is valid for the one-dimensional as well as the general three-dimensional case.

It may be worthwhile to note that the volume of the soil mass at any time is related to the effective stress in the soil at that time and not to the total stress. In other words, compressibility is a function of the effective stress. The application of a total stress increment merely creates a transient flow situation and induces consolidation through expulsion of pore water and increases in effective stress through a decrease in excess pore water pressure.

7.2.2 Compressibility and Consolidation Test-Oedometer

The apparatus developed by Terzaghi for the determination of compressibility characteristics including the time-rate of compression is called the *Oedometer*. It was later improved by A. Casagrande and G. Gilboy and referred to as the *Consolidometer*.

The consolidometer device is shown schematically in Fig. 7.1.

There are two types: The fixed ring type and the floating ring type. In the fixed ring type, the top porous plate alone is permitted to move downwards for compressing the specimen. But, in the floating ring type, both the top and bottom porous plates are free to move to compress the soil sample. Direct measurement of the permeability of the sample at any stage of the test is possible only with the fixed ring type. However, the effect of side friction on the soil sample is smaller in the floating type, while lateral confinement of the sample is available in both to simulate a soil mass *in-situ*.

240 Geotechnical Engineering

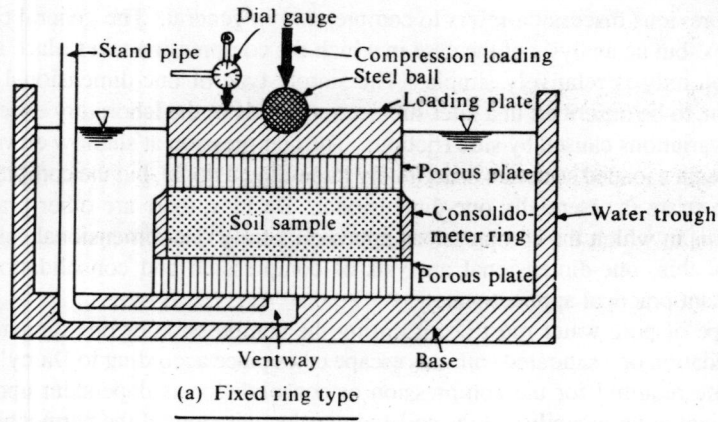

(a) Fixed ring type

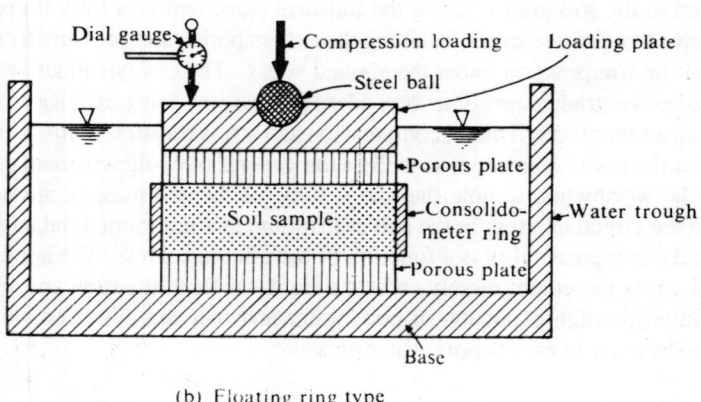

(b) Floating ring type

Fig. 7.1 Schematic of consolidometer

The consolidation test consists in placing a representative undisturbed sample of the soil in a consolidometer ring, subjecting the sample to normal stress in predetermined stress increments through a loading machine and during each stress increment, observing the reduction in the height of the sample at different elapsed times after the application of the load. The test is standardised with regard to the pattern of increasing the stress and the duration of time for each stress increment. Thus the total compression and the time-rate of compression for each stress increment may be determined. The data permits the study of the compressibility and consolidation characteristics of the soil.

The time-rate of volume change differs significantly for cohesionless soils and cohesive soils. Cohesionless soils experience compression relatively quickly, often instantaneously, after the load is imposed. But clay soils require a significant period before full compression occurs under an applied loading. Relating the time-rate of compression with compression is consolidation. Laboratory compression tests are seldom performed on cohesionless soils for two reasons: first, undisturbed soil samples cannot be obtained and secondly, the settlement is rapid, eliminating post-construction problems of settlement. If volume change or settlement characteristics are needed, these are obtained indirectly from *in-situ* density and density index and other correlations.

The following procedure is recommended by the ISI for the consolidation test [IS:2720 (Part XV), 1965]:

The specimen shall be 60 mm in diameter and 20 mm thick. The specimen shall be prepared either from undisturbed samples or from compacted representative samples. The specimen shall be trimmed carefully so that the disturbance is minimum. The orientation of the sample in the consolidometer ring must correspond to the orientation likely to exist in the field.

The porous stones shall be saturated by boiling in distilled water for at least 15 minutes. Filter papers are placed above and below the sample and porous stones are placed above and below these. The loading block shall be positioned centrally on the top porous stone.

This assembly shall be mounted on the loading frame such that the load is applied axially. In the case of the lever loading system, the apparatus shall be properly counterbalanced. The lever system shall be such that no horizontal force is imposed on the specimen at any stage during testing and should ensure the verticality of all loads applied to the specimen. Weights of known magnitude may be hung on the lever system. The holder with the dial gauge to record the progressive vertical compression of the specimen under load, shall then be screwed in place. The dial gauge shall be adjusted allowing a sufficient margin for the swelling of the soil, if any. The system shall be connected to a water reservoir with the water level being at about the same level as the soil specimen and the water allowed to flow through and saturate the sample.

An initial setting load of 50 g/cm^2 (5 kN/m^2), which may be as low as 25 g/cm^2 (2.5 kN/m^2) for very soft soils, shall be applied until there is no change in the dial gauge reading for two consecutive hours or for a maximum of 24 hours. A normal load to give the desired pressure intensity shall be applied to the soil, a stopwatch being started simultaneously with loading. The dial gauge reading shall be recorded after various intervals of time – 0.25, 1, 2.25, 4, 6.25, 9, 12.25, 16, 20.25, 25, 36, 49, 64, 81, 100, 121, 144, 169, 196, 225, 256, 289, 324, 361, 400, 500, 600, and 1440 minutes.*

The dial gauge readings are noted until 90% consolidation is reached. Thereafter, occasional observations shall be continued. For soils which have slow primary consolidation, loads should act for at least 24 hours and in extreme cases or where secondary consolidation must be evaluated, much longer.

---

*The significance as well as convenience of choosing the time intervals as perfect squares will be understood later on, after the reader goes through the Sub-section 7.7.1 – " the square root of Time Fitting Method".

At the end of the period specified, the load intensity on the soil specimen is doubled.* Dial and time readings shall be taken as earlier. Then successive load increments shall be applied and the observations repeated for each load till the specimen has been loaded to the desired intensity. The usual sequence of loading is 0.1, 0.2, 0.4, 0.8, 1.6, 3.2 and 6.4 kg/cm^2 (10, 20, 40, 80, 160, 320 and 640 kN/m^2). Smaller increments may be desirable for very soft soil samples. Alternatively, 6, 12, 25, 50, 100 and 200 per cent of the maximum field loading may be used. An alternative loading or reloading schedule may be employed that reproduces the construction stress changes, obtains better definition of some part of the stress-void ratio curve, or aids in interpreting the field behaviour of the soil.

After the last load has been on for the required period, the load should be decreased to 1/4 the value of the last load and allowed to stand for 24 hours. No time-dial readings are normally necessary during the rebound, unless information on swelling is required. The load shall be further reduced in steps of one-fourth the previous intensity till an intensity of 0.1 kg/m^2 (10 kN/m^2) is reached. If data for repeated loading is desired, the load intensity may now be increased in steps of double the immediately preceding value and the observations repeated.

Throughout the test, the container shall be kept filled with water in order to prevent desiccation and to provide water for rebound expansion. After the final reading has been taken for 0.1 kg/cm^2 (10 kN/m^2) the load shall be reduced to the initial setting load, kept for 24 hours and the final reading of the dial gauge noted.

When the observations are completed, the assembly shall be quickly dismantled, the excess surface water on the specimen is carefully removed by blotting and the ring with the consolidated soil specimen weighed. The soil shall then be dried to constant weight in an oven maintained at 105° to 110°C and the dry weight recorded.

7.2.3 Presentation and Analysis of Compression Test Data

There are several ways in which the data from a laboratory compression test may be presented and analysed.

The consolidation is rapid at first, but the rate gradually decreases. After a time, the dial reading becomes practically steady, and the soil sample may be assumed to have reached a condition of equilibrium. For the common size of the soil sample, this condition is generally attained in about twenty-four hours, although, theoretically speaking, the time required for complete consolidation is infinite. This variation of compression or the dial gauge reading with time may be plotted for each one of the stress increments. Figure 7.2 depicts a typical time versus compression curve.

A curve of this type may be transformed in a certain manner and used for a specific purpose as will be indicated in Sec. 7.7.

---

*The significance of this procedure will be understood after the reader goes through Sub-section 7.2.6 – "Normally consolidated soil and overconsolidated soil". The objective is to see that the soil sample is normally consolidated throughout the test.

Compressibility and Consolidation of Soils 243

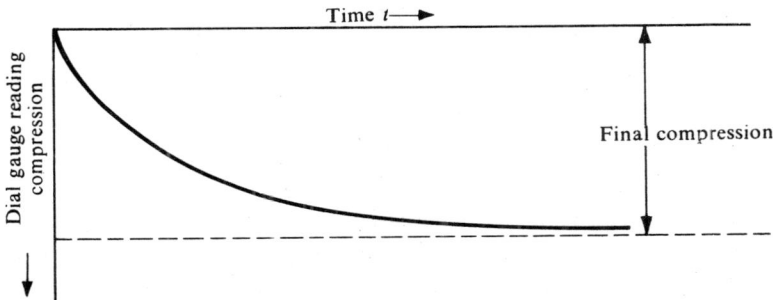

Fig. 7.2 Typical time-compression curve for a stress increment on clay

The time-compression curves for consecutive increments of stress appear somewhat as shown in Fig. 7.3:

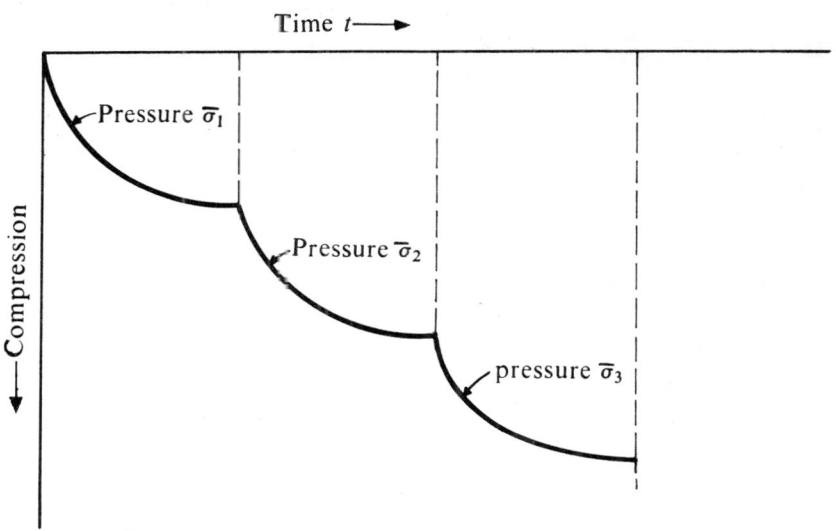

Fig. 7.3 Time-compression curve for successive increments of stress

Since compression is due to decrease in void spaces of the soil, it is commonly indicated as a change in the void ratio. Therefore, the final stress-strain relationships, are presented in the form of a graph between the pressure and void ratio, with a point on the curve for the final condition of each pressure increment.

Accurate determination of the void ratio is essential and may be made as follows:

$$e = \frac{V}{V_s} - 1$$

$$V_s = \frac{W_s}{G \cdot \gamma_w}$$

$$V = A.H$$

Here, A = area of cross-section of the sample;
H = height of the sample at any stage of the test;
W_s = weight of solids or dry soil, obtained by drying and weighing the sample at the end of the test;
G = specific gravity of solids, found separately for the soil sample.

At any stage of the test, the height of the sample may be obtained by deducting the reduction in thickness, got from dial gauge readings, from the initial thickness which will be the same as the internal height of the consolidometer.

Alternatively, the void ratio at any stage may be computed as follows:

The void ratio at the end of test may be obtained as $e = w.G$, where w is the water content at the end of the test.

$$V = A.H = V_s(1+e)$$

If ΔH is the change in H and the corresponding change in void ratio is Δe,
$A . \Delta H = V_s \cdot \Delta e$

Dividing one by the other,

$$\frac{\Delta H}{H} = \frac{\Delta e}{(1+e)}$$

$$\therefore \quad \Delta e = [(1+e)/H] \cdot \Delta H \qquad \ldots \text{(Eq. 7.1)}$$

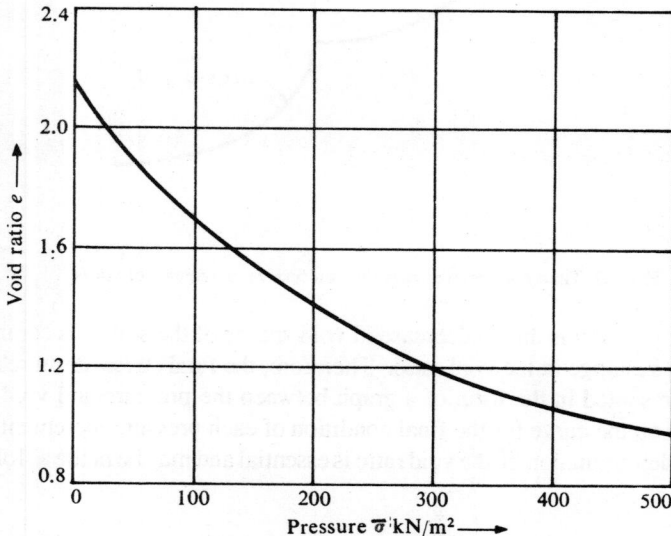

Fig. 7.4 Pressure-void ratio relationship

Compressibility and Consolidation of Soils 245

Working backwards from the known value of the final void ratio, the void ratio corresponding to each pressure may be computed. A typical pressure-void ratio curve is shown in Fig. 7.4.

The slope of this curve at any point is defined as the coefficient of compressibility, a_v. Mathematically speaking,

$$a_v = -\frac{\Delta e}{\Delta \overline{\sigma}} \qquad \ldots \text{(Eq. 7.2)}$$

The negative sign indicates that as the pressure increases, the void ratio decreases. (Alternatively, the curve may be approximated to a straight line between this point and another later point of pressure and its slope may be taken as a_v). It is difficult

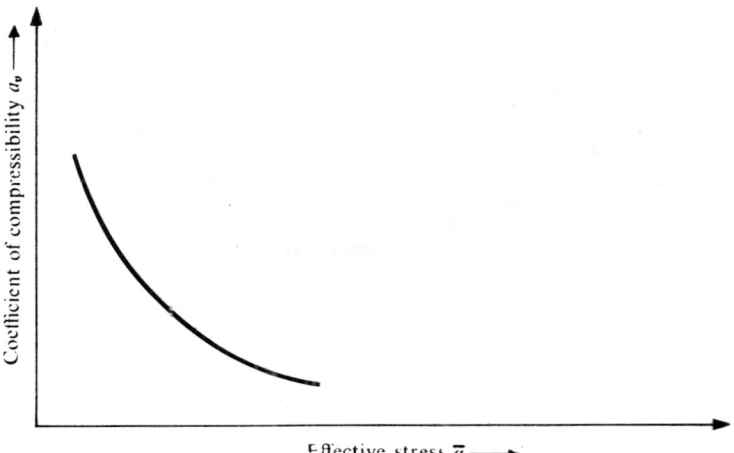

Fig. 7.5 Compressibility – a function of effective stress

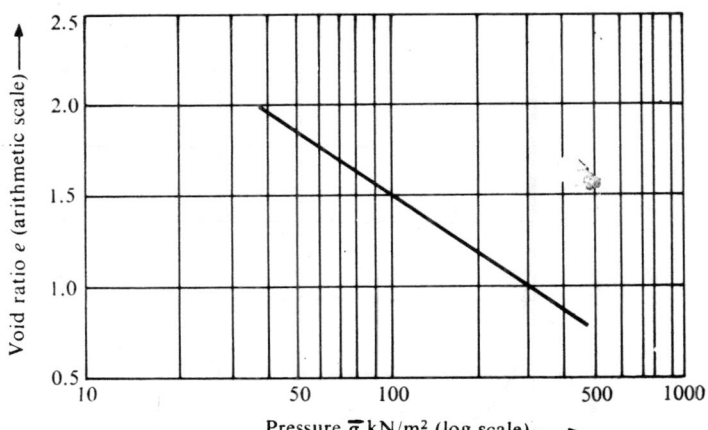

Fig. 7.6 Pressure-void ratio relationship (Semilog co-ordinates)

to use a_v in a mathematical analysis, because of the constantly changing slope of the curve. This leads us to the fact that compressibility is a function of the effective stress as shown in Fig. 7.5.

If the void-ratio is plotted versus the logarithm of the pressure, the data will plot approximately as a straight line (or as a series of straight lines, as described later), as shown in Fig. 7.6. In this form the test data are more adaptable to analytical use.

7.2.4 Compressibility of Sands

The pressure-void ratio relationship for a typical sand under one-dimensional compression is shown in Fig. 7.7. A typical time-compression curve for an increment of stress is shown in Fig. 7.8.

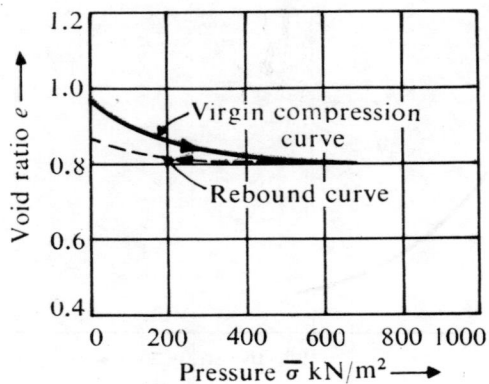

Fig. 7.7 Pressure void-ratio relationship for a typical sand

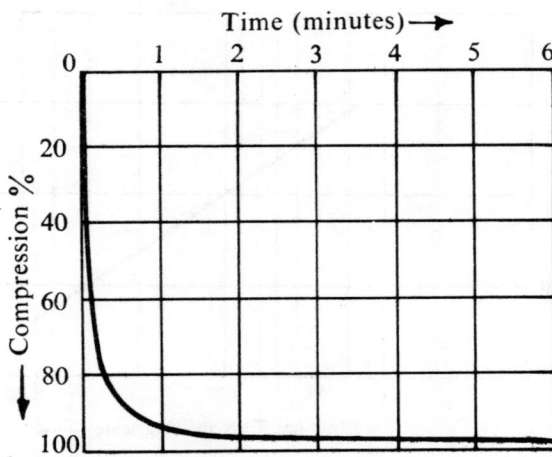

Fig. 7.8 Typical time-compression curve for a sand

Compressibility and Consolidation of Soils 247

It is observed that although there is some rebound on release of pressure, it is never cent per cent; as such, the pressure-void ratio curves for initial loading and unloading, which are respectively referred to as the 'Virgin Compression Curve' and 'Rebound Curve', will be somewhat different from each other. It is also observed that not much of reduction in void ratio occurs in the sands, indicating that their compressibility is relatively very low.

It is also observed from the time-compression curve that the major part of the compression takes place almost instantaneously. In about one minute about 95% of the compression has occurred in this particular case.

The time-lag during compression is largely of a frictional nature in the case of sands. In clean sands, it is about the same whether it is saturated or dry. Upon application of an increment of load, a successive irregular, localised building up and breaking down of stresses in groups of grains occur. A continuous rearrangement of particle positions occurs; the time-lag in reaching the final state is referred to as the frictional lag.

7.2.5 Compressibility and Consolidation of Clays

A typical pressure versus void ratio curve for a clay to natural pressure scale is shown in Fig. 7.9 and to the logarithmic pressure scale in Fig. 7.10. A typical time-compression relationship for an increment of stress for a clay has already been shown in Fig. 7.2. The virgin compression curve and the rebound curve, covering one cycle of loading and unloading, are being presented in Figs. 7.9 and 7.10.

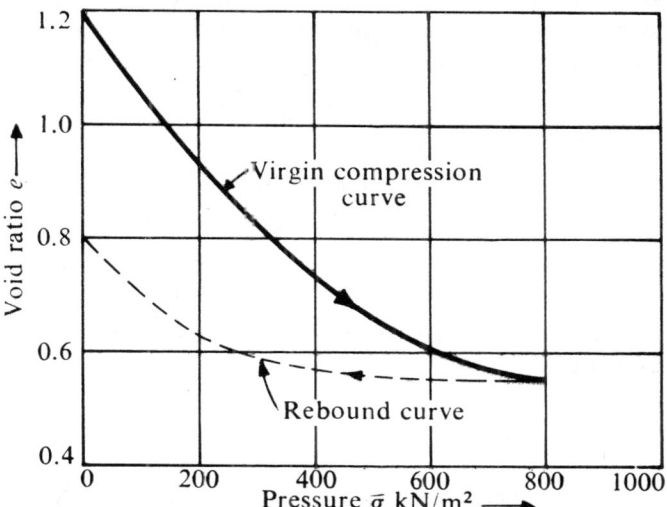

Fig. 7.9 Pressure-void ratio relationship for a typical clay
(Natural or arithmetic scale)

248 *Geotechnical Engineering*

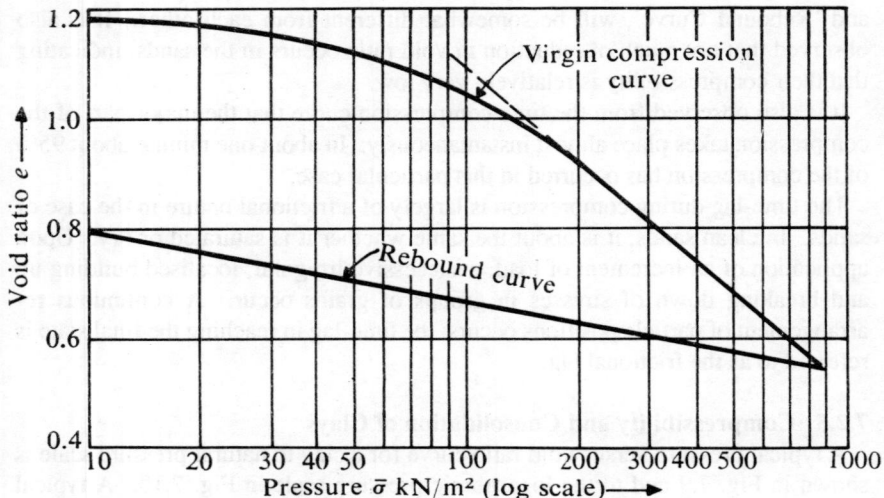

Fig. 7.10 Pressure-void ratio relationship for a typical clay
(Pressure to logarithmic scale)

It is clear that a clay shows greater compressibility than a sand for the same pressure range. It is also clear that the rebound on release of pressure during unloading is much less. The second mode of semi-log plotting yields straight lines in certain zones of loading and unloading.

In the semi-logarithmic plot, it can be seen that the virgin compression curve in this case approximates a straight line from about 200 kN/m² pressure. The equation of this straight line portion may be written in the following form:

$$e = e_0 - C_c \log_{10} \frac{\bar{\sigma}}{\bar{\sigma}_0} \qquad \text{...(Eq. 7.3)}$$

where e corresponds to $\bar{\sigma}$ and e_0 corresponds to $\bar{\sigma}_0$. The value arbitrarily chosen for $\bar{\sigma}_0$ is 100 kN/m², usually (1 kg/cm²), although the straight line has to be produced backward to reach this pressure.

The numerical value of the slope of this straight line, C_c, which is obviously negative in view of the decreasing void ratio for increasing pressure, is called the 'Compression index':

$$C_c = \frac{(e - e_0)}{\log_{10} \frac{\bar{\sigma}}{\bar{\sigma}_0}} \qquad \text{...(Eq. 7.4)}$$

The rebound curve obtained during unloading may be similarly expressed with C_e designating what is called the 'Expansion index':

Compressibility and Consolidation of Soils 249

$$e = e_0 - C_e \log_{10} \frac{\bar{\sigma}}{\bar{\sigma}_0} \qquad \ldots(\text{Eq. 7.5})$$

If, after complete removal of all loads, the sample is reloaded with the same series of loads as in the initial cycle, a different curve, called the 'recompression curve' is obtained. It is shown in Figs. 7.11 and 7.12, with the pressure to arithmetic scale and to logarithmic scale respectively. Some of the volume change due to external loading is permanent. The difference in void ratios attained at any pressure

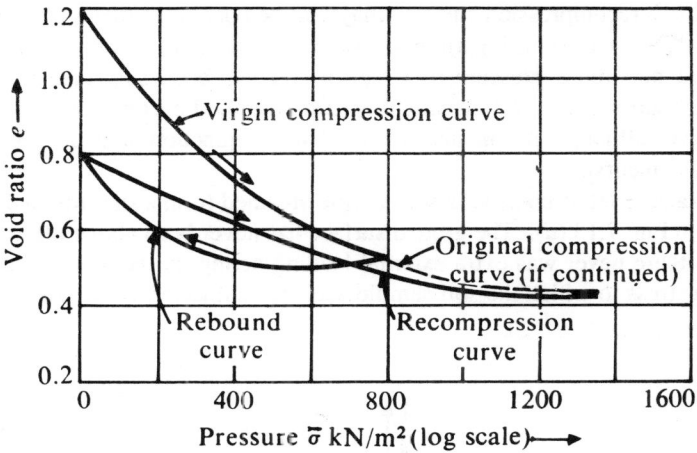

Fig. 7.11 Virgin compression, rebound and recompression curves for a clay (Arithmetic scale)

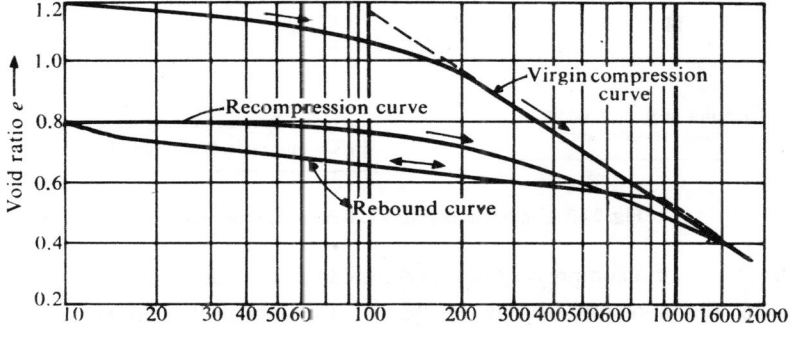

Fig. 7.12 Virgin compression, rebound and recompression curves for a clay (Pressure to logarithmic scale)

between the virgin curve and the recompression curve is predominant at lower pressures and gets decreased gradually with increasing pressure. The two curves are almost the same at the pressure from which the original rebound was made to occur during unloading. The recompression curve is less steep than the virgin curve.

It may be noted from Fig. 7.12 that the curvature of the virgin compression curve at pressures smaller than about 200 kN/m² resembles the curvature of the recompression curve at pressures smaller than about 800 kN/m² from which the rebound occurred. This resemblance indicates that the specimen was probably subjected to a pressure of about 150 to 200 kN/m² at some time before its removal from the ground. Therefore, the initial curved portion of the so-called virgin curve can be visualised as a recompression curve; it may also be concluded that a convex curvature on this type of semi-logarithmic plot always indicates recompression.

This past maximum pressure to which a soil has been subjected is called "Preconsolidation pressure"; usually this term is applied in conjunction with the virgin curve, although it can also be used in conjunction with a laboratory recompression curve.

As an example let us consider a soil sample obtained from a site from a depth z as shown in Fig. 7.13 (a). The ground surface has never been above the existing level, and there never was extra external loading acting on the area. Thus, the maximum stress to which the soil sample was ever subjected is the current overburden pressure $\bar{\sigma}_{v_o} (= \gamma' \cdot z)$.

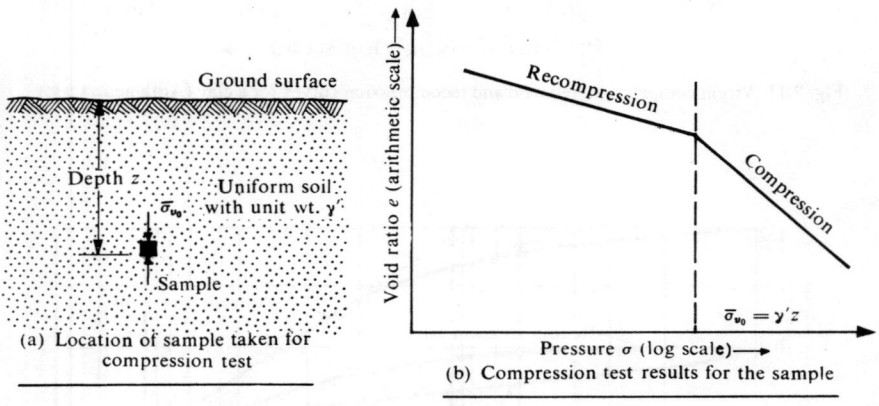

Fig. 7.13 Conditions applying to compression test sample

The results of a compression test performed on this sample are shown in Fig. 7.13 (b). For laboratory loading less than $\bar{\sigma}_{v_o}$, the slope of the compression curve is less than it is for loads greater than $\bar{\sigma}_{v_o}$, since, in so far as this soil is concerned, it represents a reloading or recompression. Thus, the portion of the curve prior to pressure $\bar{\sigma}_{v_0}$ represents a recompression curve, while that at greater pressures than $\bar{\sigma}_{v_0}$ represents the virgin compression curve.

It is obvious that a change in the slope of the compression curve occurs when the previous maximum pressure ever imposed on to the soil is exceeded. If the ground surface had at some time in past history been above the existing surface and had been eroded away, or if any other external load acted earlier and got released, $\bar{\sigma}_{v_0}$, the existing over-burden pressure, would not be the maximum pressure ever imposed on the sample. If this greatest past pressure is $\bar{\sigma}_{v\max}$, greater than $\bar{\sigma}_{v_0}$, compression test results would be as shown in Fig. 7.14.

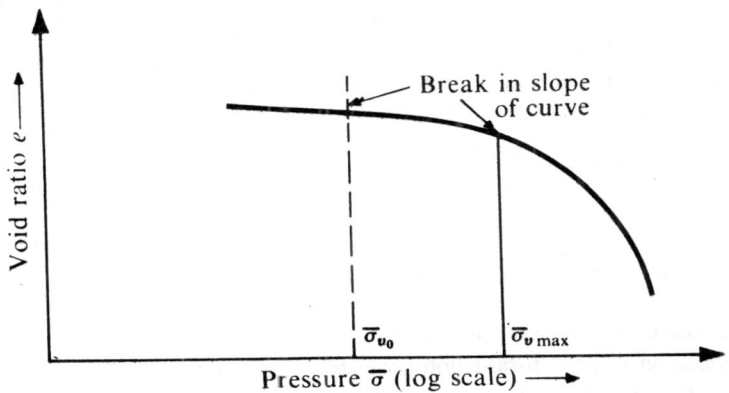

Fig. 7.14 Compression test results where past pressure exceeds present overburden pressure

7.2.6 Normally Consolidated Soil and Overconsolidated Soil

In view of the marked difference in the compressibility behaviour of clay soils which are being loaded for the first time since their origin in relation to the behaviour of clay soils which are being reloaded after initial loading and unloading, as depicted in Figs. 7.11 and 7.12, it becomes imperative that one should know the stress history of the soil to predict its compressibility behaviour. A soil for which the existing effective stress is the maximum to which it has ever been subjected in its stress history, is said to be 'normally consolidated'. The straight portion of the virgin compression curve shown in Fig. 7.12 corresponds to such a situation. For a particular change in pressure, there will be a significant change in void ratio, leading to substantial settlement in a practical foundation.

A soil is said to be 'overconsolidated' if the present effective stress in it has been exceeded sometime during its stress history. The curved portion of the virgin compression curve in Fig. 7.12 prior to the straight line portion corresponds to such a situation. An overconsolidated soil is also said to be a 'pre-compressed' soil. In this state of the soil, the change in void ratio corresponding to a certain change in pressure is relatively less and settlements due to the application of pressures of such order, which keep the soil in an overconsolidated condition, is considered insignificant. Thus the compressibility of a soil in an overconsolidated condition is much less than that for the same soil in a normally consolidated condition.

A soil which is not fully consolidated under the existing overburden pressure is said to be 'underconsolidated'.

It is worthwhile to note that these terms indicate the state or condition of a soil in relation to the pressures, present and past, and are not any special types.

A number of agencies in nature transform normally consolidated clays to overconsolidated or precompressed ones. For example, geological agencies such as glaciers apply pressures on advancing and unload on receding. Human agencies such as engineers load through construction and unload through demolition of structures. Environmental agencies such as climatic factors cause loading and unloading through ground-water movements and the phenomenon of capillarity.

A quantitative measure of the degree of overconsolidation is what is known as the 'Overconsolidation Ratio', OCR. It is defined as follows:

$$\text{OCR} = \frac{\text{Maximum effective stress to which the soil has been subjected in its stress history}}{\text{Existing effective stress in the soil}}$$

...(Eq. 7.6)

Thus, the maximum OCR of normally consolidated soil equals 1.

In this connection, it is of considerable engineering interest to be able to determine the past maximum effective stress that an overconsolidated clay in nature has experienced on its preconsolidation pressure. This would enable an engineer to know at what stress level the soil will exhibit the relatively higher compressibility characteristics of a normally consolidated clay.

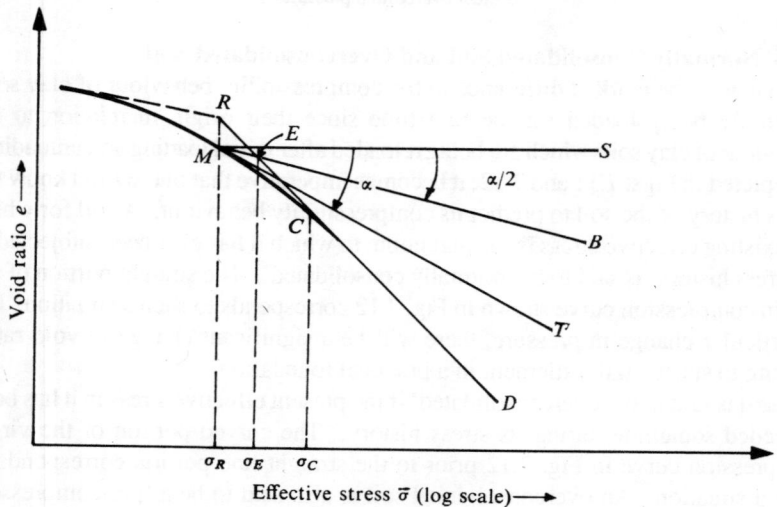

Fig. 7.15 A. Casagrande's procedure for determining preconsolidation pressure

A. Casagrande (1936) proposed a geometrical technique to evaluate past maximum effective stress or preconsolidation pressure from the e versus $\log \bar{\sigma}$ plot obtained by loading a sample in the laboratory. This technique is illustrated in Fig. 7.15.

The steps in the geometrical construction are:
1. The point of maximum curvature **M** on the curved portion of the e vs. $\log \bar{\sigma}$ plot is located.
2. A horizontal line **MS** is drawn through **M**.
3. A tangent **MT** to the curved portion is drawn through **M**.
4. The angle **SMT** is bisected, **MB** being the bisector.
5. The straight portion **DC** of the plot is extended backward to meet **MB** in **E**.
6. The pressure corresponding to the point **E**, $\bar{\sigma}_E$, is the most probable past maximum effective stress or the preconsolidation pressure.

Sometimes the lower and upper bounds for the preconsolidation pressure are also mentioned. If the tangent to the initial recompression portion and the straight line portion of the virgin curve **DC** meet at **R**, the pressure $\bar{\sigma}_R$ corresponding to **R** is said to be the minimum preconsolidation pressure, while that corresponding to **C**, $\bar{\sigma}_c$, is said to be the maximum preconsolidation pressure.

7.2.7 Time-Lags During the Compression of Clay

Considerable time is required for the full compression to occur under a given increment of stress for a clay soil. This is a well-known characteristic of clays. A typical time-compression curve for clay has already been presented in Fig. 7.2. Although it may not take more than twenty-four hours for the full compression to occur for a laboratory sample, it may take a number of years in the case of a field deposit of clay. This is the reason for settlements continuing to occur at an appreciable rate after many years for buildings founded above thick clay strata, although, generally speaking, the rate should be steadily decreasing with time.

Two phenomena are responsible for this time-lag. The first is due to the low permeability of clays and consequent time required for the escape of pore water. This is called the "hydrodynamic lag". The second is due to the plastic action in absorbed water near grain-to-grain contacts, which does not allow quick transmission of the applied stress to the grains and the effective stress to reach a constant value. This is known as the "plastic lag". The frictional lag in sands may be thought of as a simple form of plastic lag.

The theory of one-dimensional consolidation of Terzaghi, presented in section 7.4, does not recognise the existence of plastic lag, although it presents a good understanding of the hydrodynamic lag and consequent rates of settlement. This may be the reason for the prediction on the basis of the Terzaghi theory going wrong once in a while.

7.2.8 Compressibility of Field Deposits

The compressibility characteristics are usually found by performing an oedometer test in the laboratory on a "so-called" undisturbed sample of clay or

254 *Geotechnical Engineering*

on a remoulded sample of the same clay. The pressure-void ratio diagrams for these will be invariably different. This difference is attributed to the inevitable disturbance caused during remoulding. Extending this logic, the disturbance caused during sampling in the field, as also that caused during the transfer of the sample from the sampling tube into the consolidation cell, would naturally alter the compressibility characteristics of the field deposit of clay. Also, depending on the depth of sampling, a certain "stress release" occurs in the field sample by the time it is tested. For these reasons it is natural to expect that the compressibility characteristics of the so-called undisturbed samples do not reflect the true characteristics of the field deposits of clay. Extending the logic of comparison, the true compressibility of field deposit would be somewhat greater than that displayed by laboratory samples. If this difference is not recognised, we would be erring on the wrong side in so far as settlement computations are concerned.

A typical set of pressure-void ratio curves for undisturbed and remoulded samples of clay in relation to that which may be anticipated for the corresponding field deposit is given in Fig. 7.16.

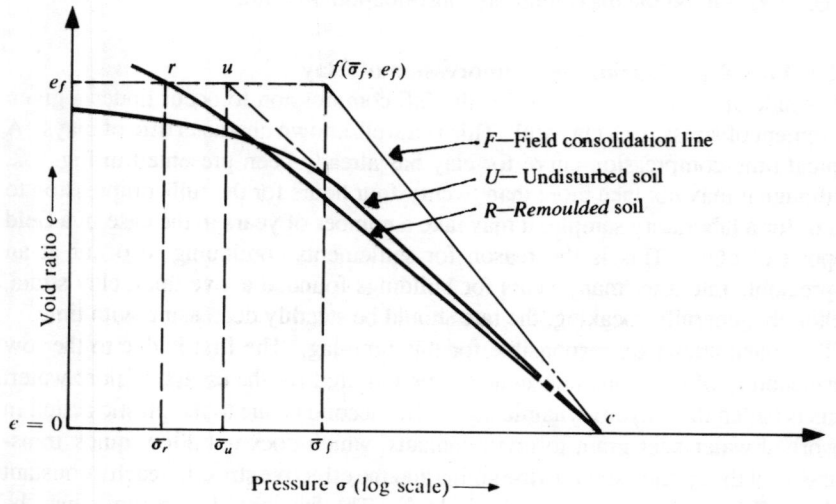

Fig. 7.16 Compression curves for undisturbed and remoulded samples for a field deposit of clay

Let a sample of clay be taken from a depth z from the ground surface. For the overburden pressure $\bar{\sigma}_f$ on it, the void ratio of the sample was, say, e_f. The point f on the plot represents these values, the pressure being plotted to the logarithmic scale. Let e_f and f be joined by a dotted line. The moment the undisturbed sample is taken out from the ground it gets freed of the overburden pressure, the water content and void ratio remaining the same. The compression curve for this sample will be curved until the pressure reaches $\bar{\sigma}_f$ and, later on, will be a straight line, as shown by the plot U. If the sample is remoulded at the same water content, the compression curve obtained will be as depicted by the plot R. If $\bar{\sigma}_r$ is the pressure corresponding to the void ratio e_f on this plot, it will be observed that this plot will be almost a straight line at pressures greater than $\bar{\sigma}_r$. If the plot U is produced downwards to meet the pressure axis at c, the straight portion of the plot R also would almost pass through c, if produced. It is, therefore, logical to assume that the compression curve corresponding to the consolidation of the field deposit, in-situ, is also a straight line tending to pass through c. Since the point f, representing the original conditions, should also lie on this line, the curve F, that is, $e_f fc$ represents the compression curve for the consolidation of the field deposit. The portion fc is referred to as the "field consolidation line".

Let the straight portion of plot U be produced backwards and upwards to meet $e_f f$ line in u and let the corresponding pressure be $\bar{\sigma}_u$. $\bar{\sigma}_u$ will be less than $\bar{\sigma}_f$ for all clays except extra-sensitive ones. The ratio $\bar{\sigma}_u/\bar{\sigma}_f$ indicates the degree of disturbance during sampling. An average value for this ratio is 0.5.

Terzaghi and Peck (1948) recommend that the field consolidation line F be taken as the basis of settlement computations. The reconstruction of this line is possible by procedures suggested by some workers, e.g., Schmertmann (1955).

7.2.9 Relationship between Compressibility and Liquid Limit

A.W. Skempton and his associates have established a relationship between the compressibility of a clay, as indicated by its compression index, and the liquid limit, by conducting experiments with clays from various parts of the world. The relationship was found to be linear as shown in Fig. 7.17.

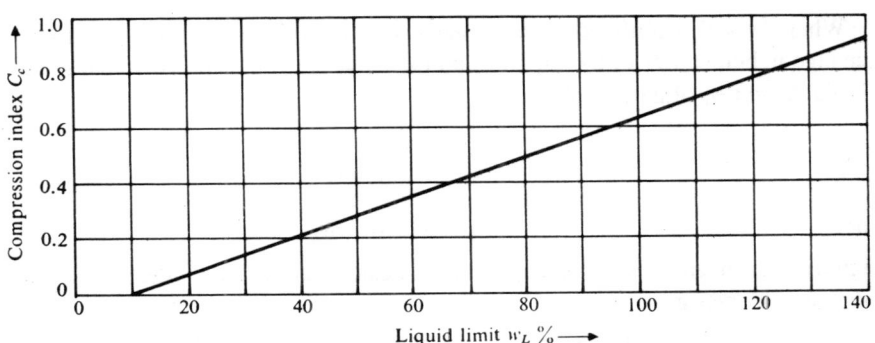

Fig. 7.17 Relationship between compression index and liquid limit for remoulded clays (After Skempton)

The equation of this straight line may be approximately written as:
$$C_c = 0.007 (w_L - 10) \quad \ldots(\text{Eq. 7.7})$$
w_L being the liquid limit in per cent.

It has also been established that the compression index of field deposits of clays of low and medium sensitivity is about 1.30 times that of their value in the remoulded state. Therefore, we may write for consolidation of field deposits of clay:

$$C_c = 0.009 \, (w_L - 10) \qquad \text{...(Eq. 7.8)}$$

This equation is observed to give a satisfactory estimate of the settlement of structures founded on clay deposits of low and medium sensitivity.

These two equations are, as reported by Terzaghi and Peck (1948), based on the work of Skempton and his associates.

7.2.10 Modulus of Volume Change and Consolidation Settlement

The 'modulus of volume change,is defined as the change in volume of a soil per unit initial volume due to a unit increase in effective stress. It is also called the 'coefficient of volume change' or 'coefficient of volume compressibility' and is denoted by the symbol, m_v.

$$m_v = -\frac{\Delta e}{(1+e_0)} \cdot \frac{1}{\Delta \overline{\sigma}} \qquad \text{...(Eq. 7.9)}$$

Δe represents the change in void ratio and represents the change in volume of the saturated soil occurring through expulsion of pore water, and $(1+e_0)$ represents initial volume, both for unit volume of solids.

But we know from Eq. 7.2 that

$-\dfrac{\Delta e}{\Delta \overline{\sigma}} = a_v$, the coefficient of compressibility.

$$\therefore \quad m_v = \frac{a_v}{(1+e_0)} \qquad \text{...(Eq. 7.10)}$$

When the soil is confined laterally, the change in volume is proportional to the change in height, ΔH of the sample, and the initial volume is proportional to the initial height H_0 of the sample.

$$\therefore \quad m_v = -\frac{\Delta H}{H_0} \cdot \frac{1}{\Delta \overline{\sigma}}$$

or
$$\Delta H = m_v \cdot H_0 \cdot \Delta \overline{\sigma} \qquad \text{...(Eq. 7.11)}$$

ignoring the negative sign which merely indicates that the height decreases with increase in pressure.

Thus, the consolidation settlement, S_c, of a clay for full compression under a pressure increment $\Delta \overline{\sigma}$, is given by Eq. 7.11.

This is under the assumption that $\Delta \overline{\sigma}$ is transmitted uniformly over the thickness. However, it is found that $\Delta \overline{\sigma}$ decreases with depth non-linearly. In such cases, the consolidation settlement may be obtained as:

$$S_c = \int_0^H m_v \cdot \Delta \overline{\sigma} \cdot dz \qquad \text{...(Eq. 7.12)}$$

Compressibility and Consolidation of Soils

This integration may be performed numerically by dividing the stratum of height H into thin layers and considering $\Delta\bar{\sigma}$ for the mid-height of the layer as being applicable for the thin layer. The total settlement of the layer of height H will be given by the sum of settlements of individual layers.

The consolidation settlement S_c, may also be put in a different, but more common form, as follows:

$$m_v = \frac{e}{(1+e_0)} \cdot \frac{1}{\Delta\bar{\sigma}}, \text{ ignoring sign.}$$

$$\frac{\Delta H}{H_0} = \frac{\Delta e}{(1+e_0)}$$

$$S_c = \Delta H = \frac{\Delta e}{(1+e_0)} \cdot H_0$$

Substituting for Δe in terms of the compression index, C_c, from Eq. 7.4, recognising $(e - e_0)$ as Δe, we have:

$$S_c = \Delta H = H_0 \cdot \frac{C_c}{(1+e_0)} \cdot \log_{10} \frac{\bar{\sigma}}{\bar{\sigma}_0} \quad \text{...(Eq. 7.13)}$$

$\bar{\sigma}_0 = $ preconsolidation pressure.

or *Normally Consolidated*

$$S_c = H_0 \cdot \frac{C_c}{(1+e_0)} \cdot \log_{10} \left(\frac{\bar{\sigma}_0 + \Delta\bar{\sigma}}{\bar{\sigma}_0} \right) \quad \text{...(Eq. 7.14)}$$

This is the famous equation for computing the ultimate or total settlement of a clay layer occurring due to the consolidation process under the influence of a given effective stress increment.

7.3 A MECHANISTIC MODEL FOR CONSOLIDATION

The process of consolidation, and the Terzaghi theory to be presented in section 7.4, can be better understood only if an important simplifying assumption is explained and appreciated.

The pressure-void ratio relationship for the increment of pressure under question is taken to be linear as shown in Fig. 7.18, when both the variables are plotted to the natural or arithmetic scale. It is further assumed that this linear relationship holds under all conditions, with no variation because of time effects or any other factor. If there were no plastic lag in clay, this assumption would have been acceptable; however, clays are highly plastic.

The process of consolidation may be explained on the basis of this simplifying assumption as follows:

Let the soil sample be in equilibrium under the pressure $\bar{\sigma}_1$ throughout its depth, at the void ratio e_1. Immediately on application of the higher pressure $\bar{\sigma}_2$, the void ratio is e_1 only. The pressure $\bar{\sigma}_2$ cannot be effective within the soil until the void ratio becomes e_2, and the effective pressure is still $\bar{\sigma}_1$. The increase in pressure, $(\bar{\sigma}_2 - \bar{\sigma}_1)$ tends to produce a strain $(e_1 - e_2)$. On account of hydrodynamic lag, this

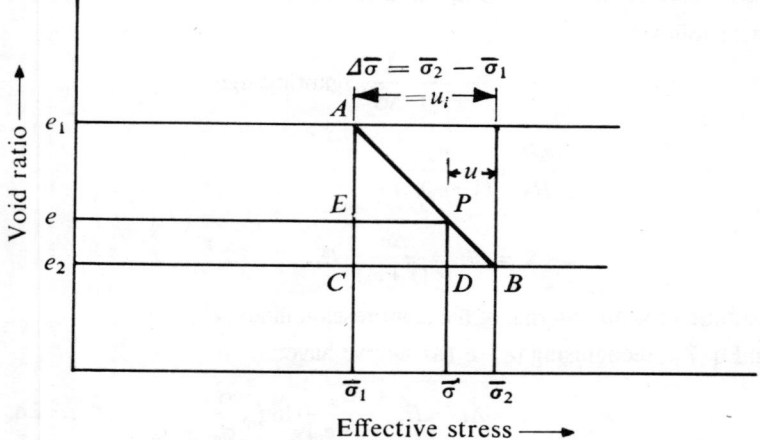

Fig. 7.18 Idealised pressure-void ratio relationship

cannot take place at once. Thus, there is only one possibility – the increase in pressure is carried by the pore water, with the pressure in the soil skeleton still being $\bar{\sigma}_1$. This increase in pressure in the pore water produced by transient conditions as given above, is called 'hydrostatic excess pressure', u. The initial value, u_i, for this is $(\bar{\sigma}_2 - \bar{\sigma}_1)$.

If the samples were to be hermetically sealed, permitting no escape of water, the conditions mentioned above would continue indefinitely. But, in the laboratory oedometer sample, the porous stone disks tend to promptly eliminate the hydrostatic excess pressure at the top and bottom of the sample creating a high gradient of pressure and consequent rapid drainage. Gradually the void ratio decreases, as the hydrostatic excess pressure dissipates and the effective or intergranular pressure increases; this process is in a more advance state near the drainage ends at the top and bottom than at the centre of the sample. The sample is said to be "consolidating" under the stress increase $(\bar{\sigma}_2 - \bar{\sigma}_1)$. This continues until the void ratio at all points becomes e_2. Theoretically, no more water is forced out when the hydrostatic excess pressure becomes zero; the effective pressure in the soil skeleton is $\bar{\sigma}_2$, and the sample is said to have been "consolidated" under the pressure $\bar{\sigma}_2$. It should be noted that "Consolidation" is a relative term, which refers to the degree to which that the gradual process has advanced, and does not refer to the stiffness of the material.

In fact, a quantitative idea of the consolidation or the 'Degree of consolidation' may be obtained by what is called the 'Consolidation ratio', U_z.

$$U_z = \frac{(e_1 - e)}{(e_1 - e_2)} \qquad \text{...(Eq. 7.15)}$$

Compressibility and Consolidation of Soils 259

with reference to Fig. 7.18; this is the fundamental definition for U_z.

Usually it is expressed as per cent and is referred to as the 'Per cent consolidation'. It can be shown from Fig. 7.18 that U_z may also be written as follows:

$$U_z = \frac{e_1 - e}{e_1 - e_2} = \left(\frac{\overline{\sigma} - \overline{\sigma}_1}{\overline{\sigma}_2 - \overline{\sigma}_1}\right) = 1 - \frac{u}{u_i} \qquad \ldots\text{(Eq. 7.16)}$$

in view of the relationship $\overline{\sigma}_2 = \overline{\sigma}_1 + u_i = \overline{\sigma} + u$, from the same figure.

A mechanistic model for the phenomenon of consolidation was given by Taylor (1948), by which the process can be better understood. This model, with slight modifications, is presented in Fig. 7.19 and is explained below:

A spring of initial height H_i is surrounded by water in a cylinder. The spring is analogous to the soil skeleton and the water to the pore water. The cylinder is fitted with a piston of area A through which a certain load may be transmitted to be the system representing a saturated soil. The piston, in turn, is fitted with a vent, and a valve by which the vent may be opened or closed.

Referring to Fig. 7.19(a), let a load P be applied on the piston. Let us assume that the valve of the vent is open and no flow is occurring. This indicates that the system is in equilibrium under the total stress P/A which is fully borne by the spring, the pressure in the water being zero.

Referring to Fig. 7.19(b), let us apply an increment of load δP to the piston, the valve being kept closed. Since no water is allowed to flow out, the piston cannot move downwards and compress the spring; therefore, the spring carries the earlier stress of P/A, while the water is forced to carry the additional stress of $\delta P/A$ imposed on the system, the sum counteracting the total stress imposed. This additional stress $\delta P/A$ in the water is known as the hydrostatic excess pressure.

Referring to Fig. 7.19(c), let us open the valve and start reckoning time from that instant. Water just starts to flow under the pressure gradient between it and the atmosphere seeking to return to its equilibrium or atmospheric pressure. The excess pore pressure begins to diminish, the spring starts getting compressed as the piston descends consequent to expulsion of pore water. It is just the beginning of transient flow, simulating the phenomenon of consolidation. The openness of the valve is analogous to the permeability of soil.

Referring to Fig. 7.19(d), flow has occurred to the extent of dissipating 50% of the excess pore pressure. The pore water pressure at this instant is half the initial value, i.e., $\frac{1}{2}(\delta P/A)$. This causes a corresponding increase in the stress in the spring of $\frac{1}{2}(\delta P/A)$, the total stress remaining constant at $[(P/A) + (\delta P/A)]$. This stage refers to that of "50% consolidation".

Referring to Fig. 7.19(e), the final equilibrium condition is reached when the transient flow situation ceases to exist, consequent to the complete dissipation of the pore water pressure. The spring compresses to a final height $H_f < H_i$, carrying the total stress of $(P + \delta P)/A$, all by itself, since the excess pore water pressure has

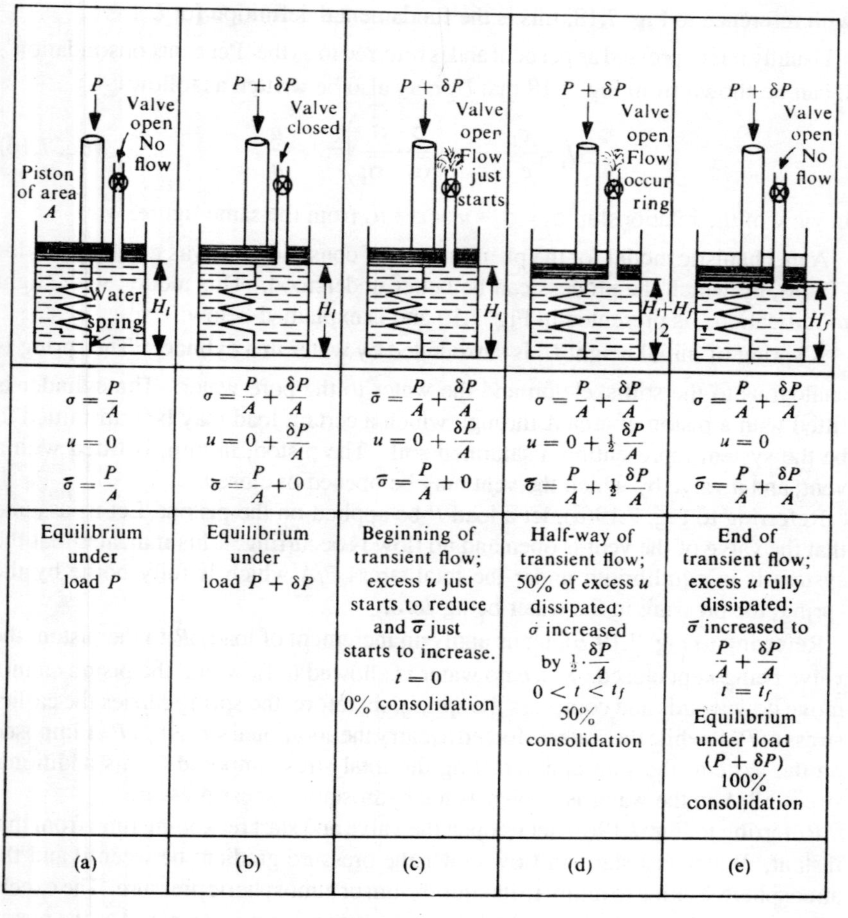

260 Geotechnical Engineering

(a)	(b)	(c)	(d)	(e)
P, Valve open, No flow	$P + \delta P$, Valve closed	$P + \delta P$, Valve open, Flow just starts	$P + \delta P$, Valve open, Flow occurring	$P + \delta P$, Valve open, No flow
$\sigma = \dfrac{P}{A}$	$\sigma = \dfrac{P}{A} + \dfrac{\delta P}{A}$	$\sigma = \dfrac{P}{A} + \dfrac{\delta P}{A}$	$\sigma = \dfrac{P}{A} + \dfrac{\delta P}{A}$	$\sigma = \dfrac{P}{A} + \dfrac{\delta P}{A}$
$u = 0$	$u = 0 + \dfrac{\delta P}{A}$	$u = 0 + \dfrac{\delta P}{A}$	$u = 0 + \tfrac{1}{2}\dfrac{\delta P}{A}$	$u = 0$
$\bar{\sigma} = \dfrac{P}{A}$	$\bar{\sigma} = \dfrac{P}{A} + 0$	$\bar{\sigma} = \dfrac{P}{A} + 0$	$\bar{\sigma} = \dfrac{P}{A} + \tfrac{1}{2}\dfrac{\delta P}{A}$	$\bar{\sigma} = \dfrac{P}{A} + \dfrac{\delta P}{A}$
Equilibrium under load P	Equilibrium under load $P + \delta P$	Beginning of transient flow; excess u just starts to reduce and $\bar{\sigma}$ just starts to increase. $t = 0$ 0% consolidation	Half-way of transient flow; 50% of excess u dissipated; $\bar{\sigma}$ increased by $\tfrac{1}{2}\cdot\dfrac{\delta P}{A}$ $0 < t < t_f$ 50% consolidation	End of transient flow; excess u fully dissipated; $\bar{\sigma}$ increased to $\dfrac{P}{A} + \dfrac{\delta P}{A}$ $t = t_f$ Equilibrium under load $(P + \delta P)$ 100% consolidation

Fig. 7.19 A mechanistic model for consolidation
(adapted from Taylor, 1948)

been reduced to zero, the pressure in it having equalled the atmospheric. The system has reached the equilibrium condition under the load $(P + \delta P)$. This represents "100% consolidation" under the applied load or stress increment. We may say that the "soil" has been consolidated to an effective stress of $(P + \delta P)/A$.

In this mechanistic model, the compressible soil skeleton is characterised by the spring and the pore water by the water in the cylinder. The more compressible the soil, the longer the time required for consolidation; the more permeable the soil, the shorter the time required.

There is one important aspect in which this analogy fails to simulate consolidation of a soil. It is that the pressure conditions are the same throughout the height of the cylinder, whereas the consolidation of a soil begins near the drainage surfaces and gradually progresses inward. It may be noted that soil consolidates only when effective stress increases; that is to say, the volume change behaviour of a soil is a function of the effective stress and not the total stress.

Similar arguments may be applied to the expansion characteristics under the decrease of load.

An alternative mechanical analogy to the consolidation process is shown in Fig. 7.20.

A cylinder is fitted with a number of pistons connected by springs to one another. Each of the compartments thus formed is connected to the atmosphere with the aid of standpipes. The cylinder is full of water and is considered to be airtight. The pistons are provided with perforations through which water can move from one compartment to another. The topmost piston is fitted with valves which may open or close to the atmosphere. It is assumed that any pressure applied to the top piston gets transmitted undiminished to the water and springs.

Initially, the cylinder is full of water and weights of the pistons are balanced by the springs; the water is at atmospheric pressure and the valves may be open. The water level stands at the elevation **PP** in the standpipes as shown. The valves are now closed, the water level continuing to remain at **PP**. An increment of pressure $\Delta\sigma$ is applied on the top piston. It will be observed that the water level rises instantaneously in all the stand pipes to an elevation **QQ**, above **PP** by a height $h = \Delta\sigma/\gamma_w$. Let all the valves be opened simultaneously with the application of the pressure increment, the time being reckoned from that instant. The height of the springs remains unchanged at that instant and the applied increment of pressure is fully taken up by water as the hydrostatic excess pressure over and above the atmospheric. An equal rise of water in all the standpipes indicates that the hydrostatic excess pressure is the same in all compartments immediately after application of pressure. As time elapses, the water level in the pipes starts falling, the pistons move downwards gradually and water comes out through the open valves. At any time $t = t_1$, the water pressure in the first compartment is least and that in the last or the bottommost is highest, as indicated by the water levels in the standpipes. The variation of hydrostatic excess pressure at various points in the depth of the cylinder, as shown by the dotted lines, varies with time. Ultimately, the hydrostatic excess pressure reduces to zero in all compartments, the water levels in the standpipes reaching elevation **PP**; this, theoretically speaking, is supposed to happen after the lapse of infinite time. As the hydrostatic excess pressure decreases in each compartment, the springs in each compartment experience a corresponding pressure and get compressed. For example, at time $t = t_1$, the hydrostatic excess pressure in the first compartment is given by the head **PJ**; the pressure taken by the springs is indicated by the head **JQ**, the sum of the two at all times being equivalent to the applied pressure increment; that is to say, it is analogous to the effective stress principle: $\sigma = \bar{\sigma} + u$, the pressure transferred to the springs being analogous to intergranular or effective stress in a saturated soil, and the hydrostatic excess pressure to the neutral pressure or excess pore water pressure.

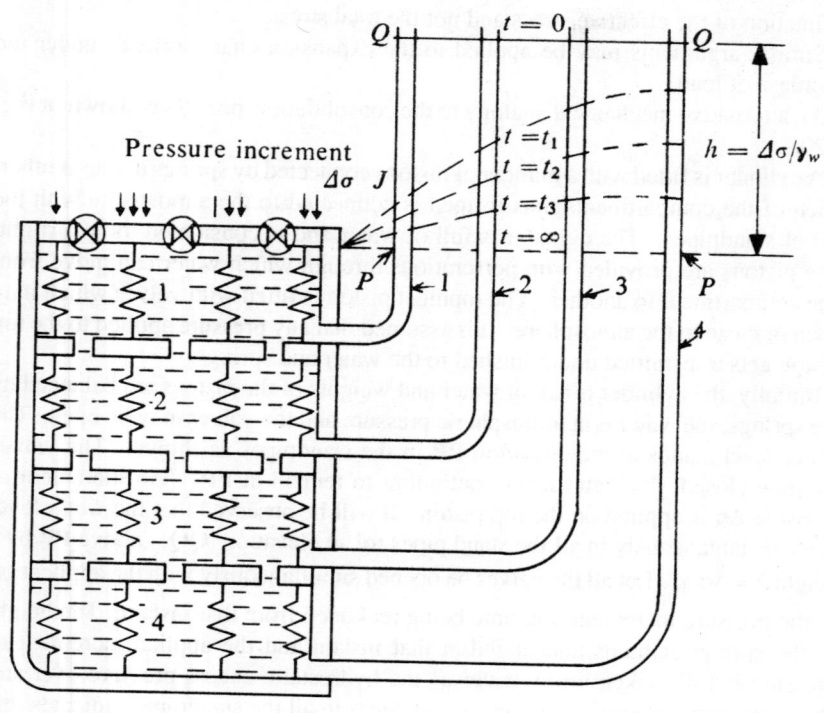

Fig. 7.20 Mechanical analogy to consolidation process

Since water is permitted to escape only at one end, it is similar to the case of a single drainage face for a consolidating clay sample. The distribution of hydrostatic excess pressure will be symmetrical about mid-depth for the situation of a double drainage face, the maximum occurring at mid-depth and the minimum or zero values occurring at the drainage faces.

7.4 TERZAGHI'S THEORY OF ONE-DIMENSIONAL CONSOLIDATION

Terzaghi (1925) advanced his theory of one-dimensional consolidation based upon the following assumptions, the mathematical implications being given in parentheses:

(1) The soil is homogeneous (k_z is independent of z).
(2) The soil is completely saturated ($S = 100\%$).
(3) The soil grains and water are virtually incompressible (γ_w is constant and volume change of soil is only due to change in void ratio).

(4) The behaviour of infinitesimal masses in regard to expulsion of pore water and consequent consolidation is no different from that of larger representative masses (Principles of calculus may be applied).
(5) The compression is one-dimensional (u varies with z only).
(6) The flow of water in the soil voids is one-dimensional, Darcy's law being valid.

$$\left(\frac{\partial v_x}{\partial x} = \frac{\partial v_y}{\partial y} = 0 \text{ and } v_z = k_z \cdot \frac{\partial h}{\partial z}\right).$$

Also, flow occurs on account of hydrostatic excess pressure ($h = u/\gamma_w$).

(7) Certain soil properties such as permeability and modulus of volume change are constant; these actually vary somewhat with pressure. (k and m_v are independent of pressure).
(8) The pressure versus void ratio relationship is taken to be the idealised one, as shown in Fig. 7.18 (a_v is constant).
(9) Hydrodynamic lag alone is considered and plastic lag is ignored, although it is known to exist. (The effect of k alone is considered on the rate of expulsion of pore water).

The first three assumptions represent conditions that do not vary significantly from actual conditions. The fourth assumption is purely of academic interest and is stated because the differential equations used in the derivation treat only infinitesimal distances. It has no significance for the laboratory soil sample or for the field soil deposit. The fifth assumption is certainly valid for deeper strata in the field owing to lateral confinement and is also reasonably valid for an oedometer sample. The sixth assumption regarding flow of pore water being one-dimensional may be taken to be valid for the laboratory sample, while its applicability to a field situation should be checked. However, the validity of Darcy's law for flow of pore water is unquestionable.

The seventh assumption may introduce certain errors in view of the fact that certain soil properties which enter into the theory vary somewhat with pressure but the errors are considered to be of minor importance.

The eighth and ninth assumptions lead to the limited validity of the theory. The only justification for the use of the eighth assumption is that, otherwise, the analysis becomes unduly complex. The ninth assumption is necessitated because it is not possible to take the plastic lag into account in this theory. These two assumptions also may be considered to introduce some errors.

Now let us see the derivation of Terzaghi's theory with respect to the laboratory oedometer sample with double drainage as shown in Fig. 7.21:

Let us consider a layer of unit area of cross-section and of elementary thickness dz at depth z from the pervious boundary. Let the increment of pressure applied be $\Delta\sigma$. Immediately on application of the pressure increment, pore water starts to flow towards the drainage faces. Let ∂h be the head lost between the two faces of this elementary layer, corresponding to a decrease of hydrostatic excess pressure ∂u.

264 *Geotechnical Engineering*

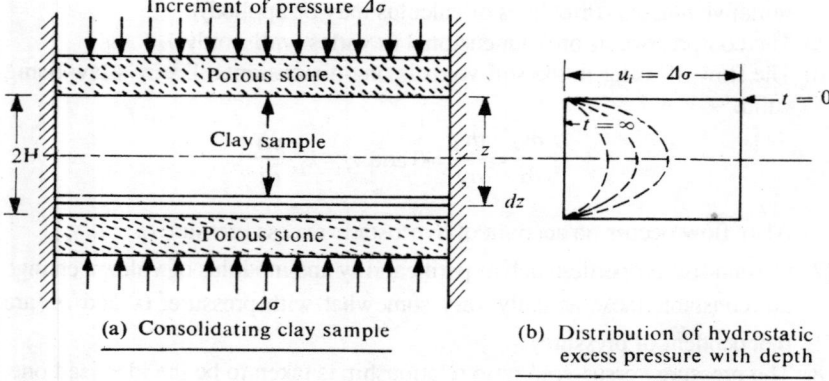

(a) Consolidating clay sample

(b) Distribution of hydrostatic excess pressure with depth

Fig. 7.21 Consolidation of a clay sample with double drainage

Equation 6.2, for flow of water through soil, holds here also.

$$k_x \cdot \frac{\partial^2 h}{\partial x^2} + k_z \frac{\partial^2 h}{\partial z^2} = \frac{1}{(1+e)} \left[e \cdot \frac{\partial s}{\partial t} + S \cdot \frac{\partial e}{\partial t} \right] \qquad \text{...(Eq. 6.2)}$$

For one-dimensional flow situation, this reduces to:

$$k_z \cdot \frac{\partial^2 h}{\partial z^2} = \frac{1}{(1+e)} \left[e \cdot \frac{\partial s}{\partial t} + S \cdot \frac{\partial e}{\partial t} \right]$$

During the process of consolidation, the degree of saturation is taken to remain constant at 100%, while void ratio changes causing reduction in volume and dissipation of excess hydrostatic pressure through expulsion of pore water; that is, $S = 100\%$ or unity, and $\frac{\partial S}{\partial t} = 0$.

$$\therefore \quad k_z \cdot \frac{\partial^2 h}{\partial z^2} = -\frac{1}{(1+e)} \cdot \frac{\partial e}{\partial t} = -\frac{\partial}{\partial t} \left[\frac{e}{1+e} \right],$$

negative sign denoting decrease of e for increase of h.

Since volume decrease can be due to a decrease in the void ratio only as the pore water and soil grains are virtually incompressible, $\frac{\partial}{\partial t}\left(\frac{e}{1+e}\right)$ represents time-rate of volume change per unit volume.

The flow is only due to the hydrostatic excess pressure,

$h = \frac{u}{\gamma_w}$, where γ_w = unit weight of water.

$$\therefore \quad \frac{k}{\gamma_w} \cdot \frac{\partial^2 u}{\partial z^2} = -\frac{\partial V}{\partial t} \qquad \text{...(Eq. 7.17)}$$

(This can also be considered as the continuity equation for a non-zero net out-flow, while Laplace's equation represents inflow being equal to out-flow).

Here k is the permeability of soil in the direction of flow, and ∂V represents the change in volume per unit volume. The change in hydrostatic excess pressure, ∂u, changes the intergranular or effective stress by the same magnitude, the total stress remaining constant.

The change in volume per unit volume, ∂V, may be written, as per the definition of the modulus of volume change, m_v;

$\partial V = m_v \cdot \partial \bar{\sigma} = - m_v \cdot \partial u$, since an increase $\partial \bar{\sigma}$ represents a decrease ∂u.

Differentiating both sides with respect to time,

$$\frac{\partial V}{\partial t} = - m_v \cdot \frac{\partial u}{\partial t} \qquad \text{...(Eq. 7.18)}$$

From Eqs. 7.17 and 7.18, we have:

$$\frac{\partial u}{\partial t} = \frac{k}{\gamma_w \cdot m_v} \cdot \frac{\partial^2 u}{\partial z^2}$$

This is written as:

$$\frac{\partial u}{\partial t} = c_v \cdot \frac{\partial^2 u}{\partial z^2} \qquad \text{...(Eq. 7.19)}$$

where $c_v = \dfrac{k}{\gamma_w \cdot m_v}$

c_v is known as the "Coefficient of consolidation". u represents the hydrostatic excess pressure at a depth z from the drainage face at time t from the start of the process of consolidation.

The coefficient of consolidation may also be written in terms of the coefficient of compressibility:

$$c_v = \frac{k}{\gamma_w m_v} = \frac{k(1+e_0)}{a_v \gamma_w} \qquad \text{...(Eq. 7.20)}$$

Equation 7.19 is the basic differential equation of consolidation according to Terzaghi's theory of one-dimensional consolidation. The coefficient of consolidation combines the effect of permeability and compressibility characteristics on volume change during consolidation. Its units can be shown to be mm²/s or $L^2 T^{-1}$.

The initial hydrostatic excess pressure, u_i, is equal to the increment of pressure $\Delta \sigma$, and is the same throughout the depth of the sample, immediately on application of the pressure, and is shown by the heavy line in Fig. 7.21(b). The horizontal portion of the heavy line indicates the fact that, at the drainage face, the hydrostatic excess pressure instantly reduces to zero, theoretically speaking. Further, the hydrostatic excess pressure would get fully dissipated throughout the depth of the

266 *Geotechnical Engineering*

sample only after the lapse of infinite time*, as indicated by the heavy vertical line on the left of the figure. At any other instant of time, the hydrostatic excess pressure will be maximum at the farthest point in the depth from the drainage faces, that is, at the middle and it is zero at the top and bottom. The distribution of the hydrostatic excess pressure with depth is sinusoidal at other instants of time, as shown by dotted lines. These curves are called "Isochrones".

Aliter

With reference to Fig. 7.21, the hydraulic gradient i_1 at depth z $\Big\} = \dfrac{\partial h}{\partial z} = \dfrac{1}{\gamma_w} \cdot \dfrac{\partial u}{\partial z}$

Hydraulic gradient i_2 at depth $(z + dz)$ $\Big\} = \dfrac{1}{\gamma_w} \left(\dfrac{\partial u}{\partial z} + \dfrac{\partial^2 u}{\partial z^2} \cdot dz \right)$

Rate of inflow per unit area = Velocity at depth $z = k \cdot i_1$, by Darcy's law.

Rate of outflow per unit area = velocity at $(z + dz) = k \cdot i_2$

Water lost per unit time = $k(i_2 - i_1) = \dfrac{k}{\gamma_w} \cdot \dfrac{\partial^2 u}{\partial z^2} \cdot dz$

This should be the same as the time-rate of volume decrease. Volumetric strain = $m_v \cdot \Delta \overline{\sigma} = - m_v \partial(\sigma - u)$, by definition of the modulus of volume change, m_v.
(The negative sign denotes decrease in volume with increase in pressure).
∴ Change of volume = $- m_v \partial(\sigma - u) \cdot dz$,
since the elementary layer of thickness dz and unit cross-sectional area is considered.

Time-rate of change of volume $= - m_v \cdot \dfrac{\partial}{\partial t}(\sigma - u) \cdot dz$

$\dfrac{\partial \sigma}{\partial t} = 0$, since σ is constant.

∴ Time-rate of change of volume $= + m_v \cdot \dfrac{\partial u}{\partial t} \cdot dz$

Equating this to water lost per unit time,

$$\dfrac{k}{\gamma_w} \cdot \dfrac{\partial^2 u}{\partial z^2} \cdot dz = m_v \cdot \dfrac{\partial u}{\partial t} \cdot dz$$

---

*As the process of consolidation is in progress, the hydrostatic excess pressure causing flow decreases, which, in turn, slows down the rate of flow. This, again, reduces the rate of dissipation of pore water pressure, and so on. This results in an asymptotic relation between time and excess pore pressure. Therefore, mathematically speaking, it takes infinite time for 100% consolidation. Fortunately, it takes finite time for 99% or even 99.9% consolidation. This is good enough from the point of view of engineering accuracy.

or

$$\frac{\partial u}{\partial t} = c_v \cdot \frac{\partial^2 u}{\partial z^2} \qquad \text{...(Eq. 7.19)}$$

where

$$c_v = \frac{k}{m_v \gamma_w} = \frac{k(1+e)}{a_v \cdot \gamma_w} \qquad \text{...(Eq. 7.20)}$$

7.5 SOLUTION OF TERZAGHI'S EQUATION FOR ONE-DIMENSIONAL CONSOLIDATION

Terzaghi solved the differential equation (Eq. 7.19) for a set of boundary conditions which have utility in solving numerous engineering problems and presented the results in graphical form using dimensional parameters.

The following are the boundary conditions:
(1) There is drainage at the top of the sample: At $z = 0$, $u = 0$, for all t.
(2) There is drainage at the bottom of the sample: At $z = 2H$, $u = 0$, for all t.
(3) The initial hydrostatic excess pressure u_i is equal to the pressure increment, $\Delta \sigma$ $u = u_i = \Delta \sigma$, at $t = 0$.

Terzaghi chose to consider this situation where $u = u_i$ initially *throughout the depth*, although solutions are possible when u_i varies with depth in any specified manner. The thickness of the sample is designated by $2H$, the distance H thus being the length of the longest drainage path, i.e., maximum distance water has to travel to reach a drainage face because of the existence of two drainage faces. (In the case of only one drainage face, this will be equal to the total thickness of the clay layer).

The general solution for the above set of boundary conditions has been obtained on the basis of separation of variables and Fourier Series expansion and is as follows:

$$u = f(z, t) = \sum_{n=1}^{\infty} \left(\frac{1}{H} \int_0^{2H} u_i \sin \frac{n \pi z}{2H} \cdot dz \right) \left(\sin \frac{n \pi z}{2H} \right) e^{-n^2 \pi^2 c_v t / 4H^2}$$

...(Eq. 7.21)

This solution enables the hydrostatic excess u to be computed for a soil mass under any initial system of stress u_i, at any depth z, and at any time t.

In particular, if u_i is considered constant with respect to depth, this equation reduces to

$$u = \sum_{n=1}^{\infty} \left[\frac{2u_i}{n \pi} (1 - \cos n \pi) \left(\sin \frac{n \pi z}{2H} \right) e^{-n^2 \pi^2 c_v t / 4H^2} \right]$$

...(Eq. 7.22)

When n is even, $(1 - \cos n\pi)$ vanishes; when n is odd, this factor becomes 2. Therefore it is convenient to replace n by $(2m + 1)$, m being an integer. Thus, we have

$$u = \sum_{m=0}^{\infty} \frac{4u_i}{(2m+1)\pi} \left[\sin \frac{(2m+1)\pi z}{2H} \right] e^{-(2m+1)^2 \pi^2 c_v t/4H^2}$$

...(Eq. 7.23)

It is convenient to use the symbol M to represent $(\pi/2)(2m + 1)$, which occurs frequently:

$$u = \sum_{m=0}^{\infty} \frac{2u_i}{M} \left(\sin \frac{Mz}{H} \right) e^{-M^2 c_v t/H^2} \qquad \text{...(Eq. 7.24)}$$

Three-dimensionless parameters are introduced for convenience in presenting the results in a form usable in practice. The first is z/H, relating to the location of the point at which consolidation is considered, H being the maximum length of the drainage path. The second is the consolidation ratio, U_z, defined in sec. 7.3, to indicate the extent of dissipation of the hydrostatic excess pressure in relation to the initial value:

$$U_z = (u_i - u)/u_i = \left(1 - \frac{u}{u_i}\right) \qquad \text{...(Eq. 7.16)}$$

The subscript z is significant, since the extent of dissipation of excess pore water pressure is different for different locations, except at the beginning and the end of the consolidation process.

The third dimensionless parameter, relating to time, and called 'Time-factor', T, is defined as follows:

$$T = \frac{c_v t}{H^2} \qquad \text{...(Eq. 7.25)}$$

where c_v is the coefficient of consolidation,

H is the length drainage path,

and $\quad t$ is the elapsed time from the start of consolidation process.

In the context of consolidation process at a particular site, c_v and H are constants, and the time factor is directly proportional to time.

Introducing the time factor into Eq. 7.24, we have

$$u = \sum_{m=0}^{\infty} \frac{2u_i}{M} \left(\sin \frac{Mz}{H} \right) e^{-M^2 T} \qquad \text{...(Eq. 7.26)}$$

Introducing the consolidation ratio, U_z, we have:

$$U_z = 1 - \frac{u}{u_i} = 1 - \sum_{m=0}^{\infty} \frac{2}{M} \left(\sin \frac{Mz}{H} \right) \cdot e^{-M^2 T} \qquad \text{...(Eq. 7.27)}$$

It may be shown that if there exists a single drainage face only for the layer, one of the boundary conditions gets modified as $\frac{\partial u}{\partial z} = 0$ at the impermeable boundary.

Noting the fact that the maximum drainage path in this case is the total thickness

Compressibility and Consolidation of Soils 269

of the layer itself and designating the latter as H, instead of $2H$, we shall arrive at the same solution as indicated by Eq. 7.27. That is to say, the effect of double drainage or single drainage may be easily accounted for by substituting for H in the solution the length of the maximum drainage path.

The average degree of consolidation over the depth of the stratum at any time during the consolidation process may be determined as follows:

The average initial hydrostatic excess pressure may be written as:

$$\frac{1}{2H}\int_0^{2H} u_i \cdot dz$$

Similarly, the average hydrostatic excess pressure at any time t during consolidation is

$$\frac{1}{2H}\int_0^{2H} u \cdot dz$$

The average consolidation ratio U is the average value of $U_z (= 1 - u/u_i)$ over the depth of the stratum. It may be written as

$$U = 1 - \frac{\int_0^{2H} u\, dz}{\int_0^{2H} u_i \cdot dz} \qquad \ldots\text{(Eq. 7.28)}$$

Substituting for u from Eq. 7.26, we have

$$U = 1 - \sum_{m=0}^{\infty} \frac{2\int_0^{2H} u_i \cdot \sin\frac{Mz}{H} \cdot dz}{M \int_0^{2H} u_i \cdot dz} \cdot e^{-M^2 T} \qquad \ldots\text{(Eq. 7.29)}$$

In the special case of constant initial hydrostatic excess pressure, this reduces to

$$^*U = 1 - \sum_{m=0}^{\infty} \frac{2}{M^2} \cdot e^{-M^2 T} \qquad \ldots\text{(Eq. 7.30)}$$

A numerically equivalent procedure may also be employed for arriving at the average degree of consolidation from the graphical modes of presentation of results as indicated in the following section.

---

*The following approximate expressions have been found to yield values for T with good degree of precision:

When $U < 60\%$, $T = (\pi/4)U^2$

When $U > 60\%$, $T = -0.9332 \log_{10}(1 - U) - 0.0851$.

270 Geotechnical Engineering

7.6 GRAPHICAL PRESENTATION OF CONSOLIDATION RELATIONSHIPS

One-dimensional consolidation, subject to the condition of constant initial hydrostatic excess pressure, is the type of consolidation that is of major interest. It applies in the laboratory consolidation tests and is usually assumed, although it generally is not strictly applicable, in the cases of consolidation in the field. Equations 7.27 and 7.30 give the final results of the mathematical solution for this case.

A graphical presentation of the results indicated by Eq. 7.27 is given in Fig. 7.22. By assigning different values of z/H and T, different values of U_z are solved and plotted to obtain the family of curves shown. The tedious computations involved in this will no longer be required in view of the utility of the chart.

Figure 7.22 presents an excellent pictorial idea of the process of consolidation in an especially instructive manner. At the start of the process, $t = 0$ and $T = 0$, and U_z is zero for all depths. The heavy vertical line representing $U_z = 0$ indicates that the process of dissipation of excess pore pressure has yet to begin. It is seen that consolidation proceeds most rapidly at the drainage faces and least rapidly at the middle of the layer for double drainage conditions. (For single drainage

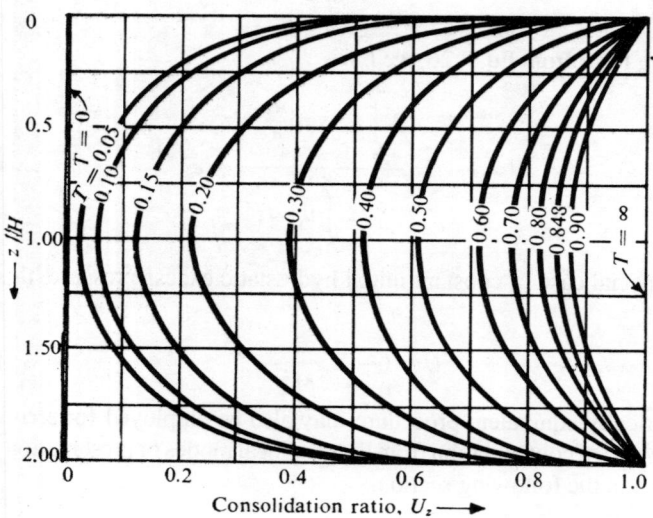

Fig. 7.22 Graphical solution for consolidation equation

conditions, consolidation proceeds least rapidly at the impermeable surface). At any finite time factor, the consolidation ratio is 1 at drainage faces and is minimum at the middle of the layer. For example, for $T = 0.20$: $U_z = 0.23$ at $z/H = 1$; $U_z = 0.46$ at $z/H = 0.5$ and 1.5; and $U_z = 0.70$ at $z/H = 0.25$ and 1.75. This indicates that at a depth of one-eighth of the layer, consolidation is 70% complete; at a depth of one-fourth of the layer, consolidation is 46% complete; while, at the middle of the layer, consolidation is just 23% complete. The distribution is somewhat parabolic in shape. As time elapses and the time factor increases, the per cent consolidation at every point increases. Finally, after a lapse of theoretically infinite time, consolidation is 100% complete at all depths, the hydrostatic excess pressure is zero as all applied pressure is carried by the soil grains.

Figure 7.22 does not depict how much consolidation occurs *as a whole* in the entire stratum. This information is of primary concern to the geotechnical engineer and may be deduced from Fig. 7.22 by the following procedure:

The relation between U_z and z/H for a time factor, $T = 0.848$, is reproduced in Fig. 7.23:

Average degree of consolidation at this time factor is the average abscissa for the entire depth and is therefore given by the shaded area from $T = 0$ to $T = 0.848$ divided by the total area from $T = 0$ to $T = \infty$; this is because the abscissa from $T = 0$ to $T = 0.848$ is the consolidation completed at $T = 0.848$, while that from $T = 0$ to $T = \infty$ represents complete or 100% consolidation.

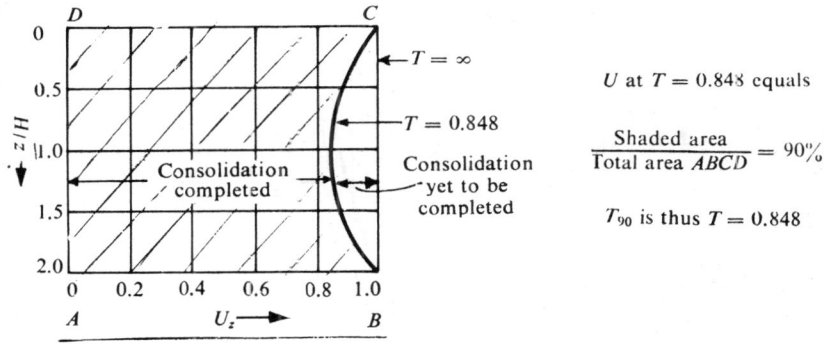

Fig. 7.23 Average consolidation at time factor 0.848

In this case the ratio of the shaded area to the total area is found to be 90%. Thus, the time factor corresponding to an average degree of consolidation of 90%, denoted by T_{90}, is 0.848.

If this exercise is repeated for different time factors, the relation between average degree of consolidation and time factor can be established as shown by curve I in Fig. 7.24. Alternatively, the curve could have been obtained by the direct application of Eq. 7.30.

In this figure, the relationship is also given in a few cases wherein the initial hydrostatic excess pressure is not constant with depth. Equation 7.29 must be applied and the corresponding mathematical expression for u_i substituted and the indicated integrations performed. Three examples–I(b), II, and III–of variable u_i are presented. It is interesting to note that the results would be identical for all linear variations. Curve II is for a sinusoidal variation of initial hydrostatic excess.

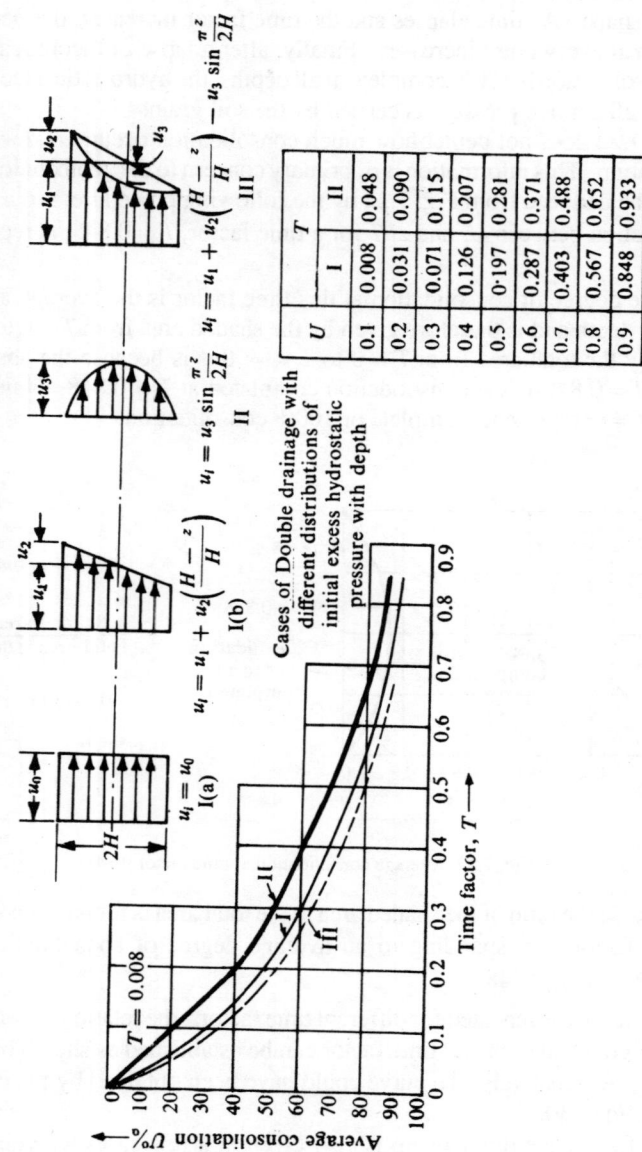

Fig. 7.24 Average degree of consolidation versus time factor (After Taylor, 1948)

Case III is a particular combination of linear and sinusoidal variations. Actual field cases may be closely approximated by such combinations. It is interesting to note, once again, that curve III is not very different from the curve I of constant initial hydrostatic excess. This is the reason for the generally accepted conclusion that curve I is an adequate representation of typical cases in nature.

All these curves are applicable to the conditions of double drainage.

There are many clay strata in nature in which drainage is at the top surface only, the bottom surface being in contact with impervious rock. All such occurrences of single drainage may be considered to be the upper-half of a case of double drainage, the other half being a fictitious mirror image. The drainage path in this case is the thickness of the stratum itself and so, $2H$ must be substituted for H in the equations for double drainage conditions.

If the hydrostatic excess pressure is constant throughout the depth, the solution for the single drainage conditions will be the same as that for the corresponding double drainage case; that is to say, curve I of Fig. 7.24 will apply. For other distributions of hydrostatic excess pressure, the results will be different. For triangular distributions of hydrostatic excess pressure indicated in Fig. 7.25, the values of the time factor for different degrees of consolidation are shown:

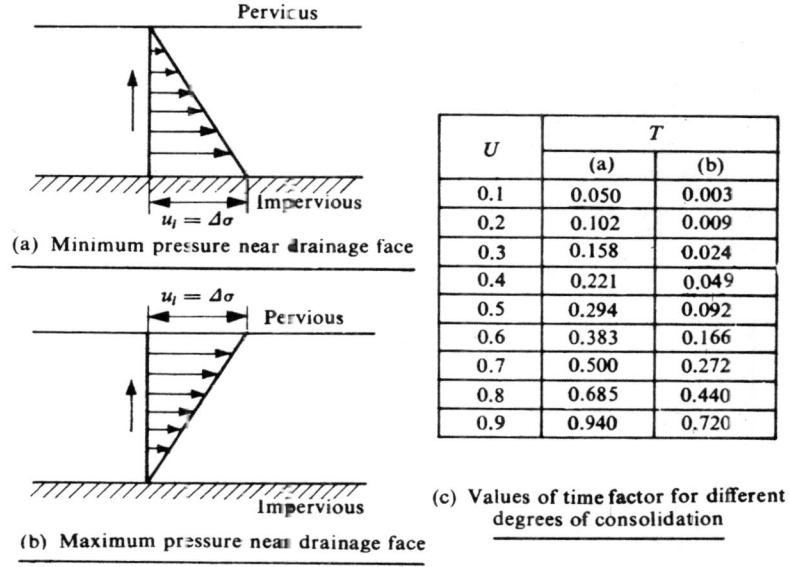

U	T	
	(a)	(b)
0.1	0.050	0.003
0.2	0.102	0.009
0.3	0.158	0.024
0.4	0.221	0.049
0.5	0.294	0.092
0.6	0.383	0.166
0.7	0.500	0.272
0.8	0.685	0.440
0.9	0.940	0.720

(a) Minimum pressure near drainage face

(b) Maximum pressure near drainage face

(c) Values of time factor for different degrees of consolidation

Fig. 7.25 Single drainage condition—triangular distributions of initial hydrostatic excess pressure

Unless otherwise stated, it is the average consolidation ratio for a stratum that is referred to.

7.7 EVALUATION OF COEFFICIENT OF CONSOLIDATION FROM OEDOMETER TEST DATA

The coefficient of consolidation, C_v, in any stress range of interest, may be evaluated from its definition given by Eq. 7.20, by experimentally determining the parameters k, a_v and e_0 for the stress range under consideration. k may be got from a permeability test conducted on the oedometer sample itself, after complete consolidation under the particular stress increment. a_v and e_0 may be obtained from the oedometer test data, by plotting the $e - \bar{\sigma}$ curve. However, Eq. 7.20 is rarely used for the determination of c_v. Instead, c_v is evaluated from the consolidation test data by the use of characteristics of the theoretical relationship between the time factor T, and the degree of consolidation, U. These methods are known as 'fitting methods', as one tries to fit in the characteristics of the theoretical curve with the experimental or laboratory curve. In this context, it is pertinent to note the striking similarity between curve I of Fig. 7.24 and the typical time-compression curve for clays given in Fig. 7.2.

The more generally used fitting methods are the following:
(a) The square root of time fitting method
(b) The logarithm of time fitting method

These two methods will be presented in the following sub-sections.

7.7.1 The Square Root of Time Fitting Method

This method has been devised by D.W. Taylor (1948). The coefficient of consolidation is the soil property that controls the time-rate or speed of consolidation under a load-increment. The relation between the sample thickness and elapsed time since the application of the load-increment is obtainable from an oedometer test and is somewhat as shown in Fig. 7.26 for a typical load-increment.

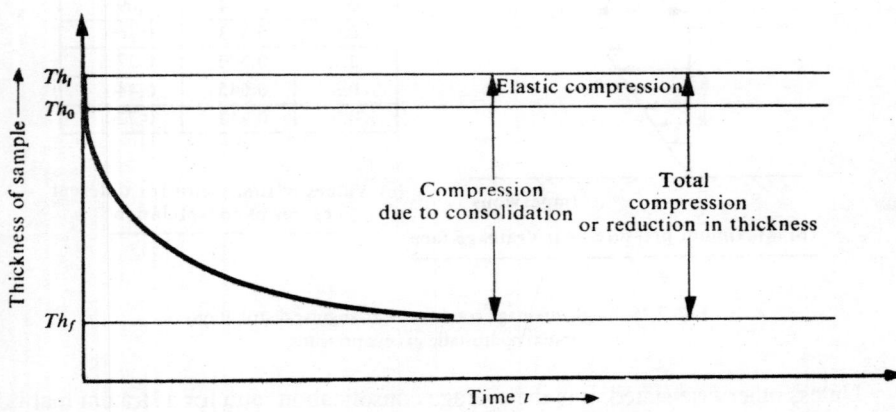

Fig. 7.26 Time versus reduction in sample thickness for a load-increment

This figure depicts change in sample thickness with time essentially due to consolidation; only the elastic compression which occurs almost instantaneously on application of load increment is shown. The effect of prolonged compression that occurs after 100% dissipation of excess pore pressure is not shown or is ignored; this effect is known as 'Secondary consolidation', which is briefly presented in the following section. The curves of Figs. 7.26 and 7.24 bear striking similarity; in fact, one should expect it if Terzaghi's theory is to be valid for the phenomenon of consolidation. This similarity becomes more apparent if the curves are plotted with square root of time/time factor as the function, as shown in Fig. 7.27(a) and (b).

The theoretical curve on the square root plot is a straight line up to about 60% consolidation with a gentle concave upwards curve thereafter. If another straight line, shown dotted, is drawn such that the abscissae of this line are 1.15 times those of the straight line portion of the theoretical curve, it can be shown to cut the theoretical curve at 90% consolidation. This may be established from the values of T at various values of U given in Fig. 7.24 for case I; that is, the value of $\sqrt{T}$ at 90% consolidation is 1.15 times the abscissa of an extension of the straight line portion of the U versus $\sqrt{T}$ relation. This property is used for 'fitting' the theoretical curve to the laboratory curve.

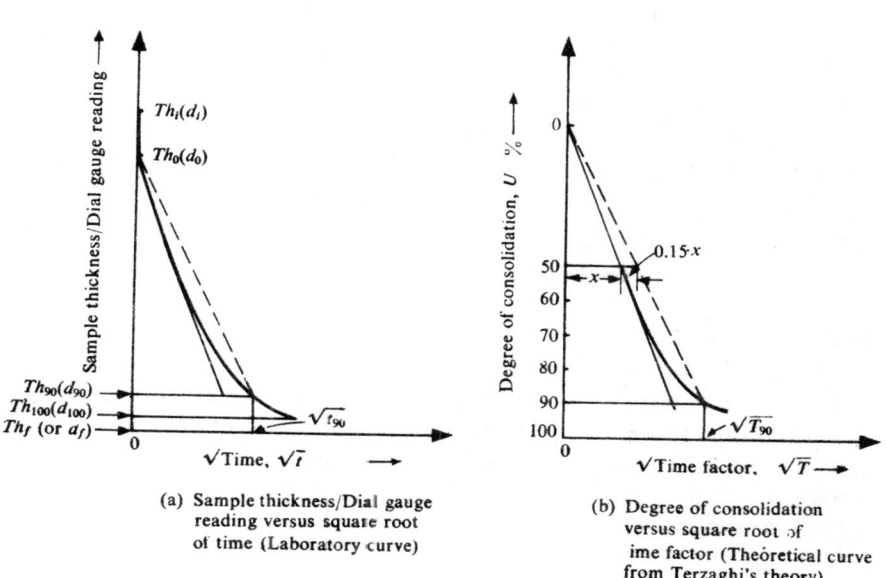

(a) Sample thickness/Dial gauge reading versus square root of time (Laboratory curve)

(b) Degree of consolidation versus square root of ime factor (Theoretical curve from Terzaghi's theory)

Fig. 7.27 Square root of time fitting method (After Taylor, 1948)

The laboratory curve shows a sudden initial compression, called 'elastic compression' which may be partly due to compression of gas in the pores. The corrected zero point at zero time is obtained by extending the straight line portion of the laboratory plot backward to meet the axis showing the sample thickness/dial gauge reading. The so-called 'primary compression' or 'primary consolidation' is reckoned from this corrected zero. A dashed line is constructed from the corrected zero such that its abscissae are 1.15 times those of the straight line portion of the laboratory plot. The intersection of the dashed line with the laboratory plot identifies the point representing 90% consolidation in the sample. The time corresponding to this can be read off from the laboratory plot. The point corresponding to 100% primary consolidation may be easily extrapolated on this plot.

The coefficient of consolidation, c_v, may be obtained from

$$c_v = \frac{T_{90} H^2}{t_{90}} \qquad \ldots \text{(Eq. 7.31)}$$

where t_{90} is read off from Fig. 7.27(a)

T_{90} is 0.848 from Terzaghi's theory

H is the drainage path, which may be taken as half the thickness of the sample for double drainage conditions, or as $(Th_0 + Th_f)/4$ in terms of the sample thickness (Fig. 7.26).

The primary compression is that from Th_0 to Th_{100}/d_0 to d_{100} in terms of sample thickness/dial gauge reading; the total compression is that from Th_i to Th_f/d_i to d_f. The ratio of primary compression to total compression is called the 'Primary Compression ratio''.

Thus, the total compression in a loading increment of a laboratory test has three parts. The part from Th_i to Th_0/d_i to d_0 is instantaneous elastic compression; that from Th_0 to Th_{100}/d_0 to d_{100} is primary compression; and that from Th_{100} to Th_f/d_{100} to d_f is secondary compression. The secondary compression may be as much as 20% or more in a number of cases.

7.7.2 The Logarithm of Time Fitting Method

This method was devised by A. Casagrande and R.E. Fadum (1939). The point corresponding to 100 per cent consolidation curve is plotted on a semi-logarithmic scale, with time factor on a logarithmic scale and degree of consolidation on arithmetic scale, the intersection of the tangent and asymptote is at the ordinate of 100% consolidation. A comparison of the theoretical and laboratory plots in this regard is shown in Figs. 7.28(a) and (b).

Since the early portion of the curve is known to approximate a parabola, the corrected zero point may be located as follows: The difference in ordinates between two points with times in the ratio of 4 to 1 is marked off; then a distance equal to this difference may be stepped off above the upper points to obtain the corrected zero point. This point may be checked by more trials, with different pairs of points on the curve.

Compressibility and Consolidation of Soils 277

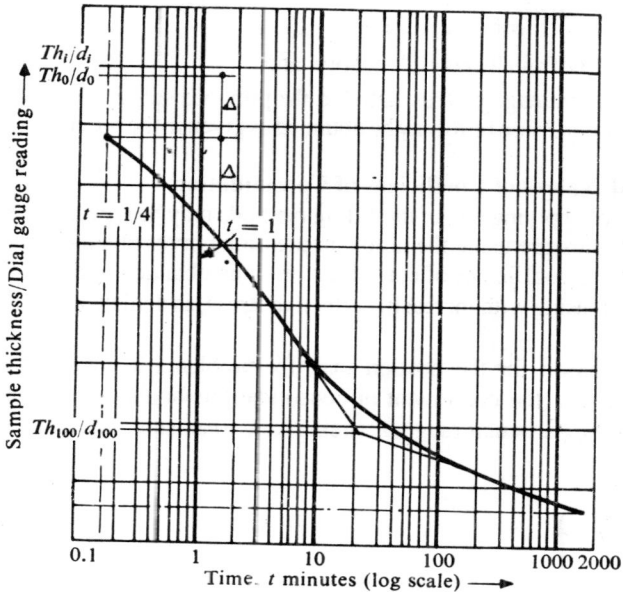

(a) Sample thickness/Dial gauge reading
versus logarithm of time (Laboratory curve)

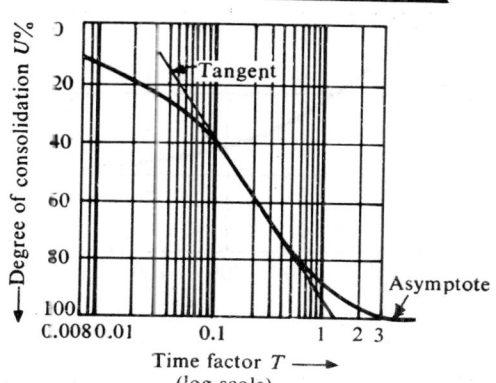

(b) Degree of consolidation versus
logarithm of time factor
(Theoretical curve from Terzaghi's theory)

Fig. 7.28 Logarithm of time fitting method
(After A. Casagrande, 1939)

After the zero and 100% primary compression points are located, the point corresponding to 50% consolidation and its time may easily be obtained and the coefficient of consolidation computed from:

278 Geotechnical Engineering

$$C_v = \frac{T_{50}H^2}{t_{50}} \qquad \text{...(Eq. 7.32)}$$

where t_{50} is read off from Fig. 7.28(a)

$T_{50} = 0.197$ from Terzaghi's theory, and

H is the drainage path as stated in the previous subsection.

The primary compression ratio may be obtained as given in the previous subsection.

7.7.3 Typical Values of Coefficient of Consolidation

The process of applying one of the fitting methods may be repeated for different increments of pressure using the time-compression curves obtained in each case. The values of the coefficient of consolidation thus obtained will be found to be essentially decreasing with increasing effective stress, as depicted typically in Fig. 7.29.

This is the reason for the caution that, for problems in the field involving settlement analysis, the coefficient of consolidation should be evaluated in the laboratory for the particular range of stress likely to exist in the field.

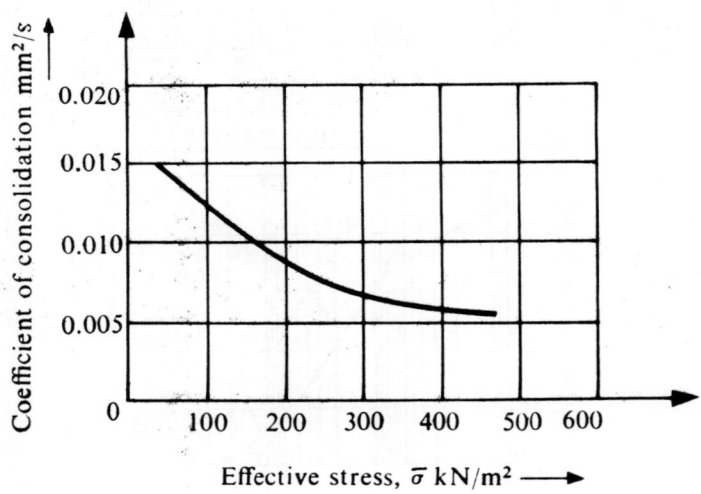

Fig. 7.29 Variation of coefficient of consolidation with effective stress

The range of values for C_v is rather wide—5×10^{-4} mm^2/s to 2×10^{-2} mm^2/s. Further, it is also found that the value of C_v decreases as the liquid limit of the clay increases. This should be expected since, in general, clays of increasing plasticity should be requiring more time for a particular degree of consolidation, as is evident from Eq. 7.25.

The value of C_v is useful in the determination of the time required for a finite percentage of consolidation to occur. Usually 90 to 95% consolidation time may be treated to be that required for ultimate settlement.

Compressibility and Consolidation of Soils 279

It is interesting to note that a consolidation test provides an indirect way of obtaining the coefficient of permeability of a clay by applying Eq. 7.20, after evaluating the coefficient of consolidation by one of the available fitting methods.

7.8 SECONDARY CONSOLIDATION

The time-settlement curve for a cohesive soil has three distinct parts as illustrated in Fig. 7.30:

When the hydrostatic excess pressure is fully dissipated, no more consolidation should be expected. However, in practice, the decrease in void ratio continues, though very slowly, for a long time after this stage, called 'Primary Consolidation'. The effect or the phenomenon of continued consolidation after the complete dissipation of excess pore water pressure is termed 'Secondary Consolidation' and the resulting compression is called 'Secondary Compression'. During this stage, plastic readjustment of clay platelets takes place and other effects as well as colloidal-chemical processes and surface phenomena such as induced electrokinetic potentials occur. These are, by their very nature, very slow.

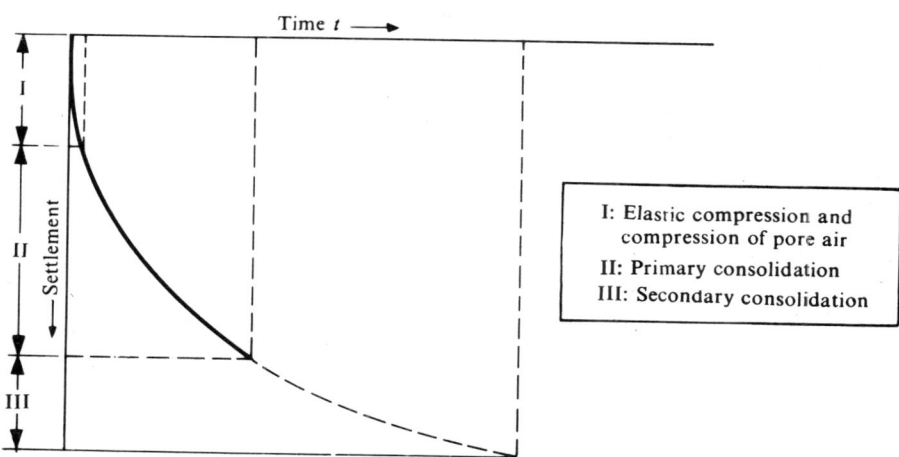

Fig. 7.30 Time-settlement curve for a cohesive soil

I: Elastic compression and compression of pore air
II: Primary consolidation
III: Secondary consolidation

Secondary consolidation is believed to come into play even in the range of primary consolidation, although its magnitude is small because of the existence of a plastic lag right from the beginning of loading. However, it is almost impossible to separate this component from the primary compression. Since dissipation of excess pore pressure is not the criterion here, Terzaghi's theory is inapplicable to secondary consolidation. The fact that experimental time-compression curves are

280 Geotechnical Engineering

in agreement with Terzaghi's theoretical curve only up to about 60% consolidation is, in itself, an indication of the manifestation of secondary consolidation even during the stage of primary consolidation.

Secondary consolidation of mineral soils is usually negligible but it may be considerable in the case of organic soils due to their colloidal nature. This may constitute a substantial part of total compression in the case of organic soils, micaceous soils, loosely deposited clays, etc. A possible disintegration of clay particles is also mentioned as one of the reasons for this phenomenon. Secondary compression is usually assumed to be proportional to the logarithm of time. A more detailed discussion is out of scope of the present text.

7.9 ILLUSTRATIVE EXAMPLES

Example 7.1: In a consolidation test the following results have been obtained. When the load was changed from 50 kN/m² to 100 kN/m², the void ratio changed from 0.70 to 0.65. Determine the coefficient of volume decrease, m_v, and the compression index, C_c.

(S.V.U.—B.Tech., (Part-time)—Sep., 1982)

$e_0 = 0.70$ $\bar{\sigma}_0 = 50$ kN/m²

$e_1 = 0.65$ $\bar{\sigma} = 100$ kN/m²

Coefficient of compressibility, $a_v = \dfrac{\Delta e}{\Delta \bar{\sigma}}$, ignoring sign.

$$= \dfrac{(0.70 - 0.65)}{(100 - 50)} \text{ m}^2/\text{kN} = 0.05/50 \text{ m}^2/\text{kN} = 0.001 \text{ m}^2/\text{kN}.$$

Modulus of volume change, or coefficient of volume decrease,

$$m_v = \dfrac{a_v}{(1+e_0)} = \dfrac{0.001}{(1+0.70)} = \dfrac{0.001}{1.7} \text{ m}^2/\text{kN}.$$

$$= 5.88 \times 10^{-4} \text{ m}^2/\text{kN}$$

Compression index, $C_c = \dfrac{\Delta e}{\Delta(\log \bar{\sigma})} = \dfrac{(0.70 - 0.65)}{(\log_{10} 100 - \log_{10} 50)}$

$$= \dfrac{0.05}{\log_{10}\frac{100}{50}} = \dfrac{0.05}{\log_{10} 2} = \dfrac{0.050}{0.301}$$

$$= \mathbf{0.166}$$

Example 7.2: A sand fill compacted to a bulk density of 18.84 kN/m³ is to be placed on a compressible saturated marsh deposit 3.5 m thick. The height of the sand fill is to be 3 m. If the volume compressibility m_v of the deposit is 7×10^{-4} m²/kN, estimate the final settlement of the fill.

(S.V.U.—B.E., (N.R.)—March-April, 1966)

Ht. of sand fill = 3 m
Bulk unit weight of fill = 18.84 kN/m³
Increment of the pressure on top of marsh deposit $\Delta\bar{\sigma}$ = 3 × 18.84
$$= 56.52 \text{ kN/m}^2$$
Thickness of marsh deposit, H_0 = 3.5 m
Volume compressibility $m_v = 7 \times 10^{-4} \text{ m}^2/\text{kN}$
Final settlement of the marsh deposit, ΔH
$$= m_v \cdot H_0 \cdot \Delta\bar{\sigma}$$
$$= 7 \times 10^{-4} \times 3500 \times 56.52 \text{ mm}$$
$$= 138.5 \text{ mm}$$

Example 7.3: The following results were obtained from a consolidation test:
Initial height of sample H_i = 2.5 cm
Height of solid particles H_s = 1.25 cm

Pressure in kg/cm²	Dial reading in cm
0.00	0.000
0.13	0.000
0.27	0.004
0.54	0.016
1.08	0.044
2.14	0.104
4.80	0.218
9.60	0.340
15.00	0.420

Plot the pressure-void ratio curve and determine (*a*) the compression index and (*b*) the preconsolidation pressure.

(S.V.U.—B.E.,(R.R.)—Dec., 1968)

Initial void ratio, $e_0 = \dfrac{(H_i - H_s)}{H_s} = \dfrac{(2.50 - 1.25)}{1.25} = \dfrac{1.25}{1.25} = 1.000$

The heights of sample and the void ratios at the end of each pressure increments are tabulated below:

Pressure in kg/cm²	Dial reading in cm	Height of sample, H in cm	Void ratio
0.00	0.000	2.500	1.000
0.13	0.000	2.500	1.000
0.27	0.004	2.496	0.997
054	0.016	2.484	0.987
1.08	0.044	2.456	0.965
2.14	0.104	2.396	0.917
4.80	0.218	2.282	0.826
9.60	0.340	2.160	0.728
15.00	0.420	2.080	0.664

282 *Geotechnical Engineering*

Now, the pressure-void ratio curve, is drawn on a semi-logarithmic scale, the pressure being represented on logarithmic scale as shown in Fig. 7.31.
From the diagram shown in Fig. 7.31.
(a) $C_c = \Delta e / \log(\bar{\sigma}_2/\bar{\sigma}_1)$
   = Δe, if one logarithmic cycle of pressure is chosen as the base.
   = **0.303**, in this case.
(b) The pre-consolidation pressure by A. Casagrande's method
   = **1.8 kg/cm²**

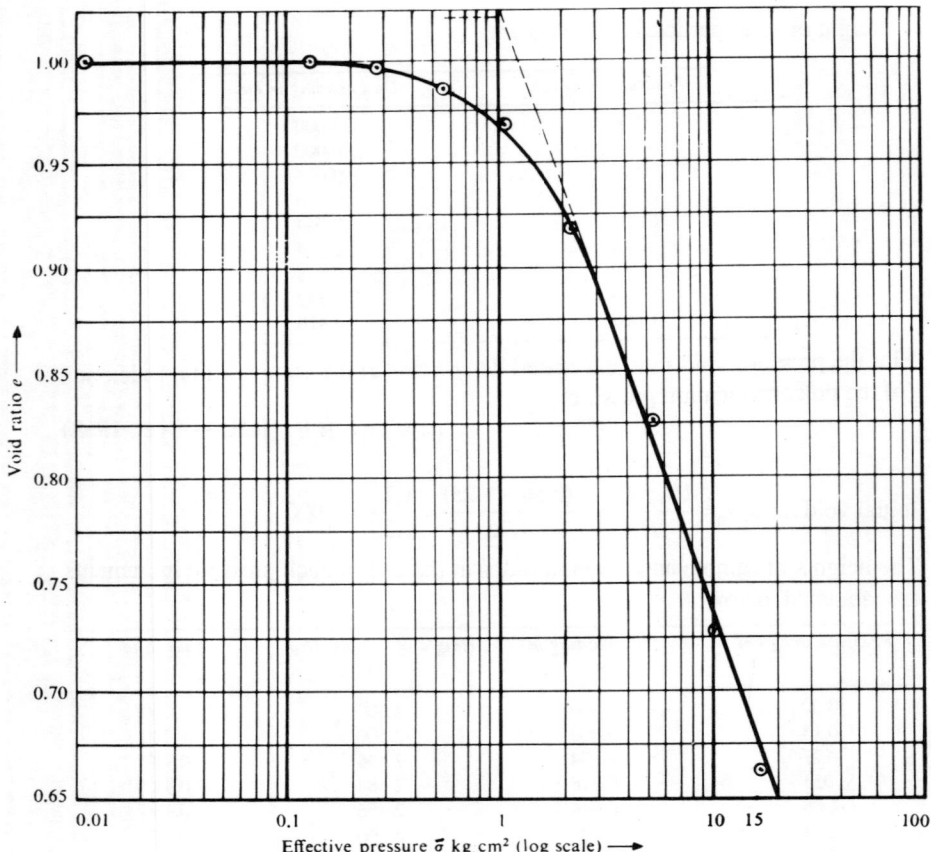

Fig. 7.31 Pressure-void ratio diagram (Example 7.3)
(pressure to logarithmic scale)

Compressibility and Consolidation of Soils 283

Example 7.4: A layer of soft clay is 6 m thick and lies under a newly constructed building. The weight of sand overlying the clayey layer produces a pressure of 2.6 kg/cm² and the new construction increases the pressure by 1.0 kg/cm². If the compression index is 0.5, compute the settlement. Water content is 40% and specific gravity of grains is 2.65

(S.V.U.—B.E., (R.R.)—Dec., 1976)

Initial pressure, $\bar{\sigma}_0 = 2.5$ kg/cm²
Increment of pressure, $\Delta\bar{\sigma} = 1.0$ kg/cm²
Thickness of clay layer, $H = 6$ m $= 600$ cm.
Compression index, $C_c = 0.5$
Water content, $w = 40\%$
Specific gravity of grains, $G = 2.65$
Void ratio, $e = wG$, (since the soil is saturated) $= 0.40 \times 2.65 = 1.06$
This is taken as the initial void ratio, e_0.
Consolidation settlement,

$$S = \frac{H \cdot C_c}{(1+e_0)} \log_{10}\left(\frac{\bar{\sigma}_0 + \Delta\bar{\sigma}}{\bar{\sigma}_0}\right)$$

$$= \frac{600 \times 0.5}{(1+1.06)} \log_{10}\left(\frac{2.5+1.0}{2.5}\right) \text{ cm}$$

$$= \frac{300}{2.06} \log_{10}\left(\frac{3.5}{2.5}\right) \text{ cm}$$

$$= 21.3 \text{ cm.}$$

Example 7.5: The settlement analysis (based on the assumption of the clay layer draining from top and bottom surfaces) for a proposed structure shows 2.5 cm of settlement in four years and an ultimate settlement of 10 cm. However, detailed sub-surface investigation reveals that there will be no drainage at the bottom. For this situation, determine the ultimate settlement and the time required for 2.5 cm settlement.

(S.V.U.—B.E., (R.R.)—Nov., 1973)

The ultimate settlement is not affected by the nature of drainage, whether it is one-way or two-way.
Hence, the ultimate settlement = 10 cm.
However, the time-rate of settlement depends upon the nature of drainage.
Settlement in four years = 2.5 cm.

$$T = \frac{C_v t}{H^2}$$

$$U = 2.5/10.0 = 25\%$$

284 *Geotechnical Engineering*

Since the settlement is the same, $U\%$ is the same;
hence, the time-factor is the same.

$$\therefore \quad T/C_v = t/H^2 = \text{Constant}.$$

or

$$\frac{t_2}{H_2^2} = \frac{t_1}{H_1^2},$$

t_2 and H_2 referring to double drainage, and t_1 and H_1 referring to single drainage.

The drainage path for single drainage is the thickness of the layer itself, while that for double drainage is half the thickness.

$$\therefore \quad H_1 = 2H_2$$

$$\therefore \quad \frac{t_2}{H_2^2} = \frac{t_1}{4H_2^2},$$

$$\therefore \quad t_1 = 4t_2 = 4 \times 4 \text{ yrs} = \textbf{16 yrs.}$$

Example 7.6: There is a bed of compressible clay of 4 m thickness with pervious sand on top and impervious rock at the bottom. In a consolidation test on an undisturbed specimen of clay from this deposit 90% settlement was reached in 4 hours. The specimen was 20 mm thick. Estimate the time in years for the building founded over this deposit to reach 90% of its final settlement.

(S.V.U.—B.E., (R.R.)—Sept., 1978)

This is a case of one-way drainage in the field.

$\therefore$ Drainage path for the field deposit, $H_f = 4$ m = 4000 mm. In the laboratory consolidation test, commonly it is a case of two-way drainage.

$\therefore$ Drainage path for the laboratory sample, $H_l = 20/2 = 10$ mm

Time for 90% settlement of laboratory sample = 4 hrs.
Time factor for 90% settlement, $T_{90} = 0.848$

$$\therefore \quad T_{90} = \frac{C_v t_{90_f}}{H_f} = \frac{C_v t_{90_l}}{H_l}$$

or

$$\frac{t_{90_f}}{H_f} = \frac{t_{90_l}}{H_l}$$

$$\therefore \quad T_{90f} = \frac{t_{90_l}}{H_l} \times H_f = \frac{4 \times 4000}{10} \text{ hrs.}$$

$$= \frac{4 \times 400}{24 \times 365} \text{ years}$$

$$= 0.18265 \text{ yrs} \left(66\frac{2}{3} \textbf{ days}\right).$$

Example 7.7: The void ratio of clay A decreased from 0.572 to 0.505 under a change in pressure from 120 to 180 kg/m². The void ratio of clay B decreased from 0.612 to 0.597 under the same increment of pressure. The thickness of sample A was 1.5 times that of B. Nevertheless the time required for 50% consolidation was three times longer for sample B than for sample A. What is the ratio of the coefficient of permeability of A to that of B?

(S.V.U.—B.E., (N.R.)—Sep., 1967)

Clay A
$e_0 = 0.572$
$e_1 = 0.505$
$\bar{\sigma}_0 = 120$ kN/m²
$\bar{\sigma}_1 = 180$ kN/m²

$$a_{v_A} = \frac{\Delta e}{\Delta \sigma} = \frac{0.067}{60} \text{ m}^2/\text{kN}$$

$$m_{v_A} = \frac{0.067}{6}/(1+0.572)$$

$$= 7.10 \times 10^{-4} \text{ m}^2/\text{kN}$$

Clay B
$e_0 = 0.612$
$e_1 = 0.597$
$\bar{\sigma}_0 = 120$ kN/m²
$\bar{\sigma}_1 = 180$ kN/m²

$$a_{v_B} = \frac{\Delta e}{\Delta \sigma} = \frac{0.015}{60} \text{ m}^2/\text{kN}$$

$$m_{v_B} = \frac{0.015}{6}/(1+0.612)$$

$$= 1.55 \times 10^{-4} \text{ m}^2/\text{kN}$$

$H_A/H_B = 1.5$ and $t_{50_B}/t_{50_A} = 3$

$$T_{50} = C_v t_{50}/H^2$$

$$\therefore \quad T_{50} = \frac{C_{v_A} \cdot t_{50_A}}{H_A^2} = \frac{C_{v_B} \cdot t_{50_B}}{H_B^2}$$

$$\frac{C_{v_A}}{C_{v_B}} = \frac{t_{50_B}}{t_{50_A}} \cdot \frac{H_A^2}{H_B^2} = 3 \times (1.5)^2 = 6.75$$

But $C_v = k/m_v \gamma_w$

or $k = C_v m_v \cdot \gamma_w$

$$\therefore \quad k_A/k_B = \frac{C_{v_A} \cdot m_{v_A}}{C_{v_B} \cdot m_{v_B}} = 6.75 \times \frac{7.10 \times 10^{-4}}{1.55 \times 10^{-4}} = 30.92 \approx 31$$

Example 7.8: A saturated soil has a compression index of 0.25. Its void ratio at a stress of 10 kN/m² is 2.02 and its permeability is 3.4×10^{-7} mm/s. Compute:
(i) Change in void ratio if the stress is increased to 19 kN/m²;

(ii) Settlement in (i) if the soil stratum is 5 m thick; and
(iii) Time required for 40% consolidation if drainage is one-way.
                              (S.V.U.—B.Tech., (Part-time)—Apr., 1982)

Compression index, $C_c = 0.25$

$$e_0 = 2.02 \quad \bar{\sigma}_0 = 10 \text{ kN/m}^2 \quad k = 3.4 \times 10^{-7} \text{ mm/s}$$

$$\bar{\sigma}_1 = 19 \text{ kN/m}^2$$

(i) $$C_c = \frac{\Delta e}{\log_{10}(\bar{\sigma}_1/\bar{\sigma}_0)} \qquad \therefore 0.25 = \frac{\Delta e}{\log_{10}(19/10)}$$

$$\therefore \quad \Delta e = 0.25 \log_{10}(1.9) \approx \mathbf{0.07}$$

or Void ratio at a stress of 19 kN/m² = 2.02 − 0.07 = 1.95

$$a_v = \Delta e / \Delta \bar{\sigma} = 0.07/9 = 0.00778 \text{ m}^2/\text{kN}$$

$$m_v = a_v/(1+e_0) = 0.00778/(1+2.02) = 2.575 \times 10^{-3} \text{ m}^2/\text{kN}$$

(ii) Thickness of soil stratum, $H = 5$ m.

$$\text{Settlement, } S = \frac{H \cdot C_c}{(1+e_0)} \log_{10}\left(\frac{\bar{\sigma}_0 + \Delta\bar{\sigma}}{\bar{\sigma}_0}\right) = \frac{H \cdot C_c}{(1+e_0)} \log_{10}\left(\frac{\bar{\sigma}_1}{\bar{\sigma}_0}\right)$$

$$= \frac{5 \times 1000 \times 0.25}{(1+2.02)} \log_{10}(19/10) \text{ mm} \approx \mathbf{115.4 \text{ mm}}$$

(iii) If drainage is one way, drainage path, H = thickness of stratum = 5 m

$$T_{40} = \frac{C_v t_{40}}{H^2}; \quad T_{40} = (\pi/4)U^2 = (\pi/4) \times (0.40)^2 = 0.04 = 0.125664$$

$$C_v = k/m_v \cdot \gamma_w$$

$$= \frac{3.4 \times 10^{-7} \times 10^{-3} \times 9 \times (1+2.02)}{2.575 \times 10^{-3} \times 9.81} \text{ m}^2/\text{s} = 3.66 \times 10^{-7} \text{ m}^2/\text{s}.$$

$$\therefore \quad t_{40} = \frac{T_{40} \cdot H^2}{C_v}$$

$$= \frac{0.125664 \times 5 \times 5}{3.66 \times 10^{-7} \times 60 \times 60 \times 24} \text{ days}$$

$$\approx \mathbf{99.35 \text{ days}.}$$

Example 7.9: (*a*) The soil profile at a building site consists of dense sand up to 2 m depth, normally loaded soft clay from 2 m to 6 m depth, and stiff impervious rock below 6 m depth. The ground-water table is at 0.40 m depth below ground

Compressibility and Consolidation of Soils 287

level. The sand has a density of 1.85 t/m³ above water table and 1.90 t/m³ below it. For the clay, natural water content is 50%, liquid limit is 65% and grain specific gravity is 2.65. Calculate the probable ultimate settlement resulting from a uniformly distributed surface load of 4.0 t/m² applied over an extensive area of the site.

(b) In a laboratory consolidation test with porous discs on either side of the soil sample, the 25 mm thick sample took 81 minutes for 90% primary compression. Calculate the value of coefficient of consolidation for the sample.

(S.V.U.—Four year B.Tech.,—April, 1983)

(a) The soil profile is as shown in Fig. 7.32:

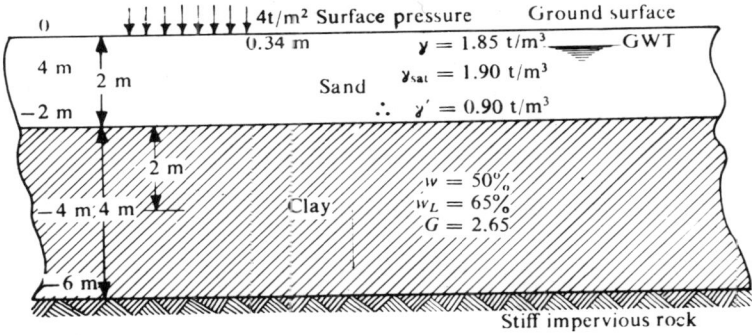

Fig. 7.32 Soil profile at a building site (Example 7.9)

For the clay stratum:

$$w = 50\%$$
$$G = 2.65$$

Since it is saturated, $e = w.G = 0.50 \times 2.65 = 1.325$

This is the initial void ratio, e_0.

$$\gamma_{sat} = \frac{(G+e)}{(1+e)} \cdot \gamma_w = \frac{(2.65+1.325)}{(1+1.325)} \times 1 \text{ t/m}^3 = \frac{3.975}{2.325} \text{ t/m}^3$$

$$= 1.71 \text{ t/m}^3$$

$$\gamma' = (\gamma_{sat} - \gamma_w) = 0.71 \text{ t/m}^3$$

Initial effective overburden pressure at the middle of the clay layer:

$$\bar{\sigma}_0 = (0.4 \times 1.85 + 1.6 \times 0.90 + 2 \times 0.71) \text{ t/m}^2$$

$$= 3.60 \text{ t/m}^2$$

Let us assume that the applied surface pressure of 4.0 t/m² gets transmitted to the middle of the clay layer undiminished.

$$\therefore \quad \Delta\bar{\sigma} = 4.0 \text{ t/m}^2$$

The compression index, C_c may be taken as:

$$C_c = 0.009 (w_L - 10) \qquad \ldots(Eq. 7.8)$$

$$\therefore \quad C_c = 0.009 (65 - 10) = 0.495$$

The consolidation settlement, S, is given by:

$$S = \frac{H \cdot C_c}{(1+e_0)} \log_{10}\left(\frac{\overline{\sigma}_0 + \Delta\overline{\sigma}}{\overline{\sigma}_0}\right)$$

$$= \frac{400 \times 0.495}{(1+1.325)} \log_{10} \frac{(3.6+4.0)}{3.6} \text{ cm}$$

$$\approx 27.66 \text{ cm}.$$

(b) Thickness of the laboratory sample = 25 mm.

Since it is two-way drainage with porous discs on either side, the drainage path, $H = 25/2 = 12.5$ mm.

Time for 90% primary compression, $t_{90} = 81$ minutes.

Time factor, T_{90}, for $U = 90\%$ is known to be 0.848.

(Alternatively, $T = -0.9332 \log_{10}(1-U) - 0.0851$

$$= -0.9332 \log_{10} 0.10 - 0.0851$$

$$= 0.9332 - 0.0851 = 0.8481)$$

$$\therefore \quad T_{90} = \frac{C_v t_{90}}{H^2}$$

Coefficient of consolidation

$$C_v = \frac{T_{90} \cdot H^2}{t_{90}} = \frac{0.848 \times 1^2 \cdot 25}{81} \text{ cm}^2/\text{min}.$$

$$= \frac{0.848 \times 1^2 \cdot 25}{81 \times 60} \text{ cm}^2/\text{s}$$

$$= 2.726 \times 10^{-4} \text{ cm}^2/\text{s}$$

SUMMARY OF MAIN POINTS

1. Specifically, the compressibility of a soil depends on the structural arrangement of the soil particles. Grain-shape also influences this aspect.
2. Consolidation means expulsion of pore water from a saturated soil; it is relevant to clays and is a function of the effective stress rather than the total stress.

3. Oedometer or consolidometer is the device used for investigating the compressibility characteristics of a soil in the laboratory—both the total compression and its time-rate under specific pressure.
4. In a sand, the total or ultimate compression under the influence of a stress increment occurs very fast, almost instantaneously; however, in a clay it takes at least 24 hours or even much more.
5. The slope of the straight line portion of pressure (logarithmic scale) and void ratio (natural scale) is called the 'compression index'.
6. Preconsolidation pressure or past maximum pressure is important since compression is very little until this pressure is reached. The concept is applicable primarily to field deposits but can also be applied to laboratory samples.
7. 'Normally consolidated' refers to a condition wherein the existing effective stress is the maximum which the soil has ever been subjected to in its stress history; 'overconsolidated', on the other hand, refers to a condition wherein the present effective stress is smaller than the past maximum pressure.
8. The compression index, C_c, is related to the liquid limit, LL, as established by Skempton and his associates.
9. The total consolidation settlement is given by:

$$S_c = \frac{H_0 \cdot C_c}{(1+e_0)} \log_{10}\left(\frac{\bar{\sigma}_0 + \Delta\bar{\sigma}}{\bar{\sigma}_0}\right)$$

10. Terzaghi's one-dimensional consolidation theory, expressed mathematically, states:

$$\frac{\partial u}{\partial t} = C_v \cdot \frac{\partial^2 u}{\partial z^2}, \text{ where } c_v = \frac{k}{m_v \gamma_w}, \text{ coefficient of consolidation.}$$

11. Fitting methods are those used to compare laboratory time-compression curves with the theoretical curve, with a view to evaluating the coefficient of consolidation.
12. Plastic readjustment of clay platelets and certain colloidal chemical processes lead to continued consolidation even after cent per cent dissipation of excess pore water pressure, which is termed 'secondary consolidation'; however, it is very slow and negligible compared to primary consolidation.

REFERENCES

1. A. Casagrande: *The Determination of the Pre-consolidation Load and its Practical Significance,* Proceedings, First International Conference on Soil Mechanics and Foundation Engineering, Cambridge, Massachusetts, USA, June, 1936.
2. A. Casagrande and R.E. Fadum: *Notes on Soil Testing for Engineering Purposes,* Graduate School of Engineering—Harvard University, Cambridge, Massachusetts, USA, Soil Mechanics series No. 8, 1939-40.

3. IS:2720 (Part XV) 1965: *Methods of Test for Soils—Determination of consolidation Properties.*
4. A.R. Jumikis: *Soil Mechanics,* D. Van Nostrand Co., Princeton, NJ, USA, 1962.
5. D.F. McCarthy: *Essentials of Soil Mechanics and Foundations,* Reston Publishing Company, Reston, Va, USA, 1977.
6. J.M. Schmertmann: *The Undisturbed Consolidation of Clay,* Trans. ASCE, Vol. 120, 1955.
7. G.N. Smith: *Essentials of Soil Mechanics for Civil and Mining Engineers,* Third edition Metric, Crosby Lockwood Staples, London, 1974.
8. M.G. Spangler: *Soil Engineering,* International Textbook Company, Scranton, USA, 1951.
9. D.W. Taylor: *Fundamentals of Soil Mechanics,* John Wiley & Sons, Inc., New York, 1948.
10. K. Terzaghi: *Erdbaumechanik auf bodenphysikalicher Grundlage,* Leipzig und Wien, Franz Deuticke, 1925.
11. K. Terzaghi and R.B. Peck: *Soil Mechanics in Engineering Practice,* John Wiley & Sons, Inc., NY, 1948.

QUESTIONS AND PROBLEMS

7.1 Write short notes on the following:
   (a) Log fitting method for evaluation of C_v from laboratory consolidation test.
   (b) Precompression in clays.
   (S.V.U.—Four year B.Tech.—Apr., 1983)
7.2 (a) State the assumptions made in Terzaghi's theory of one-dimensional consolidation.
   (S.V.U.—B.E., (N.R.)—March, 1966)
   (b) Define the terms 'Compression index', coefficient of consolidation', and 'coefficient of compressibility', and indicate their units and symbols.
   (S.V.U.—B.Tech., (Part-time)—May, 1983, B.E., (R.R.)—May, 1971)
7.3 Define 'preconsolidation pressure'. In what ways is its determination important in soil engineering practice? Describe a suitable procedure for determining the preconsolidation pressure.
   (S.V.U.—B.Tech., (Part-time) Apr., 1982 & Sep., 1982, B.E., (R.R.)—May, 1975)
7.4 Explain with neat sketches:
   (i) The influence of load-increment ratio on time-settlement curve.
   (ii) Terzaghi's assumptions.
   (S.V.U.—B.Tech., (Part-time)—Apr., 1982)
7.5 Differentiate between 'compaction' and 'consolidation'.
   (S.V.U.—B.Tech., (Part-time)—June, 1981, B.E., (R.R)—Feb., 1976)

7.6 (a) Define (i) Compression Index, (ii) Coefficient of volume decrease, (iii) Coefficient of consolidation and (iv) Per cent consolidation.
(S.V.U.—B.Tech., (Part-time)—June, 1981)
(b) Describe a suitable method of determining the compression index of a soil.
(S.V.U.—B.Tech., (Part-time)—June, 1981, B.E., (R.R.)—Nov., 1973)

7.7 Explain what is meant by normally consolidated clay stratum and over-consolidated clay stratum. Sketch typical results of consolidation test data to a suitable plot relating the void ratio and consolidation pressure in each case and show how preconsolidation can be estimated.
(S.V.U.—B.E., (R.R.)—Sep., 1978)

7.8 (a) Distinguish between normally consolidated and over consolidated soils.
(b) Explain in detail any one method for determining the coefficient of consolidation of a soil.
(S.V.U.—B.E., (R.R.)—Feb., 1976, Nov., 1973)

7.9 Obtain the differential equation defining the one-dimensional consolidation as given by Terzaghi, listing the various assumptions.
(S.V.U.—B.E., (R.R.)—May, 1975, Nov., 1974, Nov., 1973, May, 1971, Dec., 1970, Nov., 1969)

7.10 Define and distinguish between coefficient of volume compressibility and coefficient of consolidation.
Describe clearly one method of computing coefficient of consolidation, given oedometer test data.
(S.V.U.—B.E., (R.R.)—Nov., 1973)

7.11 (a) Draw a typical time-consolidation curve for an increment of load and show the process of consolidation.
(b) Explain why is it necessary to double the load in a consolidation test.
(S.V.U.—B.E., (R.R.)—May, 1970, B.E., (N.R.)—May, 1969)

7.12 Explain with suitable analogy Terzaghi's theory of one-dimensional consolidation of soils.
(S.V.U.—B.E., (R.R.)—Dec., 1968)

7.13 (a) Derive the expression for the total settlement of a normally consolidated saturated clay subjected to an additional pressure.
(b) Indicate the nature of e-log $\bar{\sigma}$ curve that is got in a laboratory test on (i) undisturbed normally consolidated clay, (ii) Undisturbed overconsolidated clay.
(S.V.U.—B.E., (N.R.)—Sep., 1968)

7.14 Derive the equation $\Delta H = \int H_i \dfrac{C_c}{(1+e_0)} \log_{10} \dfrac{(p_0 + \Delta p)}{p_0}$ for consolidation of clays.
(S.V.U.—B.E., (R.R.)—May, 1969)

7.15 Explain "Secondary Consolidation".
(S.V.U.—B.E., (R.R.)—May, 1970, May, 1971, Four year B.Tech.—June, 1982)

7.16 Determine the amount of settlement given the following data:
Thickness of compressible medium = 3 m
Coefficient of volume decrease = 0.02 cm²/kg.
Pressure increment at the centre of the compressible medium = 0.75 kg/cm².

(S.V.U.—B.Tech. (Part-time)—June, 1981)

7.17 The subsurface consists of 6 m of sandy soil (γ = 1.84 g/cc) underlain by a deposit of clay (γ = 1.94 g/cc). The water table is at 4.2 m below the ground surface. Given the information (in the following table) from a consolidation test of an undisturbed clay sample obtained from a depth of 9.0 m from the ground surface, find C_c and preconsolidation pressure. Explain with reason whether this is a normally or overconsolidated clay.

Pressure, kg/cm²	0	0.25	0.5	1.0	2.0	4.0	8.0	16.0
Void ratio	0.960	0.950	0.942	0.932	0.901	0.870	0.710	0.540

(S.V.U.—B.E., (R.R.)—Nov., 1975)

7.18 The co-ordinates of two points on a straight-line section of a semi-logarithmic plot of compression diagram are: e_1 = 2.50, $\bar{\sigma}_1$ = 150 kN/m²; and e_2 = 1.75, $\bar{\sigma}_2$ = 600 kN/m². Calculate the compression index.

7.19 The void ratio of a clay is 1.56, and its compression index is found to be 0.8 at the pressure 180 kN/m². What will be the void ratio if the pressure is increased to 240 kN/m² ?

7.20 The compression diagram for a precompressed clay indicates that it had been compressed under a pressure of 240 kN/m². The compression index is 0.9; its void ratio under a pressure of 180 kN/m² is 1.5. Estimate the void ratio if the pressure is increased to 360 kN/m².

7.21 In a clay stratum below the water table, the pore pressure is 36 kN/m² at a depth of 3 m. Is the clay fully consolidated under the existing pressure? Explain.

7.22 A saturated clay specimen is subjected to a pressure of 240 kN/m². After the lapse of a time, it is determined that the pore pressure in the specimen is 72 kN/m². What is the degree of consolidation?

7.23 A compressible stratum is 6 m thick and its void ratio is 1.70. If the final void ratio after the construction of a building is expected to be 1.61, what will be the probable ultimate settlement of the building?

7.24 The total anticipated settlement due to consolidation of a clay layer under a certain pressure is 150 mm. If 45 mm of settlement has occurred in 9 months, what is the expected settlement in 18 months?

7.25 A stratum of a clay 5 m thick is sandwiched between highly permeable sand strata. A sample of this clay, 25 mm thick, experienced 50% of ultimate settlement in 12 minutes after the application of a certain pressure. How long will it take for a building proposed to be constructed at this site, and which is expected to increase the pressure to a value comparable to that applied in the laboratory test, to settle 50% of the ultimate value?

7.26 A 30 mm thick oedometer sample of clay reached 30% consolidation in 15 minutes with drainage at top and bottom. How long would it take the clay layer from which this sample was obtained to reach 60% consolidation? The clay layer had one-way drainage and was 6 m. thick.

7.27 A clay stratum is 4.5 m thick and rests on a rock surface. The coefficient of consolidation of a sample of this clay was found to be 4.5×10^{-8} m²/s in the laboratory. Determine probable period of time required for the clay stratum to undergo 50% of the ultimate settlement expected under a certain increment of pressure.

7.28 A saturated clay layer 4 m thick is located 6 m below GL. The void ratio of the clay is 1.0. When a raft foundation is located at 2 m below GL, the stresses at the top and bottom of the clay layer increased by 1.5 and 1.0 kg/cm² respectively. Estimate the consolidation settlement if the coefficient of compressibility is 0.02 cm²/kg.

(S.V.U.—B.Tech., (Part-time)—May, 1983)

7.29 In a consolidation test the following data was obtained:
Void ratio of the soil = 0.75
Specific gravity of the soil = 2.62
Compression Index = 0.1

Determine the settlement of a footing resting on the saturated soil with properties as given above. The thickness of the compressible soil is 3 m. The increase in pressure at the centre of the layer is 60 kN/m². The pre-consolidation pressure is 50 kN/m².

If the coefficient of consolidation is 2×10^{-7} m²/s, determine the time in days for 90% consolidation. Assume one-way drainage.

(S.V.U.—B.Tech., (Part-time)—Apr., 1982)

7.30 A clay layer 5 m thick has double drainage. It was consolidated under a load of 127.5 kN/m². The load is increased to 197.5 kN/m². The coefficient of volume compressibility is 5.79×10^{-4} m²/kN and value of $k = 1.6 \times 10^{-8}$ m/min. Find total settlement and settlement at 50% consolidation. If the test sample is 2 cm thick and attains 100% consolidation in 24 hours, what is the time taken for 100% consolidation in the actual layer?

(S.V.U.—Four year B.Tech.,—June, 1982)

7.31 The thickness of a saturated specimen of clay under a consolidation pressure of 1 kg/cm² is 24 mm and its water content is 20%. On increase of the consolidation pressure to 2 kg/cm², the specimen thickness decreases by 3 mm. Determine the compression index for the soil if the specific gravity of the soil grains is 2.70.

(S.V.U.—B.Tech., (Part-time)—Sept., 1983)

7.32 If a representative clay specimen 20 mm thick, under double drainage, took 121 minutes for 90% primary compression, estimate the time required for 50% primary compression of a field layer 2 m thick, bounded by impervious boundary at the bottom and sand at the top.

(S.V.U.—B.Tech., (Part-time)—Sept., 1983)

7.33 A bed of sand 12 m thick is underlain by a compressible stratum of normally loaded clay, 6 m thick. The water table is at a depth of 5 m below the ground level. The bulk densities of sand above and below the water table are 17.5 kN/m^3 and 20.5 kN/m^3 respectively. The clay has a natural water content of 40% and LL of 45%. G = 2.75. Estimate the probable final settlement if the average increment in pressure due to a footing is 100 kN/m^2.

(S.V.U.—B.E. (Part-time)—Apr., 1982)

8
SHEARING STRENGTH OF SOILS

8.1 INTRODUCTION

'Shearing Strength' of a soil is perhaps the most important of its engineering properties. This is because all stability analyses in the field of geotechnical engineering, whether they relate to foundations, slopes of cuts or earth dams, involve a basic knowledge of this engineering property of the soil. 'Shearing strength' or merely 'Shear strength', may be defined as the resistance to shearing stresses and a consequent tendency for shear deformation.

Shearing strength of a soil is the most difficult to comprehend in view of the multitude of factors known to affect it. A lot of maturity and skill may be required on the part of the engineer in interpreting the results of the laboratory tests for application to the conditions in the field.

Basically speaking, a soil derives its shearing strength from the following:
(1) Resistance due to the interlocking of particles.
(2) Frictional resistance between the individual soil grains, which may be sliding friction, rolling friction, or both.
(3) Adhesion between soil particles or 'cohesion'.

Granular soils or sands may derive their shear strength from the first two sources, while cohesive soils or clays may derive their shear strength from the second and third sources. Highly plastic clays, however, may exhibit the third source alone for their shearing strength. Most natural soil deposits are partly cohesive and partly granular and as such, may fall into the second of the three categories just mentioned, from the point of view of shearing strength.

The shear strength of a soil cannot be tabulated in codes of practice since a soil can significantly exhibit different shear strengths under different field and engineering conditions.

8.2 FRICTION

'Friction' is the primary source of shearing strength in most natural soils. Hence, a few important aspects of the concept of frictional resistance need to be considered.

8.2.1 Friction between Solid Bodies
When two solid bodies are in contact with each other, the frictional resistance available is dependent upon the normal force between the two and an intrinsic

property known as the 'Coefficient of friction'. The coefficient of friction, in turn, depends upon the nature and the condition of the surfaces in contact. This is so even when a solid body rests on a rigid surface, as shown in Fig. 8.1.

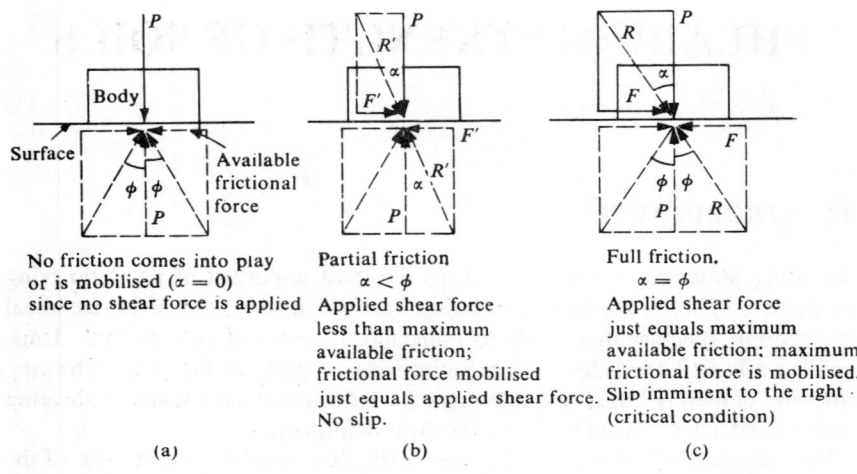

Fig. 8.1 Friction between a solid and a rigid surface

The available frictional resistance F when a normal force P is acting is related to P as follows:

$$F = P \cdot \mu = P \cdot \tan \phi \qquad \text{...(Eq. 8.1)}$$

Here μ is called the 'Coefficient of friction' and ϕ is known as the 'Angle of friction'.

The characteristics μ and ϕ are properties of the materials in contact and they are independent of the applied forces and are fairly constant. The available frictional resistance F does not come into play or get mobilised unless it is required to resist an applied shearing force.

In Fig. 8.1(a), no frictional resistance is mobilised because there is no applied shearing force. The normal force exerted by the solid body on the rigid surface is resisted by an equal force by way of reaction from the rigid surface.

In Fig. 8.1(b), a small magnitude of shearing force is applied. This causes a resultant force R' to be acting at an angle α with respect to the normal to the rigid surface. This angle is called the 'Angle of Obliquity' and is dependent only upon the applied forces. To resist this applied shearing force R', an equal magnitude of the available frictional resistance is mobilised. Since F' is less than the maximum available frictional resistance F, the angle of obliquity α is less than ϕ. In this case there is equilibrium and there is no slip between the body and the surface.

In Fig. 8.1(c), a shearing force equal to the maximum available frictional resistance is applied. The entire frictional resistance available will get mobilised now to resist the applied force. Angles α and ϕ are equal; slip or sliding to the right is incipient or imminent. If the applied shearing force is reversed in direction, the direction of imminent slip also gets reversed. The frictional force, as is easily understood, tends to oppose motion.

The friction angle is the limiting value of obliquity; the criterion of slip is therefore an angle of obliquity equal to the friction angle. The condition of incipient slip for solid bodies in contact may be expressed as follows:

$$F/P = \tan \phi = \mu \qquad \ldots(\text{Eq. 8.2})$$

For solid bodies which are in contact but which have no adhesion between them, the term 'friction' is synonymous with the terms 'shearing strength' and 'maximum shearing resistance'. In most natural soils friction represents only a part of the shearing strength, although an important part, but other phenomena contribute to the shearing strength, particularly in fine-grained soils.

8.2.2 Internal Friction within Granular Soil Masses

In granular or cohesionless soil masses, the resistance to sliding on any plane through the point within the mass is similar to that discussed in the previous sub-section; the friction angle in this case is called the 'angle of internal friction'. However, the frictional resistance in granular soil masses is rather more complex than that between solid bodies, since the nature of the resistance is partly sliding friction and partly rolling friction. Further, a phenomenon known as 'interlocking' is also supposed to contribute to the shearing resistance of such soil masses, as part of the frictional resistance.

The angle of internal friction which is a limiting angle of obliquity and hence the primary criterion for slip or failure to occur on a certain plane, varies appreciably for a given sand with the density index, since the degree of interlocking is known to be directly dependent upon the density. This angle also varies somewhat with the normal stress. However, the angle of internal friction is mostly considered constant, since it is almost so for a given sand at a given density.

Since failure or slip within a soil mass cannot be restricted to any specific plane, it is necessary to understand the relationships that exist between the stresses on different planes passing through a point, as a prerequisite for further consideration of shearing strength of soils.

8.3 PRINCIPAL PLANES AND PRINCIPAL STRESSES—MOHR'S CIRCLE

At a point in a stressed material, every plane will be subjected, in general, to a normal or direct stress and a shearing stress. In the field of geotechnical engineering, compressive direct stresses are usually considered positive, while tensile stresses are considered negative.

A 'Principal plane' is defined as a plane on which the stress is wholly normal, or one which does not carry shearing stress. From mechanics, it is known that there exists three principal planes at any point in a stressed material. The normal stresses acting on these principal planes are known as the 'principal stresses'. The

298 Geotechnical Engineering

three principal planes are to be mutually perpendicular. In the order of decreasing magnitude the principal stresses are designated the 'major principal stress', the 'intermediate principal stress' and the 'minor principal stress', the corresponding principal planes being designated exactly in the same manner. It can be shown that satisfactory solutions may be obtained for many problems in the field of geotechnical engineering by two-dimensional analysis, the intermediate principal stress being commonly ignored.

Let us consider an element of soil whose sides are chosen as the principal planes, the major and the minor, as shown in Fig. 8.2(a):

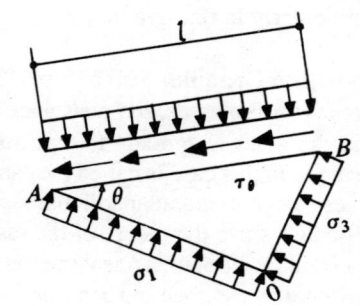

(a) Stress system

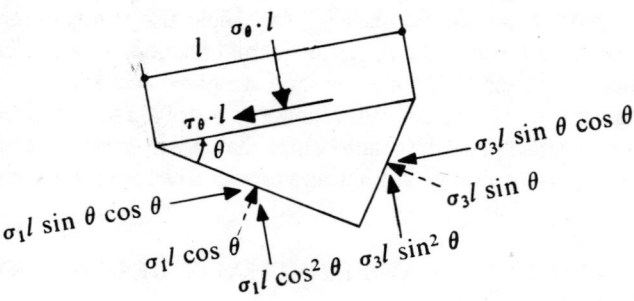

(b) Force system

Fig. 8.2 Stresses on a plane inclined to the principal planes

Let O be any point in the stressed medium and OA and OB be the major and minor principal planes, with the corresponding principal stresses σ_1 and σ_3. The plane of the figure is the intermediate principal plane. Let it be required to determine the stress conditions on a plane normal to the figure, and inclined at an angle θ to the major principal plane, considered positive when measured counter-clockwise.

If the stress conditions are uniform, the size of the element is immaterial. If the stresses are varying, the element must be infinitesimal in size, so that the variation of stress along a side need not be considered.

Let us consider the element to be of unit thickness perpendicular to the plane of the figure, AB being l. The forces on the sides of the element are shown dotted and their components parallel and perpendicular to AB are shown by full lines. Considering the equilibrium of the element and resolving all forces in the directions parallel and perpendicular to AB, the following equations may be obtained:

$$\sigma_\theta = \sigma_1 \cos^2 \theta + \sigma_3 \sin^2 \theta = \sigma_3 + (\sigma_1 - \sigma_3) \cos^2 \theta$$

$$= \frac{(\sigma_1 + \sigma_3)}{2} + \frac{(\sigma_1 - \sigma_3)}{2} \cdot \cos 2\theta \qquad \ldots \text{(Eq. 8.3)}$$

$$\tau_\theta = \frac{(\sigma_1 - \sigma_3)}{2} \sin 2\theta \qquad \ldots \text{(Eq. 8.4)}$$

Thus it may be noted that the normal and shearing stresses on any plane which is normal to the intermediate principal plane may be expressed in terms of σ_1, σ_3, and θ.

Otto Mohr (1882) represented these results graphically in a circle diagram, which is called Mohr's circle. Normal stresses are represented as abscissae and shear stresses as ordinates. If the coordinates σ_θ and τ_θ represented by Eqs. 8.3 and 8.4

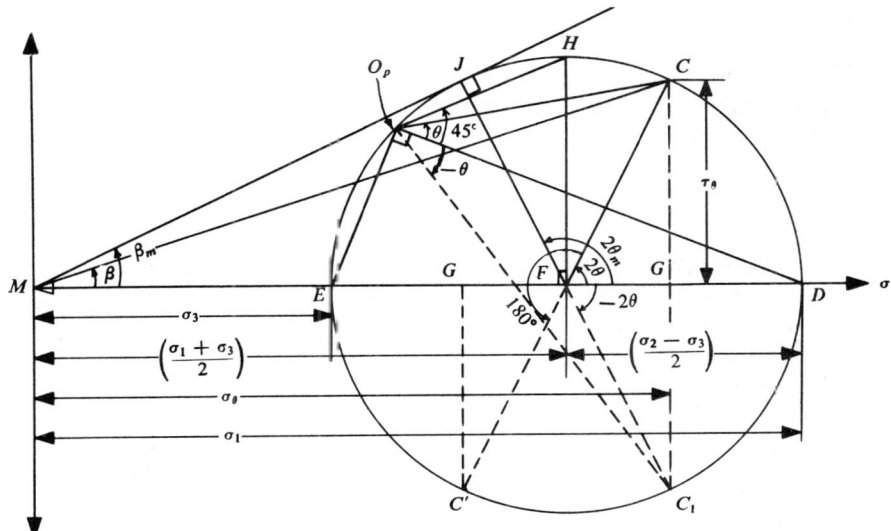

Fig. 8.3 Mohr's circle for the stress conditions illustrated in Fig. 8.2

are plotted for all possible values of θ, the locus is a circle as shown in Fig. 8.3. This circle has its centre on the axis and cuts it at values σ_3 and σ_1. This circle is known as the Mohr's circle.

The Mohr's circle diagram provides excellent means of visualisation of the orientation of different planes. Let a line be drawn parallel to the major principal plane through D, the coordinate of which is the major principal stress. The intersection of this line with the Mohr's circle, O_p, is called the 'Origin of planes'. If a line parallel to the minor principal plane is drawn through E, the co-ordinate of which is the minor principal stress, it will also be observed to pass through O_p; the angle between these two lines is a right angle from the properties of the circle. Likewise, it can be shown that any line through O_p, parallel to any arbitrarily chosen plane, intersects the Mohr's circle at a point the co-ordinates of which represent the normal and shear stresses on that plane. Thus the stresses on the plane represented by AB in Fig. 8.2(a), may be obtained by drawing O_pC parallel to AB, that is, at an angle θ with respect to O_pD, the major principal plane, and measuring off the co-ordinates of C, namely σ_θ and τ_θ.

Since angle $CO_pD = \theta$, angle $CFD = 2\theta$, from the properties of the circle. From the geometry of the figure, the co-ordinates of the point C, are established as follows:

$$\sigma_\theta = MG = MF + FG$$

$$= \frac{(\sigma_1 + \sigma_3)}{2} + \frac{(\sigma_1 - \sigma_3)}{2} \cdot \cos 2\theta$$

$$\tau_\theta = CG = \frac{(\sigma_1 - \sigma_3)}{2} \cdot \sin 2\theta$$

These are the same as in Eqs. 8.3 and 8.4, which prove our statement.

In the special case where the major and minor principal planes are vertical and horizontal respectively, or vice-versa, the origin of planes will be D or E, as the case may be. In other words, it will lie on the σ-axis.

A few important basic facts and relationships may be directly obtained from the Mohr's circle:

1. The only planes free from shear are the given sides of the element which are the principal planes. The stresses on these are the greatest and smallest normal stresses.
2. The maximum or principal shearing stress is equal to the radius of the Mohr's circle, and it occurs on planes inclined at 45° to the principal planes.

$$\tau_{max} = (\sigma_1 - \sigma_3)/2 \qquad \text{...(Eq. 8.5)}$$

3. The normal stresses on planes of maximum shear are equal to each other and is equal to half the sum of the principal stresses.

$$\sigma_C = (\sigma_1 + \sigma_3)/2 \qquad \text{...(Eq. 8.6)}$$

Shearing Strength of Soils 301

4. Shearing stresses on planes at right angles to each other are numerically equal and are of an opposite sign. These are called conjugate shearing stresses.
5. The sum of the normal stresses on mutually perpendicular planes is a constant $(MG' + MG = 2MF = \sigma_1 + \sigma_3)$. If we designate the normal stress on a plane perpendicular to the plane on which it is σ_θ as $\sigma_{\theta'}$:

$$\sigma_\theta + \sigma_{\theta'} = \sigma_1 + \sigma_3 \qquad \text{...(Eq. 8.7)}$$

Of the two stresses σ_θ and $\sigma_{\theta'}$, the one which makes the smaller angle with σ_1 is the greater of the two.

6. The resultant stress, σ_r, on any plane is $\sqrt{\sigma_\theta^2 + \tau_\theta^2}$ and has an obliquity, β, which is equal to $\tan^{-1}(\tau_\theta/\sigma_\theta)$.

$$\sigma_r = \sqrt{\sigma_\theta^2 + \tau_\theta^2} \qquad \text{...(Eq. 8.8)}$$

$$\beta = \tan^{-1}(\tau_\theta/\sigma_\theta) \qquad \text{...(Eq. 8.9)}$$

7. Stresses on conjugate planes, that is, planes which are equally inclined in different directions with respect to a principal plane are equal. (This is indicated by the co-ordinates of C and C_1 in Fig. 8.3).
8. When the principal stresses are equal to each other, the radius of the Mohr's circle becomes zero, which means that shear stresses vanish on all planes. Such a point is called an isotropic point.
9. The maximum angle of obliquity, β_m, occurs on a plane inclined at $\theta_{cr}\left(=45° + \dfrac{\beta_m}{2}\right)$ with respect to the major principal plane.

$$\theta_{cr} = 45° + \frac{\beta_m}{2} \qquad \text{...(Eq. 8.10)}$$

This may be obtained by drawing a line which passes through the origin and is tangential to the Mohr's circle. The co-ordinates of the point of tangency are the stresses on the plane of maximum obliquity; the shear stress on this plane is obviously less than the principal or maximum shear stress.

On the plane of principal shear the obliquity is slightly smaller than β_m. It is the plane of maximum obliquity which is most liable to failure and not the plane of maximum shear, since the criterion of slip is limiting obliquity. When β_m approaches and equals the angle of internal friction, ϕ, of the soil, failure will become incipient.

Mohr's circle affords an easy means of obtaining all important relationships. The following are a few such relationships:

$$\sin \beta_m = \left(\frac{\sigma_1 - \sigma_3}{\sigma_1 + \sigma_3}\right) \qquad \text{...(Eq. 8.11)}$$

$$\sigma_1/\sigma_3 = \left(\frac{1 + \sin \beta_m}{1 - \sin \beta_m}\right) \qquad \text{...(Eq. 8.12)}$$

302 Geotechnical Engineering

σ for the plane of maximum obliquity,

$$\sigma_{cr} = \sigma_3(1 + \sin \beta_m) \qquad \ldots(\text{Eq. 8.13})$$

In case the normal and shearing stresses on two mutually perpendicular planes are known, the principal planes and principal stresses may be determined with the aid of the Mohr's circle diagram, as shown in Fig. 8.4. The shearing stresses on two mutually perpendicular planes are equal in magnitude by the principle of complementary shear.

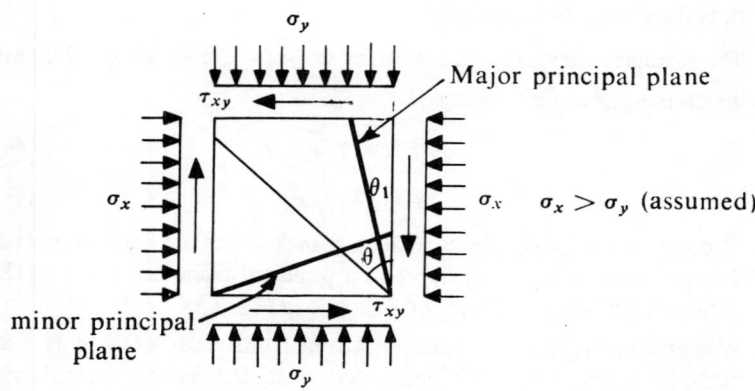

(a) General two-dimensional stress system

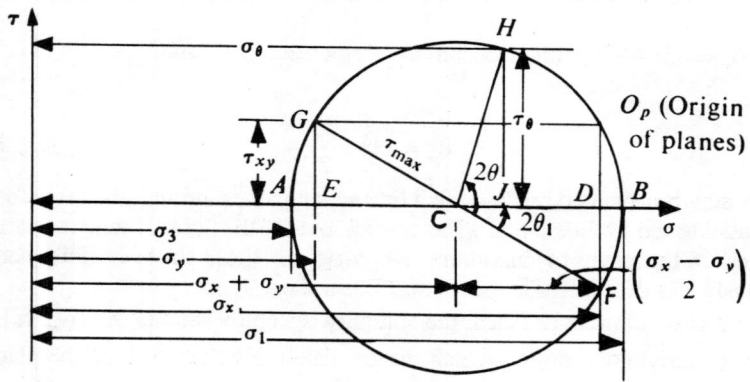

Fig. 8.4 Determination of principal planes and principal stresses from Mohr's circle

Figure 8.4(a) shows an element subjected to a general two-dimensional stress system, normal stresses σ_x and σ_y on mutually perpendicular planes and shear stresses τ_{xy} on these planes, as indicated. Figure 8.4(b) shows the corresponding Mohr's circle, the construction of which is obvious.

From a consideration of the equilibrium of a portion of the element, the normal and shearing stress components, σ_θ and τ_θ, respectively, on a plane inclined at an angle θ, measured counter-clockwise with respect to the plane on which σ_x acts, may be obtained as follows:

$$\sigma_\theta = \frac{(\sigma_x + \sigma_y)}{2} + \frac{(\sigma_x - \sigma_y)}{2} \cdot \cos 2\theta + \tau_{xy} \sin 2\theta \qquad \text{...(Eq. 8.14)}$$

$$\tau_\theta = \frac{(\sigma_x - \sigma_y)}{2} \cdot \sin 2\theta - \tau_{xy} \cdot \cos 2\theta \qquad \text{...(Eq. 8.15)}$$

Squaring Eqs. 8.14 and 8.15 and adding,

$$\left[\sigma_\theta - \frac{(\sigma_x + \sigma_y)}{2}\right]^2 + \tau_\theta^2 = \left(\frac{\sigma_x - \sigma_y}{2}\right)^2 + \tau_{xy}^2 \qquad \text{...(Eq. 8.16)}$$

This represents a circle with centre $\left[\frac{(\sigma_x + \sigma_y)}{2}, 0\right]$, and radius

$$\sqrt{\left(\frac{\sigma_x - \sigma_y}{2}\right)^2 + \tau_{xy}^2}.$$

Once the Mohr's circle is constructed, the principal stresses σ_1 and σ_3, and the orientation of the principal planes may be obtained from the diagram.

The shearing stress is to be plotted upward or downward according as it is positive or negative. It is common to take a shear stress which tends to rotate the element counter-clockwise, positive.

It may be noted that the same Mohr's circle and hence the same principal stresses are obtained, irrespective of how the shear stresses are plotted. (The centre of the Mohr's circle, C, is the mid-point of DE, with the co-ordinates $\left(\frac{\sigma_x + \sigma_y}{2}\right)$ and 0;

the radius of the circle is CG), the co-ordinates of G being σ_y and τ_{xy}.

The following relationships are also easily obtained:

$$\sigma_1 = \left(\frac{\sigma_x + \sigma_y}{2}\right) + \frac{1}{2}\sqrt{(\sigma_x - \sigma_y)^2 + 4\tau_{xy}^2} \qquad \text{...(Eq. 8.17)}$$

$$\sigma_3 = \left(\frac{\sigma_x + \sigma_y}{2}\right) - \frac{1}{2}\sqrt{(\sigma_x - \sigma_y)^2 + 4\tau_{xy}^2} \qquad \text{...(Eq. 8.18)}$$

$$\tan 2\theta_{1,3} = 2\tau_{xy}/(\sigma_x - \sigma_y) \qquad \text{...(Eq. 8.19)}$$

$$\tau_{max} = \frac{1}{2}\sqrt{(\sigma_x - \sigma_y)^2 + 4\tau_{xy}^2} \qquad \text{...(Eq. 8.20)}$$

Invariably, the vertical stress will be the major principal stress and the horizontal one the minor principal stress in geotechnical engineering situations.

8.4 STRENGTH THEORIES FOR SOILS

A number of theories have been propounded for explaining the shearing strength of soils. Of all such theories, the Mohr's strength theory and the Mohr-Coulomb theory, a generalisation and modification of the Coulomb's equation, meet the requirements for application to a soil in an admirable manner.

8.4.1 Mohr's Strength Theory

We have seen that the shearing stress may be expressed as $\tau = \sigma \tan \beta$ on any plane, where β is the angle of obliquity. If the obliquity angle is the maximum or has limiting value ϕ, the shearing stress is also at its limiting value and it is called the shearing strength, s. For a cohesionless soil the shearing strength may be expressed as:

$$s = \sigma \tan \phi \qquad \ldots \text{(Eq. 8.21)}$$

If the angle of internal friction ϕ is assumed to be a constant, the shearing strength may be represented by a pair of straight lines at inclinations of $+\phi$ and $-\phi$ with the σ-axis and passing through the origin of the Mohr's circle diagram. A line of this type is called a Mohr envelope. The Mohr envelopes for a cohesionless soil, as shown in Fig. 8.5, are the straight lines OA and OA'.

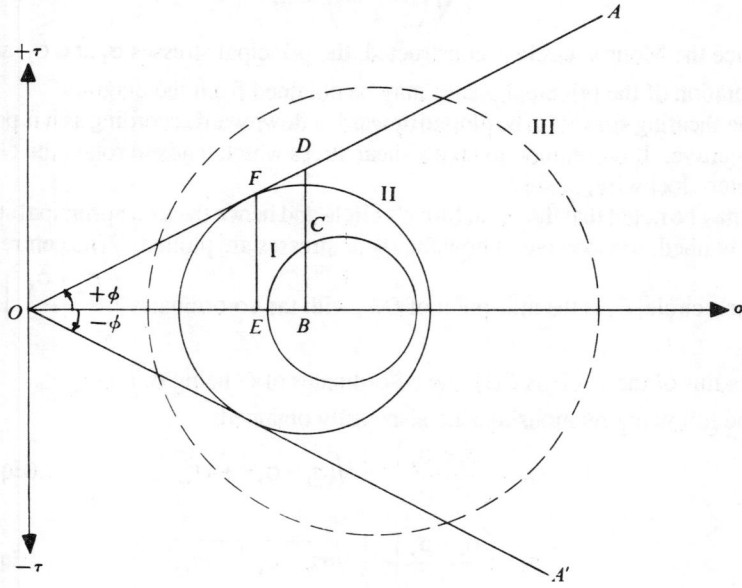

Fig. 8.5 Mohr's strength theory—Mohr envelopes for cohesionless soil

If the stress conditions at a point are represented by Mohr's circle I, the shear stress on any plane through the point is less than the shearing strength, as indicated by the line BCD; BC represents the shear stress on a plane on which the normal stress is given by OD. BD, representing the shearing strength for this normal stress, is greater than BC.

The stress conditions represented by the Mohr's Circle II, which is tangential to the Mohr's envelope at F, are such that the shearing stress, EF, on the plane of maximum obliquity is equal to the shearing strength. Failure is incipient on this plane and will occur unless the normal stress on the critical plane increases.

It may be noted that it would be impossible to apply the stress conditions represented by Mohr's circle III (dashed) to this soil sample, since failure would have occurred even by the time the shear stress on the critical plane equals the shearing strength available on that plane, thus eliminating the possibility of the shear stress exceeding the shearing strength.

The Mohr's strength theory, or theory of failure or rupture, may thus be stated as follows: The stress condition given by any Mohr's circle falling within the Mohr's envelope represents a condition of stability, while the condition given by any Mohr's circle tangent to the Mohr's envelope indicates incipient failure on the plane relating to the point of tangency. The Mohr's envelope may be treated to be a property of the material and independent of the imposed stresses. Also, the Mohr's circle of stress depends only upon the imposed stresses and has nothing to do with the nature and properties of the material.

To emphasise that the stresses in Eq. 8.21 are those on the plane on which failure is incipient, we add the subscript f to σ:

$$s = \sigma_f \tan \phi \qquad \ldots(\text{Eq. 8.22})$$

It is possible to express the strength in terms of normal stress on any plane, with the aid of the Mohr's circle of stress. Some common relationships are:

$$\sigma_f = \sigma_3(1 + \sin \phi) = \sigma_1(1 - \sin \phi)$$

$$= \left(\frac{\sigma_1 - \sigma_3}{2}\right) \cdot \frac{\cos^2 \phi}{\sin \phi} \qquad \ldots(\text{Eq. 8.23})$$

$$s = \sigma_f \tan \phi = \sigma_3 \tan \phi \, (1 + \sin \phi)$$

$$= \sigma_1 \tan \phi \, (1 - \sin \phi) = \left(\frac{\sigma_1 - \sigma_3}{2}\right) \cdot \cos \phi \qquad \ldots(\text{Eq. 8.24})$$

The primary assumptions in the Mohr's strength theory are that the intermediate principal stress has no influence on the strength and that the strength is dependent only upon the normal stress on the plane of maximum obliquity. However, the shearing strength, in fact, does depend to a small extent upon the intermediate principal stress, density, speed of application of shear, and so on. But the Mohr theory explains satisfactorily the strength concept in soils and hence is in vogue.

It may also be noted that the Mohr envelope will not be a straight line but is actually slightly curved since the angle of internal friction is known to decrease slightly with increase in stress.

8.4.2 Mohr-Coulomb Theory

The Mohr-Coulomb theory of shearing strength of a soil, first propounded by Coulomb (1976) and later generalised by Mohr, is the most commonly used concept. The functional relationship between the normal stress on any plane and the shearing strength available on that plane was assumed to be linear by Coulomb; thus the following is usually known as Coulomb's law:

306 Geotechnical Engineering

$$s = c + \sigma \tan \phi \qquad \ldots(\text{Eq. 8.25})$$

where c and ϕ are empirical parameters, known as the 'apparent cohesion' and 'angle of shearing resistance' (or angle of internal friction), respectively. These are better visualised as 'parameters' and not as absolute properties of a soil since they are known to vary with water content, conditions of testing such as speed of shear and drainage conditions, and a number of other factors besides the type of soil.

Coloumb's law is merely a mathematical equation of the failure envelope shown in Fig. 8.6(a); Mohr's generalisation of the failure envelope as a curve which becomes flatter with increasing normal stress is shown in Fig. 8.6(b).

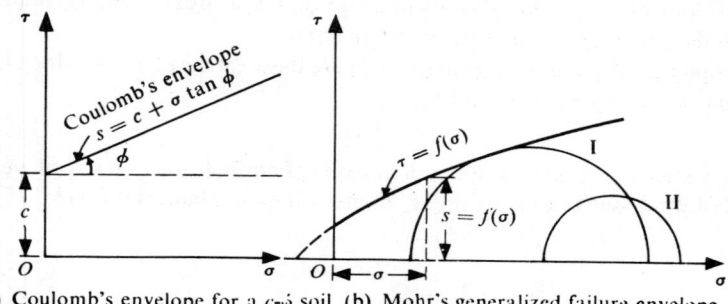

(a) Coulomb's envelope for a c-ϕ soil (b) Mohr's generalized failure envelope

Fig. 8.6 Mohr-Coulomb Theory—failure envelopes

The envelopes are called 'strength envelopes' or 'failure envelopes'. The meaning of an envelope has already been given in the previous section; if the normal and shear stress components on a plane plot on to the failure envelope, failure is supposed to be incipient and if the stresses plot below the envelope, the condition represents stability. And, it is impossible that these plot above the envelope, since failure should have occurred previously.

Coulomb's law is also written as follows to indicate that the stress condition refers to that on the plane of failure:

$$s = c + \sigma_f \tan \phi \qquad \ldots(\text{Eq. 8.26})$$

In a different way, it can be said that the Mohr's circle of stress relating to a given stress condition would represent, incipient failure condition if it just touches or is tangent to the strength or failure envelope (circle I); otherwise, it would wholly lie below the envelopes as shown in circle II, Fig. 8.6(b).

The Coulomb envelope in special cases may take the shapes given in Fig. 8.7(a) and (b); for a purely cohesionless or granular soil or a pure sand, it would be as shown in Fig. 8.7(a) and for a purely cohesive soil or a pure clay, it would be as shown in Fig. 8.7(b).

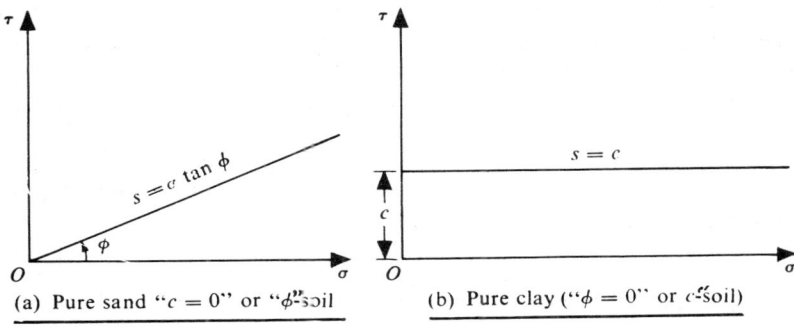

Fig. 8.7 Coulomb envelopes for pure sand and for pure clay

8.5 SHEARING STRENGTH—A FUNCTION OF EFFECTIVE STRESS

Equation 8.26 apparently indicates that the shearing strength of a soil is governed by the *total* normal stress on the failure plane. However, according to Terzaghi, it is the effective stress on the failure plane that governs the shearing strength and not the total stress.

It may be expected intuitively that the denser a soil, the greater the shearing strength. It has been learnt in chapter seven that a soil deposit becomes densest under any given pressure after the occurrence of complete consolidation and consequent dissipation of pore water pressure. Thus, complete consolidation, dependent upon the dissipation of pore water pressure and hence upon the increase in the effective stress, leads to increase in the shearing strength of a soil. In other words, it is the effective stress in the case of a saturated soil and not the total stress which is relevant to the mobilisation of shearing stress.

Further, the density of a soil increases when subjected to shearing action, drainage being allowed simultaneously. Therefore, even if two soils are equally dense on having been consolidated to the same effective stress, they will exhibit different shearing strengths if drainage is permitted during shear for one, while it is not for the other.

These ideas lead to a statement that "the strength of a soil is a unique function of the effective stress acting on the failure plane".

Equation 8.26 may now be modified to read:

$$s = c' + \bar{\sigma}_f \tan \phi' \qquad \text{...(Eq. 8.27)}$$

where c' and ϕ' are called the effective cohesion and effective angle of internal friction, respectively, since they are based on the effective normal stress on the failure plane. Collectively, they are called 'effective stress parameters', while c and ϕ of Eq. 8.26 are called "total stress parameters".

More about this differentiation and other related concepts will be seen in later sections.

8.6 HVORSLEV'S TRUE SHEAR PARAMETERS

Hvorslev (1960), based on his experimental work on remoulded cohesive soils, proposed that the shearing strength, s, can be represented by the following general equation, irrespective of the stress history of the soil:

$$s = f(e_f, \bar{\sigma}_f) \qquad \text{...(Eq. 8.28)}$$

where $f(e_f, \bar{\sigma}_f) =$ a function of the void ratio, e_f, at failure, and the effective normal stress on the failure plane, at failure.

This may be written more explicitly as follows:

$$s = c_e + \bar{\sigma}_f \tan \phi_e \qquad \text{...(Eq. 8.29)}$$

where $c_e =$ 'true cohesion' or effective cohesion, and $\phi_e =$ 'true angle of internal friction' or effective friction angle.

The true angle of internal friction is found to be practically constant. However, the true cohesion is found to be dependent upon the water content or void ratio of the soil at failure. In fact, it is found to be directly proportional to the 'equivalent' consolidation pressure or the pressure from the virgin impression curve corresponding to the void ratio at failure.

That is to say:

$$c_e = K \cdot \bar{\sigma}_e \qquad \text{...(Eq. 8.30)}$$

where $c_e =$ true cohesion,

$\bar{\sigma}_e =$ equivalent consolidation pressure,

and $K =$ constant of proportionality, called "cohesion factor" or "coefficient of cohesion".

In view of this, Eq. 8.29 may be rewritten as:

$$s = K \cdot \bar{\sigma}_e + \bar{\sigma}_f \tan \phi_e \qquad \text{...(Eq. 8.31)}$$

Unlike the Coulomb parameters, c and ϕ, the parameters K and ϕ_e are constant for a soil, irrespective of its stress history and other conditions. Thus, these parameters are known as Hvorslev's true shear parameters.

Ordinarily, the Coulomb parameters are sufficient for practical application provided, the field conditions such as stress history are properly simulated during the laboratory evaluation of these parameters; however, evaluation of Hvorslev's true shear parameters is an essential feature of fundamental research in the field of shearing strength of remoulded clays.

8.7 TYPES OF SHEAR TESTS BASED ON DRAINAGE CONDITIONS

Before considering various methods of conducting shearing strength tests on a soil, it is necessary to consider the possible drainage conditions before and during the tests since the results are significantly affected by these.

A cohesionless or a coarse-grained soil may be tested for shearing strength either in the dry condition or in the saturated condition. A cohesive or fine-grained soil is usually tested in the saturated condition. Depending upon whether drainage is permitted before and during the test, shear tests on such saturated soils are classified as follows:

Unconsolidated undrained test

Drainage is not permitted at any stage of the test, that is, either before the test during the application of the normal stress or during the test when the shear stress is applied. Hence no time is allowed for dissipation of pore water pressure and consequent consolidation of the soil; also, no significant volume changes are expected. Usually, 5 to 10 minutes may be adequate for the whole test, because of the shortness of drainage path. However, undrained tests are often performed only on soils of low permeability.

This is the most unfavourable condition which might occur in geotechnical engineering practice and hence is simulated in shear testing. Since a relatively small time is allowed for the testing till failure, it is also called the 'Quick test'. It is designated UU, Q, or Q_u test.

Consolidated undrained test

Drainage is permitted fully in this type of test during the application of the normal stress and no drainage is permitted during the application of the shear stress. Thus volume changes do not take place during shear and excess pore pressure develops. Usually, after the soil is consolidated under the applied normal stress to the desired degree, 5 to 10 minutes may be adequate for the test.

This test is also called 'consolidated quick test' and is designated CU or Q_c test. These conditions are also common in geotechnical engineering practice.

Drained test

Drainage is permitted fully before and during the test, at every stage. The soil is consolidated under the applied normal stress and is tested for shear by applying the shear stress also very slowly while drainage is permitted at every stage. Practically no excess pore pressure develops at any stage and volume changes take place. It may require 4 to 6 weeks to complete a single test of this kind in the case of cohesive soils, although not so much time is required in the case of cohesionless soils as the latter drain off quickly.

This test is seldom conducted on cohesive soils except for purposes of research. It is also called the 'Slow Test' or 'consolidated slow test' and is designated CD, S, or S_c test.

The shear parameters c and ϕ vary with the type of test or drainage conditions. The suffixes u, cu, and d are used for the parameters obtained from the UU-, CU- and CD-tests respectively.

The choice as to which of these tests is to be used depends upon the types of soil and the problem on hand. For problems of short-term stability of foundations, excavations and earth dams UU-tests are appropriate. For problems of long-term stability, either CU-test or CD-tests are appropriate, depending upon the drainage conditions in the field. For a more detailed exposition of these and other related aspects, the reader is referred to *"The relevance of triaxial test to the solution of stability problems"* by A.W. Bishop and L. Bjerrum—Proc. ASCE Res. Conf. Shear strength of Cohesive soils, Colorado, USA, 1960.

A fuller discussion of the nature of results obtained from these various types of tests and the choice of test conditions with a view to simulating field conditions, is postponed to a later section.

8.8 SHEARING STRENGTH TESTS

Determination of shearing strength of a soil involves the plotting of failure envelopes and evaluation of the shear strength parameters for the necessary conditions. The following tests are available for this purpose:

Laboratory tests
1. Direct Shear Test
2. Triaxial Compression Test
3. Unconfined Compression Test
4. Laboratory Vane Shear Test
5. Torsion Test
6. Ring Shear Tests

Field tests
1. Vane Shear Test
2. Penetration Test

The first three tests among the laboratory tests are very commonly used, while the fourth is gaining popularity owing to its simplicity. The fifth and sixth are mostly used for research purposes and hence are not dealt with here.

The principle of the field vane test is the same as that of the laboratory vane shear test, except that the apparatus is bigger in size for convenience of field use. The penetration test involves the measurement of resistance of a soil to penetration of a cone or a cylinder, as an indication of the shearing strength. This procedure is indirect and rather empirical in nature although correlations are possible. The field tests are also not considered here. The details of the test procedures are available in the relevant I.S. codes or any book on laboratory testing, such as Lambe (1951).

8.8.1 Direct Shear Test

The direct shear device, also called the 'shear box apparatus', essentially consists of a brass box, split horizontally at mid-height of the soil specimen, as shown schematically in Fig. 8.8. The soil is gripped in perforated metal grilles, behind which porous discs can be placed if required to allow the specimen to drain. For undrained tests, metal plates and solid metal grilles may be used. The usual plan size of the specimen is 60 mm square; but a larger size such as 300 mm square or even more, is employed for testing larger size granular material such as gravel. The minimum thickness or height of the specimen is 20 mm.

After the sample to be tested is placed in the apparatus or shear box, a normal load which is vertical is applied to the top of the sample by means of a loading yoke and weights. Since the shear plane is predetermined as the horizontal plane, this becomes the normal stress on the failure plane, which is kept constant throughout the test. A shearing force is applied to the upper-half of the box, which is zero initially and is increased until the specimen fails.

Two types of application of shear are possible – one in which the shear stress is controlled and the other in which the shear strain is controlled. The principles of these two types of devices are illustrated schematically in Fig. 8.8(b) and (c), respectively. In the stress-controlled type, the shear stress, which is the controlled variable, may be applied at a constant rate or more commonly in equal increments by means of calibrated weights hung from a hanger attached to a wire passing over a pulley. Each increment of shearing force is applied and held constant, until the shearing deformation ceases. The shear displacement is measured with the aid of a dial guage attached to the side of the box. In the strain-controlled type, the shear displacement is applied at a constant rate by means of a screw operated manually or by motor. With this type of test the shearing force necessary to overcome the resistance within the soil is automatically developed. This shearing force is measured with the aid of a proving ring—a steel ring that has been carefully machined, balanced and calibrated. The deflection of the annular ring is measured with the aid of a dial gauge set inside the ring, the causative force being got for any displacement by means of the calibration chart supplied by the manufacturer. The shear displacement is measured again with the aid of another dial gauge attached to the side of the box.

In both cases, a dial gauge attached to the plunger, through which the normal load is applied, will enable one to determine the changes in the thickness of the soil sample which will help in the computation of volume changes of the sample, if any. The strain-controlled type is very widely used. The strain is taken as the ratio of the shear displacement to the thickness of the sample. The proving ring readings may be taken at fixed displacements or even at fixed intervals of time as the rate of strain is made constant by an electric motor. A sudden drop in the proving ring reading or a levelling-off in successive readings, indicates shear failure of the soil specimen.

The shear strain may be plotted against the shear stress; it may be plotted versus the ratio of the shearing stress to normal stress; and it may also be plotted versus volume change. Each plot may yield information useful in one way or the other. The stresses may be obtained from the forces by dividing them by the area of cross-section of the sample.

The stress-conditions on the failure plane and the corresponding Mohr's circle for direct shear test are shown in Fig. 8.9(a) and (b) respectively.

The failure plane is predetermined as the horizontal plane here. Several specimens are tested under different normal loads and the results plotted to obtain failure envelopes.

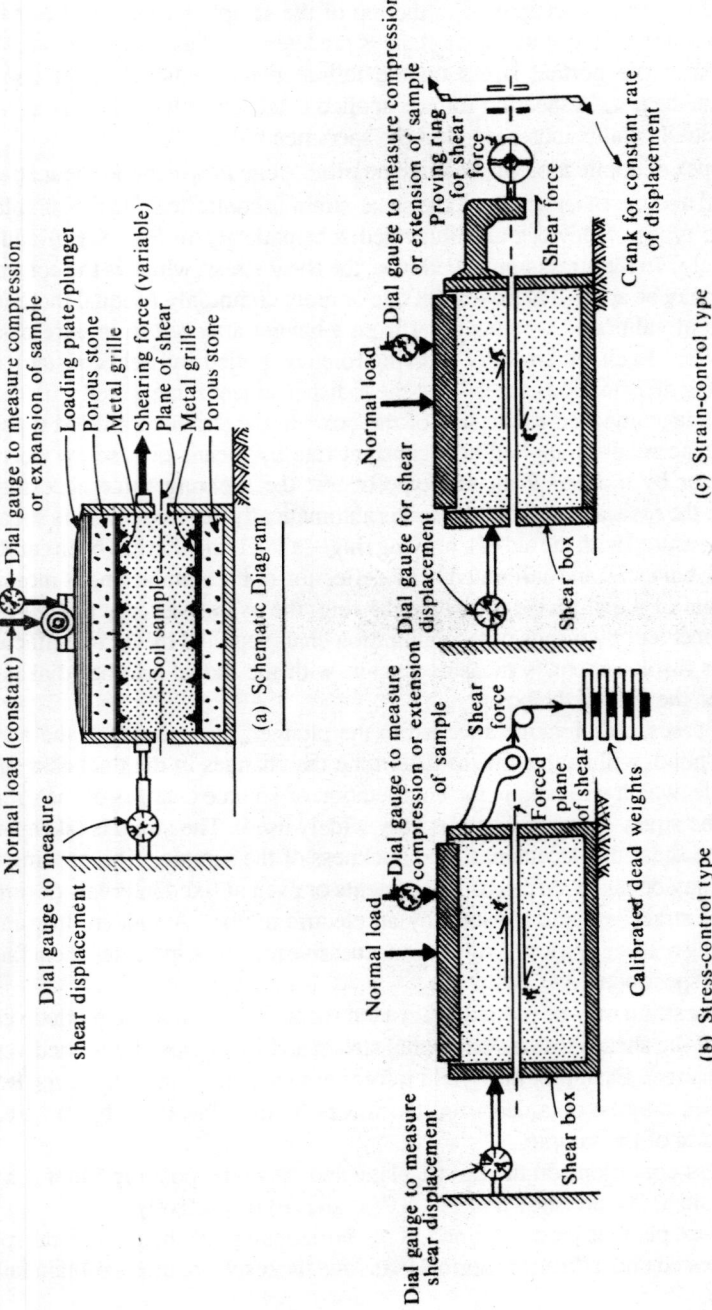

Fig. 8.8 Direct shear device

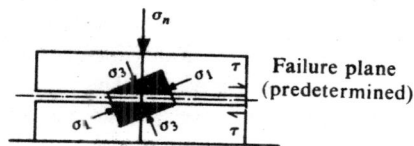

(a) Conditions of stress in the shear box

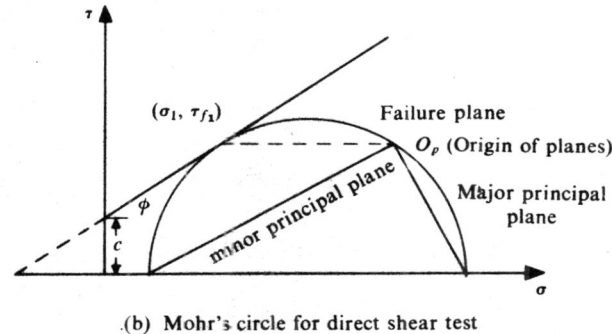

(b) Mohr's circle for direct shear test

Fig. 8.9 Mohr's circle representation of stress conditions in direct shear test

The direct shear test is a relatively simple test. Quick drainage, i.e., quick dissipation of pore pressures is possible since the thickness of the specimen is small. However, the test suffers from the following inherent disadvantages, which limit its application.

1. The stress conditions are complex primarily because of the non-uniform distribution of normal and shear stresses on the plane.
2. There is virtually no control of the drainage of the soil specimen as the water content of a saturated soil changes rapidly with stress.
3. The area of the sliding surface at failure will be less than the original area of the soil specimen and strictly speaking, this should be accounted for.
4. The ridges of the metal gratings embedded on the top and bottom of the specimen, causes distortion of the specimen to some degree.
5. The effect of lateral restraint by the side walls of the shear box is likely to affect the results.
6. The failure plane is predetermined and this may not be the weakest plane. In fact, this is the most important limitation of the direct shear test.

8.8.2 Triaxial Compression Test

The triaxial compression test, introduced by Casagrande and Terzaghi in 1936, is by far the most popular and extensively used shearing strength test, both for field application and for purposes of research. As the name itself suggests, the soil specimen is subjected to three compressive stresses in mutually perpendicular directions, one of the three stresses being increased until the specimen fails in shear. Usually a cylindrical specimen with a height equal to twice its diameter is used. The desired three-dimensional stress system is achieved by an initial application of all-round fluid pressure or confining pressure through water. While this confining pressure is kept constant throughout the test, axial or vertical loading is increased gradually and at a uniform rate. The axial stress thus constitutes the major principal stress and the confining pressure acts in the other two principal directions, the intermediate and minor principal stresses being equal to the confining pressure. The principle is shown in Fig. 8.10.

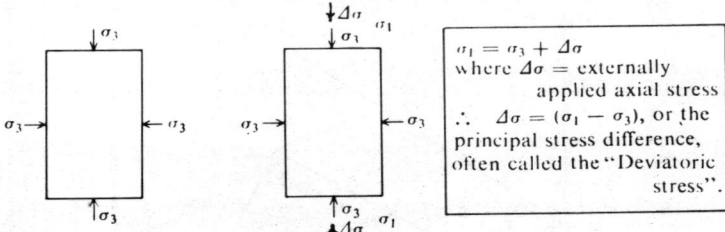

(a) Initially, upon application of all-round fluid pressure, or confining pressure

(b) After application of external axial stress in addition to the confining pressure, held constant until failure

$\sigma_1 = \sigma_3 + \Delta\sigma$
where $\Delta\sigma$ = externally applied axial stress
$\therefore \Delta\sigma = (\sigma_1 - \sigma_3)$, or the principal stress difference, often called the "Deviatoric stress".

Fig. 8.10 Principle and stress conditions of triaxial compression test

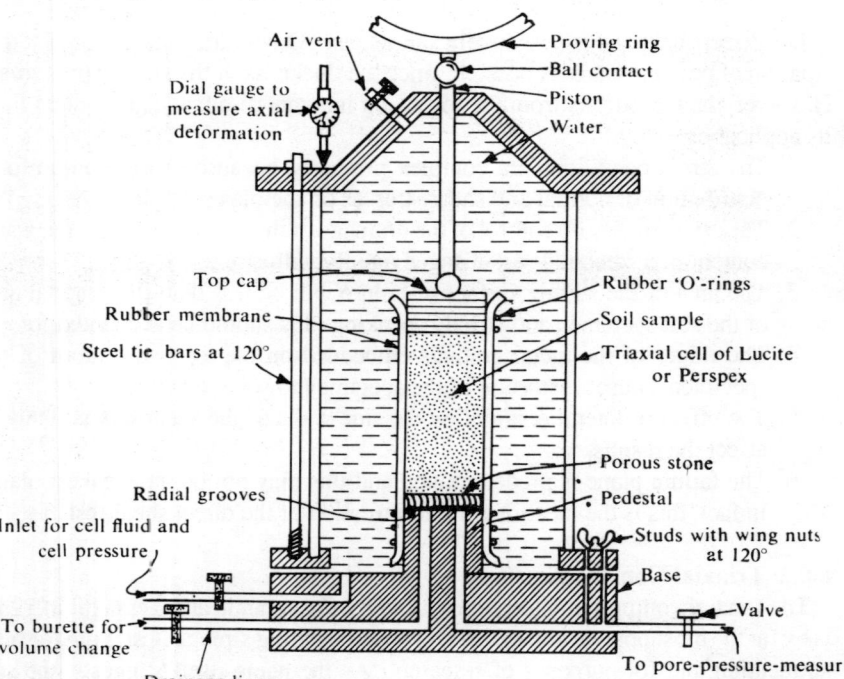

Fig. 8.11 Triaxial cell with accessories

Shearing Strength of Soils 315

The apparatus, consists of a lucite or perspex cylindrical cell, called 'triaxial cell', with appropriate arrangements for an inlet of cell fluid and application of pressure by means of a compressor, outlet of pore water from the specimen if it is desired to permit drainage which otherwise may serve as pore pressure connection and axial loading through a piston and loading cap, as shown in Fig. 8.11.

The assembly may be placed on the base of a motorised loading frame with a proving ring made to bear on the loading piston for the purpose of measuring the axial load at any stage of the test.

Test procedure

The essential steps in the conduct of the test are as follows:
(i) A saturated porous stone is placed on the pedestal and the cylindrical soil specimen is placed on it.
(ii) The specimen is enveloped by a rubber membrane to isolate it from the water with which the cell is to be filled later; it is sealed with the pedestal and top cap by rubber "O" rings.
(iii) The cell is filled with water and pressure is applied to the water, which in turn is transmitted to the soil specimen all-round and at top. This pressure is called 'cell pressure', 'chamber pressure' or 'confining pressure'.
(iv) Additional axial stress is applied while keeping the cell pressure constant. This introduces shearing stresses on all planes except the horizontal and vertical planes, on which the major, minor and intermediate principal stresses act, the last two being equal to the cell pressure on account of axial symmetry.
(v) The additional axial stress is continuously increased until failure of the specimen occurs. (What constitutes failure is often a question of definition and may be different for different kinds of soils. This aspect would be discussed later on).

A number of observations may be made during a triaxial compression test regarding the physical changes occurring in the soil specimen:
(a) As the cell pressure is applied, pore water pressure develops in the specimen, which can be measured with the help of a pore pressure measuring apparatus, such as Bishop's pore pressure device (Bishop, 1960), connected to the pore pressure line, after closing the valve of the drainage line.
(b) If the pore pressure is to be dissipated, the pore water line is closed, the drainage line opened and connected to a burette. The volume decrease of the specimen due to consolidation is indicated by the water drained into the burette.
(c) The axial strain associated with the application of additional axial stress can be measured by means of a dial gauge, set to record the downward movement of the loading piston.
(d) Upon application of the additional axial stress, some pore pressure develops. It may be measured with the pore pressure device, after the drainage line is closed. On the other hand, if it is desired that any pore pressure developed be allowed to be dissipated, the pore water line is closed and the drainage line opened as stated previously.
(e) The cell pressure is measured and kept constant during the course of the test.

(f) The additional axial stress applied is also measured with the aid of a proving ring and dial gauge.

Thus the entire triaxial test may be visualised in two important stages:

(i) The specimen is placed in the triaxial cell and cell pressure is applied, during the first stage.

(ii) The additional axial stress is applied and is continuously increased to cause a shear failure, the potential failure plane being that with maximum obliquity during the second stage.

Area correction for the determination of additional axial stress or deviatoric stress

The additional axial load applied at any stage of the test can be determined from the proving ring reading. During the application of the load, the specimen undergoes axial compression and horizontal expansion to some extent. Little error is expected to creep in if the volume is supposed to remain constant, although the area of cross-section varies as axial strain increases. The assumption is perfectly valid if the test is conducted under undrained conditions, but, for drained conditions, the exact relationship is somewhat different.

If A_0, h_0 and V_0 are the initial area of cross-section, height and volume of the soil specimen respectively, and if A, h, and V are the corresponding values at any stage of the test, the corresponding changes in the values being designated ΔA, Δh, and ΔV, then

$$A(h_0 + \Delta h) = V = V_0 + \Delta V$$

$$\therefore \quad A = \frac{V_0 + \Delta V}{h_0 + \Delta h}$$

But, for axial compression, Δh is known to be negative.

$$\therefore \quad A = \frac{V_0 + \Delta V}{h_0 - \Delta h} = \frac{V_0\left(1 + \frac{\Delta V}{V_0}\right)}{h_0\left(1 - \frac{\Delta h}{h_0}\right)} = \frac{A_0\left(1 + \frac{\Delta V}{V_0}\right)}{(1 - \varepsilon_a)},$$

since the axial strain, $\varepsilon_a = \Delta h / h_0$.

For an undrained test, $A = \dfrac{A_0}{(1 - \varepsilon_a)}$,

since $\Delta V = 0$. ...(Eq. 8.32)

This is called the 'Area correction' and $\dfrac{1}{(1 - \varepsilon_a)}$ is the correction factor.

A more accurate expression for the corrected area is given by

$$A = \frac{A_0}{(1-\varepsilon_a)} \cdot \left(1 + \frac{\Delta V}{V_0}\right) = \frac{V_0 + \Delta V}{(h_0 - \Delta h)} \quad \ldots \text{(Eq. 8.33)}$$

Once the corrected area is determined, the additional axial stress or the deviator stress, $\Delta \sigma$, is obtained as

$$\Delta\sigma = \sigma_1 - \sigma_3 = \frac{\text{Axial load (from proving ring reading)}}{\text{Corrected area}}$$

The cell pressure or the confining pressure, σ_c, itself being the minor principal stress, σ_3, this is constant for one test; however, the major principal stress, σ_1, goes on increasing until failure.

$$\sigma_1 = \sigma_3 + \Delta\sigma \qquad \ldots \text{(Eq. 8.34)}$$

Mohr's circle for triaxial test

The stress conditions in a triaxial test may be represented by a Mohr's circle, at any stage of the test, as well as at failure, as shown in Fig. 8.12:

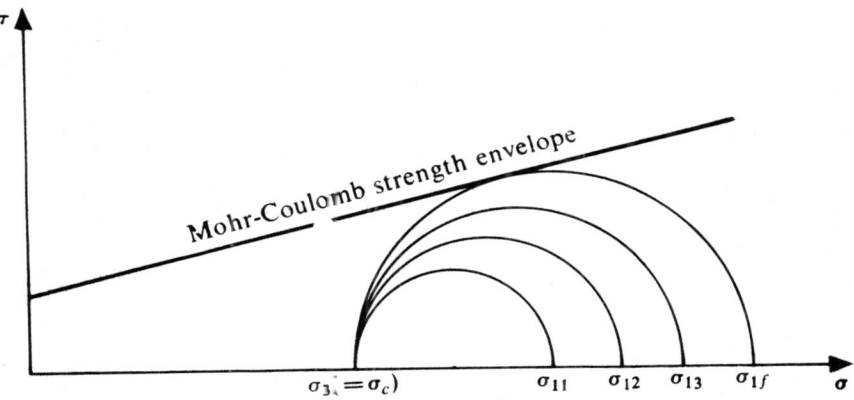

Fig. 8.12 Mohr's circles during triaxial test

The cell pressure, σ_c, which is also the minor principal stress is constant and σ_{11}, $\sigma_{12}, \sigma_{13}, \ldots \sigma_{1f}$ are the major principal stresses at different stages of loading and at failure. The Mohr's circle at failure will be tangential to the Mohr-Coulomb strength envelope, while those at intermediate stages will be lying wholly below it. The Mohr's circle at failure for one particular value of cell pressure will be as shown in Fig. 8.13.

The Mohr's circles at failure for one particular cell pressure are shown for the three typical cases of a general $c - \phi$ soil, a ϕ-soil and a c-soil in Figs. 8.13(*a*), (*b*), and (*c*) respectively.

With reference to Fig. 8.13(*a*), the relationship between the major and minor principal stresses at failure may be established from the geometry of the Mohr's circle, as follows:

From ΔDCG,

$$2\alpha = 90° + \phi$$

$$\alpha = 45° + \phi/2$$

318 *Geotechnical Engineering*

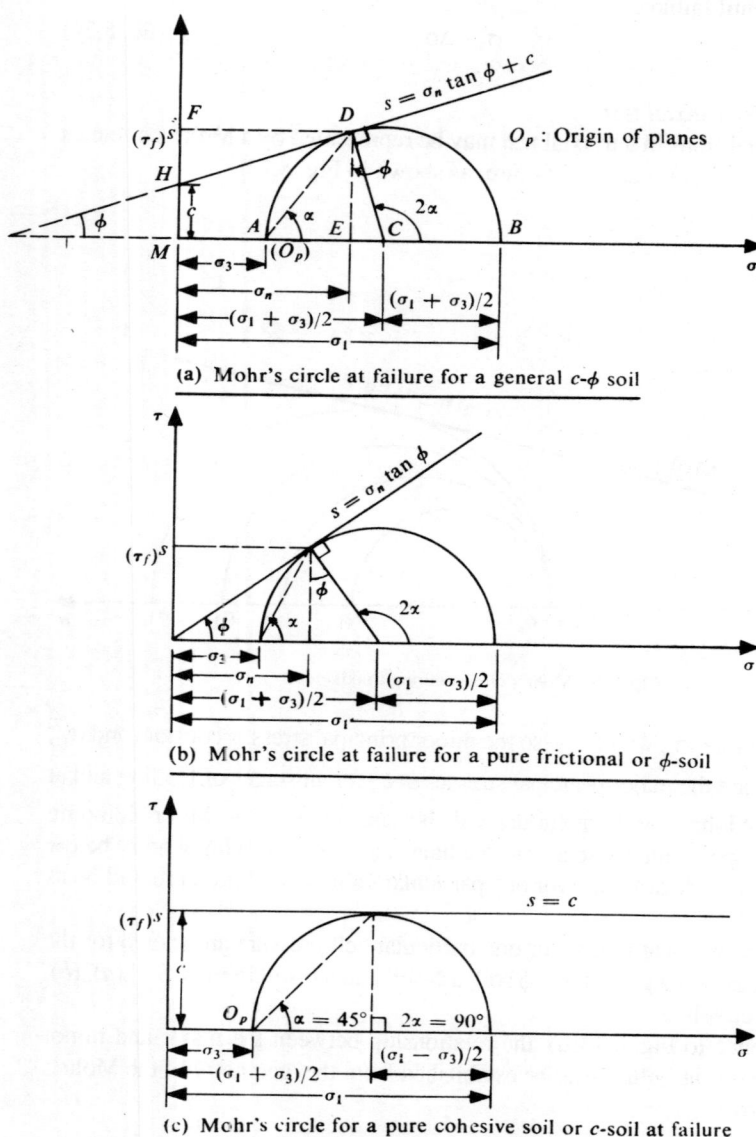

Fig. 8.13 Mohr's circle at failure for one particular cell pressure for triaxial test.

Again from ΔDCG

$$\sin\phi = DC/GC = DC/(DM + MC) = \frac{(\sigma_1 - \sigma_3)/2}{c\cot\phi + (\sigma_1 + \sigma_3)/2}$$

$$= \frac{(\sigma_1 - \sigma_3)}{2c\cot\phi + (\sigma_1 + \sigma_3)}$$

$$\therefore \quad (\sigma_1 - \sigma_3) = 2c\cos\phi + (\sigma_1 + \sigma_3)\sin\phi \quad \text{...(Eq. 8.35)}$$

or $\quad \sigma_1(1 - \sin\phi) = \sigma_3(1 + \sin\phi) + 2c \cdot \cos\phi$

$$\therefore \quad \sigma_1 = \frac{\sigma_3(1 + \sin\phi)}{(1 - \sin\phi)} + \frac{2c\cos\phi}{(1 - \sin\phi)}$$

or $\quad \sigma_1 = \sigma_3 \tan^2(45° + \phi/2) + 2c\tan(45° + \phi/2) \quad \text{...(Eq. 8.36)}$

or $\quad \sigma_1 = \sigma_3 \tan^2\alpha + 2c\tan\alpha \quad \text{...(Eq. 8.37)}$

This is also written as

$$\sigma_1 = \sigma_3 N_\phi + 2c\sqrt{N_\phi} \quad \text{...(Eq. 8.38)}$$

where, $\quad N_\phi = \tan^2\alpha = \tan^2(45° + \phi/2) \quad \text{...(Eq. 8.39)}$

Equation 8.36 or Eqs. 8.38 and 3.39 define the relationship between the principal stresses at failure. This state of stress is defined as 'Plastic equilibrium condition', when failure is imminent.

From one test, a set of σ_1 and σ_3 is known; however, it can be seen from Eq. 8.36, that at least two such sets are necessary to evaluate the parameters c and ϕ. Conventionally, three or more such sets are used from a corresponding number of tests.

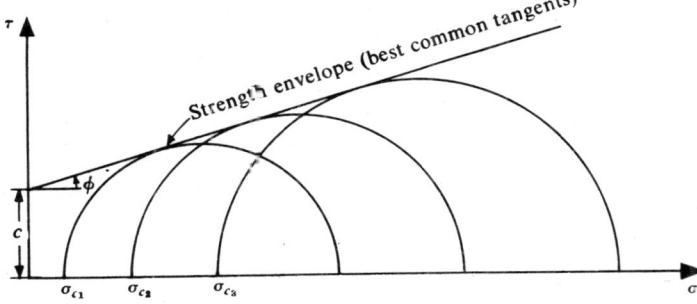

Fig. 8.14 Mohr's circles for triaxial tests with different cell pressures and strength envelope

320 Geotechnical Engineering

The usual procedure is to plot the Mohr's circles for a number of tests and take the best common tangent to the circles as the strength envelope. A small curvature occurs in the strength envelope of most soils, but since this effect is slight, the envelope for all practical purposes, may be taken as a straight line. The intercept of the strength envelope on the τ-axis gives the cohesion and the angle of slope of this line with σ-axis gives the angle of internal friction, as shown in Fig. 8.14.

Lambe and Whitman (1969) advocate a modified procedure to obtain the failure envelope, as a function of $(\sigma_1 + \sigma_3)/2$ and $(\sigma_1 - \sigma_3)/2$.

Equation 8.35 may be rewritten as follows:

$$(\sigma_1 - \sigma_3)/2 = d + \frac{(\sigma_1 + \sigma_3)}{2} \cdot \tan \psi \qquad \ldots \text{(Eq. 8.40)}$$

Here,
$$\tan \psi = \sin \phi \qquad \ldots \text{(Eq. 8.41)}$$

and
$$d = c \cos \phi \qquad \ldots \text{(Eq. 8.42)}$$

Equation 8.40 indicates a linear relationship between $(\sigma_1 + \sigma_3)/2$ and $(\sigma_1 - \sigma_3)/2$ and may be plotted from the results of a series of triaxial compression tests, as shown in Fig. 8.15.

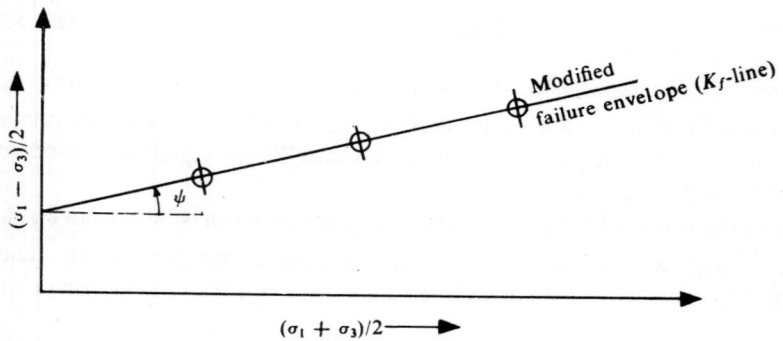

Fig. 8.15 Alternative procedure of evaluating shear strength parameters
(After Lambe and Whitman, 1969)

The best straight line is fitted to the data so that the averaging of the inevitable scatter of the experimental results is automatically taken care of. Once the values, d and ψ are obtained, c and ϕ may be computed by using Eqs. 8.41 and 8.42.

Graphical presentation of data from triaxial compression tests

The following are the usual set of graphs plotted making use of data from triaxial compression tests:

(i) Major principal stress versus % axial strain
(ii) σ_1/σ_3 versus % axial strain
(iii) % Volumetric strain versus % axial strain
(iv) Major principal stress versus volume change
(v) Mohr's circles at failure for each set of soil samples tested, from which the shear strength parameters may be evaluated.

A host of other useful information may be obtained from the data gathered from the trixial compression test if it is properly presented.

Types of failure of a triaxial compression test specimen

A triaxial compression test specimen may exhibit a particular pattern or shape as failure is reached, depending upon the nature of the soil and its condition, as illustrated in Fig. 8.16.

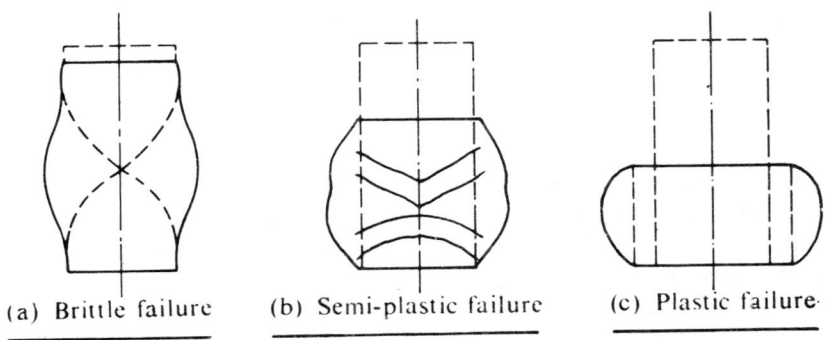

(a) Brittle failure   (b) Semi-plastic failure   (c) Plastic failure

Fig. 8.16 Failure patterns in triaxial compression tests

The first type is a brittle failure with well-defined shear plane, the second type is semi-plastic failure showing shear cones and some lateral bulging, and the third type is plastic failure with well-expressed lateral bulging.

In the case of plastic failure, the strain goes on increasing slowly at a reduced rate with increasing stress, with no specific stage to pin-point failure. In such a case, failure is assumed to have taken place when the strain reaches an arbitrary value such as 20%.

Merits of triaxial compression test

The following are the significant points of merit of triaxial compression test:
(1) Failure occurs along the weakest plane unlike along the predetermined plane in the case of direct shear test.
(2) The stress distribution on the failure plane is much more uniform than it is in the direct shear test; the failure is not also progressive, but the shear strength is mobilised all at once. Of course, the effect of end restraint for the sample is considered to be a disadvantage; however, this may not have pronounced effect on the results since the conditions are more uniform to the desired degree near the middle of the height of the sample where failure usually occurs.
(3) Complete control of the drainage conditions is possible with the triaxial compression test; this would enable one to simulate the field conditions better.

322   *Geotechnical Engineering*

(4) The possibility to vary the cell pressure or confining pressure also affords another means to simulate the field conditions for the sample, so that the results are more meaningfully interpreted.
(5) Precise measurements of pore water pressure and volume changes during the test are possible.
(6) The state of stress within the specimen is known on all planes and not only on a predetermined failure plane as it is with direct shear tests.
(7) The state of stress on any plane is capable of being determined not only at failure but also at any earlier stage.
(8) Special tests such as extension tests are also possible to be conducted with the triaxial testing apparatus.
(9) It provides an ingenious and a symmetrical three-dimensional stress system better suited to simulate field conditions.

8.8.3 Unconfined Compression Test

This is a special case of a triaxial compression test; the confining pressure being zero. A cylindrical soil specimen, usually of the same standard size as that for the triaxial compression, is loaded axially by a compressive force until failure takes place. Since the specimen is laterally unconfined, the test is known as 'unconfined compression test'. No rubber membrane is necessary to encase the specimen. The axial or vertical compressive stress is the major principal stress and the other two principal stresses are zero.

This test may be conducted on undisturbed or remoulded cohesive soils. It cannot be conducted on coarse-grained soils such as sands and gravels as these cannot stand without lateral support. Also the test is essentially a quick or undrained one because it is assumed that there is no loss of moisture during the test, which is performed fairly fast. Owing to its simplicity, it is often used as a field test, besides being used in the laboratory. The failure plane is not predetermined and failure takes place along the weakest plane.

The test specimen is loaded through a calibrated spring by a simple manually operated screw jack at the top of the machine. Different springs with stiffness values ranging from 2 to 20 N/mm may be used to test soils of varying strengths. The graph of load versus deformation is traced directly on a sheet of paper by means of an autographic recording arm. For any vertical or axial strain, the corrected area can be computed, assuming no change in volume. The axial stress is got by dividing the load by the corrected area. The apparatus is shown in Fig. 8.17.

The specimen is placed between two metal cones attached to two horizontal plates, the upper plate being fixed and the lower one sliding on vertical rods. The spring is supported by a plate and a screw on either side. The plate is capable of being raised by turning a handle so as to apply a compressive load on the soil specimen.

The stress-strain diagram is plotted autographically. The vertical movement of the pen relative to the chart is equal to the extension of the spring, and hence, is

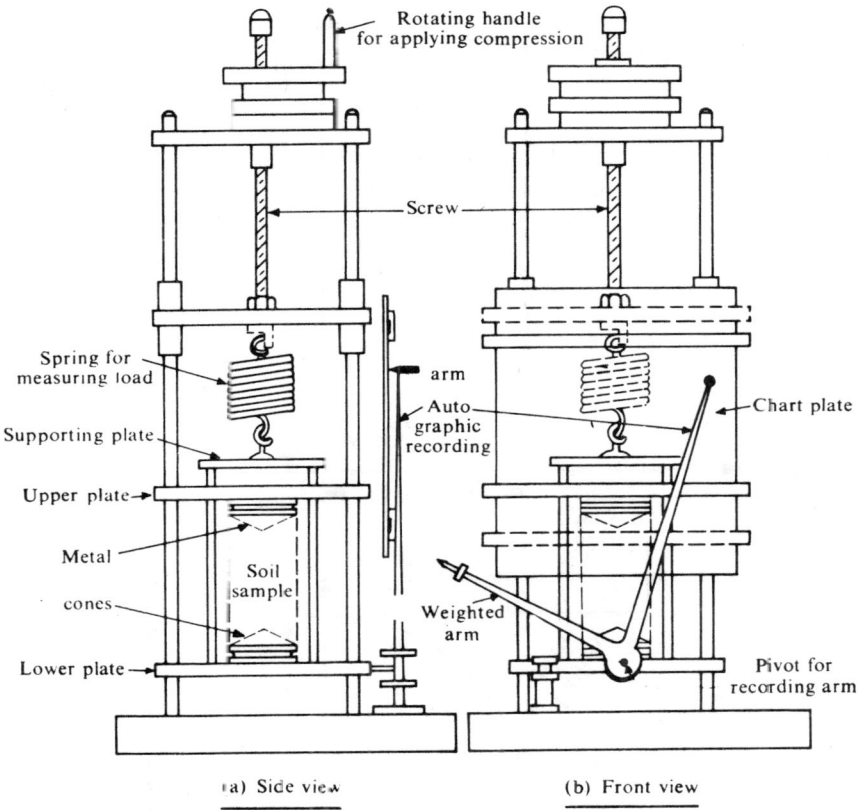

Fig. 8.17 Unconfined compression apparatus

proportional to the load. As the lower plate moves upwards, the upper one swings sideways, the weighted arms bearing on a stop. The lateral movement of the pen is thus proportional to the axial strain of the soil specimen. The area of cross-section increases as the specimen gets compressed. A transparent calibrated mask is used to read the stress direct from the chart.

Alternatively, a loading frame with proving ring and a dial gauge for measuring the axial compression of the specimen may also be used. The maximum compressive stress is that at the peak of the stress-strain curve. If the peak is not well-defined, an arbitrary strain value such as 20% is taken to represent failure.

Mohr's circle for unconfined compression test

The Mohr's circles for the unconfined compression test are shown in Fig. 8.18. From Eq. 8.36, recognising that $\sigma_3 = 0$

$$\sigma_1 = 2c \tan (45° + \phi/2) \qquad \ldots(\text{Eq. 8.43})$$

324 *Geotechnical Engineering*

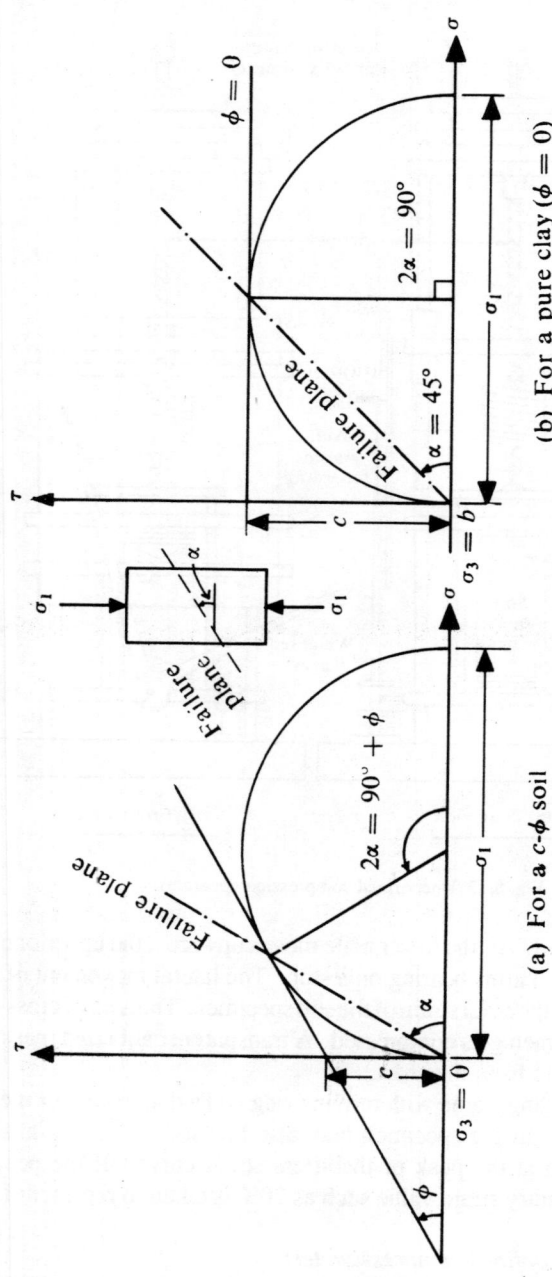

Fig. 8.18 Mohr's circles for unconfined compression test

The two unknowns – c and ϕ – cannot be solved since any number of unconfined compression tests would give only one value for σ_1. Therefore, the unconfined compression test is mostly found useful in the determination of the shearing strength of saturated clays for which ϕ is negligible or zero, under undrained conditions. In such a case, Eq. 8.43 reduces to

$$\sigma_1 = \phi_u = 2c \qquad \ldots(\text{Eq. 8.44})$$

where, ϕ_u is the unconfined compression strength.

Thus, the shearing strength or cohesion value for a saturated clay from unconfined compression test is taken to be half the unconfined compression strength.

8.8.4 Vane Shear Test

If suitable undisturbed or remoulded samples cannot be got for conducting triaxial or unconfined compression tests, the shear strength is determined by a device called the Shear Vane.

The vane shear test may also be conducted in the laboratory. The laboratory shear vane will be usually smaller in size as compared to the field vane.

The shear vane usually consists of four steel plates welded orthogonally to a steel rod, as shown in Fig. 8.19.

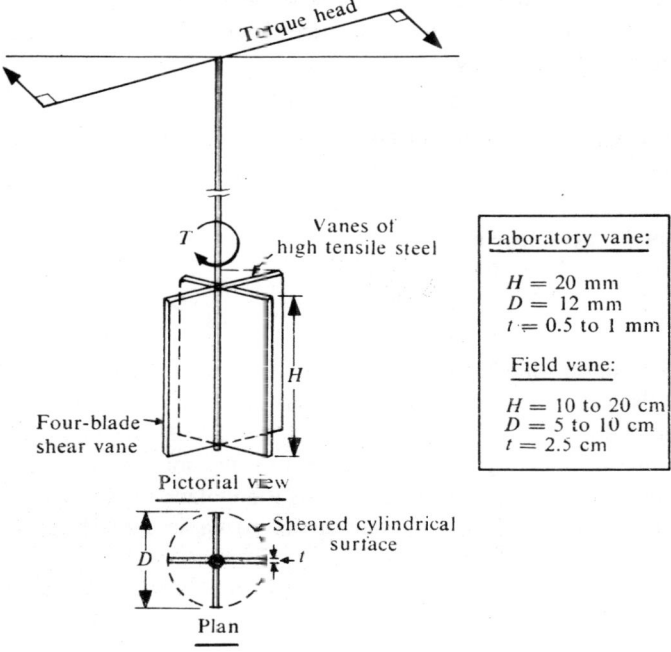

Fig. 8.19 Laboratory shear vane

The applied torque is measured by a calibrated torsion spring, the angle of twist being read on a special gauge. A uniform rotation of about 1° per minute is used. The vane is forced into the soil specimen or into the undisturbed soil at the bottom of a bore-hole in a gentle manner and torque is applied. The torque is computed by multiplying the angle of twist by the spring constant.

The shear strength s of the clay is given by

$$s = \frac{T}{\pi D^2(H/2 + D/6)} \qquad \ldots(\text{Eq. 8.45})$$

if both the top and bottom of the vane partake in shearing the soil.

Here, T = torque
 D = diameter of the vane
 H = height of the vane

If only one end of the vane partakes in shearing the soil, then

$$s = \frac{T}{\pi D^2(H/2 + D/12)} \qquad \ldots(\text{Eq. 8.46})$$

Equation 8.45 may be derived as follows :

The shearing resistance is mobilised at failure along a cylindrical surface of diameter D, the diameter of the vane, as also at the two circular faces at top and bottom.

The shearing force at the cylindrical surface = $\pi D.H.s$, where s is the shearing strength of the soil. The moment of this force about the axis of the vane contributes to the torque and is given by

$$\pi DH.s.D/2 \text{ or } \pi sH.D^2/2$$

For the circular faces at top or bottom, considering the shearing strength of a ring of thickness dr at a radius r, the elementary torque is

$$(2\pi r\, dr)\cdot s.r$$

and the total for one face is

$$\int_0^{D/2} 2\pi s r^2 dr = \frac{2\pi s}{3}\cdot\frac{D^3}{8} = \frac{\pi s}{12}\cdot D^3$$

If we add these contributions considering both the top and bottom faces and equate to the torque T at failure, we get Eq. 8.45, and if only one face is considered, we get Eq. 8.46.

Regarding the shearing stress distribution on the soil cylinder, it is assumed uniform on the cylindrical surface but it is triangular over the shear end faces, varying from zero at the axis of the vane device, to maximum at the edge, as shown in Fig. 8.20.

The vane shear test is particularly suited for soft clays and sensitive clays for which suitable cylindrical specimens cannot be easily prepared.

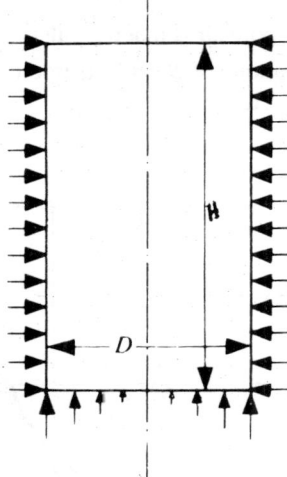

Fig. 8.20 Shearing distribution on the sides and faces of soil cylinder in the vane shear test

8.9 PORE PRESSURE PARAMETERS

Pore water pressures play an important role in determining the strength of soil. The change in pore water pressure due to change in applied stress is characterised by dimensionless coefficients, called 'Pore pressure coefficients' or 'Pore pressure parameters' A and B. These parameters have been proposed by Prof. A.W. Skempton (Skempton, 1954) and are now universally accepted.

In an undrained triaxial compression test, pore water pressures develop in the first stage of application of cell pressure or confining pressure, as also in the second stage of application of additional axial stress or deviator stress.

The ratio of the pore water pressure developed to the applied confining pressure is called the B-parameter:

$$B = \frac{\Delta u_c}{\Delta \sigma_c} = \frac{\Delta u_c}{\Delta \sigma_3} \qquad \ldots\text{(Eq. 8.47)}$$

Since no drainage is permitted, the decrease in volume of soil skeleton is equal to that in the volume of pore water. Using this and the principles of theory of elasticity, it can be shown that

$$B = \frac{1}{1 + n \cdot \frac{C_v}{C_c}} \qquad \ldots\text{(Eq. 8.48)}$$

where C_v and C_c represent the volume compressibilities (change in volume per unit volume per unit pressure increase) of pore water and soil respectively and n is the porosity.

For a saturated soil C_c is very much greater than C_v, and B is very nearly unity; for a dry soil C_v, the value for pore air is much greater than C_c and B is practically negligible or zero. The variation of B with degree of saturation, found experimentally, is shown in Fig. 8.21.

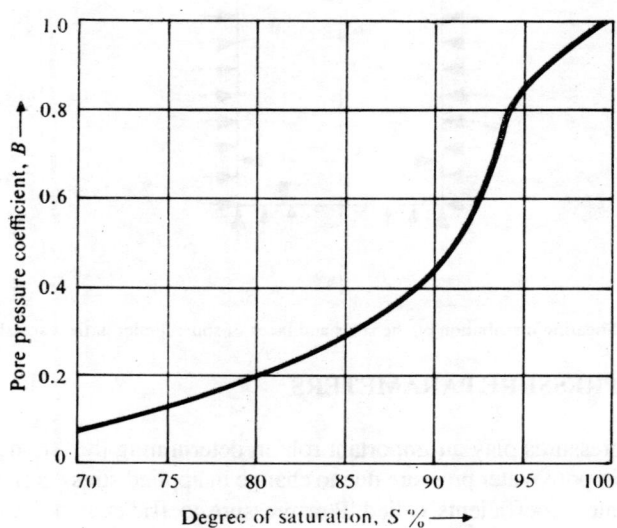

Fig. 8.21 Variation of B-factor with degree of saturation

Pore water pressures develop during the application of the deviator stress also in a triaxial compression test; the pore pressure coefficient or parameter A is defined as follows:

$$A = \frac{\Delta u_d}{(\Delta\sigma_1 - \Delta\sigma_3)} \qquad \text{...(Eq. 8.49)}$$

where Δu_2 = pore pressure developed due to an increase of deviator stress $(\Delta\sigma_1 - \Delta\sigma_3)$

The A-factor or parameter is not a constant. It varies with the soil, its stress history and the applied deviator stress. Its value can be specified at failure or maximum deviator stress or at any other desired stage of the test. A-factor varies also with the initial density index in the case of sands and with over-consolidation ratio in the case of clays. Its variation with over-consolidation ratio, as given by Bishop and Henkel (1962), is shown in Fig. 8.22.

The general expression for the pore water pressure developed and changes in applied stresses is as follows:

$$\Delta u = B\{\Delta\sigma_3 + A(\Delta\sigma_1 - \Delta\sigma_3)\} \qquad \text{...(Eq. 8.50)}$$

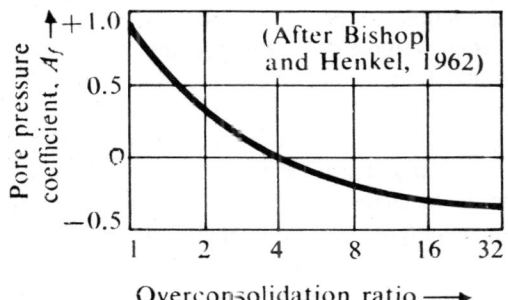

Fig. 8.22 Variation of A-factor at failure with over-consolidation ratio

A for a perfectly elastic material may be shown to be 1/3. This may also be written in the form:

$$\Delta u = B \cdot \Delta\sigma_3 + \overline{A}(\Delta\sigma_1 - \Delta\sigma_3) \qquad \text{...(Eq. 8.51)}$$

where $\overline{A} = A \cdot B$.

If Δu is considered to be the sum of two components Δu_d and Δu_c,

$$\Delta u_c = B \cdot \Delta\sigma_3$$

and
$$\Delta u_d = \overline{A}(\Delta\sigma_1 - \Delta\sigma_3)$$

For the conventional triaxial test at constant cell pressure, during the application of the deviator stress, $\Delta\sigma_3 = 0$ and $\Delta\sigma_1 = (\sigma_1 - \sigma_3)$. Taking B as unity for full saturation, Eq. 8.50 for this case of UU-test will reduce to

$$\Delta u = \Delta\sigma_3 + A(\sigma_1 - \sigma_3) \qquad \text{...(Eq. 8.52)}$$

$\overline{A}$ and hence A can be easily determined from the conventional triaxial compression test of UU type.

For CU tests where drainage is permitted during the application of cell pressure, Δu_c is zero, and the corresponding value of Δu is given by

$$\Delta u = A(\sigma_1 - \sigma_3) \qquad \text{...(Eq. 8.53)}$$

A-factor may be as high as 2 to 3 for saturated fine sand in loose condition, and as low as -0.5 for heavily preconsolidated clay.

8.10 SHEARING CHARACTERISTICS OF SANDS

The shearing strength in sand may be said to consist of two parts, the internal frictional resistance between grains, which is a combination of rolling and sliding friction and another part known as 'interlocking'. Interlocking, which means locking of one particle by the adjacent ones, resisting movements, contributes a large portion of the shearing strength in dense sands, while it does not occur in loose sands. The Mohr strength theory is not invalidated by the occurrence of interlocking. The Mohr envelopes merely show larger ordinates and steeper slopes for dense soils than for loose ones.

The angle of internal friction is a measure of the resistance of the soil to sliding along a plane. This varies with the density of packing, characterised by density index, particle shape and roughness and particle size distribution. Its value increases with the density index, with the angularity and roughness of particles and also with better gradation. This is influenced to some extent by the normal pressure on the plane of shear and also the rate of application of shear.

The 'angle of repose' is the angle to the horizontal at which a heap of dry sand, poured freely from a small height, will stand without support. It is approximately the same as the angle of friction in the loose state.

Some clean sands exhibit slight cohesion under certain conditions of moisture content, owing to capillary tension in the water contained in the voids. Since this is small and may disappear with change in water content, it should not be relied upon for shear strength. On the other hand, even small percentages of silt and clay in a sand give it cohesive properties which may be sufficiently large so as to merit consideration.

Unless drainage is deliberately prevented, a shear test on a sand will be a drained one as the high value of permeability makes consolidation and drainage virtually instantaneous. A sand can be tested either in the dry or in the saturated condition. If it is dry, there will be no pore water pressures and if it is saturated, the pore water pressure will be zero due to quick drainage. In either case, the intergranular pressure will be equal to the applied stress. However, there may be certain situations in which significant pore pressures are developed, at least temporarily, in sands. For example, during earth-quakes, heavy blasting and operation of vibratory equipment, instantaneous pore pressures are likely to develop due to large shocks or dynamic loads. These may lead to the phenomenon of 'liquefaction' or sudden and total loss of shearing strength, which is a grave situation of lack of stability.

Further discussion of shear characteristics of sands is presented in the following sub-sections.

8.10.1 Stress-strain Behaviour of Sands

The stress-strain behaviour of sands is dependent to a large extent on the initial density of packing, as characterised by the density index. This is represented in Fig. 8.23.

It can be observed from Fig. 8.23(a), the shear stress (in the case of direct shear tests) or deviator stress (in the case of triaxial compression tests) builds up gradually for an initially loose sand, while for an initially dense sand, it reaches a peak value and decreases at greater values of shear/axial strain to an ultimate value comparable to that for an initially loose specimen. The behaviour of a medium-dense sand is intermediate to that of a loose sand and a dense sand. Intuitively, it should be expected that the denser a sand is, the stronger it is. The hatched portion represents the additional strength due to the phenomenon of interlocking in the case of dense sands.

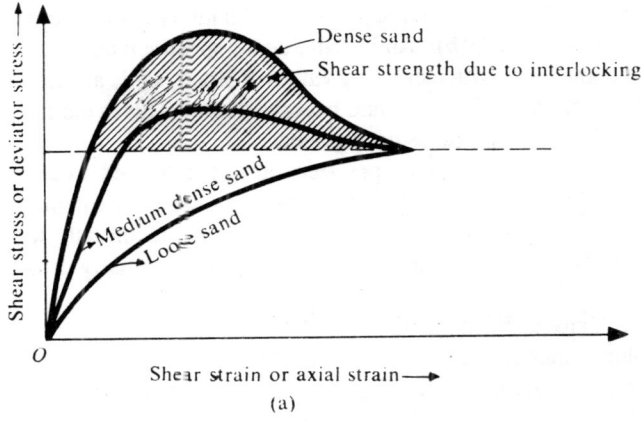

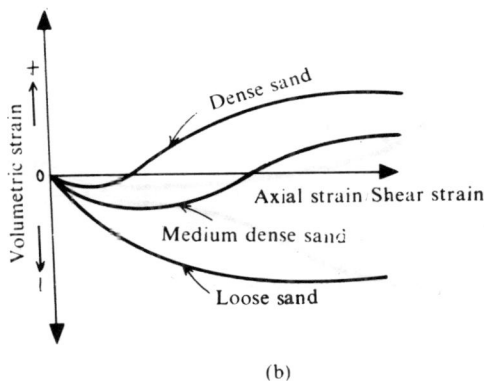

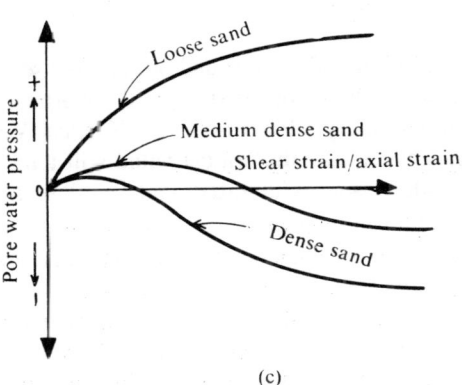

Fig. 8.23 Stress-strain characteristics of sands

The volume change characteristics of sands is another interesting feature, as depicted in Fig. 8.23(*b*). An initially dense specimen tends to increase in volume and become loose with increasing values of strain, while an initially loose specimen tends to decrease in volume and become dense. This is explained in terms of the rearrangement of particles during shear.

The changes in pore water pressure during undrained shear, which is rather not very common owing to high permeability of sands, are depicted in Fig. 8.23(*c*). Positive pore pressures develop in the case of an initially loose specimen and negative pore pressures develop in the case of an initially dense specimen.

8.10.2 Critical Void Ratio

Volume change characteristics depend upon various factors such as the particle size, particle shape and distribution, principal stresses, previous stress history and significantly on density index. Volume changes, expressed in terms of the void ratio versus shear strain are typically as shown in Fig. 8.24.

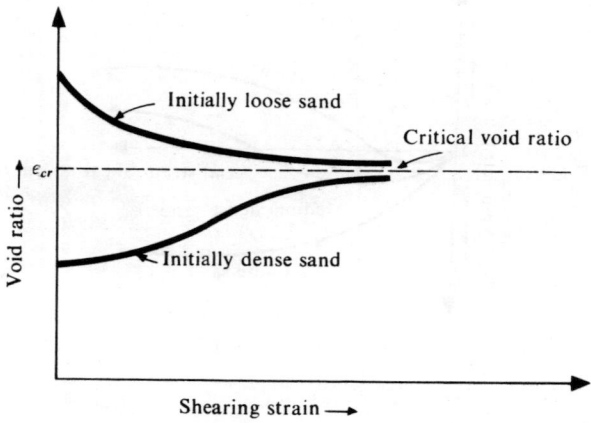

Fig. 8.24 Effect of initial density on changes in void ratio

At large strains both initially loose and initially dense specimens attain nearly the same void ratio, at which further strain will not produce any volume changes. Such a void ratio is usually referred to as the 'Critical Void Ratio'. Sands with initial void ratio greater than the critical value will tend to decrease in volume during shearing, while sands with initial void ratio less than the critical will tend to increase in volume.

The critical void ratio is dependent upon the cell pressure (in the case of triaxial compression tests) or effective normal pressure (in the case of direct shear tests), besides a few other particle characteristics. It bears a reciprocal relationship with pressure. The value of critical void ratio under a given set of conditions may be determined by plotting the volume changes versus void ratio. The value for which the volume change is zero is the critical one.

8.10.3 Shearing Strength of Sands

The shearing strength of cohesionless soils has been established to depend primarily upon the angle of internal friction which itself is dependent upon a number of factors including the normal pressure on the failure plane. The nature of the results of the shear tests will be influenced by the type of test—direct shear or triaxial compression, by the fact whether the sand is saturated or dry and also by the nature of stresses considered—total or effective.

Each direct shear test is usually conducted under a certain normal stress. Each stress-strain diagram therefore reflects the behaviour of a specimen under a particular normal stress. A number of specimens are tested under different normal stresses. It is to be noted that only the effective normal stress is capable of mobilising shear strength. The results when plotted appear as shown in Fig. 8.25.

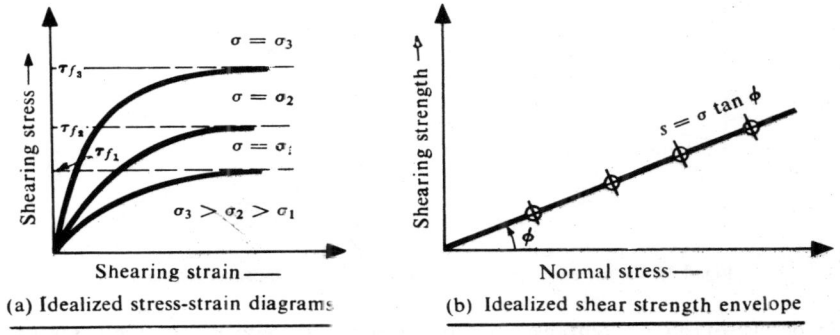

Fig. 8.25 Shear characteristics of sands from direct shear tests

It may be observed from Fig. 8.25(a) that the greater the effective normal pressure during shear, the greater is the shearing stress at failure or shearing strength. The shear strength plotted against effective normal pressure gives the Coulomb strength envelope as a straight line, passing through the origin and inclined at the angle of internal friction to the normal stress axis. It is shown in Fig. 8.25(b). The failure envelope obtained from ultimate shear strength values is assumed to pass through the origin for dry cohesionless soils. The same is true even for saturated sands if the plot is made in terms of effective stresses. In the case of dense sands, the values of ϕ obtained by plotting peak strength values will be somewhat greater than those from ultimate strength values.

Ultimate values of ϕ may range from 29 to 35° and peak values from 32 to 45° for sands. The values of ϕ selected for use in practical problems should be related to soil strains expected. If soil deformation is limited, using the peak value for ϕ would be justified. If the deformation is relatively large, ultimate value of ϕ should be used.

If the sand is moist, the failure envelope does not pass through the origin as shown in Fig. 8.26. The intercept on the shear stress axis is referred to as the 'apparent cohesion', attributed to factors such as surface tension of the moisture films on the grains. The extra strength would be lost if the soil were to dry out or to become saturated or submerged. For this reason, the extra shear strength attributed to apparent cohesion is neglected in practice.

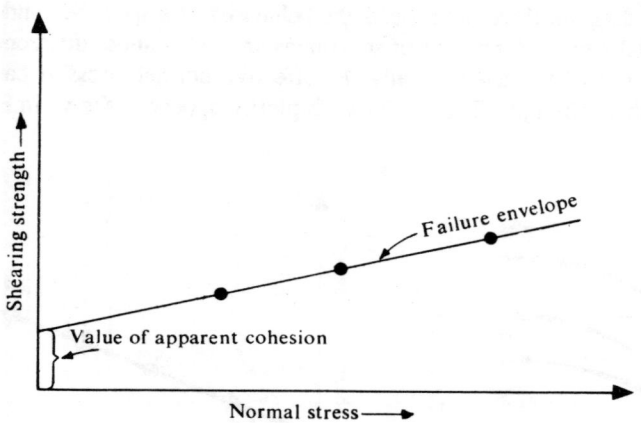

Fig. 8.26 Failure envelope for moist sand indicating apparent cohesion

In the case of triaxial compression tests, different tests with different cell pressure are to be conducted to evaluate the shearing strength and the angle of internal friction. In each test, the axial normal stress is gradually increased keeping the cell pressure constant, until failure occurs. The value of ϕ is obtained by plotting the Mohr Circles and the corresponding Mohr's envelope.

The failure envelope obtained from a series of drained triaxial compression tests on saturated sand specimens initially at the same density index is approximately a straight line passing through the origin, as shown in Fig. 8.27.

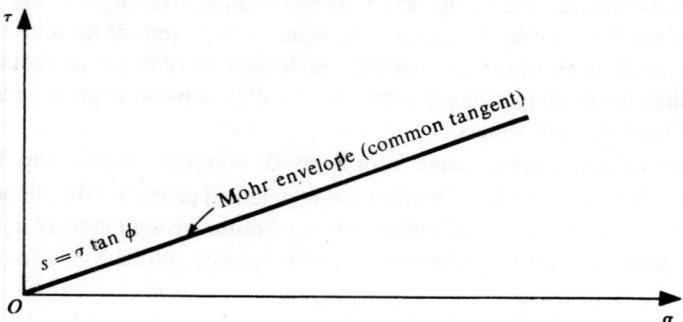

Fig. 8.27 Drained triaxial compression tests on saturated sand

Similar results are obtained when undrained triaxial compression tests are conducted with pore pressure measurements on saturated sand samples and Mohr's circles are plotted in terms of effective stresses. However, if Mohr's circles are plotted in terms of total stresses, the shape of envelopes will be similar to those for a purely cohesive soil. The failure envelope will be approximately horizontal with an intercept on the shearing stress axis, indicating the so-called 'apparent cohesion', as shown in Fig. 8.28.

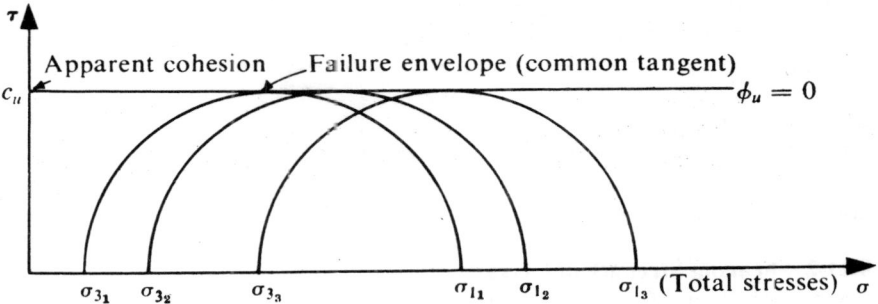

Fig. 8.28 Undrained triaxial compression tests on saturated sands (total stresses)

8.11 SHEARING CHARACTERISTICS OF CLAYS

The understanding of the fundamentals of shearing strength is much more important in the case of cohesive soils or clays in view of their troublesome nature with regard to stability. In fact, the most complex physical property of clays is the shearing strength, as it is dependent on a multitude of inter-related factors. One of the most difficult tasks is to interpret results of laboratory shearing strength tests to the shearing strength of natural clay deposits.

8.11.1 Source and Nature of Shearing Strength of Clays

Cohesion

This is a characteristic of true clay. This is sometimes referred to as no-load shear strength and is responsible for the strength of unconfined specimens. Cohesion in clays is a proerty which varies considerably with consistency. Cohesion therefore varies with both the type of clay and condition of clay. It is a kind of surface attraction among particles.

Adhesion

Whereas cohesion is the mutual attraction of two different parts of a clay mass to each other, clay often also exhibits the property of 'adhesion', which is a pro-

336 *Geotechnical Engineering*

pensity to adhere to other materials at a common surface. This has no relation to normal pressure. This is of particular interest in relation to the supporting capacity of friction piling in clays and to the lateral pressures on retaining walls.

Viscous friction

Solid friction effects are of relatively minor importance and the effects of viscous friction are quite pronounced. The laws of viscous friction are, in general, opposite to those of solid friction. The total frictional resistance is independent of normal force, but varies directly with the contact area. It varies with some power of the relative velocity of adjacent layers of fluid or with the rate of shearing. The well-established fact that the strength of saturated clays varies with consistency also is in accord with the concept that strength is due to viscous rather than solid friction.

Tensile strength

In varying degrees and for different periods of time, many clays are capable of developing a certain amount of tensile strength. This may affect the magnitude of normal stresses on failure planes.

8.11.2 Shearing Strength of Clays

Shear behaviour of clays is influenced by the fact whether the clay is normally consolidated or overconsolidated, by the fact whether it is undisturbed or remoulded, by the drainage conditions during testing, consistency of the clay, by certain structural effects, by the type of test and by the type and rate of strain. The following discussion relates to the shearing strength of saturated clays which are in a normally consolidated state; the modifications that may be expected in case the clay is in an overconsolidated state are indicated at the appropriate places.

Unconsolidated undrained tests

It is difficult, if not impossible, to utilise the concept of effective stress in connection with the shearing strength of saturated clays. It is difficult to imagine that any substantial part of the normal stress is transmitted through particle contacts when grain-to-grain contacts are relatively infrequent or when the solid phase is weak in itself. For this reason, it is common practice to consider only total stresses in the case of saturated clays. The results of unconsolidated undrained tests in direct shear are indicated in Fig. 8.29.

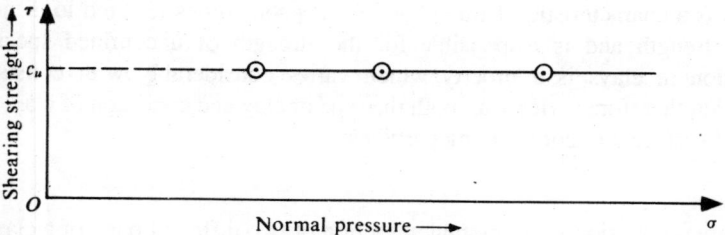

Fig. 8.29 Unconsolidated undrained tests in direct shear on saturated clays

Shearing Strength of Soils

It is seen that the total normal pressure does not influence the shearing strength of a saturated clay from undrained tests; the intercept of the horizontal plot on the shear strength axis gives the cohesion c_u. The strength of a clay is often reported simply in terms of unit cohesion, regardless of the overburden pressure.

The results of such tests in triaxial compression are indicated in Fig. 8.30.

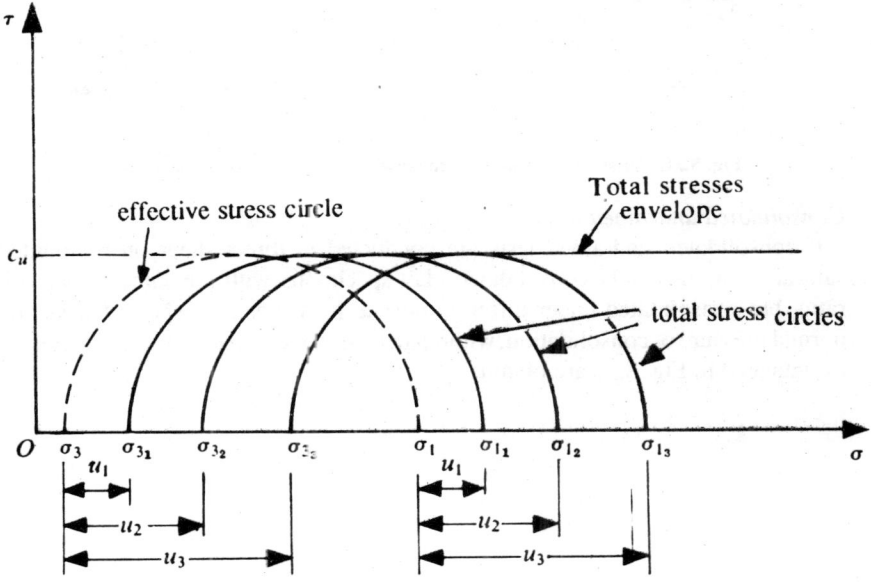

Fig. 8.30 Unconsolidated undrained tests in triaxial compression on saturated clays

Since drainage is not permitted both during the application of cell pressure and during the application of deviator stress (or additional axial stress), the increase in cell pressure or axial stress automatically increases the pore water pressure by an equal magnitude, the effective stress remaining constant. In view of this, the diameter of the effective stress circle will be the same as that of the total stress circles with mere lateral shifts. The total stress envelope is thus a horizontal line, the intercept on the shearing strength axis being cohesion c_u, and ϕ_u being zero. It may also be easily understood that the effective stress envelope cannot be obtained from these tests since only one circle will be obtained for all tests. Consolidated undrained or drained tests may be used for this purpose. Pore pressurements are not usually made in the unconsolidated undrained tests as they are not useful.

It is common knowledge that the shear strength of clay varies widely with its consistency, the shear strength being negligible when the water content is at liquid limit. This is reflected in Fig. 8.31.

The shearing strength of partially saturated clays is a more complex phenomenon and, hence, is considered outside the scope of the present work.

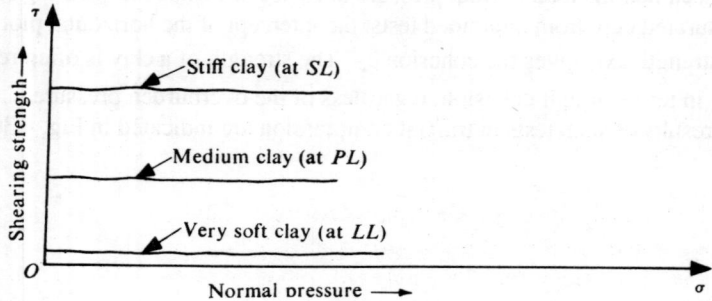

Fig. 8.31 Variation of shearing strength with consistency of saturated clays

Consolidated undrained tests

If consolidated undrained tests are conducted in direct shear on remoulded, saturated and normally consolidated clay specimens with the same initial void ratio, but consolidated under different normal pressures, and sheared under the normal pressure of consolidation, without permitting drainage during shear, results as indicated in Fig. 8.32 are obtained.

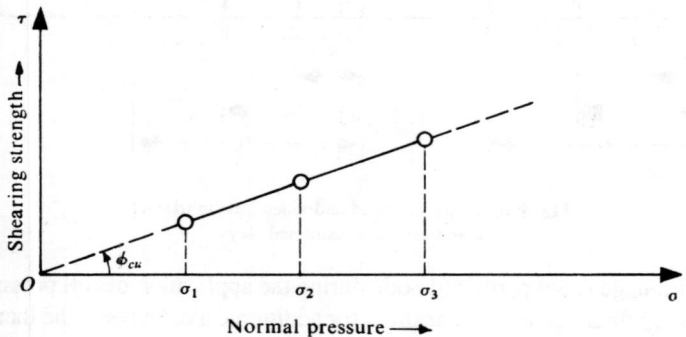

Fig. 8.32 Consolidated undrained tests in direct shear on remoulded, saturated, and normally consolidated clay
(consolidated and sheared under normal pressures σ_1, σ_2, and σ_3)

It is observed that the shear strength is proportional to the normal pressure. The strength envelope passes through the origin, giving an angle of shearing resistance ϕ_{cu}.

However, it is fallacious to assume that the shear strength is related to the normal pressure during the application of shear. This may be demonstrated by consolidating all the samples under one particular pressure and testing them in shear under a different pressure. In such a case the results will appear somewhat as shown in Fig. 8.33.

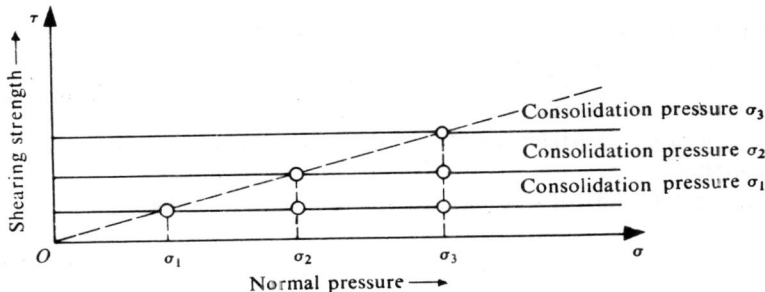

Fig. 8.33 Consolidated undrained tests in direct shear on remoulded, saturated, and normally consolidated clay (consolidated under normal pressures σ_1, σ_2, and σ_3, and sheared under different normal pressures)

It is observed that the shearing strength is independent of the normal pressure during shear but is dependent only on the normal pressure during consolidation or consolidation pressure. The process of preconsolidation may thus be viewed simply as a method of changing the consistency of the clay, the strength at a given consistency being practically independent of normal pressure during shear.

Similarly, consolidated undrained tests may be conducted in triaxial compression by either of the following procedures:

(i) The specimens of saturated, remoulded, and normally consolidated clay are consolidated under different cell pressures and sheared, without permitting drainage, under a cell pressure equal to the consolidation pressure. This approach is more commonly used.

(ii) The specimens are consolidated under the same cell pressure σ_c, and then sheared under undrained conditions with different cell pressures by increasing the axial stress; different series of these tests may be performed with different values of cell pressure for consolidation, which will be constant for any one series, as stated above.

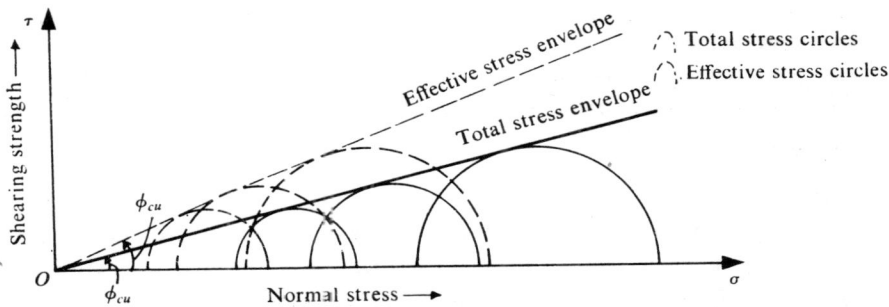

Fig. 8.34 Consolidated undrained tests in triaxial compression on remoulded, saturated, and normally consolidated clay (consolidated under different cell pressures and sheared undrained under the same cell pressures)

The results from the first method appear somewhat as shown in Fig. 8.34; total stress envelopes as well as effective stress envelopes are shown.

The failure envelopes pass through the origin, giving $c_{cu} = c'_{cu} = 0$, and values of ϕ_{cu} and ϕ'_{cu} such that $\phi'_{cu} > \phi_{cu}$. If the tests are conducted starting with a very low consolidation pressure, the initial portion of the envelope is usually curved and shows a cohesion intercept. The straight portion when extended passes through the origin.

An overconsolidated clay shows an apparent cohesion; the equation for shear strength is:

$$s = c_{cu} + (\sigma - \sigma_c) \tan \phi_{cu} \qquad \ldots(\text{Eq. 8.54})$$

The corresponding equation for a normally consolidated clay is:

$$s = \sigma \tan \phi_{cu} \qquad \ldots(\text{Eq. 8.55})$$

Here, σ_c is the consolidation pressure and σ is the applied normal pressure.

The envelope is generally curved up to the preconsolidation pressure and shows a cohesion intercept. The corresponding equations for shear strength in terms of effective stresses are written with primes.

The effect of preconsolidation is to reduce the value of A-parameter and thus cause higher strength. At higher values of over-consolidation ratio, A-factor may be even negative; the effective stress circles will then get shifted to the right of the total stress circles instead of to the left. This gives lower value of effective apparent cohesion and higher value of effective angle of shearing resistance than those of total stress values.

The results from the second method appear somewhat as shown in Fig. 8.35.

The results indicate that, for a particular series, the deviator stress at failure is independent of the cell pressure. The failure envelope will be horizontal for each series, the apparent cohesion c_{cu} being different for different series; the angle ϕ_{cu} is zero, as indicated by Fig. 8.35(a), (b) and (c). The greater the effective consolidation pressure, the greater is the apparent cohesion. This is indicated in Fig. 8.35(d). If the clay is over-consolidated, the consolidation pressure versus apparent cohesion curve will show a discontinuity at the pressure corresponding to the preconsolidation pressure; below this pressure, the relationship is non-linear and will show an intercept at zero pressure and, above this pressure, it is linear. If the clay is normally consolidated for all the consolidation pressures used in the tests, this relationship will be a straight line, which, when produced backwards, will pass through the origin.

Drained tests

The specimen is first consolidated under a certain cell pressure and is then sheared sufficiently slowly so that no pore pressures are allowed to develop at any stage. The effective stresses will be the same as the total stresses. The results will be similar to those obtained from the consolidated undrained tests, with the same modifications as for a clay in an overconsolidated condition, as shown in Fig. 8.36.

Shearing Strength of Soils 341

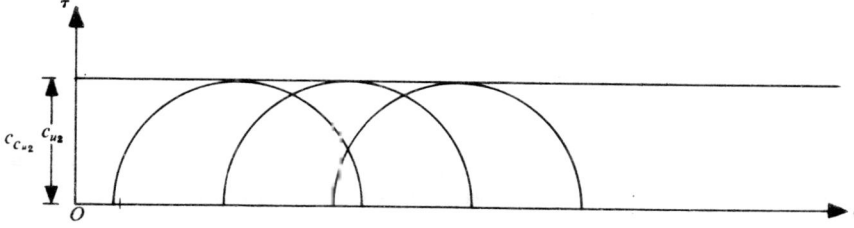

(a) Failure envelope for the first series with consolidation pressure

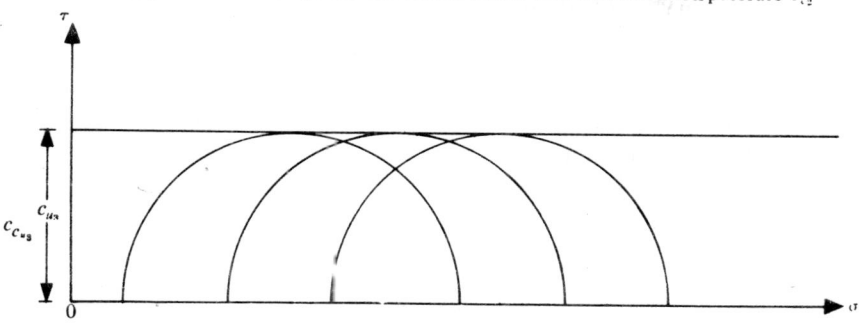

(b) Failure envelope for the second series with consolidation pressure σ_{c_2}

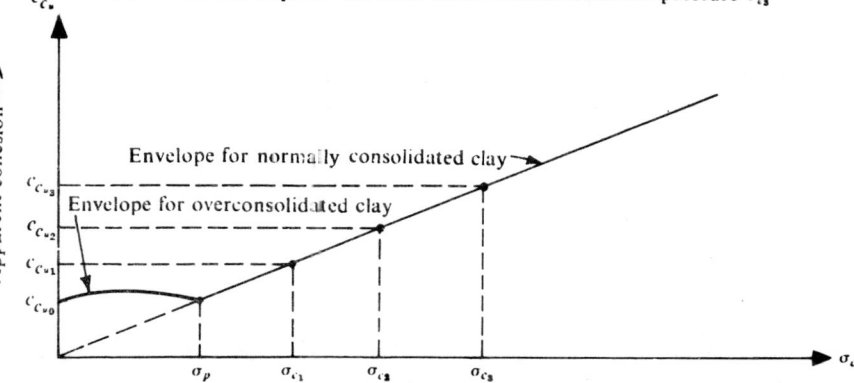

(c) Failure envelope for the third series with consolidation pressure σ_{c_3}

(d) Variation of apparent cohesion with consolidation pressure

Fig. 8.35 Consolidated undrained tests in triaxial compression on remoulded, saturated, and normally consolidated clay consolidated under a particular cell pressure and sheared undrained under cell pressures different from consolidated pressures

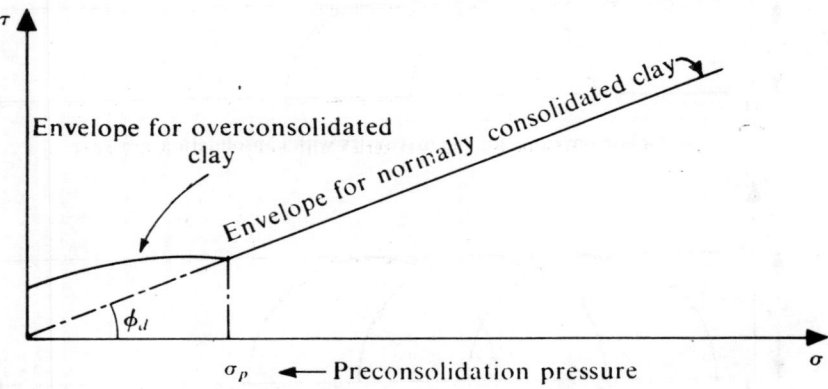

Fig. 8.36 Drained tests in triaxial compression on a remoulded saturated clay sheared under cell pressure equal to the consolidation pressure

Stress-strain behaviour of clays

The stress-strain behaviour of clays is primarily dependent upon whether the clay is in a normally consolidated state or in an overconsolidated state. The stress-strain relationships for a normally consolidated clay and those for an overconsolidated clay are shown in Figs. 8.37 and 8.38 respectively.

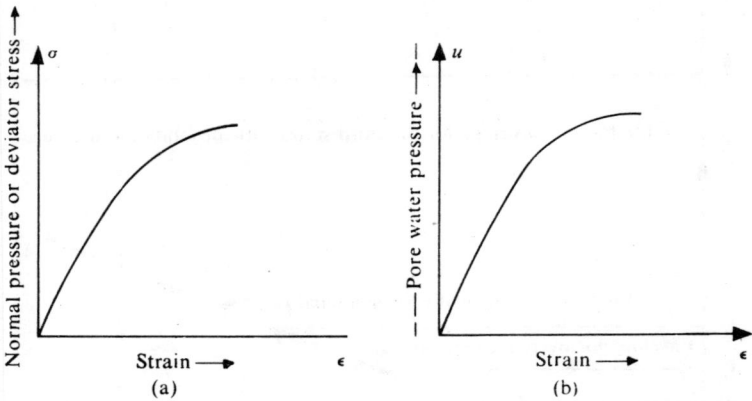

Fig. 8.37 Stress-strain relationships for a normally consolidated clay

The behaviour of a normally consolidated clay is somewhat similar to that of a loose sand and that of an overconsolidated clay is similar to that of a dense sand. In the case of plastic nature of stress-strain relationship with no specific failure point, an arbitrary strain of 15 to 20% is considered to be representative of failure condition.

Shearing Strength of Soils 343

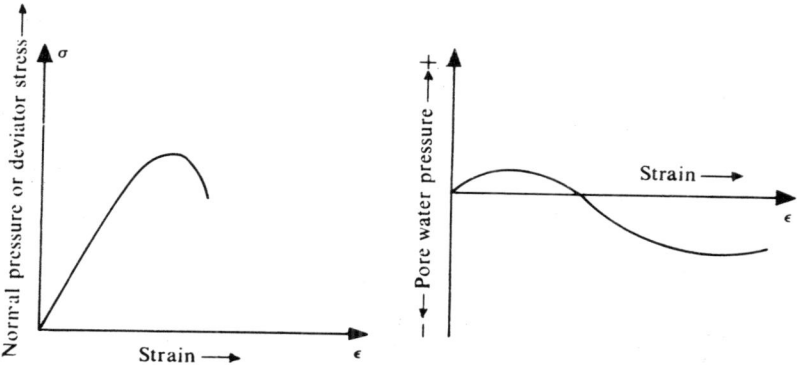

Fig. 8.38 Stress-strain relationships for an overconsolidated clay

Effect of rate and nature of shear strain

Clays are often sensitive to the rate and manner of shearing. Usually standard rates of shearing are adopted for proper comparison. A strain of about 0.10 to 0.15 cm/min., is considered standard in strain-controlled direct shear. However, it is not common that strain is controlled in nature or in construction operations.

It is observed that shear strength increases somewhat with increased rates of strain. If the loading is not at a uniform rate but is effected in increments, much greater shearing resistance is developed; however, the failure in such a case is observed to occur rather suddenly. The increase in shear strength could be as much as 25% with increase in rate of strain from a very slow rate; this increase would be as high as 100% or more if the loading is by increments.

If there is interruption of strain, the shear stress could decrease steadily by a creep in saturated clays; but in the case of sands, this will not have any significant effect on shearing stress.

Also, greater shearing displacements are associated with smaller rates of shearing strain and vice versa. This is also in contrast to the behaviour of sand for which these factors do not appear to materially affect the results.

Sensitivity of clays

If the strength of an undisturbed sample of clay is measured and its strength is again measured after remoulding at the same water content to the same dry density, a reduction in strength is often observed. This is an important phenomenon which is quantitatively characterised by 'Sensitivity', defined as follows:

$$\text{Sensitivity, } S_t = \frac{\text{Unconfined compression strength, undisturbed}}{\text{Unconfined compression strength, remoulded}}$$

...(Eq. 8.56)

A comparison of stress-strain curves for a sensitive clay in the undisturbed and remoulded states is shown in Fig. 8.39.

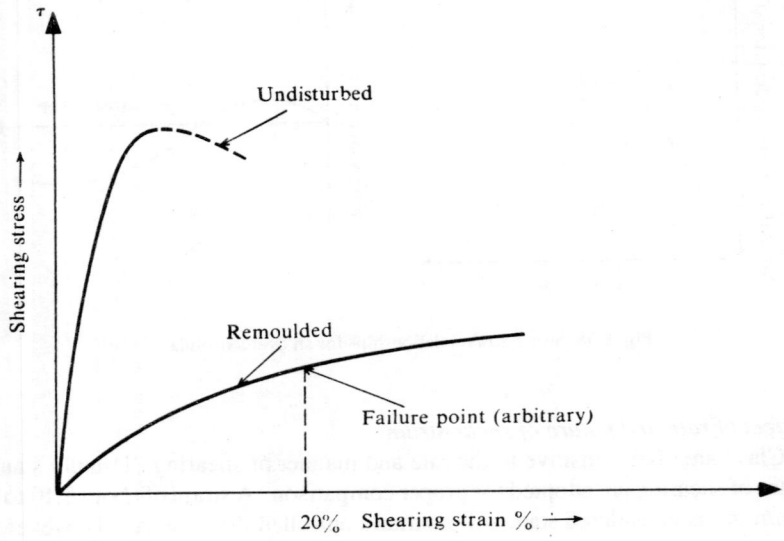

Fig. 8.39 Stress-strain curves for a sensitive clay in the undisturbed and remoulded states

Sensitivity classification is given in the table below:

Table 4.1 Sensitivity classification of clays (Smith, 1974)

Sensitivity S_t	Classification
1	Insensitive
1 – 2	Low
2 – 4	Medium
4 – 8	Sensitive
8 – 16	Extra-sensitive
Greater than 16	Quick (S_t can be even up to 150)

Overconsolidated clays are rarely sensitive, although some quick clays have been found to be overconsolidated.

8.12 ILLUSTRATIVE EXAMPLES

Example 8.1: The stresses at failure on the failure plane in a cohesionless soil mass were: Shear stress = 4 kN/m²; normal stress = 10 kN/m². Determine the resultant stress on the failure plane, the angle of internal friction of the soil and the angle of inclination of the failure plane to the major principal plane.

Resultant stress $= \sqrt{\sigma^2 + \tau^2}$

$$= \sqrt{10^2 + 4^2} = \mathbf{10.77 \text{ kN/m}^2}$$

$\tan \phi = \tau/\sigma = 4/10 = 0.4$

$\phi = \mathbf{21°48'}$

$\theta = 45° + \phi/2 = 45° + \dfrac{21°48'}{2} = \mathbf{55°54'}$

Graphical solution (Fig. 8.40):

The procedure is first to draw the σ- and τ-axes from an origin O and then, to a suitable scale, set-off point D with coordinates $(10, 4)$. Joining O to D, the strength envelope is got. The Mohr Circle should be tangential to OD at D. DC is drawn perpendicular to OD to cut OX in C, which is the centre of the circle. With C as the centre and CD as radius, the circle is completed to cut OX in A and B.

By scaling, the resultant stress $= OD = 10.8$ kN/m^2.
With protractor, $\phi = 22°$ and $\theta = 55°53'$
We also observe that $\sigma_3 = OA = 7.25$ kN/m^2 and $\sigma_1 = OB = 15.9$ kN/m^2

Example 8.2: Clean and dry sand samples were tested in a large shear box, 25 cm × 25 cm and the following results were obtained:

Normal load (kN)	5	10	15
Peak shear load (kN)	5	10	15
Ultimate shear load (kN)	2.9	5.8	8.7

Determine the angle of shearing resistance of the sand in the dense and loose states.

The value of ϕ obtained from the peak stress represents the angle of shearing resistance of the sand in its initial compacted state; that from the ultimate stress corresponds to the sand when loosened by the shearing action.

The area of the shear box $= 25 \times 25 = 625$ cm^2.
$= 0.0625$ m^2.
Normal stress in the first test $= 5/0.0625$ kN/m$^2 = 80$ kN/m^2
Similarly the other normal stresses and shear stresses are obtained by dividing by the area of the box and are as follows in kN/m^2:

Normal stress, σ	80	160	240
Peak shear stress, τ_{max}	80	160	240
Ultimate shear stress, τ_f	46.4	92.8	139.2

346 *Geotechnical Engineering*

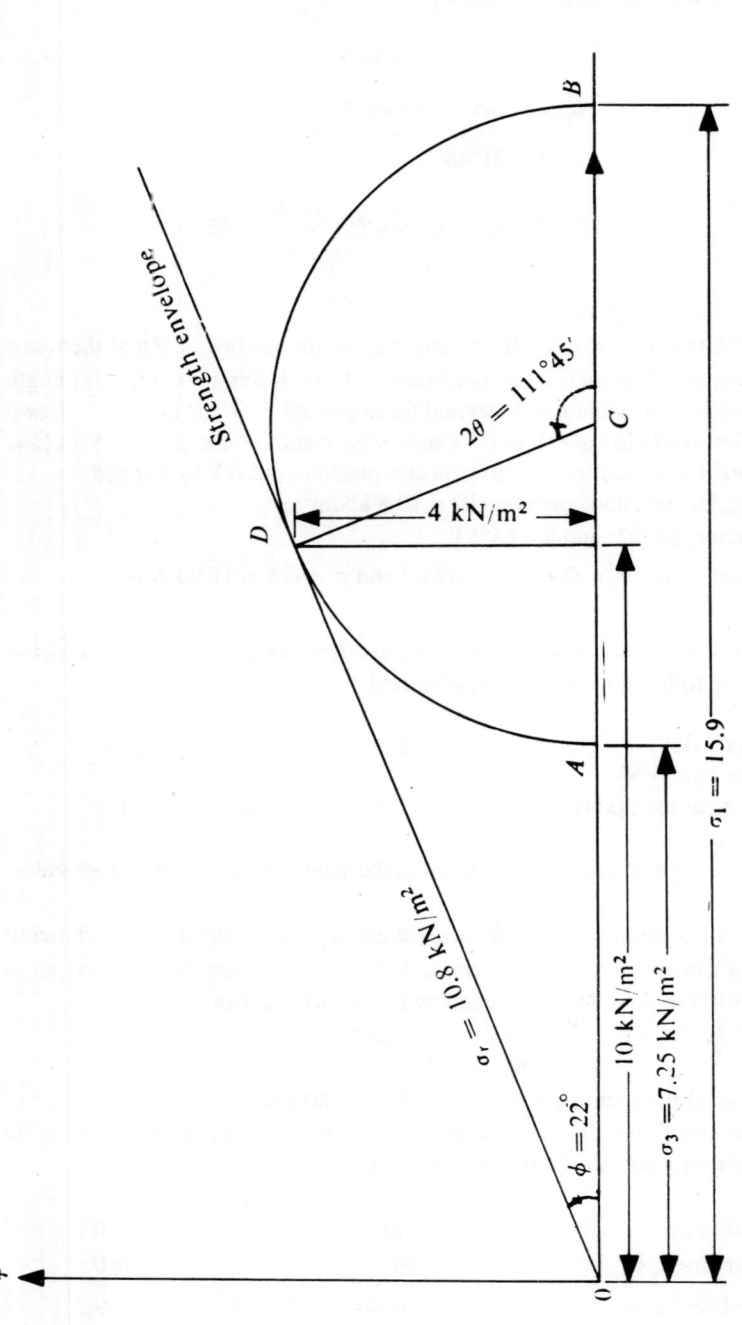

Fig. 8.40 Mohr's circle (Ex. 8.1)

Since more than one set of values are available, graphical method is better:

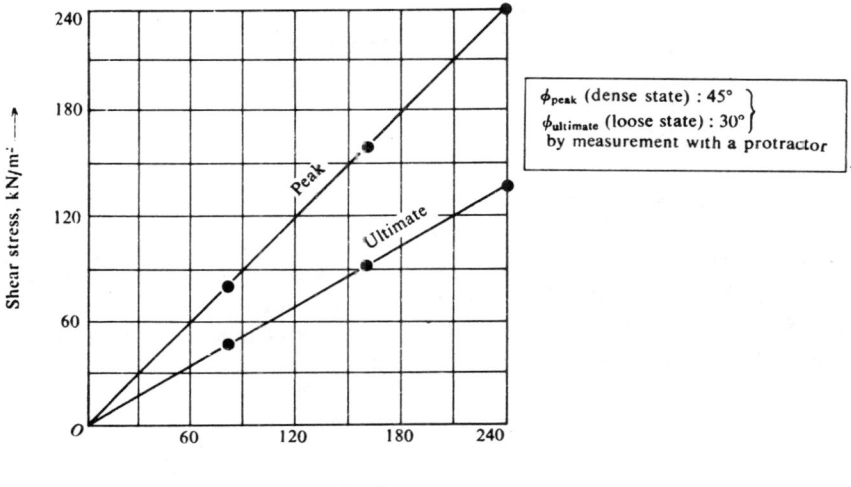

Fig. 8.41 Failure envelopes (Ex. 8.2)

Example 8.3: Calculate the potential shear strength on a horizontal plane at a depth of 3 m below the surface in a formation of cohesionless soil when the water table is at a depth of 3.5 m. The degree of saturation may be taken as 0.5 on the average. Void ratio = 0.50; grain specific gravity = 2.70; angle of internal friction = 30°. What will be the modified value of shear strength if the water table reaches the ground surface?

(S.V.U—B.E., (R.R.)—Feb., 1976)

Effective unit weight $\gamma' = \dfrac{(G-1)}{(1+e)} \cdot \gamma_w$

$= \dfrac{(2.70-1)}{(1+0.5)} \times 1000 \text{ kg/m}^3 = 1133 \text{ kg/m}^3$

Unit weight, γ, at 50% saturation

$= \dfrac{(G+S_e)}{(1+e)} \cdot \gamma_w = \dfrac{(2.70+0.5 \times 0.5)}{(1+0.5)} \times 1000 \text{ kg/m}^3$

$= 1960.7 \text{ kg/m}^3$

(a) When the water table is at 3.5 m below the surface:
Normal stress at 3 m depth, $\sigma = 1967 \times 3 = 5900 \text{ kg/m}^2$

Shear strength, $s = \sigma \cdot \tan \phi$ for a sand
$$= 5900 \tan 30° = \mathbf{3400 \text{ kg/m}^2} \text{ (nearly)}.$$
(b) When water table reaches the ground surface:
Effective Normal stress at 3 m depth
$$\overline{\sigma} = \gamma' \cdot h = 1133 \times 3 = 3400 \text{ kg/m}^2$$

Shear strength, $s = \overline{\sigma} \tan \phi$

$$= 3400 \tan 30°$$

$$= \mathbf{1960 \text{ kg/m}^2} \text{ (nearly)}.$$

Example 8.4: The following data were obtained in a direct shear test. Normal pressure = 20 kN/m², tangential pressure = 16 kN/m². Angle of internal friction = 20°, cohesion = 8 kN/m². Represent the data by Mohr's Circle and compute the principal stresses and the direction of the principal planes.

(S.V.U.—B.E., (N.R.)—May, 1969)

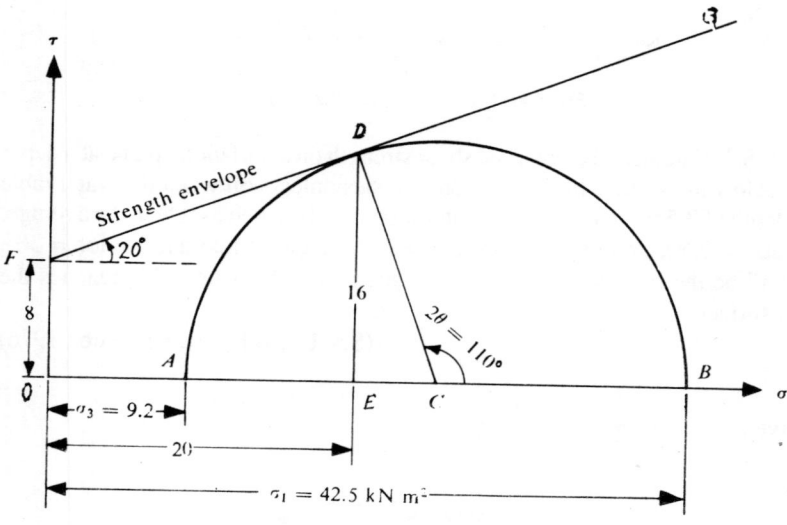

Fig. 8.42 Mohr's Circle (Ex. 8.4)

The strength envelope FG is located since both c and ϕ are given. Point D is set-off with co-ordinates (20, 16) with respect to the origin O; it should fall on the envelope. (In this case, there appears to be slight discrepancy in the data). DC is drawn perpendicular to FD to meet the σ-axis in C. With C as centre and CD as radius, the Mohr's circle is completed. The principal stresses σ_3 (OA) and σ_1 (OB) are scaled off and found to be **9.2 kN/m²** and **42.5 kN/m²**. Angle BCD is measured and found to be 110°. Hence the major principal plane is inclined at 55° (clockwise) and the minor principal plane at 35° (counter clockwise) to the plane of shear (horizontal plane, in this case).

Analytical solution:

$$\sigma_1 = \sigma_3 N_\phi + 2c\sqrt{N_\phi}$$

$$N_\phi = \tan^2(45° + \phi/2) = \tan^2 55° = 2.04$$

$$\sigma_1 = 2.04\,\sigma_3 + 2\times 8 \times \tan 55° = 2.04\sigma_3 + 22.88 \qquad \ldots(1)$$

$$\sigma_n = \sigma_1 \cos^2 55° + \sigma_3 \sin^2 55° = 20$$

$$0.33\,\sigma_1 + 0.67\,\sigma_3 = 20 \qquad \ldots(2)$$

Solving, $\sigma_1 = 42.5$ kN/m² and $\sigma_3 = 9.2$ kN/m², as obtained graphically.

Example 8.5: The following results were obtained in a shear box test. Determine the angle of shearing resistance and cohesion intercept:

Normal stress (kg/cm²)	1	2	3
Shear stress (kg/cm²)	1.30	1.85	2.40

(S.V.U.—B.Tech. (Part-time)— June, 1981)

The normal and shear stresses on the failure plane are plotted as shown:

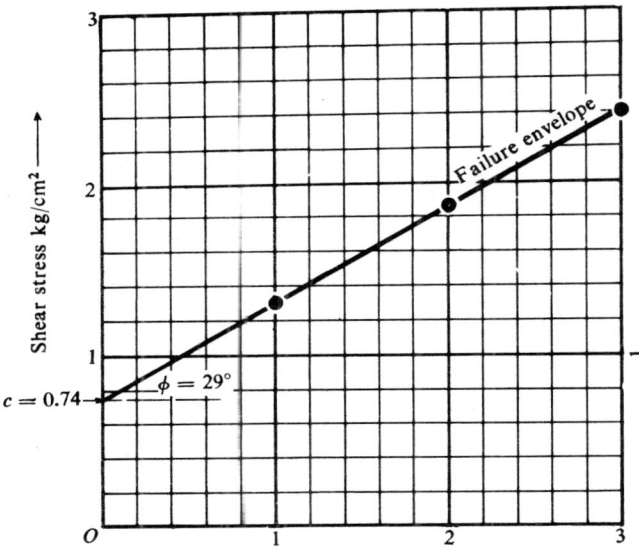

Fig. 8.43 Failure envelope (Ex. 8.5)

350 Geotechnical Engineering

The intercept on the shear stress axis is cohesion, c, and the angle of inclination of the failure envelope with the normal stress axis of the angle of shearing resistance, ϕ.

From Fig. 8.43,
$$c = 0.74 \text{ kg/cm}^2$$
$$\phi = 29°$$

Example 8.6: A series of shear tests were performed on a soil. Each test was carried out until the sample sheared and the principal stresses for each test were:

Test No.	(kN/m²)	(kN/m²)
1	200	600
2	300	900
3	400	1200

Plot the Mohr's circles and hence determine the strength envelope and angle of internal friction of the soil.

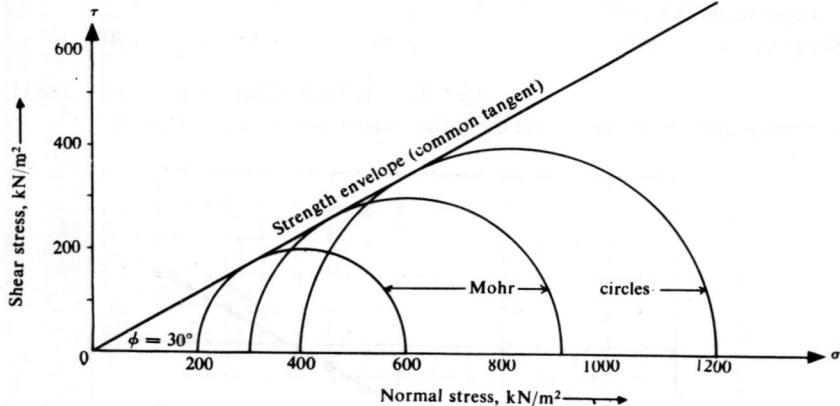

Fig. 8.44 Mohr's circles and strength envelopes (Ex. 8.6)

The data indicate that the tests are triaxial compression tests; the Mohr's circles are plotted with $(\sigma_1 - \sigma_3)$ as diameter and the strength envelope is obtained as the common tangent.

The angle of internal friction is found to be **30°**, by measurement with a protractor from Fig. 8.44.

Example 8.7: A particular soil failed under a major principal stress of 300 kN/m² with a corresponding minor principal stress of 100 kN/m². If, for the same soil, the minor principal stress had been 200 kN/m², determine what the major principal stress would have been if
(a) $\phi = 30°$ and (b) $\phi = 0°$.

Graphical solution:

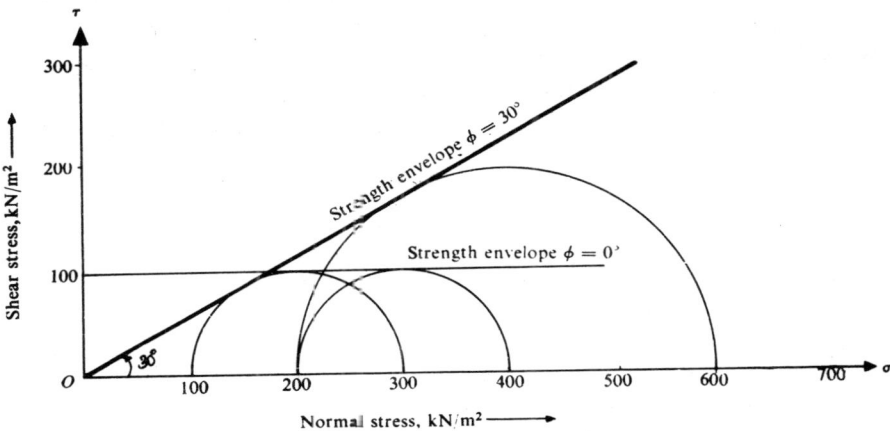

Fig. 8.45 Mohr's circle and strength envelope (Ex. 8.7)

The Mohr circle of stress is drawn to which the strength envelope will be tangential; the envelopes for $\phi = 0°$ and $\phi = 30°$ are drawn. Two stress circles, each starting at a minor principal stress value of 200 kN/m², one tangential to $\phi = 0°$ envelope, and the other tangential to $\phi = 30°$ envelope are drawn.

The corresponding major principal stresses are scaled off as **400 kg/m² and 600 kN/m²**.

Analytical solution:
(a) $\phi = 30°$;

$$\sigma_3 = 100 \text{ kN/m}^2 \quad \sigma_1 = 300 \text{ kN/m}^2$$

$$\frac{\sigma_3}{\sigma_1} = \frac{1 - \sin \phi}{1 + \sin \phi} = \frac{1 - \sin 30°}{1 + \sin 30°} = 1/3$$

The given stress circle will be tangential to the strength envelope with $\phi = 30°$.
With $\sigma_3 = 200$ kN/m², $\sigma_1 = 3 \times 200 = $ **600 kN/m²**,
if the circle is to be tangential to the strength envelope $\phi = 30°$ passing through the origin.

(b) $\phi = 0°$;

If the given stress circle has to be tangential to the strength envelope $\phi = 0°$, the envelope has to be drawn with $c = \tau = 100$ kN/m². The deviator stress will then be 200 kN/m², irrespective of the minor principal stress.
Hence $\sigma_1 = 200 + 200 = $ **400 kN/m²** for $\sigma_3 = 200$ kN/m².

352 *Geotechnical Engineering*

Example 8.8: The stresses acting on the plane of maximum shearing stress through a given point in sand are as follows: total normal stress = 250 kN/m²; pore-water pressure = 88.5 kN/m²; shearing stress = 85 kN/m². Failure is occurring in the region surrounding the point. Determine the major and minor principal effective stresses, the normal effective stress and the shearing stress on the plane of failure and the friction angle of the sand. Define clearly the terms 'plane of maximum shearing stress' and 'plane of failure' in relation to the Mohr's rupture diagram.
(S.V.U.—B.E., (R.R.)—Nov., 1973)

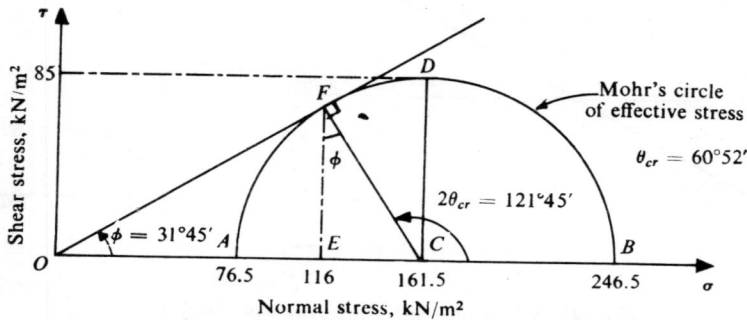

Fig. 8.46 Mohr's circle of effective stresses (Ex. 8.8)

Total normal stress = 250 kN/m²
Pore water pressure = 88.5 kN/m²
Effective normal stress on the plane of maximum shear = (250 − 88.5) = 161.5 kN/m²
Max. Shear stress = 85 kN/m²

Graphical solution:
 The normal stress on the plane of maximum shear stress is plotted as OC to a suitable scale; CD is plotted perpendicular to $\bar{\sigma}$-axis as the maximum shear stress. With C as centre and CD as radius, the Mohr's circle is established. A tangent drawn to the circle from the origin O establishes the strength envelope. The foot of the perpendicular E from the point of tangency F is located. The principal effective stresses and the stresses on the plane of failure are scaled-off. The angle of internal friction is measured with a protractor.

The results are: Major effective principal stress = (OB) = **246.5 kN/m²**
 Minor effective principal stress (OA) = **76.5 kN/m²**
Angle of internal friction, ϕ (angle FOB) = **31°45′**
Normal effective stress on plane of failure (DE) = **116 kN/m²**
Shearing stress on the plane of failure (EF) = **72 kN/m²**

Analytical solution:
$$\left(\frac{\bar{\sigma}_1 + \bar{\sigma}_3}{2}\right) = \text{Normal stress on the plane of maximum shear} = 161.5 \quad \ldots(1)$$

$$\left(\frac{\bar{\sigma}_1 - \bar{\sigma}_3}{2}\right) = \text{Maximum shear stress} = 85 \qquad ...(2)$$

Solving (1) and (2), $\bar{\sigma}_1$ = **246.5 kN/m²** (Major principal effective stress)

$\bar{\sigma}_3$ = **76.5 kN/m²** (Minor principal effective stress)

$$\sin \phi = \frac{(\bar{\sigma}_1 - \bar{\sigma}_3)/2}{(\bar{\sigma}_1 + \bar{\sigma}_3)/2} = \frac{85}{161.5} = 0.526$$

∴ Angle of internal friction ϕ = **31°45′** nearly.

Normal stress on the failure plane

$$= \left(\frac{\bar{\sigma}_1 + \bar{\sigma}_3}{2}\right) - \left(\frac{\bar{\sigma}_1 - \bar{\sigma}_3}{2}\right) \cdot \sin \phi$$

$$= 161.5 - \frac{85 \times 85}{161.5} = 116.76 \text{ kN/m}^2$$

Shear stress on the failure plane

$$= \left(\frac{\bar{\sigma}_1 - \bar{\sigma}_3}{2}\right) \cdot \cos \phi$$

$$= 85 \times \cos 31°45′ = 72.27 \text{ kN/m}^2$$

The answers from a graphical approach compare very well with those from the analytical approach. The planes of maximum shear, i.e., the planes on which the shearing stress is the maximum, are inclined at 45° with the principal planes.

The failure plane i.e., the plane on which the resultant has max. obliquity, is inclined at $(45° + \phi/2)$ or **61°52′** (Counterclockwise) with the major principal plane. These observations are confirmed from the Mohr's circle of stress.

Example 8.9: In an unconfined compression test, a sample of sandy clay 8 cm long and 4 cm in diameter fails under a load of 12 kg at 10% strain. Compute the shearing resistance taking into account the effect of change in cross-section of the sample.

(S.V.U.— B.Tech. (Part-time)—May, 1983)

Size of specimen = 4 cm dia. × 8 cm long.

Initial area of cross-section = $(\pi/4) \times 4^2 = 4\pi$ cm².

Area of of cross-section at failure $= \dfrac{A_0}{(1-\varepsilon)}$

$$= \frac{4\pi}{(1-0.10)} = 4\pi/0.9 = 40\pi/9 \text{ cm}^2$$

Load at failure = 12 kg.

Axial stress at failure $= \dfrac{12 \times 9}{40\pi}$ kg/cm²

$= 2.7/\pi$ kg/cm²

$= 0.86$ kg/cm²

Shear stress at failure $= \dfrac{1}{2} \times 0.86$ kg/cm²

$= \mathbf{0.43}$ **kg/cm²**

The corresponding Mohr's circle is shown in Fig. 8.47.

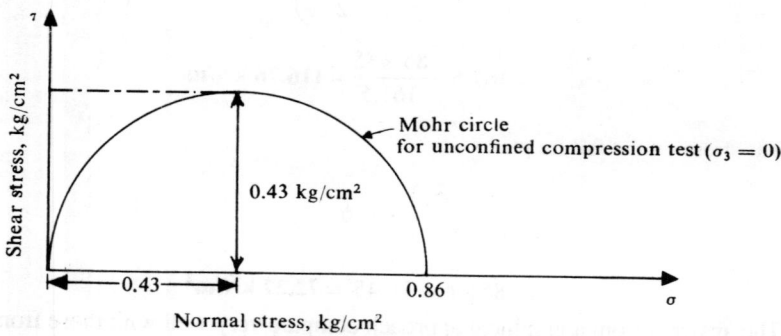

Fig. 8.47 Mohr's circle for unconfined compression test (Ex. 8.9)

Example 8.10: A cylindrical specimen of a saturated soil fails under an axial stress 150 kN/m² in an unconfined compression test. The failure plane makes an angle of 52° with the horizontal. Calculate the cohesion and angle of internal friction of the soil.

Analytical solution:

The angle of the failure plane with respect to the plane on which the major principal (axial) stress acts is:

$$\theta_{cr} = 45° + \phi/2 = 52°$$

$$\therefore \quad \phi/2 = 7° \text{ or } \phi = \mathbf{14°}$$

$$\sigma_1 = 150 \text{ kN/m}^2 \quad \sigma_3 = 0$$

$$\sigma_1 = \sigma_3 N_\phi + 2c\sqrt{N_\phi}$$

where $N_\phi = \tan^2(45° + \phi/2) = \tan^2 52°$

$\sqrt{N_\phi} = \tan 52°$

∴ $150 = 0 + 2 \times c \tan 52°$

∴   Cohesion, $c = 75/\tan 52° = \mathbf{58.6 \text{ kN/m}^2}$

Graphical solution:

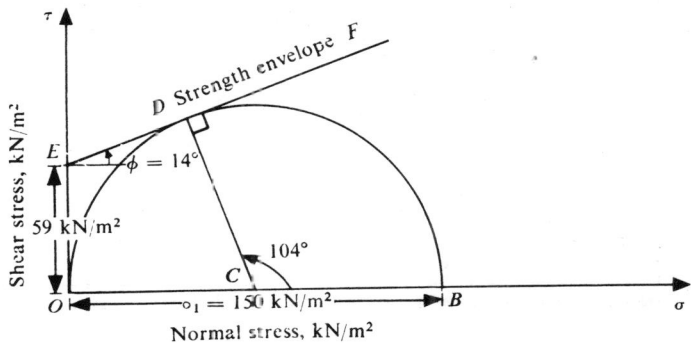

Fig. 8.48 Mohr's circle and strength envelope (Ex. 8.10)

The axial stress is plotted to a suitable scale as *OB*. With *OB* as diameter, the Mohr's circle is established. At the centre *C*, angle *ACD* is set-off as $2 \times 52°$ or $104°$ to cut the circle in *D*. A tangent to the circle at *D* establishes the strength envelope. The intercept of this on the τ-axis gives the cohesion *c* as 59 kN/m² and the angle of slope of this line with horizontal gives φ as 14°. These values compare very well with those from the analytical approach.

Example 8.11: In a triaxial shear test conducted on a soil sample having a cohesion of 12 kN/m² and angle of shearing resistance of 36°, the cell pressure was 200 kN/m². Determine the value of the deviator stress at failure.

(S.V.U.—B.E., (R.R)—Nov., 1974)

The strength envelope is drawn through *E* on the τ-axis, *OE* being equal to $C = 12$ kN/m² to a convenient scale, at an angle φ = 36° with the σ-axis. The cell pressure, $\sigma_3 = 200$ kN/m² is plotted as *OA*. With centre on the σ-axis, a circle is drawn to pass through *A* and be tangential to the envelope, by trial and error. *AC* is scaled-off, *C* being the centre of the Mohr's circle, which is $(\sigma_1 - \sigma_3)/2$. The deviator stress is double this value. In this case the result is **616 kN/m²**.

Analytical solution :

$$c = 12 \text{ kN/m}^2$$

$$\phi = 36°$$

$$\sigma_3 = 200 \text{ kN/m}^2$$

$$\sigma_1 = \sigma_3 N_\phi + 2c\sqrt{N_\phi}$$

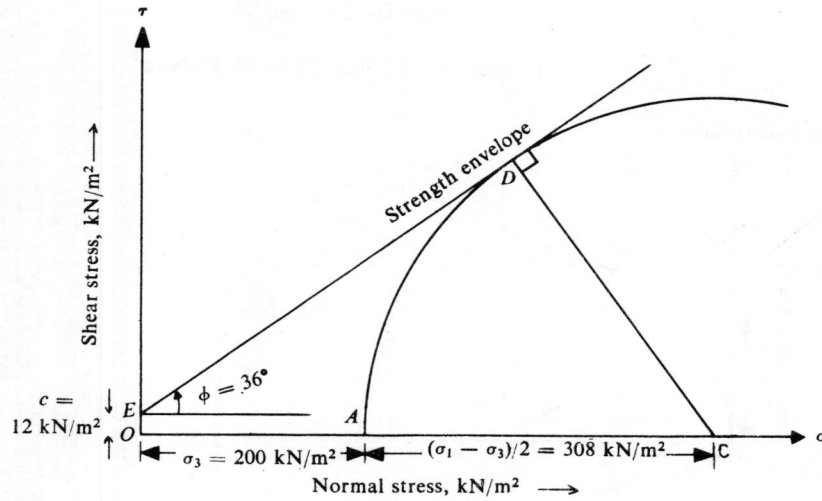

Fig. 8.49 Mohr's circle for triaxial test (Ex. 8.11)

where $N_\phi = \tan^2(45° + \phi/2)$.

$$N_\phi = \tan^2(45° + 18°) = \tan^2 63° = 3.8518$$

$$\sqrt{N_\phi} = \tan 63° = 1.9626$$

$$\therefore \quad \sigma_1 = 200 \times 3.8518 + 2 \times 12 \times 1.9626 = 817.5 \text{ kN/m}^2$$

Deviator stress = $\sigma_1 - \sigma_3 = (817.5 - 200)$ kN/m² = 617.5 kN/m²

The result from the graphical solution agrees well with this value.

Example 8.12: A triaxial compression test on a cohesive sample cylindrical in shape yields the following effective stresses:

    Major principal stress   ...   8 MN/m²
    Minor principal stress   ...   2 MN/m²

Angle of inclination of rupture plane is 60° to the horizontal. Present the above data, by means of a Mohr's circle of stress diagram. Find the cohesion and angle of internal friction.

                                      (S.V.U.—Four-year B.Tech.—June, 1982)

The minor and major principal stresses are plotted as *OA* and *OB* to a convenient scale on the σ-axis. The mid-point of *AB* is located as *C*. With *C* as centre and *CA* or *CB* as radius, the Mohr's stress circle is drawn. Angle *BCD* is plotted as $2\theta_{cr}$ or $2 \times 60° = 120°$ to cut the circle in *D*. A tangent to the circle drawn at *D* (perpendicular to *CD*) gives the strength envelope. The intercept of this envelope, on the τ-axis gives the cohesion, *c*, and the inclination of the envelope with σ-axis gives the angle of internal friction, φ.

Graphical solution :

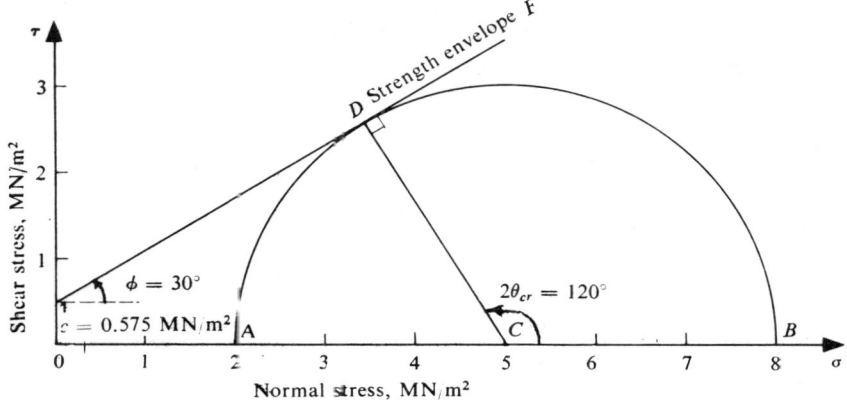

Fig. 8.50 Mohr's circle and strength envelope (Ex. 8.12)

The results obtained graphically are: $c = 0.575$ MN/m^2;
$$\phi = 30°$$

Analytical method:
$\sigma_1 = 8$ MN/m^2 and $\sigma_3 = 2$ MN/m^2 $\theta_{cr} = 60°$

$\theta_{cr} = 45° + \phi/2 = 60°$, whence $\phi = 30°$

$N_\phi = \tan^2(45° + \phi/2) = \tan^2 60° = 3$ $\sqrt{N_\phi} = \sqrt{3}$

$\sigma_1 = \sigma_3 N_\phi + 2c\sqrt{N_\phi}$

∴ $8 = 2 \times 3 + 2 \times c\sqrt{3}$, whence, $c = 1/\sqrt{3} = 0.577$ MN/m^2

The results obtained graphically show excellent agreement with these values.

Example 8.13: A sample of dry sand is subjected to a triaxial test. The angle of internal friction is 37 degrees. If the minor principal stress is 200 kN/m^2, at what value of major principal stress will the soil fail?

(S.V.U.—B.E., (R.R.)—May, 1970)

Analytical method:

$$\phi = 37°$$

$$\sigma_3 = 200 \text{ kN/m}^2$$

For dry sand, $c = 0$.

$$\sigma_1 = \sigma_3 N_\phi + 2c\sqrt{N_\phi}$$

$$= \sigma_3 N_\phi, \text{ since } c = 0$$

358 Geotechnical Engineering

$$N_\phi = \tan^2(45° + \phi/2) = \tan^2(45° + 18°30') = \tan^2 63°30' = 4.0228$$

$$\therefore \quad \sigma_1 = \sigma_3 N_\phi = 200 \times 4.0228 \text{ kN/m}^2$$

Major principal stress, $\sigma_1 = \mathbf{804.56 \text{ kN/m}^2}$

Graphical method:

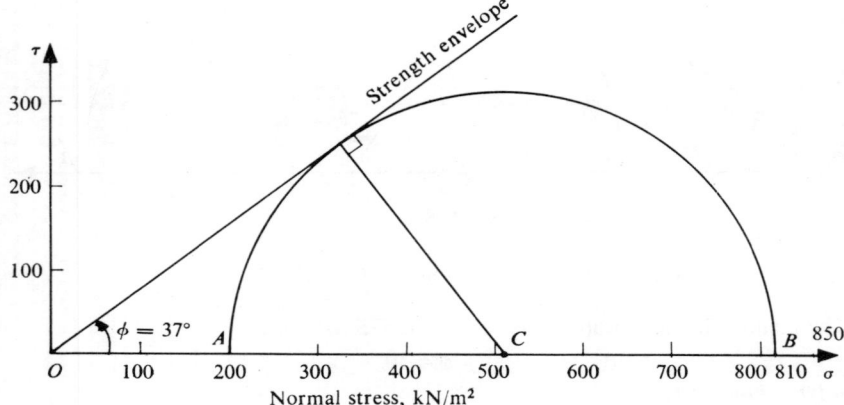

Fig. 8.51 Mohr's circle and strength envelope (Ex. 8.13)

The strength envelope is drawn at 37° to σ-axis, through the origin. The minor principal stress 200 kN/m² is plotted as OA on the σ-axis to a convenient scale. With the centre on the σ-axis, draw a circle to pass through A and be tangential to the strength envelope by trial and error. If the circle cuts σ-axis at B also, OB is scaled-off to give the major principal stress, σ_1.

The result in this case is 810 kN/m² which compares favourably with the analytical value.

Example 8.14: In a drained triaxial compression test, a saturated specimen of cohesionless sand fails under a deviator stress of 5.35 kg/cm² when the cell pressure is 1.5 kg/cm². Find the effective angle of shearing resistance of sand and the approximate inclination of the failure plane to the horizontal. Graphical method is allowed.

(S.V.U.—B.E., (R.R.)—Nov., 1972)

The cell pressure will be the minor principal stress and the major principal stress will be got by adding the deviator stress to it. These principal stresses are plotted as OA and OB to a convenient scale on the σ-axis. C, the mid-point of AB, is the centre of the circle. The Mohr circle is completed with radius as CA or CB. Since this is pure sand, the strength envelope is drawn as the tangent to the circle passing through the origin. Angles DOC and angle BCD are measured with a protractor to give ϕ and $2\theta_{cr}$, respectively. The values in this case are:

Graphical method:

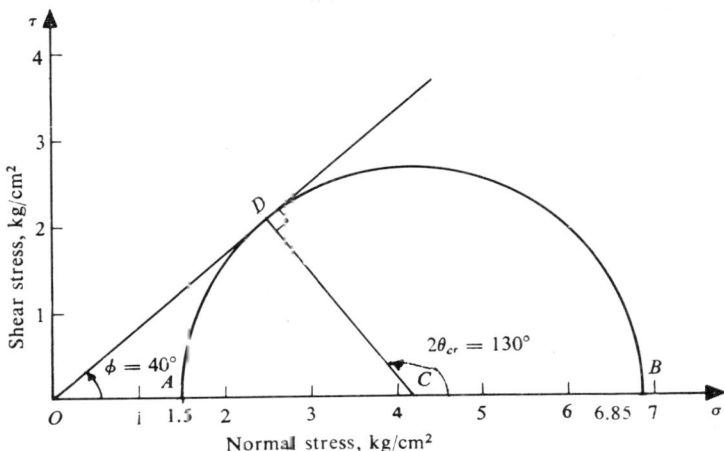

Fig. 8.52 Mohr's circle and strength envelope (Ex. 8.14)

$$\phi = 40°;\ \theta_{cr} = 65° \text{ with the horizontal.}$$

Analytical method:

$$\sigma_3 = 1.5 \text{ kg/cm}^2$$

$$(\sigma_1 - \sigma_3) = 5.35 \text{ kg/cm}^2$$

$$\therefore \quad \sigma_1 = 6.85 \text{ kg/cm}^2$$

$$\sigma_1 = \sigma_3 N_\phi, \text{ where } N_\phi = \tan^2(45° + \phi/2)$$

$$\therefore \quad N_\phi = \sigma_1/\sigma_3 = \frac{6.85}{1.50} = 4.57$$

$$(45° + \phi/2) = 64°55'$$

$$\therefore \quad \phi/2 = 19°55'$$

or $$\phi = 39°50'$$

Hence, $\theta_{cr} = (45° + \phi/2) = \mathbf{64°55'}$.

The graphical values compare very well with these results.

Example 8.15: The shearing resistance of a soil is determined by the equation $s = c' + \sigma' \tan \phi'$. Two drained triaxial tests are performed on the material. In the first test the all-round pressure is 200 kN/m² and failure occurs at an added axial stress of 600 kN/m². In the second test all-round pressure is 350 kN/m² and failure occurs at an added axial stress of 1050 kN/m². What values of c' and ϕ' correspond to these results?

(S.V.U.—B.E., (R.R.)—Nov., 1973)

Graphical method:

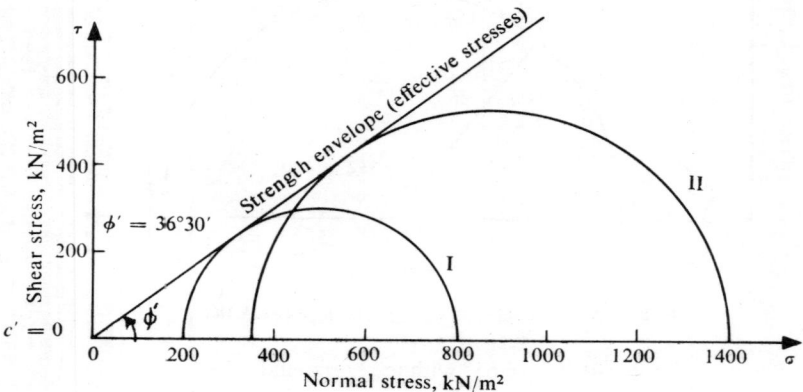

Fig. 8.53 Mohr's circle for effective stress (Ex. 8.15)

Since the cell pressures (σ_3) and the added axial stresses ($\sigma_1 - \sigma_3$) are known, σ_1-values are also obtained by addition. The Mohr's circles for the two tests are drawn. The common tangent to the two circles is seen to pass very nearly through the origin and is sketched. The inclination of this line, which is the strength envelope in terms of effective stresses, with the σ-axis is the effective friction angle. The value of c' is zero; and the value of ϕ', as measured with a protractor, is **36°30′**.

Analytical method:

Since the tests are drained tests, we may assume $c' = 0$. On this basis, we may obtain N_ϕ.

From both tests, $N_\phi = \sigma_1/\sigma_3 = 4$

$$\therefore \quad \sqrt{N_\phi} = \tan(45° + \phi/2) = 2$$

or
$$45° + \phi'/2 = 63°25', \quad \phi'/2 = 18°25'$$

$$\therefore \quad \phi' = 36°50'.$$

The graphical result compares favourably with this value.

Example 8.16: The following data relate to a triaxial compression tests performed on a soil sample:

Test No.	Chamber pressure	Max. deviator stress	Pore pressure at maximum deviator stress
1	80 kN/m^2	175 kN/m^2	45 kN/m^2
2	150 "	240 "	50 "
3	210 "	300 "	60 "

Determine the total and effective stress parameters of the soil.

(S.V.U.—Four year B. Tech.—April, 1983)

Graphical solution:

(a) Total stresses:			(b) Effective stresses = (Total stress − pore pressure)		
S.No.			S.No.		
1	255	80	1	210	35
2	390	150	2	340	100
3	510	210	3	450	150

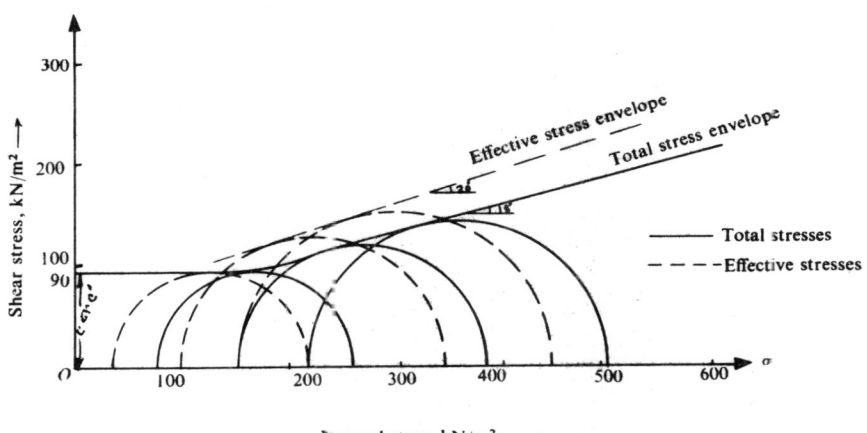

Fig. 8.54 Effective stress and total stress envelopes (Ex. 8.16)

Total stress parameters:

$$c = 90 \text{ kN/m}^2; \phi = 18°$$

Effective stress parameters:

$$c' = 90 \text{ kN/m}^2; \phi' = 20°$$

Example 8.17: Given the following data from a consolidated undrained test with pore water pressure measurement, determine the total and effective stress parameters:

σ_3	100 kN/m²	200 kN/m²
$(\sigma_1 - \sigma_3)_f$	156 kN/m²	198 kN/m²
u_f	58 kN/m²	138 kN/m²

(S.V.U.—B.Tech. (Part-time)—Sept., 1982)

(a) Total stresses: (b) Effective stresses:

σ_3 100 kN/m²   200 kN/m² $\bar{\sigma}_3$ (100 – 58) = 42 kN/m²   62 kN/m²

σ_1 (100 + 156) = 256 kN/m² 398 kN/m² $\bar{\sigma}_1$ (256 – 58) = 198 kN/m² 260 kN/m²

These principal stresses are used to draw the corresponding Mohr's circles (Fig. 8.54).

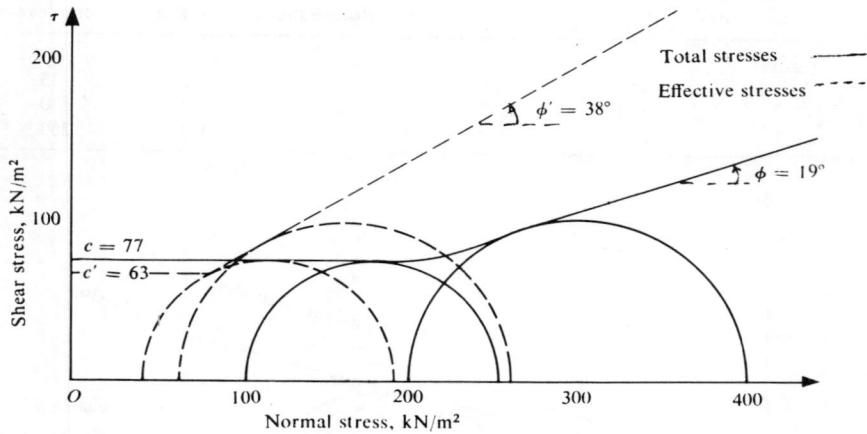

Fig. 8.55 Effective stress and total stresses envelopes (Ex. 8.17)

Total stress parameters:
$$c = 77 \text{ kN/m}^2; \phi = 19°$$

Effective stress parameters:
$$c' = 63 \text{ kN/m}^2; \phi' = 38°$$

The clay should have been overconsolidated.

Example 8.18: A thin layer of silt exists at a depth of 18 m below the surface of the ground. The soil above this level has an average dry density of 1.53 Mg/m³ and an average water content of 36%. The water table is almost at the surface. Tests on undisturbed samples of the silt indicate the following values:

$$c_u = 45 \text{ kN/m}^2; \phi_u = 180°; c' = 36 \text{ kN/m}^2; \phi' = 27°$$

Estimate the shearing resistance of the silt on a horizontal plane, (a) when the shear stress builds up rapidly and (b) when the shear stress builds up very slowly.

Bulk unit weight, $\gamma = \gamma_d(1+w)$

$$= 1.53 \times 1.36 = 2.081 \text{ Mg/m}^3$$

Submerged unit weight, $\gamma' = 2.081 - 1.0 = 1.081 \text{ Mg/m}^3$

Total normal pressure at 18 m depth $= 2.081 \times 9.81 \times 18$

$$\sigma = 367.5 \text{ kN/m}^2$$

Effective pressure at 18 m depth $= 1.081 \times 9.81 \times 18$

$$\bar{\sigma} = 190.9 \text{ kN/m}^2$$

(a) For rapid build-up, the properties for the undrained state and total pressure are to be used:

$$s = c_u + \sigma \tan \phi_u$$

Shear strength $= 45 + 367.5 \tan 18°$

$$= 164.4 \text{ kN/m}^2$$

(b) For slow build-up, the effective stress properties and effective pressure are to be used:

$$s = c' + \bar{\sigma} \tan \phi'$$

Shear strength $= 36 + 190.9 \tan 27°$

$$= 133.3 \text{ kN/m}^2$$

Example 8.19: A vane, 10.8 cm long, 7.2 cm in diameter, was pressed into a soft clay at the bottom of a bore hole. Torque was applied and the value at failure was 45 Nm. Find the shear strength of the clay on a horizontal plane.

$$T = c\pi \left(\frac{D^2 H}{2} + \frac{D^3}{6} \right)$$

for both ends of the vane shear device partaking in shear.

$$45/1000 = c\pi \left(\frac{702^2 \times 10.8}{2} + \frac{7.2^3}{6} \right) \times \frac{1}{100 \times 100 \times 100}$$

$$c = \frac{45 \times 100 \times 100 \times 100}{1000 \left(\frac{7.2^2 \times 10.8}{2} + \frac{7.2^3}{6} \right)} \text{ kN/m}^2$$

$$\approx 42 \text{ kN/m}^2$$

The shear strength of the clay (cohesion) is **42 kN/m²**, nearly.

SUMMARY OF MAIN POINTS

1. Shearing strength of a soil is defined as the resistance to shearing stresses; it is perhaps the most important engineering property and also the most difficult to comprehend in view of the multitude of factors affecting it.
2. Interlocking, friction, and cohesion between soil grains are the important phenomena from which a soil derives its shearing strength.
3. The Mohr's stress circle from which the state of stress on any plane as well as the principal stresses may be obtained, is a versatile tool useful for the solution of problems in shearing strength.
4. According to the Mohr's strength theory and the Mohr-Coulomb theory, if the Mohr's stress circle corresponding to the existing state of stress at a point in a soil touches the failure envelope, failure will be imminent; if it is within the envelope, the strength mobilised is lower than the ultimate strength and the soil is safe.
5. According to the conditions of drainage, shearing strength tests may be classified as the unconsolidated undrained (quick), consolidated-undrained (consolidated quick), and drained (slow) tests; these tend to simulate certain conditions obtaining in field situations.
6. Direct shear, triaxial compression and the unconfined compression tests are the more important of laboratory shear strength tests; triaxial compression test is the most versatile test, capable of simulating many field situations. Unconfined compression test is a simple special case of the triaxial compression test. Field vane and penetration tests are commonly used for field tests.
7. $(\sigma_1 - \sigma_3)/2$ is plotted, against $(\sigma_1 + \sigma_3)/2$ to give Lambe and Whitman's k_f-line or modified failure envelope.
8. The change in pore pressure due to change in applied stress is characterised by dimensionless coefficients, called Skempton's pore pressure parameters A and B.
9. The behaviour of dense sand and of loose sand in shear differ significantly from each other, especially in respect of volume change behaviour. The void ratio at which no volume change occurs in shear is called the 'critical void ratio'. Apparent cohesion is exhibited by saturated sand in UU tests.
10. Cohesion, adhesion, and viscous friction are the sources of shear strength for clays. In UU-tests, cohesion is exhibited with $\phi = 0$; in CU-tests, ϕ alone is exhibited and the strength is independent of the normal pressure during shear, but is dependent only on the consolidation pressure. Strength envelopes may be shown in terms of total stresses as well as in terms of effective stresses; the friction angle from the latter is always greater than that from the former, indicating increase in strength upon drainage.
11. The shear behaviour of overconsolidated clay is different from that of normally consolidated clay; the strength envelope for the former will be much flatter than that for the latter.

12. Sensitivity of a clay is an index of the loss of strength or disturbance of the structure; quantitatively, it is the ratio of the unconfined compression strength values in the undisturbed and in the remoulded states.

REFERENCES

1. Alam Singh and B.C. Punmia: *Soil Mechanics and Foundations*, Standard Book-house, Delhi-6, 1970.
2. A.W. Bishop: *The Measurement of Pore Pressure in the Triaxial Test*, conference on Pore Pressure and Suction in soils, London: Butterworths, 1960.
3. A.W. Bishop and D.J. Henkel: *The Measurement of Soil Properties in the Triaxial Test*, Edward Arnold Ltd., London, 1962.
4. C.A. Coulomb: *Essai sur une application des règles de maximis et minimis à quelques problems de statique relatifs à l'architecture, Memoires de la mathèmatique et de physique*, présentés a l'Academic Royale des Sciences, par divers savants, et lius dans ses Assemblèes, Paris, De L'Imprimerie Royale, 1776.
5. M.J. Hvorslev: *The Physical Components of the Shear Strength of Saturated Clays*, ASCE Research conference on Shear strength of cohesive soils, Boulder, Colarado USA, 1960.
6. A.R. Jumikis: *Soil Mechanics*, D.Van Nostrand Co., Princeton, NJ, USA, 1962.
7. T.W. Lambe: *Soil Testing for Engineers*, John Wiley and Sons, Inc., New York, 1951.
8. T.W. Lambe and R.V. Whitman: *Soil Mechanics*, John Wiley and Sons Inc., New York, 1969.
9. D.F. McCarthy: *Essentials of Soil Mechanics and Foundations*, Reston Publishing Company, Reston, Va., USA, 1977.
10. Otto Mohr: *Über die Darstellung des Spannunge zustandes und das Deformationzustandes eines Körper Elements*, Zivilingenieur, 1882.
11. Otto Mohr: *Technische Mechanik*, Wilhelm Ernst und Sohn, Berlin, 1906.
12. V.N.S. Murthy: *Soil Mechanics and Foundation Engineering*, Dhanpat Rai and Sons, Delhi-6, 2nd ed., 1977.
13. S.B. Sehgal: *A Textbook of Soil Mechanics*, Metropolitan Book Co. Pvt. Ltd., Delhi, 1967.
14. A.W. Skempton: *The Pore Pressure Coefficients A and B*, Geotechnique, Vol.4., 1954.
15. G.N. Smith: *Essentials of Soil Mechanics for Civil and Mining Engineers*, Third Edition Metric, Crosby Lockwood Staples, London, 1974.
16. M.G. Spangler: *Soil Engineering*, International Textbook Company, Scranton, USA, 1951.
17. D.W. Taylor: *Fundamentals of Soil Mechanics*, John Wiley and Sons, Inc., New York, 1948.

18. K. Terzaghi: *The Shearing Resistance of Saturated Soils and the Angle between Planes of Shear*, Proceedings, First International Conference of Soil Mechanics and Foundation Engineering, Cambridge, Mass., USA., 1936.

QUESTIONS AND PROBLEMS

8.1 Explain the principle of the direct shear test. What are the advantages of this test? What are its limitations?
(S.V.U.—Four year B.Tech.—April, 1983)

8.2 What are the advantages and disadvantages of a triaxial compression test? Briefly explain how you conduct the test and compute the shear parameters for the soil from the test data.
(S.V.U.—B.Tech.(Part-time)—May, 1983), (B.E., (R.R.)—Sep., 1978)

8.3 Differentiate between unconsolidated undrained test and a drained test. Under what conditions are these test results used for design purposes?
(S.V.C.—B.Tech. (Part-time)—Sep., 1982)

8.4 Write brief critical notes on:
(a) Mohr's Circle
(B.Tech. (Part-time)—Sep., 1982)
(b) Unconfined compression Test
(B.Tech. (Part-time)—June, 1981)
(c) Triaxial test and its merits
(B.Tech. (R.R.)—Feb., 1976)

8.5 (a) Explain the basic differences between a box shear test and a triaxial shear test for soils.
(b) Differentiate between shear strength parameters obtained from total and effective stress considerations.
(S.V.U.—B.Tech. (Part-time)—Apr., 1982)

8.6 (a) Explain the Mohr-Coulomb strength envelope.
(b) Sketch the stress-strain relationship for dense and loose sand.
(S.V.U.—B.Tech. (Part-time)—Apr., 1982)

8.7 What are the three standard triaxial shear tests with respect to drainage conditions? Explain with reasons the situations for which each test is to be preferred.
(S.V.U.—B.Tech. (Part-time)—June, 1981, B.E. (R.R.)—Nov., 1974)

8.8 Explain the shear characteristics of sand and normally loaded clays.
(S.V.U. (B.E.) (R.R.)—May, 1971)

8.9 (a) What is the effect of pore pressure in strength of soils?
(b) Explain Coulomb's law for shearing strength of soils and its modification by Terzaghi.
(S.V.U.—B.E. (N.R.)—May, 1969)

8.10 A granular soil is subjected to a minor principal stress of 200 kN/m^2. If the angle of internal friction is 30°, determine the inclination of the plane of failure with respect to the direction of the major principal stress. What are the stresses on the plane of failure and the maximum shear stress induced?

8.11 Samples of compacted, clean, dry sand were tested in a shear box, 6 cm × 6 cm, and the following observations were recorded:

Normal load	(N):	100	200	300
Peak shear load	(N):	90	180	270
Ultimate shear load	(N):	75	150	225

Determine the angle of shearing resistance in (a) the dense state and in (b) the loose state.

8.12 An embankment consists of clay fill for which $c' = 25$ kN/m² and $\phi = 27°$ (from consolidated-undrained tests with pore-pressure measurement). The average bulk unit-weight of the fill is 2 Mg/m³. Estimate the shear-strength of the material on a horizontal plane at a point 20 m below the surface of the embankment, if the pore pressure at this point is 180 kN/m² as shown by a piezometer.

8.13 The following data are from a direct shear test on an undisturbed soil sample. Represent the data by a Mohr Circle and compute the principal stresses and direction of principal planes:
Normal pressure = 16.2 kN/m²; Tangential pressure = 14.4 kN/m²,
Angle of internal friction = 24°; $c = 7.2$ kN/m².

(S.V.U.—B.E. (N.R.)—Sept., 1967)

8.14 From a direct shear test on an undisturbed soil sample, the following data have been obtained. Evaluate the undrained strength parameters by plotting the results. Also draw a Mohr Circle corresponding to the second test. Hence determine the major and minor principal total stresses for the second test.

Normal stress (kg/cm²)	...	0.70	0.96	1.14
Shear stress (kg/cm²)	...	1.38	1.56	1.70

(S.V.U.—B.E. (R.R.)—May, 1975)

8.15 Two samples of a soil were subjected to shear tests. The results were as follows:

Test No.	σ_3 (kN/m²)	σ_1 (kN/m²)
1	100	240
2	300	630

In a further sample of the same soil was tested under a minor principal stress of 200 kN/m², what value of major principal stress can be expected at failure?

8.16 A shear box test carried out on a soil sample gave:

Test No.	Vertical stress (kN/m²)	Horizontal shear stress (kN/m²)
1	100	80
2	200	144
3	300	216

Determine the magnitude of the major and minor principal stresses at failure when the vertical stress on the sample was 200 kN/m². Determine also the inclination to the horizontal of these stresses.

368 Geotechnical Engineering

8.17 A series of undrained shear box tests (area of box = 360 mm^2) were carried out on a soil with the following results:

Normal load (N)	Shear force at failure (N)
90	70
180	90
270	117

(i) Determine the cohesion and angle of friction of the soil with respect to total stresses.

(ii) If a 30 mm diameter, 72 mm long sample of the same soil was tested in a triaxial machine, with a cell pressure 270 kN/m^2 what would be the additional axial load at failure if the sample shortened by 6.3 mm?

(iii) If a further sample of the soil was tested in an unconfined compression apparatus, at what value of compressive stress would failure be expected?

8.18 A cylindrical specimen of saturated clay, 4.5 cm in diameter, and 9 cm long, is tested in an unconfined compression apparatus. Find the cohesion if the specimen fails at an axial load of 45 kg. The change in length of the specimen at failure is 9 mm.

8.19 A cylindrical specimen of a saturated soil fails at an axial stress of 180 kN/m^2 in an unconfined compression test. The failure plane makes an angle of 54° with the horizontal. What are the cohesion and angle of internal friction of the soil?

8.20 The following results were obtained from an undrained triaxial test on a soil:

Cell pressure kN/m^2	Additional axial stress at failure (kN/m^2)
200	690
400	840
600	990

Determine the cohesion and angle of internal friction of the soil with respect to total stresses.

8.21 A triaxial compression test on a cohesive soil sample of cylindrical shape yielded the following results:
Major principal stress ... 10 kg/cm^2
Minor principal stress ... 2.5 kg/cm^2
If the angle of inclination of the rupture plane to the horizontal is 60°, determine the cohesion and angle of internal friction by drawing Mohr circle or by calculation.

(S.V.U.—B.E. (R.R.)—May, 1969)

8.22 A sample of dry sand is subjected to a triaxial test. The angle of internal friction is 36°. If the cell pressure is 180 kN/m^2, at what value of deviator stress will the soil fail?

8.23 In a drained triaxial compression test, a saturated specimen of cohesionless sand fails at a deviator stress of 450 kN/m^2 when the cell pressure was 135 kN/m^2. Find the effective angle of shearing resistance of sand and the angle of inclination of the failure plane with the horizontal.

8.24 An undisturbed soil sample, 100 mm in diameter and 200 mm high, was tested in a triaxial machine. The sample failed at an additional axial load of 3 kN with a vertical deformation of 20 mm. The failure plane was inclined at 50° to the horizontal and the cell pressure was 300 kN/m². Determine, from Mohr's circle, the total stress parameters. A further sample of the soil was tested in a shear box under the same drainage conditions as used for the triaxial test. If the area of the box was 3600 mm², and the normal load was 540 N what would have been the failure shear stress?

8.25 Pore pressure measurements were made during undrained triaxial tests on samples of compacted fill material from an earth dam after saturating them in the laboratory. The results were as follows:

Property measured (t/m²)	I Test	II Test
Lateral pressure (σ_3)	15	45
Total vertical pressure (σ_1)	40	100
Pore water pressure (u)	3	12.5

Determine the apparent cohesion and the angle of shearing resistance as referred to (i) total stress and (ii) effective stress.

(S.V.U.—B.E. (R.R.)—Sept., 1978)

8.26 The following results were obtained from a series of undrained triaxial tests carried out on undisturbed samples of soil.

Cell pressure (kN/m²)	Additional axial load at failure (N)
200	270
400	330
600	390

Each sample, originally 36 mm diameter and 72 mm high, had a vertical deformation of 5.4 mm. Determine total stress parameters.

8.27 For a normally consolidated insensitive clay $\phi_{CID} = 30°$. Deviator stress at failure of the same soil is 2.5 kg/cm² in *UU* test. If Skempton's A-parameter at failure is 0.62, find out ϕ_{CIU} for this soil. Also, find out the confining pressure during the consolidation stage.

(S.V.U.—B.Tech. (Part-time)—April, 1982)

9
STABILITY OF EARTH SLOPES

9.1 INTRODUCTION

Earth slopes may be found in nature or may be man-made. These are invariably required in the construction of highways, railways, earth dams and river-training works. The stability of these earth slopes is, therefore, of concern to the geotechnical engineer, since failure entails loss of life and property.

The failure of an earth slope involves a 'slide'. Gravitational forces and forces due to seepage of water in the soil mass, progressive disintegration of the structure of the soil mass and excavation near the base are among the chief reasons for the failure of earth slopes. Slides and consequent failure of earth slopes can occur slowly or suddenly.

The slides that occurred during the construction of the Panama canal, connecting the Atlantic and the Pacific Oceans and during the construction of railways in Sweden spurred the geotechnical engineers all over the world into a lot of research on various aspects of the stability of earth slopes. Swedish engineers were in the forefront in this regard.

Determination of the potential failure surface and the forces tending to cause slip and those tending to restore or stabilise the mass of earth are the essential steps in the stability analysis of earth slopes and the available margin of safety. The soil mass is assumed to be homogeneous. It is also assumed that it is possible to compute the seepage forces from the flow net and the shearing strength of the soil from the Mohr-Coulomb theory.

The slope may be an 'infinite' one or a 'finite' one. An infinite slope represents the surface of a semi-infinite inclined soil mass; obviously, such a slope is rather hypothetical in nature. A slope of a finite extent, bounded by a top surface is said to be finite. Slopes involving a cohesive-frictional soil are most common; however, the case of purely cohesionless soils is also treated as a useful introduction to the treatment of $c - \phi$ soils.

9.2 INFINITE SLOPES

An *'infinite slope'* is one which represents the boundary surface of a semi-infinite soil mass inclined to the horizontal. In practice, if the height of the slope is very large, one may consider it as an infinite one. It is assumed that the soil is homogeneous in its properties. If different strata are present the strata boundaries are assumed to be parallel to the surface. Failure tends to occur only along a plane parallel to the surface. The stability analysis for such slopes is relatively simple and it is dealt with for the cases of purely cohesionless soil, purely cohesive soil and cohesive-frictional soil; the cases in which seepage forces under steady seepage and rapid drawdown occur are also considered for a purely cohesionless soil.

9.2.1 Infinite Slope in Cohesionless Soil

Let us consider an infinite slope in cohesionless soil, inclined at an angle β to the horizontal, as shown in Fig. 9.1.

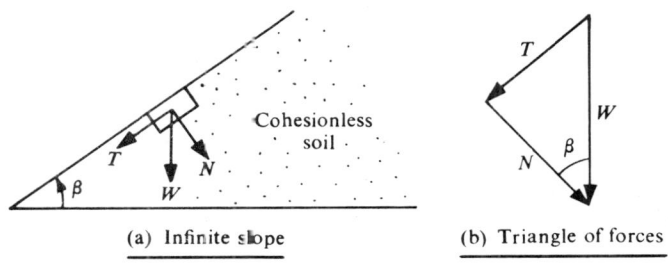

(a) Infinite slope (b) Triangle of forces

Fig. 9.1 Infinite slope in a cohesionless soil

If the weight of an element of the soil mass at the surface is W, the components of W parallel to and perpendicular to the surface of the slope are $T = W \sin \beta$ and $N = W \cos \beta$ respectively. The maximum force restraining the sliding action of T is the shear resistance that could be mobilised by the normal component N. For a cohesionless soil, this is given by $N \tan \phi$ or $W \cos \beta \tan \phi$, where ϕ is the angle of internal friction.

The factor of safety F against sliding or failure is given by:

$$F = \frac{\text{Restraining force}}{\text{Sliding force}} = \frac{W \cos \beta \tan \phi}{W \sin \beta} = \frac{\tan \phi}{\tan \beta} \qquad \text{...(Eq. 9.1)}$$

For limiting equilibrium ($F = 1$),

$$\tan \beta = \tan \phi$$

or
$$\beta = \phi.$$

Thus, the maximum inclination of an infinite slope in a cohesionless soil for stability is equal to the angle of internal friction of the soil. It is interesting to note that the stability is affected neither by the unit weight of the soil nor by the water content, provided seepage forces do not enter into the picture.

Purely granular soils are infrequent as most soils possess some cohesion, but a study of the former affords useful introductory ideas to the treatment of cohesive-frictional soils which are of most frequent occurrence in nature.

Even if a vertical element extending to a finite depth is considered, similar situations exist and the factor of safety against slippage on a plane parallel to the surface at that depth is crucial. In terms of the shearing stresses and the shearing strength as defined by the Mohr-Coulomb envelope, the limit angle of inclination for stability of the slope may be indicated as in Fig. 9.2.

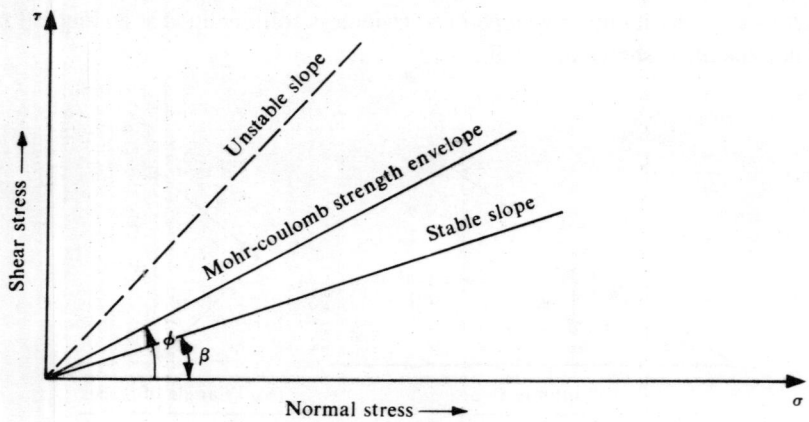

Fig. 9.2 Relation between strength envelope and angle of slope

Rapid drawdown in a slope in cohesionless soil

When the water level in a river or reservoir recedes, say after floods or after a drawdown, the water in the slope of the embankment may not fall as rapidly as that in the river or the reservoir, depending upon the permeability of the soil. This gives rise to a condition commonly known as "sudden or rapid drawdown". The effect of this is that seepage occurs from the high water level in the slope to the lower water level of the river. A flow net can be drawn for this condition and the excess hydrostatic head at any point within the slope can be determined.

Let us consider an element within the slope as shown in Fig. 9.3.

Let the weight of the element be W. Let the excess pore water pressure induced by seepage be u at the base of the element. Let the length of the element perpendicular to the plane of the figure be unity.

$$\text{Normal stress } \sigma_n = \frac{\text{Normal component } N \text{ of weight } W}{l},$$

l being the width of the element parallel to the surface.

$$\therefore \quad \sigma_n = \frac{W \cos \beta}{l} = \frac{W \cos^2 \beta}{b}, \text{ since } l = b/\cos \beta$$

$$\sigma_n = \frac{\gamma z\, b \cos^2 \beta}{b} = \gamma z \cos^2 \beta \qquad \ldots\text{(Eq. 9.2)}$$

Stability of Earth Slopes 373

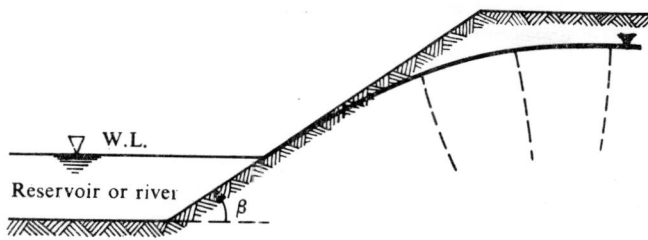

(a) Earth slope subjected to rapid drawdown

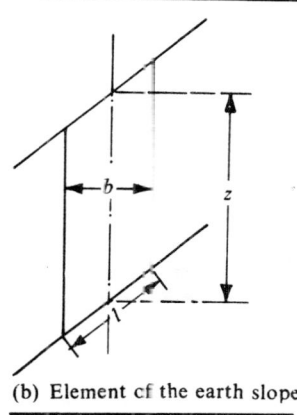

(b) Element of the earth slope

Fig. 9.3 Rapid drawdown in a slope in cohesionless soil

Effective normal stress $\bar{\sigma}_n = (\sigma_n - u)$

$$= (\gamma z \cos^2 \beta - u)$$

(γ is the average unit weight of the slice, which is usually considered saturated.)

Shear stress $\tau = \dfrac{W \sin \beta}{l} = \gamma z \sin \beta \cos \beta$ ...(Eq. 9.3)

Shear strength of soil $= \bar{\sigma}_n \tan \phi$

$$= (\gamma z \cos^2 \beta - u) \tan \phi$$

Factor of safety against slippage, $F = \dfrac{\text{Shear strength}}{\text{Shear stress}}$

$$\therefore \quad F = \dfrac{(\gamma z \cos^2 \beta - u) \tan \phi}{\gamma z \sin \beta \cos \beta}$$

$$= [(1/\tan \beta) - (u/\gamma z \sin \beta \cos \beta)] \cdot \tan \phi$$

$$= \left(1 - \frac{u}{\gamma z \cos^2 \beta}\right) \frac{\tan \phi}{\tan \beta}$$

or
$$F = \left(1 - \frac{r_u}{\cos^2 \beta}\right) \cdot \frac{\tan \phi}{\tan \beta} \qquad \text{...(Eq. 9.4)}$$

where
$$r_u = u/\gamma z \qquad \text{...(Eq. 9.5)}$$

r_u is called the 'pore pressure ratio'.

Flow parallel to the surface and at the surface of a slope in cohesionless soil

If there is a flow parallel to the surface and at the surface of a slope in the cohesionless soil, the flow net is very simple and is depicted in Fig. 9.4.

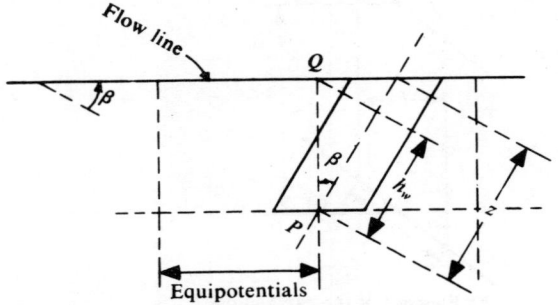

Fig. 9.4 Flow parallel to the surface and at the surface

The excess pore water pressure at the centre P of the base of the element, similar to the previous case, expressed as a head, is represented by the height h_w. From the figure, $PQ = z \cos \beta$, $h_w = PQ \cdot \cos \beta$

$$\therefore \quad h_w = z \cdot \cos^2 \beta$$

The excess pore water pressure $u = \gamma_w h_w = \gamma_w z \cos^2 \beta$

$$\therefore \quad r_u = u/\gamma \cdot z = \frac{\gamma_w z \cos^2 \beta}{\gamma z} = (\gamma_w/\gamma) \cdot \cos^2 \beta \qquad \text{...(Eq. 9.6)}$$

The factor of safety against slippage may be written as:

$$F = \left(1 - \frac{\gamma_w}{\gamma}\right) \frac{\tan \phi}{\tan \beta} = \left(\frac{\gamma - \gamma_w}{\gamma}\right) \frac{\tan \phi}{\tan \beta} = \frac{\gamma'}{\gamma_{sat}} \cdot \frac{\tan \phi}{\tan \beta} \qquad \text{...(Eq. 9.7)}$$

9.2.2. Infinite Slope in a Purely Cohesive Soil

Let us consider an infinite slope in purely cohesive soil as shown in Fig. 9.5.

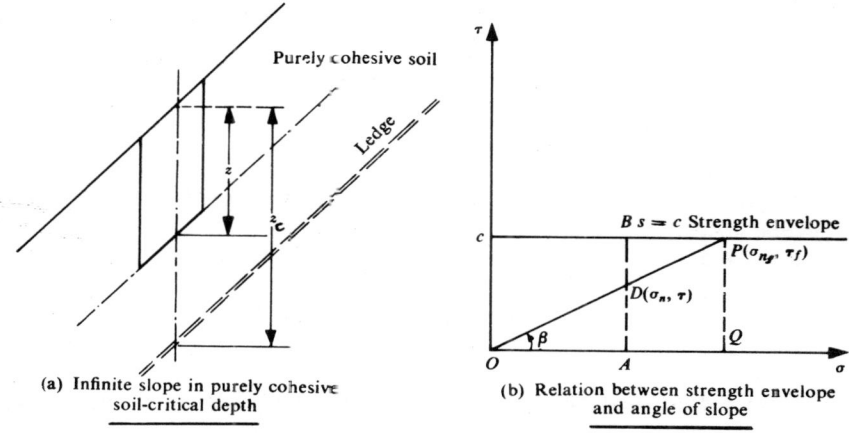

(a) Infinite slope in purely cohesive soil-critical depth

(b) Relation between strength envelope and angle of slope

Fig. 9.5 Infinite slope in a purely cohesive soil

For a particular depth z, the values of the normal and shear stresses at the base of the element are given by Eqs. 9.2 and 9.3, i.e.,

$$\sigma_n = \gamma \cdot z \cos^2 \beta$$

and
$$\tau = \gamma \cdot z \sin \beta \cdot \cos \beta$$

If these are represented as co-ordinates on a $\sigma - \tau$ plot, point D is obtained. This should lie on a line through origin O inclined at the angle of slope β, since $\dfrac{\tau}{\sigma_n} = \tan \beta$. If this point D lies below the Coulomb strength envelope, $s = c$ for the purely cohesive soil, the slope will be stable.

The factor of safety against slippage will be $\dfrac{AB}{AD}$, at a depth z from the surface.

$$\therefore \quad F = c/\tau = \dfrac{c}{\gamma z \sin \beta \cos \beta} \qquad \ldots\text{(Eq. 9.8)}$$

If the line OD is extended it will meet the horizontal strength envelope at a point, say P, the foot of the perpendicular from P on to σ-axis being Q. The point P represents a stress condition for a different depth, greater than z. At this point the shearing stress at the base of the element equals the shearing strength of the soil; that is to say, failure is incipient at this depth. In other words, the slope will be stable only up to a maximum depth z_c, called the critical depth, at which the shearing stress reaches the value of the shearing strength of the soil, which is merely c in this case, as it is a purely cohesive soil. A ledge or some other material with a sufficiently large strength exists below the soil of critical depth.

376 Geotechnical Engineering

The critical depth z_c can be evaluated by equating F to unity.
From Eq. 9.8,

$$1 = \frac{c}{\gamma z_c \sin \beta \cos \beta}$$

or
$$z_c = \frac{c}{\gamma \sin \beta \cos \beta} \qquad \text{...(Eq. 9.9)}$$

Thus for a given value of β, z_c is proportional to cohesion and inversely proportional to the unit weight.

From Eq. 9.9,

$$\frac{c}{\gamma \cdot z_c} = \sin \beta \cos \beta \qquad \text{...(Eq. 9.10)}$$

$\left(\dfrac{c}{\gamma \cdot z_c}\right)$ is a dimensionless quantity and is called the 'stability number', it is designated by S_n.

By combining Eqs. 9.8 and 9.10, we get $F = z_c/z$ ...(Eq. 9.11)

Thus, the factor of safety with respect to cohesion is the same as that with respect to depth. The stability number concept facilitates the preparation of charts and tables for slope stability analysis in more complex situations, especially in the case of finite slopes to be dealt with later.

9.2.3 Infinite Slope in Cohesive-Frictional Soil

Let us consider an infinite slope in a cohesive-frictional soil as shown in Fig. 9.6.

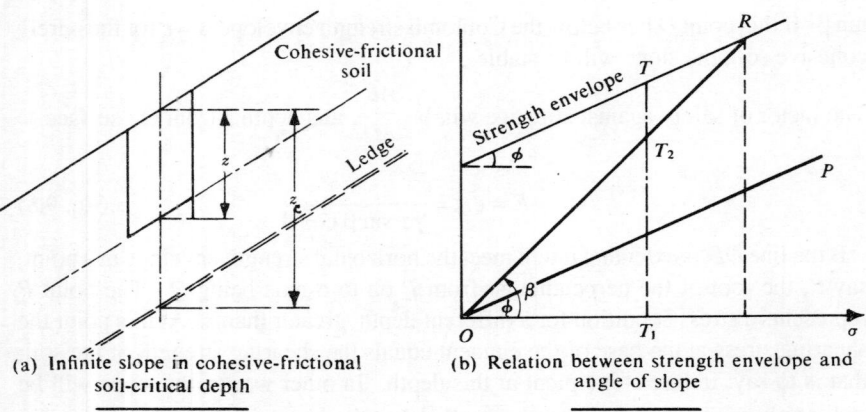

(a) Infinite slope in cohesive-frictional soil-critical depth

(b) Relation between strength envelope and angle of slope

Fig. 9.6 Infinite slope in a cohesive-frictional soil

Stability of Earth Slopes

It is obvious that for a slope with an angle of inclination less than or equal to ϕ, the shearing stress will be less than the shearing strength for any depth, as represented by the line OP; the slope will be stable irrespective of the depth in that case. If the slope is inclined at an angle β greater than ϕ, it cuts the strength envelope at some point such as R. The point R represents the state of stress at a certain depth at which the shearing stress equals the shearing strength and hence denotes incipient failure. For any depth less than that represented by R, the shearing stress will be less than the shearing strength and hence the slope remains stable.

For example, the depth z corresponding to the point T_2 is stable for the slope angle $\beta > \phi$. However, if $\beta > \phi$, the slope can be stable only up to a limited depth, which is known as the critical depth z_c; the state of stress at this depth is represented by R, as already stated.

The equation of the strength envelope is given by:
$$s = c + \sigma_n \tan \phi$$

At failure, $s = \tau_f = c + \sigma_{n_f} \tan \phi$

But $\sigma_{n_f} = \gamma \cdot z_c \cdot \cos^2 \beta$

and $\tau_f = \gamma z_c \cdot \sin \beta \cos \beta$, from Eqs. 9.2 and 9.3.

$$\therefore \quad \gamma z_c \cdot \sin \beta \cos \beta = c + \gamma z_c \cdot \cos^2 \beta \cdot \tan \phi$$

$$z_c \gamma \cos \beta (\sin \beta - \cos \beta \tan \phi) = c$$

$$\therefore \quad z_c = (c/\gamma) \cdot 1/[\cos^2 \beta (\tan \beta - \tan \phi)] \quad \ldots \text{(Eq. 9.12)}$$

Thus the critical depth is proportional to cohesion, for particular values of β and ϕ.

From Eq. 9.12,
$$\frac{c}{\gamma z_c} = \cos^2 \beta (\tan \beta - \tan \phi) \quad \ldots \text{(Eq. 9.13)}$$

The quantity $\dfrac{c}{\gamma z_c}$ is called the stability number S_n.

For any depth z less than z_c, the factor of safety
$$F = \frac{\text{Shearing strength}}{\text{Shearing stress}}$$

$$\therefore \quad F = \frac{c + \gamma z \cos^2 \beta \cdot \tan \phi}{\gamma z \cos \beta \cdot \sin \beta} \quad \ldots \text{(Eq. 9.14)}$$

Since the factor of safety F_c with respect to cohesion,
$$F_c = \frac{c_m}{c}, \text{ where } c_m = \text{mobilised cohesion, at depth } z,$$

378 *Geotechnical Engineering*

$$S_n = c/\gamma z_c = c_m/\gamma z = c/F_c \cdot \gamma z = \cos^2 \beta (\tan \beta - \tan \phi) \qquad \ldots \text{(Eq. 9.15)}$$

From Eqs. 9.13 and 9.15,

$$F_c = \frac{z_c}{z}$$

This is the same as Eq. 9.11, as for a purely cohesive soil.

This is based on the assumption that the frictional resistance of the soil is fully developed. The actual factor of safety should be based on the simultaneous development of cohesion and friction.

If there is a seepage parallel to the ground surface throughout the entire mass of soil, it can be shown that:

$$\frac{c}{\gamma z} = \cos^2 \beta \left(\tan \beta - \frac{\gamma'}{\gamma} \cdot \tan \phi \right) \qquad \ldots \text{(Eq. 9.16)}$$

since effective stress alone is capable of mobilising shearing strength.

9.3 FINITE SLOPES

A 'finite slope', as has already been defined, is one with a base and a top surface, the height being limited. The inclined faces of earth dams, embankments, excavations and the like are all finite slopes. Thus, the stability analysis of such slopes is of vital importance to the geotechnical engineering profession.

Investigation of the stability of finite slopes involves the following steps according to the commonly adopted procedure:
(a) Assuming a possible slip surface,
(b) studying the equilibrium of the forces acting on this surface, and

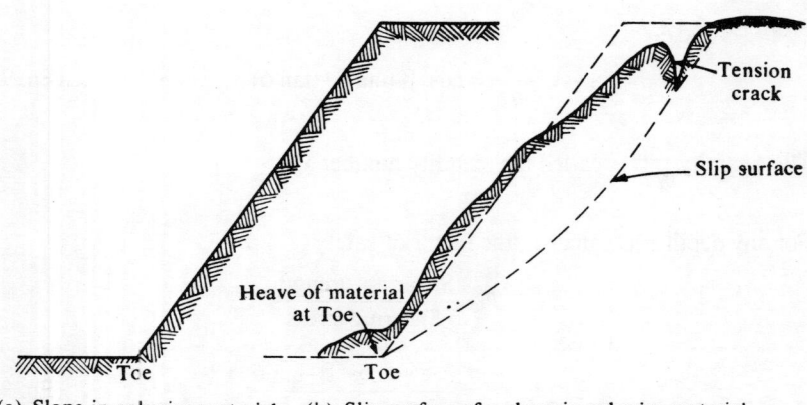

(a) Slope in cohesive material   (b) Slip surface of a slope in cohesive material

Fig. 9.7 Typical characteristics of the rotational slip in a cohesive soil

(c) repeating the process until the worst slip surface, that is, the one with minimum margin of safety, is found.

Failure of finite slopes in cohesive or cohesive-frictional soils tends to occur by rotation, the slip surface approximating to the arc of a circle as shown in Fig. 9.7.

The following important methods will be considered:
- (i) Total stress analysis for purely cohesive soil
- (ii) Total stress analysis for cohesive-frictional soil—the Swedish method of slices
- (iii) Effective stress analysis for conditions of steady seepage, rapid drawdown and immediately after construction.
- (iv) Effective stress analysis by Bishop's method
- (v) Friction circle method
- (vi) Taylor's method

9.3.1 Total Stress Analysis for a Purely Cohesive Soil

Analysis based on total stresses, also called '$\phi = 0$ analysis', gives the stability of an embankment immediately after its construction. It is assumed that the soil has had no time to drain and the shear strength parameters used relate to the undrained strength with respect to total stresses. These may be obtained from either unconfined compression test or an undrained triaxial test without pore pressure measurements.

Let AB be a trial slip surface (a circular arc of radius r) as shown in Fig. 9.8.

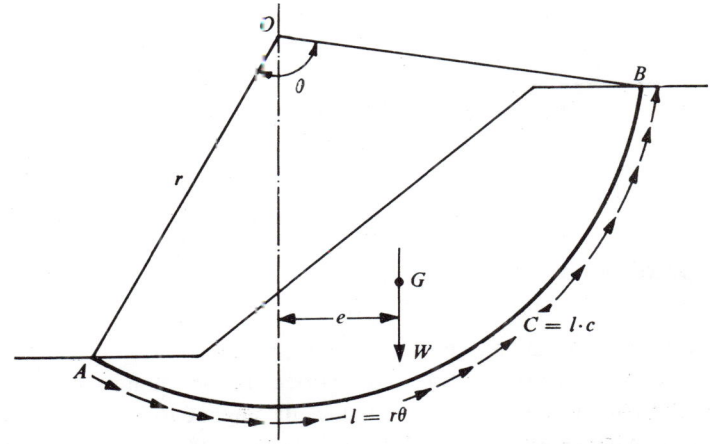

Fig. 9.8 Total stress analysis for a purely cohesive soil

Let W be the weight of the soil within the slip surface and G the position of its centre of gravity. The shearing strength of the soil is c, since $\phi = 0°$.

Taking moments about 0, the centre of rotation:
$W.e = c.l.r = c.r.\theta.r = cr^2.\theta$, for equilibrium (incipient failure).

$$\text{Factor of safety, } F = \frac{\text{Restraining moment}}{\text{Sliding moment}} = \frac{c.r^2.\theta}{W \cdot e} \qquad \text{...(Eq. 9.17)}$$

Here $W.e$ is dependent on the cohesion mobilised which will be less than the maximum cohesion of soil.

The exact position of G is not required and it is only necessary to ascertain the position of the line of action of W. This may be got by dividing the sector into a set of vertical slices and taking moments of area of these slices about any convenient vertical axis.

Effect of Tension Cracks

When slip is imminent in a cohesive soil there will always develop a tension crack at the top surface of the slope along which no shear resistance can develop, as indicated in Fig. 9.9.

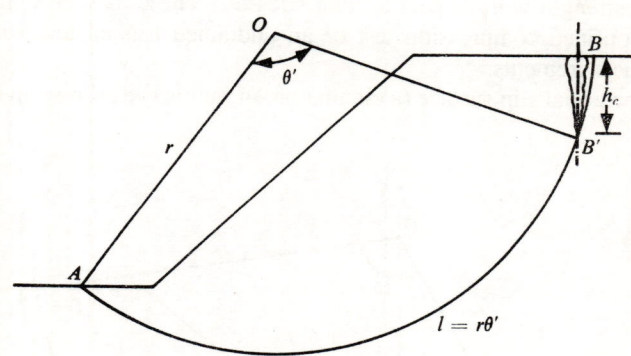

Fig. 9.9 Effect of tension crack in a purely cohesive soil

The depth of tension crack is given by:
$$h_c = 2c/\gamma \qquad \text{...(Eq. 9.18)}$$
(The concept and derivation of this is given in Ch. 13).

The effect of the tension crack is to shorten the arc along which shearing resistance gets mobilised to AB' and to reduce the angle θ to θ'.

In computing the factor of safety F against slippage, θ' is to be used instead of θ, and the full weight W of the soil within the sliding surface AB to compensate for any water pressure that may be exerted, if the crack gets filled with rain water.

The Swedish Method of slices for a cohesive-frictional soil

If the soil is not purely cohesive but is a cohesive-frictional one or if it is partially saturated, the undrained strength envelope shows both c and ϕ values. The total stress analysis can be adapted to this case by dividing the area within the slip circle into a number of vertical slices, as shown in Fig. 9.10.

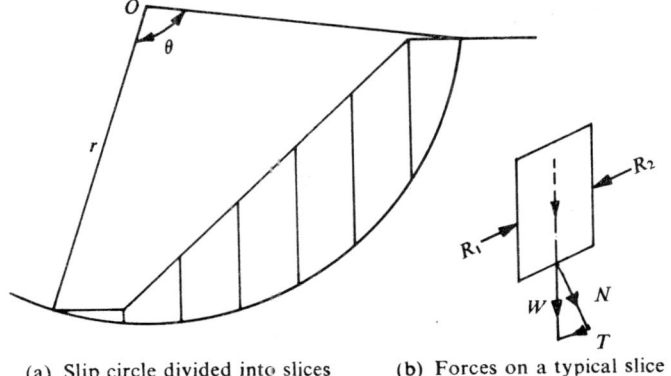

(a) Slip circle divided into slices   (b) Forces on a typical slice

Fig. 9.10 Swedish method of slices for a cohesive-frictional soil

The forces on a typical slice are given in Fig. 9.10(b). The reactions R_1 and R_2 at the sides of the slice are assumed equal. The weight W of the slice is set-off at the base of the slice to a convenient scale. The known directions of its normal component N and the tangential component T are drawn to complete the vector triangle. The values of N and T are scaled-off.

Taking moments about the centre of rotation,

Sliding moment = $r\Sigma T$ (reckoned positive if clockwise)

Restoring moment = $r(cr\theta + \Sigma N \tan \phi)$ (reckoned positive if counterclockwise)

$$\therefore \quad \text{Factor of Safety } F = \frac{(cr\theta + \Sigma N \tan \phi)}{\Sigma T} \quad \ldots\text{(Eq. 9.19)}$$

The effect of a tension crack can be allowed for as in the previous case. But the depth of the crack in this case is given by:

$$h_c = (2c/\gamma) \cdot \tan(45° + \phi/2) \quad \ldots\text{(Eq. 9.20)}$$

The tangential components of a few slices at the base may cause restoring moments. In that case these may be considered negative, thus reducing the denominator in Eq. 9.19. Alternatively these may be added to the numerator.

The values of W, ΣN, and ΣT may be found conveniently in a number of ways.

For example, the values for all slices may be tabulated as follows and summed up:

Slice No.	Area m²	Weight W (kN)	Normal component N (kN)	Tangential components T (kN)
1.				
2.				
3.				
.			Sum, $\Sigma N = \ldots$ kN	$\Sigma T = \ldots$ kN

382 Geotechnical Engineering

Another approach is to draw the N-curve and T-curve, showing the variation of N- and T-values for the various slices with the breadth of the slice as the base, as shown in Fig. 9.11.

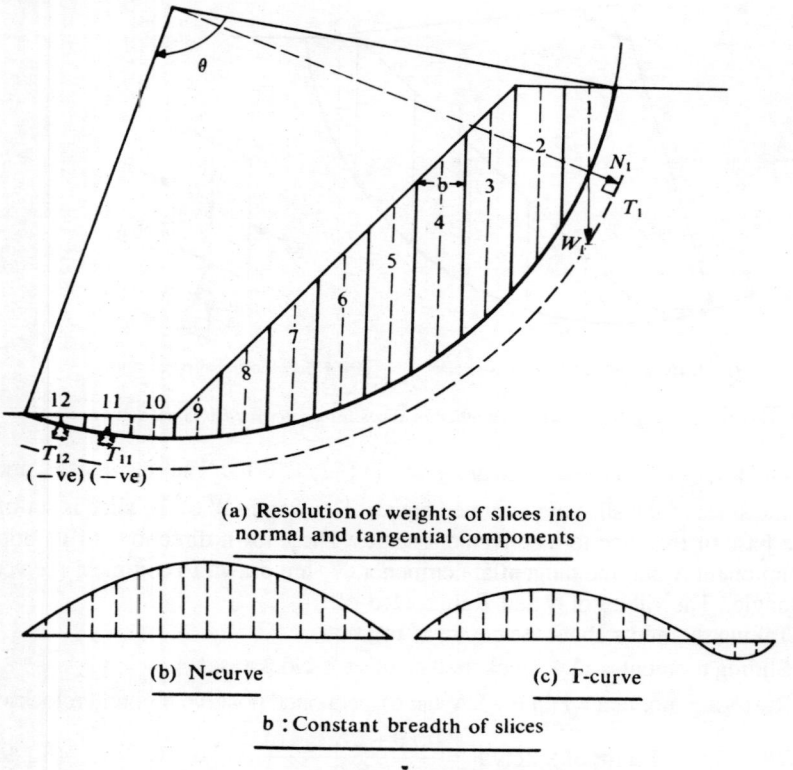

Fig. 9.11 Determination of ΣN and ΣT in the Swedish method of slices

If the areas under the N- and T-curves are found out by a planimeter or otherwise and divided by the constant breadth of the slices, relatively accurate values of ΣN and ΣT will be obtained. The weights of the respective slices can be considered to be approximately proportional to the mid-ordinates and the scale can be easily determined.

The Swedish method of slices is a general approach which is equally applicable to homogeneous soils, stratified deposits, partially submerged cases and non-uniform slopes. Seepage effects also can be considered.

Location of the most critical circle

The centre of the most critical circle can be found only by trial and error. A number of slip circles are to be analysed and the minimum factor of safety finally obtained. One of the procedures suggested for this is shown in Fig. 9.12.

Stability of Earth Slopes 383

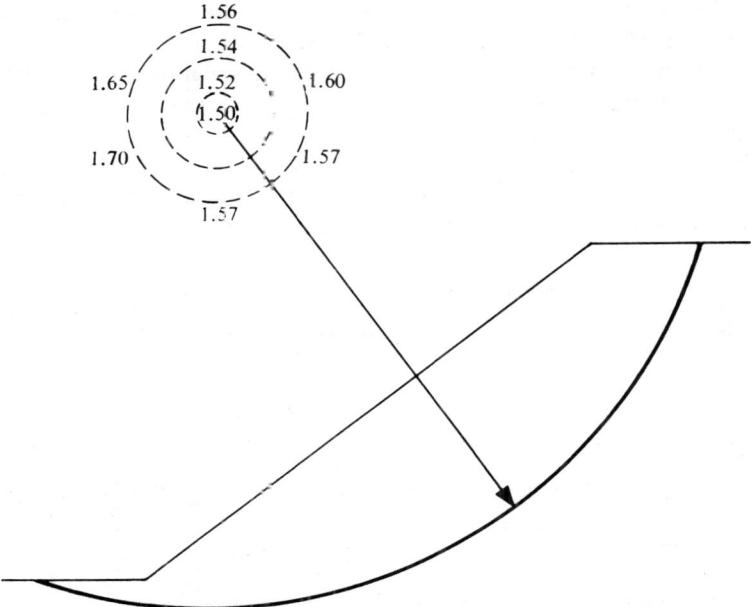

Fig. 9.12 Location of centre of critical circle by contours of factor of safety

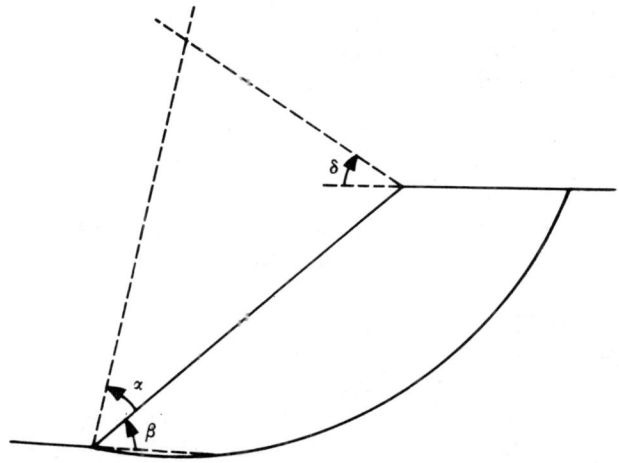

Fig. 9.13 Centre of critical circle – Fellenius' procedure

The centre of each trial circle is plotted and the value of the corresponding factor of safety marked near it. After analysing a number of such trial circles, contours of the factor of safety may be drawn. These will be nearly elliptical. The centre of these ellipses indicates the most probable centre for the critical circle. It may be noted that the value of the safety factor is more sensitive to horizontal movement of the centre of the circle than to vertical movement.

If the slope is made out of homogeneous cohesive soil it is possible to determine directly the centre of the critical circle by a procedure given by Fellenius (1936) which is indicated in Fig. 9.13.

The centre of the critical circle is the intersection of two lines set off from the base and top of the slope at angles α and δ respectively. The values given by Fellenius for α and δ for different values of slope angle β are set out in Table 9.1.

Table 9.1 Fellenius' values for α and δ for the different values of β

S. No.	Slope	Angle of slope β	Angle α at base	Angle δ at top
1	1 : 5	11°.32	25°	37°
2	1 : 3	18°.43	25°	35°
3	1 : 2	26°.57	25°	35°
4	1 : 1.50	33°.79	26°	35°
5	1 : 1	45°	28°	37°
6	1 : 0.58	60°	29°	40°

This procedure is not applicable in its original form to cohesive-frictional soils; however, Jumikis (1962) modified it to be applicable to $c - \phi$ soil, provided they are homogeneous. The modified procedure is shown in Fig. 9.14.

The centre of the Fellenius' circle, O_1, is fixed as given earlier. Then a point P is fixed such that it is $2H$ below the top of the slope and $4.5H$ horizontally from the toe of the slope, H being the critical height of the slope. The centre of the critical circle O, lies on the line PO_1 produced beyond O_1. The distance O_1O increases with the angle of internal friction. After a few trials with centres lying on PO_1 produced, the critical circle is located as the one which gives the minimum factor of safety.

These procedures become less reliable for non-homogeneous conditions such as irregular slope or the existence of pore water pressures.

Typical failure surfaces

A study of the various types of slip that can occur is helpful in determining a reasonable position of the centre of a trial slip circle.

The following information relating to homogeneous soils is relevant.

For soils with $\phi \not< 3°$, the critical slip circle is invariably through the toe. It is so irrespective of the value of ϕ for inclination of slopes exceeding 53°. However, when there is a stiff layer at the base of the slope, the slip circle will be tangential to it.

For cohesive soils with a small friction angle the slip circle tends to be deeper and may extend in front of the toe. If there is a stiff layer below the base, the depth of the slip circle would be limited by this stiff layer.

These various possibilities are illustrated in Fig. 9.15.

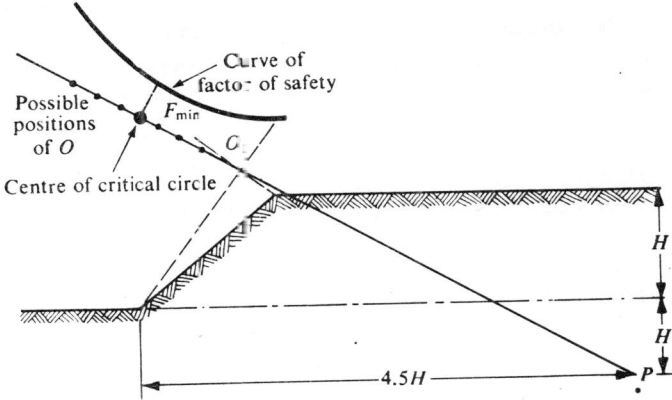

Fig. 9.14 Fellenius' procedure, modified by Jumikis for a $c - \phi$ soil for the centre of the critical circle

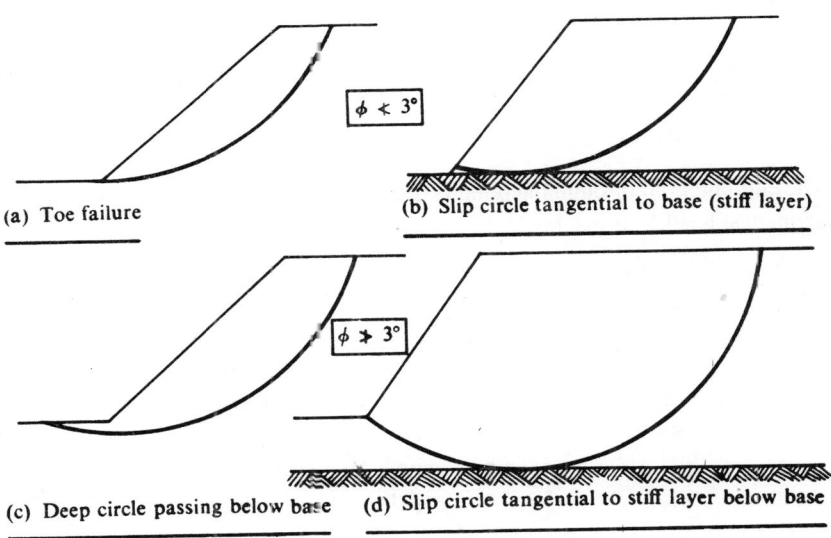

Fig. 9.15 Types of slip surfaces

9.3.2 Effective Stress Analysis

Total stress analysis is applicable for the analysis of stability of a slope soon after construction under undrained conditions. If pore water pressures exist in an embankment under certain conditions of drainage or seepage, an analysis in terms of effective stress is considered appropriate; in fact, this method is applicable at any stage of drainage – from no drainage to full drainage – or for any value of the pore pressure ratio, r_u. Depending upon the situation, the pore water pressure may depend upon the ground water level within the embankment or the flow net pattern owing to the impounded water. It may also depend upon the magnitude of the applied stresses, for example, during rapid construction of an earth dam or embankment.

Steady seepage

The case when steady seepage is occurring at the maximum possible rate through an earth dam or an embankment is considered as the critical condition for the stability of the downstream slope.

The pore pressure ratio r_u can be easily obtained from the flow net for this case, as shown in Fig. 9.16.

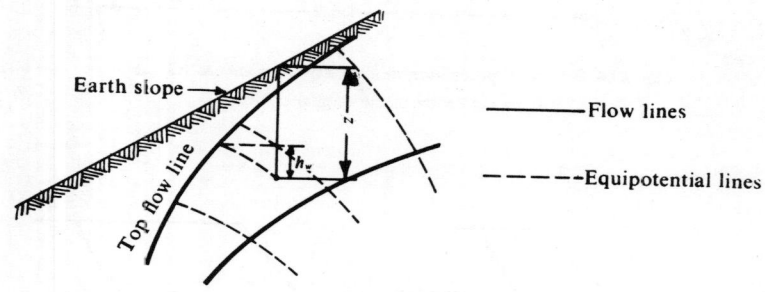

Fig. 9.16 Determination of pore water pressure from flow net

The equipotential through the point is traced up to the top of the flow net, so that piezometric head h_w is established.

Since $u = \gamma_w h_w$, $r_u = \dfrac{\gamma_w h_w}{\gamma z}$ ...(Eq. 9.21)

The effect of the pore water pressure is to reduce the effective stress and thereby reduce the stability because the shear strength mobilised would be decreased.

The factor of safety F may be written as:

$$F = \frac{c'r\theta + \tan\phi' \Sigma(N-U)}{\Sigma T}$$ ...(Eq. 9.22)

Here c' and ϕ' are the shear parameters based on effective stress analysis, which may be got from drained tests in the laboratory.

ΣU = Total force because of pore pressure on the surface.

ΣU can be obtained by obtaining values of u at the points of inter-section of the slip surface with equipotentials as stated earlier by Eq. 9.21 and showing the respective values as ordinates normal to the slip surface and getting the area of this U-diagram, similar to the N- and T-curves.

Stability of Earth Slopes 387

In the absence of a flow net, the approximate value of F may be given by:
$$F = \frac{c'r\theta + \tan\phi' \Sigma N'}{\Sigma T} \qquad \ldots(\text{Eq. 9.23})$$

Here the normal components N' of the weights of slices have to be obtained using effective or buoyant unit weight γ' and the tangential components T using the saturated unit weight.

A value of F ranging from 1.25 to 1.50 is considered to be satisfactory for an earth slope. For economic reasons, a value greater than 1.50 is not desired. Hence, $F = 1.50$ may be considered to be necessary as well as sufficient.

Rapid drawdown

For the upstream slope of an embankment or an earth dam, the case of rapid or sudden drawdown represents the critical condition since the seepage force in that condition adds to the sliding moment while it reduces the shear resistance mobilised by decreasing the effective stress. The effect of rapid drawdown on slope stability depends very much on the opportunity for drainage at the base. If the base material is pervious the flow pattern tends to be downwards, which is conducive to stability; otherwise, the seepage forces may create more unfavourable conditions with respect to stability. The pore water pressure along the slip surface can be determined from the flow net. Referring to Fig. 9.17, the pore water pressure u may be written as follows:

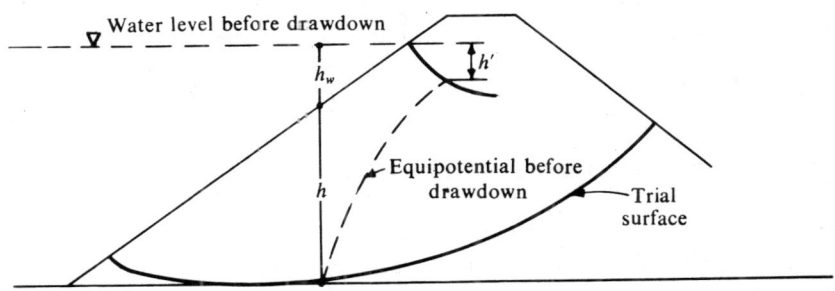

Fig. 9.17 Upstream slope subjected to rapid drawdown

$$u = u_o + \Delta u$$

$$u_o = \gamma_w(h_w + h - h') \qquad \ldots(\text{Eq. 9.24})$$

$\Delta u = \overline{B} \cdot \Delta \sigma_1 = \overline{B} \cdot (-\gamma_w h_w)$, $\overline{B}$ being defined as $\dfrac{\Delta u}{\Delta \sigma_1}$.

Here $\overline{B}$ is commonly assumed as unity.

$$\therefore \quad \Delta u = -\gamma_w h_w \qquad \ldots(\text{Eq. 9.25})$$

$$\therefore \quad u = \gamma_w(h - h') \qquad \ldots(\text{Eq. 9.26})$$

Equation 9.22 may now be used to determine the factor of safety, F. If a flow net is not available, the approximate value of F may be got from Eq. 9.23.

Immediately after construction

When an earth dam or an embankment is constructed rather rapidly, excess pore pressures are likely to develop which affect the factor of safety. Assuming the initial pore pressure to be negligible, the pore pressure at any stage is the change in pore pressure, Δu.

But $\Delta u = B[\Delta\sigma_3 + A(\Delta\sigma_1 - \Delta\sigma_3)]$

from Skempton's concept of the pore pressure parameters.

$$\therefore \quad \frac{\Delta u}{\Delta\sigma_1} = \overline{B} = B\left[\frac{\Delta\sigma_3}{\Delta\sigma_1} + A\left(1 - \frac{\Delta\sigma_3}{\Delta\sigma_1}\right)\right] \quad \ldots\text{(Eq. 9.27)}$$

$$r_u = u/\gamma z$$

$$= \Delta u/\gamma z$$

$$= \frac{\overline{B} \cdot \Delta\sigma_1}{\gamma z}$$

$\Delta\sigma_1$ may be taken to be nearly equal to the weight of the material above the point, or γz.

$$\therefore \quad r_u = \overline{B} \quad \ldots\text{(Eq. 9.28)}$$

The pore pressure coefficient $\overline{B}$ may be determined from a triaxial test in which the sample is subjected to increases in the principal stresses $\Delta\sigma_1$ and $\Delta\sigma_3$ of magnitudes expected in the field. The resulting pore pressure is measured and $\overline{B}$ is obtained.

Once an idea of pore pressures is got, the factor of safety immediately after construction may be obtained in the usual manner.

Effective stress analysis by Bishop's method

Bishop (1955) gave an effective stress analysis of which he took into account, at least partially, the effect of the forces on the vertical sides of the slices in the Swedish method.

Figure 9.18 illustrates a trial failure surface and all the forces on a vertical slice which tend to keep it in equilibrium.

Let R_n and R_{n+1} be the reactions on the vertical sides of the slice under consideration.

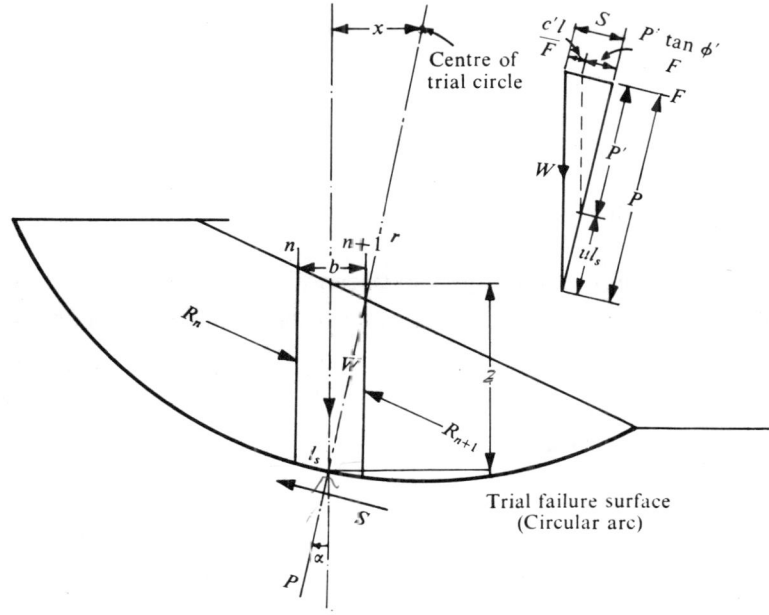

Fig. 9.18 Bishop's procedure for effective stress analysis of slope stability

Let the other forces on the slice be:
 W : weight of slice.
 P : Total normal force acting on the base of the slice.
 S : Shearing resistance acting at the base of slice.
Also, let b : breadth of slice
 l_s : length of slice along the curved surface at the base.
 z : height of the slice
 x : horizontal distance of the centre of the slice from the centre of the trial slip circle.
 α : angle between P and the vertical.
The shear resistance (stress) mobilised is:
$$\tau = \frac{c' + (\sigma_n - u)\tan\phi'}{F}$$
Total normal stress σ_n on the base of the slice = P/l_s
$$\therefore \quad \tau = \frac{1}{F}[c' + \{(P/l_s) - u\}\tan\phi']$$
Shearing force acting on the base of the slice, $S = \tau l_s$.
For equilibrium,
 Sliding moment = Restoring moment

or
$$\Sigma W \cdot x = \Sigma S \cdot r = \Sigma \tau \cdot l_s \cdot r$$

$$= \frac{r}{F} \Sigma [c'l_s + (P - ul_s) \tan \phi']$$

$$\therefore \quad F = \frac{r}{\Sigma W x} \Sigma [c'l_s + (P - ul_s) \tan \phi'] \quad \ldots\text{(Eq. 9.29)}$$

Let the normal effective force, $P' = (P - ul_s)$
Resolving the forces vertically,
$$W = P \cos \alpha + S \sin \alpha \quad \ldots\text{(Eq. 9.30)}$$
(The vertical components of R_n and R_{n+1} are taken to be equal and hence to be nullifying each other, the error from this assumption being considered negligible).
Here $P = P' + ul_s$

$$S = 1/F(c'l_s + P' \tan \phi')$$

Substituting these values of P and S in Eq. 9.30, we have:
$$W = (P' + ul_s) \cos \alpha + \frac{(c'l_s + P' \tan \phi') \sin \alpha}{F}$$

$$= P' \left(\cos \alpha + \frac{\tan \phi'}{F} \sin \alpha \right) + l_s \{u \cos \alpha + (c'/F) \sin \alpha\}$$

$$\therefore \quad P' = \frac{W - l_s \{u \cos \alpha + (c'/F) \sin \alpha\}}{\cos \alpha + \frac{\tan \phi' \sin \alpha}{F}} \quad \ldots\text{(Eq. 9.31)}$$

Substituting this value of P' for $(P - ul_s)$ in Eq. 9.29, we get
$$F = \frac{r}{\Sigma W x} \Sigma \left[\frac{c'l_s + \left(W - ul_s \cos \alpha - \frac{c'l_s}{F} \sin \alpha\right) \tan \phi'}{\left(\cos \alpha + \frac{\tan \phi' \sin \alpha}{F}\right)} \right] \quad \ldots\text{(Eq. 9.32)}$$

Here $x = r \sin \alpha$

$b = l_s \cos \alpha$

$$\frac{ub}{W} = \frac{u}{\gamma z} = r_u$$

Substituting these into Eq. 9.32,
$$F = \frac{1}{\Sigma W \sin \alpha} \Sigma \left[\{c'b + W(1 - r_u) \tan \phi'\} \frac{\sec \alpha}{\left(1 + \frac{\tan \phi' \tan \alpha}{F}\right)} \right] \quad \ldots\text{(Eq. 9.33)}$$

Since this equation contains F on both sides, the solution should be one by trial and error.

Bishop and Morgenstern (1960) evolved stability coefficients m and n, which depend upon $c'/\gamma H$, ϕ', and cot β. In terms of these coefficients,
$$F = m - n r_u \quad \ldots\text{(Eq. 9.34)}$$

Stability of Earth Slopes

m is the factor of safety with respect to total stresses and n is a coefficient representing the effect of the pore pressures on the factor of safety. Bishop and Morgenstern prepared charts of m and n for sets of $c'/\gamma H$ values and for different slope angles.

If the effect of forces R_n and R_{n+1} is completely ignored, the only vertical force acting on the slice is W.

Hence $P = W \cos \alpha$

$$\therefore \quad F = \frac{r}{\Sigma W x} \Sigma [c' l_s + (W \cos \alpha - u l_s) \tan \phi']$$

$$= \frac{1}{\Sigma W \sin \alpha} \Sigma [c' l_s + (W \cos \alpha - u l_s) \tan \phi'] \quad \ldots\text{(Eq. 9.35)}$$

since $r \sin \alpha = x$.

If u is expressed in terms of pore pressure ratio r_u,

$$u = r_u \gamma \cdot z = r_u \cdot \frac{W}{b}$$

But $\quad b = l_s \cos \alpha$

$$\therefore \quad u = \frac{r_u W}{l_s \cos \alpha} = \frac{r_u W}{l_s} \cdot \sec \alpha$$

$$\therefore \quad F = \frac{1}{\Sigma W \sin \alpha} \Sigma [c' l_s + W(\cos \alpha - r_u \sec \alpha) \tan \phi'] \quad \ldots\text{(Eq. 9.36)}$$

This is nothing but the Eq. 9.22, obtained by the method of slices and adapted to the case of steady seepage, pore pressure effects being taken into account. This approximate approach is the conventional one while Eq. 9.33 represents the vigorous approach.

9.3.3 Friction Circle Method

The friction circle method is based on the fact that the resultant reaction between the two portions of the soil mass into which the trial slip circle divides the slope will be tangential to a concentric smaller circle of radius $r \sin \phi$, since the obliquity of the resultant at failure is the angle of internal friction, ϕ. (This, of course, implies the assumption that friction is mobilised in full). This can be understood from Fig. 9.19.

This smaller circle is called the 'friction circle' or 'ϕ-circle'.

The forces acting on the sliding wedge are:
(i) weight W of the wedge of soil
(ii) reaction R due to frictional resistance, and
(iii) cohesive force C_m mobilised along the slip surface. These are shown in Fig. 9.20.

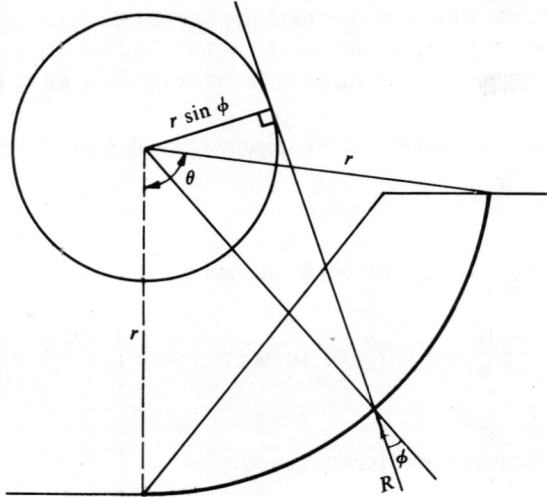

Fig. 9.19 Concept of friction circle

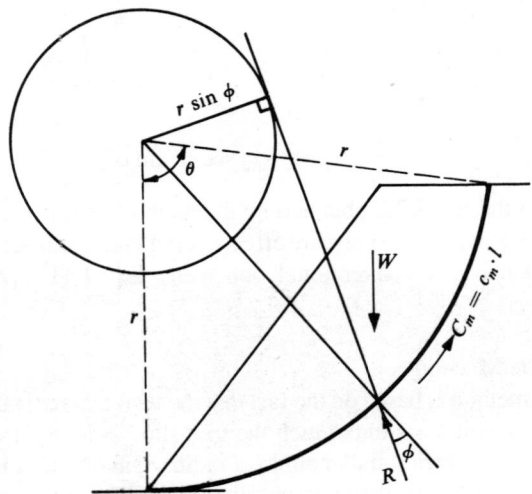

Fig. 9.20 Forces on the sliding wedge

The total reaction R, strictly speaking, will not be exactly tangential to the friction circle, but will pass at a slightly greater radial distance than $r \sin \phi$ from the centre of the circular arc. Thus, it can be considered as being tangential to modified friction circle of radius $kr \sin \phi$ where k is a constant greater than unity, the value of which is supposed to depend upon the central angle θ and the nature of the distribution of the intergranular pressure along the sliding surface. This concept will be clear if the reactions from small finite lengths of the arc are considered as shown in Fig. 9.21, the total value of R being obtained as the vector sum of such small values.

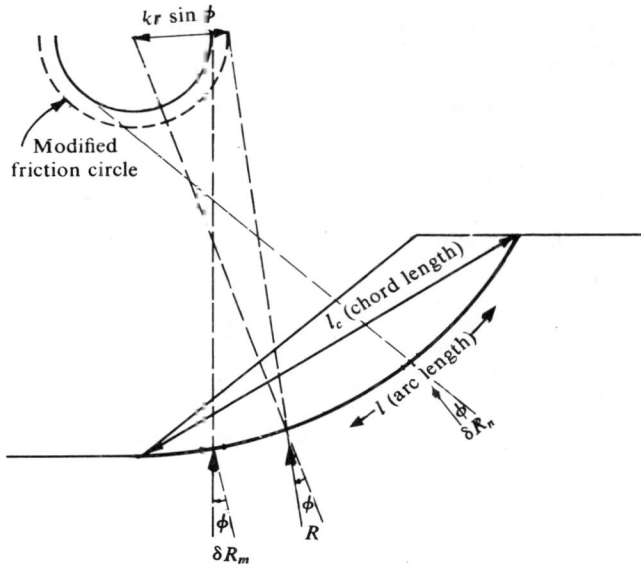

Fig. 9.21 Resultant frictional force–tangential to modified friction circle

The value of k may be obtained from Fig. 9.22.

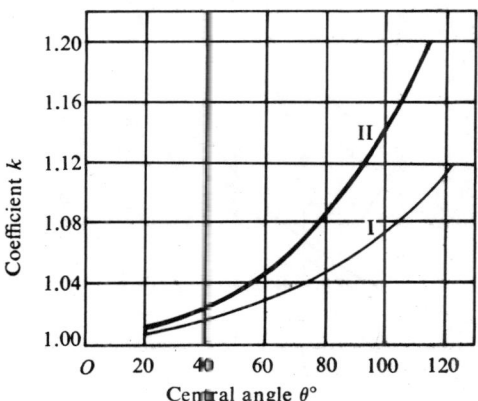

[*Note*: (1) The relationship I is valid for sinusoidal variation of intergranular pressure with zero values at the ends of the arc, which is considered nearer the actual distribution. (2) The relationship II is valid for uniform pressure distribution.]

Fig. 9.22 Central angle versus coefficient k for the modified friction circle

394 *Geotechnical Engineering*

Similarly the resultant mobilised cohesive force C_m can be located by equating its moments and the cohesive forces from elementary or finite lengths, into which the whole arc may be divided, about the centre. If C_m is the mobilised unit cohesion, the total mobilised cohesive force all along the arc is $c_m \cdot l$; but the resultant total cohesive force c_m can be shown to be $c_m \cdot l_c$ only where l_c is the chord length since the resultant of an infinite number of small vectors along the arc is the vector along the chord. Putting it in another way, the components parallel to the chord add up to one another while those perpendicular to the chord cancel out on the whole.

Thus, if a is the lever arm of the total cohesive force mobilised, C_m, from the centre of the circle,

$$C_m \cdot a = c_m \cdot l_c \cdot a = c_m \cdot l \cdot r$$

$$a = \frac{1}{l_c} \cdot r \qquad \qquad \ldots(\text{Eq. 9.37})$$

It may be noted that the line of action of C_m does not depend upon the value of C_m.

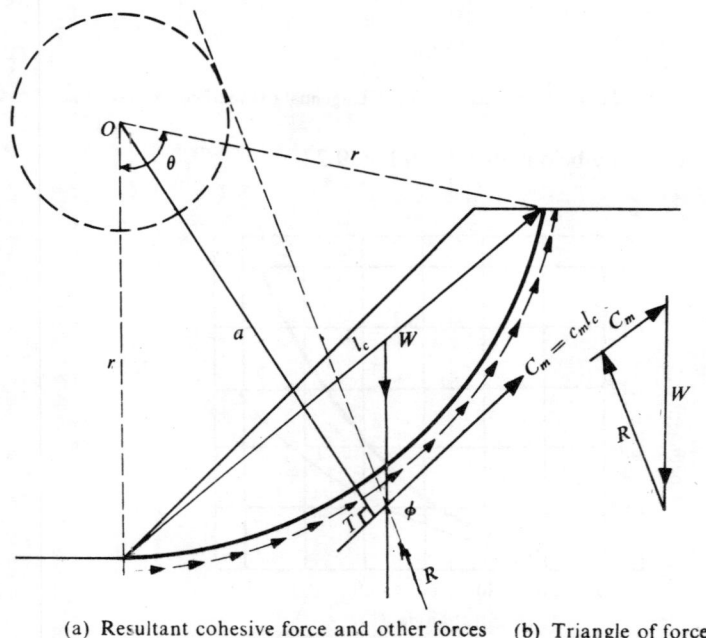

(a) Resultant cohesive force and other forces   (b) Triangle of forces

Fig. 9.23 Location of the resultant cohesive force and triangle of forces

Figure 9.23 illustrates these points clearly in addition to showing all the forces and the corresponding triangle of forces.

Stability of Earth Slopes

The lines of action of W and C_m are located first. A tangent is drawn to the modified friction circle from the point of intersection of W and C_m, to give the direction of R. Now the triangle of forces may be completed as shown in Fig. 9.23(b). drawing W to a suitable scale.

The factor of safety with respect to cohesion, assuming that friction is mobilised in full, is given by:

$$F_c = c/c_m \qquad \ldots(\text{Eq. 9.38})$$

The factor of safety with respect to friction, assuming that cohesion is mobilised in full, is given by:

$$F_\phi = \frac{\tan \phi'}{\tan \phi_m} \qquad \ldots(\text{Eq. 9.39})$$

where ϕ' and ϕ_m are the effective friction angle and mobilised friction angle. If the factor of safety with respect to the total shear strength F_s is required, ϕ_m is to be chosen such that F_c and F_ϕ are equal. This is common-sense and may also be established mathematically:

$$F_s = s/\tau \qquad \ldots(\text{Eq. 9.40})$$

where $s = c' + \bar{\sigma} \tan \phi'$ (shear strength) ...(Eq. 9.41)

and $\tau = c_m + \bar{\sigma} \tan \phi_m$ (shear strength mobilised) ...(Eq. 9.42)

If there were to be neutral pressure due to submergence, it will add to the actuating force as shown in Fig. 9.24.

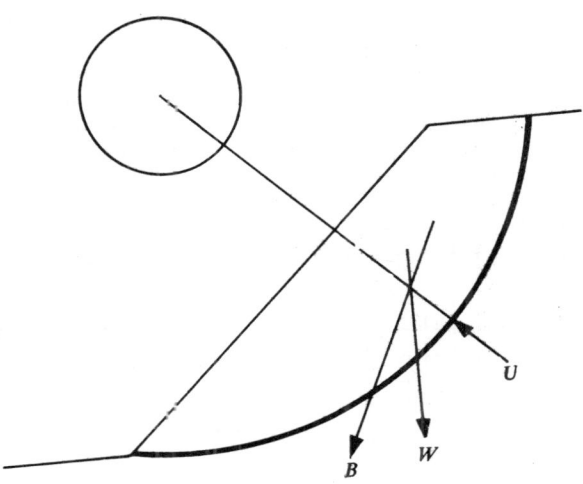

Fig. 9.24 Effect of neutral pressure on the stability of a slope

396 *Geotechnical Engineering*

The line of action of U will pass through the centre of the circle.
The resultant of W and U is the actuating force B.
The triangle of forces will consist of the forces B, C_m and R.

9.3.4 Taylor's Method

For slopes made from two different soils the ratio $c_m/\gamma H$ has been shown to be the same for each slope provided that the two soils have the same angle of friction. This ratio is known as the *'stability number'* and is designated by the symbol, N.

$$\therefore \quad N = C_m/\gamma H \qquad \ldots(\text{Eq. 9.43})$$

where $N =$ stability number (same as S_n of Eq. 9.15)
$c_m =$ Unit cohesion mobilised (with respect to total stress)
$\gamma =$ Unit weight of soil
and $H =$ Vertical height of the slope (Similar to z of Eq. 9.15).

Taylor (1948) prepared two charts relating the stability number to the angle of slope, based on the friction circle method and an analytical approach. The first is for the general case of a $c - \phi$ soil with the angle of slope less than 53°, as shown in Fig. 9.25. The second is for a soil with $\phi = 0$ and a layer of rock or stiff material at a depth DH below the top of the embankment, as shown in Fig. 9.26. Here, D is known as the depth factor; depending upon its value, the slip circle will pass through the toe or will emerge at a distance nH in front of the toe (the value of n may be obtained from the curves). Theoretically, the critical arc in such cases extends to an infinite depth (slope angle being less than 53°), however, it is limited to the hard stratum. For $\phi = 0$ and a slope angle greater than 53°, the first chart is to be used.

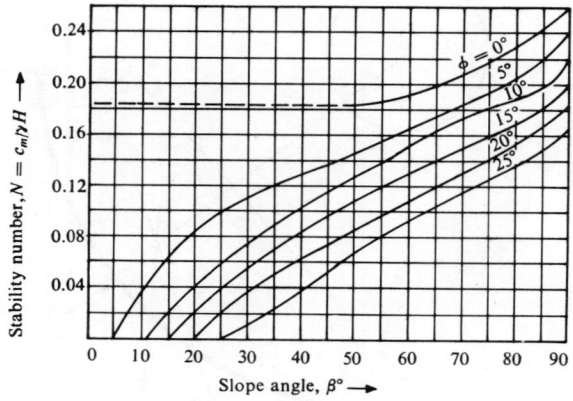

Fig. 9.25 Taylor's charts for slope stability (After Taylor, 1948)
(for $\phi = 0°$ and $\beta < 53°$, use Fig. 9.26)

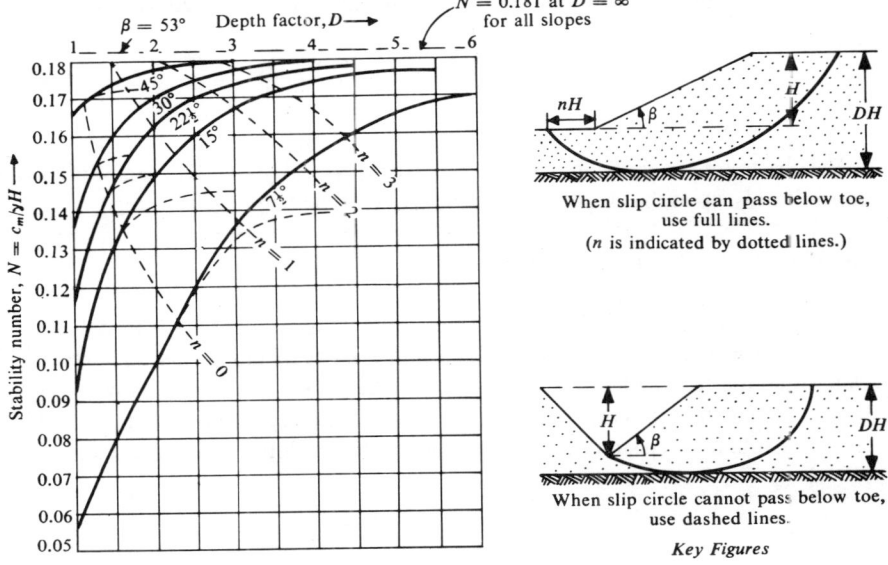

Fig. 9.26 Taylor's chart for slopes with depth limitation
(After Taylor, 1948) (for β > 53°, use Fig. 9.25)

(**Note:** For $\phi = 0°$ and $\beta = 90°$, $N = 0.26$. So the maximum unsupported height of a vertical-cut in pure clay is $c_u/\gamma N$ or $4c/\gamma$ nearly).

The use of the charts is almost self-explanatory. For example, the first chart may be used in one of the two following ways, depending upon the nature of the problem on hand:

1. If the slope angle and mobilised friction angle are known, the stability number can be obtained. Knowing unit weight and vertical height of the slope, the mobilised cohesion can be got.

   The factor of safety may be evaluated as the ratio of the effective cohesional strength to the mobilised unit cohesion.

2. Knowing the height of the slope, unit weight of the earth material constituting the slope and the desired factor of safety, the stability number can be evaluated. The slope angle can be found from the chart against the permissible angle of internal friction.

If the slope is submerged, the effective unit weight γ' instead of γ is to be used.

For the case of sudden drawdown, the saturated unit weight γ_{sat} is to be used for γ; in addition, a reduced value of ϕ, ϕ_w, should be used, where:

$$\phi_w = (\gamma'/\gamma_{sat}) \times \phi \qquad \text{...(Eq. 9.44)}$$

398 Geotechnical Engineering

Taylor's charts are based on the assumption of full mobilisation of friction, that is, these give the factor of safety with respect to cohesion.

This is all right for a purely cohesive soil; but, in the case of a $c - \phi$ soil, where the factor of safety F_s with respect to shearing strength is desired, ϕ_m should be used for ϕ:

$$\tan \phi_m = \tan \phi / F_s \qquad \ldots(\text{Eq. 9.45})$$

$$(\text{Also } \phi_m \approx \phi/F_s)$$

The charts are not applicable for a purely frictional soil ($c = 0$). The stability then depends only upon the slope angle, irrespective of the height of the slope.

9.4 ILLUSTRATIVE EXAMPLES

Example 9.1: Figure 9.27 shows the details of an embankment made of cohesive soil with $\phi = 0$ and $c = 30$ kN/m². The unit weight of the soil is 18.9 kN/m³. Determine the factor of safety against sliding along the trial circle shown. The weight of the sliding mass is 360 kN acting at an eccentricity of 5.0 m from the centre of rotation. Assume that no tension crack develops. The central angle is 70°.

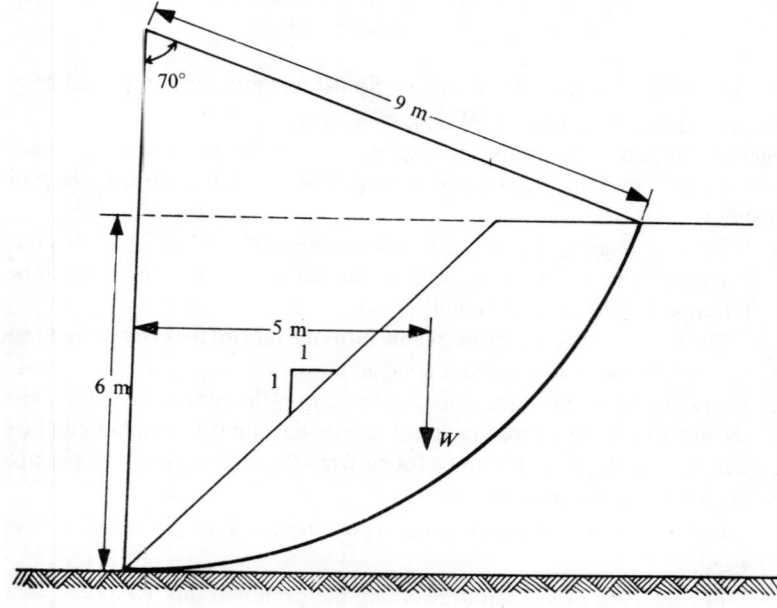

Fig. 9.27 Trial slip circle (Ex. 9.1)

Sliding moment = $360 \times 5 = 1800$ kNm

Restoring moment = $c \cdot r^2 \theta = 30 \times 9^2 \times \dfrac{70}{180} \times \pi = 2970$ kNm

Factor of safety against sliding,

$$F = 2970/1800 = 1.65$$

Example 9.2: A cutting is made 10 m deep with sides sloping at 8 : 5 in a clay soil having a mean undrained strength of 5 t/m² and a mean bulk density of 1900 kg/m³. Determine the factor of safety under immediate (undrained) conditions given the following details of the impending failure circular surface: The centre of rotation lies vertically above the middle of the slope. Radius of failure arc = 16.5 m. The deepest portion of the failure surface is 2.5 m below the bottom surface of the cut (i.e., the centre of rotation is 4 m above the top surface of the cut). Allowance is to be made for tension cracks developing to a depth of 3.5 m from surface. Assume that there is no external pressure on the face of the slope.

(S.V.U.—B.E., (R.R.)—Sep., 1978)

The data are shown in Fig. 9.28.

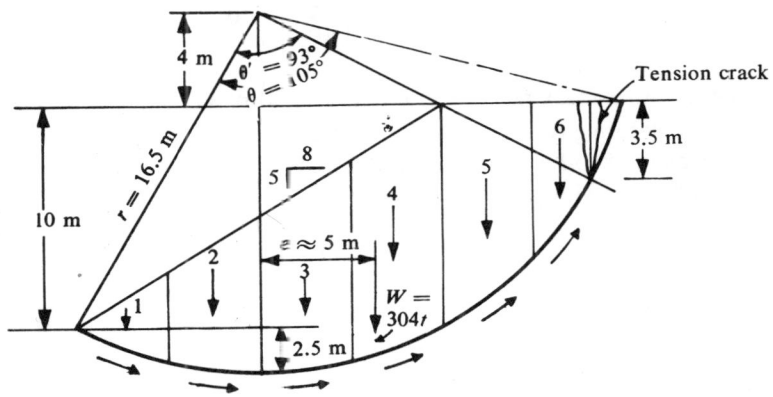

Fig. 9.28 Trial failure surface (Ex. 9.2)

Mean undrained strength = 5 t/m², ∴ $c = 5$ t/m²

Factor of safety $F_c = cr^2 \theta'/W \cdot e$

$= 5 \times 16.5^2 \times (93/180) \times \pi \times 1/(304 \times 5)$

$= \mathbf{1.45}$

(**Note**: Here, W and e are obtained by dividing the sliding mass into six slices as shown in Fig. 9.28 and by taking moments of the weights of these about the centre of rotation).

Example 9.3: An embankment 10 m high is inclined at an angle of 36° to the horizontal. A stability analysis by the method of slices gives the following forces per running meter:

$$\Sigma \text{ Shearing forces} = 450 \text{ kN}$$
$$\Sigma \text{ Normal forces} = 900 \text{ kN}$$
$$\Sigma \text{ Neutral forces} = 216 \text{ kN}$$

The length of the failure arc is 27 m. Laboratory tests on the soil indicate the effective values c' and ϕ' as 20 kN/m² and 18° respectively.

Determine the factor of safety of the slope with respect to (*a*) shearing strength and (*b*) cohesion.

(*a*) Factor of safety with respect to shearing strength

$$F_s = \frac{c'r\theta + \{\Sigma(N-U)\} \tan \phi'}{\Sigma T}$$

$$= \frac{20 \times 27 + (900 - 216) \tan 18°}{450} = 1.70$$

(*b*) Factor of safety with respect to cohesion

$$F_c = \frac{c'r\theta}{\Sigma T}$$

$$= \frac{20 \times 27}{450} = 1.20$$

Example 9.4: An embankment is inclined at an angle of 35° and its height is 15 m. The angle of shearing resistance is 15° and the cohesion intercept is 200 kN/m². The unit weight of soil is 18.0 kN/m³. If Taylor's stability number is 0.06, find the factor of safety with respect to cohesion.

(S.V.U.—B.Tech. (Part-time)—Apr., 1982)

$\beta = 35°$
$H = 15$ m
$\phi = 15°$
$c = 200$ kN/m²
$\gamma = 18.0$ kN/m³
Taylor's stability Number $N = 0.06$

Since

$$N = \frac{c_m}{\gamma H}$$

∴

$$0.06 = \frac{c_m}{18 \times 15}$$

∴ Mobilised cohesion,

$$c_m = 0.06 \times 18 \times 15 \text{ kN/m}^2$$

$$= 16.2 \text{ kN/m}^2$$

Cohesive strength $c = 200$ kN/m²

∴ Factor of safety with respect to cohesion:

$$F_c = \frac{c}{c_m} = \frac{200}{16.2} \approx 12.3$$

Example 9.5: An embankment has a slope of 30° to the horizontal. The properties of the soil are: $c = 30$ kN/m², $\phi = 20°$, $\gamma = 18$ kN/m³. The height of the embankment is 27 m. Using Taylor's charts, determine the factor of safety of the slope.

From Taylor's charts, it will be seen that a slope with $\theta = 20°$ and $\beta = 30°$ has a stability number 0.025. That is to say, if the factor of safety with respect to friction were to be unity (implying full mobilisation of friction), the mobilised cohesion required will be found from

$$N = \frac{c_m}{\gamma H}$$

$$0.025 = \frac{c_m}{18 \times 27}$$

∴ $c_m = 18 \times 27 \times 0.025 = 12.15$ kN/m²

∴ Factor of safety with respect to cohesion

$$F_c = c/c_m = 30/12.15 = 2.47$$

But the factor of safety F_s against shearing strength is more appropriate:

$$F_s = \frac{c + \sigma \tan \phi}{\tau}$$

F may be found by successive approximations as follows:

Let us try $F = 1.5$

$\dfrac{\tan \phi}{F} = 0.364/1.5 = 0.24267 \approx$ tangent of angle $13\dfrac{2°}{3}$

With this value of ϕ, the new value N from charts is found to be 0.055.

$$c = 0.055 \times 18 \times 27 = 26.73$$

∴ F with respect to $c = 30/26.73 = 1.12$

Try $F = 1.3$

$\dfrac{\tan \phi}{F} = 0.364/1.3 = 0.280 =$ Tangent of $15\dfrac{2°}{3}$

From the charts, new value of $N = 0.045$

$$c = 0.045 \times 18 \times 27 = 21.87$$

∴ $F_c = 30/21.87 = 1.37$

402 Geotechnical Engineering

Let us try $F = 1.35$

$$\frac{\tan \phi}{F} = 0.364/1.35 = 0.27 = \text{tangent of angle } 15°.1$$

From the charts, the new value of $N = 0.046$

$$c = 0.046 \times 18 \times 27 = 22.356$$

$$\therefore \quad F_c = 30/22.356 = 1.342 \, (= F_\phi)$$

This is not very different from the assumed value.
∴ The factor of safety for the slope = **1.35**

Example 9.6: A cutting is to be made in clay for which the cohesion is 0.35 kg/cm² and $\phi = 0°$. The density of the soil is 2 g/cm³. Find the maximum depth for a cutting of side slope $1\frac{1}{2}$ to 1 if the factor of safety is to be 1.5. Take the stability number for a $1\frac{1}{2}$ to 1 slope and $\phi = 0°$ as 0.17.

(S.V.U.—B.E., (N.R.)—Apr., 1966)

$c = 0.35$ kg/cm² $\qquad \phi = 0°$

$\gamma = 2$ g/cm³ $\qquad N = 0.17$

$\qquad\qquad\qquad F_c = 1.5$

$\therefore \qquad\qquad c_m = c/F_c = 0.35/1.5 = 0.7/3 \text{ kg/cm}^2$

But $\qquad N = \dfrac{c_m}{\gamma H}$

$\therefore \qquad 0.17 = \dfrac{c_m \times 1000}{2 \times H} = \dfrac{0.7 \times 1000}{3 \times 2 \times H}$

$\therefore \qquad H = \dfrac{0.7 \times 1000}{6 \times 0.17} \text{ cm}$

$\qquad\qquad = 7/1.02 \text{ m} = \textbf{6.86 m.}$

Example 9.7: A cut 9 m deep is to be made in a clay with a unit weight of 18 kN/m³ and a cohesion of 27 kN/m². A hard stratum exists at a depth of 18 m below the ground surface. Determine from Taylor's charts if a 30° slope is safe. If a factor of safety of 1.50 is desired, what is a safe angle of slope?

Depth factor $D = 18/9 = 2$
From Taylor's charts,
$\qquad$ for $\qquad D = 2; \beta = 30°$

$\qquad\qquad N = 0.172$

Stability of Earth Slopes

$$0.172 = \frac{c_m}{18 \times 9}$$

$$c_m = 0.172 \times 18 \times 9 = 27.86 \text{ kN/m}^2$$

$$c = 27 \text{ kN/m}^2$$

$$F_c = c/c_m = 27/27.86 = 0.97$$

The proposed slope is therefore not safe.

For
$$F_c = 1.50$$

$$c_m = c/F_c = 27/1.5 = 18 \text{ kN/m}^2$$

$$N = 18/18 \times 9 = 1/9 = 0.11$$

For
$$D = 2.0, \text{ and } N = 0.11,$$

from Taylor's charts,
we have $\beta = 8°$

∴ Safe angle of slope is 8°.

Example 9.8: A canal is to be excavated through a soil with $c = 15$ kN/m², $\phi = 20°$, $e = 0.9$ and $G = 2.67$. The side slope is 1 in 1. The depth of the canal is 6 m. Determine the factor of safety with respect to cohesion when the canal runs full. What will be the factor of safety if the canal is rapidly emptied?

$$\gamma_{sat} = \left(\frac{G+e}{1+e}\right) \cdot \gamma_w = \left(\frac{2.67+0.90}{1+0.90}\right) \times 9.81 \text{ kN/m}^3$$

$$= \frac{3.57}{1.90} \times 9.81 \text{ kN/m}^3 = 18.43 \text{ kN/m}^3$$

$$\gamma' = \gamma_{sat} - \gamma_w = 8.62 \text{ kN/m}^3$$

$$\beta = 45°, \phi = 20°.$$

(a) Submerged condition:
  From Taylor's charts, for these values of β and ϕ, the stability number N is found to be 0.06.

$$\therefore \quad 0.06 = \frac{c_m}{\gamma' \cdot H} = \frac{c_m}{8.62 \times 6}$$

$$c_m = 8.62 \times 6 \times 0.06 \text{ kN/m}^2 = 3.10 \text{ kN/m}^2.$$

Factor of safety with respect to cohesion, $F_c = c/c_m = 15/3.10 = \mathbf{4.84}$

(b) Rapid drawdown condition:

$$\phi_w = (\gamma'/\gamma_{sat}) \times \phi = (8.62/18.43) \times 20° = 9.35°$$

For $\beta = 45°$ and $\phi = 9.35°$, Taylor's stability number from charts is found to be 0.114.

$$\therefore \quad 0.114 = \frac{C_m}{\gamma_{sat} H} = \frac{C_m}{18.43 \times 6}$$

$$C_m = 0.114 \times 18.43 \times 6 \text{ kN/m}^2 = 12.60 \text{ kN/m}^2$$

Factor of safety with respect to cohesion $F_c = c/c_m = \frac{15.0}{12.6} \approx 1.2$

(**Note:** The critical nature of a rapid drawdown should now be apparent).

Example 9.9: The cross-section of an earth dam on an impermeable base is shown in Fig. 9.29. The stability of the downstream slope is to be investigated using the slip circle shown. Given:

$$\gamma_{sat} = 19.5 \text{ kN/m}^3$$

$$c' = 9 \text{ kN/m}^2$$

$$\phi' = 27°$$

$$r = 9 \text{ m}$$

$$\theta = 88°$$

For this circle determine the factor of safety by the conventional approach, as well as the rigorous one.

The following r_u values obtained from a flow net and weights of slices are given:

Slice No.	1	2	3	4	5
r_u	0.360	0.420	0.375	0.300	0.075
W (kN)	42	114	150	162	75

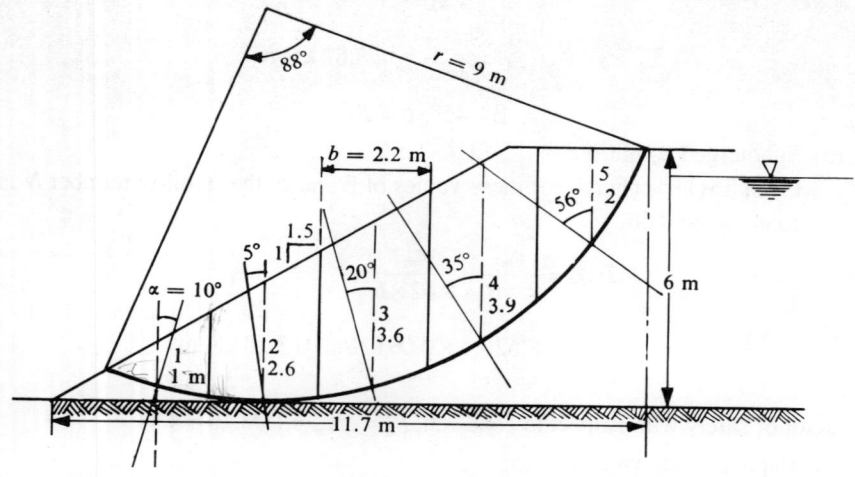

Fig. 9.29 Stability analysis by the conventional approach (Ex. 9.9)

Conventional approach:

The calculations are best set out in a tabular form as shown below for the conventional approach:

Table 9.2 Stability of downstream slope of an earth dam—conventional approach

Slice No.	$z(m)$	$b(m)$	W (kN)	$\alpha°$	$\cos\alpha$	$\sec\alpha$	$\cos\alpha - r_u \sec\alpha$	$W(\cos\alpha - r_u \sec\alpha)\tan\phi$	$\sin\alpha$	$W \sin\alpha$
1	1.0	2.2	42	−10	0.985	1.015	0.620	13.27	−0.174	−7.31
2	2.6	2.2	114	5	0.996	1.004	0.574	33.34	0.087	9.92
3	3.6	2.2	150	20	0.940	1.063	0.541	41.35	0.342	51.30
4	3.9	2.2	162	35	0.819	1.208	0.457	37.72	0.574	92.99
5	2.2	2.2	75	56	0.559	1.788	0.425	16.24	0.829	62.18
								$\Sigma = 141.92$		$\Sigma = 209.08$

$\theta = 88° \therefore c'r\theta = 9 \times 9 \times 88\pi/180 = 124.41$ kN

Factor of safety $F = \dfrac{(124.41 + 141.92)}{209.08} = \mathbf{1.274}$

(b) Rigorous approach:

The calculations for the rigorous approach are set out in the tabular form (Table 9.3).

F with the first approximation $= \dfrac{294.45}{209.08} = 1.41$

Columns (2) and (3) are recalculated with an F value of **1.4.**

F with the second approximation $= \dfrac{290.55}{209.08} = 1.39$,

which is very near the assumed value of F, i.e. 1.4

Thus, the factor of safety may be taken as **1.4** by the rigorous approach.

Table 9.3 Stability of downstream slope of an earth dam—Rigorous approach

Slice No.	$z(m)$	$b(m)$	W (kN)	$\alpha°$	$\sin\alpha$	$W\sin\alpha$	$c'\bar{b}$	$W(1-r_u)\tan\phi'$	$c'b+W(1-r_u)\tan\phi'$ (1)	$\sec\alpha$	$\tan\alpha$	$\dfrac{\sec\alpha}{1+\frac{\tan\phi'\tan\alpha}{F}}$ (2)		$(1)\times(2)=(3)$	
												$F=1.5$	$F=1.4$	$F=1.5$	$F=1.4$
1	1.0	2.2	42	−10	−0.174	−7.31	19.8	13.7	33.5	1.015	−0.176	1.08	1.084	36.18	36.32
2	2.6	2.2	114	5	0.087	9.92	19.8	33.7	53.5	1.004	0.087	0.98	0.973	52.43	52.05
3	3.6	2.2	150	20	0.342	51.30	19.8	47.8	67.6	1.063	0.364	0.95	0.939	64.22	63.48
4	3.9	2.2	162	35	0.574	92.99	19.8	57.8	77.6	1.208	0.700	0.98	0.963	76.05	74.73
5	2.2	2.2	75	56	0.829	62.18	19.8	35.3	55.1	1.788	1.483	1.19	1.161	65.57	63.97
						Σ=209.08								Σ=294.45	Σ=290.55

SUMMARY OF MAIN POINTS

1. Earth slopes may be classified as infinite slopes and finite slopes; practically speaking, a slope with a large height is treated as an infinite one.
2. The critical angle of slope of an infinite earth slope in cohesionless soil is equal to the angle of internal friction.
3. For an infinite slope in cohesive soil, the critical depth, z_c, is related to the angle of slope, β; the stability number, S_n, defined as $(c/\gamma \cdot z_c)$ equals $\sin \beta \cos \beta$.

   For an infinite slope in a cohesive frictional soil, the stability number, S_n, equals $\cos^2 \beta \, (\tan \beta - \tan \phi)$. In both these cases, the factor of safety is z_c/z, where z is the actual height.
4. For steady seepage and rapid drawdown conditions, both total stress analysis and effective stress analysis may be performed. Bishop's approach is considered more rational for the latter.
5. The factor of safety, F, as per the total stress analysis for a purely cohesive soil is given by $F = \dfrac{cr^2\theta}{W \cdot e}$, with respect to a trial slip circle of radius r with a central angle θ. The least of such values is the factor of safety for the slope. The effect of a tension crack is to reduce the value of F.
6. The factor of safety, as per the Swedish method of slices for a cohesive-frictional soil, is given by $F = \dfrac{cr\theta + \Sigma N \tan \phi}{\Sigma T}$.
7. Fellenius' procedure is useful for the location of the most critical circle. The type of failure surface is partly dependent upon ϕ-value. If there exists a hard stratum at or near the base of the slope, the slip circle is taken to be tangential to it.
8. The friction circle method is based upon the premise that the resultant reaction along a slip surface is tangential to a circle of radius $r \sin \phi$, where r is the radius of the slip circle.

   The factors of safety with respect to cohesion and with respect to friction are $F_c = c/c_m$ and $F_\phi = \dfrac{\tan \phi'}{\tan \phi_m}$, respectively, where c_m and ϕ_m are mobilised values.
9. Taylor's stability number N is defined as $c_m/\gamma H$; the procedure is based on the friction circle method and is an analytical approach. The results are embodied in Taylor's design charts which may be used for determining the factor of safety of a slope or for designing the height for a desired safety factor.

REFERENCES

1. Alam Singh and B.C. Punmia: *Soil Mechanics and Foundations*, Standard Book House, Delhi-6, 1970.
2. A.W. Bishop: *The Use of Slip Circle in the Stability Analysis of Earth Slopes*, Geotechnique, Vol. 5, 1955.
3. A.W. Bishop and N.R. Morgenstern: *Stability Coefficients for Earth Slopes*, Geotechnique, Vol. 10, 1960.
4. W. Fellenius: *Calculation of the Stability of Earth Dams*, Transactions, 2nd congress on Large Dams, Washington, D.C. 1936.
5. A.R. Jumikis: *Soil Mechanics*, D. Van Nostrand Co., Princeton, NJ, USA, 1962.
6. T.W. Lambe and R.V. Whitman: *Soil Mechanics*, John Wiley and Sons., Inc., New York, 1969.
7. D.F. McCarthy: *Essentials of Soil Mechanics and Foundations*, Reston Publishing Company, Reston, Va, U.S.A., 1977.
8. V.N.S. Murthy: *Soil Mechanics and Foundation Engineering*, Dhanpat Rai and Sons, Delhi-6, 2nd edition, 1977.
9. S.B. Sehgal: *A Text Book of Soil Mechanics*, Metropolitan Book Co. Pvt. Ltd., Delhi, 1967.
10. G.N. Smith: *Essentials of Soil Mechanics for Civil and Mining Engineers*, Third Edition Metric, Crosby Lockwood Staples, London, 1974.
11. M.G. Spangler: *Soil Engineering*, International Textbook Company, Scranton, USA, 1951.
12. D.W. Taylor: *Fundamentals of Soil Mechanics*, John Wiley and Sons, Inc., New York, 1948.

QUESTIONS AND PROBLEMS

9.1 Explain the method of slices for stability analysis of slopes. How can steady seepage be accounted for in this method?
(S.V.U.—Four-year B.Tech. April, 1983 B.E., (R.R.)—Feb., 1976)

9.2 Write the expressions for the factor of safety using the method of slices when the slope of a homogeneous earth dam is dry and when fully submerged. Assume the soil to possess both cohesion and friction.
(S.V.U.—B.Tech. (Part-time)—May, 1983)

9.3 Write brief critical notes on 'Taylor's Stability Number'.
(S.V.U.—B.Tech. (Part-time)—Sep., 1982, April, '82, June, '81; Four-year B.Tech—June, 1982)

9.4 Write critical notes on the friction circle method of analysing the stability of slopes.
(S.V.U.—B.Tech. (Part-time)—April, 1982, B.E., (R.R.)—Sept., 1978, Nov., 1972, Dec., '71, Dec., '70, May, 1970)

Stability of Earth Slopes 409

9.5 Give the step by step procedure for analysing the stability of the upstream slope of an earth dam by the Swedish method of slices. Bring out the effect of sudden drawdown on the stability of the slope.

(S.V.U.—Four-year B.Tech.—June, 1982)

9.6 (a) Describe a suitable method of stability analysis of slopes in (i) purely saturated cohesive soil, (ii) cohesionless sand.
(b) Under what conditions (i) a base failure and (ii) a toe failure are expected? Explain.
(c) Critically discuss the basic assumptions made in the stability analysis of slopes.

(S.V.U.—B.Tech. (Part-time)—June, 1981)

9.7 A 40-degree clay slope has a height of 5 m. Assuming a toe circle failure starting 1 m from the edge of the slope (at the top), calculate the shear strength required for the soil for a factor of safety of 1.5.
(Hint: Assume $\gamma = 19.6$ kN/m^3. Also since the existence of the hard layer is not mentioned, take Taylor's N as 0.1817)

(S.V.U.—B.E., (R.R.)—Nov., 1969)

9.8 The unit weight of a soil of a 30° slope is 1.75 g/cm^3. The shear parameters c and ϕ for the soil are 0.10 kg/cm^2 and 20° respectively. Given that the height of the slope is 12 m and the stability number obtained from the charts for the given slope and angle of internal friction is 0.025, compute the factor of safety.

(S.V.U.—B.Tech. (Part-time)—May, 1983)

9.9 What is the maximum depth to which a trench of vertical sides can be excavated in a clay stratum with $c = 0.5$ kg/cm^2 and $\gamma = 1.6$ g/cm^2? Assume the clay to be saturated.

(S.V.U.—B.E., (N.R.)—Sep., 1967)

9.10 A cutting is to be made in a soil with a slope of 30° to the horizontal and a depth of 15 m. The properties of the soil are: $c = 25$ kN/m^2, $\phi = 15°$, and $\gamma = 19.1°$ kN/m^3. Determine the factor of safety of the slope against slip, assuming friction and cohesion to be mobilised to the same proportion of their ultimate values.

9.11 An earth dam of height 20 m is constructed of soil of which the properties are: $\gamma = 20$ kN/m^2, $c = 45$ kN/m^2, and $\phi = 20°$. The side slopes are inclined at 30° to the horizontal. Find the factor of safety immediately after drawdown.

9.12 A cutting of depth 10.5 m is to be made in a soil for which the density is 18 kN/m^3 and cohesion is 39 kN/m^2. There is a hard stratum under the clay at 12.5 m below the original ground surface. Assuming $\phi = 0°$ and allowing for a factor of safety of 1.5, find the slope of the cutting.

10
STRESS DISTRIBUTION IN SOIL

10.1 INTRODUCTION

Stress in soil is caused by the first or both of the following:
(a) Self-Weight of soil.
(b) Structural loads, applied at or below the surface.

Many problems in foundation engineering require a study of the transmission and distribution of stresses in large and extensive masses of soil. Some examples are wheel loads transmitted through embankments to culverts, foundation pressures transmitted to soil strata below footings, pressures from isolated footings transmitted to retaining walls, and wheel loads transmitted through stabilised soil pavements to sub-grades below. In such cases, the stresses are transmitted in all downward and lateral directions.

Estimation of vertical stresses at any point in a soil mass due to external loading is essential to the prediction of settlements of buildings, bridges and embankments. The theory of elasticity, which gives primarily the interrelationships of stresses and strains (Timoshenko and Goodier, 1951), has been the basis for the determination of stresses in a soil mass. According to the elastic theory, constant ratios exist between stresses and strains. For the theory to be applicable, the real requirement is not that the material necessarily be elastic but that there must be constant ratios between stresses and the corresponding strains.

It is known that, only at relatively small magnitudes of stresses, the proportionality between strains and stresses exists in the case of soil. Fortunately, the order of magnitudes of stresses transmitted into soil from structural loadings is also small and hence the application of the elastic theory for determination of stress distribution in soil gives reasonably valid results.

The most widely used theories regarding distribution of stress in soil are those of Boussinesq and Westergaard. They have developed first for point loads and later, the values for point load have been integrated to give stresses below uniform strip loads, uniformly loaded circular and rectangular areas.

The vertical stress in soil owing to its self-weight, also called 'geostatic stress' (already dealt with in Chapter 5), is given by:

$$\sigma_z = \gamma \cdot z \qquad \ldots(\text{Eq. 10.1})$$

where σ_z = vertical stress in soil at depth z below the surface due to its self-weight,
and γ = unit weight of soil.

If there are imposed structural loadings also on the soil, the resultant stress may be obtained by adding algebraically the stress due to self-weight and stress transmitted due to structural loadings.

10.2 POINT LOAD

Although a point load or a concentrated load is, strictly speaking, hypothetical in nature, consideration of it serves a useful purpose in arriving at the solutions for more complex loadings in practice.

The most fundamental of the solutions of stress distribution in soil is that for a point load applied at the surface. Boussinesq and Wastergaard have given the solution with different assumptions regarding the soil medium. These solutions which form the basis for further work in this regard and other pertinent topics will be dealt with in the following sub-sections.

10.2.1 Boussinesq's Solution

Boussinesq (1885) has given the solution for the stresses caused by the application of a point load at the surface of a homogeneous, elastic, isotropic and semi-infinite medium, with the aid of the mathematical theory of elasticity. (A semi-infinite medium is one bounded by a horizontal boundary plane, which is the ground surface for soil medium).

The following is an exhaustive list of assumptions made by Boussinesq in the derivation of his theory:

(i) The soil medium is an elastic, homogeneous, isotropic, and semi-infinite medium, which extends infinitely in all directions from a level surface. (Homogeneity indicates identical properties at all points in identical directions, while isotropy indicates identical elastic properties in all directions at a point).
(ii) The medium obeys Hooke's law.
(iii) The self-weight of the soil is ignored.
(iv) The soil is initially unstressed.
(v) The change in volume of the soil upon application of the loads on to it is neglected.
(vi) The top surface of the medium is free of shear stress and is subjected to only the point load at a specified location.
(vii) Continuity of stress is considered to exist in the medium.
(viii) The stresses are distributed symmetrically with respect to Z-axis.

The notation with regard to the stress components and the co-ordinate system is as shown in Fig. 10.1.

In Fig. 10.1(*a*), the origin of co-ordinates is taken as the point of application of the load Q and the location of any point A in the soil mass is specified by the co-ordinates x, y, and z. The stresses acting at point A on planes normal to the co-ordinate axes are shown in Fig. 10.1(*b*). σ's are the normal stresses on the planes normal to the co-ordinate axes; τ's are the shearing stresses. The first subscript of τ denotes the axis normal to which the plane containing the shear

412 *Geotechnical Engineering*

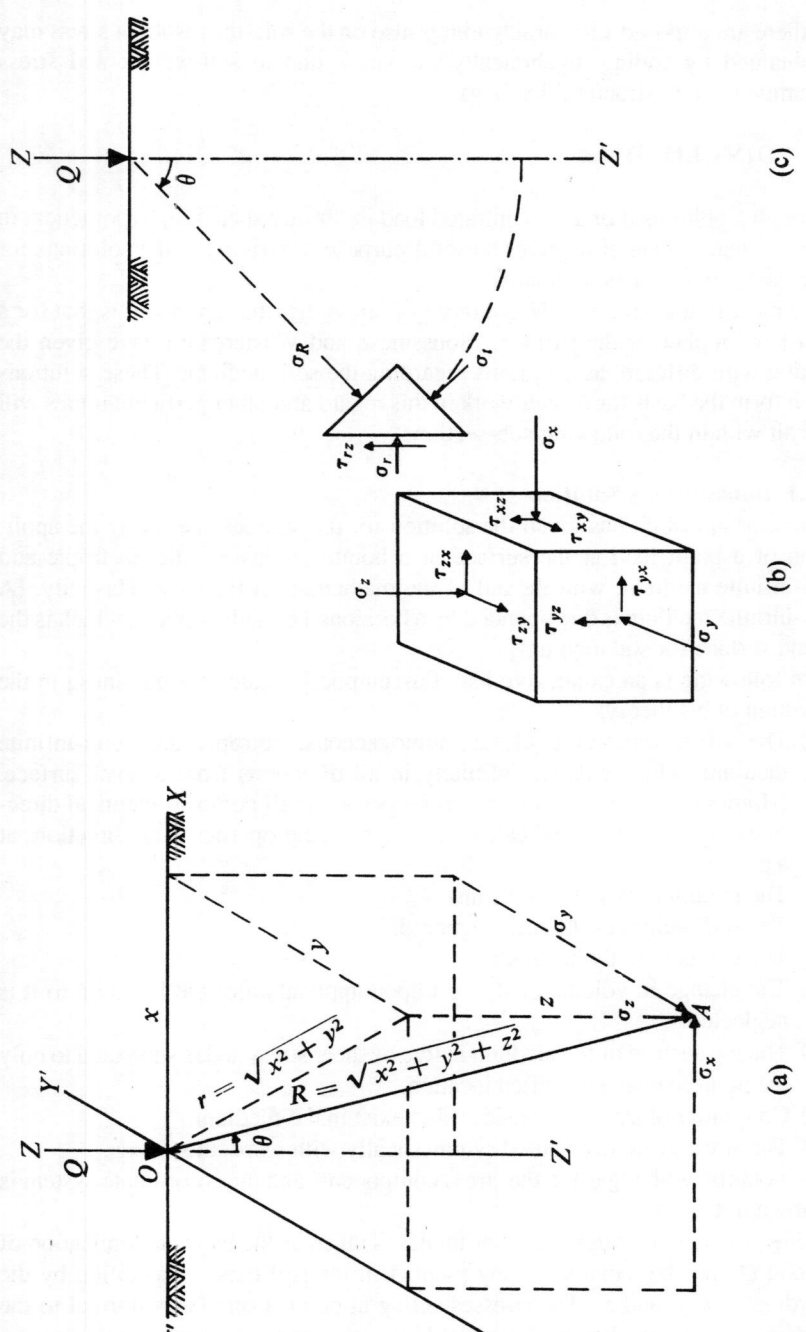

Fig. 10.1 Notation for Boussinesq's analysis

stress is, and the second subscript indicates direction of the axis parallel to which the shear stress acts. In Fig. 10.1(c), the cylindrical co-ordinates and the corresponding normal stresses—radial stress σ_r, tangential stress σ_t, and the shear stress τ_{rz}—are shown; σ_z is another principal stress in the cylindrical co-ordinates; the polar radial stress σ_R is also shown.

The Boussinesq equations are as follows:

$$\sigma_z = \frac{3Q}{2\pi} \cdot \frac{z^3}{R^5} \qquad \ldots\text{(Eq. 10.2a)}$$

$$= \frac{3Q}{2\pi} \cdot \frac{\cos^5 \theta}{z^2} \qquad \ldots\text{(Eq. 10.2b)}$$

$$= \frac{3Q}{2\pi} \cdot \frac{z^3}{(r^2+z^2)^{5/2}} \qquad \ldots\text{(Eq. 10.2c)}$$

$$= \frac{3Q}{2\pi z^2} \left[\frac{1}{1+(r/z)^2}\right]^{5/2} \qquad \ldots\text{(Eq. 10.2d)}$$

$$\sigma_x = \frac{Q}{2\pi}\left[\frac{3x^2 z}{R^5} - (1-2\nu)\left\{\frac{x^2-y^2}{Rr^2(R+z)} + \frac{y^2 z}{R^3 r^2}\right\}\right] \qquad \ldots\text{(Eq. 10.3)}$$

$$\sigma_y = \frac{Q}{2\pi}\left[\frac{3y^2 z}{R^5} - (1-2\nu)\left\{\frac{y^2-x^2}{Rr^2(R+z)} + \frac{x^2 z}{R^3 r^2}\right\}\right] \qquad \ldots\text{(Eq. 10.4)}$$

$$\sigma_R = \frac{3Q}{2\pi} \cdot \frac{\cos\theta}{R^2} \qquad ..\text{(Eq. 10.5)}$$

$$\sigma_r = \frac{Q}{2\pi}\left[\frac{3zr^2}{R^2} - \frac{(1-2\nu)}{R(R+z)}\right] \qquad \ldots\text{(Eq. 10.6a)}$$

$$= \frac{Q}{2\pi}\left[\frac{3zr^2}{(r^2+z^2)^{5/2}} - \frac{(1-2\nu)}{r^2+z^2+z\sqrt{r^2+z^2}}\right] \qquad \ldots\text{(Eq. 10.6b)}$$

$$= \frac{Q}{2\pi z^2}\left[3\sin^2\theta \cos^3\theta - \frac{(1-2\nu)\cos^2\theta}{(1+\cos\theta)}\right] \qquad \ldots\text{(Eq. 10.6c)}$$

$$\sigma_t = -\frac{Q}{2\pi z^2}(1-2\nu)\left[\cos^3\theta - \frac{\cos^2\theta}{1+\cos\theta}\right] \qquad \ldots\text{(Eq. 10.7a)}$$

$$= \frac{-Q}{2\pi}(1-2\nu)\left[\frac{z}{(r^2+z^2)^{3/2}} - \frac{1}{r^2+z^2+z\sqrt{r^2+z^2}}\right] \qquad \ldots\text{(Eq. 10.7b)}$$

$$\tau_{rz} = \frac{3Q}{2\pi} \cdot \frac{rz^2}{R^5} \qquad \ldots\text{(Eq. 10.8a)}$$

$$= \frac{3Qr}{2\pi z^3}\left[\frac{1}{1+(r/z)^2}\right]^{5/2} \qquad \ldots\text{(Eq. 10.8b)}$$

$$= (3Q/2\pi z^2) \cdot (\sin\theta \cos^4\theta) \qquad \ldots\text{(Eq. 10.8c)}$$

Here ν is 'Poisson's ratio' of the soil medium.

A geotechnical engineer must understand the assumptions on which these formulae are based, in order to be able to identify those problems to which they are directly applicable and those in which some modifications are necessary. There is usually no need for one to understand the advanced mathematical procedures by which the solution was obtained. For proofs, the reader is referred to Timoshenko and Goodier (1951) and Jumikis (1962).

Some modern methods of settlement analysis, such as those proposed by Lambe (1964, 1967), necessitate determining the increments of both major and minor principal stresses; however, in most foundation problems it is only necessary to be acquainted with the increase in vertical stresses (for settlement analysis) and the increase in shear stresses (for shear strength analysis).

Equation 10.2(d) may be rewritten in the form:

$$\sigma_z = K_B \cdot \frac{Q}{z^2} \qquad \ldots\text{(Eq. 10.9)}$$

where K_B, Boussinesq's influence factor, is given by:

$$K_B = \frac{(3/2\pi)}{[1+(r/z)^2]^{5/2}} \qquad \ldots\text{(Eq. 10.10)}$$

The influence factor is a function of r/z as shown in Fig. 10.2.

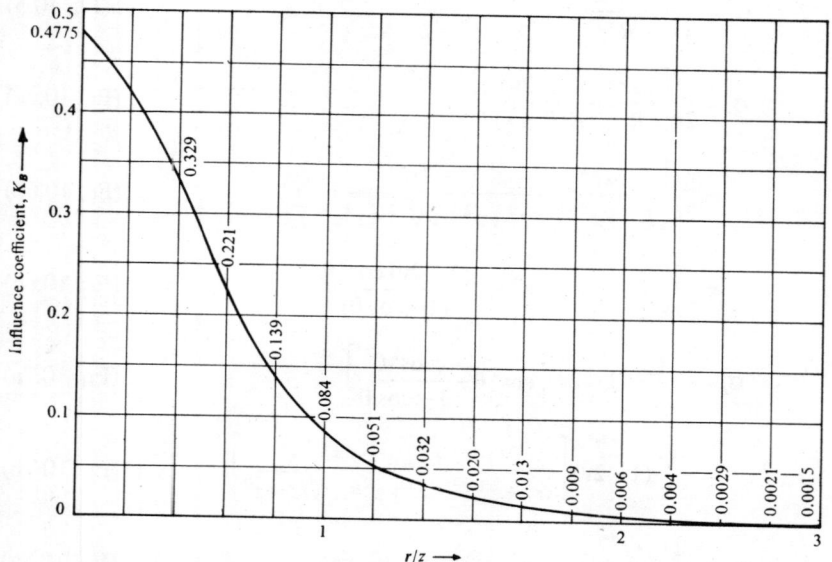

Fig. 10.2 Influence coefficients for vertical stress due to concentrated load
(After Boussinesq, 1885)

Gilboy (1933) has prepared a table of Boussinesq's influence coefficients for a large range of values of r/z. (K_B is as low as 0.0001 for r/z value 6.15). It is interesting to note that the influence factors for shearing stress, τ_{rz}, can be found by multiplying the K_B-values for σ_z by the r/z-ratio. The intensity of vertical stress, directly below the point load, on its axis of loading, is given by:

$$\sigma_z = \frac{0.4775\,Q}{z^2} \qquad \ldots \text{(Eq. 10.11)}$$

Poisson's ratio v, for the soil enters the equations for σ_x, σ_y, σ_r, and σ_t. For elastic materials v ranges from 0 to 0.5. (For cork, v is nearly zero and for clay soil it is nearly the maximum of 0.5). For a material for which v approaches 0.5, the volume change is negligible on loading; then it is said to be practically incompressible. Poisson's ratio for a soil is a highly tenuous property and one which is very difficult to determine. However, it has been found that it is closer to the upper limit of 0.5 than it is to zero.

If the value of 0.5 is taken for v for soil, the equations for σ_x, σ_y, σ_r, and σ_t get simplified as follows:

$$\sigma_x = \frac{3Q}{2\pi} \cdot \frac{x^2 z}{R^5} \qquad \ldots \text{(Eq. 10.12a)}$$

$$\sigma_y = \frac{3Q}{2\pi} \cdot \frac{y^2 z}{R^5} \qquad \ldots \text{(Eq. 10.12b)}$$

$$\sigma_r = \frac{3Q}{2\pi} \cdot \frac{r^2 z}{R^2} \qquad \ldots \text{(Eq. 10.12c)}$$

$$\sigma_t = 0 \qquad \ldots \text{(Eq. 10.12d)}$$

10.2.2 Pressure Distribution

It is possible to calculate the following pressure distributions by Eq. 10.2(d) of Boussinesq and present them graphically:

(i) Vertical stress distribution on a horizontal plane, at a depth z below the ground surface.
(ii) Vertical stress distribution along a vertical line, at a distance r from the line of action of the single concentrated load.

Vertical stress distribution on a horizontal plane

The vertical stress on a horizontal plane at depth z is given by:

$$\sigma_z = K_B \cdot \frac{Q}{z^2}, \; z \text{ being a specified depth.}$$

For several assumed values of r, r/z is calculated and K_B is found for each, the value of σ_z is then computed. For $r = 0$, σ_z is the maximum of $0.4775\,Q/z^2$; for $r = 2z$, it is only about 1.8% of the maximum, and for $r = 3z$, it is just 0.3% of the maximum. The distribution is as shown in Fig. 10.3 and Table 10.1.

416 Geotechnical Engineering

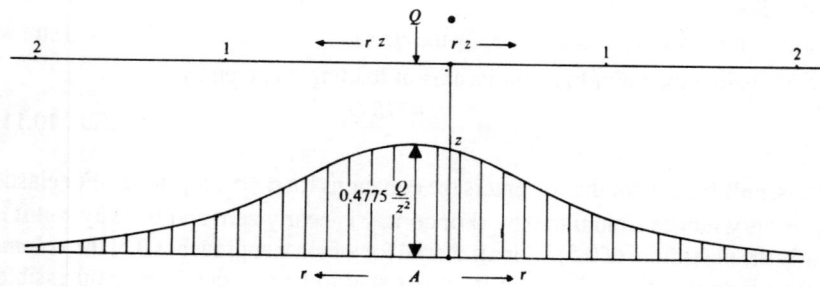

Fig. 10.3 Vertical stress distribution on a horizontal plane at depth z (Boussinesq's)

Theoretically, the stress approaches zero at infinity, although practically speaking, it reaches a negligible value at a short finite distance. The maximum pressure ordinate is relatively high at shallow elevations and it decreases with increasing depth. In other words, the bell-shaped figure flattens out with increasing depth.

If Q is taken as unity, this diagram becomes what is known as the 'Influence Diagram' for the vertical stress at A. With the aid of such a diagram, it is possible to determine the vertical stress at point A due to the combined effect of a number of concentrated loads at different radial distances from A, which will be the summation of the products of each of the loads and the ordinates of this diagram under each load.

Table 10.1 Variation of vertical stress with radial distance at a specified depth ($z = 1$ unit, say)

r	r/z	K_B	σ_z
0	0	0.4775	0.4775 Q
0.25	0.25	0.4103	0.4103 Q
0.50	0.50	0.2733	0.2733 Q
0.75	0.75	0.1565	0.1565 Q
1.00	1.00	0.0844	0.0844 Q
1.25	1.25	0.0454	0.0454 Q
1.50	1.50	0.0251	0.0251 Q
1.75	1.75	0.0144	0.0144 Q
2.00	2.00	0.0085	0.0085 Q

Vertical stress distribution along a vertical line

The variation of vertical stress with depth at a constant radial distance from the axis of the load may be shown by horizontal ordinates as in Fig. 10.4.

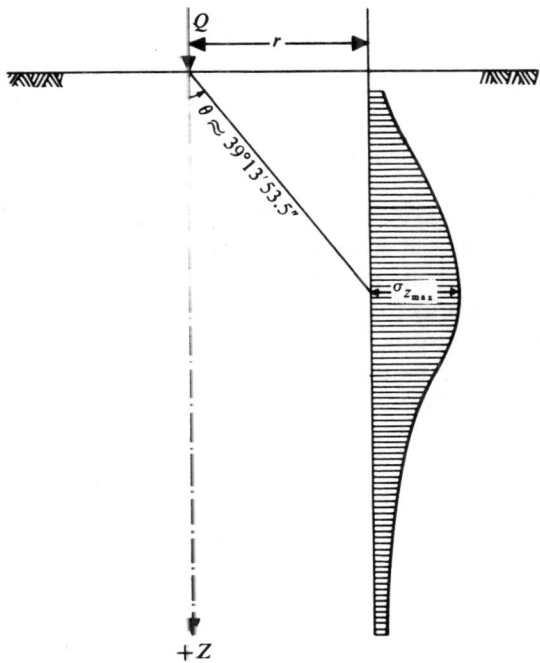

Fig. 10.4 Vertical stress distribution along a vertical line at radial distance r

As z increases, r/z decreases for a constant value of r. As r/z decreases K_B-value in the equation for σ_z increases, but since z^2 is involved in the denominator of the expression for σ_z, its value first increases with depth, attains a maximum value, and then decreases with further increase in depth. It can be shown that the maximum value of σ_z occurs when the angle θ made by the polar ray attains a value 39°13'53.5", corresponding to a value of $\sqrt{2/3}$ or 0.817 for r/z; the maximum value σ_z is then 0.0888 Q. This value decreases rapidly with depth; for $r/z = 0.1$, the value is just 0.0047 Q.

The values are tabulated for convenience as shown below:

Table 10.2 Variation of vertical stress with depth at constant values of r (Say $r = 1$ unit)

Depth z (Units)	r/z	K_B	K_B/z^2	σ_z
0	∞	--	--	Indeterminate
0.5	2.0	0.0085	0.0340	0.0340 Q
1	1.0	0.0844	0.0844	0.0844 Q
2	0.5	0.2733	0.0683	0.0683 Q
5	0.2	0.4329	0.0173	0.0173 Q
10	0.1	0.4657	0.0047	0.0047 Q

10.2.3 Stress Isobar or Pressure Bulb Concept

An *'isobar'* is a stress contour or a line which connects all points below the ground surface at which the vertical pressure is the same. In fact, an isobar is a spatial curved surface and resembles a bulb in shape; this is because the vertical pressure at all points in a horizontal plane at equal radial distances from the load is the same. Thus, the stress isobar is also called the 'bulb of pressure' or simply the 'pressure bulb'. The vertical pressure at each point on the pressure bulb is the same.

Pressures at points inside the bulb are greater than that at a point on the surface of the bulb; and pressures at points outside the bulb are smaller than that value. Any number of pressure bulbs may be drawn for any applied load, since each one corresponds to an arbitrarily chosen value of stress. A system of isobars indicates the decrease in stress intensity from the inner to the outer ones and reminds one of an 'Onion bulb'. Hence the term 'pressure bulb'. An isobar diagram, consisting of a system of isobars appears somewhat as shown in Fig. 10.5:

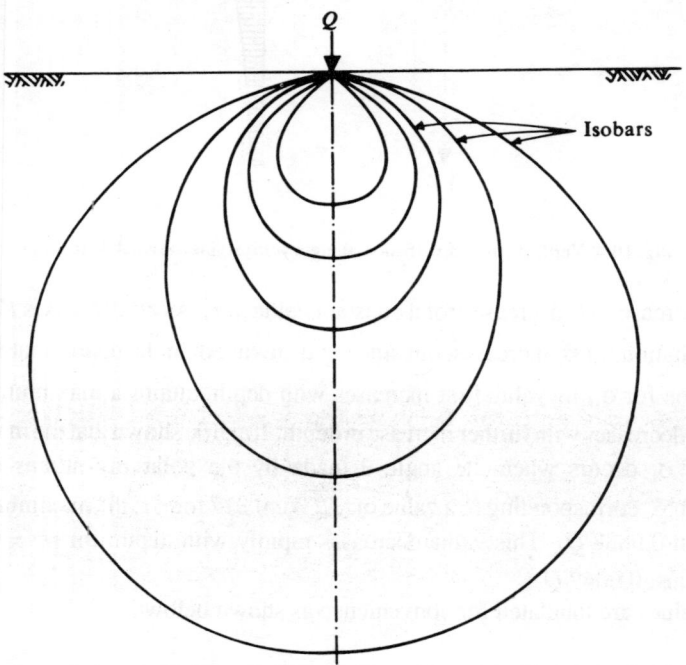

Fig. 10.5 Isobar diagram (A system of pressure bulbs for a point load—Boussinesq's)

The procedure for plotting an isobar is as follows:
Let it be required to plot an isobar for which $\sigma_z = 0.1\ Q$ per unit area (10% isobar):
From Eq. 10.9,

$$K_B = \frac{\sigma_z \cdot z^2}{Q} = \frac{0.1 Q \cdot z^2}{Q} = 0.1 z^2$$

Assuming various values for z, the corresponding K_B-values are computed; for these values of K_B, the corresponding r/z-values are obtained; and, for the assumed values of z, r-values are got.

It is obvious that, for the same value of r on any side of the z-axis, or line of action of the point load, the value of σ_z is the same; hence the isobar is symmetrical with respect to this axis.

When $r = 0$, $K_B = 0.4775$; the isobar crosses the line of action of the load at a depth of:

$$z = \sqrt{K_B/0.1} = \sqrt{\frac{0.4775}{0.1}} = \sqrt{4.775} = 2.185 \text{ units.}$$

The calculations are best performed in the form of a table as given below:

Table 10.3 Data for isobar of $\sigma_z = 0.1\,Q$ per unit area

Depth z (units)	Influence Coefficients K_B	r/z	r (units)	σ_z
0.5	0.0250	1.501	0.750	$0.1\,Q$
1.0	0.1000	0.932	0.932	$0.1\,Q$
1.5	0.2550	0.593	0.890	$0.1\,Q$
2.0	0.4000	0.271	0.542	$0.1\,Q$
2.185	0.4775	0	0	$0.1\,Q$

In general, isobars are not circular curves. Rather, their shape approaches that of the lemmiscate.

10.2.4 Westergaard's Solution

Natural clay strata have thin lenses of coarser material within them; this accentuates the non-isotropic condition commonly encountered in sedimentary soils, which is the primary reason for resistance to lateral strain in such cases.

Westergaard (1938) has obtained an elastic solution for stress distribution in soil under a point load based on conditions analogous to the extreme condition of this type. The material is assumed to be laterally reinforced by numerous, closely spaced horizontal sheets of negligible thickness but of infinite rigidity, which prevent the medium from undergoing lateral strain; this may be viewed as representative of an extreme case of non-isotropic condition.

The vertical stress σ_z caused by a point load, as obtained by Westergaard, is given by:

$$\sigma_z = \frac{Q}{z^2} \cdot \frac{\frac{1}{2\pi}\sqrt{\frac{1-2\nu}{2-2\nu}}}{\left[\left(\frac{1-2\nu}{2-2\nu}\right)+\left(\frac{r}{z}\right)^2\right]^{3/2}} \qquad \ldots\text{(Eq. 10.13)}$$

The symbols have the same meaning as in the case of Boussinesq's solution; ν is Poisson's ratio for the medium, and may be taken to be zero for large lateral restraint. (This gives, in fact, the flattest curve for stress distribution, as shown in Fig. 10.6, a flat curve being the logical shape for a case of large lateral restraint). Then the equation for σ_z reduces to:

$$\sigma_z = \frac{Q}{z^2} \cdot \frac{1/\pi}{[1+2(r/z)^2]^{3/2}} \qquad \ldots\text{(Eq. 10.14)}$$

or

$$\sigma_z = K_W \cdot \frac{Q}{z^2} \qquad \ldots\text{(Eq. 10.15)}$$

where

$$K_W = \frac{1/\pi}{[1+2(r/z)^2]^{3/2}} \qquad \ldots\text{(Eq. 10.16)}$$

K_W is Westergaard's influence coefficient, the variation of which with (r/z) is shown in Fig. 10.6; for comparison, the variation of K_B of Eq. 10.10 is also super-imposed:

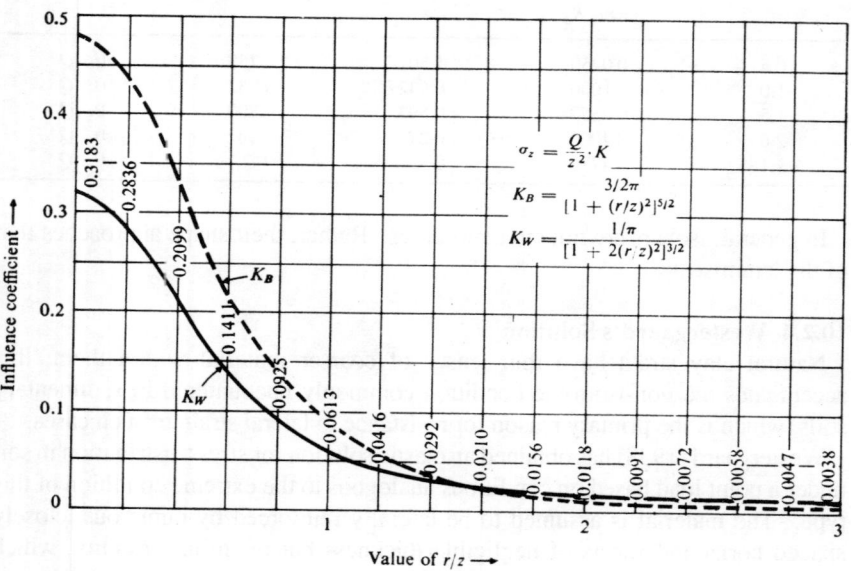

Fig. 10.6 Influence coefficients for vertical stress due to concentrated load (Westergaard's and Boussinesq's solutions) (After Taylor, 1948)

For cases of point loads with r/z less than about 0.8, Westergaard's stress values, assuming ν to be zero, are approximately equal to two-thirds of Boussinesq's stress values. For r/z of about 1.5, both solutions give identical values of stresses.

This is also reflected in the comparison of vertical stress distribution on a horizontal plane at a specified depth from Boussinesq's and Westergaard's solutions, as shown in Fig. 10.7 below:

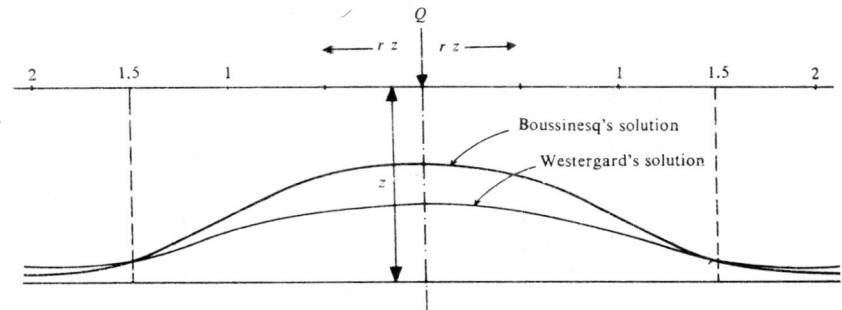

Fig. 10.7 Vertical stress distribution on a horizontal plane at specified depth—comparison between Boussinesq's and Westergaard's solutions

10.3 LINE LOAD

Let a load, uniformly distributed along a line, of intensity q' per unit length of a straight line of infinite extension, act on the surface of a semi-infinite elastic medium. Such a loading produces a state of plane strain; that is, the strains and stresses in all planes normal to the line of the loading are identical and it is adequate to consider the conditions in one such plane as in Fig. 10.8(a). Let the y-axis be directed along the line of loading as shown in Fig. 10.8(b).

Let us consider a small length dy of the line load as shown; the equivalent point load is $q' \cdot dy$ and, the vertical stress at A due to this load is given by:

$$d\sigma_z = \frac{3(q' \, dy)z^3}{2\pi r^5} = \frac{3q' \cdot z^3 dy}{2\pi(x^2 + y^2 + z^2)^{5/2}}$$

The vertical stress σ_z at A due to the infinite length of line load may be obtained by integrating the equation for $d\sigma_z$ with respect to the variable y within the limits $-\infty$ and $+\infty$.

$$\therefore \quad \sigma_z = \int_{-\infty}^{\infty} \frac{3q'z^3 dy}{2\pi(x^2 + y^2 + z^2)^{5/2}} = 2\int_0^{\infty} \frac{3q'z^3 dy}{2\pi(x^2 + y^2 + z^2)^{5/2}}$$

or

$$\sigma_z = \frac{2q'z^3}{\pi(x^2 + z^2)^2} = \frac{2q'}{\pi z} \cdot \frac{1}{[1 + (x/z)^2]^2} \qquad \ldots(\text{Eq. 10.17})$$

422 *Geotechnical Engineering*

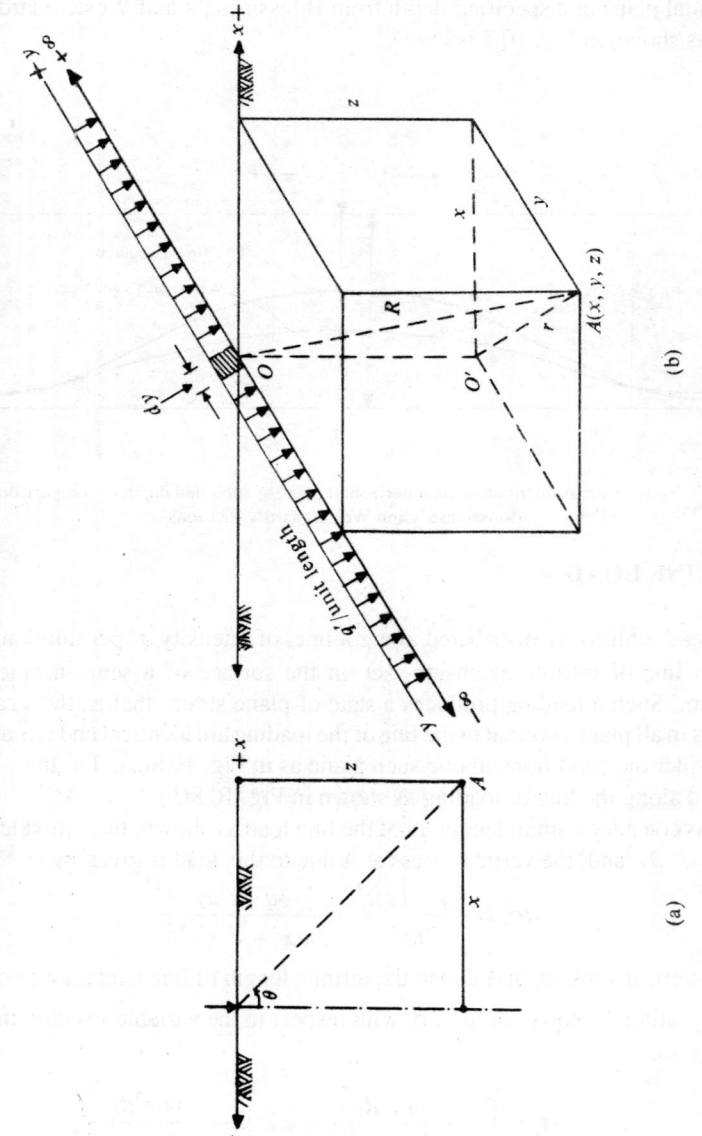

Fig. 10.8 Line load acting on the surface of semi-infinite elastic soil medium

Equation 10.17 may be written in either of the two forms:

$$\sigma_z = \frac{q'}{z} \cdot K_I \qquad \ldots(Eq.\ 10.18)$$

where
$$K_I = \frac{(2/\pi)}{[1+(x/z)^2]^2} \qquad \ldots(Eq.\ 10.19)$$

K_I being the influence coefficient for line load using Boussinesq's theory; or

$$\sigma_z = \frac{2q'}{\pi z} \cdot \cos^4 \theta \qquad \ldots(Eq.\ 10.20)$$

since the Y-co-ordinate may be taken as zero for any position of the point relative to the line load, in view of the infinite extension of the latter in either direction.

σ_x and τ_{xz} may be also derived and shown to be:

$$\sigma_x = \frac{2}{\pi} \cdot \frac{q'}{z} \cdot \cos^2 \theta \cdot \sin^2 \theta \qquad \ldots(Eq.\ 10.21)$$

and
$$\tau_{xz} = \frac{2}{\pi} \cdot \frac{q'}{z} \cdot \cos^3 \theta \cdot \sin \theta \qquad \ldots(Eq.\ 10.22)$$

If the point A is situated vertically below the line load, at a depth z, we have $x = 0$, and hence the vertical stress is then given by:

$$\sigma_z = \frac{2}{\pi} \cdot \frac{q'}{z} \qquad \ldots(Eq.\ 10.23)$$

10.4 STRIP LOAD

Let a uniform load of intensity q per unit area be acting on a strip of infinite length and a constant width B (= $2b$) as shown in Fig. 10.9.

The previous case of line load may be applied and integration used to obtain the stresses at any point such as A due to the strip load.

Considering the effect of a small width dx of the strip load and taking it as a line load of intensity ($q.\ dx$) per unit length in the Y-direction, the value of $d\sigma_z$ at A is given by:

$$d\sigma_z = \frac{2}{\pi} \cdot \frac{(q \cdot dx)}{z} \cos^4 \theta$$

Since $x = z \tan \theta$ and $dx = z \sec^2 \theta \cdot d\theta$,

$$d\sigma_z = \frac{2}{\pi} \cdot \frac{q \cdot z \sec^2 \theta \cdot \cos^4 \theta\ d\theta}{z} = \frac{2q}{\pi} \cos^2 \theta \cdot d\theta$$

Integrating within the limits, θ_1 and θ_2 for θ, we get

$$\sigma_z = \frac{q}{\pi}[\theta + \sin \theta \cos \theta]_{\theta_1}^{\theta_2} \qquad \ldots(Eq.\ 10.24)$$

424 *Geotechnical Engineering*

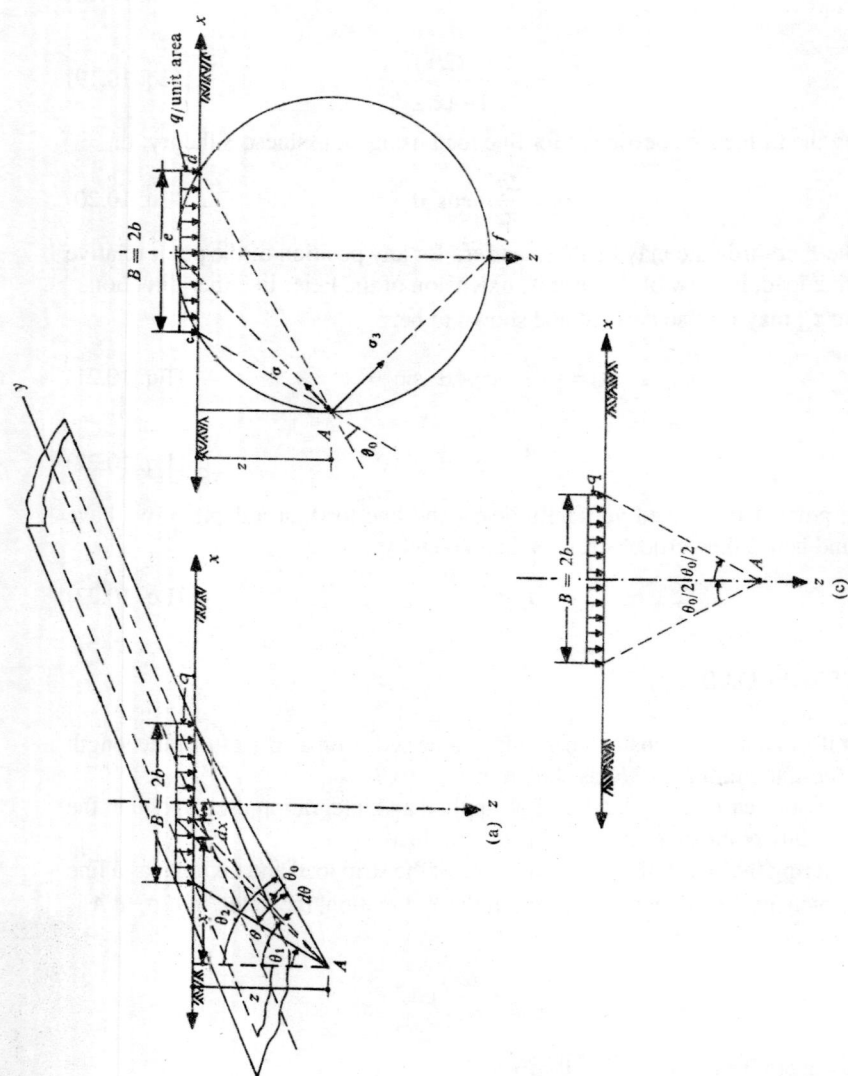

Fig. 10.9 Strip load of infinite length acting at the surface of a semi-infinite elastic soil medium (After Terzaghi, 1943)

or
$$\sigma_z = \frac{q}{\pi}[(\theta_2 - \theta_1) + (\sin\theta_2 \cos\theta_2 - \sin\theta_1 \cos\theta_1)]$$

$$= \frac{q}{\pi}\left[(\theta_2 - \theta_1) + \frac{1}{2}(\sin 2\theta_2 - \sin 2\theta_1)\right]$$

$$\therefore \quad \sigma_z = \frac{q}{\pi}[(\theta_2 - \theta_1) + \cos(\theta_2 + \theta_1)\cdot \sin(\theta_2 - \theta_1)] \quad \ldots\text{(Eq. 10.25)}$$

Similarly, starting from Eqs. 10.21 and 10.22, and adopting precisely the same procedure, one arrives at the following:

$$\sigma_x = \frac{q}{\pi}[(\theta - \sin\theta \cos\theta)]_{\theta_1}^{\theta_2} \quad \ldots\text{(Eq. 10.26)}$$

or
$$\sigma_x = \frac{q}{\pi}[(\theta_2 - \theta_1) - (\sin\theta_2 \cos\theta_2 - \sin\theta_1 \cos\theta_1)]$$

$$= \frac{q}{\pi}\left[(\theta_2 - \theta_1) - \frac{1}{2}(\sin 2\theta_2 - \sin 2\theta_1)\right]$$

$$\therefore \quad \sigma x = \frac{q}{\pi}[(\theta_2 - \theta_1) - \cos(\theta_2 + \theta_1)\cdot \sin(\theta_2 - \theta_1)] \quad \ldots\text{(Eq. 10.27)}$$

$$\tau_{xz} = \frac{q}{\pi}[\sin^2\theta]_{\epsilon_1}^{\epsilon_2} \quad \ldots\text{(Eq. 10.28)}$$

$$= \frac{q}{\pi}(\sin^2\theta_2 - \sin^2\theta_1)$$

$$\therefore \quad \tau_{xz} = \frac{q}{\pi}[\sin(\theta_2 + \theta_1)\cdot \sin(\theta_2 - \theta_1)] \quad \ldots\text{(Eq. 10.29)}$$

The corresponding principal stresses may be established as:

$$\sigma_1 = \frac{q}{\pi}(\theta_0 + \sin\theta_0) \quad \ldots\text{(Eq. 10.30)}$$

and
$$\sigma_3 = \frac{q}{\pi}(\theta_0 - \sin\theta_0) \quad \ldots\text{(Eq. 10.31)}$$

where $\theta_0 = \theta_2 - \theta_1$ [See Fig. 10.9(a)].

According to these equations the principal stresses for a given value of q depend solely on the value of θ_0; hence for every point on a circle through c, d, and A [Fig. 10.9(b)] the principal stresses have the same intensity. It can also be shown that the principal stresses at every point on the circle cdA pass through the points e and f respectively. These two points are located at the intersection between the circle and the plane of symmetry of the loaded strip.

The special case when A lies on the plane of symmetry of the loaded strip is shown in Fig. 10.9(c). The vertical stress σ_z and the horizontal stress σ_x themselves will be the principal stresses since τ_{xy} reduces to zero in view of $(\theta_2 + \theta_1)$ being zero, in this case. Hence, substituting $\theta_2 = +\dfrac{\theta_0}{2}$ and $\theta_1 = -\dfrac{\theta_0}{2}$ in equations 10.25 and 10.27, we have:

$$\sigma_z = \sigma_1 = \frac{q}{\pi}(\theta_0 + \sin\theta_0) \qquad \text{...(Eq. 10.32)}$$

$$\sigma_x = \sigma_3 = \frac{q}{\pi}(\theta_0 - \sin\theta_0) \qquad \text{...(Eq. 10.33)}$$

The vertical stresses at different depths below the centre of a uniform load of intensity q and width B are as follows:

Table 10.4 Vertical stress under centre of strip load

Depth z	σ_z
0.1 B	0.997 q
0.2 B	0.977 q
0.50 B	0.818 q
B	0.550 q
2 B	0.306 q
5 B	0.126 q
10 B	0.064 q

A few typical pressure bulbs for this case of strip loading are shown in Fig. 10.10.

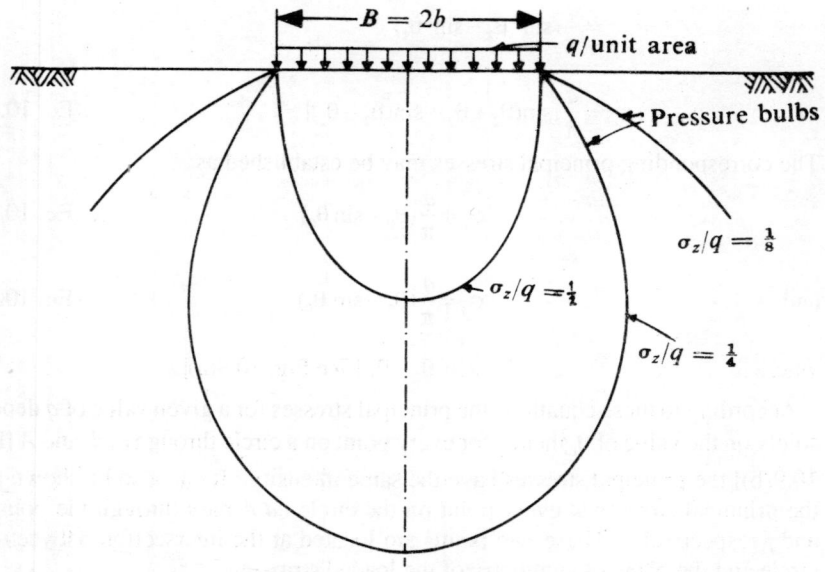

Fig. 10.10 Pressure bulbs for strip load of infinite length
(After Terzaghi, 1943)

Stress Distribution in Soil

Since the principal stresses are known from equations 10.30 and 10.31, the maximum shear stress τ_{max} may be obtained as:

$$\tau_{max} = \frac{(\sigma_1 - \sigma_3)}{2} = \left(\frac{q}{\pi}\right) \cdot \sin \theta_0 \qquad \ldots(Eq.\ 10.34)$$

This will attain its highest value when $\theta_0 = 90°$, which equals q/π.

$$\therefore \quad \tau_{absolute\ maximum} = \frac{q}{\pi} \qquad \ldots(Eq.\ 10.35)$$

This value, it is easily understood, occurs at points lying on a semi-circle of diameter equal to the width of the strip, B. Hence the maximum shear stress under the centre of a continuous strip occurs at a depth of $B/2$ beneath the centre.

The knowledge of shear stresses may not be important in normal foundation design procedure, but Jürgenson (1934) obtained the solution for this case. Pressure bulbs of shear stress as obtained by him are shown in Fig. 10.11.

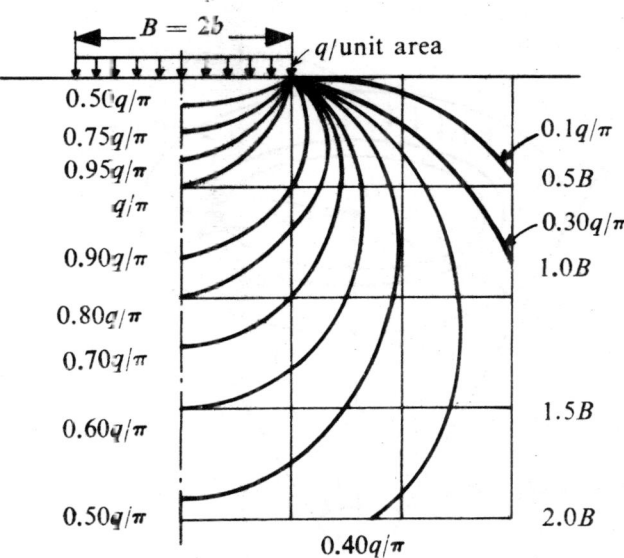

Fig. 10.11 Pressure bulbs of shear stress under strip load
(After Jürgenson, 1934)

10.5 UNIFORM LOAD ON CIRCULAR AREA

This problem may arise in connection with settlement studies of structures on circular foundations, such as gasoline tanks, grain elevators, and storage bins.

The Boussinesq equation for the vertical stress due to a point load can be extended to find the vertical stress at any point beneath the centre of a uniformly loaded circular area. Let the circular area of radius a be loaded uniformly with q per unit area as shown in Fig. 10.12.

Let us consider an elementary ring of radius r and thickness dr of the loaded area. This ring may be imagined to be further divided into elemental areas, each δA; the load from such an elemental area is $q \cdot \delta A$. The vertical stress $\delta \sigma_z$ at point A, at a depth z below the centre of the loaded area, is given by:

$$\delta \sigma_z = \frac{3(q \cdot \delta A)}{2\pi} \cdot \frac{z^3}{(r^2 + z^2)^{5/2}}$$

The stress $d\sigma_z$ due to the entire ring is given by:

$$d\sigma_z = \frac{3q}{2\pi}(\Sigma \delta A) \cdot \frac{z^3}{(r^2 + z^2)^{5/2}} = \frac{3q}{2\pi} \times (2\pi r\, dr) \cdot \frac{z^3}{(r^2 + z^2)^{5/2}}$$

$$\therefore \quad d\sigma_z = \frac{3qz^3 \cdot r\, dr}{(r^2 + z^2)^{5/2}}$$

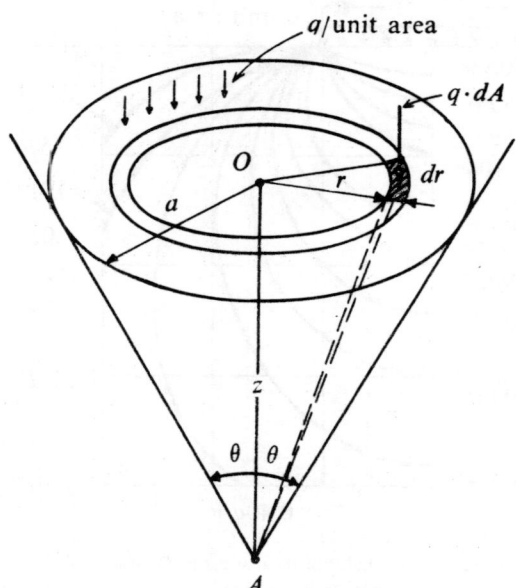

Fig. 10.12 Uniform load over circular area

The total vertical stress σ_z at A due to entire loaded area is obtained by integrating $d\sigma_z$ within the limits $r = 0$ to $r = a$.

$$\therefore \quad \sigma_z = 3qz^3 \int_{r=0}^{r=a} \frac{r\, dr}{(r^2 + z^2)^{5/2}}$$

Setting $r^2 + z^2 = R^2$, $r\, dr = R \cdot dR$, the limits for R will be z and $(a^2 + z^2)^{1/2}$.

$$\therefore \quad \sigma_z = 3qz^3 \int_{R=z}^{R=(a^2+z^2)^{1/2}} \frac{dR}{R^4}$$

$$= qz^3 \left[\frac{1}{z^3} - \frac{1}{(a^2+z^2)^{3/2}} \right]$$

$$\therefore \quad \sigma_z = q \left[1 - \frac{1}{\{1+(a/z)^2\}^{3/2}} \right] \quad \ldots(\text{Eq. 10.36})$$

This may be written as:

$$\sigma_z = q \cdot K_{B_C} \quad \ldots(\text{Eq. 10.37})$$

where K_{B_C} = Boussinesq influence coefficient for uniform load on circular area,

and,

$$K_{B_C} = \left[1 - \frac{1}{\{1+(a/z)^2\}^{3/2}} \right] \quad \ldots(\text{Eq. 10.38})$$

If θ is the angle made by OA with the tangent to the periphery of the loaded area from A,

$$\sigma_z = q(1 - \cos^3 \theta) \quad \ldots(\text{Eq. 10.39})$$

The vertical stress at a point not lying on the vertical axis through the centre of the loaded area may also be found; but it requires the evaluation of a more difficult integral.

Spangler (1951) gives the influence coefficients for both cases, the values for the case where the point lies directly beneath the centre of the loaded area being those in the column for $\frac{r}{a} = 0$ (Table 10.5).

Table 10.5 Influence coefficients for vertical stress due to uniform load on a circular area (After Spangler, 1951)

$\frac{z}{a}$	\multicolumn{9}{c}{r/a}								
	0	0.5	1.0	1.5	2.0	2.5	3.0	3.5	4.0
0.5	0.911	0.840	0.418	0.060	0.010	0.003	0.000	0.000	0.000
1.0	0.646	0.560	0.335	0.125	0.043	0.016	0.007	0.003	0.000
1.5	0.424	0.374	0.256	0.137	0.064	0.029	0.013	0.007	0.002
2.0	0.284	0.258	0.194	0.127	0.073	0.041	0.022	0.012	0.006
2.5	0.200	0.186	0.150	0.109	0.073	0.044	0.028	0.017	0.011
3.0	0.146	0.137	0.117	0.091	0.066	0.045	0.031	0.022	0.015
4.0	0.087	0.083	0.076	0.061	0.052	0.041	0.031	0.024	0.018
5.0	0.057	0.056	0.052	0.045	0.039	0.033	0.027	0.022	0.018
10.0	0.015	0.014	0.014	0.013	0.013	0.013	0.012	0.012	0.011

Pressure bulbs or isobar patterns for vertical stresses and shear stresses, as presented by Jürgenson (1934), are shown in Figs. 10.13(a) and (b).

430 Geotechnical Engineering

If Westergaard's theory is to be used, equation 10.14 may be integrated to obtain the vertical stress at a point beneath the centre of a uniformly loaded circular area, which is given by:

$$\sigma_z = q\left[1 - \frac{1}{\sqrt{1 + (a/\eta z)^2}}\right] \quad \ldots(\text{Eq. 10.40})$$

where $\eta = \sqrt{\dfrac{1-2\nu}{2-2\nu}}$, ν being Poisson's ratio.

for $\nu = 0$, $\qquad \eta = \dfrac{1}{\sqrt{2}} = 0.707$

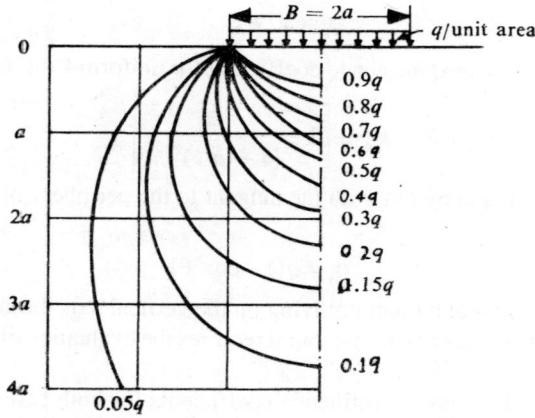

(a) For vertical stress

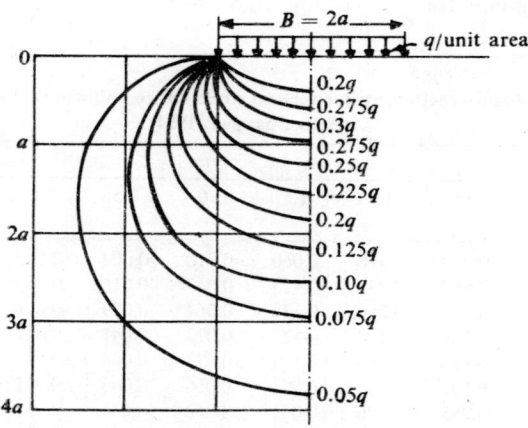

(b) For shear stress

Fig. 10.13 Pressure bulbs for vertical and shear stress due to uniform load on a circular area
(After Jürgenson, 1934)

10.6 UNIFORM LOAD ON RECTANGULAR AREA

The more common shape of a loaded area in foundation engineering practice is a rectangle, especially in the case of buildings. Applying the principle of integration, one can obtain the vertical stress at a point at a certain depth below the centre or a corner of a uniformly loaded rectangular area, based either on Boussinesq's or on Westergaard's solution for a point load.

10.6.1 Uniform Load on Rectangular Area based on Boussinesq's Theory

Newmark (1935) has derived an expression for the vertical stress at a point below the corner of a rectangular area loaded uniformly as shown in Fig. 10.14.

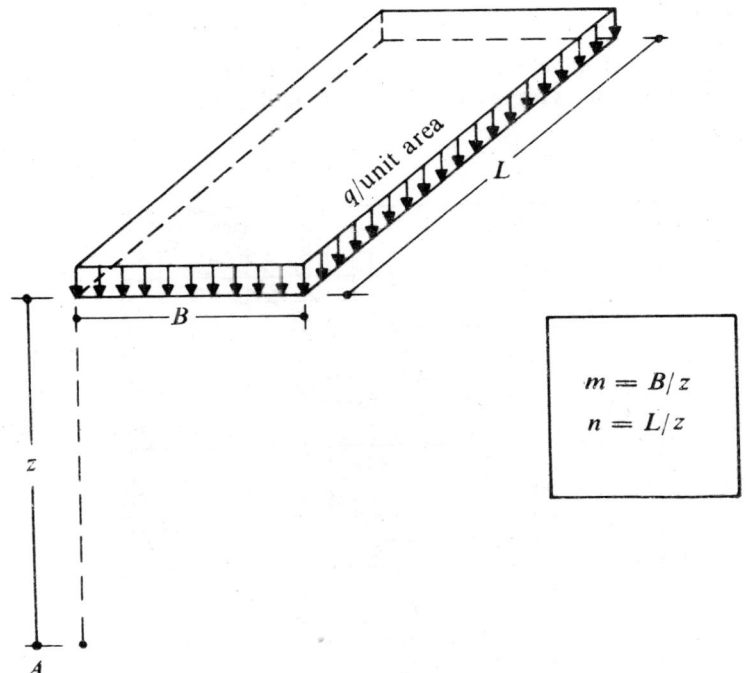

Fig. 10.14 Vertical stress at the corner of a uniformly loaded rectangular area

The following are the two popular forms of Newmark's equation for σ_z:

$$\sigma_z = \frac{q}{4\pi}\left[\left(\frac{2mn\sqrt{m^2+n^2+1}}{(m^2+n^2+1+m^2n^2)}\right)\left(\frac{m^2+n^2+2}{m^2+n^2+1}\right)+\sin^{-1}\left(\frac{2mn\sqrt{m^2+n^2+1}}{m^2+n^2+1+m^2n^2}\right)\right] \quad \text{...(Eq. 10.41)}$$

$$\sigma_z = \frac{q}{4\pi}\left[\frac{(2mn\sqrt{m^2+n^2+1})}{(m^2+n^2+1+m^2n^2)}\cdot\frac{(m^2+n^2+2)}{(m^2+n^2+1)}+\tan^{-1}\frac{2mn\sqrt{m^2+n^2+1}}{(m^2+n^2+1-m^2n^2)}\right] \quad \text{...(Eq. 10.42)}$$

where $m = B/z$ and $n = L/z$.

The second term within the brackets is an angle in radians. It is of interest to note that the above expressions do not contain the dimension z; thus, for any magnitude of z, the underground stress depends only on the ratios m and n and the surface load intensity. Since these equations are symmetrical in m and n, the values of m and n are interchangeable.

Equation 10.42 may be written in the form:

$$\sigma_z = q \cdot I_\sigma \qquad \ldots \text{(Eq. 10.43)}$$

where I_σ = Influence value

$$= (1/4\pi)\left[\left(\frac{2mn\sqrt{m^2+n^2+1}}{m^2+n^2+1+m^2n^2}\right) \cdot \left(\frac{m^2+n^2+2}{m^2+n^2+1}\right) + \tan^{-1}\left(\frac{2mn\sqrt{m^2+n^2+1}}{m^2+n^2+1-m^2n^2}\right)\right]$$

...(Eq. 10.44)

Based on this equation, Fadum (1941) has prepared a chart for the influence values for sets of values for m and n, as shown in Fig. 10.15.

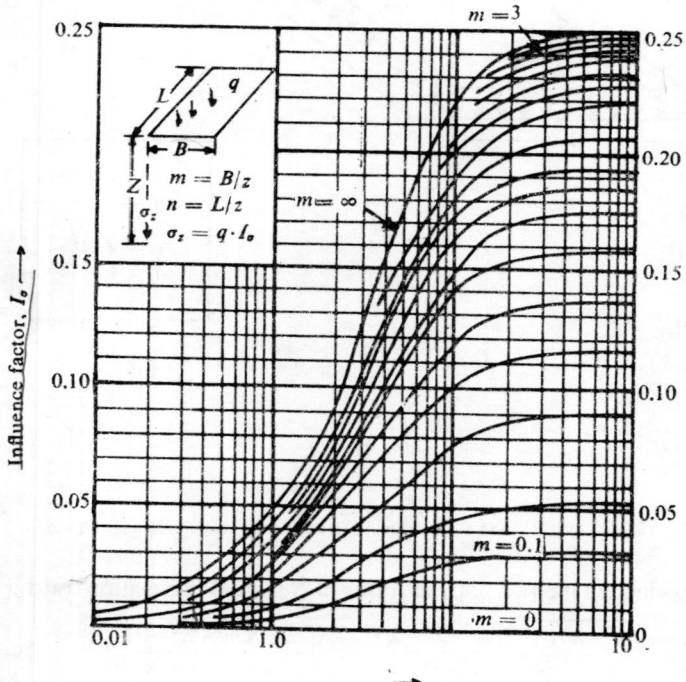

Fig. 10.15 Influence factors for vertical stress beneath a corner
of a uniformly loaded rectangular area
(After Fadum, 1941)

Steinbrenner (1934) has given another form of chart for this purpose, which is shown in Fig. 10.16, plotting the influence values I_σ on the horizontal axis and $1/m$ (= z/B) on the vertical axis, for different values of L/B (= n/m).

The vertical stress at a point beneath the centre of a uniformly loaded rectangular area may be found using the influence value for a corner by the principle of superposition, dividing the rectangle into four equal parts by lines parallel to the sides and passing through the centre. In fact, if we derive the expression for the vertical stress beneath the centre of the rectangle, we may obtain that beneath a

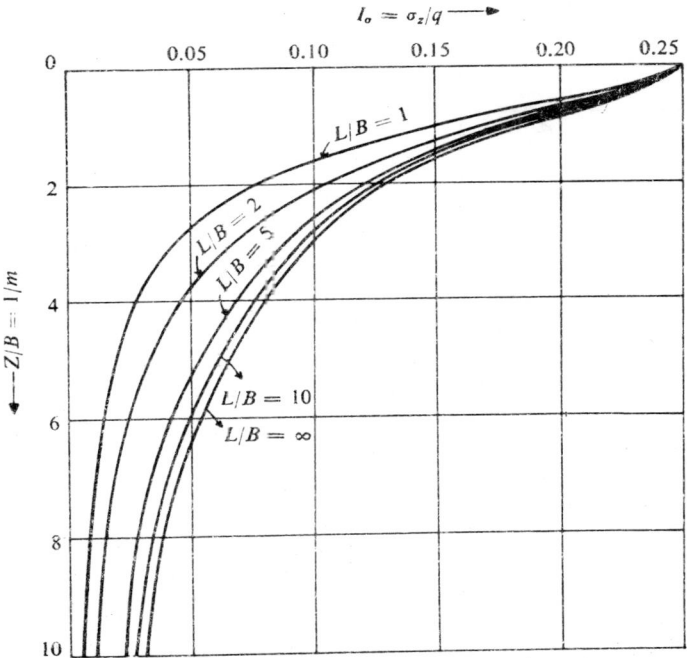

Fig. 10.16 Influence values for vertical stress at the corner of a uniformly loaded rectangular area (After Steinbrenner, 1934)

corner due to load from one fourth of this area by just dividing it by four, by the same principle of superposition; the formula may be generalised by making the necessary modifications in respect of the reduced dimensions of the rectangle.

The principle of superposition may be conveniently employed to compute the stress beneath any point either inside or outside a uniformly loaded rectangular area. This is illustrated as follows:

Let the point A at which the vertical stress is required be at a depth z beneath A', inside the uniformly loaded rectangular area $PQRS$ as in Fig. 10.17(a).

Imagine TU and VW parallel to the sides and passing through A'. σ_z at A is given by:

$$\sigma_z = q\left(I_{\sigma_\mathrm{I}} + I_{\sigma_\mathrm{II}} + I_{\sigma_\mathrm{III}} + I_{\sigma_\mathrm{IV}}\right) \qquad \text{...(Eq. 10.45)}$$

where $I_{\sigma_I}, I_{\sigma_{II}}$ ..., are the influence factors for the stress at A due to the rectangular areas I, II, ..., by the principle of superposition, since A' happens to be a corner for these areas.

If the point A is beneath A', outside the uniformly loaded rectangular area PQRS, as in Fig. 10.17(b), imagine PQ, SR, PS and QR to be extended such that PWA'T is a rectangle and U and V' are points on its sides, as shown.

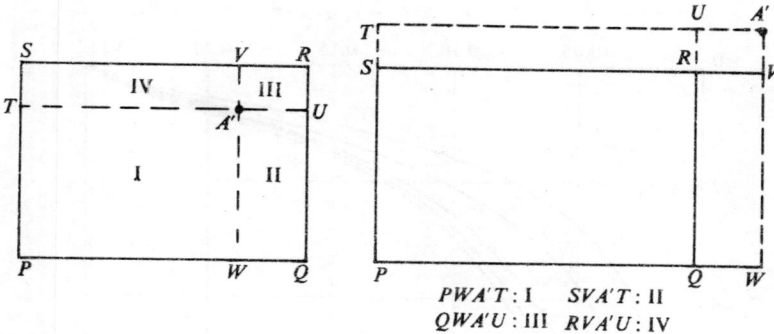

(a) Point inside the loaded area   (b) Point outside the loaded area

Fig. 10.17 Stress at a point other than under a corner of a rectangular area

Then σ_z at A is given by:

$$\sigma_z = q(I_{\sigma_I} - I_{\sigma_{II}} - I_{\sigma_{III}} - I_{\sigma_{IV}}) \qquad \text{...(Eq. 10.46)}$$

where, $I_{\sigma_I}, I_{\sigma_{II}}$ ..., are the influence factors for the stress at A due to the rectangular areas designated I, II, ..., by the principle of superposition. (Since area IV is deducted twice, its influence has to be added once).

10.6.2 Uniform Load on Rectangular Area based on Westergaard's Theory

If the soil conditions correspond to those assumed in Westergaard's theory, that is, the soil consists of very thin horizontal sheets of infinite rigidity, which prevent the occurrence of any lateral strain and the vertical stress at a point below a corner of a uniformly loaded rectangular area then it may be obtained by integration of the stress due to a point load under similar conditions, and shown to be:

$$\sigma_z = (q/2\pi)\left[\cot^{-1}\sqrt{\left(\frac{1-2\nu}{2-2\nu}\right)\left(\frac{1}{m^2}+\frac{1}{n^2}\right)+\left(\frac{1-2\nu}{2-2\nu}\right)^2 \cdot \frac{1}{m^2 n^2}}\right] \qquad \text{...(Eq. 10.47)}$$

If the Poisson's ratio, ν, is taken as zero, this reduces to:

$$\sigma_z = (q/2\pi)\left[\cot^{-1}\sqrt{\frac{1}{2m^2}+\frac{1}{2n^2}+\frac{1}{4m^2 n^2}}\right] \qquad \text{...(Eq. 10.48)}$$

The notation is the same as that for equations 10.41 and 10.42. For this case, if σ_z is written as:

$$\sigma_z = q \cdot I_{\sigma_w} \quad \ldots \text{(Eq. 10.49)}$$

where I_{σ_w} = Influence coefficient for vertical stress at the corner of a uniformly loaded rectangular area from Westergaard's Theory.

Taylor (1948) has given a chart for the determination of I_{σ_w}, for different values of m and n. It is obvious that m and n are interchangeable.

For m and n values less than unity, I_{σ_w} is about two-thirds of I_σ based on Boussinesq's theory. In such cases, sound judgement is called for regarding which theory is more appropriate for the particular conditions of the soil medium.

10.7 UNIFORM LOAD ON IRREGULAR AREAS—NEWMARK'S CHART

It may not be possible to use Fadum's influence coefficients or chart for irregularly shaped loaded areas. Newmark (1942) devised a simple, graphical procedure for computing the vertical stress in the interior of a soil medium, loaded by uniformly distributed, vertical load at the surface. The chart devised by him for this purpose is called an 'Influence Chart'. This is applicable to a semi-infinite, homogeneous, isotropic and elastic soil mass (and not for a stratified soil).

The vertical stress underneath the centre of a uniformly loaded circular area has been shown to be:

$$\sigma_z = q \left[1 - \frac{1}{\{1 + (a/z)^2\}^{3/2}} \right] \quad \ldots \text{(Eq. 10.36)}$$

where a = radius of the loaded area, z = depth at which the vertical stress is required, and q = intensity of the uniform load. This equation may be rewritten in the form:

$$\frac{a}{z} = \sqrt{\left(1 - \frac{\sigma_z}{q}\right)^{-2/3} - 1} \quad \ldots \text{(Eq. 10.50)}$$

Here (a/z) may be interpreted as relative sizes or radii of circular-loaded areas required to cause particular values of the ratio of the vertical stress to the intensity of the uniform loading applied.

If a series of values is assigned for the ratio σ_z/q, such as 0, 0.1, 0.2, ..., 0.9, and 1.00, a corresponding set of values for the relative radii, a/z, may be obtained. If a particular depth is specified, then a series of concentric circles can be drawn. Since the first has a zero radius and the eleventh has infinite radius, in practice, only nine circles are drawn. Each ring or annular space causes a stress of $q/10$ at a point beneath the centre at the specified depth z, since the number of annular spaces (c) is ten.

The relative radii may be tabulated as shown below:

436 Geotechnical Engineering

Table 10.6 Relative radii for Newmark's influence chart

S. No. of circle	σ_z/q	Relative radii a/z	Number of influence meshes per ring
1	0.0	0.000	...
2	0.1	0.270	20
3	0.2	0.400	20
4	0.3	0.518	20
5	0.4	0.637	20
6	0.5	0.766	20
7	0.6	0.918	20
8	0.7	1.110	20
9	0.8	1.387	20
10	0.9	1.908	20
11	1.0	∞	...

From this table it can be seen that the widths of the annular slices or rings are greater the farther away they are from the centre. The circle for an influence of 1.0 has an infinitely large radius. Now let us assume that a set of equally spaced rays, say s in number, is drawn emanating from the centre of the circles, thus dividing each annular area into s sectors, and the total area into cs sectors. If the usual value of 20 is adopted for s, the total number of sectors in this case will be 10×20 or 200. Each sector will cause a vertical stress of 1/200th of the total value at the centre at the specified depth and is referred to as a 'mesh' or an 'influence unit'. The value 1/200 or 0.005 is said to be the 'influence value' (or 'influence factor') for the chart. Each mesh may thus be understood to represent an influence area.

The construction of Newmark's influence chart, as this is usually called, may be given somewhat as follows:

For the specified depth z (say, 10 m), the radii of the circles, a, are calculated from the relative radii of Table 10.6 (2.70 m, 4.00 m, 5.18 m, ... and so on.). The circles are then drawn to a convenient scale (say, 1 cm = 2 m). A suitable number of uniformly spaced rays (say, 20) is drawn, emanating from the centre of the circles. The resulting diagram will appear as shown in Fig. 10.18; on it is drawn a vertical line ON, representing the depth z to the scale used in drawing the circles (if the scale used is 1 cm = 2 m, ON will be 5 cm). The influence value for this chart will be $\frac{1}{10 \times 20}$ or 0.005. The diagram can be used for other values of the depth z by simply assuming that the scale to which it is drawn alters; thus, if z is to be 5 m the line ON now represents 5 m and the scale is therefore 1 cm = 1 m (similarly, if z = 20 m, the scale becomes 1 cm = 4 m).

The operation or use of the Newmark's chart is as follows:

The chart can be used for any uniformly loaded area of whatever shape that may be. First, the loaded area is drawn on a tracing paper, using the same scale to which the distance ON on the chart represents the specified depth; the point at which the vertical stress is desired is then placed over the centre of the circles on the chart. The number of influence units encompassed by or contained in the boundaries of the loaded area are counted, including fractional units, if any; let this total equivalent number be N. The stress σ_z at the specified depth at the specified point is then given by:

$\sigma_z = I \cdot N \cdot q$, where I = influence value of the chart. ...(Eq. 10.51)

(**Note**: The stress may be found at any point which lies either inside or outside the loaded area with the aid of the chart).

Although it appears remarkably simple, Newmark's chart has also some inherent deficiencies:
1. Many loaded areas have to be drawn; alternatively, many influence charts have to be drawn.
2. For each different depth, counting of the influence meshes must be done. Considerable amount of guesswork may be required in estimating the influence units partially covered by the loaded area.

However, the primary advantage is that it can be used for loaded areas of any shape and that it is relatively rapid. This makes it attractive.

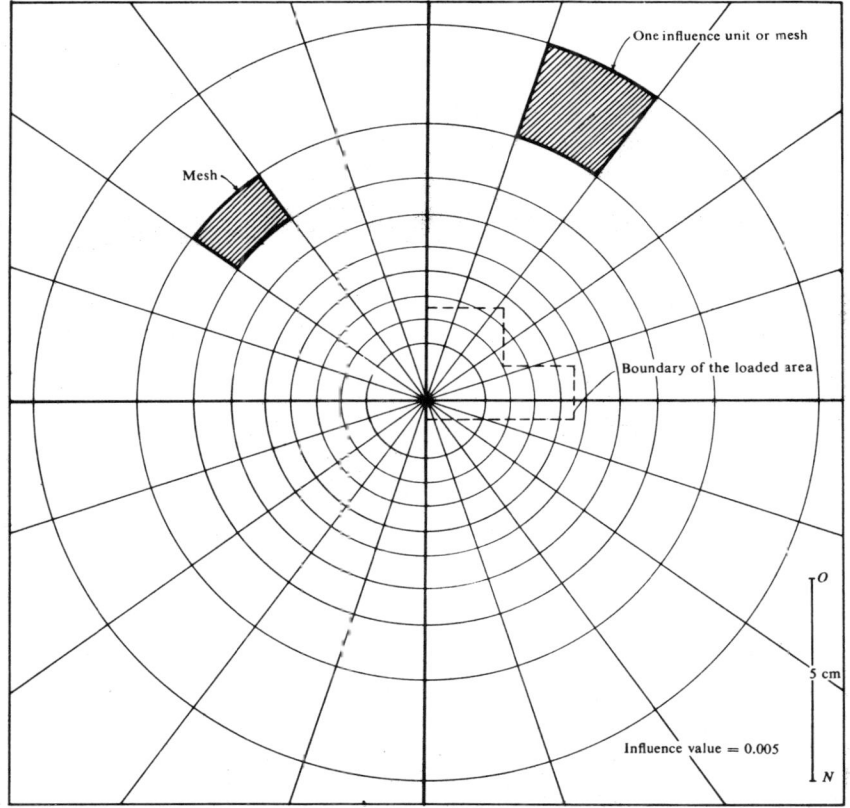

Fig. 10.18 Newmark's influence chart

10.8 APPROXIMATE METHODS

Approximate methods are used to determine the stress distribution in soil under the influence of complex loadings and/or shapes of loaded areas, saving time and labour without sacrificing accuracy to any significant degree.

Two commonly used approximate methods are given in the following subsections.

10.8.1 Equivalent Point Load Method

In this approach, the given loaded area is divided into a convenient number of smaller units and the total load from each unit is assumed to act at its centroid as a point load. The principle of superposition is then applied and the required stress at a specified point is obtained by summing up the contributions of the individual point loads from each of the units by applying the appropriate Point Load formula, such as that of Boussinesq.

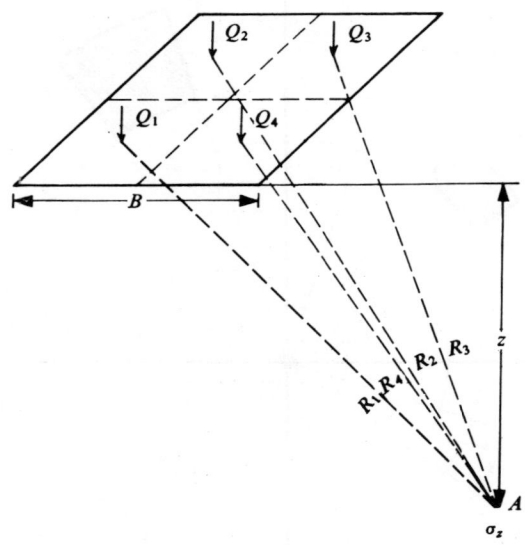

Fig. 10.19 Equivalent point load method

Referring to Fig. 10.19, if the influence values are K_{B_1}, K_{B_2}, ... for the point loads $Q_1, Q_2, ...$, for σ_z at A, we have:

$$\sigma_z = \left(Q_1 K_{B_1} + Q_2 K_{B_2} + ... \right) \qquad ...(\text{Eq. 10.51})$$

If a square area of size B is acted on by a uniform load q, the stress obtained by Newmark's influence value differs from the approximate value obtained by treating the total load of $q \cdot B^2$ to be acting at the centre. It has been established that this difference is negligible for engineering purposes if $z/B \geq 3$. This gives a hint to us that, in dividing the loaded area into smaller units, we have to remember to do it such that $z/B \geq 3$; that is to say, in relation to the specified depth, the size of any unit area should not be greater than one-third of the depth.

10.8.2 Two is to One Method

This method involves the assumption that the stresses get distributed uniformly on to areas the edges of which are obtained by taking the angle of distribution at 2 vertical to 1 horizontal ($\tan \theta = 1/2$), where θ is the angle made by the line of distribution with the vertical, as shown in Fig. 10.20.

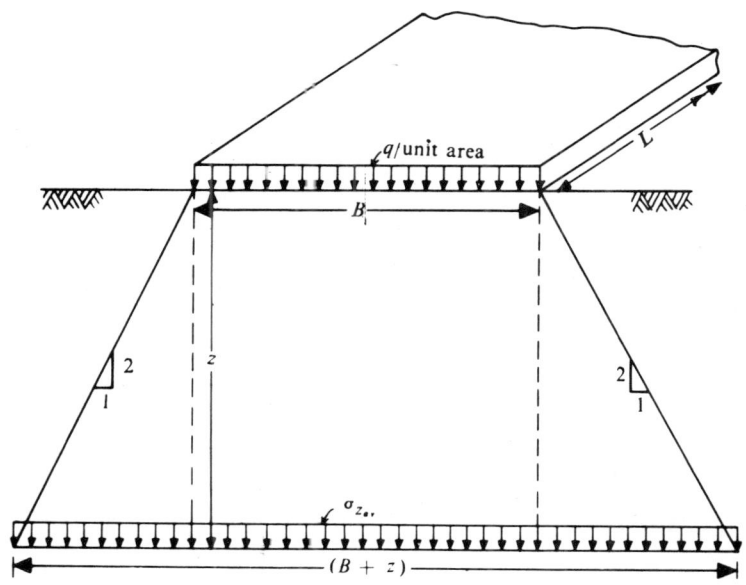

Fig. 10.20 Two is to one method

The average vertical stress at depth z is obtained as:

$$\sigma_{z_{av}} = \frac{q \cdot B \cdot L}{(B+z)(L+z)} \quad \ldots\text{(Eq. 10.52)}$$

The discrepancy between this and the accurate value of the maximum vertical stress is maximum at a value of $z/B = 0.5$, while there is no discrepancy at all at a value of $z/B \approx 2$.

10.9 ILLUSTRATIVE EXAMPLES

Example 10.1: A concentrated load of 2,250 kg acts on the surface of a homogeneous soil mass of large extent. Find the stress intensity at a depth of 15 metres and (*i*) directly under the load, and (*ii*) at a horizontal distance of 7.5 metres. Use Boussinesq's equations.

(S.V.U.—B.E., (R.R.)—Dec., 1970)

According to Boussinesq's theory,

$$\sigma_z = \frac{Q}{z^2} \cdot \frac{(3/2\pi)}{[1+(r/z)^2]^{5/2}}$$

(*i*) Directly under the load:

$$r = 0. \therefore r/z = 0$$
$$z = 15 \text{ m} \quad Q = 2{,}250 \text{ kg}$$

$$\therefore \sigma_z = \frac{2250}{15 \times 15} \cdot \frac{(3/2\pi)}{(1+0)^{5/2}}$$

$$= \mathbf{4.775 \text{ kg/m}^2}$$

(*ii*) At a horizontal distance of 7.5 metres:

$$r = 7.5 \text{ m} \quad z = 15 \text{ m}$$
$$r/z = 7.5/15 = 0.5$$

$$\therefore \sigma_z = \frac{2250}{15 \times 15} \cdot \frac{(3/2\pi)}{[(1+(0.5)^2]^{5/2}}$$

$$= \mathbf{2.733 \text{ kg/m}^2}$$

Example 10.2: A load of 1000 kN acts as a point load at the surface of a soil mass. Estimate the stress at a point 3 m below and 4 m away from the point of action of the load by Boussinesq's formula. Compare the value with the result from Westergaard's theory. (S.V.U.—B.E., (N.R.)—Sept., 1967)

Boussinesq's theory:

$$\sigma_z = \frac{Q}{z^2} \cdot \frac{(3/2\pi)}{[1+(r/z)^2]^{5/2}}$$

Here $r = 4$ m, $z = 3$ m and $Q = 1000$ kN

$$\therefore \sigma_z = \frac{1000}{3 \times 3} \cdot \frac{(3/2\pi)}{[1+(4/3)^2]^{5/2}} = 4.125 \text{ kN/m}^2$$

Westergaard's Theory:

$$\sigma_z = \frac{Q}{z^2} \cdot \frac{(1/\pi)}{[1+2(r/z)^2]^{3/2}}$$

∴ $\sigma_z = \dfrac{1000}{3 \times 3} \cdot \dfrac{(1/\pi)}{[1 + 2(4/3)^2]^{3/2}} = 3.637 \text{ kN/m}^2$

Example 10.3: A raft of size 4 m × 4 m carries a uniform load of 200 kN/m². Using the point load approximation with four equivalent point loads, calculate the stress increment at a point in the soil which is 4 m below the centre of the loaded area.

(S.V.U.—B.E., (N.R.)—March-April, 1966)

Depth below the centre O of the loaded area (raft) = 4 m. Dividing the loaded area into four equal squares of 2 m size, as shown in Fig. 10.21, the load from each small square may be taken to act through its centre.

Thus, the point loads at A, B, C and D are each:

$$200 \times 4 = 800 \text{ kN}$$

The radial distance r to O for each of the loads is $\sqrt{2}$ m,

∴ $r/z = \dfrac{\sqrt{2}}{4} = \dfrac{1}{2\sqrt{2}}$

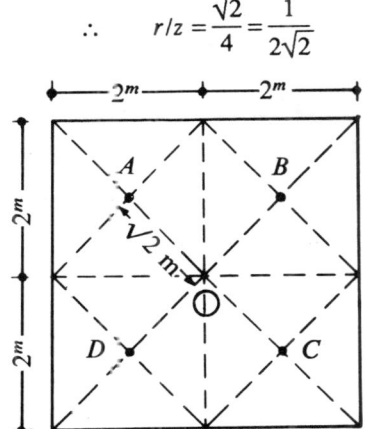

Fig 10.21 Loaded area (Ex. 10.3)

By symmetry the stress σ_z at O at 4 m depth is four times that caused by one load.

∴ $\sigma_z = \dfrac{4 \times 800}{4 \times 4} \cdot \dfrac{(3/2\pi)}{[1 + (1/2\sqrt{2})^2]^{5/2}}$

$= 71.14 \text{ kN/m}^2$

Example 10.4: A line load of 100 kN/metre run extends to a long distance. Determine the intensity of vertical stress at a point, 2 m below the surface and (*i*) directly under the line load, and (*ii*) at a distance of 2 m perpendicular to the line. Use Boussinesq's theory.

442 Geotechnical Engineering

$$q' = 100 \text{ kN/m}$$
$$z = 2 \text{ m}$$
$$\sigma_z = (q'/z) \cdot \frac{(2/\pi)}{[1+(x/z)^2]^2}$$

(i) Referring to Fig. 10.22 at point A_1,
$$x = 0$$

$$\therefore \quad \sigma_z = (q/z) \times (2/\pi) = \frac{100}{2} \times \frac{2}{\pi} \text{ kN/m}^2 = \mathbf{31.83 \text{ kN/m}^2}$$

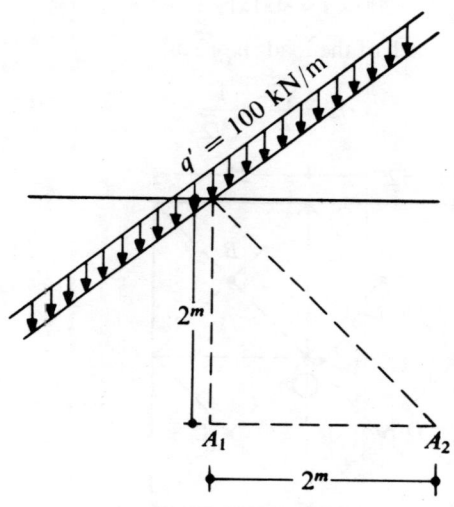

Fig. 10.22 Line load (Ex. 10.4)

(ii) $x = 2$ m at point A_2,
$$x/z = 1$$

$$\therefore \quad \sigma_z = \frac{q'}{z} \cdot \frac{(2/\pi)}{4}$$

$$= \frac{q'}{z} \cdot \frac{1}{2\pi}$$

$$= \frac{100}{2} \cdot \frac{1}{2\pi}$$

$= 25/\pi \text{ kN/m}^2$

$= 7.96 \text{ kN/m}^2$

Example 10.5: The load from a continuous footing of 1.8 metres width, which may be considered to be a strip load of considerable length, is 180 kN/m². Determine the maximum principal stress at 1.2 metres depth below the footing, if the point lies (*i*) directly below the centre of the footing, (*ii*) directly below the edge of the footing, and (*iii*) 0.6 m away from the edge of the footing. What is the maximum shear stress at each of these points? What is the absolute maximum shear stress and at what depth will it occur directly below the middle of the footing?

$B = 2b = 1.8$ m
$q = 180 \text{ kN/m}^2$
$z' = 1.2$ m
Referring to Fig. 10.23,

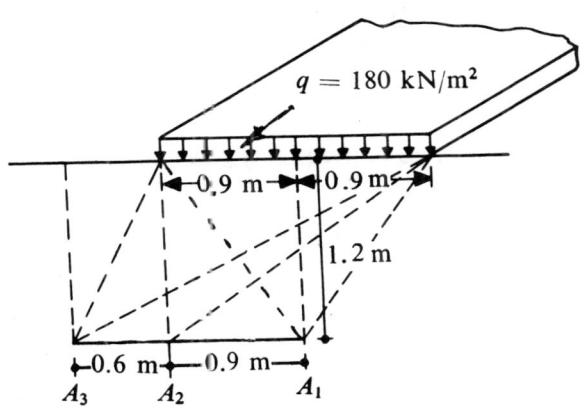

Fig. 10.23 Strip load (Ex. 10.5)

(*i*) For point A_1,

$$\frac{\theta_0}{2} = \tan^{-1}\frac{0.9}{1.2} = 35°.87 = 0.6435 \text{ rad.}$$

$\theta_0 = 1.287$ rad.

Maximum principal stress

$$\sigma_1 = (q/\pi)(\theta_0 + \sin \theta_0)$$

$$= \frac{180}{\pi}(1.287 + 0.960) = \mathbf{128.74 \text{ kN/m}^2}$$

Maximum shear stress,

$$\tau_{max} = \frac{q}{\pi} \cdot \sin \theta_0 = \frac{180}{\pi} \times 0.960 = \mathbf{55.00 \ kN/m^2}$$

(ii) for point A_2,

$$\theta_0 = \tan^{-1} \frac{1.8}{1.2} = 56°.31. = 0.9828 \ rad.$$

$$\sigma_1 = \frac{180}{\pi} (0.9828 + 0.8321) = \mathbf{104 \ kN/m^2}$$

$$\tau_{max} = \frac{180}{\pi} \cdot (0.8321) = \mathbf{47.68 \ kN/m^2}$$

(iii) for point A_3,

$$\theta_1 = \tan^{-1} \frac{0.6}{1.2} = 26°.565 = 0.464 \ rad.$$

$$\theta_2 = \tan^{-1} \frac{2.4}{1.2} = 63°.435 = 1.107 \ rad.$$

$$\theta_0 = (\theta_2 - \theta_1) = (1.107 - 0.464) \ rad. = 0.643 \ rad.$$

$$\sigma_1 = (180/\pi)(0.643 + 0.600) = \mathbf{71.22 \ kN/m^2}$$

$$\tau_{max} = \frac{180}{\pi} \times 0.6 = \mathbf{34.38 \ kN/m^2}$$

Absolute maximum shear stress = $q/\pi = 180/\pi = \mathbf{57.3 \ kN/m^2}$
This occurs at a depth $B/2$ or 0.9 m below the centre of the footing.

Example 10.6: A circular area on the surface of an elastic mass of great extent carries a uniformly distributed load of 12 t/m². The radius of the circle is 3 m. Compute the intensity of vertical pressure at a point 5 metres beneath the centre of the circle using Boussinesq's method.

(S.V.U.—B.E., (Part-Time)—Apr., 1982)

Radius 'a' of the loaded area = 3 m

$$q = 12 \ t/m^2$$

$$z = 5 \ m$$

$$z = q \left[1 - \frac{1}{\{1 + (a/z)^2\}^{3/2}} \right]$$

$$= 12 \left[1 - \frac{1}{\{1 + (3/5)^2\}^{3/2}} \right]$$

$$= 12\left[1 - \frac{1}{(34/25)^{3/2}}\right]$$

$$= 4.43 \text{ t/m}^2$$

Example 10.7: A raft of size 4 m-square carries a load of 200 kN/m². Determine the vertical stress increment at a point 4 m below the centre of the loaded area using Boussinesq's theory. Compare the result with that obtained by the equivalent point load method and with that obtained by dividing the area into four equal parts the load from each of which is assumed to act through its centre.

(*i*) Square Area:
Imagine, as in Fig. 10.24, the area to be divided into four equal squares. The stress at A will be four times the stress produced under the corner of the small square.

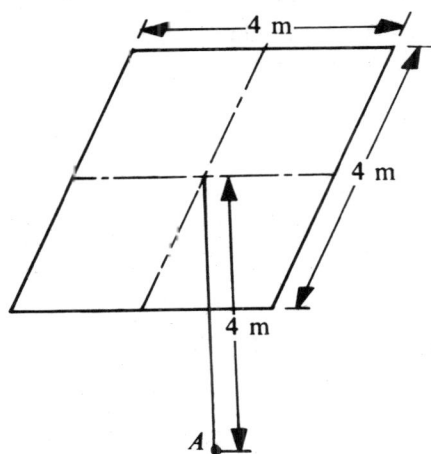

Fig. 10.24 Uniform load on square area (Ex. 10.7)

$m = 2/4 = 0.5, \quad n = 2/4 = 0.5$

$$I_\sigma = \frac{1}{4\pi}\left[\frac{2mn\sqrt{m^2+n^2+1}}{m^2+n^2+1+m^2n^2} \cdot \frac{m^2+n^2+2}{m^2+n^2+1} + \tan^{-1}\frac{2mn\sqrt{m^2+n^2+1}}{m^2+n^2+1-m^2n^2}\right]$$

$$= \frac{1}{4\pi}\left[\frac{2\times0.5\times0.5\sqrt{0.25+0.25+1}}{1.50+0.25\times0.25} \cdot \frac{0.25+0.25+2}{0.25+0.25+1} + \tan^{-1}\frac{2\times0.5\times0.5\sqrt{0.25+0.25+1}}{1.50-0.25\times0.25}\right]$$

$$= \frac{1}{4\pi}\left[\frac{0.5\sqrt{1.5}}{1.5625} \times \frac{2.50}{1.50} + \tan^{-1}\frac{0.5\sqrt{1.5}}{1.4375}\right] = 0.0840$$

(The value may be obtained from Tables or Charts also.)

$\therefore \quad \sigma_z = 4 \times 200 \times 0.084 = 67.2 \text{ kN/m}^2$

446 Geotechnical Engineering

(ii) Equivalent point load method:
$$Q = 200 \times 16 = 3200 \text{ kN}$$

$$\sigma_z = \frac{Q}{z^2} \cdot \frac{(3/2\pi)}{[1+(r/z)^2]^{5/2}} = \frac{3200}{16} \times \frac{(3/2\pi)}{1^{5/2}} = 95.5 \text{ kN/m}^2$$

(iii) Four equivalent point loads:
From Example 10.3, $\sigma_z = 71.14$ kN/m²

Thus, percentage error in the equivalent point load method
$$= \frac{(95.5 - 67.2)}{67.2} \times 100 = \mathbf{42.11}$$

Percentage error in four equivalent point loads approach
$$= \frac{(71.14 - 67.20)}{67.20} \times 100 = \mathbf{5.86}$$

Example 10.8: A rectangular foundation, 2 m × 4 m, transmits a uniform pressure of 450 kN/m² to the underlying soil. Determine the vertical stress at a depth of 1 metre below the foundation at a point within the loaded area, 1 metre away from a short edge and 0.5 metre away from a long edge. Use Boussinesq's theory.

Depth $z = 1$ m. $q = 450$ kN/m²

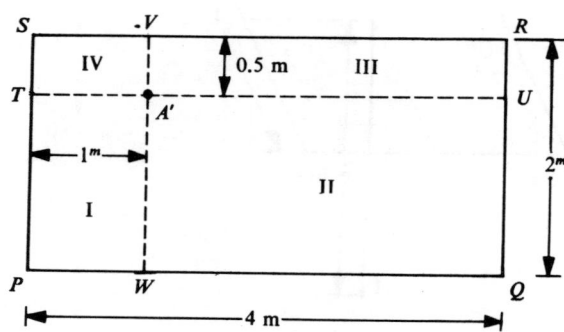

Fig. 10.25 Stress at a point inside a loaded area (Ex. 10.8)

The loaded area and the plan position of the point A' at which the vertical stress is required are shown in Fig. 10.25. The area is divided into four parts as shown, such that A' forms a corner of each.

$$\sigma_z = q[I_{\sigma_\text{I}} + I_{\sigma_\text{II}} + I_{\sigma_\text{III}} + I_{\sigma_\text{IV}}]$$

Area I: $m = 1/1 = 1$; $n = 1.5/1 = 1.5$

$$I_{\sigma_\text{I}} = \frac{1}{4\pi}\left[\frac{2mn\sqrt{m^2+n^2+1}}{m^2+n^2+1+m^2n^2} \cdot \frac{m^2+n^2+2}{m^2+n^2+1} + \tan^{-1}\frac{2mn\sqrt{m^2+n^2+1}}{m^2+n^2+1-m^2n^2}\right]$$

Stress Distribution in Soil 447

$$= \frac{1}{4\pi}\left[\frac{2\times 1\times 1.5\sqrt{1^2+1.5^2-1}}{1^2+1.5^2+1+1^2\times 1.5^2}\cdot\frac{1^2+1.5^2+2}{1^2+1.5^2+1}+\tan^{-1}\frac{2\times 1\times 1.5\sqrt{1^2+1.5^2+1}}{1^2+1.5^2+1-1^2\times 1.5^2}\right]$$

$$= 0.1936$$

Area II: $m = 1.5/1 = 1.5;\ n = 3/1 = 3$

$$I_{\sigma_{II}} = \frac{1}{4\pi}\left[\frac{2\times 1.5\times 3\sqrt{1.5^2+3^2+1}}{1.5^2+3^2+1+1.5^2\times 3^2}\cdot\frac{1.5^2+3^2+2}{1.5^2+3^2+1}+\tan^{-1}\frac{2\times 1.5\times 3\sqrt{1.5^2+3^2+1}}{1.5^2+3^2+1-1.5^2\times 3^2}\right]$$

$$= 0.2290$$

Area III: $m = 0.5/1 = 0.5;\ N = 3/1 = 3$

$$I_{\sigma_{III}} = \frac{1}{4\pi}\left[\frac{2\times 0.5\times 3\sqrt{0.5^2+3^2+1^2}}{0.5^2+3^2+1+0.5^2\times 3^2}\cdot\frac{0.5^2+3^2+2}{0.5^2+3^2+1}+\tan^{-1}\frac{2\times 0.5\times 3\sqrt{0.5^2+3^2+1}}{0.5^2+3^2+1-0.5^2\times 3^2}\right]$$

$$= 0.1368$$

Area IV: $m = 0.5/1 = 0.5;\ n = 1/1 = 1$

$$I_{\sigma_{IV}} = \frac{1}{4\pi}\left[\frac{2\times 0.5\times 1\sqrt{0.5^2+1^2+1}}{0.5^2+1^2+1+0.5^2\times 1^2}\cdot\frac{0.5^2+1^2+2}{0.5^2+1^2+1}+\tan^{-1}\frac{2\times 0.5\times 1\sqrt{0.5^2+1^2+1}}{0.5^2+1^2+1-0.5^2\times 1^2}\right]$$

$$= 0.1202$$

$\therefore\quad \sigma_z = 450(0.1936+0.2290+0.1368+0.1202)$

$$= 305.8\ \text{kN/m}^2$$

Example 10.9: A rectangular foundation $2\ \text{m}\times 3\ \text{m}$, transmits a pressure of 360 kN/m² to the underlying soil. Determine the vertical stress at a point 1 metre vertically below a point lying outside the loaded area, 1 metre away from a short edge and 0.5 metre away from a long edge. Use Boussinesq's theory.

$z = 1\ \text{m};\ q = 360\ \text{kN/m}^2$

since the point at which the stress is required is outside the loaded area, rectangles are imagined as snown in Fig. 10.26, so as to make A' a corner of all the concerned rectangles. With the notation of Fig. 10.26,

$$\sigma_z = q\left(I_{\sigma_I}-I_{\sigma_{II}}-I_{\sigma_{III}}-I_{\sigma_{IV}}\right)$$

Area I: $m = 2.5/1 = 2.5;\ n = 4/1 = 4$

$$I_{\sigma_I} = \frac{1}{4\pi}\left[\frac{2mn\sqrt{m^2+n^2+1}}{m^2+n^2+1+m^2n^2}\cdot\frac{m^2+n^2+2}{m^2+n^2+1}+\tan^{-1}\frac{2mn\sqrt{m^2+n^2+1}}{m^2+n^2+1-m^2n^2}\right]$$

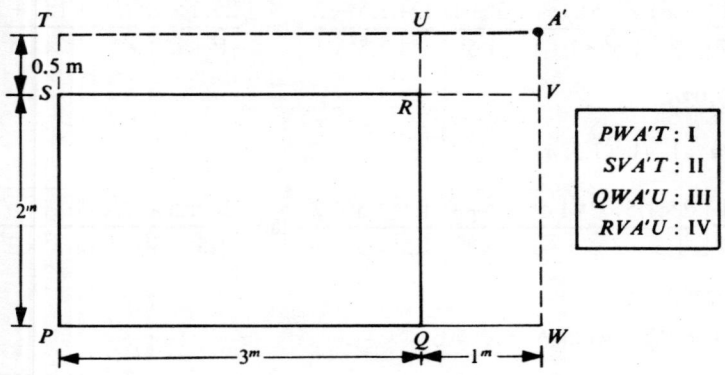

Fig. 10.26 Stress at a point outside loaded area (Ex. 10.9)

$$= \frac{1}{4\pi}\left[\frac{2\times2.5\times4\sqrt{2.5^2+4^2+1}}{2.5^2+4^2+1+2.5^2\times4^2}\cdot\frac{2.5^2+4^2+2}{2.5^2+4^2+1}+\tan^{-1}\frac{2\times2.5\times4\sqrt{2.5^2+4^2+1}}{2.5^2+4^2+1-2.5^2\times4^2}\right]$$

$= 0.2434$

Area II: $m = 0.5/1 = 0.5; n = 4/1 = 4$

$$I_{\sigma_{II}} = \frac{1}{4\pi}\left[\frac{2\times0.5\times4\sqrt{0.5^2+4^2+1}}{0.5^2+4^2+1+0.5^2\times4^2}\cdot\frac{0.5^2+4^2+2}{0.5^2+4^2+1}+\tan^{-1}\frac{2\times0.5\times4\sqrt{0.5^2+4^2+1}}{0.5^2+4^2+1-0.5^2\times4^2}\right]$$

$= 0.1372$

Area III: $m = 1/1 = 1; n = 2.5/1 = 2.5$

$$I_{\sigma_{III}} = \frac{1}{4\pi}\left[\frac{2\times1\times2.5\sqrt{1^2+2.5^2+1}}{1^2+2.5^2+1+1^2\times2.5^2}\cdot\frac{1^2+2.5^2+2}{1^2+2.5^2+1}+\tan^{-1}\frac{2\times1\times2.5\sqrt{1^2+2.5^2+1}}{1^2+2.5^2+1-1^2\times2.5^2}\right]$$

$= 0.2024$

Area IV : $m = 0.5/1 = 0.5; n = 1/1 = 1$

$$I_{\sigma_{IV}} = \frac{1}{4\pi}\left[\frac{2\times0.5\times1\sqrt{0.5^2+1^2+1}}{0.5^2+1^2+1+0.5^2\times1^2}\cdot\frac{0.5^2+1^2+2}{0.5^2+1^2+1}+\tan^{-1}\frac{2\times0.5\times1\sqrt{0.5^2+1^2+1}}{0.5^2+1^2+1-0.5^2\times1^2}\right]$$

$= 0.1202$

$\therefore \quad \sigma_z = 360(0.2434 - 0.1372 - 0.2024 + 0.1202)$

$\qquad\qquad\qquad\qquad = \mathbf{8.64\ kN/m^2}$

SUMMARY OF MAIN POINTS

1. When the surface of a soil mass is level and its unit weight constant with depth, the vertical geostatic stress increases linearly with depth, the constant of proportionality being the unit weight itself.
2. The Boussinesq solution for point load is the most popular and is applicable to a homogeneous, isotropic and elastic semi-infinite medium, which obeys Hooke's law within the range of stresses considered.
3. The Westergaard solution is applicable to sedimentary soil deposits with negligible lateral strain.
4. The stress isobar or pressure bulb concept is very useful in geotechnical engineering practice, especially in the determination of the soil mass contributing to the settlement of a structure.
5. More complicated loadings such as line load, strip load, circular loaded area, and rectangular loaded area may be dealt with by integration of the stresses due to point load.
6. The rectangular loaded area is the most common type in foundation engineering; the stress due to it may be evaluated by Fadum's or Steinbrenner's charts for influence values, or by Newmark's formula for a point beneath a corner.
7. The stress at a point inside or outside the loaded area may be conveniently determined by forming rectangles, true or hypothetical, for which the point forms a corner and by applying the principle of superposition appropriately.
8. Newmark's influence chart may be conveniently used in the case of irregular areas (it is, of course, applicable to regular areas also).
9. Approximate methods such as the equivalent point load approach yield reasonably satisfactory results under certain conditions.

REFERENCES

1. J.V. Boussinesq: *Application des potentials à l'etude de l'equilibre et du mouvement des solids elastiques*, Paris, Gauthier—Villars, 1885.
2. P.L. Capper, W.F. Cassie and J.D. Geddes: *Problems in Engineering Soils*, S.I. Edition, E & F.N. Spon Ltd., London, 1971.
3. R.E. Fadum: *Influence values for Vertical stresses in a semi-infinite solid due to surface Loads*, Graduate School of Engineering, Harvard University (Unpublished), 1941.
4. G. Gilboy: *Soil Mechanics Research*, Transactions, ASCE, 1933.
5. L. Jürgenson: *The Application of Theories of Elasticity to Foundation Problems*, Journal of Boston Society of Civil Engineers, July, 1934.
6. A.R. Jumikis: *Soil Mechanics*, D. Van Nostrand Company Inc., Princeton, NJ, USA, 1962.
7. T.W. Lambe: *Methods of Estimating Settlement*, Proc. of the ASCE Settlement Conference, North Western University, Evanston, Illinois, June, 1964.

8. T.W. Lambe: *Shallow Foundations on Clay*, Proc. of a Symposium on Bearing Capacity and Settlement of Foundations, Duke University, Durham, NC, USA, 1967.
9. D.F. McCarthy: *Essentials of Soil Mechanics and Foundations*, Reston Publishing Company, Reston, Va, USA, 1977.
10. N.M. Newmark: *Simplified computation of Vertical pressures in Elastic Foundations*, Engineering Experiment Station Circular No. 24, University of Illinois, 1935.
11. N.M. Newmark: *Influence Charts for Computation of Stresses in Elastic Foundations*, Engineering Experiment Station Bulletin Series No. 338, University of Illinois, Nov., 1942.
12. G.N. Smith: *Essentials of Soil Mechanics for Civil and Mining Engineers*, Third Edition, Metric, Crosby Lockwood Staples, London, 1974.
13. M.G. Spangler: *Soil Engineering*, International Textbook Company, Scranton, USA, 1951.
14. W. Steinbrenner: *Tafeln zur Setzungsberechnung*, Die Strasse, Vol. 1, 1934.
15. D.W. Taylor: *Fundamentals of Soil Mechanics*, John Wiley and Sons, Inc., New York, USA, 1948.
16. K. Terzaghi: *Theoretical Soil Mechanics*, John Wiley and Sons, Inc., New York, USA, 1943.
17. S. Timoshenko and J.N. Goodier: *Theory of Elasticity*, McGraw-Hill Book Company, Inc., 1951.
18. H.M. Westergaard: *A problem of Elasticity Suggested by a Problem in Soil Mechanics: Soft Material Reinforced by Numerous Strong Horizontal Sheets*, Contributions to the Mechanics of Solids, Stephen Timoshenko 60th Anniversary Volume, New York, Macmillan, 1938.

QUESTIONS AND PROBLEMS

10.1 State the basic requirements to be satisfied for the validity of Boussinesq equation for stress distribution.

(S.V.U.—B.E., (N.R.)—Sep., 1967)

10.2 (a) State Boussinesq's equation for vertical stress at a point due to a load on the surface of an elastic medium.

(b) Using Boussinesq's expression, derive the expression for vertical stress at depth h under the centre of a circular area of radius a loaded uniformly with a load q at the surface of the mass of soil.

(S.V.U.—B.E., (R.R.)—Dec., 1968)

10.3 (a) Explain the concept of 'Pressure Bulb' in soils.

(b) Derive the principle of construction of Newmark's chart and explain its use.

(S.V.U.—B.E., (R.R.)—Nov., 1969)

10.4 (a) Explain stress distribution in soils for concentrated loads by Boussinesq's equation.

Stress Distribution in Soil 451

(b) What do you understand by 'Pressure bulb'? Illustrate with sketches.
(S.V.U.—B.E., (R.R.)—May, 1971)

10.5 Write a brief critical note on 'Newmark's influence chart'.
(S.V.U.—B.E., (R.R.)—Nov., 1973, May, 1975 & Feb., 1976)
Four-year B.Tech.—Dec., 1982
Four-year B.Tech.—Apr., 1983
B.Tech.—(Part-time)—Sept., 1982

10.6 Write a brief critical note on 'the concept of pressure bulb and its use in soil engineering practice'.
(S.V.U.—B.E., (R.R.)—Nov., 1974, Nov., 1975)

10.7 What are the basic assumptions in Boussinesq's theory of stress distribution in soils? Show the vertical stress distribution on a horizontal plane at a given depth and also the vertical stress distribution with depth. What is a 'Pressure Bulb'?
(S.V.U.—B.E., (R.R.)—Feb., 1976)

10.8 Explain in detail the construction of Newmark's chart with an influence value of 0.002.
Explain Boussinesq's equation for vertical stress within an earth mass.
(S.V.U.—Four-year—B.Tech., Oct., 1982)

10.9 Derive as per Boussinesq's theory, expressions for vertical stress at any point in a soil mass due to
(i) line load on the surface, and
(ii) strip load on the surface
State the assumptions.
(S.V.U.—B.Tech. (Part-Time)—Sept., (1983)

10.10 Find the vertical pressure at a point 4 metres directly below a 2 tonnes point load acting at a horizontal ground surface. Use Boussinesq's equations.
(S.V.U.—B.E., (R.R.)—Dec., 1971)

10.11 A 2.5 tonne point load acts on the surface of a horizontal ground. Find the intensity of vertical pressure at 6 m directly below the load. Use Boussinesq's equation.
(S.V.U.—B.E., (R.R.)—June, 1972)

10.12 A reinforced concrete water tank of size 6 m × 6 m and resting on ground surface carries a uniformly distributed load of 20 tonnes/m^2. Estimate the maximum vertical pressure at a depth of 12 metres vertically below the centre of the base.
(S.V.U.—B.E. (Part-time)—Dec., 1971)

10.13 A line load of 90 kN/metre run extends to a long distance. Determine the intensity of vertical stress at a point 1.5 metres below the surface (i) directly under the line load and (ii) at a distance 1 m perpendicular to the line. Use Boussinesq's theory.

10.14 A strip load of considerable length and 1.5 m width transmits a pressure of 150 kN/m² to the underlying soil. Determine the maximum principal stress at 0.75 m depth below the footing, if the point lies (*i*) directly below the centre of the footing, and (*ii*) directly below the edge of the footing. What is the absolute maximum shear stress and where does it occur ?

10.15 A circular footing of 1.5 m radius transmits a uniform pressure of 90 kN/m². Calculate the vertical stress at a point 1.5 m directly beneath its centre.

10.16 A rectangular area 4 m × 6 m carries a uniformly distributed load of 10 t/m² at the ground surface. Estimate the vertical pressure at a depth of 6 m vertically below the centre and also below a corner of the loaded area. Compare the results with those obtained by an equivalent point load method and also by dividing the loaded area into four equal parts and treating the load from each as a point load.

(S.V.U.—B.Tech. (Part-Time)—Sept., 1983)

10.17 A 4.5 m square foundation exerts a uniform pressure of 180 kN/m² on a soil. Determine the vertical stress increment at a point 3 m below the foundation and 3.75 m from its centre along one of the axes of symmetry.

10.18 The plan of a foundation is shown in Fig. 10.27(a). The uniform pressure on the soil is 45 kN/m². Determine the vertical stress increment due to the foundation at a depth of 4 m below the point *A*.

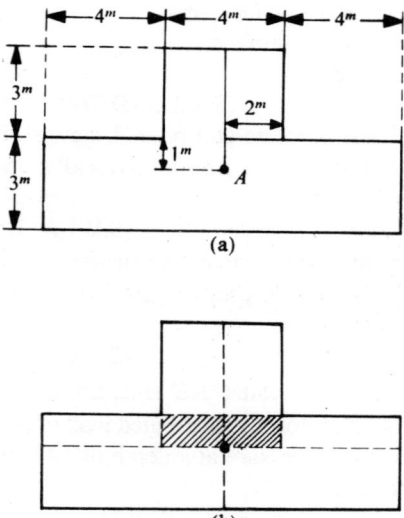

Fig. 10.27 Plan of loaded area (Prob. 10.18)

(*Hint*: In order to obtain a set of rectangles whose corners meet at a point, a part of the area is sometimes included twice and later a correction is applied. For this problem, the area must be divided into six rectangles, as shown in Fig. 10.27(b). The effect of the shaded portion is included twice and must therefore be subtracted once).

11
SETTLEMENT ANALYSIS

11.1 INTRODUCTION

Foundations of all structures have to be placed on soil. The structure may undergo settlement depending upon the characteristics such as compressibility of the strata of soil on which it is founded. Thus the term 'settlement' indicates the sinking of a structure due to the compression and deformation of the underlying soil. Clay strata often need a very long time—a number of years—to get fully consolidated under the loads from the structure. The settlement of any loose strata of cohesionless soil occurs relatively fast. Thus, there are two aspects—the total settlement and the time-rate of settlement—which need consideration.

If it can be assumed that the expulsion of water necessary for the consolidation of the compressible clay strata takes place only in the vertical direction, Terzaghi's theory of one-dimensional consolidation may be used for the determination of total settlement and also the time-rate of settlement.

Depending upon the location of the compressible strata in the soil profile relative to the ground surface, only a part of the stress transmitted to the soil at foundation level may be transmitted to these strata as stress increments causing consolidation. The theories of stress-distribution in soil have to be applied appropriately for this purpose. The vertical stress due to applied loading gets dissipated fast with respect to depth and becomes negligible below a certain depth. If the compressible strata lie below such depth, their compression or consolidation does not contribute to the settlement of the structure in any significant manner.

There is the other aspect of whether a structure is likely to undergo 'uniform settlement' or 'differential settlement'. Uniform settlement or equal settlement under different points of the structure does not cause much harm to the structural stability of the structure. However, differential settlement or different magnitudes of settlement at different points underneath a structure—especially a rigid structure—is likely to cause supplementary stress and thereby cause harmful effects such as cracking, permanent and irreparable damage, and ultimate yield and failure of the structure. As such, differential settlement must be guarded against.

11.2 DATA FOR SETTLEMENT ANALYSIS

A procedure for the computation of anticipated settlements is called *'Settlement analysis'*. This analysis may be divided into three parts. The first part consists of obtaining the soil profile, which gives an idea of the depths of various characteristic zones of soil at the site of the structure, as also the relevant properties of soil such as initial void ratio, grain specific gravity, water content, and the consolidation and compressibility characteristics. The second part consists of the analysis of the transmission of stresses to the subsurface strata, using a theory such as Boussinesq's for stress distribution in soil. The final part consists of the final settlement predictions based on concepts from the theory of consolidation and data from the first and second parts.

11.2.1 Soil Profile and Soil Properties

The thicknesses of the various strata of soil in the area in which the structure is to be built have to be ascertained carefully and presented in the form of a soil profile. Sufficient number of borings should be made for this purpose and boring logs prepared from the data. The location of the water table and water-bearing strata are also to be determined. In case the boring data show some irregularities in the various soil strata, an average idealised profile is chosen such that it is free of horizontal variation. Adequate boring data and good judgement in the interpretation are the prime requisites for good analysis. Usually five bore holes—one at each corner and the fifth at the centre of the site are adequate. The depth should not be less than five times the shorter dimension of the foundation. A relatively simple but typical profile is shown in Fig. 11.1.

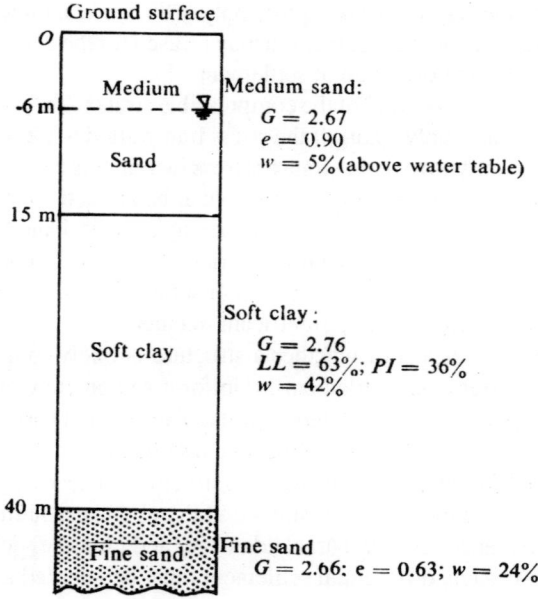

Fig. 11.1 Typical soil profile

The special feature of this profile is that a clay stratum with no horizontal variation is sandwiched by sand strata.

The soil properties of the compressible strata, especially clay strata, are also to be ascertained by taking samples at different depths. If the thickness of a compressible stratum is considerable, the properties of the soil taken from different depths are obtained and if there is significant variation, their average values are considered for the analysis.

The existing void ratio is computed from the grain specific gravity and water content, on the assumption of complete saturation, which is invariably valid for relatively deep strata below the water table. The consolidation characteristics are determined for representative soil samples; the compression index and the coefficient of consolidation are evaluated and these are utilised in the settlement computation. As far as possible undisturbed samples must be used for this purpose.

11.2.2 Stresses in Subsoil Before and After Loading

The initial and final values of the intergranular pressure, i.e., the effective stresses in the subsoil before and after loading from the structure, have to be evaluated, since these are necessary for the computation of the settlement.

When there is no horizontal variation in the strata, the total vertical stress at any depth below ground surface is dependent only on the unit weight of the overlying material. The total stress, the neutral pressure and the effective stress at the mid-depth of the compressible stratum may be computed and used. However, if the thickness of the stratum is large, these values must be got at least at the top, middle and the bottom of the stratum and averaged.

The values of neutral pressures and hence the intergranular pressures depend upon the conditions prior to the loading from the structure. The four possible basic conditions are:

1. The simple static condition
2. The residual hydrostatic excess condition
3. The artesian condition
4. The precompressed condition

Sometimes combinations of these cases may also occur.

The simple static condition

This is the simplest case and the one commonly expected, the neutral pressure at any depth being equal to the unit weight of water multiplied by the depth below the free water surface. It may be noted that the neutral pressure need not be determined in this case since the intergranular pressure may be obtained by using the submerged unit weight for all the zones which are below the water table.

The residual hydrostatic excess condition

A condition of partial consolidation under the preloading overburden exists if part of the overburden has been recently placed. This kind of situation exists in made-up soil or deltaic deposits. The hydrostatic excess pressure would have been only partially dissipated in the compressible clay stratum. The remaining excess pressure is referred to as the 'residual hydrostatic excess pressure'.

Any structure built above such a stratum must eventually undergo not only the settlement caused by its own weight but also the settlement inherent in the residual hydrostatic excess pressure.

The artesian condition

The water pressure at the top of the clay layer normally depends only on the elevation of the ground water surface, but the neutral pressure at the bottom of the clay stratum may depend upon very different conditions. A pervious stratum below the clay stratum may extend to a distant high ground. In such a case the water above it will cause a pressure at the bottom of the clay stratum which is referred to as the 'artesian condition'. If the artesian pressure below the clay remains constant, it will cause only slight increase in the intergranular pressures and has no tendency to cause consolidation. However, decrease in artesian pressure, say, due to driving of artesian wells into the pervious stratum below the clay, may cause large amounts of consolidation.

It can be demonstrated that the basic consolidation relationships are not changed by the presence of artesian pressure.

The precompressed condition

Clay strata might have been subjected in past ages to loads greater than those existing at present. This precompression may have been caused in a number of ways—by the load of glaciers of past ages, by overburden which has since been removed by erosion or by the loads of buildings that have been demolished. The existence and the amount of precompression, characterised by the preconsolidation pressure, are of greater interest than the cause.

The construction of a structure on a precompressed stratum causes recompression rather than compression. It has been seen in chapter seven that recompression causes relatively smaller settlements compared to those caused in virgin compression. This aspect, therefore, has also to be understood in the settlement analysis.

The stresses in the subsoil after loading from the structure can be obtained by computing the stress increments at the desired depths under the influence of the loading from the structure; the nature as well as the magnitude of the loading are important in this regard. The concepts of stress distribution in soil and the methods of determining it, as outlined in chapter ten, would have to be applied for this purpose. The limitations of the theories and the underlying assumptions are to be carefully understood.

11.3 SETTLEMENT

The total settlement may be considered to consist of the following contributions:
- (a) Initial settlement or elastic compression.
- (b) Consolidation settlement or primary compression.
- (c) Secondary settlement or secondary compression.

11.3.1 Initial Settlement or Elastic Compression

This is also referred to as the 'distortion settlement' or 'contact settlement' and is usually taken to occur immediately on application of the foundation load. Such immediate settlement in the case of partially saturated soils is primarily due to the expulsion of gases and to the elastic compression and rearrangement of particles. In the case of saturated soils immediate settlement is considered to be the result of vertical soil compression, before any change in volume occurs.

Immediate settlement in cohesionless soils

The elastic as well as the primary compression effects occur more or less together in the case of cohesionless soils because of their high permeabilities. The resulting settlement is termed 'immediate settlement'.

The methods available for predicting this settlement are far from perfect; either the standard penetration test or the use of charts is resorted to.

Standard penetration test

This test which is popularly used for cohesionless soils is described in detail in chapter 18. The result, which is in the form of the number of blows required for causing a standard penetration under specified standard conditions, can be used to evaluate immediate settlement in a cohesionless soil (De Beer and Martens, 1957). This method has been developed for use with the Dutch Cone Penetrometer but can be adapted for the standard penetration test.

The immediate settlement, S_i, is given by:

$$S_i = \frac{H}{C_s} \cdot \log_e \left(\frac{\overline{\sigma}_0 + \Delta\overline{\sigma}}{\overline{\sigma}_0} \right) \quad \ldots \text{(Eq. 11.1)}$$

where H = thickness of the layer getting compressed,
 $\overline{\sigma}_0$ = effective overburden pressure at the centre of the layer before any excavation or application of load,
 $\Delta\overline{\sigma}$ = vertical stress increment at the centre of the layer,
and C_s = compressibility constant, given by:

$$C_s = 1.5 \frac{C_r}{\overline{\sigma}_0} \quad \ldots \text{(Eq. 11.2)}$$

 C_r being the static cone resistance (in kN/m²), and
 $\overline{\sigma}_0$ being the effective overburden pressure at the point tested.

The value of C_r obtained from the Dutch Cone penetration test must be correlated to the recorded number of blows, N, obtained from the standard penetration test. Its variation appears to be wide. According to Meigh and Nixon (1961), C_r ranged from 430 N (kN/m²) to 1930 N (kN/m²). However, C_r is more commonly taken as 400 N (kN/m²) as proposed by Meyerhof (1956).

The use of charts: The actual number of blows, N, from the standard penetration test has to be corrected, under certain circumstances to obtain N', the corrected value. Thornburn (1963) has given a set of curves to obtain N' from N. He also extended the graphical relationship given by Terzaghi and Peck (1948) between the settlement of a 305 mm square plate under a given pressure and the N'-value of the soil immediately beneath it, as shown in Fig. 11.2.

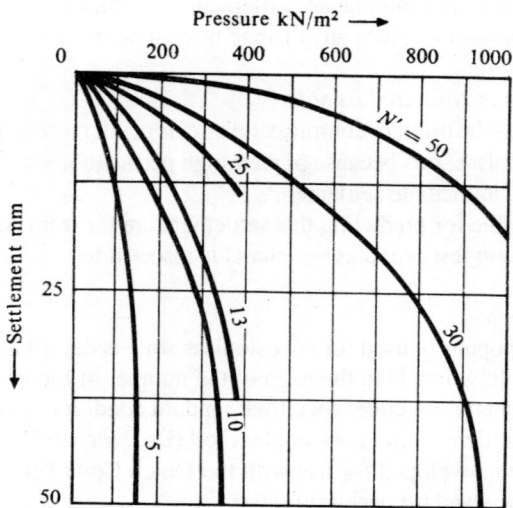

Fig. 11.2 Relationship between pressure and settlement of a 305 mm square plate, for different values of N', in cohesionless soils (After Thornburn, 1963)

This can be used for determining the settlement, S_f, of a square foundation on a deep layer of cohesionless soil by using Terzaghi and Peck's formula:

$$S_f = S_p \left(\frac{2B}{B+0.3} \right)^2 \qquad \ldots \text{(Eq. 11.3)}$$

where S_p = Settlement of a 305 mm-square plate, obtained from the chart (Fig. 11.2) and,

B = Width of foundation (metres)

The chart is applicable for deep layers only, that is, for layers of thickness not less than $4B$ below the foundation.

For rectangular foundations, a shape factor should presumably be used. It is as follows:

Table 11.1 Shape factors for rectangular foundations in cohesionless soils (After Terzaghi and Skempton)

$\frac{L}{B}$	Shape factor (flexible)	Shape factor (rigid)
1	1.00	1.00
2	1.35	1.22
3	1.57	1.31
4	1.71	1.41
5		

Note: Settlement of a rectangular foundation of width B = Settlement of square foundation of size $B \times$ shape factor.

Immediate settlement in cohesive soils ✓

If a saturated clay is loaded rapidly, excess hydrostatic pore pressures are induced; the soil gets deformed with virtually no volume change and due to low permeability of the clay little water is squeezed out of the voids. The vertical deformation due to the change in shape is the immediate settlement.

The immediate settlement of a flexible foundation, according to Terzaghi (1943), is given by:

$$S_i = q \cdot B \left(\frac{1-v^2}{E_s} \right) \cdot I_t \qquad \ldots (Eq.\ 11.4)$$

Where S_i = immediate settlement at a corner of a rectangular flexible foundation of size $L \times B$,
B = Width of the foundation,
q = Uniform pressure on the foundation,
E_s = Modulus of elasticity of the soil beneath the foundation,
v = Poisson's ratio of the soil, and
I_t = Influence Value, which is dependent on L/B

(This is analogous to I_o of chapter 10).

For a perfectly flexible square footing, the immediate settlement under its centre is twice that at its corners.

The values of I_t are tabulated below:

Table 11.2 Influence values for settlement of a corner of a flexible rectangular foundation of size $L \times B$
(After Terzaghi, 1943)

L/B	1	2	3	4	5
Influence value I_t	0.56	0.76	0.88	0.96	1.00

As in the case of computation of the vertical stress beneath any point either inside or outside a loaded area (chapter 10), the principle of superposition may be used for computing settlement by using equation 11.4; the appropriate summation of the product of B and I_t for the areas into which the total area is divided will be multiplied by $q \left(\frac{1-v^2}{E_s} \right)$.

An earth embankment may be taken as flexible and the above formula may be used to determine the immediate settlement of soil below such a construction.

Foundations are commonly more rigid than flexible and tend to cause a uniform settlement which is nearly the same as the mean value of settlement under a flexible foundation. The mean value of the settlement, S_i, for a rectangular foundation on the surface of a semi-elastic medium is given by:

$$S_i = q \cdot B \frac{(1-v^2)}{E_s} \cdot I_s \qquad \ldots\text{(Eq. 11.5)}$$

where B = width of the rectangular foundation of size $L \times B$,
q = uniform intensity of pressure,
E_s = modulus of elasticity of the soil beneath the foundation,
v = Poisson's ratio of the soil, and
I_s = influence factor which depends upon L/B.

Skempton (1951) gives the following values of I_s:

Table 11.3 Influence factors for mean value of settlement of a rectangular foundation on a semi-elastic medium (After Skempton, 1951)

L/B	Circle	1	2	5	10
Influence factor I_s	0.73	0.82	1.00	1.22	1.26

The factor $\left(\dfrac{1-v^2}{E_s}\right)(I_s)$ may be determined by conducting three or more plate load tests (chapter 14) and fitting a straight line plot for S_i versus $q \cdot B$; the slope of the plot equals this factor.

Immediate settlement of a thin clay layer

The coefficients of Tables 11.2 and 11.3 apply only to foundations on deep soil layers. A drawback of the method is that it can be applied only to a layer immediately below a foundation and extending to a great depth.

For cases when the thickness of the layer is less than $4B$, Steinbrenner (1934) prepared coefficients. His procedure was to determine the immediate settlement at the top of the layer (assuming infinite depth) and to calculate the settlement at the bottom of the layer (again assuming infinite depth below it). The difference between these two values is the actual settlement of the layer.

The immediate settlement at the corners of a rectangular foundation on an infinite layer is given by:

$$S_i = q \cdot B \left(\frac{1-v^2}{E_s}\right) \cdot I_s \qquad \ldots\text{(Eq. 11.6)}$$

The values of the influence coefficients I_s (assuming $v = 0.5$) are given in Fig. 11.3.

The principle of superposition may be used for determining the settlement underneath any point of the loaded area by dividing the area into rectangles such

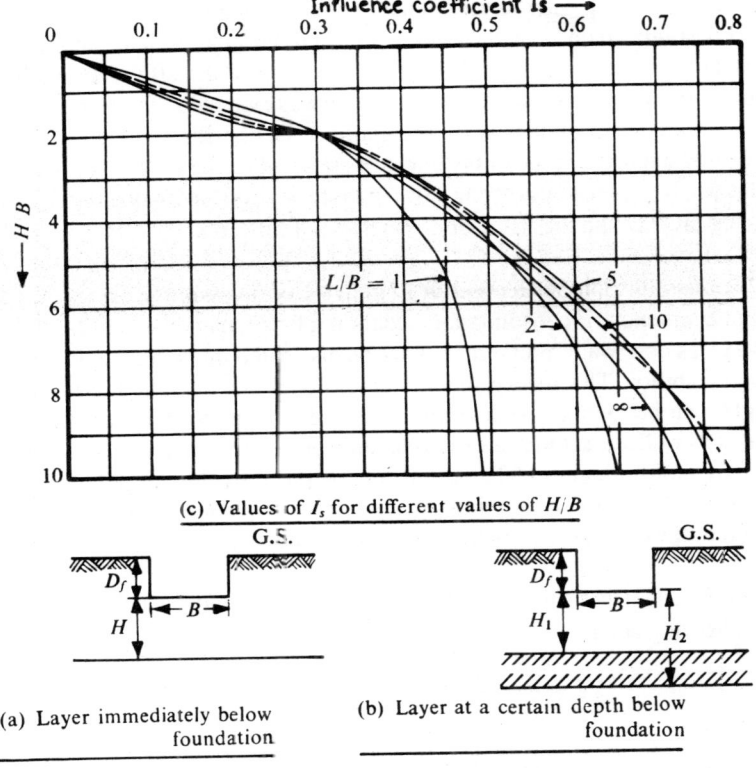

Fig. 11.3 Immediate settlement of a thin clay layer
(After Steinbrenner, 1934)

that the point forms the corner of each. The method can be extended to determine the immediate settlement of a clay layer which is located at some depth below the foundation as in Fig. 11.3(b); the settlement of a layer extending from below the foundation of thickness H_2 (using E_{s_2}), is determined first; from this value of imaginary settlement of the layer H_1 (again using E_{s_2}) is subtracted. Since this settlement is for a perfectly flexible foundation usually the value at the centre is determined and is reduced by a rigidity factor (0.8 usually) to obtain a mean value for the settlement.

Effect of depth: According to Fox (1948), the calculated settlements are more than the actual ones for deep foundations ($D_f > B$), and a reduction factor may be applied. If $D_f = B$, the reduction factor is about 0.75; it is taken as 0.50 for very deep foundations. However, most foundations are shallow. Further, in the case of foundations located at large depth, the computed settlements are, in general, small and the reduction factor is customarily not applied.

Determination of E_s: Determination of E_s, the modulus of elasticity of soil, is not simple because of the wide variety of factors influencing it. It is usually obtained from a consolidated undrained triaxial test on a representative soil sample, which is consolidated under a cell pressure approximating to the effective overburden pressure at the level from which the soil sample was extracted. The plot of deviator stress versus axial strain is never a straight line. Hence, the value must be determined at the expected value of the deviator stress when the load is applied on the foundation. If the thickness of the layer is large, it may be divided into a number of thinner layers, and the value of E_s determined for each.

11.3.2 Consolidation Settlement or Primary Compression

The phenomenon of consolidation occurs in clays (chapter seven) because the initial excess pore water pressures cannot be dissipated immediately owing to the low permeability. The theory of one-dimensional consolidation, advanced by Terzaghi, can be applied to determine the total compression or settlement of a clay layer as well as the time-rate of dissipation of excess pore pressures and hence the time-rate of settlement. The settlement computed by this procedure is known as that due to primary compression since the process of consolidation as being the dissipation of excess pore pressures alone is considered.

Total settlement: The total consolidation settlement, S_c, may be obtained from one of the following equations:

$$S_c = \frac{H \cdot C_c}{(1+e_0)} \log_{10} \left(\frac{\overline{\sigma}_0 + \Delta \overline{\sigma}}{\overline{\sigma}_0} \right) \qquad \ldots \text{(Eq. 11.7)}$$

$$S_c = m_v \cdot \Delta \overline{\sigma} \cdot H \qquad \ldots \text{(Eq. 11.8)}$$

$$S_c = \frac{\Delta e}{(1+e_0)} \cdot H \qquad \ldots \text{(Eq. 11.9)}$$

These equations and the notation have already been dealt with in chapter seven. The vertical pressure increment $\Delta \overline{\sigma}$ at the middle of the layer has to be obtained by using the theory of stress distribution in soil.

Time-rate of settlement: Time-rate of settlement is dependent, in addition to other factors, upon the drainage conditions of the clay layer. If the clay layer is sandwiched between sand layers, pore water could be drained from the top as well as from the bottom and it is said to be a case of double drainage. If drainage is possible only from either the top or the bottom, it is said to be a case of single drainage. In the former case, the settlement proceeds much more rapidly than in the latter.

The calculations are based upon the equation:

$$T = \frac{C_v t}{H^2} \qquad \ldots \text{(Eq. 11.10)}$$

Again, the use of this equation and the notation have been given in chapter seven.

A large wheel load passing on a roadway resting on a clay layer will cause immediate settlement, which is, theoretically speaking, completely recoverable after the load has passed. If the load is applied for a long time, consolidation occurs. Judgement may be necessary in deciding what portion of the superimposed load carried by a structure will be sustained long enough to cause consolidation.

In the case of foundation of finite dimensions, such as a footing resting on a thick bed of clay, lateral strains will occur and the consolidation is no longer one-dimensional. Lateral strain effects in the field may induce non-uniform pore pressures and may become one of the sources of differential settlements of a foundation.

11.3.3 Secondary Settlement or Secondary Compression

Settlement due to secondary compression is believed to occur during and mostly after the completion of primary consolidation or complete dissipation of excess pore pressure. A few theories have been advanced to explain this phenomenon, known as 'secondary consolidation', and have already been given at the end of chapter seven. In the case of organic soils and micaceous soils, the secondary compression is comparable to the primary compression; in the case of all other soils, secondary settlement is considered insignificant. Further discussion of the concept of secondary settlement, being of an advanced nature, is outside the scope of the present work.

11.4 CORRECTIONS TO COMPUTED SETTLEMENT

Certain corrections may be necessary for the computed settlement values—for example, for the effect of the construction period and for lateral strain. These and the accuracy of the computed settlements are dealt with, in brief, in the following subsections.

11.4.1 Construction Period Correction

The load from the structure has been assumed to act on the clay stratum instantaneously; but the application of the load is rather gradual as the construction proceeds. In fact, there will be a gradual stress release due to the excavation for the foundation and the net load becomes positive only after the weight of the structure exceeds that of the excavated material. No appreciable settlement occurs until this point of time. The "effective period of loading" is reckoned as the time lapse from the instant when the load becomes positive until the end of the construction; the loading diagram during this period may be taken approximately a straight line, as shown in Fig. 11.4(a). (Although some rebound and some recompression occur during the period of excavation and replacement of an equivalent load, their combined effect is considered negligible and hence is ignored).

An approximate method for the prediction of settlements during construction, advanced by Terzaghi and extended by Gilboy, is presented in Fig. 11.4(b), as given by Taylor (1948). This method is based on the assumption that at the end of construction the settlement is the same as that which would have resulted in half as much time had the entire load been active throughout.

When any specified percentage of the effective loading period has elapsed, the load acting is approximately equal to this percentage of the total load; at this time the settlement is taken as this percentage times the settlement at one-half of this time from the curve of time versus settlement under instantaneous loading.

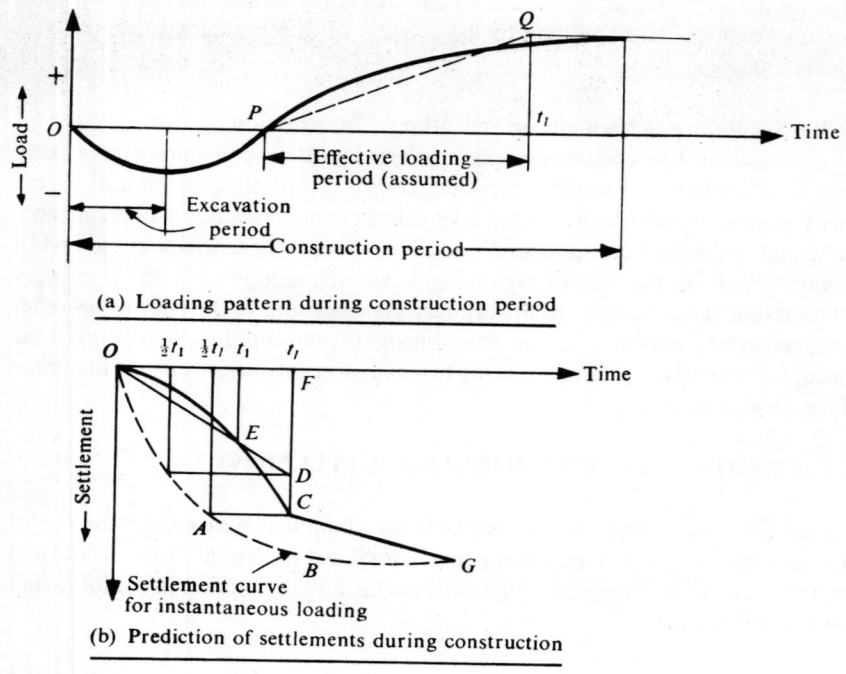

Fig. 11.4 Graphical method for determination of settlements during loading period
(After Taylor, 1948)

Let OAB be the time-settlement curve for the given case for instantaneous loading. The settlement at the end of the effective loading period t_l is equal to that at $\frac{1}{2}t_l$ on the curve OAB. Point C is obtained on the curve by projecting A horizontally on to the vertical through t_l.

For any time $t_1 < t_l$, the curve OAB shows a settlement FD at time $\frac{1}{2}t_1$. Since the load at time t_1 is t_1/t_l times the total load, the settlement at time t_1 is obtained by multiplying FD by this ratio. This is done graphically by joining O to D and projecting the vertical through t_1 to meet OD in E. Then E represents the corresponding point on the corrected time-settlement curve.

Repeating this procedure, any number of points on the curve may be obtained; the thick curve is got in this manner. Beyond point C, the curve is assumed to be the instantaneous curve AB, offset to the right by one-half of the loading period (for example, $BG = AC$). Thus, after the construction is completed, the elapsed time from the start of loading until any given settlement is reached is greater than it would be under instantaneous loading by one-half of the loading period.

11.4.2 Correction for Lateral Strain

The assumption that the soil does not undergo lateral strain made in the one-dimensional consolidation theory, is true for the oedometer sample; however, such a condition may not be true for the clay stratum in the field. This condition may still be achieved if the clay layer is thin and is sandwiched between unyielding layers of granular material or if the loaded area is large compared to the thickness of the compressible layer. Otherwise, lateral deformation takes place and the consolidation will no longer be one-dimensional, which could lead to errors in the computed settlements using the Terzaghi theory.

Skempton and Bjerrum (1957) have suggested a semi-empirical correction, based upon Skempton's pore pressure parameter A. The correction factor may be obtained from Fig. 11.5 for different H/B ratios, H being the thickness of the clay layer and B the width of the foundation.

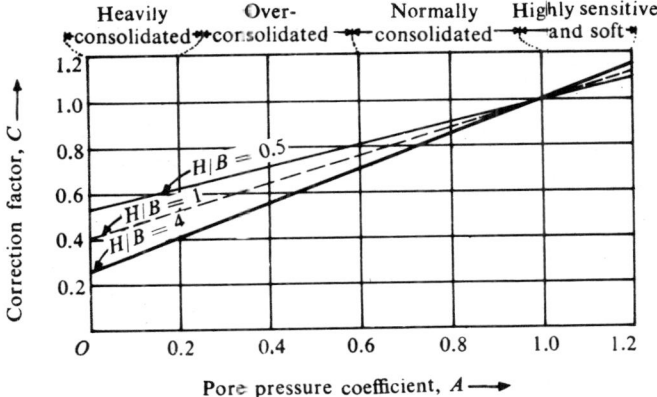

Fig. 11.5 Correction factor for the effect of lateral strain on consolidation settlement
(After Skempton and Bjerrum (1957))

The corrected consolidation settlement S_{cc} is obtained from:

$$S_{cc} = C \cdot S_c \qquad \text{(Eq. 11.11)}$$

where C is the correction factor and S_c is the computed consolidation settlement.

It may be difficult to evaluate the A-parameter accurately; in such cases it may be advisable not to apply the correction, owing to uncertainty.

11.5 FURTHER FACTORS AFFECTING SETTLEMENT

There are a few other factors which affect the settlement and which need consideration. Two of the important factors among these are the rigidity of a structure and the horizontal drainage; these two are considered in the subsections to follow.

11.5.1 Rigidity of a Structure

Considerable judgement must be used in choosing the values of load which are effective in causing settlement; the average load with respect to time must be used rather than the maximum value, especially during the early stages of settlement. However, the structural members such as the columns and footings must be designed for maximum loads.

The structural engineer determines the column loads based on the assumption that all columns undergo equal settlement. This assumption is reasonable for a large and perfectly flexible structure, such as one with timber framing and brick bearing walls, in which considerable unequal settlements can occur without causing significant changes in load distribution. However, in the case of small structures of concrete or steel framing, the settlement of any individual footing causes considerable readjustment in the load on this and the adjacent footings.

This is analogous to the settlement of supports of a continuous beam. For example, if the middle support of a three-span continuous beam settles, the reaction or the load on it gets reduced and that on the other supports gets increased correspondingly. Depending on the magnitude of the settlement, the middle support may not carry any load at all, having transferred the entire load to the other supports. Although it is possible to predict the changes in the column loads consequent to known differential settlements, the procedures are cumbersome. Thus, it is common in settlement analysis to assume column loads for equal settlement. The assumption of flexible construction is always on the safe side since it leads to greater differential settlements than actually occur. The effect of rigidity is, therefore, a desirable one, both for the building as a whole and with reference to local irregularities. In rigid construction, the start of settlement at a footing immediately transfers much of the footing load to the adjacent footings, thus greatly relieving all undesirable effects.

For a clearer understanding let us consider the two types of buildings founded on a compressible foundation, as shown in Fig. 11.6.

The flexible building shown in Fig. 11.6(a) will exert on the soil just below it a pressure distribution which is nearly uniform, shown by curve (1). This will cause a bell-shaped pressure distribution at the top of the buried compressible stratum, represented by curve (2). The settlement pattern of the surface of the stratum will then be as shown by curve (3). If the soil above is of better quality as a foundation material than the compressible stratum, the latter will be the source of practically all the settlement; the settlement pattern at foundation level is as shown by curve (4), similar to curve (3).

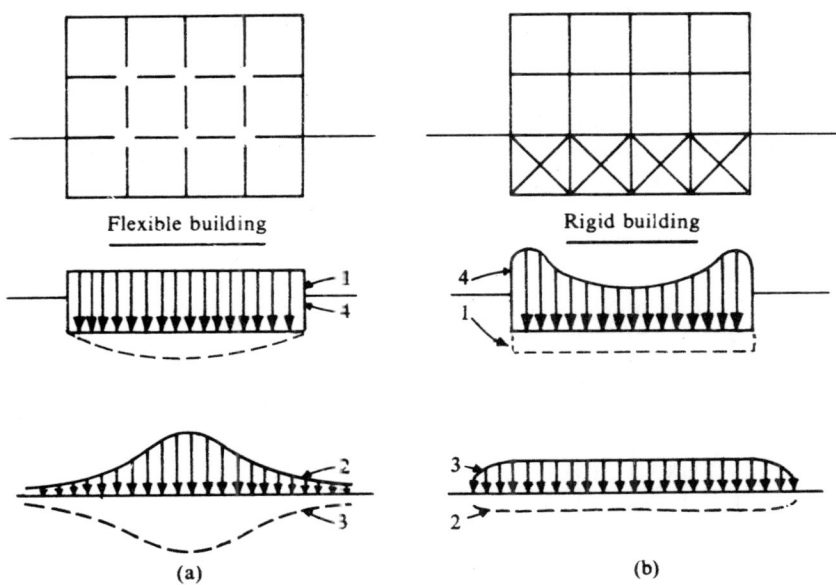

Fig. 11.6 Pressure distributions and settlement patterns for flexible and rigid structures underlain by buried compressible strata (After Taylor, 1948)

For the rigid building shown in Fig. 11.6(b), the settlement pattern is known and the curves must be considered in reversed order as compared to (a). The building must settle uniformly as shown by curve (2); the pressure must then be about uniform, as in (3). By comparing with (a), since the bell-shaped pressure distribution results from uniform pressure distribution just below the foundation, it may be deduced that in (b) the surface pressure distribution, required to cause pressure distribution at the compressible stratum shown by (3), should appear somewhat as shown by curve (4).

Thus, under a flexible structure with uniform loading, settlement at the centre is more than that at the edges, while, for a rigid structure the pressure near the edges of the loaded area is greater than that near the centre. The differential settlement in case (a) may result in cracking of the walls; in case (b) the upper storeys are not subject to distortion or cracking. But the existence of greater pressures on the outer portions of slabs in case (b) should be recognised in the design.

11.5.2 Horizontal Drainage

The hydrostatic excess pressures at the same depth may be different at different points underneath a loaded area, especially if the structure is supported on piles or columns carrying different loads. This creates horizontal flow or drainage due to the gradients in the horizontal directions.

The effect of this horizontal drainage is to accelerate the time-rate of consolidation when compared with the situation of one-dimensional flow or drainage. It has no effect on the total settlement.

The magnitude of this effect is dependent upon the size of the structure relative to the depth and thickness of the clay stratum, the effect being greatest when the plan size of the structure is small. This effect also depends greatly on the relative magnitude of the soil properties in the horizontal and vertical directions.

The effect of horizontal drainage, as presented by Gould (1946), based on his investigation with reference to a simple building, 21 m square, is shown in Fig. 11.7.

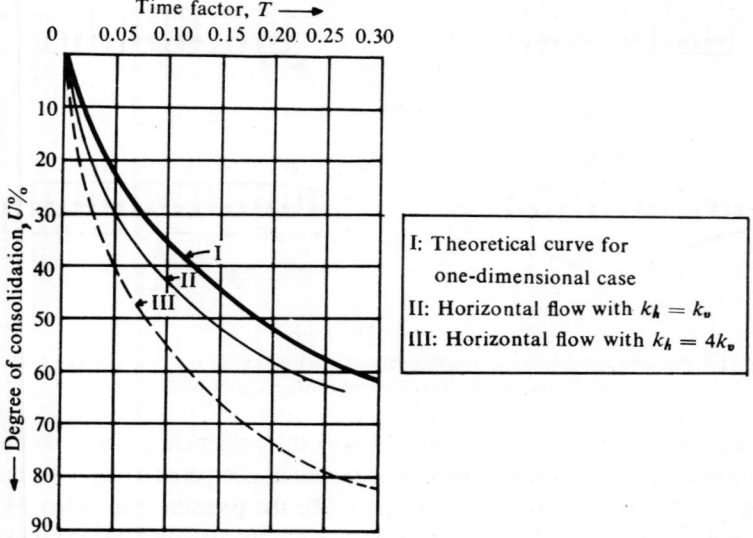

Fig. 11.7 Effect of horizontal flow on consolidation
(After Gould, 1946; as presented by Taylor, 1948)

It can be observed that the effect of horizontal drainage is to make the settlement proceed faster than in the case of one-dimensional drainage; however, this effect may not be important in many cases.

11.6 OTHER FACTORS PERTINENT TO SETTLEMENT

Three other matters pertinent to settlement—viz., accuracy of computed settlements, permissible settlements and remedial measures are dealt with in the following sub-sections.

11.6.1 Accuracy of Computed Settlemnt

The accuracy of the computed settlements is naturally dependent upon the degree or extent to which the assumptions involved in the theories made use of in the analysis are valid in any given case.

Assumptions made for the interpretation of the geological profile, especially values used for thicknesses of the strata, may lead to errors if they are incorrect. Similarly the use of soil properties obtained from partially disturbed samples, especially the consolidation characteristics may lead to errors in the estimate of total settlement as well as speed of settlement. But these are inaccuracies in data and not in the analysis, as such. The primary assumption regarding the one-dimensional nature of the compression may be valid only in the case of deeply buried clay strata; in other cases, the effects of lateral strain may be considerable.

The theories of stress distribution in soil, used in the settlement analysis, involve assumptions which may not be true in practice. For example, the assumption that soil is perfectly elastic, homogeneous and isotropic is nowhere near the facts. However, it is considered that the accuracy is not affected significantly by this erroneous assumption.

In conclusion it may be stated that settlement analyses usually give results which are at best crude estimates; however, even a crude estimate may be considered very much better than a pure guess or conjecture, which may be the only alternative.

11.6.2 Permissible Settlements

It is not easy to decide what value of settlement will have a detrimental effect on a structure. This is because uniform settlement will have little adverse effect on the structural stability, but even small differential settlement may cause trouble. There are two main ill-effects of differential settlements: (i) the architectural effect (cracking of plaster, for example) and (ii) the structural effect (redistribution of moments and shears, for example, which may ultimately lead to failure).

Many building codes and foundation authorities place restriction on differential settlement. Terzaghi and Peck (1948) specify a permissible differential settlement of 20 mm between adjacent columns and recommend that foundations on sand be designed for a total settlement of 25 mm. Skempton and MacDonald (1956) specify that the angular rotation or distortion between adjacent columns in clay should not exceed 1/300, although the total settlement may go up to 100 mm. Sowers (1957) recommends, in his discussion of the paper by Polshin and Pokar (1957) a maximum differential settlement of 1/500 for brick buildings and 1/5000 for foundations of turbogenerators. Bozozuk (1962) summarised his investigations in Ottawa as follows:

Angular rotation	Damage
1/180	None
1/120	Slight
1/90	Moderate
1/50	Severe

The I.S.I. (IS:1904-1961) recommends a permissible total settlement of 65 mm for isolated foundations on clay. 40 mm for isolated foundations on sand, 65 to 100 mm for rafts on clay and 40 to 65 mm for rafts on sand. The permissible differential settlement is 40 mm for foundations on clay and 25 mm for foundations

Table 11.4 Maximum and differential settlements of buildings

Sl. No.	Type of Structure	Isolated foundations							Raft foundations				
		Sand and hard clay			Plastic clay			Sand and hard clay			Plastic clay		
		MS mm	DS mm	AD	MS mm	DS mm	AD	MS mm	DS mm	AD	MS mm	DS mm	AD
1	For steel structure	50	0.0033L	$\frac{1}{300}$	50	0.0033L	$\frac{1}{300}$	75	0.0033L	$\frac{1}{300}$	100	0.0033L	$\frac{1}{300}$
2	For reinforced concrete structures	50	0.0015L	$\frac{1}{666}$	75	0.0015L	$\frac{1}{666}$	75	0.002L	$\frac{1}{500}$	100	0.002L	$\frac{1}{500}$
3	For plain brick walls in multi-storeyed buildings							not	likely	to be	encountered		
	(i) For $\frac{L}{H} \leq 3$	60	0.00025L	$\frac{1}{4000}$	80	0.00025L	$\frac{1}{4000}$						
	(ii) For $\frac{L}{H} > 3$	60	0.00033L	$\frac{1}{3000}$	80	0.00033L	$\frac{1}{3000}$						
4	For water towers and silos	50	0.0015L	$\frac{1}{666}$	75	0.0015L	$\frac{1}{666}$	100	0.0025L	$\frac{1}{400}$	125	0.0025L	$\frac{1}{400}$

L = Length of deflected part of wall/raft or centre-to-centre distance between columns.
H = Height of wall from foundation footing.
MS = Maximum settlement
DS = Differential settlement
AD = Angular distortion

Note: The values given in the table may be taken only as a guide and the permissible settlement and differential settlement in each case should be decided as per requirement of the designer.

on sand. The angular distortion in the case of large framed structures must not exceed 1/500 normally and 1/1000 if all kinds of minor damage also are to be prevented.

Maximum and differential settlements as specified in IS:1904-1978 "Code of Practice for structural safety of Buildings: Shallow foundations (Second revision)" are shown in Table 11.4.

Opinions on this subject vary considerably and were discussed by Rutledge (1964).

11.6.3 Remedial Measures Against Harmful Settlements

Settlement of soil is a natural phenomenon and may be considered to be unavoidable. However, a few remedial measures are possible against harmful settlements (Jumikis, 1962):

1. Removal of soft soil strata consistent with economy.
2. The use of properly designed and constructed pile foundations (chapter 16).
3. Provision for lateral restraint against lateral expulsion of soil mass from underneath the footing of a foundation.
4. Building slowly on cohesive soils to avoid lateral expansion of a soil mass and to give time for the pore water to be expelled by the surcharge load.
5. Reduction of contact pressure on the soil; more appropriately, proper adjustment between pressure, shape and size of the foundation in order to attain uniform settlements underneath the structure.
6. Preconsolidation of a building site long enough for the expected load, depending upon the tolerable settlements; alternatively, any other method of soil stabilization (chapter 17).

In order to make large settlements harmless, structures may be designed as statically determinate systems; the structures and their foundations may be designed as rigid units; and, long structrures may be subdivided into separate units.

Sometimes, structures such as bridges and water towers are supported at three points; this facilitates jacking up the structure at any support so that the structure may be raised and levelled as settlement occurs.

The need for thorough soil exploration and soil testing is obvious in the context of achieving these objectives.

11.7 SETTLEMENT RECORDS

Settlement records offer an excellent test of the accuracy of settlement analysis; as such, the maintenance of such records, wherever possible, is considered very valuable. However, different factors make it difficult to maintain such records; for example, very slow progress of settlements, winding up of construction organisations and the waning interest after the completion of construction. Careful comparison of settlement records with predicted values of settlements goes a long way in the development of better methods of analysis for future use.

The most common method of observing settlements uses periodic lines of levels and observing a representative group of reference points. Special levelling devices, such as the one described by Terzaghi (1938), may be used for more accurate records. In any case, a reliable and dependable benchmark must be available and in a locality where there is a deeply buried clay layer, benchmarks that are not disturbed by settlement are difficult to be obtained. Benchmarks are to be founded on firm ground, preferably on ledge or hard rock, for them to be satisfactory, even if this means extending to a depth of more than 30 m.

11.8 CONTACT PRESSURE AND ACTIVE ZONE FROM PRESSURE BULB CONCEPT

The concepts of contact pressure and active zone in soil based on the pressure bulb concept are relevant to settlement computation and hence are treated in the following sub-sections.

11.8.1 Contact Pressure

'Contact pressure' is the actual pressure transmitted from the foundation to the soil. It may also be looked upon as the pressure, by way of reaction, exerted by the soil on the underside of the footing or foundation. A uniformly loaded foundation will not necessarily transmit a uniform contact pressure to the soil. This is possible only if the foundation is perfectly 'flexible'; the contact pressure is uniform for a flexible foundation irrespective of the nature of the foundation soil.

If the foundation is 'rigid', the contact pressure distribution depends upon the type of the soil below the foundation as shown in Fig. 11.8.

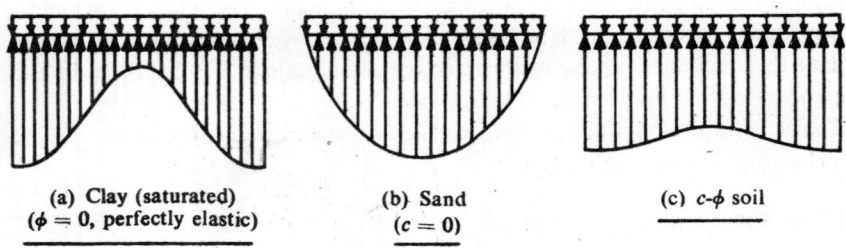

(a) Clay (saturated)
($\phi = 0$, perfectly elastic)

(b) Sand
($c = 0$)

(c) c-ϕ soil

Fig. 11.8 Contact pressure distribution under a uniformly loaded rigid foundation

On the assumption of a uniform vertical settlement of the rigid foundation, the theoretical value of the contact pressure at the edges of the foundation is found to be infinite from the theory of elasticity, in the case of perfectly elastic material such as saturated clay ($\phi = 0$). However, local yielding of the soil makes the pressure at the edges finite, as shown in Fig. 11.8(a). Under incipient failure conditions the pressure distribution, tends to be practically uniform.

For a rigid foundation, placed at the ground surface on sand ($c = 0$), the contact pressure at the edges is zero, since no resistance to shear can be mobilised for want of over-burden pressure; the pressure distribution is approximately parabolic, as shown in Fig. 11.8(b). The more the foundation is below the surface of the sand, the more the shear resistance developed at the edges due to increase in overburden pressure, and as a consequence, the contact pressure distribution tends to be more uniform.

For a general cohesive-frictional soil ($c - \phi$ soil) the contact pressure distribution will be intermediate between the extreme cases of (a) and (b), as shown in Fig. 11.8(c). Also for a foundation such as a reinforced concrete foundation which is neither perfectly flexible nor perfectly rigid, the contact pressure distribution depends on the degree of rigidity, and assumes an intermediate pattern for flexible and rigid foundation. However, in most practical cases, the assumption of uniform contact pressure distribution yields sufficiently accurate design values for moments, shears and vertical stresses, and hence is freely adopted.

11.8.2 Active Zone from Pressure Bulb Concept

Terzaghi (1936) related the bulb of pressure with the seat of settlement. Since it is possible to obtain an infinite variety of pressure bulbs for any applied pressure, one has to refer to an assumed isobar like that for $(1/n)$th of the contact pressure, q, as shown in Fig. 11.9.

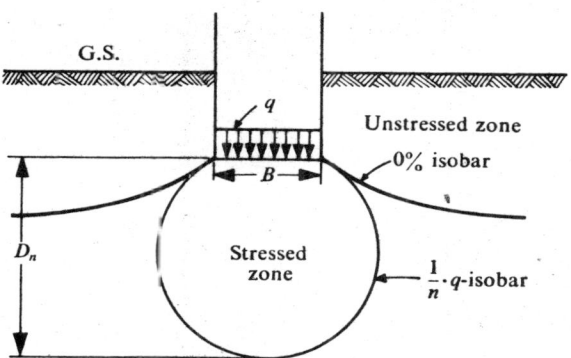

Fig. 11.9 The $(1/n).(q)$-isobar (Jumikis, 1962)

Terzaghi pointed out that the depth, D_n, of the $(1/n)$ (q)-isobar increases in direct proportion to the width of the loaded area for similar shapes of these areas:

$$\frac{D_n}{d_n} = \frac{B}{b} = \text{Constant} = f(n) \qquad \ldots\text{(Eq. 11.12)}$$

Terzaghi also observed that direct stresses are considered negligible when they are smaller than 20% of the contact stress from structural loading and that most of the settlement, nearly 80% of the total, takes place at a depth less than $D_{n=5}$. ($D_{n=5}$ is the depth of the $0.20q$-isobar). Therefore, the isobar of $0.20q$ may be taken to define the contour of the pressure bulb, which is the stressed zone within a homogeneous soil medium. The stress transmitted by the applied foundation loading on to the surface of this isobar is resisted by the shear strength of soil at this surface. The region within the $0.20q$-isobar is called by Terzaghi the "seat of settlement".

For a homogeneous, elastic, isotropic and semi-infinite soil medium, $D_{n=5} \approx 1.5B$ is considered good. (For a uniform and thick sand, $D_{n=5} < 1.5B$).

The wider the loaded area, the deeper the effect for isobars of the same intensity, as shown in Fig. 11.10.

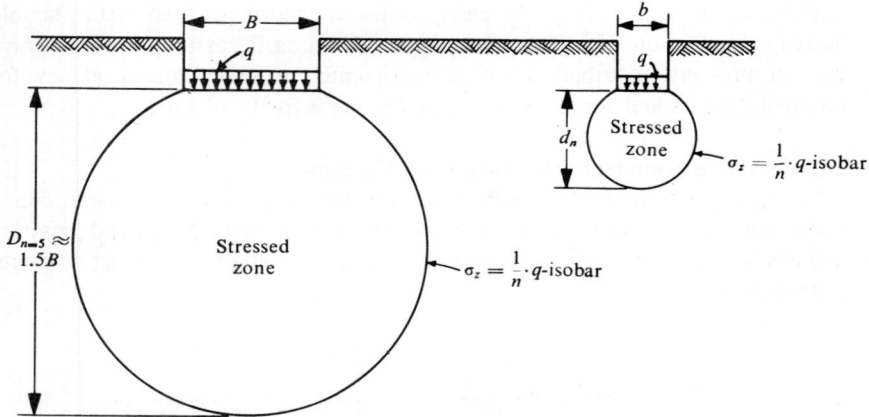

Fig. 11.10 Effect of width of foundation on depth of isobars (Jumikis, 1962)

(This will be again referred to in the plate load test in chapter 14).

The depth $D_{n=5}$, to which the $0.20q$-isobar extends below the foundation, which gives the seat of settlement, is termed the 'active zone'. The thickness of the active zone extends from the base of the foundation to that depth where the vertical stresses from the structure are 20% of the magnitude of the over-burden pressure of the soil, which contributes to most of the settlement. This is shown in Fig. 11.11.

The soil layers below the active zone are considered as being ineffective, small stresses being ignored. In other words, even if compressible strata exist below the active zone, their effect on the settlement is negligible.

It may be noted that, while the vertical stress diagram due to self-weight of soil starts with zero value at ground surface and increases linearly with depth, the stress diagram due to contact pressure caused by structural loading starts with the value of contact pressure at the base of the foundation and decreases with depth in a curvilinear fashion based on the theory used.

This concept will again be used in the determination of the depth of exploratory borings (chapter 18).

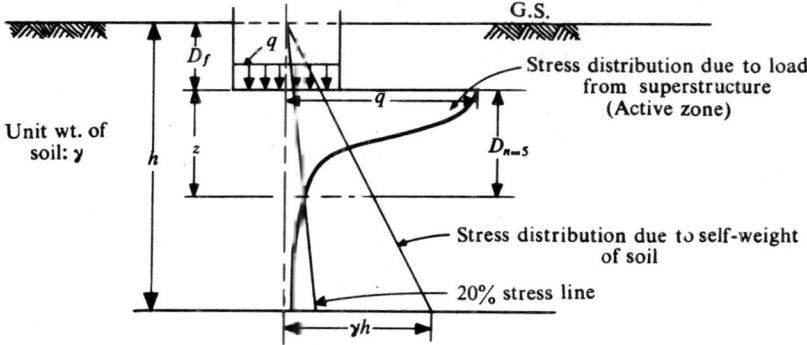

Fig. 11.11 Active zone in soil due to loading (Jumikis, 1962)

If several loaded footings are placed closely enough, the isobars of individual footings would combine and merge into one large isobar of the same intensity, as shown in Fig. 11.12.

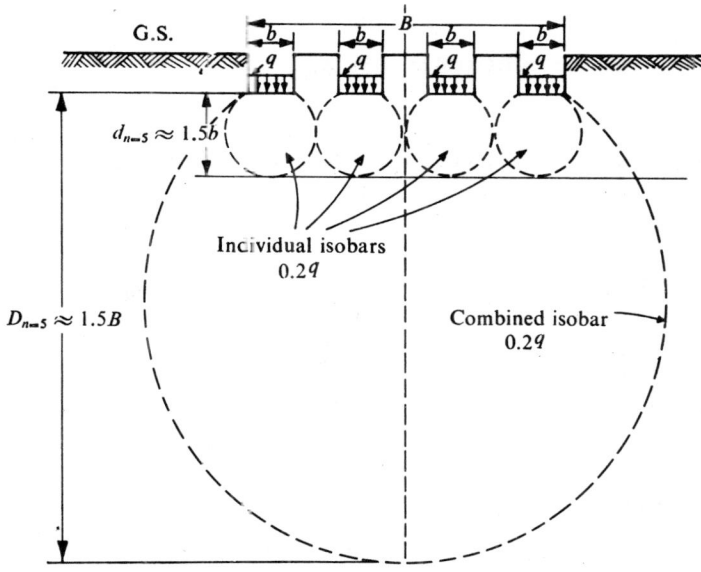

Fig. 11.12 Merging of closely-spaced isobars into one large isobar of the same intensity, reaching far deeper than the individual isobars (Jumikis, 1962)

The large isobar reaches about $D_{n=5} \approx 1.5B$ below the base of the closely spaced footings, where B is the overall width between the extreme footings.

476 Geotechnical Engineering

This concept will again be referred to in chapter 16 with regard to the settlement of a closely spaced group of friction piles.

11.9 ILLUSTRATIVE EXAMPLES

Example 11.1: A reinforced concrete foundation, of dimensions 18 m × 36 m, exerts a uniform pressure of 180 kN/m² on a soil mass, with E-value 45 MN/m². Determine the value of immediate settlement under the foundation.

The immediate settlement, S_i, is given by:

$$S_i = \frac{q \cdot B(1-v^2)}{E_s} \cdot I_s$$

$E_s = 45$ MN/m²
$q = 180$ kN/m²
$B = 18$ m
Assume $v = 0.5$
I_s for $L/B = 36/18 = 2$ is 1.00

$$\therefore \quad S_i = \frac{180 \times 18(1-0.5^2)}{45 \times 1000} \times 1.00 \text{ m}$$

$$= 0.054 \text{ m} = \textbf{54 mm}$$

Example 11.2: The plan of a proposed spoil heap is shown in Fig. 11.13(*a*). The heap will stand on a thick, soft alluvial clay with the E-value 18 MN/m². The eventual uniform bearing pressure on the soil will be about 270 kN/m². Estimate the immediate settlement under the point X at the surface of the soil.

The area is imagined to be divided into rectangles such that X forms one of the corners for each. This is as shown in Fig. 11.13(*b*).

The structure is flexible and the soil deposit is thick.

Therefore, $S_i = q \cdot B \dfrac{(1-v^2)}{E_s} \cdot I_t$

I_t being Terzaghi influence value, dependent on L/B.

By the principle of superposition,

$$S_i = q \cdot \frac{(1-v^2)}{E_s}\left(I_{t_1} B_1 + I_{t_2} \cdot B_2 + I_{t_3} \cdot B_3\right)$$

For Rectangle (1):

$$L/B = 150/50 = 3, I_{t_1} = 0.88$$

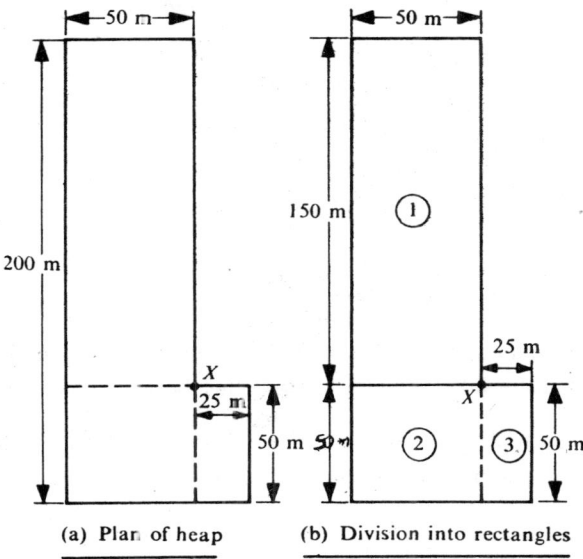

Fig. 11.13 Spoil Heap (Ex. 11.2)

For Rectangle (2) :
$$L/B = 50/50 = 1, I_{t_2} = 0.56$$

For Rectangle (3) :
$$L/B = 50/25 = 2, I_{t_3} = 0.76$$

$E_s = 18 \text{ MN/m}^2 \quad q = 270 \text{ kN/m}^2$

Assume $\nu = 0.5$

$$\therefore S_i = \frac{270 \times (1 - 0.5^2)}{18 \times 1000}(0.88 \times 50 + 0.56 \times 50 + 0.76 \times 25)$$

$$= \frac{270 \times 0.75}{18 \times 1000} \times 91 \text{ m}$$

$$\approx 1.024 \text{ m}.$$

Example 11.3: A soft, normally consolidated clay layer is 18 m thick. The natural water content is 45%. The saturated unit weight is 18 kN/m³ ; the grain specific gravity is 2.70 and the liquid limit is 63%. The vertical stress increment at the centre of the layer due to the foundation load is 9 kN/m². The ground water level is at the surface of the clay layer. Determine the settlement of the foundation.

Initial vertical effective stress at centre of layer

$$= (18 - 9.81) \times \frac{18}{2} = 73.71 \text{ kN/m}^2$$

Final effective vertical stress = 73.71 + 9.0 = 82.71 kN/m²
Initial void ratio, $e_0 = w \cdot G = 0.45 \times 2.70 = 1.215$

$$C_c = 0.009 (63 - 10) = 0.477$$

$$S_i = \frac{18 \times 0.477}{(1 + 1.215)} \cdot \log_{10} \frac{82.71}{73.71}$$

$$= 0.194 \text{ m}$$

$$= \mathbf{194 \text{ mm}}$$

This procedure may be used for rough estimate of the settlement of a small structure. For large structure, consolidation characteristics must be got from laboratory tests.

Example 11.4: A footing foundation for a water tower carries a load of 900 tonnes and is 3.6 metres square. It rests on dense sand of 9 m thickness overlying a clay

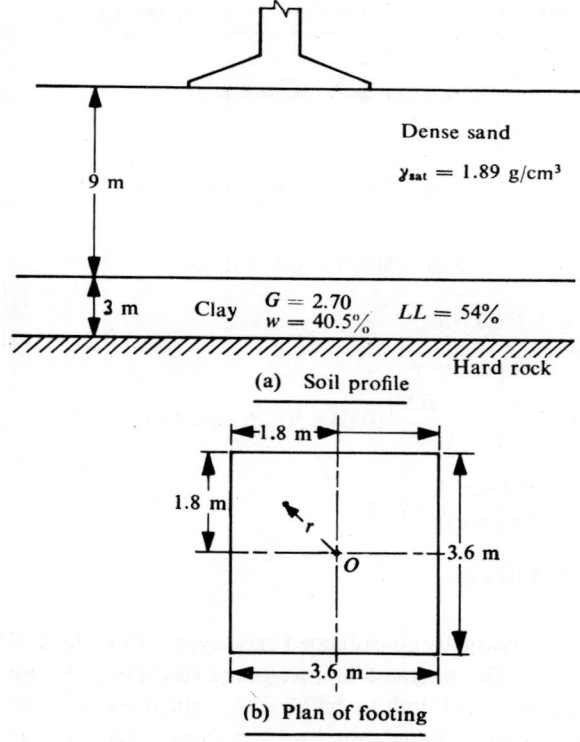

Fig. 11.14 Soil profile and plan of footing (Ex. 11.4)

layer of 3 metres depth. The clay layer overlies hard rock. Liquid limit of clay is 54%, water content 40.5%, and grain specific gravity is 2.70. The saturated unit weight of dense sand is 1.89 g/cm³. Estimate the ultimate settlement due to consolidation of the clay layer, assuming the site to be flooded.

The data area shown in Fig. 11.14.

The dimensions of the footing are more than one-third the depth at which the vertical stress is to be computed. Therefore the load may be taken as being uniformly distributed.

$$q = \frac{900}{3.6 \times 3.6} \text{ t/m}^2 = 69.44 \text{ t/m}^2$$

Since the stress below the centre of the footing is required, the area may be divided into four squares and the load from each square may be treated as a point load, acting at the centre of the square.

The radial distance of 0 from each point load,

$$r = 0.9\sqrt{2} \text{ m}$$

$$\frac{r}{z} = \frac{0.9\sqrt{2}}{(9+1.5)} = 0.1212$$

Influence factor, $K_B = \dfrac{(3/2\pi)}{\left(1+\dfrac{r^2}{z^2}\right)^{5/2}} = \dfrac{(3/2\pi)}{\{1+(0.1212)^2\}^{5/2}} = 0.460$

$$\therefore \sigma_z = 4 \times \frac{0.460}{(10.5)^2} \times \frac{900}{4} = 414/(10.5)^2 = 3.755 \text{ t/m}^2$$

Void ratio for clay $= wG = 0.405 \times 2.70 = 1.094$

γ_{sat} for clay $= \dfrac{(G+e)}{(1+e)} \cdot \gamma_w = \dfrac{2.70+1.094}{2.094} = 1.81 \text{ g/cm}^3$

γ' for dense sand $= 0.89 \text{ g/cm}^3$ γ' for clay $= 0.81 \text{ g/cm}^3$

Overburden pressure at 10.5 m depth $= (9 \times 0.89 + 1.5 \times 0.81) = 9.225 \text{ t/m}^2$

$C_c = 0.009(54 - 10) = 0.396$

Consolidation settlement, $S_c = \dfrac{300 \times 0.396}{2.094} \log_{10} \dfrac{(9.225+3.755)}{9.225} \text{ cm} \approx \mathbf{8.4 \text{ cm}}$

Example 11.5: A clay stratum of 18 metres thickness was found above a sand stratum when a boring was made. The clay was consolidated under the present overburden pressure. The hydrostatic pressure at the top of the clay stratum was found to be 5400 kg/m². Due to pumping of water from the sand stratum, the pressure in the pore water below the clay layer was reduced permanently by 5400 kg/m². If the void ratios of clay before and after pumping were 0.93 and 0.90, respectively, calculate the ultimate settlement due to pumping.

The settlement is caused due to reduction of water pressure by pumping in this case. The pressure is thus transferred to the soil grains as effective pressure, as shown in Fig. 11.15(c):

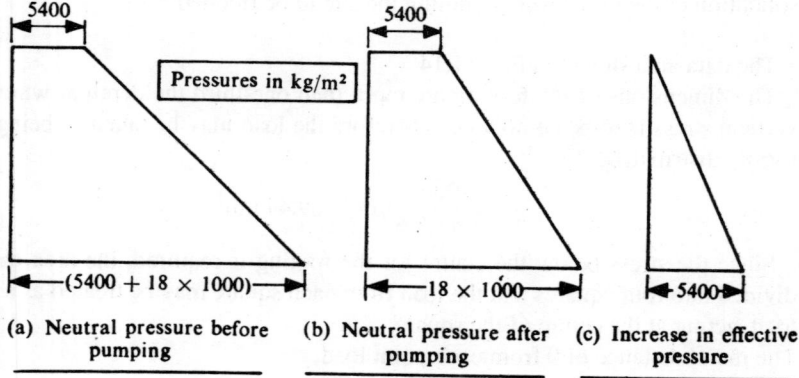

Fig. 11.15 Pressure conditions before and after pumping (Ex. 11.5)

$$\Delta\bar{\sigma} = 5400 \text{ kg/m}^2$$

$$e_0 = 0.93 \quad e_1 = 0.90 \quad \Delta e = 0.03$$

$$a_v = \frac{\Delta e}{\Delta\bar{\sigma}} = \frac{0.03}{5400} \text{ m}^2/\text{kg}$$

$$m_v = \frac{0.03}{5400} \times \frac{1}{(1+0.93)} = \frac{0.03}{5400 \times 1.93} = 2.88 \times 10^{-6} \text{ m}^2/\text{kg}$$

Consolidation settlement

$$S_c = m_v \cdot \Delta\bar{\sigma} \cdot H$$

$$= \frac{0.03}{5400 \times 1.93} \times 5400 \times 1800 \text{ cm}$$

$$\approx 28 \text{ cm}$$

∴ Ultimate settlement = **28 cm**

Example 11.6: A clay layer 24 metres thick has a saturated unit weight of 18 kN/m². Ground water level occurs at a depth of 4 metres. It is proposed to construct a reinforced concrete foundation, length 48 m and width 12 m, on the top of the layer, transmitting a uniform pressure of 180 kN/m². Determine the settlement under its centre. E for the clay is 33 MN/m² obtained from triaxial tests. Initial void ratio = 0.69. Change in void ratio = 0.02.

The details of the foundation are shown in Fig. 11.16.

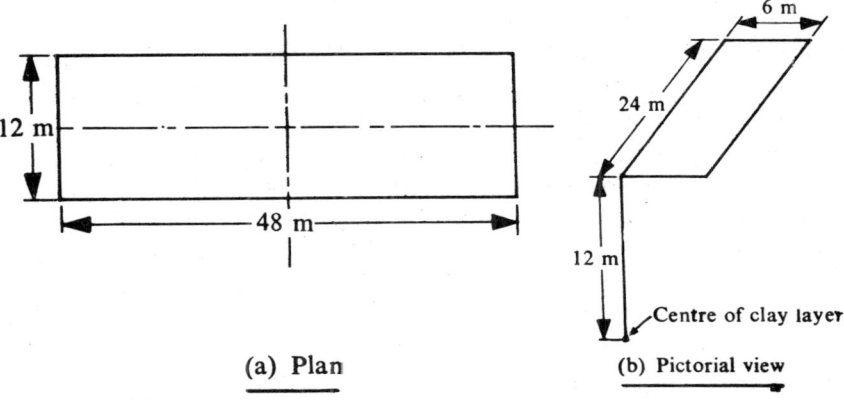

Fig. 11.16 Details of foundation (Ex. 11.6)

The vertical stress increment at the centre of the clay layer may be obtained by dividing the loaded area into four rectangles as shown.

$$m = B/z = 6/12 = 0.5$$
$$n = L/z = 24/12 = 2.0$$

Influence factor from Fadum's chart = 0.135

$$\sigma_z = 4 \times 1800 \times 0.135 = 97.2 \text{ kN/m}^2$$

Immediate settlement:

Since the thickness of the layer is less than 4B, Steinbrenner's coefficient I_s from Fig. 11.3 may be used in

$$S_i = q \cdot B \frac{(1-v^2)}{E_s} \cdot I_s \text{ and applying the principle of superposition for the four}$$

rectangles as in the case of stress.

$L/B = 24/6 = 4 \quad H/B = 24/6 = 4$ (H here is the thickness of the clay layer).
Since all four rectangles are identical,
total value of $I_s = 4 \times 0.48 = 1.92$

$$\therefore S_i = 180 \times 6 \times \frac{0.75}{33000} \times 1.92 \times 0.8 \quad \text{(assuming } v = 0.5 \text{ and rigidity factor as 0.8)}$$

$$= 0.0377 \text{ m} = 37.7 \text{ mm}$$

Consolidation settlement:

Initial effective overburden pressure at centre of clay layer
$$= 18 \times 12 - 9.81 \times 8 = 137.52 \text{ kN/m}^2$$

Consolidation settlement, $S_c = H \cdot \dfrac{\Delta e}{(1+e_0)}$

$$= 24 \times \frac{0.02}{1.69} \text{ m} = 0.284 \text{ m} = 284 \text{ mm}$$

Total settlement $S = S_i + S_c = 37.7 + 284.0$ mm = **321.7 mm**

Example 11.7: The loading period for a building extended from Feb., 1957 to Feb., 1959. In Feb., 1962, the average measured settlement was found to be 117 mm. The ultimate settlement was expected to be 360 mm. Estimate the settlement in Feb., 1967, assuming double drainage to occur. What would be this result if the measured settlement in Feb., 1962 was 153 mm instead of 117 mm?

The reckoning of time is conventionally done from mid-way through the construction or loading period. In this case,

$$S_4 = 117 \text{ mm when } t = 4 \text{ years. } S_c = 360 \text{ mm.}$$

The settlement is required at time $t = 9$ years.
Let us assume, in the first instance, that at $t = 9$ years,
U, the degree of consolidation is less than 50%.
In such a case, $U = 1.13 \sqrt{T_v}$, where T_v is the Time-factor.

$$\frac{S_4}{S_9} = \frac{U_4}{U_9} \qquad \frac{U_4}{U_9} = \sqrt{\frac{T_{v_4}}{T_{v_9}}}$$

$$\therefore \frac{S_4}{S_9} = \sqrt{\frac{t_4}{t_9}}, \text{ since } \frac{C_v}{H^2} \text{ is a constant.}$$

$$\therefore \frac{117}{S_9} = \sqrt{\frac{4}{9}} = 2/3$$

$$\therefore S_9 = (3/2) \times 117 = 175.5 \text{ mm}$$

Thus, $T_{v_9} = \dfrac{S_9}{S_c} = \dfrac{175.5}{360.0} < 50\%$. Hence the relationships used are valid.

If $S_4 = 153$ mm at $t = 4$ years,

$U_4 = 153/360 = 42.5\%$

For double drainage,
this corresponds $T_v = 0.142$, since $T_v = (\pi/4)U^2$

Then $\dfrac{C_v \times 4}{H^2} = 0.142$, or $\dfrac{C_v}{H^2} = \dfrac{0.142}{4} = 0.0355$

For $t = 9$ years, $T_{v_9} = 0.0355 \times 9 = 0.3195 \approx 0.32$

Correspondingly $U_9 = 0.632$

Hence the settlement in Feb., 1967 = 0.632×360 = **227.5 mm**

Example 11.8: A building was to be constructed on a clay stratum. Preliminary analysis indicated a settlement of 60 mm in 6 years and an ultimate settlement of 250 mm. The average increase of pressure in the clay stratum was 24 kN/m².

The following variations occurred from the assumptions used in the preliminary analysis:
(a) The loading period was 3 years, which was not considered in the preliminary analysis.
(b) Borings indicated 20% more thickness for the clay stratum than originally assumed.
(c) During construction, the water table got lowered permanently by 1 metre.

Estimate:
(i) the ultimate settlement,
(ii) the settlement at the end of the loading period, and
(iii) the settlement 2 years after completion of the building.

From preliminary analysis, the ultimate settlement is 25 cm. This will change because of the altered conditions given in (b) and (c).

Settlement S varies in direct proportion to the thickness of stratum.

∴ Modified value of ultimate settlement will be obtained by using the factor $\frac{120}{100}$

or 6/5.

Due to lowering water table, effective stress increases. In this case the increase in effective stress is vH or 9.81×1 kN/m².

Approximately, settlement varies linearly with an increase in effective stress. Therefore, the modified value on this count will be got by using the factor 33.81/24.

∴ Final value of ultimate settlement $= 250 \times \frac{6}{5} \times \frac{33.81}{24} = $ **422.6 mm**

Similarly, modified value of settlement in 6 years $= 60 \times \frac{6}{5} \times \frac{33.81}{24} = $ **101.4 mm**

But since the loading is also to be considered this settlement is supposed to occur in (6 + 3/2) or 7.5 years.

Since $\frac{S_1}{S_2} = \sqrt{\frac{t_1}{t_2}}$,

Settlement at the end of loading period, $= \frac{101.4}{\sqrt{7.5}} \times \sqrt{1.5} = $ **45.35 mm**

Similarly, settlement 2 years after completion of the building is:

$$\frac{101.4}{\sqrt{7.5}} \cdot \sqrt{3.5} = \textbf{69.27 mm}$$

Example 11.9: The plan of a proposed raft foundation 18 m × 54 m is shown in Fig. 11.17(a). The uniform pressure from the foundation is 324 kN/m². Site investigation shows that the top 6 m of subsoil is saturated coarse sand with a unit weight of 18.0 kN/m³. The ground water level occurs at 3.00 m from the top of sand. The standard penetration value of the sand taken at a depth of 4.50 m is 1.8.

484 *Geotechnical Engineering*

Below the sand there exists a clay layer of 30 m thickness ($E = 16.2$ MN/m^2, $E_{swelling} = 63$ MN/m^2). The clay rests on hard rock. Determine the total settlement under the foundation.

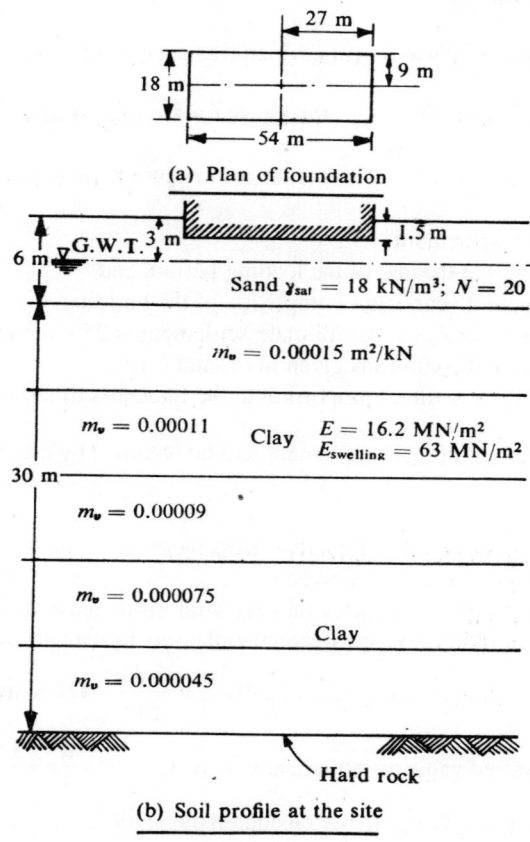

(b) Soil profile at the site

Fig. 11.17 Raft foundaion (Ex. 11.9)

Vertical stress increments:
Net pressure = 324 kN/m^2
Relief due to excavation of 1.5 m sand = $18 \times 1.5 = 27$ kN/m^2
Gross pressure = $(324 + 27) = 351$ kN/m^2.
The foundation is split into four rectangles as shown and Fadum's chart is used:

Depth (m)	B/z	L/z	I_σ	$4I_\sigma$	$\Delta\sigma_z$ (kN/m^2)
3	3	9	0.247	0.988	320
9	1	3	0.203	0.812	263
15	0.60	1.80	0.152	0.608	197
21	0.43	1.30	0.113	0.452	146
27	0.33	1.00	0.086	0.344	112
33	0.27	0.82	0.067	0.268	87

Settlement Analysis

Immediate settlement (Sand):

$$N = 18 \quad \bar{\sigma}_0 = 4.5 \times 18 - 1.5 \times 9.81 = 66.3 \text{ kN/m}^2$$

$$C_s = 1.5 \frac{C_r}{\bar{\sigma}_0}$$

But $\quad C_r = 400 N = 400 \times 18 = 7200 \text{ kN/m}^2$

$$C_s = \frac{1.5 \times 7200}{66.3} = 162.9$$

$$S_1 = \frac{H}{C_s} \cdot \log_e \left(\frac{\bar{\sigma}_0 + \Delta \bar{\sigma}}{\bar{\sigma}_0} \right)$$

$$= \frac{(6-1.5)}{162.9} \log_e \left(\frac{66.3 + 320}{66.3} \right) \text{m} = 0.0487 \text{ m} = 48.7 \text{ mm}$$

Since most part of the sand below the foundation is submerged customarily it is assumed that the settlement will be doubled.

∴ S_i for sand = $2 \times 48.7 = 97.4$ mm

Clay:
With reference to Fig. 11.3(b),

$$H_1 = 4.5 \text{ m}; H_2 = 34.5 \text{ m}$$

For H_2:

$$L/B = 27/9 = 3; \quad \frac{H_2}{B} = \frac{34.5}{9} = 3.83 \quad \therefore I_s = 0.47$$

For H_1:

$$L/B = 3; \quad \frac{H_1}{B} = \frac{4.5}{9} = 0.5, \quad \therefore I_s = 0.07$$

$$S_i = q \cdot \frac{B}{E}(1-v^2) \times 4 I_s \times \text{(rigidity factor)}$$

S_i, taking gross pressure $= \dfrac{351}{16200} \times 9 \times 0.75 \times 4(0.47 - 0.07) \times 0.8$

$$= 0.1872 \text{ m} = 187.2 \text{ mm}$$

Heave effect:

Relief pressure due to excavation = $1.5 \times 18 = 27 \text{ kN/m}^2$

$$\therefore \text{ Heave} = \frac{27}{62000} \times 9 \times 0.75 \times 4(0.47 - 0.07) \times 0.8$$

$$= 0.0037 \text{ m} = 3.7 \text{ mm}$$

Net immediate settlement in clay = $187.2 - 3.7 = 183.5$ mm

The heave effect is obviously insignificant except for great depth of excavation.

Consolidation settlement:
The clay layer is divided into five layers of 6 m thickness.

m_v	$\Delta\sigma_z$	$m_v \cdot \Delta\sigma_z \cdot H$
0.000150	263	0.2367
0.000110	197	0.1300
0.000090	146	0.0788
0.000075	112	0.0504
0.000045	87	0.0235
		0.5194 = 519.4 mm

Total settlement = (97.4 + 183.5 + 519.4) mm ≈ **800 mm**

SUMMARY OF MAIN POINTS

1. For a detailed settlement analysis, the soil profile and soil properties at the site of the structure and the stresses in the soil before and after loading are necessary.
2. The total settlement may be considered to be composed of initial settlement due to elastic compression, consolidation settlement due to primary compression and secondary settlement due to secondary compression; the latter two phenomena are restricted to cohesive soils.
3. Corrections to the computed settlement values will be necessry for the construction or loading period and for the occurrence of lateral yielding or strain; the time for assessing the time-rate of settlement is customarily reckoned from the middle of the loading period.
4. Rigidity of the structure and horizontal drainage are further factors affecting the settlement.
5. The accuracy of computed settlements is dependent upon the degree to which the inherent assumptions in the analysis are valid in a given field situation.
   Certain authorities specify permissible values for different kinds of structures and foundations, both in respect of total settlement and of differential settlement.
6. Certain remedial measures such as preconsolidation of the site and soil stabilization are possible for guarding against the occurrence of harmful settlements.
7. Settlement records are recommended to be maintained after the completion of a structure as these serve as a useful indication with regard to the accuracy of the method of analysis.
8. Contact pressure is the actual pressure transmitted from the foundation to the soil; it is uniform only for a perfectly flexible structure irrespective of the type of soil. The contact pressure distribution for a rigid structure is dependent on the nature of the soil.

9. The Terzaghi active zone in a stressed soil mass is the zone within the $0.20q$-isobar or pressure bulb; 80% of the settlement occurs due to the stress increase in this zone; the depth to which this isobar extends ($\approx 1.5B$) gives an idea of the depth to which exploratory borings should be made.

REFERENCES

1. Alam Singh and B.C. Punmia: *Soil Mechanics and Foundations*, Standard Book House, Nai Sarak, Delhi-6, 1971.
2. M. Bozozuk: *Soil shrinkage damages shallow foundations at Ottawa*, Eng. Journal, Canadian Society of Civil Engineers, 1962.
3. P.L. Capper, W.F. Cassie, and J.D. Geddes: *Problems in Engineering Soils*, S.I. Edition, E & F.N. Spon Ltd., London, 1971.
4. E. De Beer and A. Martens: *Method of computation of an upper limit for the influence of the heterogeneity of sand layers in the settlement of bridges*, Proceedings, 4th International Conference SMFE, London, 1957.
5. E.N. Fox: *The mean elastic settlement of a uniformly loaded area at a depth below the ground surface*, Proceedings, 2nd International Conference, SMFE, Rotterdam, 1948.
6. G.G. Gilboy: *Soil Mechanics Research*, Transactions, ASCE, 1933.
7. J.P. Gould: *The effect of radial flow in settlement analysis*, S.M. Thesis, Massachusetts Institute of Technology, Cambridge, Mass., U.S.A., 1946. (Unpublished).
8. IS:1904-1978: *Code of Practice for Structural Safety of Buildings: Shallow Foundations*, Second revision, ISI, New Delhi, 1978.
9. A.R. Jumikis: *Soil Mechanics*, D. Van Nostrand Co., Princeton, NJ, U.S.A., 1962.
10. A.C. Meigh and I.K. Nixon: *Comparison of in-situ tests for granular soils*, Proceedings, 5th International Conference SMFE, Paris, 1961.
11. G.G. Meyerhof: *Penetration Tests and bearing capacity of Cohesionless soils*, Proceedings, ASCE, 1956.
12. D.E. Polshin and R.A. Tokar: *Maximum Allowable Differential Settlement of Structures*, Proceedings, 4th International Conference, SMFE, London, 1957.
13. P.C. Rutledge: *Summary and closing address*, Proceedings, ASCE Settlement Conference, 1964.
14. S.B. Sehgal: *A Textbook of Soil Mechanics*, Metropolitan Book House Pvt. Ltd., Delhi-6, 1969.
15. A.W. Skempton: *The bearing capacity of clays*, Building Research Congress, UK, 1951.
16. A.W. Skempton and L. Bjerrum: *A contribution to Settlement Analysis of Foundations on Clay*, Geotechnique, 1957.
17. A.W. Skempton and R.H. MacDonald: *The allowable settlement of buildings*, Proceedings, Institution of Civil Engineers, London, 1956.
18. G.N. Smith: *Elements of Soil Mechanics for Civil and Mining Engineers*, S.I. Edition, Crosby Lockwood Staples, London, 1974.

19. G.F. Sowers: *Discussion of Maximum Allowable Differential Settlements*, Proceedings, 4th Int. Conf., SMFE, London, 1957.
20. W. Steinbrenner: *Tafeln zur Setzungsberechrung*, Schriftenreihe der strasse, Strasse, 1934.
21. D.W. Taylor: *Fundamentals of Soil Mechanics*, John Wiley & Sons, Inc., NY, USA, 1948.
22. K. Terzaghi: *The Science of Foundations*, Transactions, ASCE, 1938.
23. K. Terzaghi: *Opening discussion on Settlement of Structures*, Proceedings, first Int. Conf., SMFE, Cambridge, Mass., USA, 1936.
24. K. Terzaghi: *Settlement of Structures in Europe*, Transactions, ASCE, 1932.
25. K. Terzaghi: *Theoretical Soil Mechanics*, John Wiley & Sons, Inc., NY, USA, 1943.
26. K. Terzaghi and R.B. Peck: *Soil Mechanics in Engineering Practice*, John Wiley & Sons, Inc., NY, U.S.A., 1948.
27. S. Thornburm: *Tentative correction chart for the standard Penetration Test in non-cohesive soils*, Civil Engineering and Public Works Review, 1963.

QUESTIONS AND PROBLEMS

11.1 Write brief critical notes on 'Settlement of foundation'.
(S.V.U.—B.Tech. (Part-time)—Sept., 1983)
11.2 Write brief critical notes on 'Tolerable settlements for buildings and other structures'.
(S.V.U.—Four-year B.Tech.—Oct. & Dec., 1982)
11.3 Explain the recommended construction practices to avoid detrimental differential settlement in large structures.
(S.V.U.—B.E. (Part-time)—Dec., 1981)
11.4 Differentiate between 'total settlement' and 'differential settlement'. What are the harmful effects of differential settlement on structures? What are the possible remedial measures?
11.5 How does the construction period affect the time-rate of settlement of a structure? What is the 'effective loading period'?
11.6 (a) What is 'contact pressure'? How does it depend on the type of structure and type of soil?
(b) What is 'active zone' in soil ? Explain it with reference to the pressure bulb concept.
11.7 A reinforced concrete foundation, 20 m × 40 m, transmits a uniform pressure of 240 kN/m^2 to a soil mass, with E-value 40 kN/m^2. Determine the value of immediate settlement of the foundation.
11.8 The plan of a proposed spoil heap is shown in Fig. 11.18. The heap will stand on a thick deposit of soft clay with E-value 15 MN/m^2. The uniform pressure on the soil may be assumed as 150 kN/m^2. Estimate the immediate settlement under the point marked X at the surface of the soil.

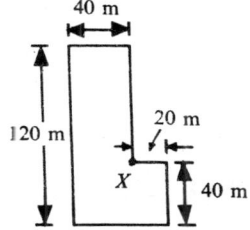

Fig. 11.18 Plan of spoil heap (Prob. 11.8)

11.9 A boring indicates the existence of a 20-metre thick clay stratum above sand. The hydrostatic pressure at the top of the clay layer is 6000 kg/m². The pore pressure at the bottom of the clay layer is reduced permanently by 6000 kg/m² by pumping. If the void ratio of clay is reduced from 1.000 to 0.975 by pumping, estimate the ultimate settlement due to this.

11.10 A clay layer 25 metres thick has a saturated unit weight of 19.2 kN/m². Ground water level occurs at a depth of 5 metres. It is proposed to construct a reinforced concrete foundation, 12.5 m × 50 m, on top of the layer, to transmit a uniform pressure of 150 kN/m².
Determine the settlement at its centre, assuming that the void ratio drops from 0.725 to 0.700 due to loading. E for the clay is 30 MN/m².

11.11 The loading period for a building extended from Aug., 1962 to Aug., 1965. The average settlement was found to be 100 mm in Aug., 1968. The ultimate settlement was expected to be 300 mm. Estimate the settlement in Aug., 1972, if there is double drainage.

11.12 Preliminary settlement analysis for a building indicated a settlement of 50 mm in 4 years and an ultimate settlement of 250 mm. The average pressure increment in the clay stratum was 30 kN/m².
If the following variations occurred in the assumptions, determine the revised value of ultimate settlement and the settlements at the end of the loading period and that at 3 years after the completion of the building.
   (i) The loading period was 2 years, which was not considered in the preliminary analysis.
   (ii) Borings indicated 25% more thickness for the clay layer than originally assumed.
   (iii) The water table got lowered permanently during construction by 1.5 metres.

12
COMPACTION OF SOIL

12.1 INTRODUCTION

'Compaction' of soil may be defined as the process by which the soil particles are artificially rearranged and packed together into a state of closer contact by mechanical means in order to decrease its porosity and thereby increase its dry density. This is usually achieved by dynamic means such as tamping, rolling, or vibration. The process of compaction involves the expulsion of air only.

In the natural location and condition, soil provides the foundation support for many structures. Besides this, soil is also extensively used as a basic material of construction for earth structures such as dams and embankments for highways and airfields. The general availability and the relatively low cost are the chief causes for using soil as construction material. Properly placed and compacted, the resulting soil mass has better strength than many natural soil formations. Such soil is referred to as a 'compacted earth fill' or a 'structural earth fill'.

For the purpose of supporting highways or buildings or for retaining water as in earth dams, the soil material must possess certain properties while in-place. These desirable features can be achieved by proper placement of an appropriate soil material. Most of these desirable qualities are associated with high unit weight (or dry density), which may be achieved by compaction.

Virtually any soil can be used for structural fill, provided it does not contain organic matter. Granular soils are capable of achieving high strength with relatively low volume changes. Properly compacted clay soils will develop relatively high strengths and low permeabilities which may be desirable features as for earth dams.

12.2 COMPACTION PHENOMENON

The process of compaction is accompanied by the expulsion of air only. In practice, soils of medium cohesion are compacted by means of rolling, while cohesionless soils are most effectively compacted by vibration. Prior to the advent of rolling equipment, earth fills were usually allowed to settle over a period of years under their own weight before the pavement or other construction was placed.

The degree of compaction of a soil is characterised by its dry density. The degree of compaction depends upon the moisture content, the amount of compactive effort or energy expended and the nature of the soil. A change in moisture content or compactive effort brings about a change in density. Thus, for compaction of soil, a certain amount of water and a certain predetermined amount of rolling are necessary.

The following are the important effects of compaction:
(*i*) Compaction increases the dry density of the soil, thus increasing its shear strength and bearing capacity through an increase in frictional characteristics;
(*ii*) Compaction decreases the tendency for settlement of soil; and,
(*iii*) Compaction brings about a low permeability of the soil.

12.3 COMPACTION TEST

To determine the soil moisture-density relationship and to evaluate a soil as to its suitability for making fills for a specific purpose, the soil is subjected to a compaction test.

Proctor (1933) showed that there exists a definite relationship between the soil moisture content and the dry density on compaction and that, for a specific amount of compaction energy used, there is a particular moisture content at which a particular soil attains its maximum dry density. Such a relationship provides a satisfactory practical approach for quality control of fill construction in the field.

12.3.1 Moisture Content—Dry Density Relationship

The relation between moisture content and dry density of a soil at a particular compaction energy or effort is shown in Fig. 12.1.

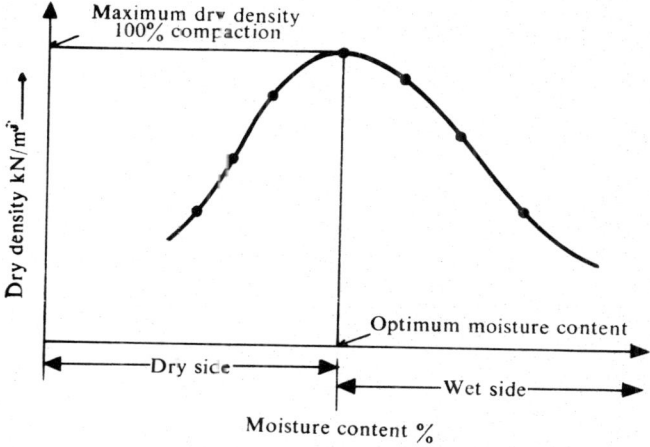

Fig. 12.1 Moisture content versus dry density at a particular compactive effort

The addition of water to a dry soil helps in bringing the solid particles together by coating them with thin films of water. At low water content, the soil is stiff and it is difficult to pack it together. As the water content is increased, water starts acting as a lubricant, the particles start coming closer due to increased workability and under a given amount of compactive effort, the soil-water-air mixture starts occupying less volume, thus effecting gradual increase in dry density. As more and more water is added, a stage is reached when the air content of the soil attains a minimum volume, thus making the dry density a maximum. The water content corresponding to this maximum dry density is called the 'optimum moisture content'. Addition of water beyond the optimum reduces the dry density because the extra water starts occupying the space which the soil could have occupied.

The curve with the peak shown in Fig. 12.1 is known as the 'moisture-content dry density curve' or the 'compactive curve'. The state at the peak is said to be that of 100% compaction at the particular compactive effort; the curve is usually of a hyperbolic form, when the points obtained from tests are smoothly joined.

The wet density and the moisture content are required in order to calculate the dry density as follows:

$$\gamma_d = \frac{\gamma}{(1+w)}, \text{ where}$$

γ_d = dry density,

γ = wet (bulk) density,

and $\qquad w$ = water content, expressed as a fraction.

11.3.2 Effect of Compactive Effort

Increase in compactive effort or the energy expended will result in an increase in the maximum dry density and a corresponding decrease in the optimum moisture content, as illustrated in Fig. 12.2.

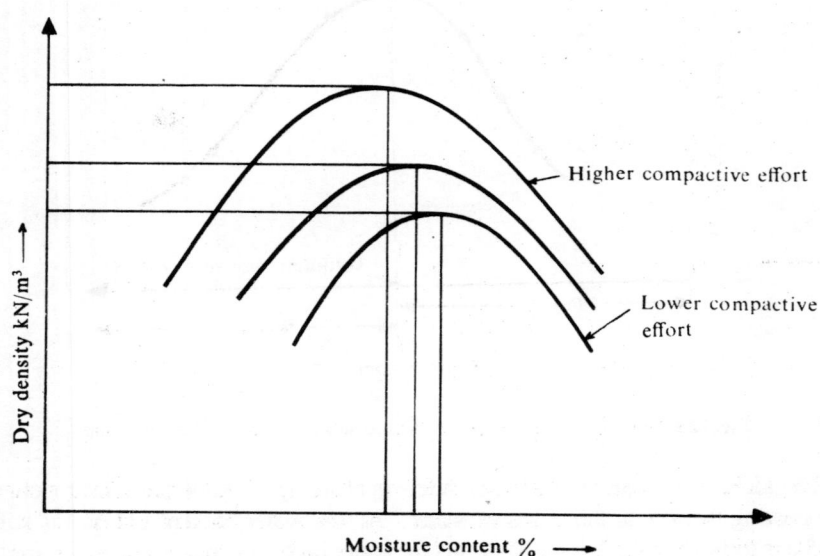

Fig. 12.2 Effect of compactive effort on compaction characteristics

Thus, for purposes of standardisation, especially in the laboratory, compaction tests are conducted at a certain specific amount of compactive effort expended in a standard manner.

12.4 SATURATION (ZERO-AIR-VOIDS) LINE

A line showing the relation between water content and dry density at a constant degree of saturation S may be established from the equation:

$$\gamma_d = \frac{G\gamma_w}{\left(1 + \frac{wG}{S}\right)}$$

Substituting $S = 95\%$, 90%, and so on, one can arrive at γ_d-values for different values of water content in %. The lines thus obtained on a plot of γ_d versus w are called 95% saturation line, 90% saturation line and so on.

If one substitutes $S = 100\%$ and plots the corresponding line, one obtains the theoretical saturation line, relating dry density with water content for a soil containing no air voids. It is said to be 'theoretical' because it can never be reached in practice as it is impossible to expel the pore air completely by compaction.

We then use

$$\gamma_d \frac{G\gamma_w}{\left(1 + \frac{wG}{100}\right)} \text{ for this situation.}$$

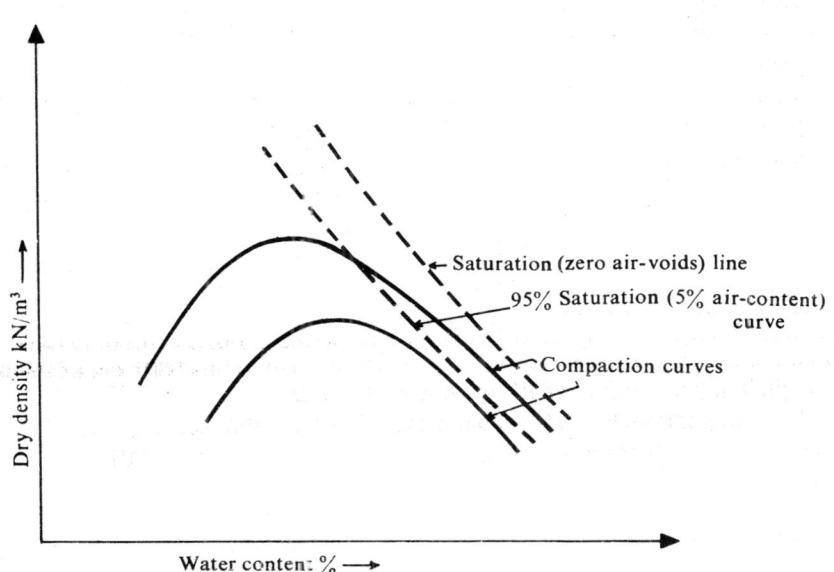

Fig. 12.3 Saturation lines superimposed on compaction curves

The saturation lines when superimposed on compaction curves give an indication of the air voids present at different points on these curves; this is shown in Fig. 12.3.

12.5 LABORATORY COMPACTION TESTS

The compaction characteristics, viz., maximum dry density and the optimum moisture content, are first determined in the laboratory. It is then specified that the unit weight achieved through compaction in the field should be a certain high percentage of the laboratory value, for quality control of the construction.

The various procedures used in the laboratory compaction tests involve application of impact loads, kneading, static loads, or vibration.

Some of the more important procedures covered are:
Standard Proctor (AASHO) Test, Modified Proctor (Modified AASHO) Test, I.S. Compaction Test, Harvard Miniature Compaction Test, Dietert Test, Abbot's Compaction Test and Jodhpur Minicompacter Test.

The primary objective of these tests is to arrive at a standard which may serve as a guide and a basis for comparison of what is achieved during compaction in the field.

12.5.1 Standard Proctor Test (AASHO Test)

This test was developed by R.R. Proctor (1933) in connection with the construction of earth dams in California (U.S.A.). The apparatus for the test consists of (*i*) a cylindrical mould of internal diameter 102 mm and an effective height of 117 mm, with a volume of 0.945 litre, (*ii*) a detachable collar of 50 mm effective height (60 mm total height), (*iii*) a detachable base plate, and (*iv*) a 50 mm diameter metal rammer of weight 2.5 kg, and a height of fall of 300 mm, moving in a metallic outer sleeve (Fig. 12.4).

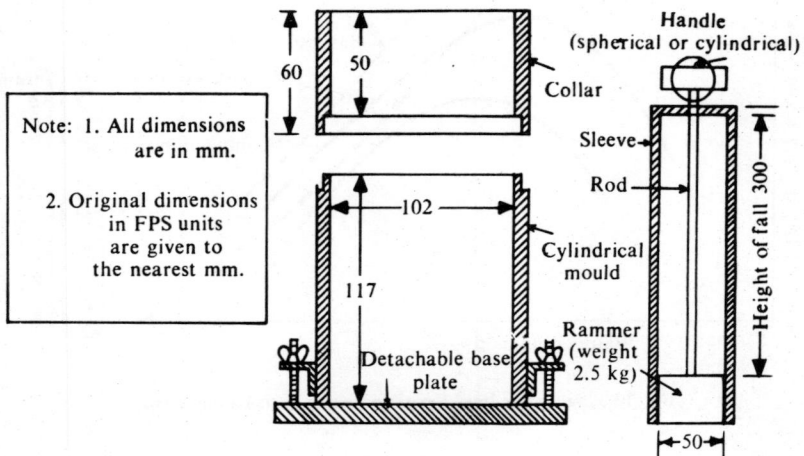

Fig. 12.4 Apparatus for standard proctor test

The test procedure consists of the following steps:
(a) About 3 kg of air-dried soil passing 20 mm sieve is taken.
(b) A reasonable amount of water is added to the soil and it is thoroughly mixed.
(c) The mould is filled with this moist soil in three equal layers to give a total compacted depth of 130 mm. Each layer is compacted by giving 25 blows with the standard rammer, pulling the rammer in the sleeve to the maximum height and then allowing it to fall freely. The position of the rammer is changed each time to distribute the compactive energy evenly to the soil. Each layer is raked with a spatula before placing fresh soil to provide proper bond.
(d) The collar is removed and the extra soil trimmed off to the top of the mould and the weight of the mould obtained. The wet weight (W) of the soil is got by subtracting the weight of the empty mould.

The bulk unit weight (γ) of the soil is obtained by dividing the wet weight of soil by the volume of the soil (V) which is the same as that of the soil.
$$\gamma = W/V$$
(e) A representative sample of the wet soil is taken and the moisture content (w %) determined in the standard manner through oven-drying.

The dry unit weight (γ_d) is obtained as
$$\gamma_d = \frac{\gamma}{\left(1 + \frac{w}{100}\right)}$$
(f) The soil is broken with hand and remixed with increased water so that the moisture content increases by 2 to 4% nearly.
(g) The test is repeated with at least six different water contents. The wet weight of the soil itself gives an indication whether the number of readings is adequate or not, because it first increases with an increase in water content, up to a certain value, and thereafter decreases. The test must be done such that this peak is established.
(h) The moisture content-dry density curve, called the 'compaction curve' is drawn.
(i) The optimum moisture content and the corresponding maximum dry unit weight are read off from the graph.

The compactive effort or energy transmitted to the soil is considered to be about 60.50 kg per 1000 cm^3 of the soil. This test has been adopted as the standard test by the AASHO (American Association of State Highway Officials) initially.

For coarse-grained soils, an initial water content of 4% and for fine-grained soils, a value of 10% are considered to be reasonable values, since the optimum value is likely to be more for the latter than that for the former.

12.5.2 Modified Proctor Test (Modified AASHO Compaction Test)

This test was developed to deliver greater compactive effort with a view to simulating the heavier compaction required for the construction of airport pavements. The mould used is almost the same as that for Standard Proctor Test but with an effective height of 127 mm. The weight of the rammer is 4.5 kg and the height of fall is 450 mm. The mould is filled in five layers, each layer being compacted with 25 blows.

The compactive energy delivered is of the order of 272.60 kgm per 1000 cm^3 of soil, which is about 4.5 times that of the Standard Proctor Test.

The moisture content-dry density relationship may be obtained by adopting a similar procedure as in the previous case. Since the compactive effort is more for this test than for the Standard Proctor Test, the compaction curve in this case may be expected to lie higher when superimposed over the curve for the latter, with the peak placed to the left.

12.5.3 Indian Standard Compaction Tests

Indian Standards specify, among the methods of test for soils, procedures for compaction tests using light compaction [IS: 2720 (Part VII)—1974] and using heavy compaction [IS: 2720 (Part VIII)—1974].

For light compaction, a 2.6 kg rammer falling through a height of 310 mm is used, while, for heavy compaction, a 4.89 kg. rammer falling through a height of 450 mm is used.

Figure 12.5 shows the details of typical mould for compaction and Fig. 12.6(*a*) and (*b*) show the details of a typical metal rammers for light compaction and for heavy compaction respectively.

A representative sample weighing about 20 kg and passing 50-mm IS Sieve of the thoroughly mixed air-dried material is taken. This is made to pass through 20-mm and 4.75-mm IS Sieves, separating the fractions retained and passing these sieves. Care should be exercised so as not to break the aggregates while pulverizing. The percentage of each fraction is determined. The fraction retained on 20-mm IS Sieve should not be used in the test. The percentages of soil coarser than 4.75-mm IS Sieve and 20-mm IS Sieve should be determined.

The ratio of fraction passing 20-mm IS Sieve and retained on 4.75-mm IS Sieve to the soil passing 4.75-mm IS Sieve should be determined. The material retained on and passing 4.75-mm IS Sieve should be mixed thoroughly to obtain about 16 to 18 kg of soil specimen. (In case the material passing 20-mm IS Sieve and retained on 4.75-mm IS Sieve is more than 20%, the ratio of such material to that passing 4.75-mm IS Sieve should be maintained for each test. If it is less than 20%, the material passing the 20-mm IS Sieve may be directly used).

Enough water should be added to the specimen to bring its moisture content to about 7% (sandy soils) and 10% (clayey soils) less than the estimated optimum moisture content. The processed soil should be kept in an airtight tin for about 18 to 20 h to ensure thorough mixing of the water with the soil.

The wet soil shall be compacted into the mould in three equal layers (five equal layers for heavy compaction), each layer being given 25 blows if the 100 mm diameter mould is used (or 56 blows if the 150-mm diameter mould is used) from the rammer weighing 2.6 kg dropping from a height of 310 mm (from the rammer weighing 4.89 kg dropping from a height of 450 mm, for heavy compaction).

The rest of the procedure and usual precautions are as for other compaction tests (for full details, *vide* the relevant IS Code).

Compaction of Soil 497

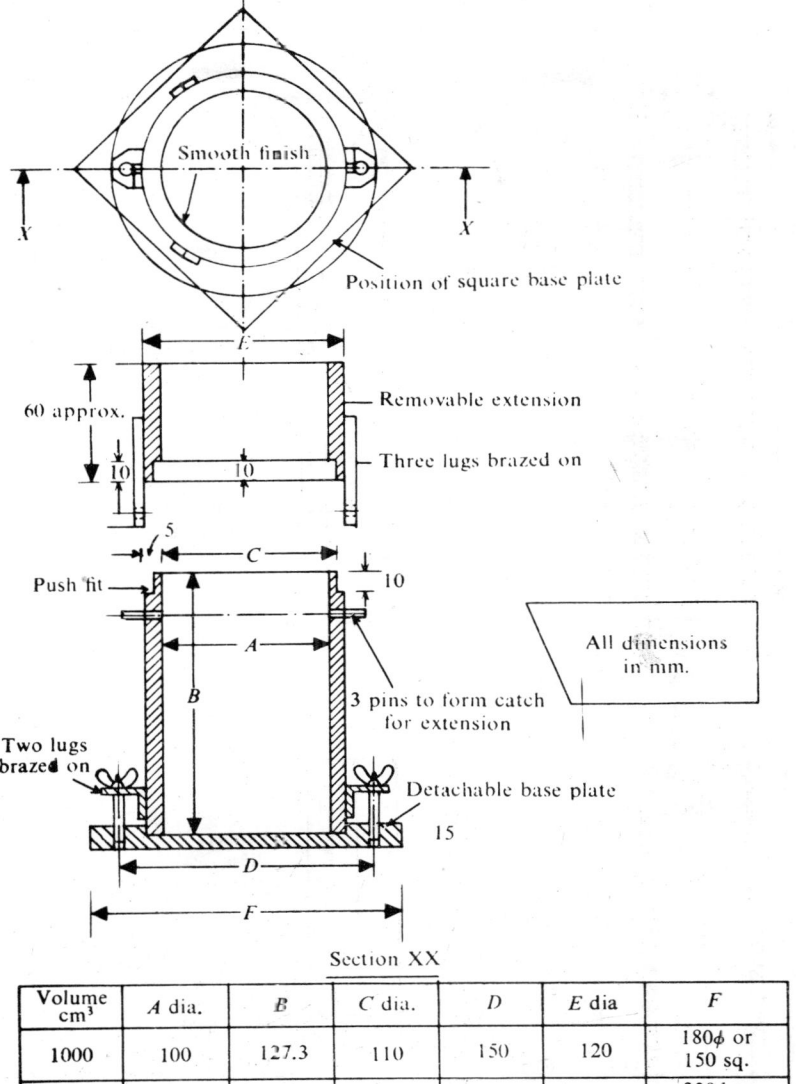

Volume cm³	A dia.	B	C dia.	D	E dia	F
1000	100	127.3	110	150	120	180φ or 150 sq.
2250	150	127.3	160	200	170	230φ or 200 sq.

Fig. 12.5 Typical mould for compaction (ISI)

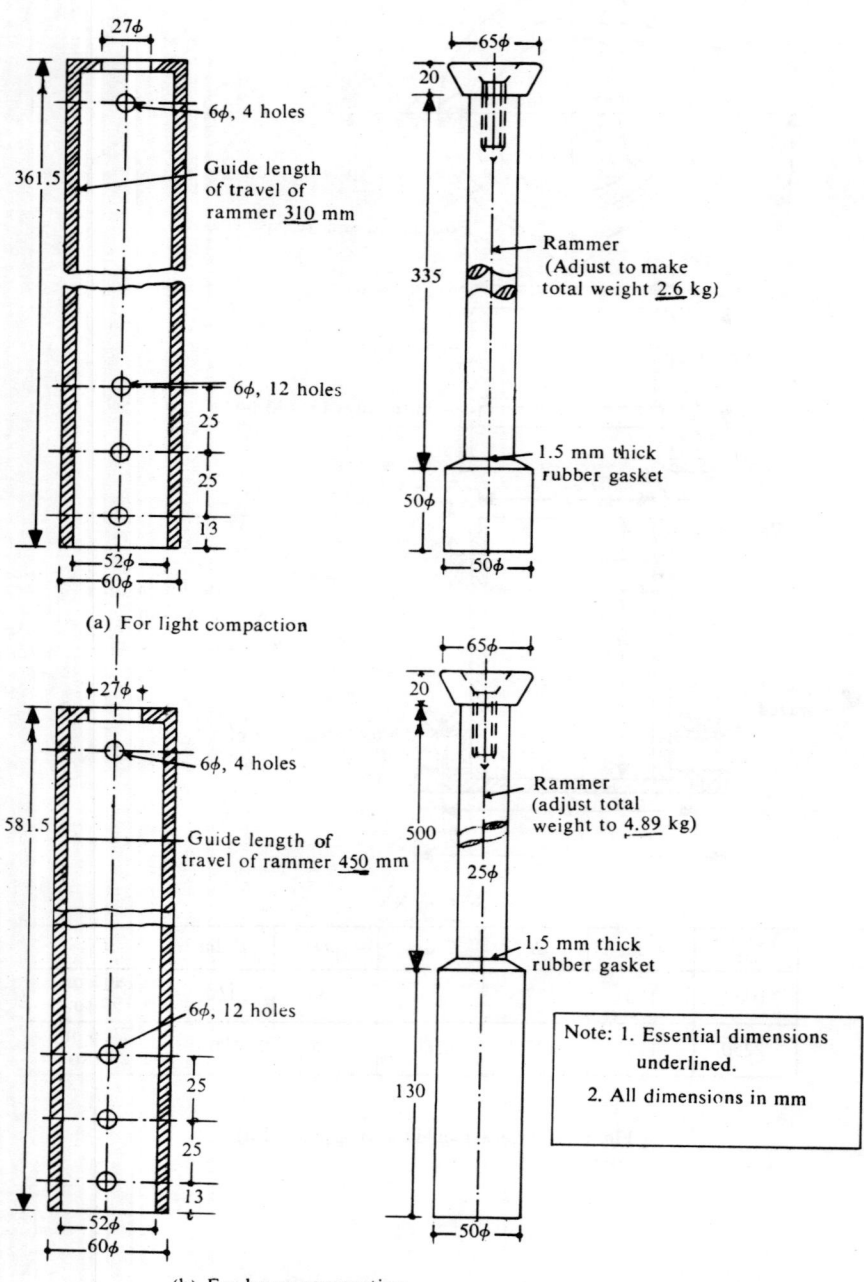

Fig. 12.6 Typical metal rammer for compaction (ISI)

Correction for oversize fraction may be applied as follows:

If the material retained on 20-mm IS Sieve (or 4.75 mm IS Sieve) has been excluded from the test, the following corrections shall be applied for getting the values of maximum dry density and optimum moisture content for the entire soil. For this purpose, the specific gravity of the portion retained and passing the 20-mm IS Sieve or the 4.75 mm IS Sieve, as the case may be, should be determined separately.

$$\text{Corrected maximum dry density} = \frac{\gamma_s \gamma_{d_{max}}}{n_1 \gamma_{d_{max}} + n_2 \cdot \gamma_s} \qquad \ldots(\text{Eq. 12.1})$$

Corrected optimum moisture content = $n_1 A_0 + n_2 w_0$ ...(12.2)

where γ_s = unit mass of oversize gravel particles in g/cm^3 (= $G \cdot \gamma_w$, where G is the specific gravity of gravel particles);

$\gamma_{d_{max}}$ = maximum dry density obtained in the test in g/cm^3;

n_1 = fraction by mass of the oversize particles in the total soil expressed as ratio;

n_2 = fraction by mass of the portion passing 20-mm IS Sieve (or 4.75 mm IS Sieve) expressed as the ratio of total soil;

A_0 = water absorption capacity of oversize material, if any, expressed as percentage of water absorbed, and

w_0 = optimum moisture content obtained in the test.

(This formula is based on the assumption that the volume of a compacted portion passing a 20-mm sieve (or a 4.75-mm sieve) is sufficient to fill the voids between the oversize particles).

12.5.4 Harvard Miniature Compaction Test

The compaction in this test is achieved by 'kneading action' of a cylindrical tamper 12.7 mm in diameter. The mould is 33.34 mm in diameter and 71.53 mm in height and has a volume of 62.4 cm^3. The tamper operates through a present compression spring so that the tamping force is controlled not to exceed a certain predetermined value. For different soils and different compactive efforts desired, the number of layers, number of tamps per layer and the tamping force may be varied.

12.5.5 Abbot's Compaction Test

A metal cylinder, 52 mm internal diameter and 400 mm effective height, with a base is used. 200 g of oven-dried soil is mixed with water and compacted in the cylindrical mould with a 50 mm diameter rammer of 2.5 kg weight falling through a height of 350 mm. The number of blows is decided by calibration with respect to Proctor's compaction or field compaction.

The height of the compacted specimen may be determined from the reading of the graduated stem of the rammer. The volume of the compacted specimen is calculated from the known cross-section and height. The wet unit weight may be obtained and also the dry unit weight from the known dry weight of the sample.

The compaction curve is obtained in the usual manner.

12.5.6 Dietert's Test

The apparatus consists of a 50.8 mm diameter mould supported on a metal base. Compaction is done by means of a piston on which a cylindrical weight of 8.165 kg operated by a cam, falls through a height of 50.8 mm.

Air dried soil weighing 150 g and passing through 2.36 mm IS Sieve, is mixed with water and compacted in the mould by application of 10 blows. The mould is inverted and another 10 blows are applied. The weight and length of the compacted soil cylinder are determined, from which the volume and bulk unit weight of the soil are obtained. The water content is determined by oven-drying. The test is repeated with different water contents and the compaction characteristics are established from a graph.

12.5.7 Jodhpur Mini-compactor Test

The Jodhpur Mini-compactor (Singh, 1965) consists of a cylindrical mould of cross-sectional area 50 cm^2 (79.8 mm diameter and 60 mm effective height) and a volume of 300 ml, and a ramming tool with a 2.5 kg drop weight (DRT which means "Dynamic Ramming Tool"), falling through 250 mm. The drop weight falls over a cylindrical base, 40 mm in diameter and 75 mm in height. The soil is compacted in two layers, each layer being given 15 blows with the ramming tool.

The compactive energy transmitted is 62.50 kgm per 1000 cm^3 of soil. Calibration tests in the laboratory (Singh and Punmia, 1965) have shown that the maximum dry unit weights and optimum moisture contents obtained from the Jodhpur mini-compactor are comparable to those obtained from the Standard Proctor Test.

12.6 IN-SITU OR FIELD COMPACTION

As indicated in Sec. 12.1, in any type of construction job which requires soil to be used as a foundation material or as a construction material, compaction in-situ or in the field is necessary.

The construction of a structural fill usually consists of two distinct operations —placing and spreading in layers and then compaction. The first part assumes greater significance in major jobs such as embankments and earth dams where the soil to be used as a construction material has to be excavated from a suitable borrow area and transported to the work site. In this phase large earth moving equipment such as self-propelled scrapers, bulldozers, graders and trucks are widely employed.

The phase of compaction may be properly accomplished by the use of appropriate equipment for compaction. The thickness of layers that can be properly compacted is known to be related to the type of soil and method or equipment of compaction. Generally speaking, granular soils can be adequately compacted in thicker layers than fine-grained soils and clays; also, for a given soil type, heavy compaction equipment is capable of compacting thicker layers than light equipment.

Although the principle of compaction in the field is relatively simple, it may turn out to be a complex process if the soil in the borrow area is not at the desired optimum moisture content for compaction. The existing moisture content is to be determined and water added, if necessary. Addition of water to the soil is normally done either during excavation or transport and rarely on the construction spot; however, water must be added before excavation in the case of clayey soils. In case the soil has more moisture content than is required for proper compaction, it has to be air-dried after excavation and compacted as soon as the desired moisture content is attained.

Soil compaction or densification can be achieved by different means such as tamping action, kneading action, vibration, and impact. Compactors operating on the tamping, kneading and impact principle are effective in the case of cohesive soils, while those operating on the kneading, tamping and vibratory principle are effective in the case of cohesionless soils.

The primary types of compaction equipment are: (*i*) rollers, (*ii*) rammers and (*iii*) vibrators. Of these, by far the most common are rollers.
Rollers are further classified as follows:
  (*a*) Smooth-wheeled rollers,
  (*b*) Pneumatic-tyred rollers,
  (*c*) Sheepsfoot rollers, and
  (*d*) Grid rollers.

Vibrators are classified as: (*a*) Vibrating drum, (*b*) Vibrating pneumatic tyre (*c*) Vibrating plate, and (*d*) Vibroflot.

The maximum dry density sought to be achieved *in-situ* is specified usually as a certain percentage of the value obtainable in the laboratory compaction test. Thus control of compaction in the field, requires the determination of *in-situ* unit weight of the compacted fill and also the moisture content.

The methods available for the determination of *in-situ* unit weight are:
  (*a*) Sand-replacement method, (*b*) Core-cutter method, (*c*) Volumenometer method, (*d*) Rubber balloon method, (*e*) Nuclear method, (*f*) Proctor plastic needle method. (All these except the last have been dealt with in Chapter 3).

Rapid methods of determination of moisture content such as the speedy moisture tester are adopted in this connection. Some of the above aspects are dealt with in the following sub-sections.

12.6.1 Types of In-situ Compaction Equipment

Certain types of *in-situ* compaction equipment are described below:

Rollers

(*a*) *Smooth-wheeled rollers:* This type imparts static compression to the soil. There may be two or three large drums; if three drums are used, two large ones in the rear and one in the front is the common pattern. The compaction pressures are relatively low because of a large contact area. This type appears to be more suitable for compacting granular base courses and paving mixtures for highway and airfield work rather than for compacting earth fill. The relatively smooth surface obtained acts as a sort of a 'seal' at the end of a day's work and drains off rain water very well. The roller is self-propelled by a diesel engine and has a weight distribution that can be altered by the addition of ballast to the rolls. The common weight is

80 kN to 100 kN (8 to 10 t), although the range may be as much as 10 kN to 200 kN (1 to 20 t). The pressure may be of the order of 300 N (30 kg) per lineal cm of the width of rear rolls. The number of passes varies with the desired compaction; usually eight passes may be adequate to achieve the equivalent of standard Proctor compaction.

(b) *Pneumatic-tyred rollers:* This type compacts primarily by kneading action. The usual form is a box or container—mounted on two axles to which pneumatic-tyred wheels are fitted; the front axle will have one wheel less than the rear and the wheels are mounted in a staggered fashion so that the entire width between the extreme wheels is covered. The weight supplied by earth ballast or other material placed in the container may range from 120 kN (12 t) to 450 kN (45 t), although an exceptionally heavy capacity of 2000 kN (200 t) may be occasionally used. Some equipment is provided with a "Wobble-wheel" effect, a design in which a slightly weaving path is tracked by the travelling wheels; this facilitates the exertion of a steady pressure on uneven ground, which is very useful in the initial stages of a fill.

The weight of the roller as well as the contact pressure is an important parameter for the performance; the latter may be varied from 0.20 to 1 N/mm^2 (2 to 10 kg/cm^2) through the adjustment of air pressure in the tyres. Although this type has originated as a towed unit, self-propelled units are also available. The number of passes required is similar to that with smooth wheeled-rollers.

This type is suitable for compacting most types of soil and has particular advantages with wet cohesive materials.

(c) *Sheepsfoot rollers:* This type of roller consists of a hollow steel drum provided with projecting studs or feet; the compaction is achieved by a combination of tamping and kneading. The drum can be filled with water or sand to provide and control the dead weight. As rolling is done, most of the roller weight is imposed through the projecting feet. (Fig. 12.7):

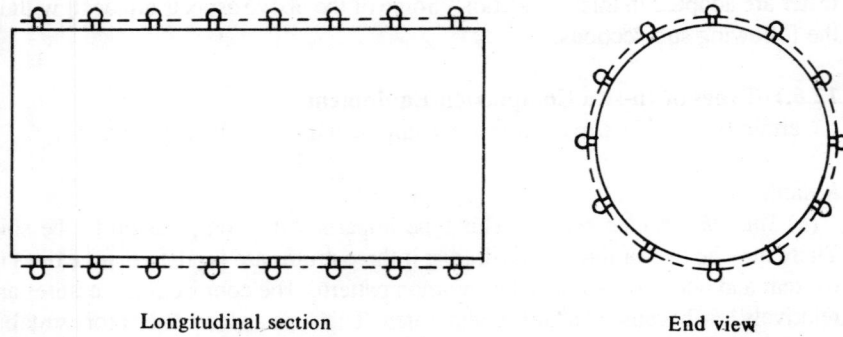

Longitudinal section   End view

Fig. 12.7 Sheepsfoot Roller

It is generally used as a towed assembly with the drums mounted either singly or in pairs; self-propelled units are also available.

Compaction of Soil 503

The feet are usually either club-shaped (100 × 75 mm) or tapered (57 × 57 mm), the number on a 50 kN (5 t) roller ranging from 64 to 88. The contact pressures of the feet may range from 700 kN/m² (7 kg/cm²) to 4200 kN/m² (42 kg/cm²) and weight per drum from 25 kN (2.5 t) to 130 kN (13 t).

Initially, the projections sink into the loose soil and compact the soil near the lowest portion of the layer. In subsequent passes with the roller, the zone of compaction continues to rise until the surface is reached, when the roller is said to "Walk-out".

The length of the studs, the contact area and the weight of roller are related to the roller performance.

This type of roller is found suitable for cohesive soils. It is unsuitable for granular soils as the studs tend to loosen these continuously. The tendency of void formation is more in soils compacted with sheepsfoot rollers.

(d) *Grid rollers:* This type consists of rolls made from 38 mm steel bars at 130 mm centres, with spaces of 90 mm square. The weight of the roller ranges from 55 kN (5.5 t) to 110 kN (11 t). This is usually a towed unit which is suitable for many types of soil including wet clays and silts.

Rammers

This type includes the dropping weight type and pneumatic and internal commission type, which are also called 'frog rammers'. They weigh up to about 1.5 kN (150 kg) and even as much as 10 kN (1 t) occasionally.

This type may be used for cohesionless soils, especially in small restricted and confined areas such as beds of drainage trenches and back fills of bridge abutments.

Vibrators

These are vibrating units of the out-of-balance weight type or the pulsating hydraulic type. Such a type is highly effective for cohesionless soils. Behind retaining walls where the soil is confined, the backfill, much deeper in thickness, may be effectively compacted by vibration type of compactors.

A few of this type are dealt with below:

(a) *Vibrating drum:* A separate motor drives an arrangement of eccentric weights so as to cause a high-frequency, low-amplitude, vertical oscillation to the drum. Smooth drums as well as sheepsfoot type of drums may be used. Layers of the order of 1 metre deep could be compacted to high densities.

(b) *Vibrating pneumatic tyre*: A separate vibrating unit is attached to the wheel axle. The ballast box is suspended separately from the axle so that it does not vibrate. A 300 mm thick layer of granular soil will be satisfactorily compacted after a few passes.

(c) *Vibrating plate:* This typically consists of a number of small plates, each of which is operated by a separate vibrating unit. These have a limited depth of effectiveness and hence are used in compacting granular base courses for highway and airfield pavements.

(d) *Vibroflot:* A method suited for compacting thick deposits of loose sandy soil is called the 'vibroflotation' process. The improvement of density is restricted to the surface zone in the case of conventional compaction equipment. The vibroflotation method first compacts deep zone in the soil and then works its way towards the surface. A cylindrical vibrator weighing about 20 kN (2 t) and approximately 400 mm in diameter and 2 m long, called the 'Vibroflot', is suspended from a crane and is jetted to the depth where compaction is to start.

The jetting consists of a water jet under pressure directed into the earth from the tip of the vibroflot; as the sand gets displaced, the vibroflot sinks into the soil. Depths up to 12 m can be reached. After the vibroflot is sunk to the desired depth, the vibrator is activated. The compaction of the soil occurs in the horizontal direction up to as much as 1.5 m outward from the vibroflot. Vibration continues as the vibroflot is slowly raised toward the surface. As this process goes on, additional sand is continually dropped into the space around the vibroflot to fill the void created. To densify the soil in a given site, locations at approximately 3-m spacings are chosen and treated with vibroflotation.

12.6.2 Control of Compaction in the Field

Control of compaction in the field consists of checking the water content in relation to the laboratory optimum moisture content and the dry unit weight achieved *in-situ* in relation to the laboratory maximum dry unit weight from a standard compaction test. Typically, each layer is tested at several random locations after it has been compacted.

Several methods are available for the determination of *in-situ* unit weight and moisture content and these have been considered in some detail in Chapter 3. The common approaches for the determination of unit weight are the core-cutter method and sand-replacement method. A faster method is what is known as the Proctor needle method, which may be used for the determination of *in-situ* unit weight as well as *in-situ* moisture content.

The required density can be specified either by 'relative compaction' (also called 'degree of compaction') or by the final air-void content. Relative compaction means the ratio of the *in-situ* dry unit weight achieved by compaction to the maximum dry unit weight obtained from an appropriate standard compaction test in the laboratory. Usually, the relative compaction of 90 to 100% (depending upon the maximum laboratory value), corresponding to about 5 to 10% air content, is specified and sought to be achieved. Typical values of dry unit weights achieved may be as high as 22.5 kN/m^3 (2250 kg/m^3) for well-graded gravel and may be as low as 14.4 kN/m^3 (1440 kg/m^3) for clays. Approximate ranges of optimum moisture content may be 6 to 10% for sands, 8 to 12% for sand-silt mixtures, 11 to 15% for silts and 13 to 21% for clays (as got from modified AASHO tests).

A variation of 5 to 10% is allowed in the field specification of dry unit weight at random locations, provided the average is about the specified value.

Proctor Needle

The Proctor needle approach given here, is an efficient and fast one for the simultaneous determination of *in-situ* unit weight and *in-situ* moisture content, it is also called 'penetration needle'. The Proctor needle apparatus is shown in Fig. 12.8.

Compaction of Soil 505

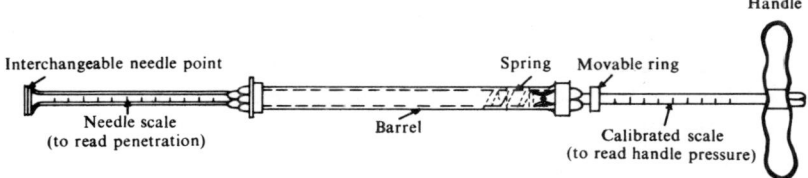

Fig. 12.8 The proctor needle

The apparatus basically consists of a needle attached to a spring-loaded plunger through a shank. An array of interchangeable needle tips is available, ranging from 6.45 to 645 mm^2, to facilitate the measurement of a wide range of penetration resistance values. A calibration of penetration against dry unit weight and water content is obtained by pushing the needle into specially prepared samples for which these values are known and noting the penetration. The penetration of the needle and the penetration resistance (load applied) may be shown on a graduated scale on the shank and the stem of handle respectively.

A sample calibration curve is shown in Fig. 12.9.

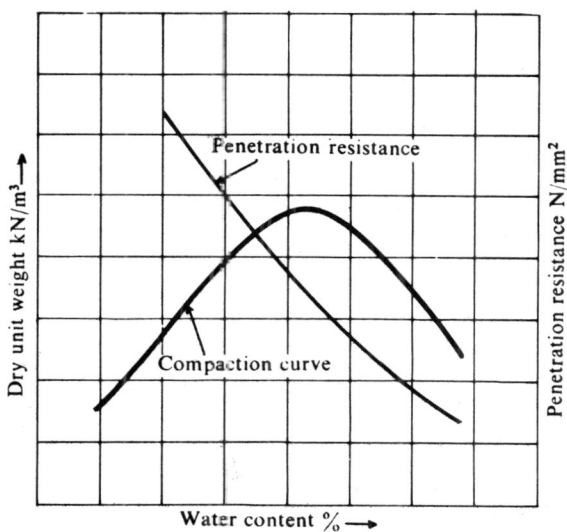

Fig. 12.9 Calibration curve for proctor needle

The procedure for the use of the Proctor 'plasticity' needle, as it is called, is obvious. The spring-loaded plunger is pressed into the compacted layer in the field with an appropriate plasticity needle. The penetration resistance is recorded for a standard depth of penetration at a standard time-rate of penetration. Against this penetration resistance, the corresponding values of water content and dry unit weight are obtained from the calibration curve.

The size of the needle to be chosen depends upon the type of soil such that the resistance to be read is neither too large nor too small.

The Tennessee Valley Authority (TVA) engineers had devised a similar device, which is called the TVA 'Penetrometer'.

12.7 COMPACTION OF SAND

The compaction characteristics of cohesionless and freely draining sands are somewhat different from those of cohesive soils.

A typical pattern of the moisture-density relationship for a cohesionless, freely-draining sand from a laboratory test will be somewhat as shown in Fig. 12.10.

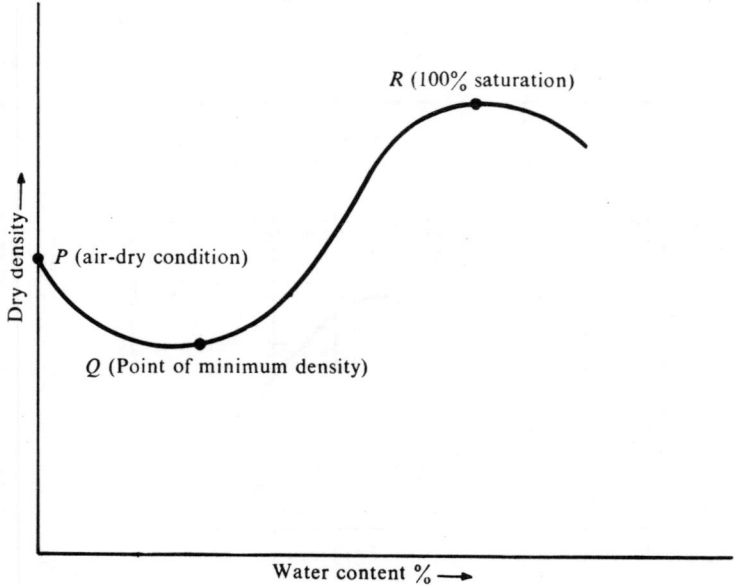

Fig. 12.10 Typical moisture-density relationship for sand

Small moisture films around the grains tend to keep them apart and can decrease the density up to a certain water content. The point Q on the curve indicates the minimum density. Later on, the apparent cohesion gets reduced as the water content increases and is destroyed ultimately at 100% saturation of the sand. Thus, the point R on the curve indicates maximum density. Thereafter, once again, the density decreases with increase in water content.

Increase of compactive effort has much less effect in the case of cohesionless soils than on cohesive soils. Vibration is considered to be the best method suitable for densifying cohesionless soils, which are either fully dry or fully saturated. This is because the stresses at the soil water menisci tend to prevent full densification. Also, relative density or density index is invariably used to indicate relative compaction or densification of sand.

12.8 COMPACTION VERSUS CONSOLIDATION

Compaction, as a phenomenon, is different from the phenomenon of consolidation of soil.

The primary differences between the two phenomena may be set out as given in Table 12.1.

Table 12.1 Compaction versus consolidation

S.No.	Compaction	Consolidation
1.	Expulsion of pore air	Expulsion of pore water
2.	Soil involved is partially saturated	Fully saturated soil
3.	Applies to cohesive as well as cohesionless soils	Applies to cohesive soils only
4.	Brought about by artificial or human agency	Brought about by application of load or by natural agencies
5.	Dynamic loading is commonly applied	Static loading is commonly applied
6.	Improves bearing power and settlement characteristics	Improves bearing power and settlement characteristics
7.	Relatively quick process	Relatively slow process
8.	Relatively complex phenomenon involving expulsion, compression, and dissolution of pore air-in water	Relatively simple phenomenon
9.	Useful primarily in embankments and earth dams	Useful as a means of improving the properties of foundation soil

12.9 ILLUSTRATIVE EXAMPLES

Example 12.1 : An earth embankment is compacted at a water content of 18% to a bulk density of 1.92 g/cm³. If the specific gravity of the sand is 2.7, find the void ratio and the degree of saturation of the compacted embankment.

(S.V.U. — B.Tech. (Part-time)—Sept, 1983)

Water content, $w = 18\%$
Bulk density, $\gamma = 1.92$ g/cm³
Specific gravity, $G = 2.7$
Dry density, $\gamma_d = \dfrac{\gamma}{(1+w)} = \dfrac{1.92}{(1+0.18)} = 1.627$ g/cm³

But $\gamma_d = \dfrac{G \cdot \gamma_w}{(1+e)}$, where $\gamma_w = 1$ g/cm^3

∴ $(1+e) = \dfrac{2.7 \times 1}{1.627} = 1.66$

Void ratio, $e = \mathbf{0.66}$

Also, $wG = S \cdot e$

∴ The degree of saturation, $S = \dfrac{wG}{e} = \dfrac{0.18 \times 2.7}{0.66}$

$= 0.7364$

∴ The degree of saturation $= \mathbf{73.64\%}$

Example 12.2 : A moist soil sample compacted into a mould of 1000 cm^3 capacity and weight 3.50 mg, weighs 5.35 kg with the mould. A representative sample of soil taken from it has an initial weight of 18.70 g and even dry weight of 16.91 g. Determine (*a*) water content, (*b*) wet density, (*c*) dry density, (*d*) void ratio and (*e*) degree of saturation of the sample.

If the soil sample is so compressed as to have all air expelled, what will be the new volume and new dry density?

(S.V.U. — Four year B.Tech. —Sept., 1983)

(*a*) Water content, $w = \dfrac{(18.70 - 16.91)}{16.91} \times 100 = \mathbf{10.58\%}$

Wet wt. of soil in the mould $= (5.35 - 3.50) = 1.85$ kg

Volume of mould $= 1000$ cm^3

(*b*) Bulk density, $\gamma = \dfrac{1.85 \times (100)^3}{1000}$ kg/m^3 $= \mathbf{1850\ kg/m^3}$

(*c*) Dry density, $\gamma_d = \dfrac{\gamma}{(1+w)} = \dfrac{1850}{(1+0.1058)} = \mathbf{1673\ kg/m^3}$

$\gamma_d = \dfrac{G\gamma_w}{(1+e)}$

Assuming a value of 2.65 for grain specific gravity,

$1673 = \dfrac{2.65 \times 1000}{(1+e)}$, since $\gamma_w = 1000$ kg/m^3.

$(1+e) = \dfrac{2.65 \times 1000}{1673} = 1.584$

(*d*) Void ratio, $e = \mathbf{0.584}$

Also, $wG = S \cdot e$

$S = \dfrac{wG}{e} = \dfrac{0.1058 \times 2.65}{0.584} = 0.48$

(e) Degree of saturation, $S = 48\%$
If air is fully expelled, the soil is fully saturated at that water content.
$\therefore wG = e = 2.65 \times 0.1058 = 0.28$

New dry density $= \dfrac{2.65 \times 1000}{(1 - 0.28)} = 2070 \text{ kg/m}^3$

Now volume $= \dfrac{(1 + 0.28)}{(1 + 0.584)} \times 1000 = 808 \text{ cm}^3$

Example 12.3 : The soil in a borrow pit has a void ratio of 0.90. A fill-in-place volume of 20,000 m³ is to be constructed with an in-place dry density of 18.84 kN/m³. If the owner of borrow area is to be compensated at Rs. 1.50 per cubic metre of excavation, determine the cost of compensation.
(S.V.U. — B.E., (R.R.)—Nov., 1975)

In-place dry density $= 18.84 \text{ kN/m}^3$
Assuming grain specific gravity as 2.70 and taking γ_w as 9.81 kN/m³,

$$18.84 = \dfrac{2.70 \times 9.81}{(1 + e_i)}$$

$$(1 + e_i) = \dfrac{2.70 \times 9.81}{18.84} = 1.406$$

$$e_i = 0.406 \quad (\text{in-place Void ratio})$$

Void-ratio of the soil in the borrow-pit,
$$e_b = 0.90$$
In-place volume of the fill, $V_i = 20,000 \text{ m}^3$
If the volume of the soil to be excavated from the borrow pit is V_b, then:

$$\dfrac{V_b}{V_i} = \dfrac{1 + e_b}{1 + e_i} = \dfrac{(1 + 0.90)}{(1 + 0.406)}$$

$$\therefore V_b = \dfrac{1.90}{1.406} \times 20,000 \text{ m}^3 = 27.027 \text{ m}^3$$

The cost of compensation to be paid to the owner of the borrow area = Rs. (1.50 × 27,027) = Rs. **40,540**.

Example 12.4 : A soil in the borrow pit is at a dry density of 17 kN/m³ with a moisture content of 10%. The soil is excavated from this pit and compacted in an embankment to a dry density of 18 kN/m³ with a moisture content of 15%. Compute the quantity of soil to be excavated from the borrow pit and the amount of water to be added for 100 m³ of compacted soil in the embankment.
(S.V.U. — B.E. (Part-time)—Apr., 1982)

510 Geotechnical Engineering

Volume of compacted soil = 100 m³
Dry density of compacted soil = 18 kN/m³
Weight of compacted dry soil = 100 × 18 = 1800 kN
This is the weight of dry soil to be excavated from the borrow pit.
Weight of wet soil to be excavated = 1800 (1 + w) = 1800 (1 + 0.10) = 1980 kN.
Wet density of soil in the borrow pit = 17 (1 + 0.10) = 18.7 kN/m³

Volume of wet soil to be excavated = $\dfrac{1980}{18.7}$ = 105.9 m³

Moisture present in the wet soil, in the borrow pit for every 100 m³ of compacted soil = 1800 × 0.10 = 180 kN
Moisture present in the compacted soil of 100 m³
$$= 1800 \times 0.15 = 270 \text{ kN}$$
Weight of water to be added for 100 m³ of compacted soil
$$= (270 - 180) \text{ kN} = 90 \text{ kN}$$
$$= \dfrac{90}{9.81} \text{ m}^3 = 9.18 \text{ kl}$$

Example 12.5 : The following data have been obtained in a standard laboratory Proctor compaction test on glacial till:

Water content %	5.02	8.81	11.25	13.05	14.40	19.25
Weight of container and compacted soil (kg)	3.580	3.730	3.932	4.000	4.007	3.907

The specific gravity of the soil particles is 2.77. The container is 944 cm³ in volume and its weight is 1.978 kg. Plot the compaction curve and determine the optimum moisture content. Also compute the void ratio and degree of saturation at optimum condition.

(S.V.U. — B.E., (R.R.)—Nov., 1973)

The dry density values are computed and shown in Table 12.2.

Table 12.2 Computation of dry density

Water Content %	5.02	8.81	11.25	13.05	14.40	19.25
Weight of container and compacted soil (kg)	3.580	3.730	3.932	4.000	4.007	3.907
Weight of container (kg)	1.978	1.978	1.978	1.978	1.978	1.978
Weight of compacted soil (kg)	1.602	1.752	1.954	2.022	2.029	1.929
Volume of container (cm³)	944	944	944	944	944	944
Bulk density of compacted soil (g/cm³)	1.697	1.856	2.070	2.142	2.149	2.043
Dry density of compacted soil (g/cm³)	1.616	1.706	1.861	1.895	1.878	1.713

The compaction curve is shown in Fig. 12.11.

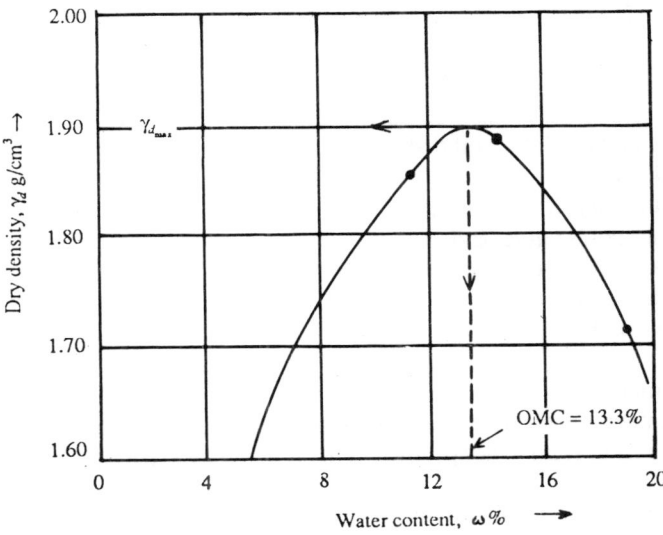

Fig. 12.11 Compaction curve (Ex. 12.5)

From the curve,
 Optimum moisture content = 13.3%
 Maximum dry density = 1.90 g/cm³
At the optimum condition, if the void ratio is e_0.

$$\gamma_{d_{max}} = \frac{G \cdot \gamma_w}{(1+e_0)}$$

$$\therefore 1.90 = \frac{2.77 \times 1}{(1+e_0)}$$

$$(1+e_0) = \frac{2.77}{1.90} = 1.46 \text{ (aprox.)}$$

$\therefore$ Void ratio = **0.46**

Since $S \cdot e = w \cdot G$,

Degree of saturation, $S = \dfrac{w \cdot G}{e}$

$$= \frac{13.3 \times 2.77}{0.46}\% = \mathbf{80.1\%}$$

Example 12.6 : Given standard soil compaction test results as follows:

Trial No.	Moisture content % by dry weight	Wet weight of compacted soil (kN/m³)
1	8.30	19.8
2	10.50	21.3
3	11.30	21.6
4	13.40	21.2
5	13.80	20.8

The specific gravity of the soil particles is 2.65.
Plot the following:
(a) Moisture-dry density curve,
(b) Zero air voids curve, and
(c) Ten per cent air content curve. (90% Saturation curve)
Determine the optimum moisture content and the corresponding maximum dry density of the soil.

(S.V.U. — B.E. (Part-time)—Dec., 1981)

Also determine the correct values of the maximum dry density and optimum moisture content if in the above test, the material retained on 20 mm Sieve, which was 9%, was eliminated. The specific gravity of these oversize particles was 2.79.

The dry density values are calculated from $\gamma_d = \dfrac{\gamma}{\left(1 + \dfrac{w}{100}\right)}$, and are shown in Table 12.3.

Table 12.3 Dry density values

Trial No.	Moisture content, w%	Wet unit weight (kN/m³)	Dry unit weight (kN/m³)
1	8.30	19.8	18.28
2	10.50	21.3	19.28
3	11.30	21.6	19.41
4	13.40	21.2	18.70
5	13.80	20.8	18.28

Since $\gamma_d = \dfrac{G \cdot \gamma_w}{(1+e)} = \dfrac{G \cdot \gamma_w}{\left(1 + \dfrac{wG}{S}\right)} = \dfrac{2.65 \times 9.81}{\left(1 + \dfrac{w \times 2.65}{S}\right)}$,

for Zero air-voids condition, $\gamma_d = \dfrac{2.65 \times 9.81}{\left(1 + \dfrac{2.65 \times w}{100}\right)}$, w being in %,

and for Ten per cent air-voids condition, $\gamma_d = \dfrac{2.65 \times 9.81}{\left(1 + \dfrac{2.65 \times w}{90}\right)}$, w being in %.

Compaction of Soil

From these equations, the following values are computed:

Water content (% w)	Dry density γ_d (kN/m³)	
	Zero air-voids condition	Ten per cent air content condition (90% saturation)
8	21.45	21.04
10	20.55	20.10
12	19.73	19.20
13	19.34	18.80
14	18.96	18.40
16	18.26	17.66

The moisture-dry density curve and the zero air-voids and ten per cent air-voids lines are shown in Fig. 12.12.

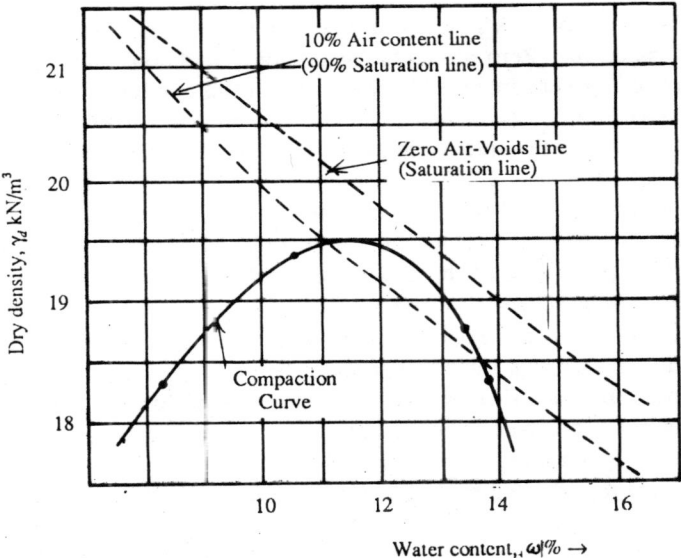

Fig. 12.12 Compaction curve and saturation lines (Ex. 12.6)

From the figure,
  Optimum moisture content = **12%**
  Maximum dry density = **19.5 kN/m³**
Correction for oversize fraction:
  $G = 2.79$ for gravel $n_1 = 0.09$ $n_2 = 0.91$

$$\text{Corrected maximum dry density} = \frac{G \cdot \gamma_w \gamma_{d_{max}}}{n_1 \gamma_{d_{max}} + n_2 G \gamma_w}$$

$$= \frac{2.79 \times 9.81 \times 19.5}{(0.09 \times 19.5 + 0.91 \times 2.79 \times 9.8)} = \mathbf{20 \text{ kN/m}^3}$$

Corrected optimum moisture content $= n_1 A_0 + n_2 w_0 = 0.91 \times 12\%$ (taking A_0 as zero)
$$= \mathbf{10.9\%}$$

SUMMARY OF MAIN POINTS

1. 'Compaction' of soil is defined as the process by which the soil grains are packed closer together by mechanical means such as tamping or vibration, in order that the dry density be increased. This involves the expulsion of pore air only, and thus differs from the phenomenon of consolidation, which involves the expulsion of pore water from a fully saturated cohesive soil under static loading.
   Compaction of a soil results in the improvement of its engineering properties such as shearing strength.
2. The moisture content-dry density relationship of a soil, as established by Proctor, is such that a peak value is reached for dry density at a moisture content, called the 'optimum' value. Increase in compactive effort will result in increased maximum dry density and decreased optimum moisture content.
3. The condition of full saturation when air in the voids is completely expelled, is called the 'Zero air-voids (Saturation) condition,' and the relationship between water content and dry density for this condition is called the 'Zero air-void line'. The compaction curve for any soil will always lie to the left of and below this line, since the air in the voids can never be fully expelled by compaction.
4. The purpose of laboratory compaction tests is to provide a guideline and a basis for control of compaction in the field, by giving a specification that a certain percentage of the maximum dry density achieved in the laboratory test shall be achieved in the construction *in-situ*.
5. Compaction is achieved *in-situ* by means of rolling or vibration, the latter being most suited for granular soils. Smooth-wheeled rollers, pneumatic-tyred rollers and sheepsfoot rollers are commonly used.
6. Control of compaction in the field involves the determination of the *in-situ* unit weight and *in-situ* moisture content.
   Sand replacement method and core-cutter method are used for the determination of *in-situ* unit weight. Quicker methods such as the use of Proctor's plasticity needle or the T.V.A. Penetrometer are used for the determination of both the *in-situ* unit weight and *in-situ* moisture content.

REFERENCES

1. Alam Singh: *Instrumentation for going metric in soil engineering testing*, Jl. Indian National Society of Soil Mechanics and Foundation Engineering, Vol.1, No. 2, New Delhi, 1965.
2. Alam Singh and B.C. Punmia: *A New Laboratory Compaction Device and its Comparison with the Proctor Test*, Highway Research News, No. 17, Highway Research Board (HRB), Washington, D.C., 1965.
3. Alam Singh and B.C. Punmia: *Soil Mechanics and Foundations*, Standard Book House, Nai Sarak, Delhi-6, 1971.

4. American Association of State Highway Officials (AASHO): Standard Laboratory Method of Test for Compaction and density of soils, *Standard Specifications for Highway Materials and Methods of Sampling and Testing*, Part-II, Methods of sampling and testing, Washington, D.C., 1942.
5. IS: 2720(Part VII)—1974 (revised): Methods of Test for soils, Part VII, *Determination of Moisture Content—Dry Density Relation Using Light Compaction*, ISI, New Delhi, 1974.
6. IS: 2720 (Part VIII)—1974 (revised): Methods of Test for soils, Part VIII, *Determination of Moisture Content—Dry Density Relation, Using Heavy Compaction*, ISI, New Delhi, 1974.
7. A.R. Jumikis: *Soil Mechanics*, D. Van Nostrand Co., Princeton, NJ, USA, 1962.
8. D.F. McCarthy: *Essentials of Soil Mechanics and Foundations*, Reston Publishing Company, Reston, Va., USA, 1977.
9. V.N.S. Murthy: *Soil Mechanics and Foundation Engineering*, Dhanpat Rai & Sons, Nai Sarak, Delhi, 2nd Ed., 1977.
10. R.R. Proctor: *Fundamental Principles of Soil Compaction*, Engineering News Record, Vol. III, nos. 9, 10, 12 and 13, 1933.
11. S.B. Sehgal: *A Textbook of Soil Mechanics*, Metropolitan Book Co., Pvt., Ltd., Delhi-6, 1967.
12. G.N. Smith: *Elements of Soil Mechanics for Civil and Mining Engineers*, SI edition, Crosby Lockwood Staples, London, 1974.

QUESTIONS AND PROBLEMS

12.1 (*a*) Discuss the effect of compaction on soil properties.
    (*b*) Write short notes on:
      (*i*) Field compaction control
      (*ii*) Methods of compaction
(S.V.U.—B.Tech. (Part-Time)—Sep., 1983)

12.2 (*a*) Derive an expression for 'zero air-void line' and draw the line for a specific gravity of 2.65.
    (*b*) What are the various factors that affect the compaction of soil in the field? How will you measure compaction in the field? Describe a method with its limitations.
(S.V.U.—Four-year B.Tech. Apr., 1983)

12.3 Draw typical compaction curves (γ_d vs. moisture content) for
    (*i*) Well-graded gravel with fines, (*ii*) Well-graded sandy clay,
    (*iii*) Silty clay and (*iv*) Highly plastic clay.
(S.V.U.—Four-year B.Tech. Dec., 1982)

12.4   (*i*) Bring out the usefulness of compaction test in the Laboratory in soil engineering practice.

(ii) By way of neat sketches explain (a) the effect of moisture content on the dry density for a constant compactive effort, (b) effect of compactive effort, on the γ_d-moisture content relationship.

(iii) Write a brief note on 'Proctor's needle'.

(iv) Write a brief note on the 'compaction in the field' bringing out the various types of rollers and their effectiveness with respect to different soil types.

(v) Write critical notes on AASHO Compaction test.

(S.V.U.—Four-year B.Tech. Oct., 1982)

12.5 Write brief notes on 'compaction' and 'consolidation' of soils differentiating the two.

(S.V.U.—B.E. Part-time—Dec., 1981)
B.E., (R.R.)—Nov., 1969.
B.E., (N.R.)—Sep., 1961.

12.6 Listing the various factors that influence the compaction of soils, show their influence with illustrative sketches of compaction curves.

(S.V.U.—B.E., (R.R.)—Sept., 1978)

12.7 What is the significance of compaction of soils? Describe how quality control is ensured in constructing an earth embankment.

(S.V.U.—B.E., (R.R.)—Dec., 1971, June, 1972, Nov., 1972—Nov., 1975)

12.8 Draw a curve showing the relation between dry density and moisture content for Standard Proctor test and indicate the salient features of the curve. Explain why soils are compacted when (i) preparing a subgrade and (ii) constructing an embankment. Describe how quality control is maintained in a rolled fill dam.

(S.V.U.—B.E., (R.R.)—May, 1975)

12.9 Explain why soils are compacted in the field. How is the degree of compaction ensured in the field (i.e., control of field compaction)? Distinguish between 'compaction' and 'consolidation' of soils. Bring out the effects of (i) moisture, (ii) compactive effort and (iii) soil type on the compaction characteristics of soils. Illustrate the answer with typical 'moisture-dry density' plots.

(S.V.U.—B.E., (R.R.)—Nov., 1973)

12.10 Describe the Proctor "Compaction Test" and give its use for construction of earth embankments.

(S.V.U.—B.E., (R.R.)—Nov., 1969)

12.11 Define "Optimum moisture content of a soil" and state on what factors it depends.

(S.V.U.—B.E., (R.R.)—Dec., 1968)

12.12 (a) What are the laws governing compaction of (i) cohesionless soils like sand and (ii) moderately cohesive soils like sandy clay?

(b) To what compaction pressure does the Standard Proctor Test correspond? Indicate how the standard proctor test can be modified to suit the compacting machinery actually used at site for the efficient compaction of embankment.

(S.V.U.—B.E., (R.R.)—Sept., 1968)

Compaction of Soil 517

12.13 The maximum dry density and optimum moisture content of a soil from standard proctor's test are 1.8 g/cm³ and 16% respectively. Compute the degree of saturation of the sample, assuming the specific gravity of soil grains as 2.70.

(S.V.U.—B.E., (R.R.)—Nov., 1973)

12.14 The wet weight of a sample is missing in a Proctor test. The oven-dry weight of this sample is 1890 g. The volume of the mould used is 1000 cm³. If the degree of saturation of this sample is 90%, determine its water content and bulk density.

12.15 A soil in the borrow pit has a water content of 11.7% and the dry density of 16.65 kN/m³. If 2,070 m³ of soil is excavated from it and compacted in an embankment at a porosity of 0.33, calculate the compacted volume of the embankment that can be constructed out of this volume of soil.

12.16 The soil from a borrow pit is at a bulk density of 17.10 kN/m³ and a water content of 12.6%. It is desired to construct an embankment with a compacted unit weight of 19.62 kN/m³ at a water content of 18%.

Determine the quantity of soil to be excavated from the borrow pit and the amount of water to be added for every 100 m³ of compacted soil in the embankment.

12.17 The following data are obtained in a compaction test.
Specific gravity = 2.65

Moisture content (%)	2	4.2	5.5	6.6	7.5	10
Wet density (g/cm³)	2.02	2.08	2.17	2.20	2.21	2.20

Determine the OMC and maximum dry density. Draw 'Zero-air-void line'.

(S.V.U.—Four year B.Tech. Dec., 1982)

12.18 The following results were obtained in a compaction test. Determine the optimum moisture content and the maximum dry density.

Test No.	1	2	3	4	5	6
Wet unit weight (kN/m³)	18.8	20.07	20.52	21.06	21.06	20.07
Water content (%)	7.4	9.7	10.5	11.5	13.1	14.4

Also find the air content at maximum dry density.

(S.V.U.—B.E., (R.R.)—Dec., 1968)

13
LATERAL EARTH PRESSURE AND STABILITY OF RETAINING WALLS

13.1 INTRODUCTION

Soil is neither a solid nor a liquid, but it exhibits some of the characteristics of both. One of the characteristics similar to that of a liquid is its tendency to exert a lateral pressure against any object in contact. This important property influences the design of retaining walls, abutments, bulkheads, sheet pile walls, basement walls and underground conduits which retain or support soil, and, as such, is of very great significance.

Retaining walls are constructed in various fields of civil engineering, such as hydraulics and irrigation structures, highways, railways, tunnels, mining and military engineering.

13.2 TYPES OF EARTH-RETAINING STRUCTURES

Earth-retaining structures may be broadly classified as retaining walls and sheet-pile walls.

Retaining walls may be further classified as:
  (i) Gravity retaining walls —usually of masonry or mass concrete.
  (ii) Cantilever walls
  (iii) Counterfort walls   } usually of reinforced concrete.
  (iv) Buttress walls

Sheet pile walls may be further classified as cantilever sheet pile walls and anchored sheet pile walls, also called 'bulkheads'.

Gravity walls depend on their weight for stability; walls up to 2 m height are invariably of this type. The other types of retaining walls, as well as sheet-pile walls, are known as 'flexible walls'. All these are shown in Fig. 13.1.

R.C. Cantilever walls have a vertical or inclined stem monolithic with a base slab. These are considered suitable up to a height of 7.5 m. A vertical or inclined stem is used in counterfort walls, supported by the base slab as well as by counterforts with which it is monolithic.

Cantilever sheet pile walls are held in the ground by the passive resistance of the soil both in front of and behind them. Anchored sheet pile wall or bulkhead is fixed at its base as a cantilever wall but supported by tie-rods near the top, sometimes using two rows of ties and properly anchored to a deadman.

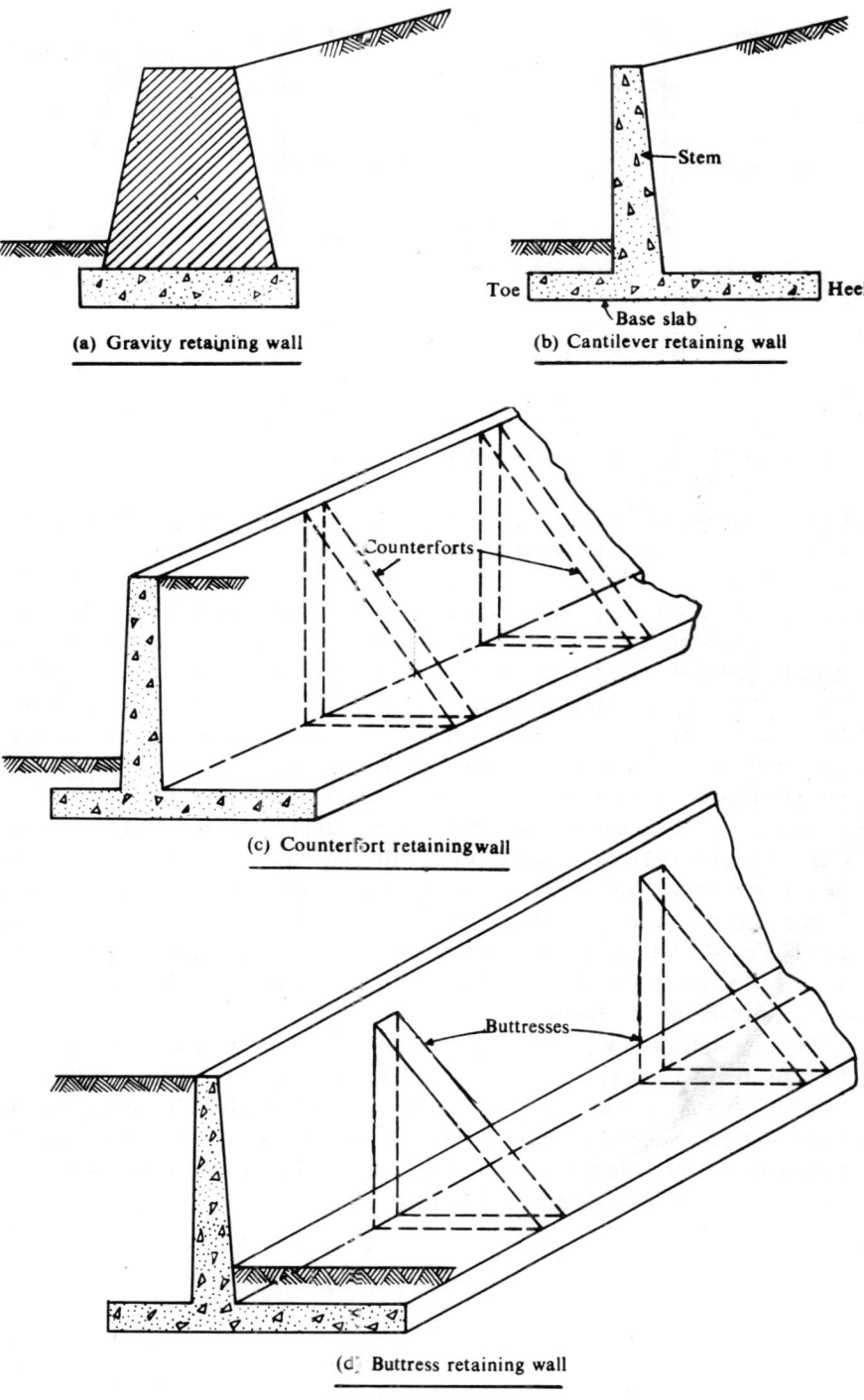

Fig. 13.1 *(continued)*

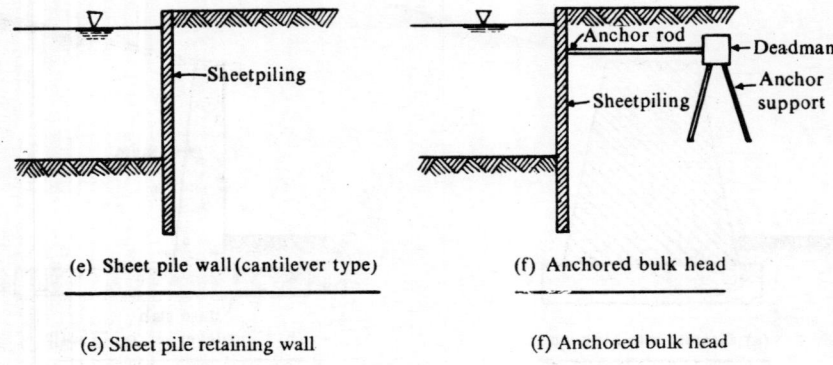

(e) Sheet pile wall (cantilever type)   (f) Anchored bulk head

(e) Sheet pile retaining wall   (f) Anchored bulk head

Fig. 13.1 Types of earth-retaining structures

13.3 LATERAL EARTH PRESSURES

Lateral earth pressure is the force exerted by the soil mass upon an earth-retaining structure, such as a retaining wall.

There are two distinct kinds of lateral earth pressure; the nature of each is to be clearly understood. First, let us consider a retaining wall which holds back a mass of soil. The soil exerts a push against the wall by virtue of its tendency to slip laterally and seek its natural slope or angle of response, thus making the wall to move slightly away from the backfilled soil mass. This kind of pressure is known as the 'active' earth pressure of the soil. The soil, being the actuating element, is considered to be active and hence the name active earth pressure. Next, let us imagine that in some manner the retaining wall is caused to move toward the soil. In such a case the retaining wall or the earth-retaining structure is the actuating element and the soil provides the resistance for maintaining stability. The pressure or resistance which soil develops in response to movement of the structure toward it is called the 'passive earth pressure', or more appropriately 'passive earth resistance' which may be very much greater than the active earth pressure. The surface over which the sheared-off soil wedge tends to slide is referred to as the surface of 'sliding' or 'rupture'.

The limiting values of both the active earth pressure and passive earth resistance for a given soil depend upon the amount of movement of the structure. In the case of active pressure, the structure tends to move away from the soil, causing strains in the soil mass, which in turn, mobilise shearing stresses; these stresses help to support the soil mass and thus tend to reduce the pressure exerted by the soil against the structure. This is indicated in Fig. 13.2.

Lateral Earth Pressure and Stability of Retaining Walls 521

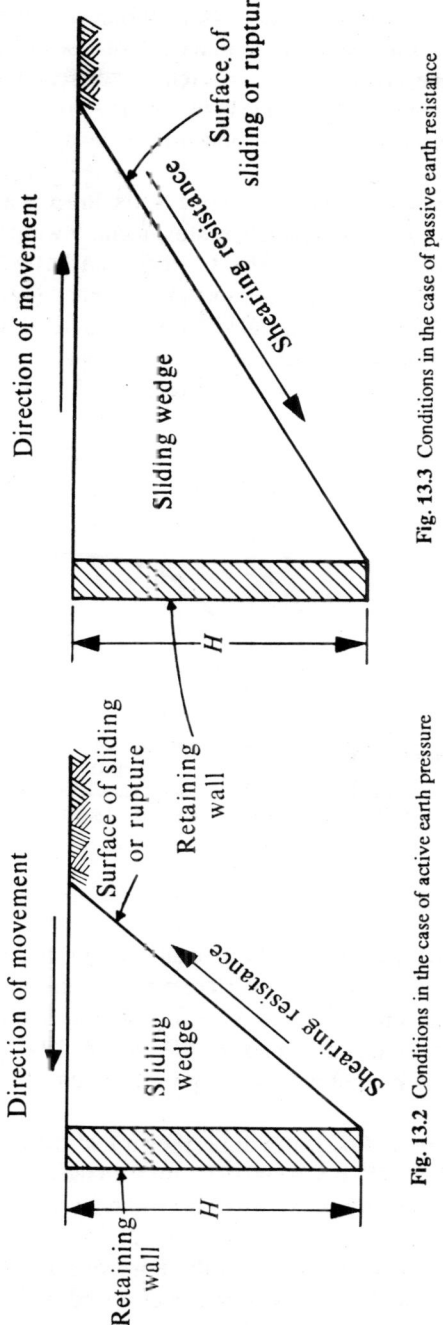

Fig. 13.3 Conditions in the case of passive earth resistance

Fig. 13.2 Conditions in the case of active earth pressure

In the case of passive earth resistance also, internal shearing stresses develop, but act in the opposite direction to those in the active case and must be overcome by the movement of the structure. This difference in direction of internal stresses accounts for the difference in magnitude between the active earth pressure and the passive earth resistance. The conditions obtaining in the passive case are indicated in Fig. 13.3.

Active pressures are accompanied by movements directed away from the soil, and passive resistances are accompanied by movements toward the soil. Logically, therefore, there must be a situation intermediate between the two when the retaining structure is perfectly stationary and does not move in either direction. The pressure which develops in this condition is called 'earth pressure at rest'. Its value is a little larger than the limiting value of active pressure, but is considerably less than the maximum passive resistance. This is indicated in Fig. 13.4.

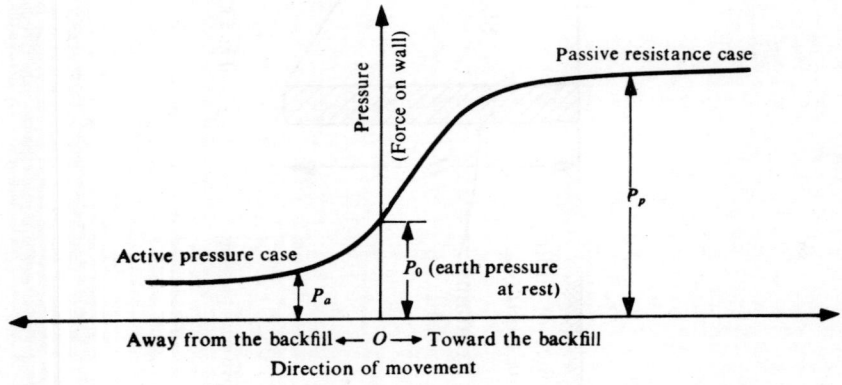

Fig. 13.4 Relation between lateral earth pressure and movement of wall

Very little movement (about 0.5% horizontal strain) is required to mobilise the active pressure; however, relatively much larger movement (about 2% of horizontal strain for dense sands and as high as 15% for loose sands) may be required to mobilise full passive resistance (Lambe and Whitman, 1969). About 50% of the passive resistance may be mobilised at a movement comparable to that required for the active case.

In a later sub-section (13.6.1), it will be shown that the failure planes will be inclined to horizontal at $(45° + \phi/2)$ and $(45° - \phi/2)$ in the active and passive cases, respectively. This means that the width of the sliding wedge at the top of the wall will be $H \cot(45° + \phi/2)$ and $H \cot(45° - \phi/2)$ for active and passive cases, respectively, H being the height of the wall. For average values of ϕ, these will be approximately $H/2$ and $2H$. The strains mentioned by Lambe and Whitman (1969) will then amount to a horizontal movement at the top of the wall of $0.0025 H$ for the active case and $0.04 H$ to $0.30 H$ for the passive case.

This agrees fairly well with Terzaghi's observation (Terzaghi, 1936) that a movement of 0.005 H of the top of the wall, or even less, is adequate for full mobilisation of active state. (In fact, Terzaghi's experiments in the 1920's indicated that even 0.001 H is adequate for this).

There are two reasons why less strain is required to reach the active condition than to reach the passive condition. First, an unloading (the active state) always involves less strain than a loading (passive state). Second, the stress change in passing to the active state is much less than the stress change in passing to the passive state. (Lambe and Whitman, 1969).

The other factors which affect the lateral earth pressure are the nature of soil —cohesive or cohesionless, porosity, water content and unit weight.

The magnitude of the total earth pressure, or to be more precise, force on the structure, is dependent on the height of the backfilled soil as also on the nature of pressure distribution along the height.

13.4 EARTH PRESSURE AT REST

Earth pressure at rest may be obtained theoretically from the theory of elasticity applied to an element of soil, remembering that the lateral strain of the element is zero. Referring to Fig. 13.5(a), the principal stresses acting on an element of soil

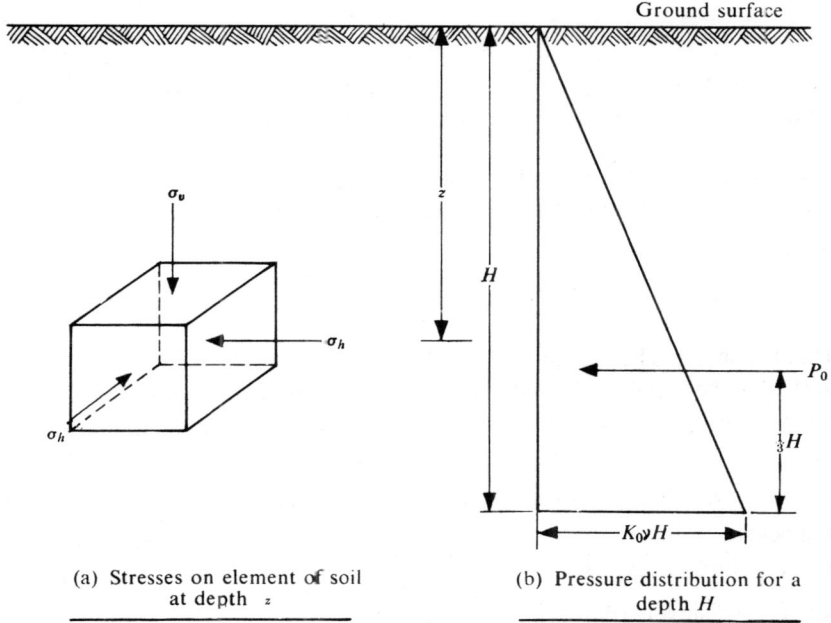

(a) Stresses on element of soil at depth z

(b) Pressure distribution for a depth H

Fig. 13.5 Stress conditions relating to earth pressure at rest

situated at a depth z from the surface in semi-infinite, elastic, homogeneous and isotropic soil mass are σ_v, σ_h and σ_h as shown. σ_v and σ_h denoting the stresses in the vertical and horizontal directions respectively.

The soil deforms vertically under its self-weight but is prevented from deforming laterally because of an infinite extent in all lateral directions. Let E_s and γ be the modulus of elasticity and Poisson's ratio of the soil respectively.

Lateral strain, $E_h = \dfrac{\sigma_h}{E_s} - \nu\left(\dfrac{\sigma_v}{E_s} + \dfrac{\sigma_h}{E_s}\right) = 0$

$$\therefore \quad \frac{\sigma_h}{\sigma_v} = \frac{\nu}{(1-\nu)} \qquad \ldots\text{(Eq. 13.1)}$$

But $\sigma_v = \gamma \cdot z$, where γ is the appropriate unit weight of the soil depending upon its condition. ...(Eq. 13.2)

$$\therefore \quad \sigma_h = \left(\frac{\nu}{1-\nu}\right) \cdot \gamma_z \qquad \ldots\text{(Eq. 13.3)}$$

Let us denote $\left(\dfrac{\nu}{1-\nu}\right)$ by K_0, which is known as the "Coefficient of earth pressure at rest" and which is the ratio of the intensity of the earth pressure at rest to the vertical stress at a specified depth.

$$K_0 = \left(\frac{\nu}{1-\nu}\right) \qquad \ldots\text{(Eq. 13.4)}$$

$$\therefore \quad \sigma_h = K_0 \cdot \gamma_z \qquad \ldots\text{(Eq. 13.5)}$$

The distribution of the earth pressure at rest with depth is obviously linear (or of hydrostatic nature) for constant soil properties such as E, ν, and γ, as shown in Fig. 13.5(b).

If a structure such as a retaining wall of height H is interposed from the surface and imagined to be held without yield, the total thrust on the wall per unit length P_0, is given by

$$P_0 = \int_0^H \sigma_h \cdot dz = \int_0^H K_0 \cdot \gamma z \cdot dz = \frac{1}{2} K_0 \cdot \gamma \cdot H^2 \qquad \text{(Eq. 13.6)}$$

This is considered to act at $(1/3)H$ above the base of the wall. As has been indicated in the previous chapter, choosing an appropriate value for the Poisson's ratio, ν, is by no means easy; this is the limitation in arriving at K_0 from equation 13.4.

Various researchers proposed empirical relationships for K_0, some of which are given below:

$K_0 = (1 - \sin\phi')$ (Jaky, 1944) ...[Eq. 13.7(a)]

$K_0 = 0.9(1 - \sin\phi')$ (Fraser, 1957) ...[Eq. 13.7(b)]

$K_0 = 0.19 + 0.233 \log I_p$ (Kenney, 1959) ...[Eq. 13.7(c)]

$K_0 = [1 + (2/3) \sin \phi'] \left(\dfrac{1 - \sin \phi}{1 + \sin \phi} \right)$ (Kezdi, 1962) ...[Eq. 13.7(d)]

$K_0 = (0.95 - \sin \phi')$ (Brooker and Ireland, 1965) ...[Eq. 13.7(e)]

ϕ' in these equations represents the effective angle of friction of the soil and I_p, the plasticity index. Brooker and Ireland (1965) recommend Jaky's equation for cohesionless soils and their own equation, given above, for cohesive soils. However, Alpan (1967) recommends Jaky's equation for cohesionless soils and Kenney's equation for cohesive soils as does Kenney (1959). Certain values of the coefficient of earth pressure at rest are suggested for different soils, based on field data, experimental evidence and experience. These are given in Table 13.1.

Table 13.1 Coefficient of earth pressure at rest

S.No.	Soil		K_0
1	Loose Sand ($e = 0.8$)		
		dry	0.64
		Saturated	0.46
2	Dense sand ($e = 0.6$)		
		dry	0.49
		saturated	0.36
3	Sand (Compacted in layers)		0.80
4	Soft clay ($I_p = 30$)		0.60
5	Hard clay ($I_p = 9$)		0.42
6	Undistrubed Silty clay ($I_p = 45$)		0.57

13.5 EARTH PRESSURE THEORIES

The magnitude of the lateral earth pressure is evaluated by the application of one or the other of the so-called 'lateral earth pressure theories' or simply 'earth pressure theories'. The problem of determining the lateral pressure against retaining walls is one of the oldest in the field of engineering. A French military engineer, Vauban, set forth certain rules for the design of revetments in 1687. Since then, several investigators have proposed many theories of earth pressure after a lot of experimental and theoretical work. Of all these theories, those given by Coulomb and Rankine stood the test of time and are usually referred to as the "Classical earth pressure theories". These theories are considered reliable in spite of some limitations and are considered basic to the problem. These theories have been developed originally to apply to cohesionless soil backfill, since this situation is considered to be more frequent in practice and since the designer will be on the safe side by neglecting cohesion. Later researchers gave necessary modifications to take into account cohesion, surcharge, submergence, and so on. Some have evolved graphical procedures to evaluate the total thrust on the retaining structure.

Although Coulomb presented his theory nearly a century earlier to Rankine's theory, Rankine's theory will be presented first due to its relative simplicity.

13.6 RANKINE'S THEORY

Rankine (1857) developed his theory of lateral earth pressure when the backfill consists of dry, cohesionless soil. The theory was later extended by Resal (1910) and Bell (1915) to be applicable to cohesive soils.

The following are the important assumptions in Rankine's theory:
  (*i*) The soil mass is semi infinite, homogeneous, dry and cohesionless.
  (*ii*) The ground surface is a plane which may be horizontal or inclined.
  (*iii*) The face of the wall in contact with the backfill is vertical and smooth. In other words, the friction between the wall and the backfill is neglected (This amounts to ignoring the presence of the wall).
  (*iv*) The wall yields about the base sufficiently for the active pressure conditions to develop; if it is the passive case that is under consideration, the wall is taken to be pushed sufficiently towards the fill for the passive resistance to be fully mobilised. (Alternatively, it is taken that the soil mass is stretched or gets compressed adequately for attaining these states, respectively. Friction between the wall and fill is supposed to reduce the active earth pressure on the wall and increase the passive resistance of the soil. Similar is the effect of cohesion of the fill soil).

Thus it is seen that, by neglecting wall friction as also cohesion of the backfill, the geotechnical engineer errs on the safe side in the computation of both the active pressure and passive resistance. Also, the fill is usually of cohesionless soil, wherever possible, from the point of view of providing proper drainage.

13.6.1 Plastic Equilibrium of Soil—Active and Passive Rankine States

A mass of soil is said to be in a state of plastic equilibrium if failure is incipient or imminent at all points within the mass. This is commonly referred to as the 'general state of plastic equilibrium' and occurs only in rare instances such as when tectonic forces act. Usually, however, failure may be imminent only in a small portion of the mass such as that produced by the yielding of a retaining structure in the soil mass adjacent to it. Such a situation is referred to as the 'local state of plastic equilibrium'.

Rankine (1857) was the first to investigate the stress conditions associated with the states of plastic equilibrium in a semi-infinite mass of homogeneous, elastic and isotropic soil mass under the influence of gravity or self-weight alone. The concept as postulated by Rankine in respect of a cohesionless soil mass is shown in Fig. 13.6.

Let us consider an element of unit area at a depth z below the horizontal ground surface. Let the unit weight of the cohesionless soil be γ. The vertical stress acting on the horizontal face of the element $\sigma_v = \gamma \cdot z$. Since any vertical plane is symmetrical with respect to the soil mass, the vertical as well as horizontal planes will be free of shear stresses. Consequently, the normal stresses acting on these planes will be principal stresses. The horizontal principal stress, σ_h, or the lateral earth pressure at rest in this case, is given by $K_0 \cdot \sigma_v$, or $K_0 \cdot \gamma \cdot z$. The element is in a state of elastic equilibrium under these stress conditions.

Lateral Earth Pressure and Stability of Retaining Walls 527

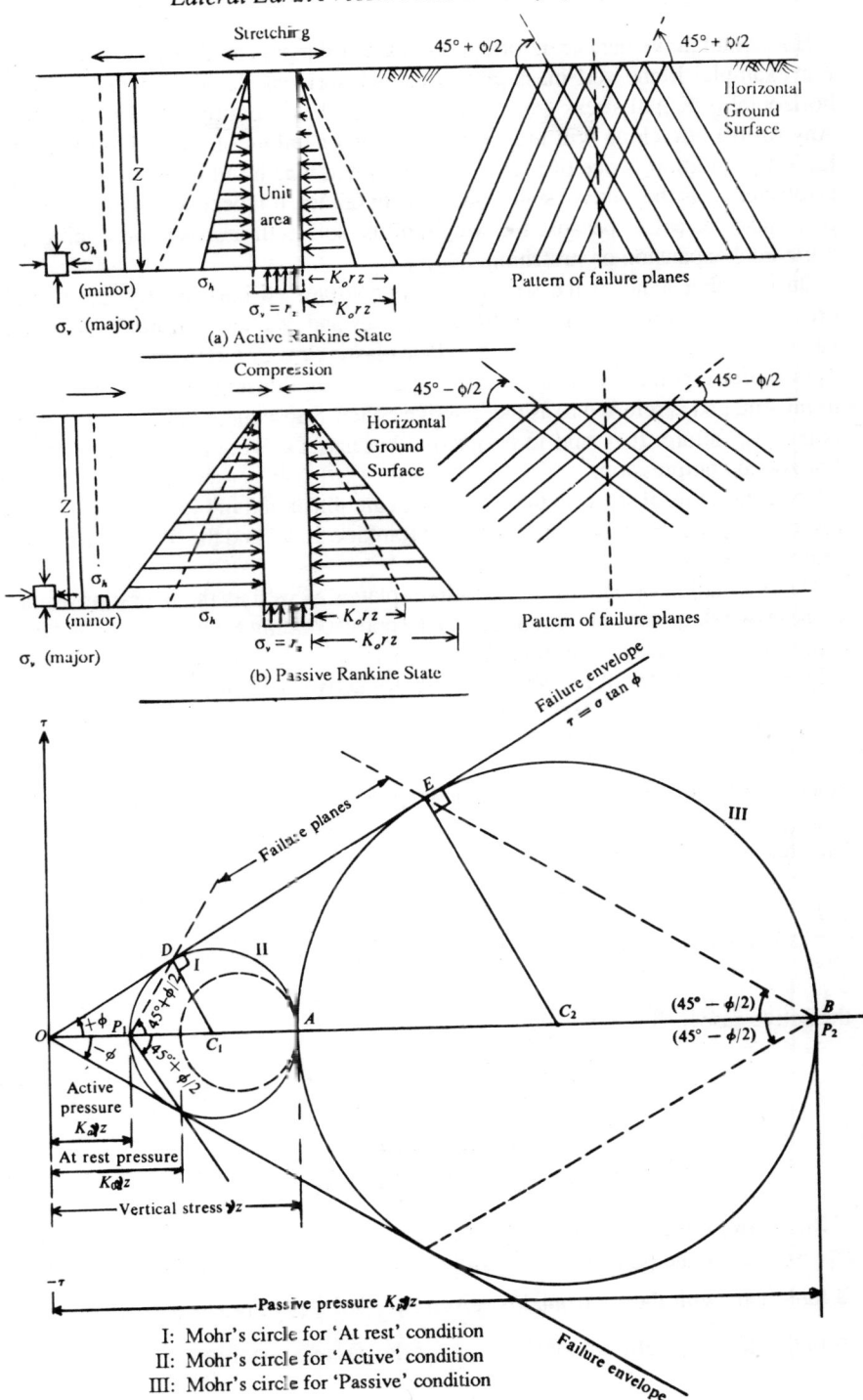

(a) Active Rankine State

(b) Passive Rankine State

I: Mohr's circle for 'At rest' condition
II: Mohr's circle for 'Active' condition
III: Mohr's circle for 'Passive' condition

(c) Mohr's stress circles and failure envelopes for active and passive states

Fig. 13.6 Rankine's states of plastic equilibrium

Horizontal movement or deformation of the soil mass can change the situation. For example, if the soil mass gets stretched horizontally, the lateral stress or horizontal principal stress gets reduced and reaches a limiting minimum value. Any further stretching will induce plastic flow or failure of the soil mass. This limiting condition is one of plastic equilibrium at which failure is imminent and is referred to as the 'active' state. Subsequent failure, if it occurs, is active failure. It is said to be active because the weight of the soil itself assists in producing the horizontal expansion or stretching.

On the other hand, if the soil mass gets compressed horizontally, the lateral pressure or horizontal principal stress increases and reaches a limiting maximum value; any further compression will induce plastic flow or failure of the soil mass. This limiting condition also is one of plastic equilibrium at which failure is imminent, and is referred to as the 'passive' state. Subsequent failure, if it occurs, is passive failure. It is said to be passive because the weight of soil resists the horizontal compression.

The conditions of stress in these two cases are illustrated in Fig. 13.6 (a) and (b) respectively and are known as the 'Active Rankine State' and the 'Passive Rankine State' respectively.

The orientation or pattern of the failure planes as well as the lateral pressures in these two states may be obtained from the corresponding Mohr's circles of stress representing the stress conditions for these two states as shown in Fig. 13.6(c).

From the geometry of the Mohr's circle, for active condition,

$$\sin\phi = \frac{DC_1}{OC_1} = \frac{(\sigma_1-\sigma_3)/2}{(\sigma_1+\sigma_3)/2} = \frac{(\sigma_1-\sigma_3)}{(\sigma_1+\sigma_3)} = \frac{(\sigma_v-\sigma_h)}{(\sigma_v+\sigma_h)}$$

since σ_v is the major principal stress and σ_h is the minor one for the active case.

This leads to $\dfrac{\sigma_h}{\sigma_v} = \dfrac{1-\sin\phi}{1+\sin\phi}$

$\dfrac{\sigma_h}{\sigma_v}$ is known as the coefficient of lateral earth pressure and is denoted by K_a for the active case.

$$\therefore \quad K_a = \frac{1-\sin\phi}{1+\sin\phi} = \tan^2\left(45° - \frac{\phi}{2}\right) \qquad \text{...(Eq. 13.8)}$$

(by trignometry.)

$AC_1D = 90° + \phi$, from $\triangle OC_1D$.

This is twice the angle made by the plane on which the stress conditions are represented by the point D on the Mohr's circle. Hence, the angle made by the failure plane with the horizontal is given by $\frac{1}{2}(90° + \phi)$ or $(45° + \phi/2)$. Similarly, from the geometry of the Mohr's circle for the passive condition,

Lateral Earth Pressure and Stability of Retaining Walls 529

$$\sin\phi = \frac{EC_2}{OC_2} = \frac{(\sigma_1-\sigma_3)/2}{(\sigma_1+\sigma_3)/2} = \frac{\sigma_1-\sigma_3}{\sigma_1+\sigma_3} = \frac{\sigma_h-\sigma_v}{\sigma_h+\sigma_v},$$

since σ_h is the major principal stress and σ_v is the minor principal stress for the passive case.

This leads to

$$\frac{\sigma_v}{\sigma_h} = \frac{1-\sin\phi}{1+\sin\phi}$$

or

$$\frac{\sigma_h}{\sigma_v} = \frac{1+\sin\phi}{1-\sin\phi}$$

$\dfrac{\sigma_h}{\sigma_v}$ is the coefficient of lateral earth pressure and is denoted by K_p for the passive case.

$$\therefore\quad K_p = \frac{1+\sin\phi}{1-\sin\phi} = \tan^2\left(45°+\frac{\phi}{2}\right) \qquad \ldots\text{(Eq. 13.9)}$$

The angle made by the failure plane with the vertical is $(45°+\phi/2)$, i.e., with the plane on which the major principal stress acts.

Thus, the angle made by the failure plane with the horizontal is $(45°-\phi/2)$ for the passive case.

The effective angle of friction, ϕ', is to be used for ϕ, if the analysis is based on effective stresses, as in the case of submerged or partially submerged backfills. These two states are the limiting states of plastic equilibrium; all the intermediate states are those of elastic equilibrium, which include 'at rest' condition.

13.6.2 Active Earth Pressure of Cohesionless Soil

Let us consider a retaining wall with a vertical back, retaining a mass of cohesionless soil, the surface of which is level with the top of the wall, as shown in Fig. 13.7(a).

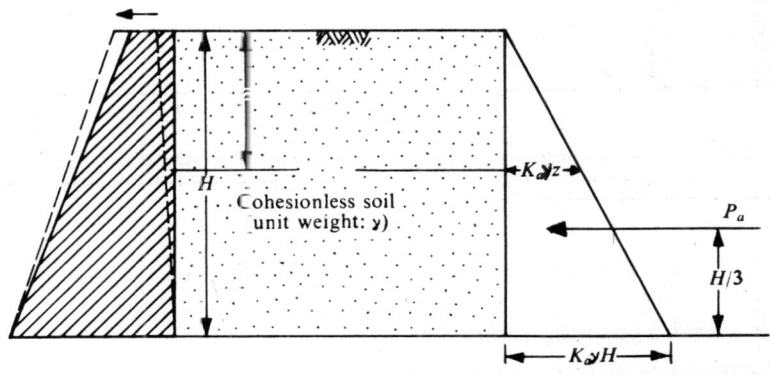

(a) Retaining wall with cohesionless backfill (b) Active pressure distribution with depth
 (moving away from the fill)

Fig. 13.7 Active earth pressure of cohesionless soil—Rankine's theory.

σ_v at a depth z below the surface $= \gamma \cdot z$

Assuming that the wall yields sufficiently for the active conditions to develop,
$$\sigma_h = K_a \cdot \sigma_v = K_a \cdot \gamma \cdot z,$$

where
$$K_a = \frac{1-\sin\phi}{1+\sin\phi} = \tan^2(45° - \phi/2)$$

The distribution of the active pressure with depth is obviously linear, as shown in Fig. 13.7(b).

For a total height of H of the wall, the total thrust P_a on the wall per unit length of the wall, is given by:

$$P_a = \frac{1}{2} K_a \gamma H^2 \qquad \text{...(Eq. 13.10)}$$

This may be taken to act at a height of $(1/3)H$ above the base as shown, through the centroid of the pressure distribution diagram.

The appropriate value of the unit weight γ should be used.

13.6.3 Passive Earth Pressure of Cohesionless Soil

Let us again consider a retaining wall with a vertical back, retaining a mass of cohesionless soil, the surface of which is level with the top of the wall, as shown in Fig. 13.8(a).

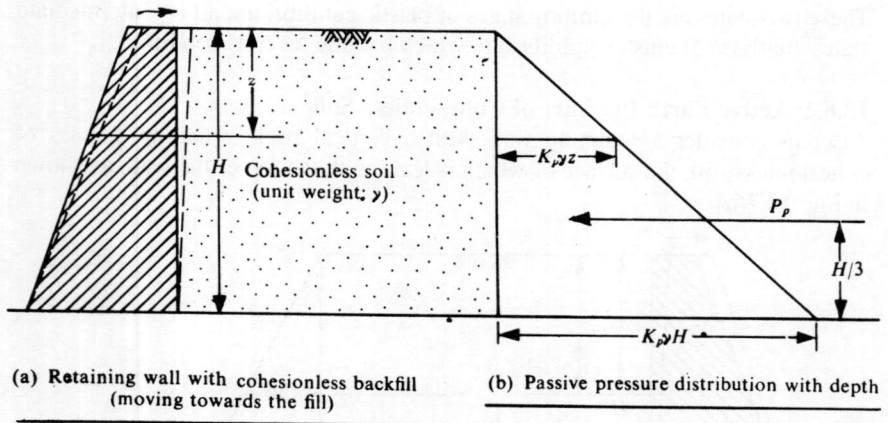

(a) Retaining wall with cohesionless backfill (moving towards the fill)

(b) Passive pressure distribution with depth

Fig. 13.8 Passive earth pressure of cohesionless soil—Rankine's theory

σ_v at a depth z below the surface $= \gamma \cdot z$

Assuming the wall moves towards the fill sufficiently to mobilise the full passive resistance,

$$\sigma_h = K_p \cdot \sigma_v = K_p \cdot \gamma \cdot z,$$

where
$$K_p = \frac{1 + \sin \phi}{1 - \sin \phi} = \tan^2(45° + \phi/2)$$

The distribution of passive pressure (resistance) with depth is obviously linear, as shown in Fig. 13.8(b).

For a total height H of the wall, the total passive thrust P_p on the wall per unit length of the wall is given by:

$$P_p = \frac{1}{2} K_p \cdot \gamma \cdot H^2 \qquad \ldots(\text{Eq. 13.11})$$

This may be taken to act at a height of $(1/3)H$ above the base as shown, through the centroid of the pressure distribution diagram. The appropriate value of γ should be used.

13.6.4 Effect of Submergence

When the backfill is fully saturated/submerged, the lateral pressure will be due to two components:
 (i) lateral earth pressure due to submerged unit weight of the backfill soil; and
 (ii) lateral pressure due to pore water.
This is shown in Fig. 13.9(a).

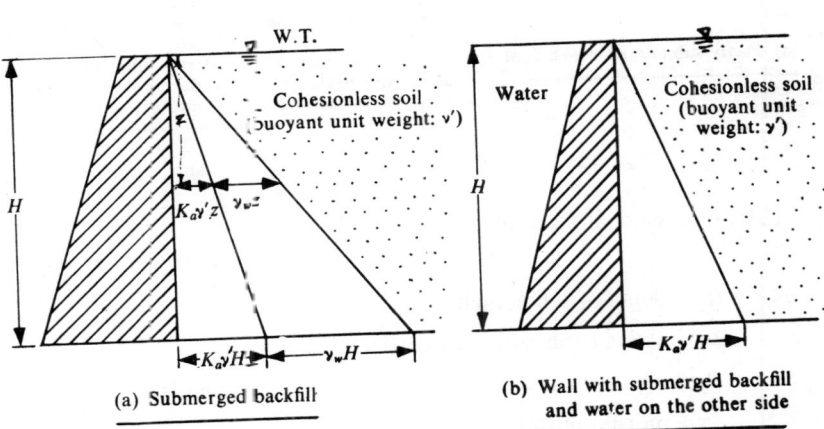

(a) Submerged backfill

(b) Wall with submerged backfill and water on the other side

Fig. 13.9 Effect of submergence on lateral earth pressure

At any depth z below the surface, the lateral pressure, σ_h, is given by:

$$\sigma_h = K_a \cdot \gamma' z + \gamma_w \cdot z$$

The pressure at the base is obtained by substituting H for z.

In case water stands to the full height of the retaining wall on the other side of the submerged backfill, as shown in Fig. 13.9(b), the net lateral pressure from the submerged backfill will be only from the first component, i.e., due to submerged unit weight of the backfill soil, as the water pressure acting on both sides will get cancelled.

In the case of passive earth pressure, the coefficient of passive earth pressure K_p, has to substituted for K_a; otherwise, the treatment will be the same.

If the backfill is submerged only to a part of its height, the backfill above the water table is considered to be moist. The lateral pressure above the water table is due to the moist unit weight of soil, and that below the water table is the sum of that due to the submerged unit weight of the soil and the water pressure. This is illustrated in Fig. 13.10(a).

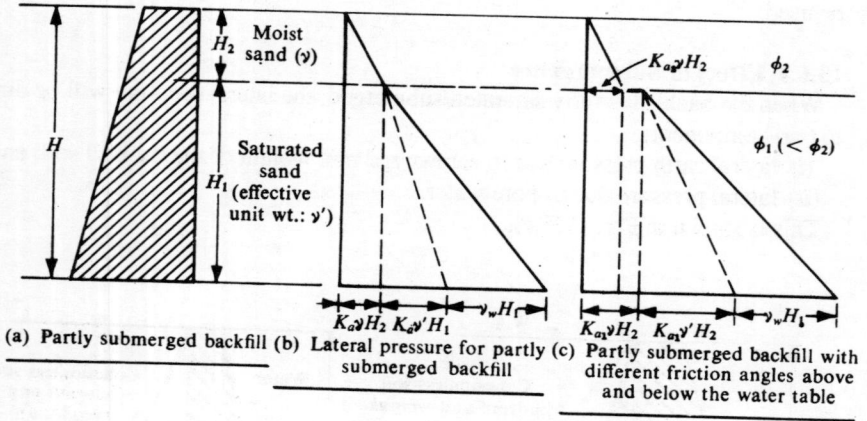

(a) Partly submerged backfill (b) Lateral pressure for partly submerged backfill (c) Partly submerged backfill with different friction angles above and below the water table

Fig. 13.10 Effect of partial submergence on lateral earth pressure

Lateral pressure at the base of wall,

$$= K_a \gamma H_2 + K_a \gamma' H_1 + \gamma_w H_1,$$ as shown in Fig. 13.10(b),

where H_1 = depth of submerged fill,

K_a = active earth pressure coefficient,

H_2 = depth of fill above water table (taken to be moist),

γ = moist unit weight, and

γ' = submerged or effective unit weight.

If the angle of internal friction below the water table is different from that above the water table (the former will usually be less than the latter), the corresponding values of K_a should be used in the respective zones. (It may be noted that K_a-values bear reciprocal relationship with φ-values while K_p-values bear direct relationship with them). At the water table, a slight but sudden increase of pressure should be expected depending upon the difference in the values of active pressure coefficients for the respective φ-values. These conditions are illustrated in Fig. 13.10(c).

13.6.5 Effect of Uniform Surcharge

The extra loading carried by a retaining structure is known as 'surcharge'. It may be a uniform load (from roadway, from stacked goods, etc.), a line load (trains running parallel to the structure), or an isolated load (say, a column footing).

Let us see the effect of a uniform surcharge on the lateral pressure acting on the retaining structure, as shown in Fig. 13.11.

In the case of a wall retaining a backfill with horizontal surface level with the top of the wall and carrying a uniform surcharge of intensity q per unit area, the

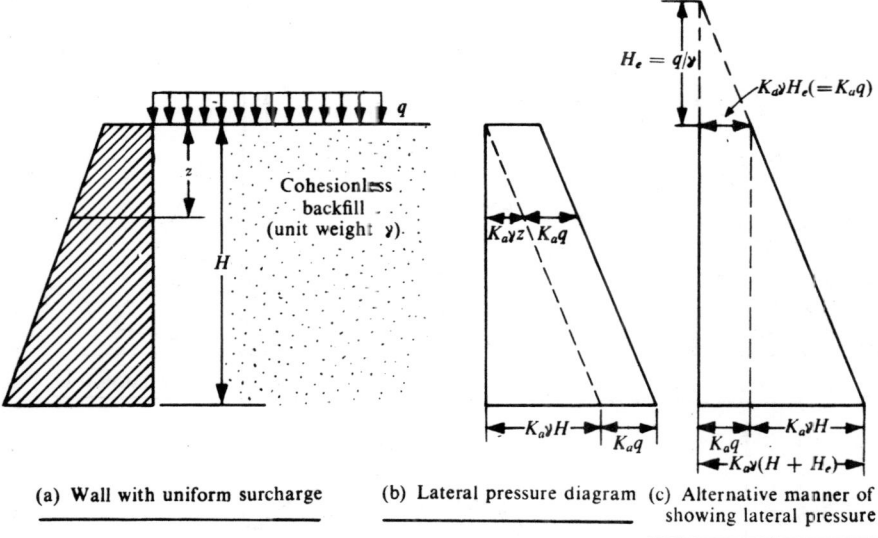

(a) Wall with uniform surcharge (b) Lateral pressure diagram (c) Alternative manner of showing lateral pressure

Fig. 13.11 Effect of uniform surcharge on lateral pressure

vertical stress at every elevation in the backfill is considered to increase by q. As such, the lateral pressure has to increase by $K_a \cdot q$.

Thus, at any depth z, $\sigma_h = K_a \gamma_z + K_a q$

Figures 13.11(b) and (c) show two different ways in which the pressure distribution may be shown. In Fig. 13.11(c), the uniform surcharge is also considered to have been converted into an equivalent height H_e, of backfill, which is easily established, as shown.

13.6.6 Effect of Inclined Surcharge—Sloping Backfill

Sometimes, the surface of the backfill will be inclined to the horizontal. This is considered to be a form of surcharge—'inclined surcharge', and the angle of inclination of the backfill with the horizontal is called the 'angle of surcharge'. Rankine's theory for this case is based on the assumption that a 'conjugate' relationship exists between the vertical pressures and lateral pressures on vertical planes within the soil adjacent to a retaining wall. It may be shown that such a conjugate relationship would hold between vertical stresses and lateral stresses on

534 Geotechnical Engineering

vertical planes within an infinite slope. Thus, it would amount to assuming that the introduction of a retaining wall into the infinite slope does not result in any changes in shearing stresses at the surface of contact between the wall and the backfill. This inherent assumption in Rankine's theory means that the effect of 'wall friction', or friction between the wall and the backfill soil is neglected.

Let us consider an element of soil of unit horizontal width at depth z below the surface of the backfill, the faces of which are parallel to the surface and to the vertical, as shown in Fig. 13.12(a).

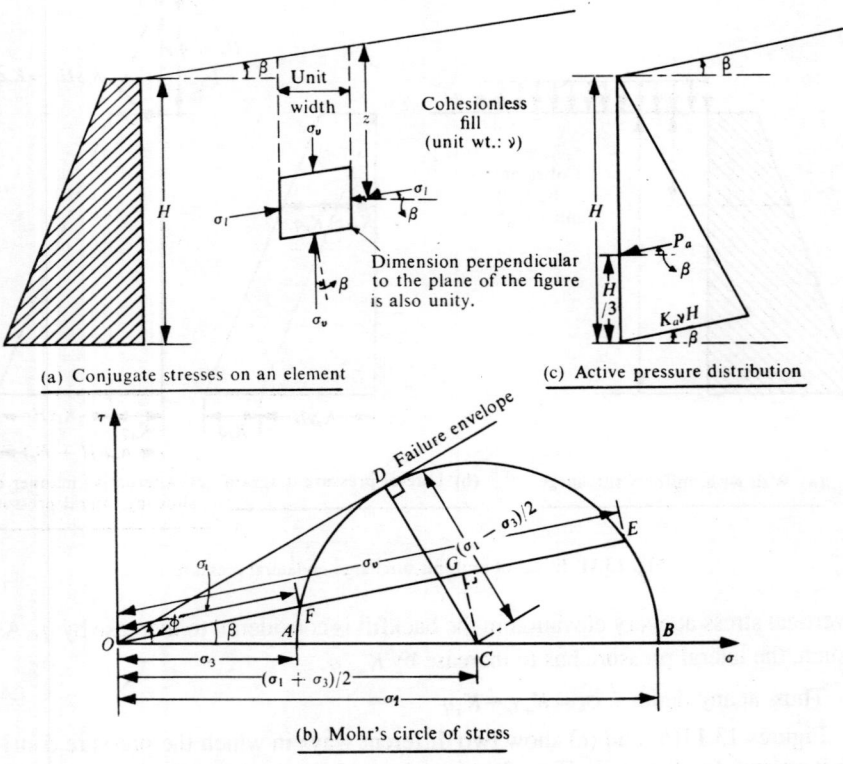

Fig. 13.12 Inclined surcharge—Rankine's theory

The vertical stress and the lateral stress on the vertical plane are each parallel to the plane of the other and, therefore, are said to be conjugate stresses. Both have obliquities equal to the angle of inclination of the slope β.

The magnitude of the vertical stress acting on the face of the element parallel to the surface can be easily obtained as follows:

The weight of the column of soil above the face = $\gamma \cdot z$

since the horizontal width is unity, the area of the parallelogram is $z \cdot 1$, and the volume of the parallelopiped is $z \cdot 1 \cdot 1$ cubic units.

This force acts on an area $\dfrac{1}{\cos \beta} \cdot 1$.

∴ The vertical stress σ_v on the face of the element parallel to the slope is:

$$\sigma_v = \frac{\gamma \cdot z}{1/\cos \beta} = \gamma \cdot z \cos \beta \qquad \ldots(\text{Eq. 13.12})$$

The conjugate nature of the lateral pressure on the vertical plane and the vertical pressure on a plane parallel to the inclined surface of the backfill may also be established from the Mohr's circle diagram of stresses, Fig. 13.12(b). It is obvious that, from the very definition of conjugate relationship, the angle of obliquity of the resultant stress should be the same for both planes. Thus, in the diagram, if a line OE is drawn at an angle β, the angle of obliquity, with the σ-axis, to cut the Mohr's circle in E and F, OE represents σ_v and OF represents σ_l, for the active case (for the passive case, it is vice versa).

Now the relationship between σ_v and σl may be derived from the geometry of the Mohr's circle, Fig. 13.12(b), as follows.

Let OD be the failure envelope inclined at ϕ to the σ-axis. Let CG be drawn perpendicular to OFE and CE, CD, and CF be joined, C being the centre of the Mohr's circle.

$$\frac{CD}{OC} = \sin \phi$$

$$\frac{(\sigma_1 - \sigma_3)/2}{(\sigma_1 + \sigma_3)/2} = \sin \phi$$

or

$$(\sigma_1 - \sigma_3) = (\sigma_1 + \sigma_3) \sin \phi \qquad \ldots(\text{Eq. 13.13})$$

$$OG = OC \cdot \cos \beta = [(\sigma_1 + \sigma_3)/2] \cos \beta$$

$$CG = OC \cdot \sin \beta = [(\sigma_1 + \sigma_3)/2] \sin \beta$$

$$FG = GE = \sqrt{CF^2 - CG^2} = \sqrt{\{(\sigma_1 - \sigma_3)/2\}^2 - \left\{\frac{(\sigma_1 + \sigma_3)}{2} \sin \beta\right\}^2}$$

$$= [(\sigma_1 + \sigma_3)/2] \sqrt{\sin^2 \phi - \sin^2 \beta}, \text{ using Eq. 13.13.}$$

Now,

$$\sigma_v = OG + GE = \frac{(\sigma_1 + \sigma_3)}{2} \cos \beta + \frac{(\sigma_1 + \sigma_3)}{2} \sqrt{\sin^2 \phi - \sin^2 \beta}$$

$$\sigma_v = \frac{(\sigma_1 + \sigma_3)}{2} (\cos \beta - \sqrt{\sin^2 \phi - \sin^2 \beta})$$

or

$$\sigma_v = \frac{(\sigma_1 + \sigma_3)}{2} (\cos \beta - \sqrt{\cos^2 \beta - \cos^2 \phi}) \qquad \ldots(\text{Eq. 13.14})$$

$$\sigma_l = OG - FG = \left(\frac{\sigma_1 + \sigma_3}{2}\right) \cos \beta - \left(\frac{\sigma_1 + \sigma_3}{2}\right) \sqrt{\sin^2 \phi - \sin^2 \beta}$$

$$\sigma_l = \left(\frac{\sigma_1 + \sigma_3}{2}\right)(\cos\beta - \sqrt{\sin^2\phi - \sin^2\beta})$$

or

$$\sigma_l = \left(\frac{\sigma_1 + \sigma_3}{2}\right)(\cos\beta - \sqrt{\cos^2\beta - \cos^2\phi}) \quad \ldots(\text{Eq. 13.15})$$

$$\frac{\sigma_l}{\sigma_v} = K = \frac{\cos\beta - \sqrt{\cos^2\beta - \cos^2\phi}}{\cos\beta + \sqrt{\cos^2\beta - \cos^2\phi}} \quad \ldots(\text{Eq. 13.16})$$

K is known as the 'Conjugate ratio'.

Using Eq. 13.12,

$$\sigma_l = \gamma z \cdot \cos\beta \cdot \left(\frac{\cos\beta - \sqrt{\cos^2\beta - \cos^2\phi}}{\cos\beta + \sqrt{\cos^2\beta - \cos^2\phi}}\right) \quad \ldots(\text{Eq. 13.17})$$

If σ_l is defined as $K_a \cdot \gamma z$ as usual,

$$K_a = \cos\beta \left(\frac{\cos\beta - \sqrt{\cos^2\beta - \cos^2\phi}}{\cos\beta + \sqrt{\cos^2\beta - \cos^2\phi}}\right) \quad \ldots(\text{Eq. 13.18})$$

K_a is the 'Rankine's Coefficient' of active earth pressure for the case inclined surcharge—sloping backfill.

The distribution of pressure with the height of the wall is linear, the pressure distribution diagram being triangular as shown in Fig. 13.12(c). The total active thrust P_a per unit length of the wall acts at $(1/3)H$ above the base of the wall and is equal to $\frac{1}{2}K_a\gamma \cdot H^2$; it acts parallel to the surface of the fill.

If the backfill is submerged, the lateral pressure due to the submerged unit weight of the backfill soil acts parallel to the surface of the backfill, while the lateral pressure due to pore water acts horizontally.

For the passive case, the right-hand sides of Eqs. 13.14 and 13.15 will represent σ_l and σ_v respectively.

The conjugate ratio, K, is given by

$$K = \frac{\cos\beta + \sqrt{\cos^2\beta - \cos^2\phi}}{\cos\beta - \sqrt{\cos^2\beta - \cos^2\phi}} \quad \ldots(\text{Eq. 13.19})$$

and the passive pressure coefficient K_p is given by

$$K_p = \cos\beta \left(\frac{\cos\beta + \sqrt{\cos^2\beta - \cos^2\phi}}{\cos\beta - \sqrt{\cos^2\beta - \cos^2\phi}}\right) \quad \ldots(\text{Eq. 13.20})$$

The total thrust or passive resistance per unit length of wall P_p is given by $\frac{1}{2}K_p\gamma \cdot H^2$, acting at $\frac{1}{3}H$ above the base of the wall, parallel to the backfill surface.

It is interesting to note that if $\beta = 0$ is substituted in Eqs. 13.18 and 13.20, we obtain Eqs. 13.8 and 13.9, respectively, for the case with the backfill surface horizontal.

13.6.7 Effect of Inclined Back of Wall

The back of a retaining wall may not always be vertical, but may occasionally be battered or inclined. In such a case, the total lateral earth pressure on an imaginary vertical surface passing through the heel of the wall is found and is combined vectorially with the weight of the soil wedge between the imaginary face and the back of the wall, to give the resultant thrust on the wall.

This procedure is applicable whether the backfill surface is horizontal or inclined, as illustrated in Fig. 13.13.

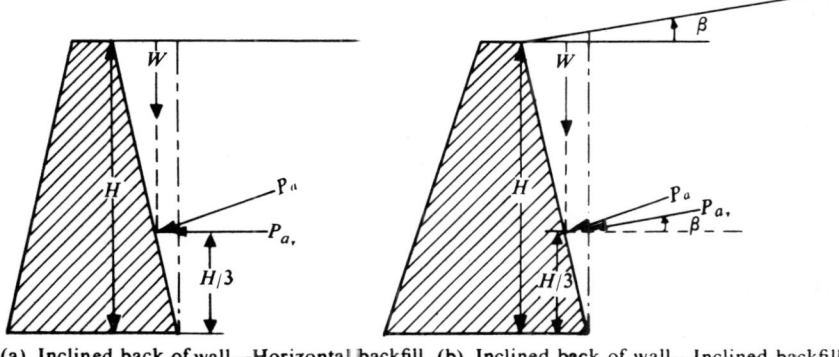

(a) Inclined back of wall—Horizontal backfill (b) Inclined back of wall—Inclined backfill

Fig. 13.13 Effect of inclined back of wall on lateral earth pressure

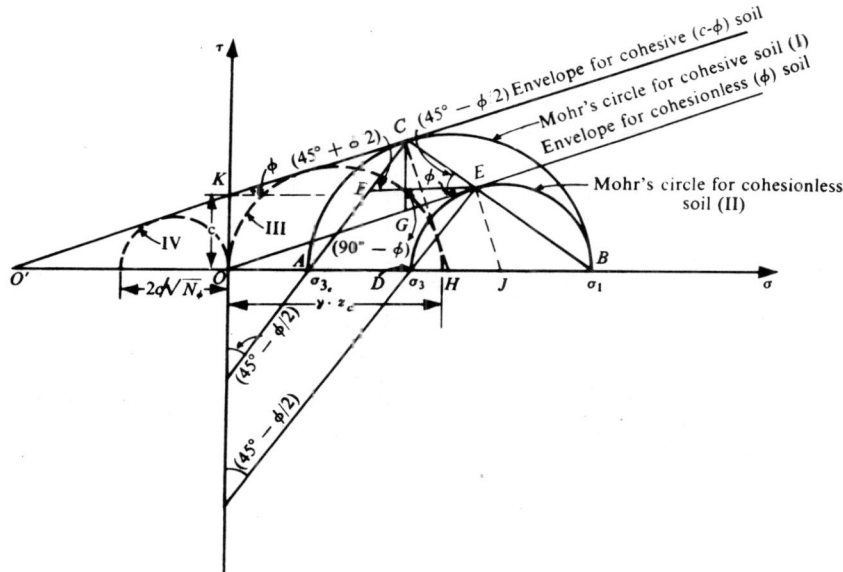

Fig. 13.14 Mohr's circles of stress for active pressure for a cohesionless soil and for a cohesive soil

538 Geotechnical Engineering

P_{a_v} is the lateral pressure on the imaginary vertical face through the heel of the wall, acting at $H/3$ above the base. The weight of the soil wedge is W, acting through its centroid. The vector sum of these is the total thrust P_a on the back of the wall.

13.6.8 Active Earth Pressure of Cohesive Soil

A cohesive soil is partially self-supporting and it will, therefore, exert a smaller pressure on a retaining wall than a cohesionless soil with the same angle of friction and density.

The Mohr's circles of stress for a cohesionless soil and for a cohesive soil for an element at a depth z for the active case are superimposed and shown in Fig. 13.14.

From the geometry of Fig. 13.14,

The difference between σ_3 and $\sigma_{3_c} = AD = EF = \dfrac{CE}{\cos(45° - \phi/2)}$

But, $\dfrac{CE}{CG} = \dfrac{CE}{c}$ and also, $\dfrac{CE}{CG} = \dfrac{\sin(90° - \phi)}{\sin(45° + \phi/2)} = 2\dfrac{\sin(45° - \phi/2)\cos(45° - \phi/2)}{\cos(45° - \phi/2)}$

$$\therefore CE = 2c \sin(45° - \phi/2)$$

Substituting, $\left(\sigma_3 - \sigma_{3_c}\right) = \dfrac{2c \sin(45° - \phi/2)}{\cos(45° - \phi/2)} = 2c \tan(45° - \phi/2)$

$$\sigma_{3_c} = \sigma_3 - 2c \tan(45° - \phi/2)$$

But, σ_3 for a cohesionless soil $= \gamma \cdot z \cdot \tan^2(45° - \phi/2)$

$$\therefore \sigma_{3_c} = \gamma \cdot z \tan^2(45° - \phi/2) - 2c \tan(45° - \phi/2) \qquad \text{...(Eq. 13.21)}$$

or $\sigma_{3_c} = \dfrac{\gamma \cdot z}{\tan^2(45° + \phi/2)} - \dfrac{2c}{\tan(45° + \phi/2)} \qquad \text{...(Eq. 13.22)}$

$$= \dfrac{\gamma z}{N_\phi} - \dfrac{2c}{\sqrt{N_\phi}} \qquad \text{...(Eq. 13.23)}$$

where $N_\phi = \tan^2(45° + \phi/2)$, called 'flow value'.

The equation for σ_{3_c}, or the lateral pressure for a cohesive soil, is known as Bell's equation.

In fact, this may be also obtained from the relationship between principal stresses expressed by Eq. 8.36 by taking $\sigma_1 = \gamma z$ and $\sigma_3 = \sigma_h$ as follows:

$$\sigma_1 = \gamma z \text{ and } \sigma_3 = \sigma_h \text{ as follows:}$$

$$\sigma_1 = \sigma_3 N_\phi + 2c \sqrt{N_\phi}$$

$$\sigma_1 = \gamma z, \ \sigma_3 = \sigma_h$$

Lateral Earth Pressure and Stability of Retaining Walls 539

$$\therefore \quad \gamma z = \sigma_h N_\phi + 2c\sqrt{N_\phi}$$

or $\quad \sigma_h = \dfrac{\gamma z}{N_\phi} - \dfrac{2c}{\sqrt{N_\phi}}$, as obtained earlier.

With the usual notation, $\dfrac{1}{N_\phi} = K_a$ for a cohesionless soil.

$$\therefore \quad \sigma_{3_c} = \sigma_h = K_a \gamma z - \dfrac{2c}{\sqrt{N_\phi}}$$

At the surface, $z = 0$ and $\sigma_h = -2c/\sqrt{N_\phi}$ ...(Eq. 13.24)

The lateral pressure distribution diagram is obtained by superimposing the diagram for the first and second terms, as shown in Fig. 13.15.

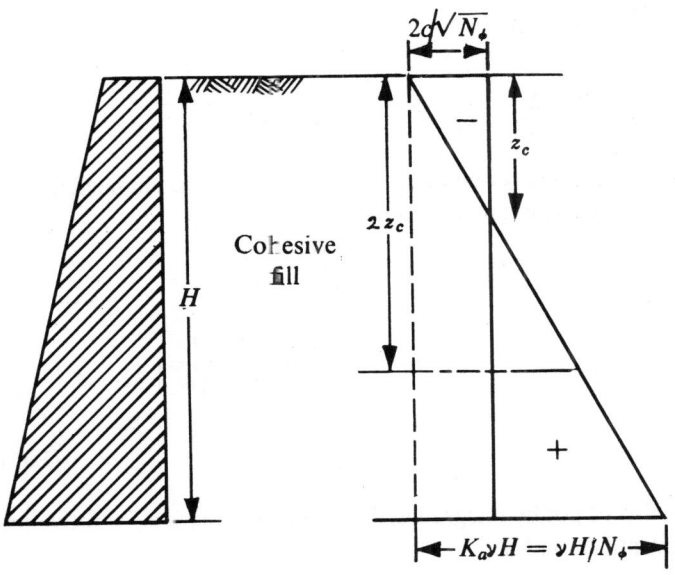

Fig. 13.15 Active pressure distribution for a cohesive soil

The negative values of active pressure up to a depth equal to half of the so-called 'critical depth' indicate suction effect or tensile stresses; however, it is well known that soils cannot withstand tensile stresses and hence, suction is unlikely to occur. Invariably, the pressure from the surface in the tension zone is ignored.

The total active thrust per unit length of the wall is obtained by considering a net area of the pressure distribution diagram or, by integrating the general expression for pressure over the entire height of the wall.

Thus,
$$P_a = \int_0^H \sigma_{3_c} \cdot dz$$
$$= \int_0^H \left(\frac{\gamma z}{N_\phi} - \frac{2c}{\sqrt{N_\phi}}\right) dz$$
$$= \left[\frac{\gamma z^2}{2N_\phi} - \frac{2c}{\sqrt{N_\phi}} \cdot z\right]_0^H$$

or
$$P_a = \frac{1}{2}\frac{\gamma H^2}{N_\phi} - \frac{2c}{\sqrt{N_\phi}} \cdot H \qquad \ldots(\text{Eq. 13.25})$$

In practice, cracks occur over the entire depth, z_c, of the tensile zone, making the backfill soil lose contact with the wall in that zone.
z_c may be got by equating σ_{3_c} to zero.

$$\sigma_{3_c} = \frac{\gamma z_c}{N_\phi} - \frac{2c}{\sqrt{N_\phi}} = 0$$

$$z_c = \frac{2c}{\sqrt{N_\phi}} \cdot \frac{N_\phi}{\gamma} = \frac{2c}{\gamma} \cdot \sqrt{N_\phi} = \frac{2c}{\gamma} \cdot \tan(45° + \phi/2) \qquad \ldots(\text{Eq. 13.26})$$

If the total active thrust per unit length of the wall is to be obtained ignoring the tensile stresses, one has to proceed as follows:

$$P_a = \frac{1}{2}(H - z_c)\left[\frac{\gamma H}{N_\phi} - \frac{2c}{\sqrt{N_\phi}}\right]$$

$$= \frac{1}{2}\left(H - \frac{2c}{\gamma} \cdot \sqrt{N_\phi}\right)\left[\frac{\gamma H}{N_\phi} - \frac{2c}{\sqrt{N_\phi}}\right]$$

or
$$P_a = \frac{1}{2}\frac{\gamma H^2}{N_\phi} - \frac{2cH}{\sqrt{N_\phi}} + \frac{2c^2}{\gamma} \qquad \ldots(\text{Eq. 13.27})$$

This may also be got by integrating the general expression for active pressure between the depths z_c and H:

$$P_a = \int_{z_c}^H \left(\frac{\gamma z}{N_\phi} - \frac{2c}{\sqrt{N_\phi}}\right) dz$$

$$= \left[\frac{\gamma z^2}{2N_\phi} - \frac{2c}{\sqrt{N_\phi}} \cdot z\right]_{z_c}^H$$

$$= \frac{\gamma}{2N_\phi}(H^2 - z_c^2) - \frac{2c}{\sqrt{N_\phi}}(H - z_c)$$

Substituting for z_c from Eq. 13.25,

$$P_a = \frac{\gamma}{2N_\phi}\left(H^2 - \frac{4c^2}{\gamma^2}\cdot N_\phi\right) - \frac{2c}{\sqrt{N_\phi}}\left(H - \frac{2c}{\gamma}\sqrt{N_\phi}\right)$$

$$= \frac{1}{2}\frac{\gamma H^2}{N_\phi} - \frac{2c^2}{\gamma} - 2\frac{cH}{\sqrt{N_\phi}} + \frac{4c^2}{\gamma}$$

or $\qquad P_a = \frac{1}{2}\frac{H^2}{N_\phi} - \frac{2cH}{\sqrt{N_\phi}} + \frac{2c^2}{\gamma}$, as obtained earlier.

For pure clay, $\phi = 0$

$$\therefore \quad P_a = \frac{1}{2}\gamma H^2 - 2cH + \frac{2c^2}{\gamma} \qquad \text{...(Eq. 13.28)}$$

This acts at $(H - z_c)/3$ above the base.

The net pressure over a depth of $2z_c$ is obviously zero. This indicates that a cohesive soil mass should be able to stand unsupported up to this depth which is known as the critical depth.

The critical depth, H_c, is given by

$$H_c = 2z_c = \frac{4c}{\gamma}\cdot\sqrt{N_\phi} \qquad \text{...(Eq. 13.29)}$$

If $\phi = 0$, $\quad H_c = \frac{4c}{\gamma} \qquad \text{...(Eq. 13.30)}$

Equations 13.24 and 13.26 may be derived from the geometry of the Mohr's circles IV and III respectively, instead of from Eq. 13.23. The proofs are let to the reader.

13.6.9 Passive Earth Pressure of Cohesive Soil

Cohesion is known to increase the passive earth resistance of a soil. This fact can be mathematically demonstrated from the relationship between the principal stresses that may be derived from the geometry of the Mohr's circle relating to the passive case for a $c - \phi$ soil, taking cognizance of the fact that $\sigma_3 = \gamma \cdot z$ and $\sigma_1 = \sigma_h$ (Fig. 13.16).

$$\sigma_1 = \sigma_3 N_\phi + 2c\sqrt{N_\phi}$$

$$\sigma_3 = \gamma z \quad \text{and} \quad \sigma_1 = \sigma_{h_c}$$

$$\therefore \quad \sigma_{1_c} = \sigma_{h_c} = \gamma z\, N_\phi + 2c\sqrt{N_\phi} \qquad \text{...(Eq. 13.31)}$$

(Here, $K_p = N_\phi$ in the usual notation).

The pressure distribution with depth is shown in Fig. 13.17.

The total passive resistance per unit length of wall is $P_P = P_P' + P_P'' = \frac{1}{2}\gamma H^2 N_\phi + 2cH\sqrt{N_\phi}$.

P_p' acts at $H/3$ and P_p'' acts at $H/2$ above the base. The location of P_p may be found by moments about the base.

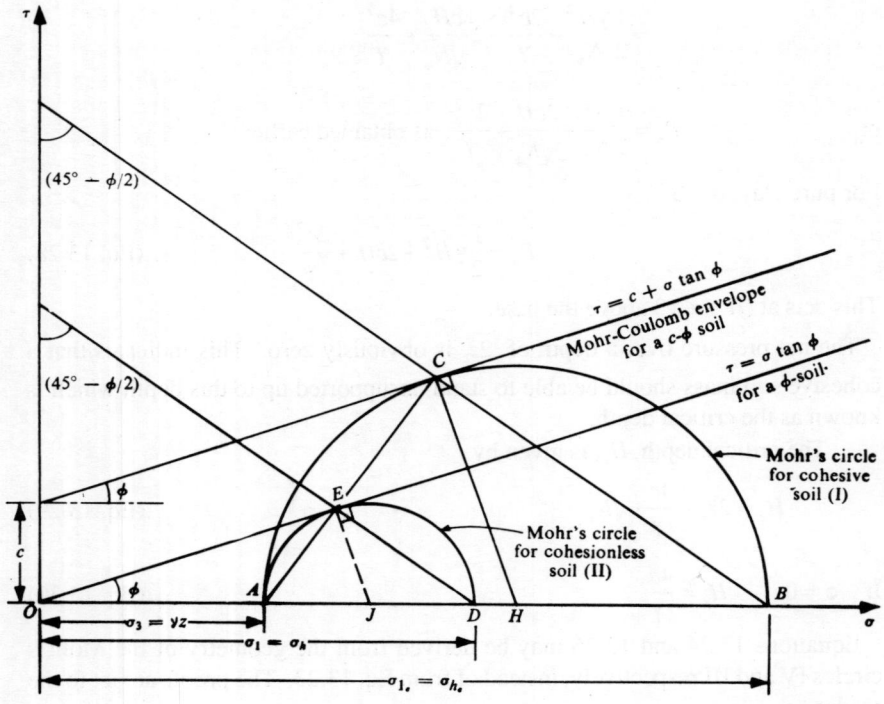

Fig. 13.16 Mohr's circles of stress for passive pressure for a cohesionless soil and for a cohesive soil

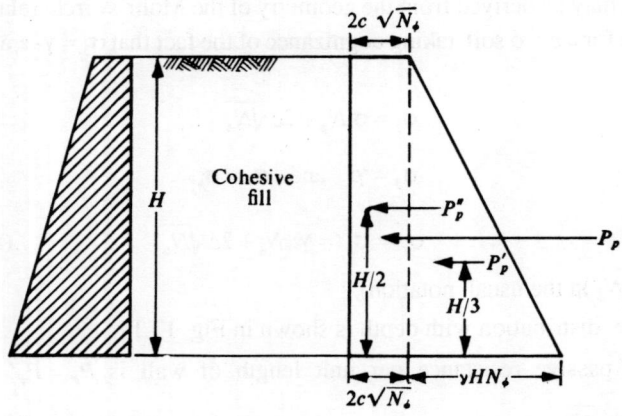

Fig. 13.17 Passive pressure distribution for the cohesive soil

13.7 COULOMB'S WEDGE THEORY

Charles Augustine Coulomb (1776), a famous French scientist and military engineer, was the first to try to give a scientific basis to the hazy and arbitrary idea existing in his time regarding lateral earth pressure on walls.

Coulomb's theory considers the soil behind the wall as a whole instead of as an element in the soil. If a wall supporting a granular soil were not to be there, the soil will slump down to its angle of repose or internal friction. It is therefore reasonable to assume that if the wall only moved forward slightly a rupture plane would develop somewhere between the wall and the surface of repose. The triangular mass of soil between this plane of failure and the back of the wall is referred to as the 'sliding wedge'. It is reasoned that, if the retaining wall were suddenly removed, the soil within the sliding wedge would slump downward. Therefore, an analysis of the forces acting on the sliding wedge at incipient failure will reveal the thrust from the lateral earth pressure which is necessary for the wall to withstand in order to hold the soil mass in place. This is why Coulomb's theory is also called the 'Wedge theory', implying the existence of a plane rupture surface. However, Coulomb recognised the possibility of the existence of a curved rupture surface, although he considered a plane surface for the sake of mathematical simplicity. In fact, it is now established that the assumption of a plane rupture surface introduces significant error in the determination of passive earth resistance, a curved rupture surface being nearer to facts, as demonstrated by experiments.

In the course of time Coulomb's theory underwent some alterations and new developments. The theory is very adaptable to graphical solution and the effects of wall friction and batter are automatically allowed for. Poncelet (1840), Culmann (1866), Rebhann (1871) and Engessor (1880) are the notable figures who contributed to further development of Coulomb's theory.

The significance of Coulomb's work may be recognised best by the fact that his ideas on earth pressure still prevail in their principal points with a few exceptions and are considered valid even today in the design of retaining walls.

13.7.1 Assumptions

The primary assumptions in Coulomb's wedge theory are as follows:
1. The backfill soil is considered to be dry, homogeneous and isotropic; it is elastically undeformable but breakable, granular material, possessing internal friction but no cohesion.
2. The rupture surface is assumed to be a plane for the sake of convenience in analysis. It passes through the heel of the wall. It is not actually a plane, but is curved and this is known to Coulomb.
3. The sliding wedge acts as a rigid body and the value of the earth pressure is obtained by considering its equilibrium.
4. The position and direction of the earth pressure are assumed to be known. The thrust acts on the back of the wall at a point one-third of the height of the wall above the base of the wall and makes an angle δ, with the normal to the back face of the wall. This is an angle of friction between the wall and backfill soil and is usually called 'wall friction'.

5. The problem of determining the earth pressure is solved, on the basis of two-dimensional case of 'plane strain'. This is to say that, the retaining wall is assumed to be of great length and all conditions of the wall and fill remain constant along the length of the wall. Thus, a unit length of the wall perpendicular to the plane of the paper is considered.
6. When the soil wedge is at incipient failure or the sliding of the wedge is impending, the theory gives two limiting values of earth pressure, the least and the greatest (active and passive) compatible with equilibrium.

The additional inherent assumptions relevant to the theory are as follows:

7. The soil forms a natural slope angle, ϕ, with the horizontal, without rupture and sliding. This is called the angle of repose and in the case of dry cohesionless soil, it is nothing but the angle of internal friction. The concept of friction was understood by Coulomb.
8. If the wall yields and the rupture of the backfill soil takes place, a soil wedge is torn off from the rest of the soil mass. In the active case, the soil wedge slides sideways and downward over the rupture surface, thus exerting a lateral pressure on the wall. In the case of passive earth resistance, the soil wedge slides sideways and upward on the rupture surface due to the forcing of the wall against the fill. These are illustrated in Fig. 13.18.
9. For a rupture plane within the soil mass, as well as between the back of the wall and the soil, Newton's law of friction is valid (that is to say, the shear force developed due to friction is the coefficient of friction times the normal force acting on the plane). This angle of friction, whose tangent is the coefficient of friction, is dependent upon the physical properties of the materials involved.
10. The friction is distributed uniformly on the rupture surface.
11. The back face of the wall is a plane.
12. The following considerations are employed for the determination of the active and passive earth pressures:

Among the infinitely large number of rupture surfaces that may be passed through the heel of the wall, the most dangerous one is that for which the active earth pressure is a maximum (the wall must resist even the greatest value to be stable).

In the case of passive earth resistance, the most dangerous rupture surface is the one for which the resistance is a minimum. The minimum force necessary to tear off the soil wedge from the soil mass when the wall is forced against the soil is thus the criterion, since failure is sure to occur at greater force. Note that this is in contrast to the minimum and maximum for active and passive cases in relation to the movement of the wall away from or towards the fill.

Also note that Coulomb's theory treats the soil mass in the sliding wedge in its entirety. The assumptions permit one to treat the problem as a statically determinate one.

Coulombs' theory is applicable to inclined wall faces, to a wall with a broken face, to a sloping backfill curved backfill surface, broken backfill surface and to concentrated or distributed surcharge loads.

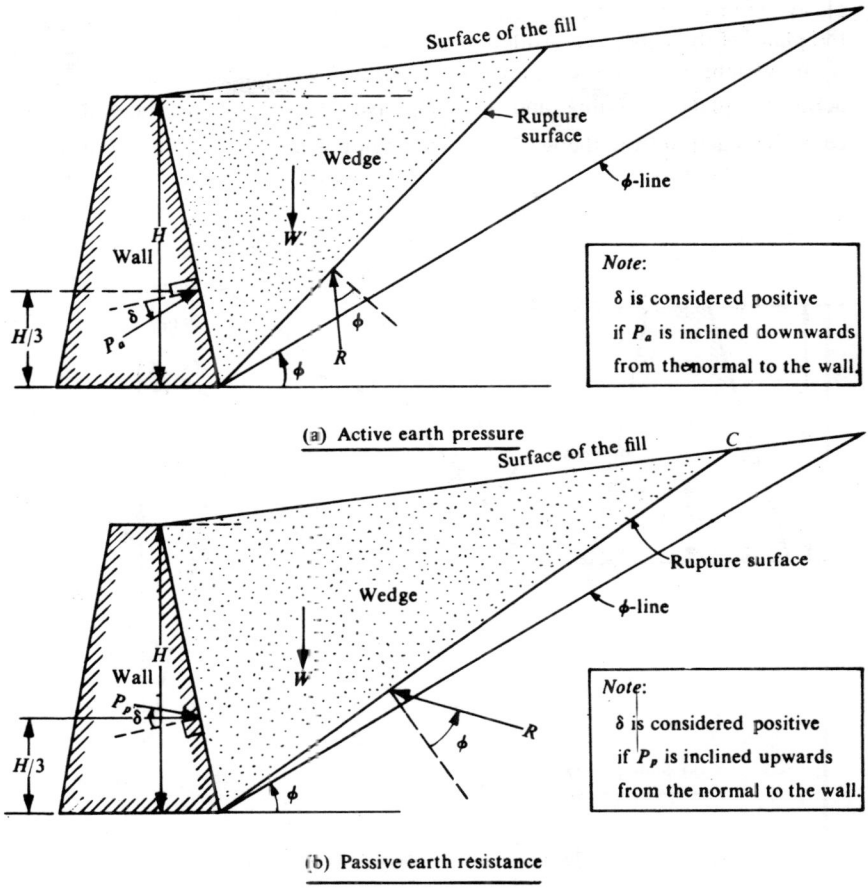

Fig. 13.18 Coulomb's theory—active and passive cases

One of the main deficiencies in Coulomb's theory is that, in general, it does not satisfy the static equilibrium condition occurring in nature. The three forces (weight of the sliding wedge, earth pressure and soil reaction on the rupture surface) acting on the sliding wedge do not meet at a common point, when the sliding surface is assumed to be planar. Even the wall friction was not originally considered but was introduced only some time later.

Regardless of this deficiency and other assumptions, the theory gives useful results in practice; however, the soil constants should be determined as accurately as possible.

13.7.2 Active Earth Pressure of Cohesionless Soil

A simple case of active earth pressure on an inclined wall face with a uniformly sloping backfill may be considered first. The backfill consists of homogeneous, elastic and isotropic cohesionless soil. A unit length of the wall perpendicular to the plane of the paper is considered. The forces acting on the sliding wedge are (*i*) W, weight of the soil contained in the sliding wedge, (*ii*) R, the soil reaction across the plane of sliding, and (*iii*) the active thrust P_a against the wall, in this case, the reaction from the wall on to the sliding wedge, as shown in Fig. 13.19.

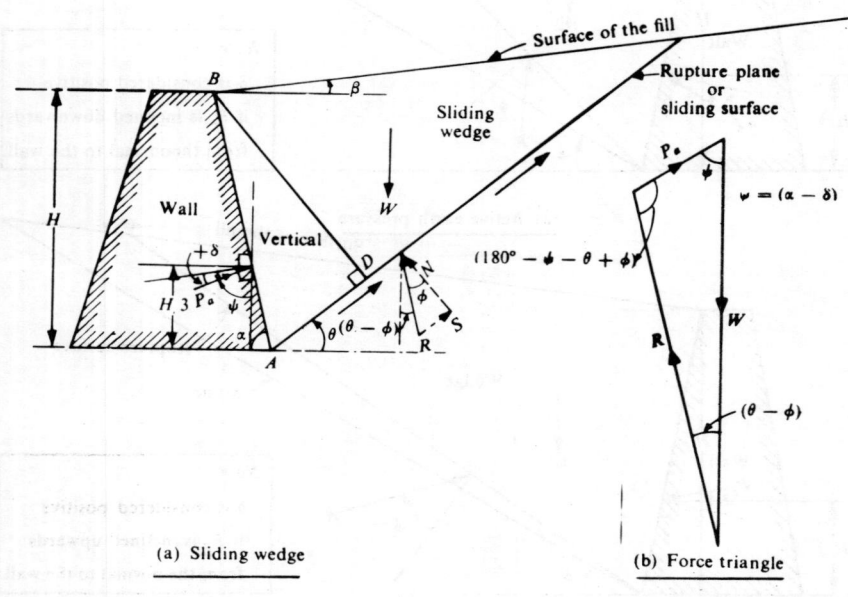

(a) Sliding wedge (b) Force triangle

Fig. 13.19 Active earth pressure of cohesionless soil—Coulomb's theory

The triangle of forces is shown in Fig. 13.19(*b*). With the nomenclature of Fig. 13.19, one may proceed as follows for the determination of the active thrust, P_a:

$$W = \gamma \,(\text{area of wedge } ABC)$$

$$\triangle ABC = \frac{1}{2} AC \cdot BD, \; BD \text{ being the altitude on to } AC.$$

$$AC = AB \cdot \frac{\sin(\alpha + \beta)}{\sin(\theta - \beta)}$$

$$BD = AB \cdot \sin(\alpha + \theta)$$

$$AB = \frac{H}{\sin \alpha}$$

Substituting and simplifying,

$$W = \gamma \frac{H^2}{2\sin^2\alpha} \cdot \sin(\theta+\alpha) \cdot \frac{\sin(\alpha+\beta)}{\sin(\theta-\beta)} \qquad \text{...(Eq. 13.3)}$$

From the triangle of forces,

$$\frac{P_a}{\sin(\theta-\phi)} = \frac{W}{\sin(180°-\psi-\theta+\phi)}$$

$$\therefore \quad P_a = W \cdot \frac{\sin(\theta-\phi)}{\sin(180°-\psi-\theta+\phi)}$$

Substituting for W,

$$P_a = \frac{1}{2} \frac{\gamma H^2}{\sin^2\alpha} \cdot \frac{\sin(\theta-\phi)}{\sin(180°-\psi-\theta+\phi)} \cdot \frac{\sin(\theta+\alpha) \cdot \sin(\alpha+\beta)}{\sin(\theta-\beta)}$$

...(Eq. 13.32)

The maximum value for P_a is obtained by equating the first derivative of P_a with respect to θ to zero;

or $\frac{\partial P_a}{\partial \theta} = 0$, and substituting the corresponding value of θ.

The value of P_a so obtained is written as

$$P_a = \frac{1}{2} \gamma \cdot H^2 \cdot \frac{\sin^2(\alpha+\phi)}{\sin^2\alpha \sin(\alpha-\delta)\left[1+\sqrt{\frac{\sin(\phi+\delta)\sin(\phi-\beta)}{\sin(\alpha-\delta)\sin(\alpha+\beta)}}\right]^2}$$

...(Eq. 13.33)

This is usually written as

$$P_a = \frac{1}{2} \gamma H^2 \cdot K_a,$$

where

$$K_a = \frac{\sin^2(\alpha+\phi)}{\sin^2\alpha \cdot \sin(\alpha-\delta)\left[1+\sqrt{\frac{\sin(\phi+\delta)\cdot\sin(\phi-\beta)}{\sin(\alpha-\delta)\sin(\alpha+\beta)}}\right]^2}$$

...(Eq. 13.34)

K_a being the coefficient of active earth pressure.

For a vertical wall retaining a horizontal backfill for which the angle of wall friction is equal to ϕ, K_a reduces to

$$K_a = \frac{\cos\phi}{(1+\sqrt{2}\sin\phi)^2} \qquad \text{...(Eq. 13.35)}$$

by substituting $\alpha = 90°, \beta = 0°,$ and $\delta = \phi$.

For a smooth vertical wall retaining a backfill with horizontal surface,
$$\alpha = 90°, \delta = 0, \text{ and } \beta = 0;$$

$$K_a = \frac{1-\sin\phi}{1+\sin\phi} = \tan^2(45°-\phi/2) = 1/N_\phi,$$

which is the same as the Rankine value.

In fact, for this simple case, one may proceed from fundamentals as follows:

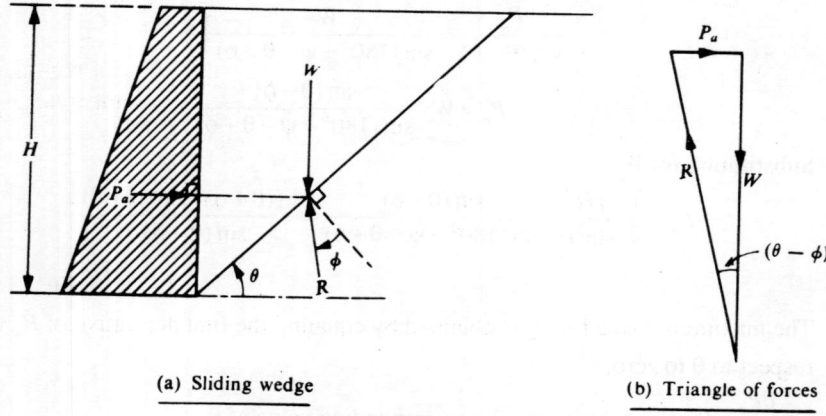

(a) Sliding wedge (b) Triangle of forces

Fig. 13.20 Active earth pressure of cohesionless soil special case: $\alpha = 90°$, $\delta = \beta = 0°$.

With reference to Fig. 13.20(b),

$$P_a = W \tan(\theta - \phi),$$

$$W = \frac{1}{2}\gamma H^2 \cdot \cot\theta$$

$$\therefore \quad P_a = \frac{1}{2}\gamma H^2 \cot\theta \tan(\theta - \phi) \qquad \text{...(Eq. 13.36)}$$

For maximum value of P_a, $\dfrac{\partial P_a}{\partial \theta} = 0$

$$\therefore \quad \frac{\partial P_a}{\partial \theta} = \frac{1}{2}\gamma H^2 \left[-\frac{\tan(\theta - \phi)}{\sin^2\theta} + \frac{\cot\theta}{\cos^2(\theta - \phi)} \right] = 0$$

or

$$\frac{-\sin(\theta - \phi)\cos(\theta - \phi) + \sin\theta \cos\theta}{\sin^2\theta \cos^2(\theta - \phi)} = 0$$

or $\sin\phi \cos(2\theta - \phi) = 0$, on simplification.

$\therefore \cos(2\theta - \phi) = 0$ or $\theta = 45° + \phi/2$

Substituting in Eq. 13.35, $P_a = \frac{1}{2}\gamma H^2 \tan^2(45° - \phi/2)$ ...(Eq. 13.37)

as obtained by substitution in the general equation.

Ironically, this approach is sometimes known as 'Rankine's method of Trial Wedges'.

A few representative values of K_a from Eq. 13.34 for certain values of ϕ, δ, α and β are shown in Table 13.2.

Table 13.2 Coefficient of active earth pressure from Coulomb's theory

δ↓ φ →	20°		30°	40°
		α = 90°, β = 0°		
0°	0.49		0.33	0.22
10°	0.45		0.32	0.21
20°	0.43		0.31	0.20
30°	...		0.30	0.20
		α = 90°, β = 10°		
0°	0.51		0.37	0.24
10°	0.52		0.35	0.23
20°	0.52		0.34	0.22
30°	....		0.33	0.22
		α = 90°, β = 20°		
0°	0.88		0.44	0.27
10°	0.90		0.43	0.26
20°	0.94		0.42	0.25
30°	....		0.42	0.25

It may be observed that the theoretical solution is thus rather complicated even for relatively simple cases. This fact has led to the development of graphical procedures for arriving at the total thrust on the wall. Poncelet (1840), Culmann (1866), Rebhann (1871), and Engessor (1880) have given efficient graphical solutions, some of which will be dealt with in the subsequent subsections.

An obvious graphical approach that suggests itself is the "Trial-Wedge method". In this method, a few trial rupture surfaces are assumed at varying inclinations, θ, with the horizontal and passing through the heel of the wall; for each trial surface the triangle of forces is completed and the value of P_a found. A $θ - P_a$ plot is made which should appear somewhat as shown in Fig. 13.21, if an adequate number of intelligently planned trial rupture surfaces are analysed.

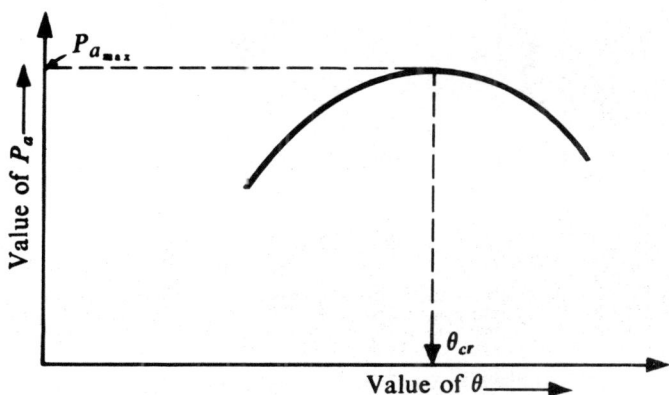

Fig. 13.21 Angle of inclination of trial rupture plane versus active thrust

The maximum value of P_a from this plot gives the anticipated total active thrust on the wall per lineal unit and the corresponding value of θ, the inclination of the most probable rupture surface.

Wall friction

At this juncture, a few comments on wall friction may be appropriate. In the active case, the outward stretching leads to a downward motion of the backfill soil relative to the wall. Such a downward shear force upon the wall is called 'positive' wall friction for the active case. This leads to the upward inclination of the active thrust exerted on the sliding wedge as shown in Fig. 13.19(a). This means that the active thrust exerted *on the wall* will be directed with an downward inclination.

In the passive case, the horizontal compression must be accompanied by an upward bulging of the soil and hence there tends to occur an upward shear on the wall. Such an upward shear on the wall is said to be 'positive' wall friction for the passive case. This leads to the downward inclination of the passive thrust exerted on the sliding wedge as shown in Fig. 13.24(a); this means that the passive resistance exerted *on the wall* will be directed with upward inclination.

In the active case wall friction is almost always positive. Sometimes, under special conditions, such as when part of the backfill soil immediately behind the wall is excavated for repair purposes and the wall is braced against the remaining earth mass of the backfill, negative wall friction might develop.

Either positive or negative wall friction may develop in the passive case. The sign of wall friction must be determined from a study of motions expected for each field situation.

Once wall friction is present, the shape of the rupture surface is curved and not plane. The nature of the surface for positive and negative values of wall friction is shown in Figs. 13.22(a) and (b), respectively.

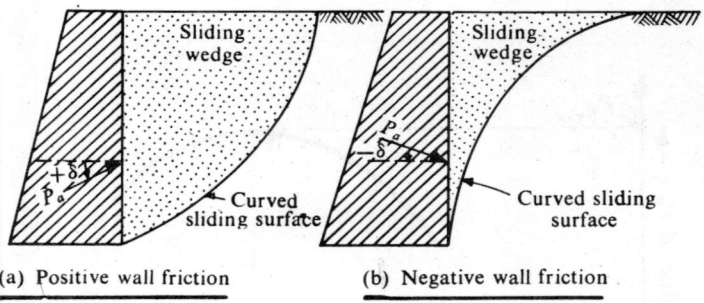

(a) Positive wall friction   (b) Negative wall friction

Fig. 13.22 Positive and negative wall friction for active case along with probable shape of sliding surface

The angle of wall friction, δ, will not be greater than ϕ; at the maximum it can equal ϕ, for a rough wall with a loose fill. For a wall with dense fill, δ will be less than ϕ. It may range from $\frac{1}{2}\phi$ to $\frac{3}{4}\phi$ in most cases; it is usually assumed as $(2/3)\phi$ in the absence of precise data.

The possibility of δ shifting from $+\phi$ to $-\phi$ in the worst case should be considered in the design of a retaining wall.

The value of K_a for the case of a vertical wall retaining a fill with a level surface, in which ϕ ranges from 20° to 40° and δ ranges from 0° to ϕ, may be obtained from the chart given in Fig. 13.23.

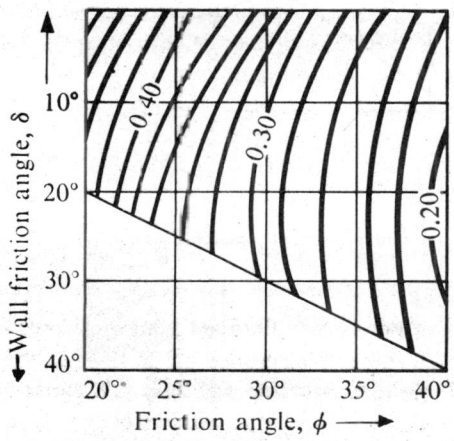

Fig. 13.23 Coefficient of active pressure as a function of wall friction

The influence of wall friction on K_a may be understood from this chart to some extent.

The assumption of plane failure in the active case of the Coulomb theory is in error by only a relatively small amount. It has been shown by Fellenius that the assumption of circular arcs for failure surfaces leads to active thrusts that generally do not exceed the corresponding values from the Coulomb theory by more than 5 per cent.

13.7.3 Passive Earth Pressure of Cohesionless Soil

The passive case differs from the active case in that the obliquity angles at the wall and on the failure plane are of opposite sign. Plane failure surface is assumed for the passive case also in the Coulomb theory but the critical plane is that for which the passive thrust is minimum. The failure plane is at a much smaller angle to the horizontal than in the active case, as shown in Fig. 13.24.

The triangle of forces is shown in Fig. 13.24(b). With the usual nomenclature, the passive resistance P_p may be determined as follows:

$$W = \frac{1}{2}\frac{\gamma H^2}{\sin^2\alpha} \cdot \sin(\theta + \alpha) \cdot \frac{\sin(\alpha + \beta)}{\sin(\theta - \beta)}, \text{ as in the active case.}$$

552 Geotechnical Engineering

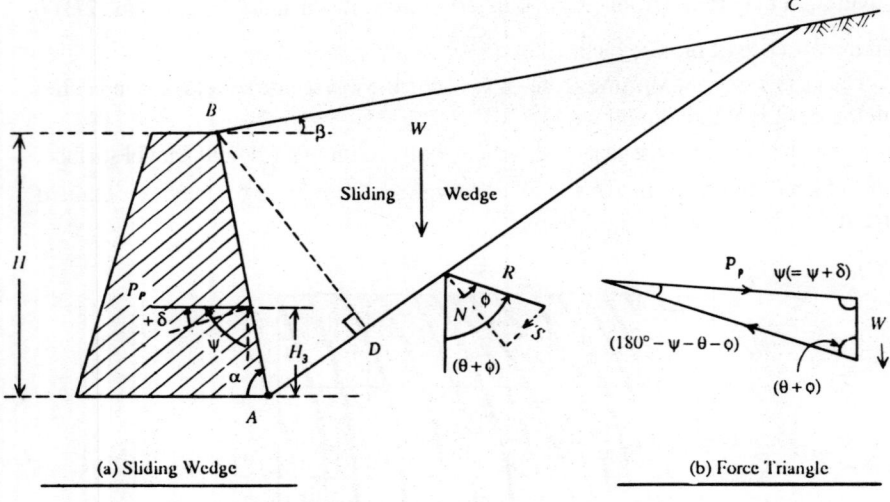

Fig. 13.24 Passive earth pressure of cohesionless soil—Coulomb's theory

From the triangle of forces

$$\frac{P_p}{\sin(\theta+\phi)} = \frac{W}{\sin(180°-\psi-\theta-\phi)}$$

$$\therefore \quad P_p = W \cdot \frac{\sin(\theta+\phi)}{\sin(180°-\psi-\theta-\phi)}$$

Substituting for W,

$$P_p = \frac{1}{2} \cdot \frac{\gamma H^2}{\sin^2 \alpha} \cdot \sin(\theta+\alpha) \cdot \frac{\sin(\alpha+\beta)}{\sin(\theta-\beta)} \cdot \frac{\sin(\theta+\phi)}{\sin(180°-\psi-\theta-\phi)}$$

...(Eq. 13.38)

The minimum value of P_p is obtained by differentiating Eq. 13.38 with respect to θ equating $\dfrac{\partial P_p}{\partial \theta}$ to zero, and substituting the corresponding value of θ.

The value of P_p so obtained may be written as

$$P_p = \frac{1}{2} \gamma H^2 \cdot K_p$$

where

$$K_p = \frac{\sin^2(\alpha-\phi)}{\sin^2 \alpha \cdot \sin(\alpha+\delta)\left[1 - \sqrt{\dfrac{\sin(\theta+\delta)\cdot\sin(\phi+\beta)}{\sin(\alpha+\delta)\cdot\sin(\alpha+\beta)}}\right]^2}$$

...(Eq. 13.39)

K_P being the coefficient of passive earth resistance.

For a vertical wall retaining a horizontal backfill and for which the wall friction is equal to ϕ,

$$\alpha = 90°, \ \beta = 0°, \text{ and } \delta = \phi, \text{ and } K_p \text{ reduces to}$$

$$K_p = \frac{\cos^2\phi}{\cos\phi\left[1 - \sqrt{\frac{2\sin\phi\cos\phi \cdot \sin\phi}{\cos\phi}}\right]^2}$$

or
$$K_p = \frac{\cos\phi}{(1 - \sqrt{2}\sin\phi)^2} \qquad \ldots\text{(Eq. 13.40)}$$

For a smooth vertical wall retaining a horizontal backfill,

$$\alpha = 90°, \ \beta = 0° \text{ and } \delta = 0°;$$

$$K_p = \frac{\cos^2\phi}{(1 - \sin\phi)^2} = \frac{1 - \sin^2\phi}{(1 - \sin\phi)^2} = \frac{(1 + \sin\phi)}{(1 - \sin\phi)} = \tan^2(45° + \phi/2) = N_\phi,$$

which is the same as the Rankine value.

For this simple case, it is possible to proceed from fundamentals, as has been shown for the active case.

[$(\theta + \phi)$ takes the place of $(\theta - \phi)$ and $(45° + \phi/2)$ that of $(45° - \phi/2)$ in the work relating to the active case.]

Coulomb's theory with plane surface of failure is valid only if the wall friction is zero in respect of passive resistance. The passive resistance obtained by plane failure surfaces is very much more than that obtained by assuming curved failure surfaces, which are nearer truth especially when wall friction is present. The error increases with increasing wall friction. This leads to errors on the unsafe side.

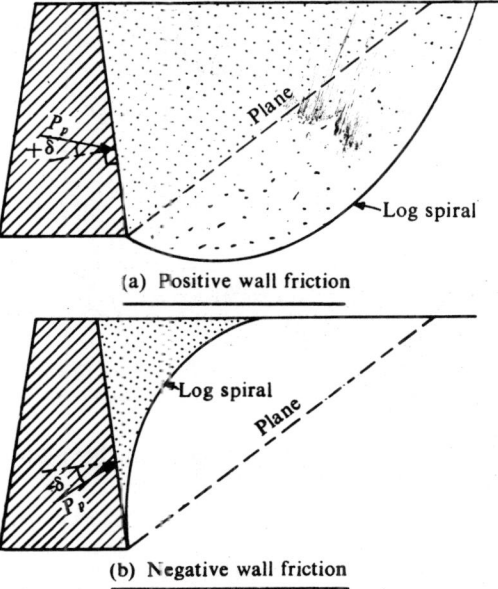

(a) Positive wall friction

(b) Negative wall friction

Fig. 13.25 Curved failure surface for estimating passive resistance

Terzaghi (1943) has presented a more rigorous type of analysis assuming curved failure surfaces (logarithmic spiral form) which resembles those shown in Fig. 13.25.

Terzaghi states that when δ is less than $(1/3)\phi$, the error introduced by assuming plane rupture surfaces instead of curved ones in estimating the passive resistance is not significant; however, when δ is greater than $(1/3)\phi$, the error is significant and hence cannot be ignored. This situation calls for the use of analysis based on curved rupture surfaces as given by Terzaghi; alternatively, charts and tables prepared by Caquot and Kerisel (1949) may be used. Extracts of such results are presented in Table 13.3 and Fig. 13.26.

Table 13.3 Passive pressure coefficient from curved failure surfaces

$\delta \downarrow \phi \rightarrow$	10°	20°	30°	40°
0°	1.42	2.04	3.00	4.60
$\phi/2$	1.56	2.60	4.80	10.40
ϕ	1.65	3.00	6.40	17.50
$-\phi$	0.72	0.58	0.54	0.52

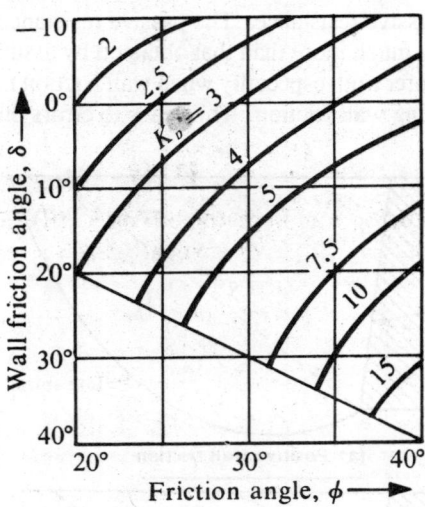

Fig. 13.26 Chart for passive pressure coefficient
(After Caquot and Kerisel, 1949)

Alternatively, Sokolovski's (1965) method may be used. This also gives essentially the same results.

The theoretical predictions regarding passive resistance with wall friction are not well confirmed by experimental evidence as those regarding active thrust and hence cannot be used with as much confidence. Tschebotarioff (1951) gives the results of a few large-scale laboratory tests in this regard.

13.7.4 Rebhann's Condition and Graphical Method

Rebhann (1871) is credited with having presented the criterion for the direct location of the failure plane assumed in the Coulomb's theory. His presentation is somewhat as follows:

Figure 13.27(a) represents a retaining wall retaining a cohesionless backfill inclined at $+\beta$ to the horizontal. Let BC be the failure plane, the position of which is to be determined.

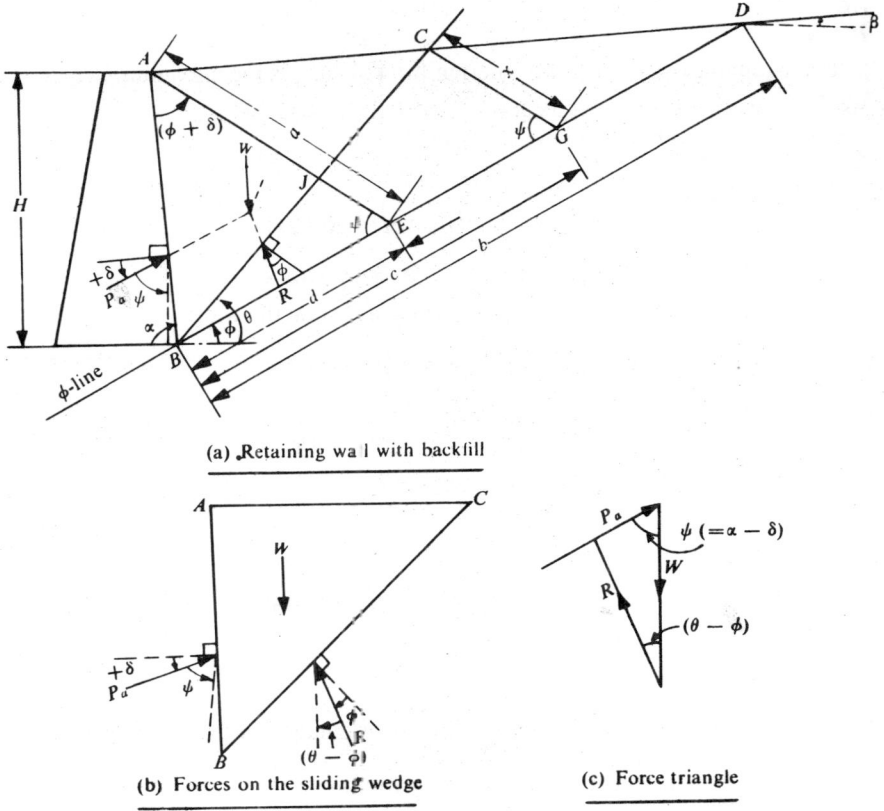

(a) Retaining wall with backfill

(b) Forces on the sliding wedge

(c) Force triangle

Fig. 13.27 Rebhann's condition for Coulomb's wedge theory—Location of failure plane for the active case

Figure 13.27 (b) represents the forces on the sliding wedge and Fig. 13.27 (c) represents the force triangle.

Let BD be a line inclined at ϕ to the horizontal through B, the heel of the wall, D being the intersection of this ϕ-line with the surface of the backfill.

The value of P_a depends upon the angle θ relating to the location of the failure plane. P_a will be zero when $\theta = \phi$, and increases with an increase in θ up to a limit, beyond which it decreases and reaches zero again when $\theta = 180° - \alpha$.

The situations when P_a is zero are both ridiculous, since in the first case, no wall is required to retain a soil mass at an angle ϕ and in the second, the failure wedge has no mass. Thus, the failure plane will lie between the ϕ-line and the back of the wall.

Let AE be drawn at an angle $(\phi + \delta)$ to the wall face AB to meet the ϕ-line in E. Let CG be drawn parallel to AE to meet the ϕ-line in G.

Let the distances be denoted as follows:
$AE = a$ $BG = c$ $CG = x$
$BD = b$ $BE = d$

It is required to determine the criterion for which P_a is the maximum, which is supposed to give the correct location of the failure surface.

Weight of the soil in the sliding wedge
$$W = \gamma \cdot (\Delta ABC)$$
$$= \gamma \cdot (\Delta ABD - \Delta BCD)$$
$$= \gamma \cdot (b/2) \cdot (\sin \psi)(a - x)$$

Value of thrust on the wedge (the same as the thrust on the wall).

$P_a = \dfrac{W \cdot x}{c}$, since ΔBCG is similar to the triangle of forces.

$$\therefore \quad P_a = \frac{\gamma bx}{2c}(a - x) \cdot \sin \psi \qquad \text{...(Eq. 13.41)}$$

$$\text{If } \frac{DG}{CG} = k, \; c = b - kx$$

$$\therefore \quad P_a = \frac{\gamma bx}{2(b - kx)} \cdot (a - x) \sin \psi$$

For the value of P_a to be a maximum, $\dfrac{\partial P_a}{\partial x} = 0$,

since x is the only value which varies with the orientation of the failure plane.

$$\therefore \quad \frac{\partial P_a}{\partial x} = (b - kx)(a - 2x) + kx(a - x) = 0$$

$$(a - x)(b - kx + kx) - x(b - kx) = 0$$

$$b(a - x) = cx \qquad \text{...(Eq.13.42)}$$

Multiplying throughout by $\frac{1}{2}\sin \psi$,

$$\tfrac{1}{2}ba \sin \psi - \tfrac{1}{2}bx \sin \psi = \tfrac{1}{2}cx \sin \psi$$

or $\Delta ABD - \Delta BCD = \Delta BCG$

or $\triangle ABC = \triangle BCG$ ...(Eq. 13.43)

This equation signifies that for EC to be the failure plane the requirement is that the area of the failure wedge ABC be equal to the area of the triangle BCG.

This is known as "Rebhann's condition", since it was demonstrated first by Rebhann in 1871.

The triangles ABC and BCG which are equal have a common base BC; hence their altitudes on to BC should be equal;
or $AJ \cdot \sin \angle AJB = CG \cdot \sin \angle BCG$ But $\angle AJB = \angle BCG$
as CG is parallel to AJ. This leads to $CG = AJ = x$; and
$$JE = a - x$$

Triangles DAE and DCG are similar.

Hence $\dfrac{(b-d)}{(b-c)} \cdot x = a$

Also, triangles BCG and BJE are similar.

Consequently, $\dfrac{d}{c} \cdot x = a - x$

Subtracting one from the another,
$$x \left(\frac{b-d}{b-c} - \frac{d}{c} \right) = x$$

Simplifying,
$$c^2 = bd$$

or $\qquad c = \sqrt{bd}$ ...(Eq. 13.44)

Thus if c is known, the position of G and hence that of the most dangerous rupture p surface, BC, can be determined and the weight of the sliding wedge, W, and the active thrust, P_a, can be calculated.

The relationship expressed by Eq. 13.44 is called the "Poncelet Rule" after Poncelet (1840). It is obvious that Rebhann's condition leads one to Poncelet's rule and the satisfaction of one of these two implies that of the other automatically.

The value of x may now be obtained from Eq. 13.42:
$$cx = b(a - x)$$

or $\qquad x = \dfrac{ab}{b+c}$ ...(Eq. 13.45)

Substituting $c = \dfrac{b}{x}(a - x)$ from Eq. 13.42 in to Eq. 13.41, one gets

$$P_a = \frac{1}{2} \gamma x^2 \cdot \sin \psi \qquad \text{...(Eq. 13.46)}$$

In summary, the Eq. $\psi = \alpha - \delta$ along with Eqs. 13.44 to 13.46, provide a sequence of steps :

558 Geotechnical Engineering

$$\psi = \alpha - \delta$$

$$c = \sqrt{bd}$$

$$x = \frac{ab}{b+c}$$

$$P_a = \frac{1}{2}\gamma x^2 \cdot \sin\psi$$

which gives an analytical procedure for the computation of the active thrust by Coulomb's wedge theory.

However, elegant graphical methods have been devised and are preferred to the analytical approach, in view of their versatility, coupled with simplicity.

The graphical method to follow is given by Poncelet and it is also sometimes known as the Rebhann's graphical method, since it is based on Rebhann's condition.

The steps involved in the graphical method are as follows, with reference to Fig. 13.28.

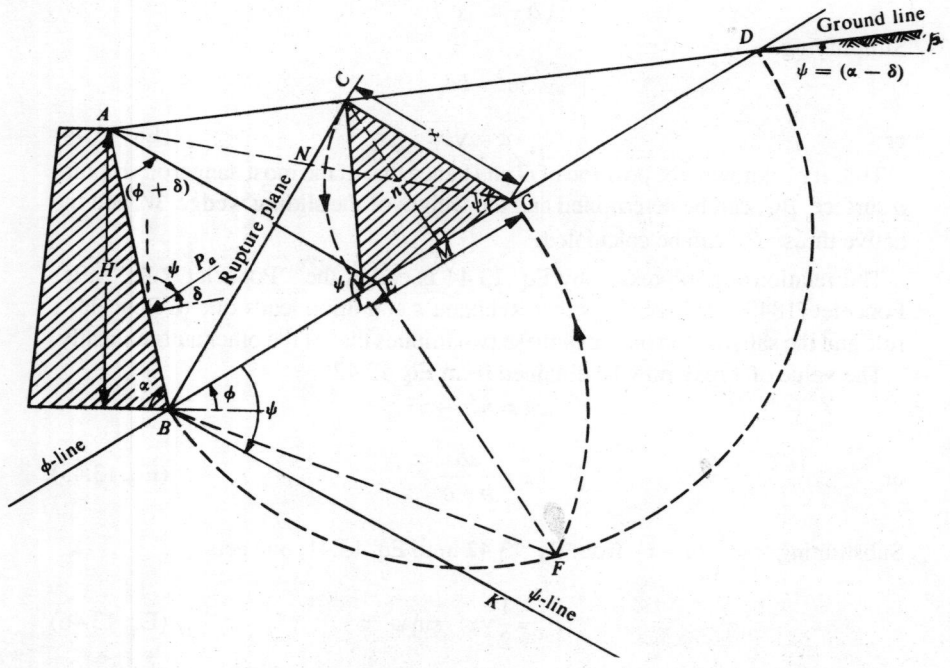

Fig. 13.28 Poncelet graphical construction for active thrust

Lateral Earth Pressure and Stability of Retaining Walls 559

(i) Let AB represent the backface of the wall and AB the backfill surface.

(ii) Draw BD inclined at ϕ with the horizontal from the heel B of the wall to meet the backfill surface in D.

(iii) Draw BK inclined at $\psi (= \alpha - \delta)$ with BD, which is the ψ-line.

(iv) Through A, draw AE parallel to the ψ-line to meet BD in E. (Alternatively, draw AE at $(\phi + \delta)$ with AB to meet BD in E).

(v) Describe a semi-circle on BD as diameter.

(vi) Erect a perpendicular to BD at E to meet the semi-circle in F.

(vii) With B as centre and BF as radius draw an arc to meet BD in G.

(viii) Through G, draw a parallel to the ψ-line to meet AD in C.

(ix) With G as centre and GC as radius draw an arc to cut BD in L; join CL and also draw a perpendicular CM from C on to LG.

BC is the required rupture surface. The criterion may be checked as follows:

Since $\triangle ABC = \triangle BCG$, and BC is their common base, their altitudes on BC must be equal; or $AN \sin \angle ANB = NG \sin \angle GNC$ that is to say $AN = NG$, since $\angle ANB = \angle GNC$. (N is intersection of AG and BC). Thus, if AN and NG are measured and found to be equal, the construction is correct.

The active thrust P_a is given by

$$P_a = \frac{1}{2} \gamma x^2 \cdot \sin \psi, \text{ where } CG = LG = x$$

$$= \gamma \cdot (\triangle CGL)$$

$$= \frac{1}{2} \gamma \cdot x \cdot n, \text{ where } n = CM, \text{ the altitude on to } LG.$$

(Incidentally, with the notation of Fig. 13.27(a), it may be easily understood that $W = \frac{1}{2} \gamma \cdot c \cdot n$. in view of Eq. 13.43).

Validity of the method

The validity of Poncelet construction may be easily demonstrated.

$$c^2 = BG^2 = BF^2 = BE^2 + EF^2$$

But $EF^2 = BE \cdot ED$, from the properties of a circle.

$$\therefore \qquad c^2 = BE^2 + BE \cdot ED$$

$$= BE(BE + ED)$$

$$= BE \cdot BD$$

or $c^2 = b \cdot d$, with the notation of Fig. 13.27(a).

This is the Poncelet rule, which implies Rebhann's condition automatically; hence the validity of the construction.

560 Geotechnical Engineering

The validity of the graphical method may be established in a different manner by deriving Eq. 13.33 for P_a from the geometry of Fig. 13.28 coupled with the notation of Fig. 13.27. However, with a view to avoiding confusion, the data required for this are given separately in Fig. 13.29.

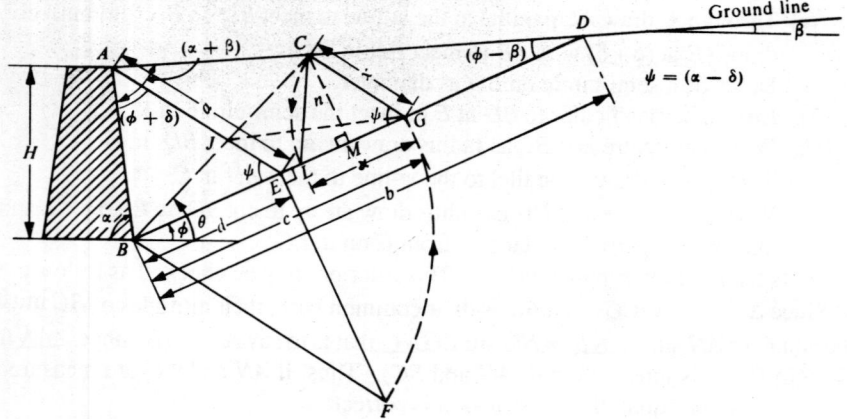

Fig. 13.29 Key figure for establishing the validity of Poncelet graphical method

$$\frac{x}{c} = \frac{\sin(\theta - \phi)}{\sin(\theta - \phi + \psi)} \qquad \ldots(\text{Eq. 13.47})$$

by applying sine rule in the triangle BCG,
where $c = \sqrt{bd}$ (the geometric mean).
For similar triangles GCD and EAD,

$$x/a = \frac{b-c}{b-d}$$

From the triangle ABE,

$$\frac{AE}{AB} = \frac{\sin(\phi + \alpha)}{\sin\psi}$$

$$\therefore \quad a = AE = AB \cdot \frac{\sin(\phi + \alpha)}{\sin\psi} = \frac{H}{\sin\alpha} \cdot \frac{\sin(\phi + \alpha)}{\sin\psi}$$

$$\therefore \quad x = \frac{a(b-c)}{(b-d)} = \frac{H}{\sin\alpha} \cdot \frac{\sin(\phi + \alpha)}{\sin\psi} \cdot \frac{(b-c)}{(b-d)} \qquad \ldots(\text{Eq. 13.48})$$

But the ratio $\dfrac{(b-c)}{(b-d)}$ may be transformed as follows:

$$\frac{(b-c)}{(b-d)} = \frac{b - \sqrt{bd}}{b-d} = \frac{1 - \sqrt{d/b}}{1 - d/b} = \frac{1}{1 + \sqrt{d/b}} \qquad \ldots(\text{Eq. 13.49})$$

From triangles ABE and ABD, by the application of the sine rule, one may obtain d/b as follows:

$$d/b = \frac{d}{AB} \cdot \frac{AB}{b} = \frac{\sin(\phi+\delta)}{\sin\psi} \cdot \frac{\sin(\phi-\beta)}{\sin(\alpha+\beta)} = \frac{\sin(\phi+\delta)\cdot\sin(\phi-\beta)}{\sin(\alpha-\delta)\cdot\sin(\alpha+\beta)}$$

...(Eq. 13.50)

Hence, substituting this in Eq. 13.49.

$$\frac{(b-c)}{(b-d)} = \frac{1}{1+\sqrt{\frac{\sin(\phi+\delta)\cdot\sin(\phi-\beta)}{\sin(\alpha-\delta)\cdot\sin(\alpha+\beta)}}}$$

...(Eq. 13.51)

Substituting this in Eq. 13.48,

$$x = \frac{H}{\sin\alpha} \cdot \frac{\sin(\phi+\alpha)}{\sin(\alpha-\delta)} \cdot \frac{1}{1+\sqrt{\frac{\sin(\phi+\delta)\cdot\sin(\phi-\beta)}{\sin(\alpha-\delta)\cdot\sin(\alpha+\beta)}}}$$

...(Eq. 13.52)

Since $P_a = \frac{1}{2}\gamma x^2 \cdot \sin\psi$ from Eq. 13.46, one obtains

$$P_a = \frac{1}{2}\gamma \cdot \sin(\alpha-\delta) \cdot \frac{H^2}{\sin^2\alpha} \cdot \frac{\sin^2(\phi+\alpha)}{\sin^2(\alpha-\delta)} \cdot \left[\frac{1}{1+\sqrt{\frac{\sin(\phi+\delta)\cdot\sin(\phi-\beta)}{\sin(\alpha-\delta)\cdot\sin(\alpha+\beta)}}}\right]^2$$

or

$$P_a = \frac{1}{2}\gamma H^2 \cdot \frac{\sin^2(\alpha+\phi)}{\sin^2\alpha \cdot \sin(\alpha-\delta)\left[1+\sqrt{\frac{\sin(\phi+\delta)\cdot\sin(\phi-\beta)}{\sin(\alpha-\delta)\cdot\sin(\alpha+\beta)}}\right]^2}$$

which is the same as Eq. 13.33 obtained previously from Coulomb's theory.
Another form for P_a is as follows (Taylor, 1848):

$$P_a = \frac{1}{2}\gamma H^2 \left[\frac{\operatorname{cosec}\alpha \cdot \sin(\alpha-\phi)}{\sqrt{\sin(\alpha-\delta)}+\sqrt{\frac{\sin(\phi+\delta)\cdot\sin(\phi-\beta)}{\sin(\alpha+\beta)}}}\right]^2$$

...(Eq.13.53)

It is interesting to note that when $\alpha = 90°$ and $\delta = \phi$, both Eqs. 13.33 and 13.53 reduce to the corresponding value obtained by using Eq. 13.17 of Rankine's theory. (Also the form for P_a expressed in Eq. 13.53 may be derived by considering the equality of the side ratios x/a and CD/AD, and those in the similar triangles BJC and BCD, JG being parallel to CD, and substituting in Eq. 13.46 for P_a.)

Special cases
(1) β is nearly equal to ϕ:
A special case of Poncelet's (Rebhann's) construction arises when β is nearly equal to ϕ so that ϕ-line and the backfill surface meet at a large distance from the wall and hence cannot be accommodated on the drawing. The following procedure may be adopted in such a case, as illustrated in Fig. 13.30.
  (i) Represent the backface of the wall AB and the backfill surface through A.
  (ii) Draw the ϕ-line through B inclined at ϕ with the horizontal.
  (iii) Draw the ψ-line BK through B inclined at ψ with the ϕ-line.
  (iv) Choose a convenient point D, on the ϕ-line, and draw the semi-circle on BD_1 as the diameter.

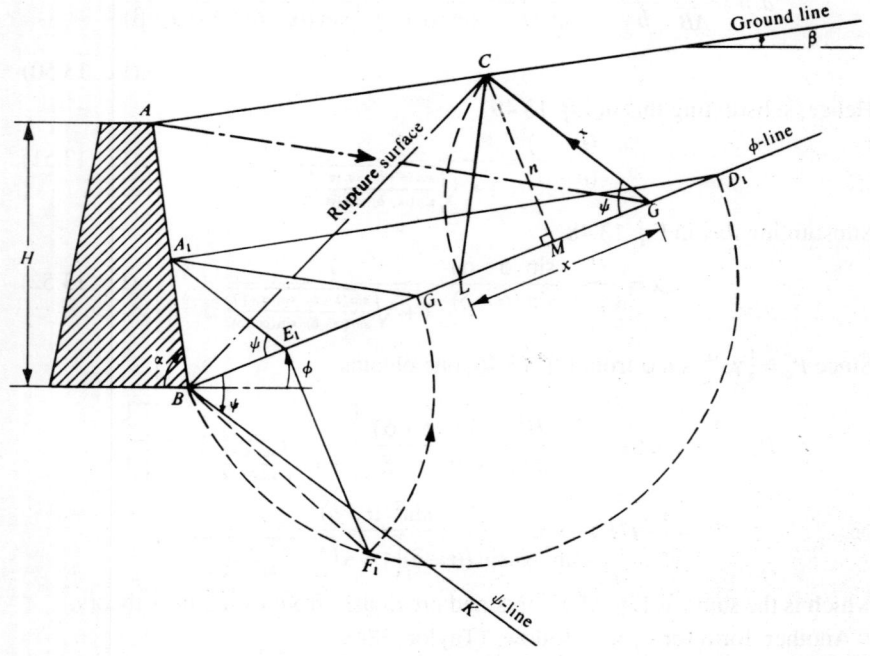

Fig. 13.30 Special case of Poncelet construction when $\beta \approx \phi$

(v) Draw D_1A_1 parallel to the backfill surface to meet the wall in A_1.
(vi) Through A_1 draw A_1E_1 parallel to the ψ-line to meet the ϕ-line in E_1.
(vii) Erect a perpendicular E_1F_1 to the ϕ-line at E_1 to meet the semi-circle in F_1.
(viii) With B as centre and BF_1 as radius, draw an arc F_1G_1 to meet the ϕ-line in G_1.
(ix) Through A draw AG parallel to A_1G_1 to meet the ϕ-line in G.
(x) Through G draw a line parallel to the ψ-line to meet the backfill surface in C.
(xi) Join GC and with G as centre and GC as radius, draw an arc to meet the ϕ-line in L.
(xii) Join CL and drop CM perpendicular to GL.

As usual, BC is the required rupture surface and P_a is given by the weight of the soil in the triangle CGL or $P_a = 1/2 \gamma x^2 \cdot \sin \psi = 1/2 \gamma x \cdot n$.

The construction is based on the principle that the location of the triangle CGL (whose weight equals P_a) gets shifted proportionately to the shift in D, the point of intersection of the backfill surface and the ϕ-line. Thus with the arbitrary location D_1 for D, AG gets shifted to A_1G_1; and this gives one the procedure required to locate the correct position of the triangle CGL, the rupture surface BC, and the active thrust P_a for this situation.

(2) β equals φ:

When β exactly equals φ, the ground line and the φ-line are parallel and will meet only at infinity. The points C and D, and the triangle CGL exist at infinity. However, the triangle CGL can be constructed anywhere between the φ-line and the ground line. The construction is shown in Fig. 13.31.

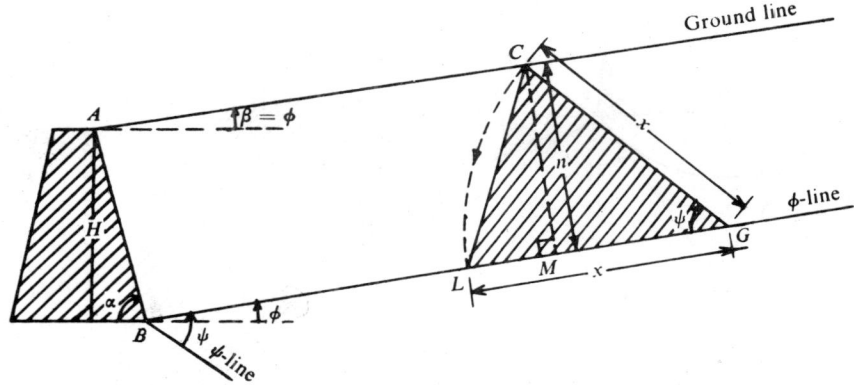

Fig. 13.31 Special case of Poncelet construction when β = φ

(i) Draw the ground line and the φ-line.
(ii) Draw the ψ-line BK through B at an angle ψ with the φ-line.
(iii) From any convenient point G on the φ-line, draw a line parallel to ψ-line to meet the ground line in C.
(iv) With G as centre and GC as radius, draw an arc to cut the line in L.
(v) Join CL and drop CM perpendicular on to LG.

The value of P_a is given by

$$P_a = \gamma(\Delta CGL) = \frac{1}{2}\gamma x^2 \sin\psi = \frac{1}{2}\gamma x\, n.$$

Poncelet construction for the determination of passive resistance

The determination of Coulomb's passive resistance graphically by the Poncelet construction is similar to that in the case of active thrust, except that the signs of the angles of internal friction of soil and wall friction have to be reversed. Graphically, this is accomplished by constructing the position line at an angle of $-(\phi + \delta)$ with the wall face, i.e., on to the opposite side of the fill, as shown in Fig. 13.32. Likewise, the φ-line is to be drawn through the heel B at an angle $(-\phi)$, i.e., below the horizontal.

(i) Draw the wall face AB and the backfill surface through A.
(ii) Draw the φ-line through B at φ below the horizontal. Let the φ-line produced meet the ground line extended backwards in D.

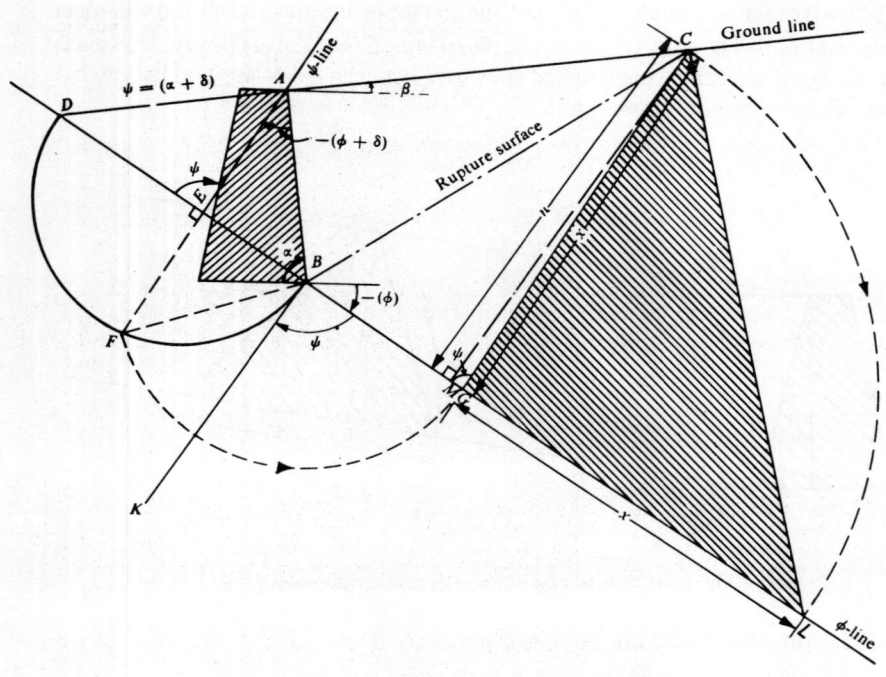

Fig. 13.32 Poncelet construction for passive resistance

(*iii*) Draw the ψ-line BK through B at $\psi\,(=\alpha+\delta)$ clockwise from the ϕ-line.
(*iv*) Through A, draw AE parallel to the ψ-line to meet BD in E. [Alternatively, draw AE through A at $(\phi+\delta)$ away from the fill to meet BD in E].
(*v*) Describe a semi-circle on BD as diameter.
(*vi*) Erect a perpendicular to BP at E to meet the semi-circle in F.
(*vii*) With B as centre and BF as radius draw an arc to meet DB produced in G.
(*viii*) Through G, draw a line parallel to the ψ-line to meet the backfill surface in C.
(*ix*) With G as centre and GC as radius draw an arc to cut DB produced beyond G in L; join CL and also draw a perpendicular CM from C on to LG.
BC is the required rupture surface.
The passive resistance P_p is given by

$$P_p = \gamma \cdot (\Delta CGL)$$

$$= \frac{1}{2}\gamma \cdot x^2 \cdot \sin\psi$$

$$= \gamma \cdot xn, \text{ as usual.}$$

13.7.5 Culmann's Graphical Method

Karl Culmann (1866) gave his own graphical method to evaluate the earth pressure from Coulomb's theory. Culmann's method permits one to determine graphically the magnitude of the earth pressure and to locate the most dangerous rupture surface according to Coulomb's wedge theory. This method has more general application than Poncelet's and is, in fact, a simplified version of the more general trial wedge method. It may be conveniently used for ground surface of any shape, for different types of surcharge loads, and for layered backfill with different unit weights for different layers.

With reference to Fig. 13.33(b), the force triangle may be imagined to be rotated clockwise through an angle $(90° - \phi)$, so as to bring the vector $\vec{W}$, parallel to the ϕ-line; in that case, the reaction, $\vec{R}$, will be parallel to the rupture surface, and the active thrust, P_a, parallel to the ψ-line.

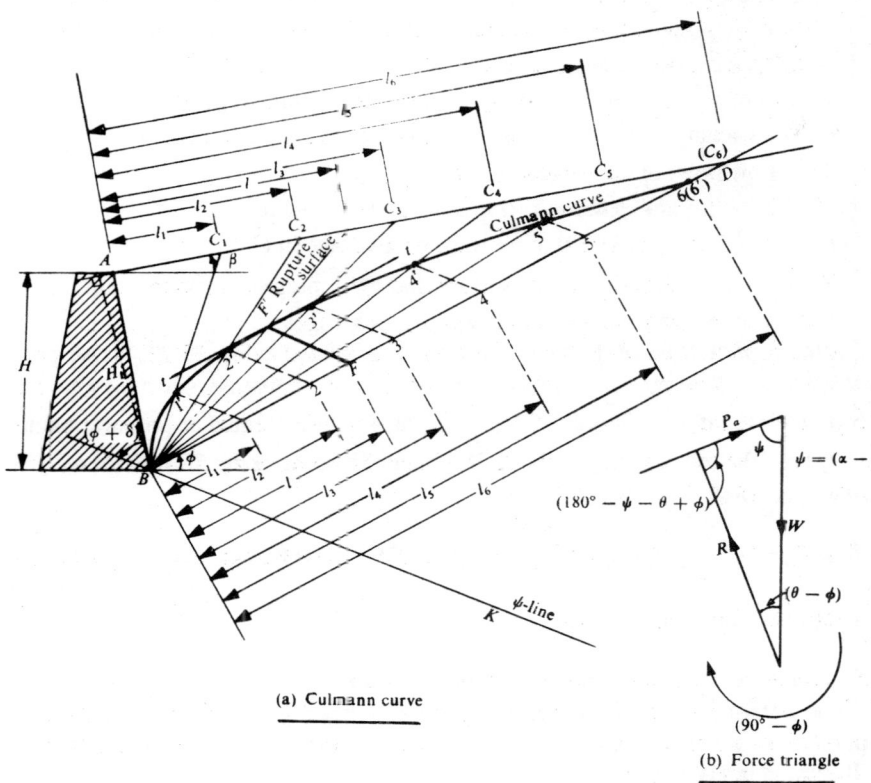

(a) Culmann curve

(b) Force triangle

Fig. 13.33 Culmann's graphical method for active thrust

Hence, if weights of the various sliding wedges arising out of arbitrarily assumed sliding surfaces are set off to a convenient force scale on the ϕ-line from the heel of the wall and if lines parallel to the ψ-line are drawn from the ends of these weight vectors to meet the respective assumed rupture lines, the force triangle for each of these sliding wedges will be complete. The end points of the active thrust vectors, when joined in a sequence, form what is known as the "Culmann-curve". The maximum value of the active thrust may be obtained from this curve by drawing a tangent parallel to the φ-line, which represents the desired active thrust, P_a. The corresponding rupture surface, which represents the most dangerous rupture surface, may be obtained by the line joining the heel of the wall to the end of the maximum pressure vector.

The steps in the construction may be set out as follows:

(*i*) Draw the ground line, φ-line, and ψ-line, and the wall face AB.

(*ii*) Choose an arbitrary failure plane BC_1. Calculate weight of the wedge ABC and plot it as B-1 to a convenient scale on the φ-line.

(*iii*) Draw 1 – 1' parallel to the ψ-line through 1 to meet BC_1 in 1'. 1' is a point on the Culmann-line.

(*iv*) Similarly, take some more failure planes BC_2, BC_3, ..., and repeat the steps (*ii*) and (*iii*) to establish points 2', 3', ...

(*v*) Join B, 1', 2', 3', etc., smoothly to obtain the Culmann curve.

(*vi*) Draw a tangent *t-t*, to the Culmann line parallel to the φ-line.
Let the point of the tangency be F'

(*vii*) Draw $F'F$ parallel to the ψ-line to meet the φ-line in F.

(*viii*) Join BF' and produce it to meet the ground line in C.

(*ix*) $BF'C$ represents the failure surface and $\overrightarrow{F'F}$ represents P_a to the same scale as that chosen to represent the weights of wedges.

If the upper surface of the backfill is a plane, as shown in Fig. 13.33, the weights of wedges will be proportional to the distances l_1, l_2 ... (bases), since they have a common-height, H_1. Thus B-1, B-2, etc..., may be made equal or proportional to l_1, l_2, etc. The sector scale may be easily obtained by comparing BF with the weight of wedge ABC.

Thus $P_a = W \cdot \dfrac{\overrightarrow{F'F}}{\overrightarrow{BF}} \dfrac{1}{2}\gamma(H_1) 1 \cdot \dfrac{\overrightarrow{F'F}}{l} = \dfrac{1}{2}\gamma H_1 (\overrightarrow{EF})$, if the bases themselves are used

to represent the weight vector.

Passive earth resistance from Culmann's approach

The determination of the passive earth resistance by Culmann's method is pursued in a similar manner as for the active earth pressure. The method is illustrated in Fig. 13.34.

Note that the φ-line is to be drawn through point B at an angle $-(\phi)$, *i.e.*, it must be drawn at an angle φ below the horizontal. On the line, the weights of the arbitrarily assumed sliding wedges are plotted to a convenient force scale. If the ground surface is plane as shown in Fig. 13.34, the weights of the wedges are proportional to the sloping distances, l_1, l_2, ..., and these distances or lengths may be plotted proportionally on the φ-line to represent the weights of the wedges. The

Lateral Earth Pressure and Stability of Retaining Walls 567

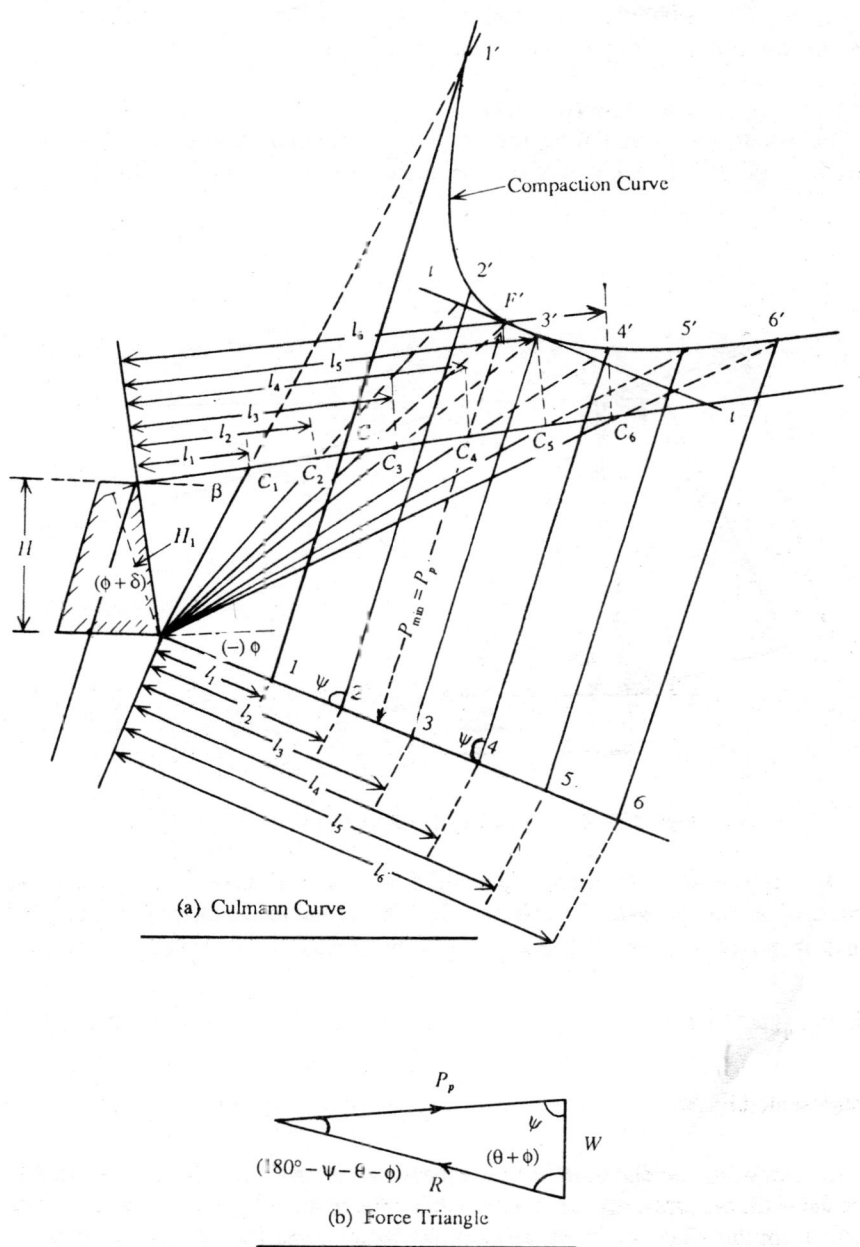

Fig. 13.34 Culmann's graphical method for passive resistance

568 *Geotechnical Engineering*

position line is drawn through A at an angle $-(\phi+\delta)$ (or to the left of the backface AB of the wall). The rest of the procedure is very much similar to that for the active case, the only difference being that the Culmann's curve will have a minimum vector which represents the passive earth resistance.

13.7.6 Break in the Backfill Surface

Sometimes the surface of the backfill may consist of a combination of two different slopes. The treatment of such a situation is illustrated in Fig. 13.35.

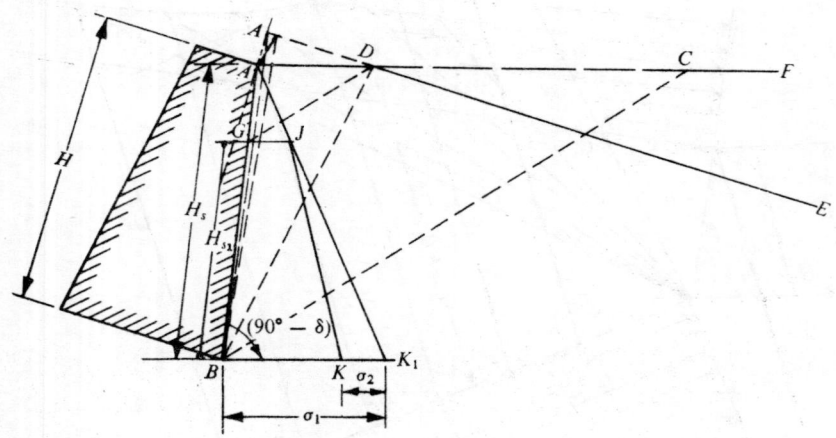

Fig. 13.35 Break in the backfill surface (After Taylor, 1948)

(*i*) Let the surface of the backfill be ADE with a break at D. Let AB represent the backface of the wall. First, ignore the line DE and locate the failure plane BC and obtain the pressure distribution AK_1B, by means of a Poncelet construction. If P_1 is the total thrust on the wall obtained from this construction $\sigma_1 = \dfrac{2P_1}{H_s}$, represented by BK_1.

(*ii*) Draw DG parallel to BC to meet the wall face in G. If AG is considered to be the wall, the pressure distribution is AJG; the break in the backfill will have no effect for this since it is to the right of the failure plane GD. However, below G the break will result in smaller pressures than those represented by the line JK_1.

(iii) The total thrust on the wall for the actual cross-section with the backfill surface ADE is obtained by another Poncelet construction. The irregular shape introduced by the break is eliminated by replacing area ADB by the area $BA'D$ as follows: The triangle BAD and $BA'D$ have a common base BD; the altitudes are made equal by drawing AA' parallel to BD, A' being on the line ED produced. The Poncelet construction with $A'B$ as the wall face and $A'E$ as the backfill surface gives the thrust P.

(iv) AK_1B represents the distribution of the thrust P_1 which is larger than the correct thrust by $(P_1 - P)$. The pressure distribution in the lower portion of the wall also may be assumed to be linear without significant loss of accuracy. Thus if $(\sigma_1 - \sigma_2)$ is the final value of the pressure at B,

$$\frac{\sigma_2}{2} \times H_{s_1} = (P_1 - P)$$

or

$$\sigma_2 = \frac{2(P_1 - P)}{H_{s_1}}$$

Thus, the final pressure distribution is given by the triangle AJG plus the trapezium $GJKB$.

The moment of the thrust P about point B may be expressed as

$$M = P_1\left(\frac{1}{3}H_s \cos\delta\right) - (P_1 - P)\left(\frac{1}{3}H_{s_1}\cos\delta\right)$$

If the total thrust alone, and not the distribution of pressure, is of interest, step (iii) is adequate. The thrust may be taken to act at approximately $\frac{1}{3}H$ above the base.

13.7.7 Effect of Uniform Surcharge and Line Load

Uniform Surcharge

Let a uniform surcharge q per unit area act on the surface of the backfill as shown in Fig. 13.36.

The effect of surcharge is to increase the intensity of vertical pressure, thereby increasing the lateral earth pressure.

Differential increase in the weight of any differential soil wedge

$$dW = -(\gamma dA + qds \cos\beta) \qquad \ldots(\text{Eq. } 13.54)$$

where dA = differential area of the differential soil wedge,
ds = differential length of the surface, and
β = the angle of surcharge.

This is under the assumption that q is the intensity of surcharge load per unit horizontal area.

The negative sign in Eq. 13.54 indicates that as θ increases, the weight of the sliding soil wedge decreases.

From the geometry of the figure,

$$dA = \frac{1}{2}H_s' \cdot ds \qquad \text{...(Eq. 13.55)}$$

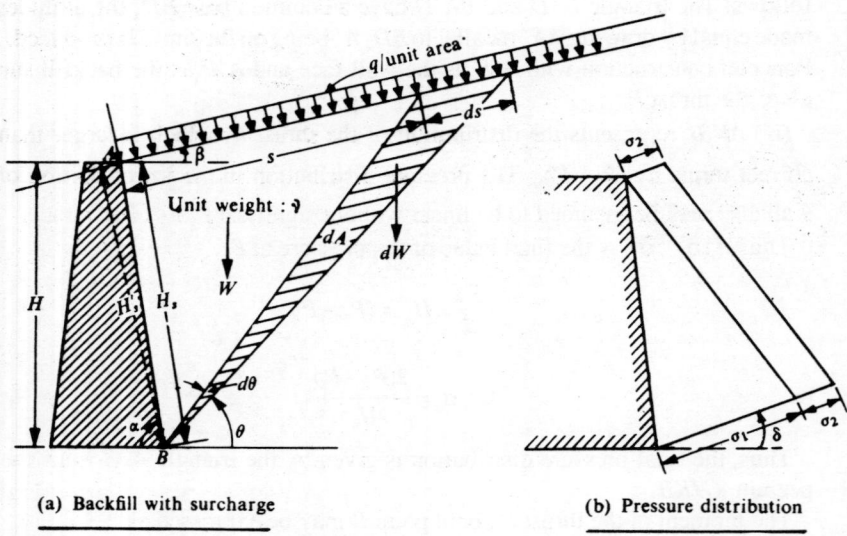

(a) Backfill with surcharge   (b) Pressure distribution

Fig. 13.36 Effect of uniform surcharge on earth pressure (Jumikis, 1962)

or
$$ds = \frac{2dA}{H_s'} \qquad \text{...(Eq. 13.56)}$$

Substituting this into Eq. 13.54, one gets

$$dW = -\left(\gamma dA + \frac{2qdA}{H_s'} \cdot \cos\beta\right) = -\left(\gamma + \frac{2q \cdot \cos\beta}{H_s'}\right) \cdot dA$$

$$= -\gamma_1 \cdot dA \qquad \text{...(Eq. 13.57)}$$

where
$$\gamma_1 = \left(\gamma + \frac{2q \cdot \cos\beta}{H_s'}\right) \qquad \text{...(Eq. 13.58)}$$

Thus, one can imagine that the effect of uniform surcharge may be taken into account by using a modified unit weight γ_1, which is given by Eq. 13.58, in the computation of the weights of trial sliding wedges in the Culmann's construction, or for γ in Eq. 13.46, if Poncelet's construction is used. It simply means that, in the force triangle, W should be taken as the weight of the trial sliding wedge plus $qs. \cos\beta$.

Equation 13.58 may also be written as follows:

$$\gamma_1 = \left[\gamma + \frac{2q \cdot \cos\beta}{H_s \sin(\alpha+\beta)}\right] \qquad \text{...(Eq. 13.59)}$$

$$\gamma_1 = \left[\gamma + \frac{2q \cdot \cos\beta}{H \sin\alpha \cdot \sin(\alpha+\beta)}\right] \qquad \text{... (Eq. 13.60)}$$

Lateral Earth Pressure and Stability of Retaining Walls

If the intensity of surcharge is specified as q per unit sloping area, Eq. 13.58 gets modified as

$$\gamma_1 = \left(\gamma + \frac{2q}{H_s'}\right) \qquad \ldots(\text{Eq. 13.61})$$

It may be shown that the location of the failure plane is not changed, as also the directions of the forces on the sliding wedge. When surcharge is added, all forces increase in the same ratio. The ratio of the additional thrust due to surcharge to that without surcharge is sometimes called the 'surcharge ratio'.

Since the expression for the modified unit weight consists of two terms—one of the unit weight of soil and the other relating to the surcharge term, the thrust may be looked upon as being composed of that without surcharge and the contribution due to the surcharge. The weight of the soil wedge and the thrust due to it are proportional to H^2, while the weight of surcharge is proportional to the surface dimension of the wedge, or to H; hence, the contribution of the surcharge to the lateral pressure is proportional to H. Thus, the lateral pressure due to surcharge is constant over the height of the wall. The distribution of the lateral pressure with depth is, therefore, as shown in Fig. 13.36(b). The pressures σ_1 and σ_2 may be obtained if the thrusts with and without discharge are determined.

If the ground surface is surcharged with different intensities q_1, q_2 etc., as shown in Fig. 13.37, the Culmann-curve may have several maximum P-values. The maximum of the several maximum values, the so-called "maximum maximorum", is then taken as the active thrust: $P_a = P_{a\,max\,max}$. This value also determines the position of the most dangerous rupture surface, BC.

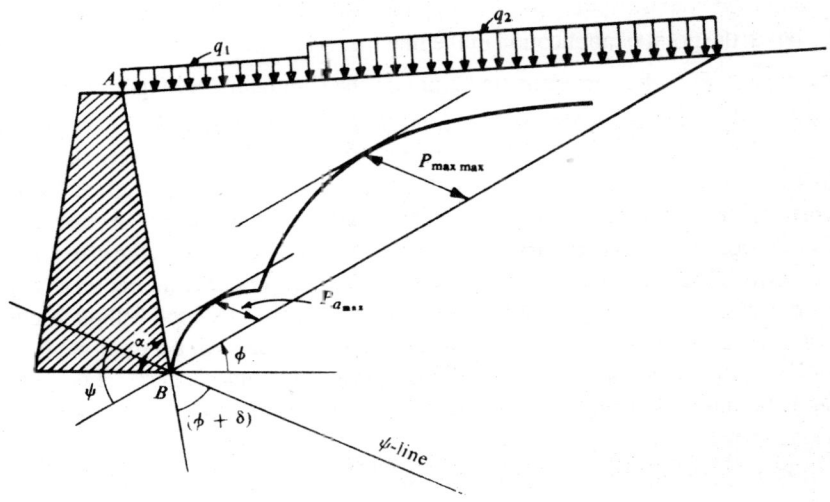

Fig. 13.37 Culmann's method for surcharges of different intensities

Line load

A railway track or a long wall of a building or a loaded wharf near a waterfront structure will constitute a line load if it runs parallel to the length of a retaining wall.

Culmann's graphical method may be adapted to take into account the effect of such a line load on the lateral earth pressure on the retaining wall, as illustrated in Fig. 13.38.

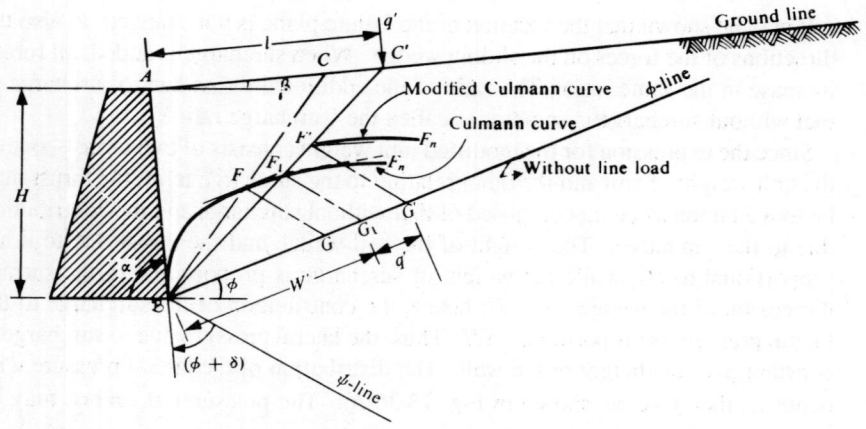

Fig. 13.38 Effect of line load on lateral earth pressure

Let AB represent the wall face and let the backfill surface or ground line be inclined to the horizontal at an angle β. Draw the ϕ-line and ψ-line through the heel, B, of the wall as shown. Using the Culmann's graphical method and ignoring the presence of the line load, obtain the Culmann's curve BFF_1F_n, the maximum ordinate GF and the failure plane BFC. Let the weight of the wedge ABC' be W', C' being the point of application of the line load. This is represented by BG_1 along the ϕ-line. F_1 is the corresponding point on the Culmann-curve ignoring the line load. If the line load is also included, the weight of the wedge ABC' will be $(W' + q')$. Letting BG' represent this increased weight, $G'F'$ is drawn parallel to the ψ-line to meet BC' in F'. For all other wedges considered to the right of the line load, the load q' should be included and the points on the Culmann-curve obtained. The 'modified' Culmann-curve obtained in this manner includes the effect of the line load. There is an abrupt increase in the lateral pressure, the increase being proportional to q'. If $G'F'$ or any other ordinate of the modified Culmann-curve is greater than GF, failure will not occur along BFC, but will occur along $BF'C'$ if $G'F'$ is the maximum ordinate, of the modified Culmann-curve. If GF is still the maximum ordinate, it means that there is no influence of the line load on the active thrust, P_a.

If $G'F'$ is the maximum ordinate, the increase in thrust is $(G'F' - GF)$, or say ΔP_a. (Invariably, the maximum ordinate of the modified Culmann-curve will be $G'F'$ indicating failure along $BF_1F'C'$, passing through the line load).

This increase in thrust ΔP_a may be obtained for several different locations of the line load, as illustrated in Fig. 13.39.

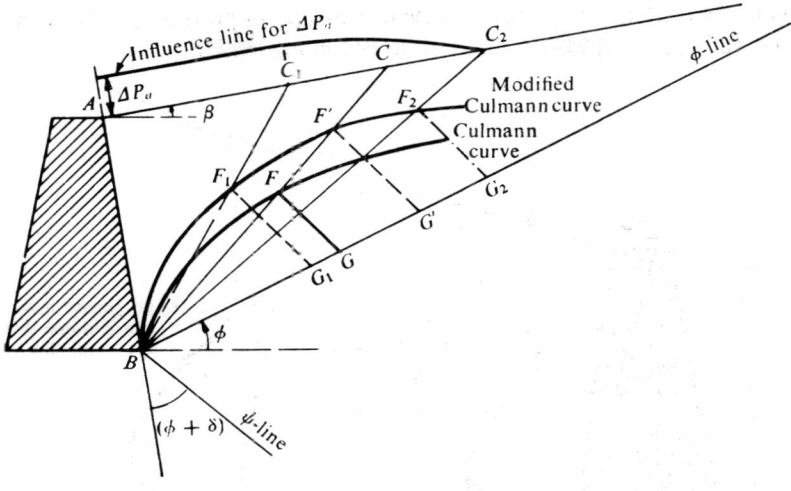

Fig. 13.39 Influence line for thrust increment due to line load

First, the Culmann-curve BFF_a is obtained ignoring the line load. Next, the modified Culmann-curve $BF_1F'F_2$ is obtained considering, the line load and adding q' to the weight of every wedge. The intercepts GF and $G'F'$ are obtained by drawing tangents parallel to the ϕ-line to the Culmann-curve and the modified Culmann-curve, respectively, these giving the greatest thrusts without and with the line load. If the tangent to the Culmann-curve at F is extended to meet the modified Culmann-curve in F_2, the intercept G_2F_2 equals GF. This implies that if the line load is placed beyond C_2, there is no effect on the lateral pressure, since $\Delta P_a = G_2F_2 - GF = 0$. For other positions of the line load between A and C_2, ΔP_a may be obtained as indicated in Fig. 13.38, and plotted as ordinates at the locations of the line load. It will be observed that ΔP_a is maximum when the load is at the face of the wall, it remains constant with positions of line load up to C_1, and then decreased gradually to zero at C_2.

This analysis is considered very useful in locating the position of a railway line or a long wall of a building on the backfill at a safe distance so that the thrust on the wall does not increase.

Alternative procedure

An alternative procedure may be adopted in case one is interested in obtaining not only the total thrust, but also the distribution of pressure across the height of the wall (Taylor, 1948). This is illustrated in Fig. 13.40.

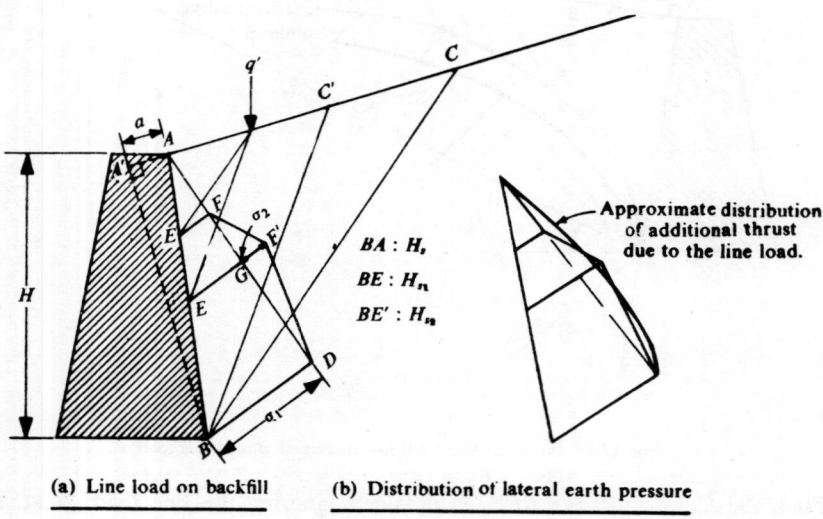

(a) Line load on backfill   (b) Distribution of lateral earth pressure

Fig. 13.40 Effect of line load on distribution of active earth pressure (After Taylor, 1948)

First, the line load is disregarded and a Poncelet construction is carried through; the failure plane BC is located and the thrust P_{a_1} is determined. The distribution of pressure ABD is obtained,
wherein
$$BD = \sigma_1 = 2P_{a_1}/H_s \qquad \ldots \text{(Eq. 13.62)}$$

Point A' is then located on the backfill surface produced, the weight of the triangle of soil being equal to q'. The distance AA' ($= a$) is given by

$$q' = \frac{1}{2}\gamma a \cdot H_s \qquad \ldots \text{(Eq. 13.63)}$$

Now another Poncelet construction, starting from point A', with *unchanged* ψ-*line*, is performed to determine the failure plane AC' and the total thrust P_{a_2}. The surcharge causes an additional thrust of $P_{a_2} - P_{a_1}$, which has a distribution that is approximately as shown in Fig. 13.40(b). This is approximated with reasonable accuracy by a broken line.

Let the lines parallel to the failure planes BC and BC', and passing through the point the action of the line load q', meet the wall face in E and E', respectively. It may be assumed that at point E, there is no effect of the line load and that the pressure is EF; that at point E' the pressure caused by the line load has its maximum value σ_2; and, that at point B there is no effect of the surcharge and that the pressure is BD.

Since $\frac{1}{2}\sigma_2$ is the average pressure added over the height of the wall DE, σ_2 is defined by the equation

$$P_{a_2} - P_{a_1} = \frac{1}{2}\sigma_2 \times H_s,$$

whence
$$\sigma_2 = \frac{2(P_{a_2} - P_{a_1})}{H_{s_1}} \qquad \ldots(\text{Eq. 13.64})$$

The use of this equation allows the completion of the pressure distribution diagram $AF\,F'D$.

The moment of this pressure diagram about the heel B of the wall, which may be required in the stability computations of the wall, is given by

$$M_B = \frac{1}{3}P_{a_1} \cdot H_s \cos\delta + \frac{1}{3}(P_{a_2} - P_{a_1})(H_{s_1} + H_{s_2})\cos\delta \qquad \ldots(\text{Eq. 13.65})$$

13.7.8 Lateral Earth Pressure of Cohesive Soil

The lateral earth pressure of cohesive soil may be obtained from the Coulomb's wedge theory; however, one should take cognisance of the tension zone near the surface of the cohesive backfill and consequent loss of contact and loss of adhesion and friction at the back of the wall and along the plane of rupture, so as to avoid getting erroneous results.

The trial wedge method may be applied to this case as illustrated in Fig. 13.41. The following five forces act on a trial wedge:
1. Weight of the wedge including the tension zone, W.
2. Cohesion along the wall face or adhesion between the wall and the fill, C_a.
3. Cohesion along the rupture plane, C.
4. Reaction on the plane of failure, R, acting at ϕ to the normal to the plane of failure.
5. Active thrust, P_a, acting at δ to the normal to the face of the wall.

The total adhesion force, C_a, is given by

$$C_a = c_a \cdot \overrightarrow{BF} \qquad \ldots(\text{Eq. 13.66})$$

where c_a is the unit adhesion between the wall and the fill, which cannot be greater than the unit cohesion, c, of the soil. c_a may be obtained from tests; however, in the absence of data, c_a may be taken as equal to c for soils with c up to 50 kN/m², c_a may be limited to 50 kN/m² for soils with c greater than this value. (Smith, 1974).

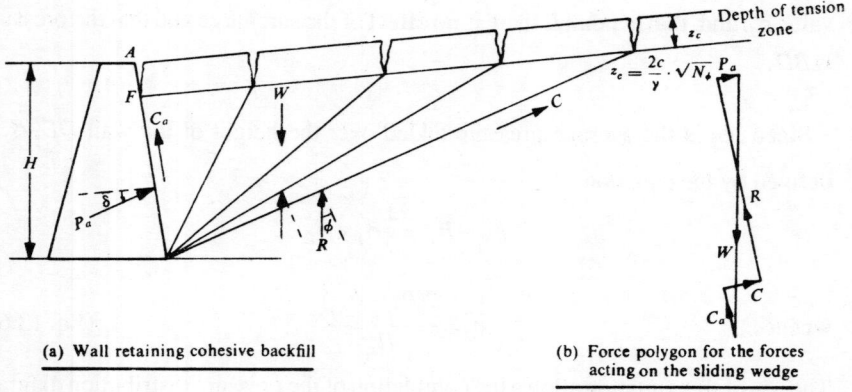

(a) Wall retaining cohesive backfill

(b) Force polygon for the forces acting on the sliding wedge

Fig. 13.41 Active earth pressure of cohesive soil—trial wedge method—Coulomb's theory

The total cohesion force, C, is given by
$$C = c \cdot \overrightarrow{BC} \qquad \text{...(Eq. 13.67)}$$
c being the unit cohesion of the fill soil and $\overrightarrow{BC}$ is the length of the rupture plane.

Thus three forces W, C_a, and C are fully known and the directions of the other two unknown forces R and P_a are known; the vector polygon may therefore be completed as shown in Fig. 13.41(b), and the value of P_a may be scaled-off.

A number of such trial wedges may be analysed and the maximum of all P_a values chosen as the active thrust. The rupture plane may also be located. The final value of the thrust on the wall is the resultant of P_a and C_a.

Culmann's method may also be adapted to suit this case, as illustrated in Fig. 13.42.

Passive earth pressure of cohesive soil

The procedure adopted to determine the active earth pressure of cohesive soil from Coulomb's theory may also be used to determine the passive earth resistance of cohesive soil.

The points of difference are that the signs of friction angles, ϕ and δ, will be reversed and the directions of C_a and C also get reversed.

Either the trial wedge approach or Culmann's approach may be used but one has also to consider the effect of the tensile zone in reducing C_a and C.

However, it must be noted that the Coulomb theory with plane rupture surfaces is not applicable to the case of passive resistance. Analysis must be carried out, strictly speaking, using curved rupture surfaces such as logarithmic spirals (Terzaghi, 1943), so as to avoid overestimation of passive resistance.

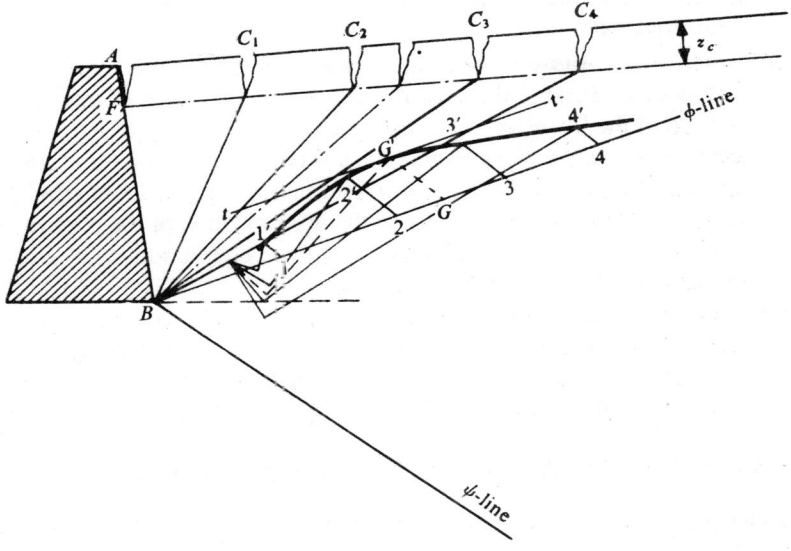

Fig. 13.42 Culmann's method adapted to allow for cohesion

13.7.9 Use of Tables and Charts for Earth Pressure

To facilitate earth pressure calculations, earth pressure coefficient tables, such as those given by Caquot and Kerisel (1956) and Jumikis (1962), may be used. These give K_a and K_p coefficients for various α, β, δ, and ϕ values, in reasonable ranges practically possible for each. Linear interpolation may be used satisfactorily for obtaining values not available directly from the tables.

For design purposes, even the use of charts may be considered all right. However, most of the charts available may have ϕ and δ as variables and consider standard common values for others, such as $\alpha = 90°$ and $\beta = 0°$. Therefore, these charts may be useful only for certain simple situations.

13.7.10 Comparison of Coulomb's Theory with Rankine's Theory

The following are the important points of comparison:
  (*i*) Coulomb considers a retaining wall and the backfill as a system; he takes into account the friction between the wall and the backfill, while Rankine does not.
  (*ii*) The backfill surface may be plane or curved in Coulomb's theory, but Rankine's allows only for a plane surface.
  (*iii*) In Coulomb's theory, the total earth thrust is first obtained and its position and direction of the earth pressure are assumed to be known; linear variation of pressure with depth is tacitly assumed and the direction is automatically obtained from the concept of wall friction. In Rankine's theory, plastic equilibrium inside a semi-infinite soil mass is considered, pressures

evaluated, a retaining wall is imagined to be interposed later, and the location and magnitude of the total earth thrust are established mathematically.

(iv) Coulomb's theory is more versatile than Rankine's in that it can take into account any shape of the backfill surface, break in the wall face or in the surface of the fill, effect of stratification of the backfill, effect of various kinds of surcharge on earth pressure, and the effects of cohesion, adhesion and wall friction. It lends itself to elegant graphical solutions and gives more reliable results, especially in the determination of the passive earth resistance; this is inspite of the fact that static equilibrium condition does not appear to be satisfied in the analysis.

(v) Rankine's theory is relatively simple and hence is more commonly used, while Coulomb's theory is more rational and versatile although cumbersome at times; therefore, the use of the latter is called for in important situations or problems.

13.8 STABILITY CONSIDERATIONS FOR RETAINING WALLS

A retaining wall is one of the most important types of retaining structures. The primary purpose of a retaining wall is to retain earth or other material at or near a vertical position. It is extensively used in a variety of situations in such fields as highway engineering, railway engineering, bridge engineering, dock and harbour engineering, irrigation engineering, land reclamation and coastal engineering.

When designing retaining structures, an engineer often needs to ensure only that total collapse or failure does not occur. Movements of several centimetres are often of no concern as long as there is assurance that larger motions will not suddenly occur. Thus the approach to the design of retaining structures generally is to analyse the conditions at collapse and to apply suitable safety factors to prevent collapse. This is known as limit design and requires analysis of limiting equilibrium conditions such as the active and the passive states, which has been the subject matter of this chapter till now.

13.8.1 Types of Retaining Walls

The common types of retaining walls have been listed in Section 13.1. Each type will be seen now in a little more detail.

Gravity retaining wall

A gravity retaining wall is typically used to form a permanent wall of an excavation whenever space requirements make it impractical to simply slope the side of the excavation. Such conditions arise, for example, when a roadway is needed immediately adjacent to an excavation. In order to construct the wall, a temporary slope is formed at the edge of the excavation, the wall is built and then backfill is dumped into the space between the wall and the temporary slope. In earlier days, masonry walls were often used. Today, most such walls are of plain concrete (Huntington, 1957).

Lateral Earth Pressure and Stability of Retaining Walls 579

The lateral earth pressure is resisted by this type of wall primarily by its weight; hence the name 'gravity type'. It is, therefore, of thicker section in contrast to a few other types. A schematic representation of a gravity retaining wall is shown in Fig. 13.43.

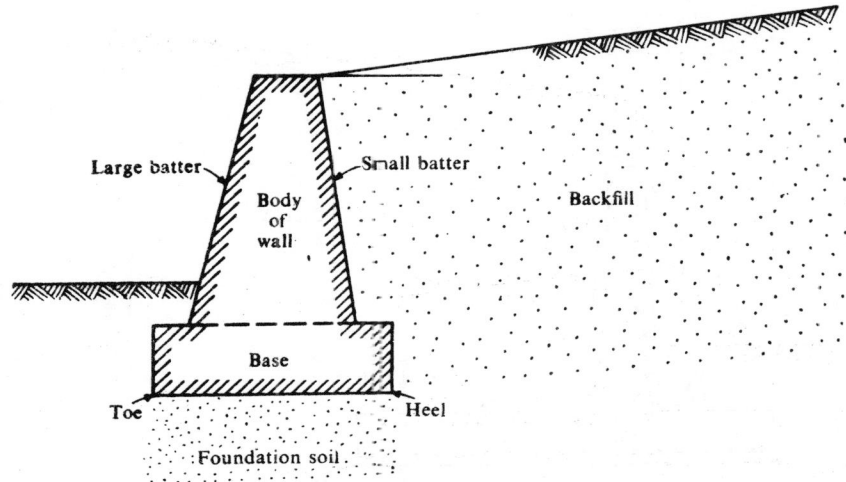

Fig. 13.43 Gravity retaining wall

Semi-gravity retaining wall

A semi-gravity retaining wall is one which resists the lateral earth pressure partly by its weight and partly by the nominal reinforcements that are provided. It is usually thinner in section as compared to the gravity type, and is shown in Fig. 13.44.

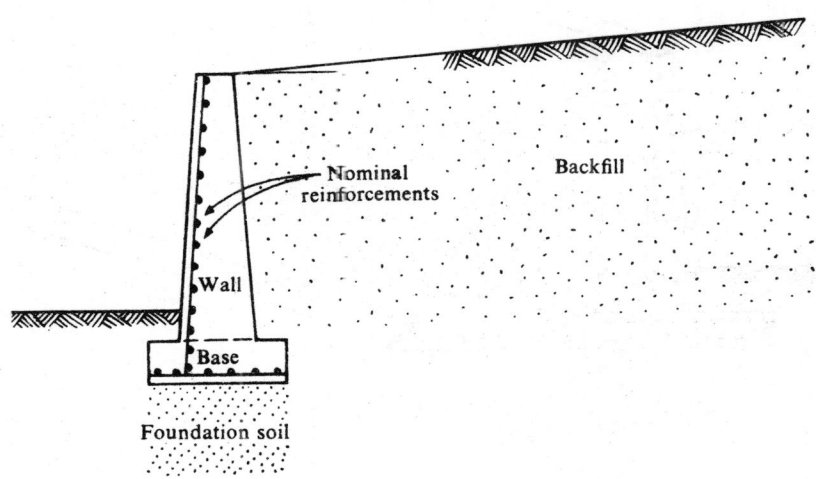

Fig. 13.44 Semi-gravity retaining wall

Cantilever retaining wall

A cantilever type of retaining wall resists the lateral earth pressure by cantilever action of the stem, toe slab and heel slab. Necessary reinforcements are provided to take care of flexural stresses, as shown in Fig. 13.45.

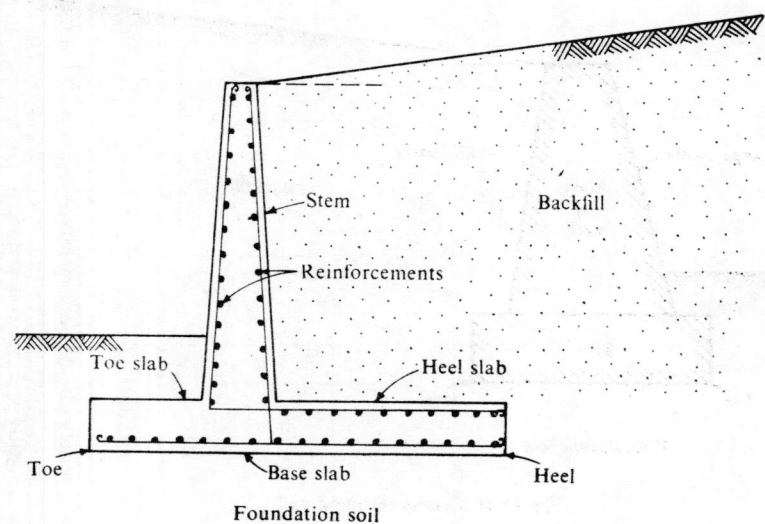

Fig. 13.45 Cantilever retaining wall

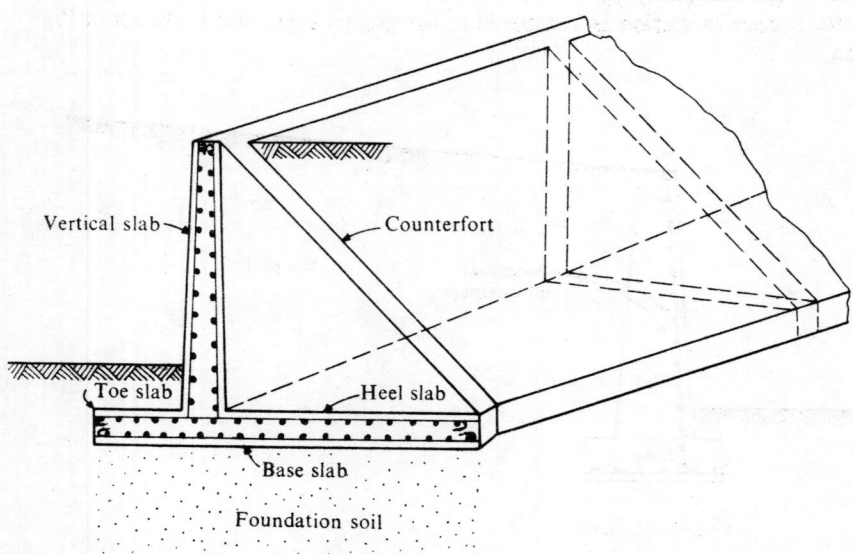

Fig. 13.46 Counterfort retaining walls

It is usually constructed in reinforced concrete and the thicknesses of the stem and base slab will be small in view of the reinforcements provided to take care of flexural stresses.

Counterfort retaining wall

This type of wall, as shown in Fig. 13.46 resists the lateral earth pressure by beam action between counterforts, which are wedge-shaped slabs. The base slab in the heel portion also resists the upward pressure of the foundation soil by beam action. This type also is constructed in reinforced concrete and is used for greater heights of the backfill.

Buttress retaining wall

This type is similar to the Counterfort type with the difference, that the counterforts, called 'buttresses' in this case, are provided on the side opposite to the fill. They are thus exposed to view and may not contribute to the elegance or aesthetic appearance. This is also constructed in reinforced concrete and may appear somewhat as shown in Fig 13.47.

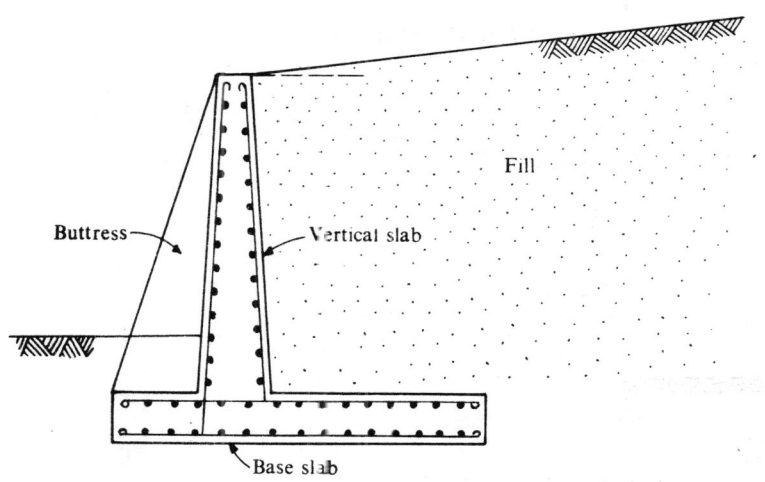

Fig. 13.47 Buttress retaining wall

Crib retaining wall

This is a box-like structure or crib made up of usually wood members with fill in between the members. The fabricated precast concrete or steel members may also form cribs. This type occupies too much of space and is used only under certain special circumstances. It appears somewhat as shown in Fig. 13.48.

13.8.2 Stability Considerations for Gravity Retaining Walls

Figure 13.49 shows in a general way the forces that act upon a gravity retaining wall. The bearing force or the reaction from the base of the wall resists the weight of the wall plus the vertical components of other forces.

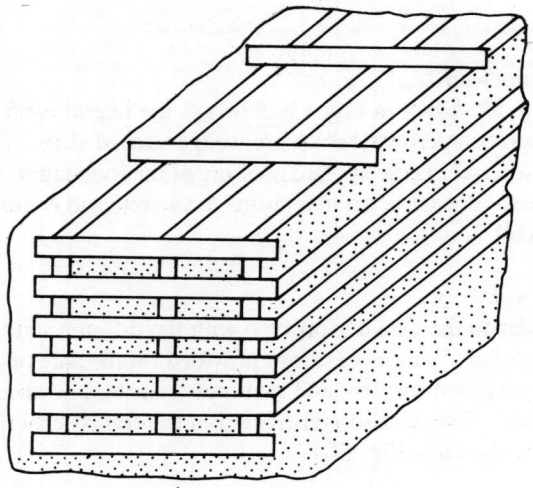

Fig. 13.48 Crib retaining wall

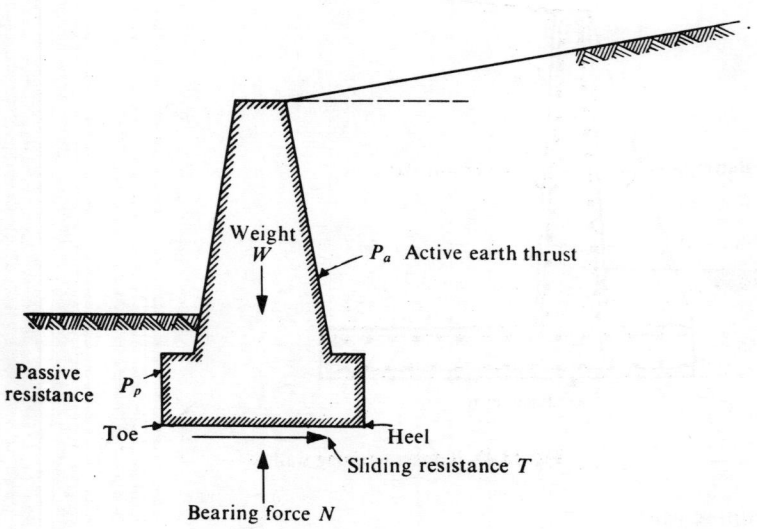

Fig. 13.49 Forces acting on a gravity retaining wall

The active thrust acts to push the wall outward. This outward motion is resisted by sliding or shear resistance along the base of the wall and by the passive resistance of the soil lying above the toe of the wall. The active thrust also tends to overturn the wall around the toe. The overturning is resisted by the weight of the wall and the vertical component of the active thrust. The weight of the wall thus acts in two ways: it resists overturning and it causes frictional sliding resistance at the base of the wall. This is the reason for calling the wall a 'gravity' retaining wall.

A gravity retaining wall, together with the retained backfill and the supporting soil, is a highly 'indeterminate' system (Lambe and Whitman, 1969). The magnitudes of the forces that act upon the wall cannot be determined from statics alone and these will be affected by the sequence of construction and backfilling operations. Hence, the design of such a wall is based on an analysis of expected forces that would exist if the wall started to fail, that is, to overturn or to slide outwards.

Considering the patterns of deformations observed from experiments, an approach to the design of gravity retaining walls may be stated. First, trial dimensions for the wall are chosen; next, the active thrust on the wall is determined under the assumption that the active pressure is fully mobilised; then the resistance offered by the weight of wall, the frictional resistance at the base of the wall, and the passive resistance, if any, at the toe of the wall, are determined. Finally, the active thrust and total resistance are compared and it is ensured that the resistance exceeds the thrust by a suitable safety factor. Consider a gravity retaining wall as shown in Fig. 13.50.

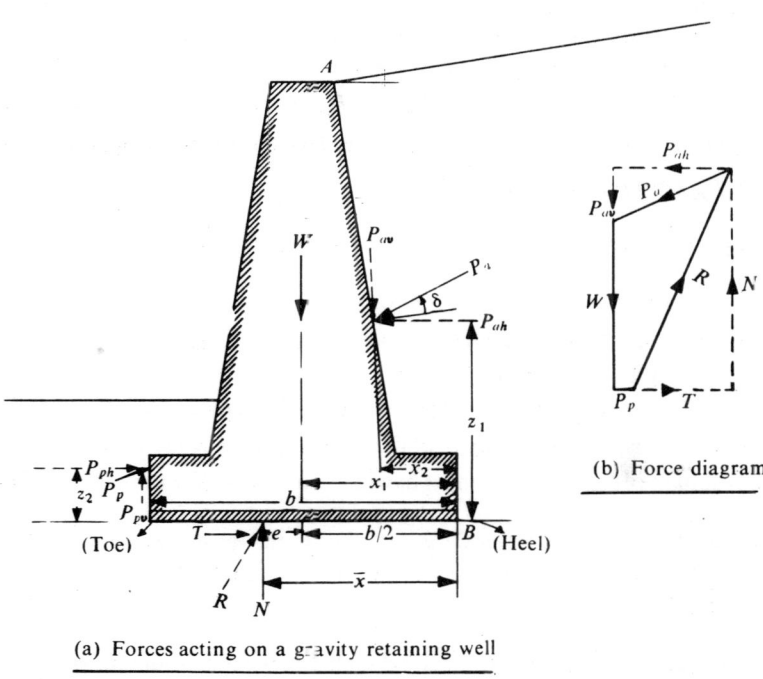

Fig. 13.50 Force system on a gravity retaining wall

Let W represent the weight of the wall per unit length perpendicular to the plane of the figure, acting through the centre of gravity of the cross-section of the wall. Let the active thrust on the wall be P_a acting at an angle δ with the normal to the back face of the wall. For convenience in design, however, P_a is resolved into its horizontal component, P_{ah} and its vertical component, P_{av}. Let R represent the reaction from the foundation soil acting on the base of the wall, which is also for convenience taken to be resolved into its horizontal component T and its vertical component N. A passive resistance P_p which is usually small, may exist on the side of the wall remote from the backfill as shown. Let its horizontal and vertical components be P_{ph} and P_{pv} respectively. P_p is often neglected in view of its relatively small magnitude.

For equilibrium of the wall under these forces, one may write

$$N = W + P_{av} - P_{pv} \qquad \text{...(Eq. 13.68)}$$

and
$$T = P_{ah} - P_{ph} \qquad \text{...(Eq. 13.69)}$$

For any arbitrarily chosen section of the wall, W, P_a and P_p may be obtained and therefore N and T may be computed.

The eccentricity e of the force N relative to the centre of the base of the wall may be computed by taking moments about B.

$$N \cdot \bar{x} = W \cdot x_1 + P_{av} x_2 + P_{ah} \cdot z_1 - P_{pv} \cdot b - P_{ph} \cdot z_2 \qquad \text{...(Eq. 13.70)}$$

(Note: If P_p strikes the body of the wall and not the base slab, the appropriate lever arm for P_{pv} with respect to B must be used).

$$\therefore \quad \bar{x} = (W x_1 + P_{av} \cdot x_2 + P_{ah} \cdot z_1 - P_{pv} \cdot b - P_{ph} \cdot z_2)/N = \frac{\Sigma M}{\Sigma V} \qquad \text{...(Eq. 13.71)}$$

$$^*e = (\bar{x} - b/2) \qquad \text{...(Eq. 13.72)}$$

Here ΣM = Algebraic sum of the moments of all the actuating forces, other than that of reaction N.

ΣV = Algebraic sum of all the vertical forces, other than T.

This simply means that the resultant of W, P_a, and P_p must be just equal and opposite to the resultant of N and T, and must have the same line of action, for equilibrium of the wall.

The problem becomes essentially one of trial; the necessary width of the base usually falls between 30% and 60% of the height of the wall.

The criteria for a satisfactory design of a gravity retaining wall may be enunciated as follows:

(a) The base width of the wall must be such that the maximum pressure exerted on the foundation soil does not exceed the safe bearing capacity of the soil.
(b) Tension should not develop anywhere in the wall.
(c) The wall must be safe against sliding; that is, the factor of safety against sliding should be adequate.

---

* If $x > b/2$, the maximum normal pressure occurs at the toe; and, if $x < b/2$, the maximum value occurs at the heel.

Lateral Earth Pressure and Stability of Retaining Walls 585

(d) The wall must be safe against overturning; that is, the factor of safety against overturning should be adequate.

For any trial value of the base width these criteria are investigated as follows:

(a) The pressure exerted by the force N on the base of the wall is a combination of direct and bending stresses owing to the eccentricity of this force with respect to the centroid of the rectangular area $b \times 1$ on which it acts. Assuming linear variation of pressure, the intensities of pressure at the toe and the heel are given by:

$$\sigma_{max} = \frac{N}{b}\left(1 + \frac{6e}{b}\right) \qquad \text{...(Eq. 13.73)}$$

$$\sigma_{min} = \frac{N}{b}\left(1 - \frac{6e}{b}\right) \qquad \text{...(Eq. 13.74)}$$

respectively.

Three different cases arise depending upon the value of e: $-e < \frac{b}{6}$, $e = \frac{b}{6}$, and $e > \frac{b}{6}$. These correspond to the situations where the resultant force (or N) strikes the base

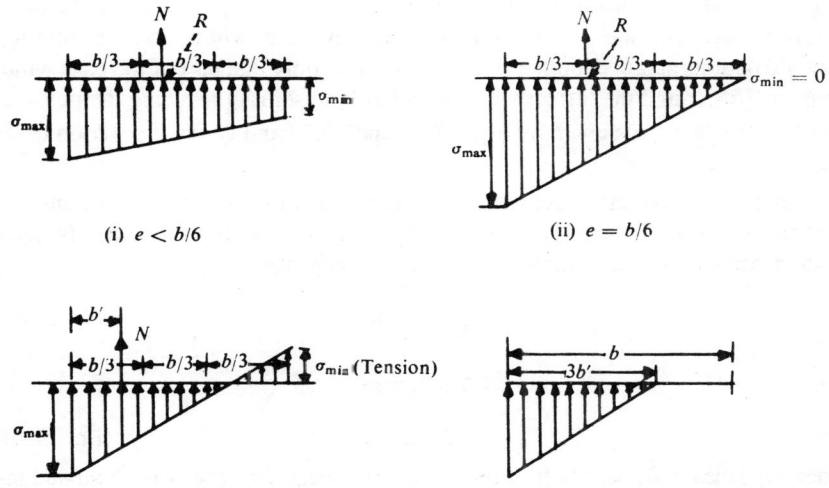

Fig. 13.51 Distributions of base pressure for different values of eccentricity of the resultant force on the base

within the 'middle-third' of the base, at the outer third-point of the base, and out of the middle-third of the base, respectively. The corresponding pressure diagrams for the base are shown in Fig. 13.51.

Equations 13.73 and 13.74 apply for the first case—$e < \dfrac{b}{6}$. For case (ii) $e = b/6$,

$$\sigma_{max} = \frac{2N}{b} \qquad \ldots\text{(Eq. 13.75)}$$

and σ_{min} is zero.

For case (iii) $e > b/6$, tension is supposed to have developed as shown.

Since soil is considered incapable of resisting any tension, the pressure is taken to be redistributed along the intact base of width $3b'$, where b' is the distance of the line of action of R (or N) from the toe. σ_{max} is then given by:

$$\sigma_{max} = \frac{2N}{3b'} \qquad \ldots\text{(Eq. 13.76)}$$

or

$$\sigma_{max} = \frac{2N}{3\left(\dfrac{b}{2} - e\right)} \qquad \ldots\text{(Eq. 13.77)}$$

$$\text{since } b' = \left(\frac{b}{2} - e\right) \qquad \ldots\text{(Eq. 13.78)}$$

σ_{max} should not be greater than the allowable bearing capacity of the soil. (More of the concept of bearing capacity will be seen in Chapter 14).

(b) The condition of no tension is also easily verified. If tension occurs, there are two choices, one is to increase the trial base width and go through the calculations again. Another is to consider that only that part of the base width equal to $3b'$ (called the 'effective' base width) is useful in resisting the pressure and recompute σ_{max} as given by Eqs. 13.76 and 13.78 and verify if criterion (a) is now satisfied.

(c) If the friction angle between the material of the base of the wall and the foundation soil is δ', the requirement of safety against sliding is that the obliquity of the reaction R be less than δ'. This may be expressed:

$$\frac{T}{N} < \tan \delta' \qquad \ldots\text{(Eq. 13.79)}$$

or

$$T < N \tan \delta'$$

or

$$T < \mu \cdot N \qquad \ldots\text{(Eq. 13.80)}$$

where μ is the coefficient of friction ($= \tan \delta'$) between the base of the wall and the foundation soil. Further, one may insist on a margin of safety by demanding a certain minimum factor of safety against sliding, η_s (greater than unity), expressed as follows:

$$\eta_s = \frac{\mu \cdot N}{T} \qquad \ldots\text{(Eq. 13.81)}$$

This means that the frictional resistance to sliding is compared with the horizontal component of the thrust, which tends to cause sliding of the wall over its base.

If passive resistance is considered, the factor of safety against sliding should be greater than two. However, more commonly, the passive resistance is ignored and it is required that the factor of safety against sliding be 1.5 or more.

(d) For the wall to be safe against overturning, the reaction R must cross the base of the wall (that is $x \not> b$). If the requirement of no tension is satisfied, complete safety against overturning is automatically assured.

The factor of safety against overturning, η_0, is expressed:

$$\eta_0 = \frac{\text{Restoring moment}}{\text{Overturning moment}} \qquad \text{...(Eq. 13.82)}$$

These moments are taken about the toe of the wall. The force P_{ah} causes an overturning moment for the wall about the toe, while the forces W, P_{av}, and P_{ph} cause a restoring moment. In this case η_0 is given by:

$$\eta_0 = \frac{W(b - x_1) + P_{av}(b - x_2) + P_{ph} z_2}{P_{ah} \cdot z_1} \qquad \text{...(Eq. 13.83)}$$

It is recommended that this value be not less than 1.5 for granular soils and 2.0 for cohesive soils, if passive pressure is ignored. However, if passive pressure is also considered, this should be more than these specified values.

13.8.3 Influence of Yield of Wall on Design

The strain conditions within the failure wedge depend upon the nature of the yield of the wall. The distribution of lateral earth pressure with depth may be shown to be highly dependent on the strain conditions within the failure wedge and hence on the nature and extent of the yield of the wall.

Figure 13.52(a) represents the case where a wall is prevented from yielding. It is then subjected to 'earth pressure at rest'. The light gridwork represents planes that are initially horizontal and planes on which slip may occur, if an active case is reached. This gridwork is used to illustrate strains that occur in cases discussed later. The earth pressure distribution in this case is known to be triangular as shown by the straight line AF in Fig. 13.52(b), the total thrust on the wall per unit length being represented by P_0 which acts at a height of $H/3$ above the base.

The wall can yield in one of two ways: either by rotation about its lower edge [Fig. 13.52(c)], or by sliding forward or translating away from the backfill [Fig. 13.52(f)]. If the wall yields sufficiently, a state of active earth pressure is reached and the thrust on the wall in both cases is about the same (P_a). However, the pressure distribution that gives this total thrust can be very different in each instance, as will be seen from the following discussion.

Let it be assumed that the wall yields by rotation about the heel B by an amount sufficient to create the active pressure conditions. During this rotation, the wedge ABC distorts in an essentially uniform manner throughout to the shape $A'BC$ of Fig. 13.52(c). The uniform distribution leads to a ϕ-obliquity condition throughout and active pressures occur on the wall over its entire height. Neglecting wall friction, the pressure distribution will appear as shown by the straight line AG, in

588 *Geotechnical Engineering*

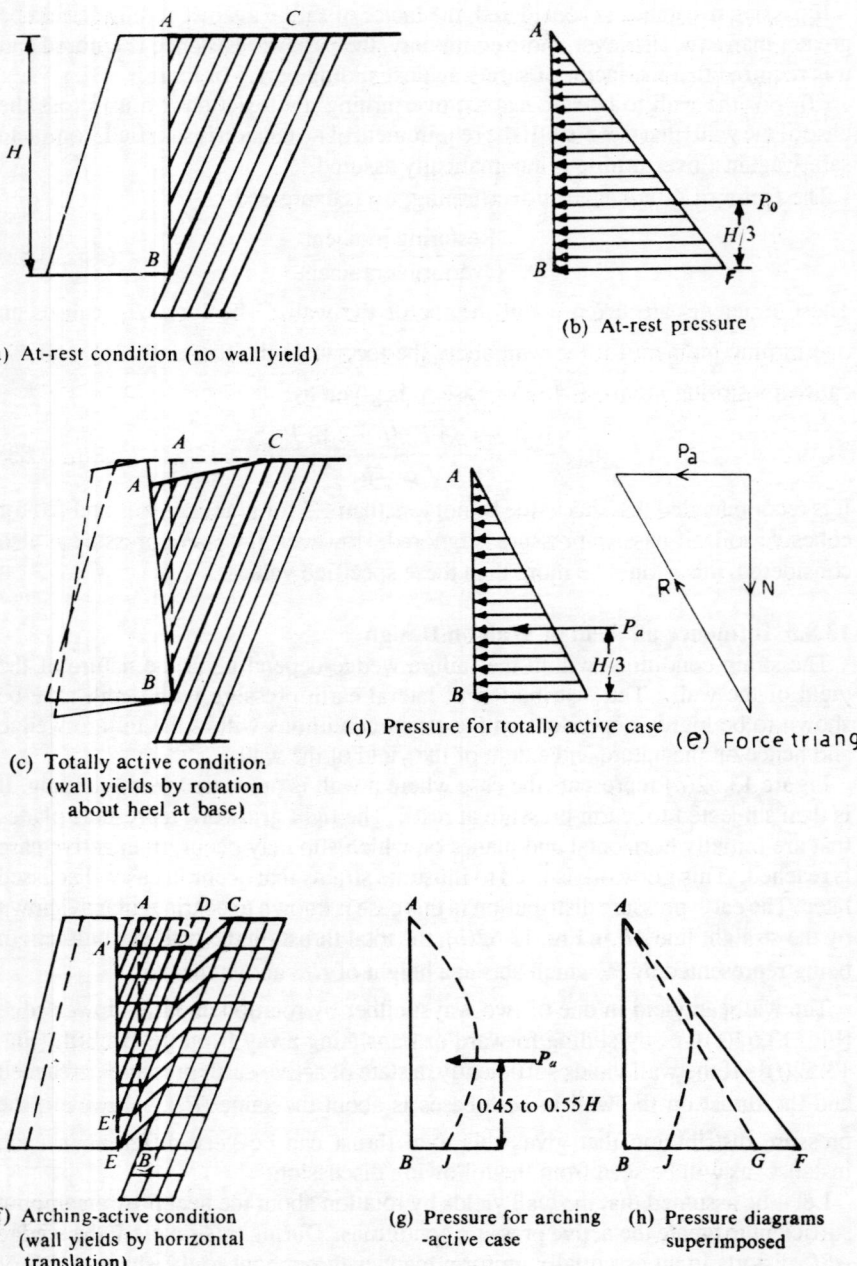

Fig. 13.52 At rest, totally active and arching—active cases (wall friction neglected) (Taylor, 1948)

Fig. 13.52(d), in a triangular shape, the pressure at any depth being less than the at-rest value. The total thrust P_a acts at $H/3$ above the base. The revised positions of the grid lines of Fig. 13.52(a) are shown in (c). Although the explanation is somewhat idealised in some respects, the general concept is essentially correct. This case is referred to as the totally active case which is the same as the active Rankine state.

Suppose, however, that the wall, starting from the at-rest condition, yields outward by horizontal translation, the face of the wall remaining vertical, until active thrust conditions are achieved. This is illustrated in Fig. 13.52(f). In this case the wedge collapses somewhat as shown in (f), and failure and the ϕ-obliquity condition occur only in a thin zone in the vicinity of line BC. The major portion of the wedge is not appreciably distorted and, therefore, the lateral pressure on the upper portion of the wall remains very much similar to that in the at-rest condition. In spite of this the total thrust on the wall in (f) is approximately the same as it was in (c). This is evident from a consideration of the force triangle shown in Fig. 13.52(e). In both cases the weight of the wedge ABC is W, which must be in equilibrium with intergranular reaction, R, on the sliding surface and the wall thrust P_a. Force W has the same magnitude and direction in (c) and (f), and the other two forces have orientations that are the same in (c) and (f). Thus the thrust P_a, representing the equilibrant of W and R, shall be essentially the same in both cases. It follows that the pressure distribution on the wall in (f) must be roughly as shown by the curved line AJ, approximating to a parabolic shape in Fig. 13.52(g). The high pressures that occur near the top of the wall and on the upper portion of the surface BC constitute an 'arching action' and has been referred to as the 'arching-active' case by Terzaghi (1936) and is described in detail by him; the conditions are described briefly, but excellently by Taylor (1948). This type of yield condition leads to a situation approximating to the wedge theory, the centre of pressure moving up to 0.45 to 0.55 H above the base.

The differences between the pressure distributions may be observed better by superimposing all in one figure as in Fig. 13.52(h); line AF represents the at-rest pressure, line AG represents the totally active pressure, and curve AJ the arching-active pressure. The total thrust in the second and third cases is the same, but is somewhat smaller than that in the first case.

Terzaghi observes

(i) If the mid-height point of the wall moves outward to a distance roughly equal to 0.05% of the wall height, an arching-active case is attained. (According to another school of thought, the top of the wall must yield about 0.10% of the wall height for this purpose). It is immaterial, whether the wall rotates or translates; however, the exact pressure distribution depends considerably on the amount of tilting of the wall.

(ii) If the top of the wall moves outward to a distance roughly equal to 0.50% of the wall height, the totally active case is attained. This criterion holds if the base of the wall either remains fixed or moves outward slightly.

Based on these concepts, the principles of design for different conditions of yield of the wall are summarised by Taylor (1948) as follows:

I. If a retaining wall with a cohesionless backfill is held rigidly in place by adjacent restraints (e.g., if it is joined to an adjacent structure), it must be designed to resist a thrust larger than the active value; for the completely restrained case it must be designed to resist the thrust relating to the at-rest condition. However, this case will not occur often, in view of the relatively small yield required to give the case given in II.

II. If a retaining wall with a cohesionless backfill is so restrained that only a small amount of yield takes place, it is likely that this movement will be sufficient to give the arching active case but not the totally active case. In this case, the assumption of a triangular pressure distribution is incorrect; the actual pressure distribution is statically indeterminate to a high degree, but is roughly parabolic.

A common example of the arching-active case is the pressure distribution on the sheeting of trenches.

III. If a retaining wall with a cohesionless backfill is not attached to any adjacent structure, it can yield considerably without harm to the structure; in such a case the totally active case is attained. (In fact, even a yield of 0.5% of the wall height is adequate for this condition). The design of such a wall on the basis of active pressure and triangular distribution of pressure is rational.

IV. In the case of a retaining wall with a cohesive backfill, the totally active case is reached as soon as the wall yields but, due to plastic flow within the clay, there is a tendency for a continuous increase in the pressure on the wall, unless the wall is permitted to yield continuously. The continuous yield, although slow, may lead to a large movement over a period of years. In such cases, one can either design for a higher pressure or design for a totally active pressure if the wall is considered capable of withstanding any movement without harmful effects; for this latter basis, which is a commonly used basis for design, the probable life of a wall with a cohesive backfill may be relatively short.

According to these principles any wall capable of yielding without detrimental results may be designed on the basis of active thrust and triangular distribution of pressure; however, the actual thrust on the wall may be more and the pressure distribution may not be triangular. This need not cause alarm to the geotechnical engineer, since any wall must have a margin of safety and will be designed to withstand thrusts greater than the calculated values. This margin of strength may prevent the wall from ever yielding sufficiently to give active conditions. Furthermore, the moment the wall is subjected to an increased thrust it merely yields a small amount, which immediately reduces the pressure. This interdependency of yield and pressure is thus a saving factor which partly takes care of any uncertainties in the theory.

13.8.4 Choice of Appropriate Earth Pressure Theory

The choice of appropriate earth pressure theory for a given situation will become easy if one remembers the various assumptions in the development of the earth pressure theories dealt with.

Lateral Earth Pressure and Stability of Retaining Walls 591

For example, if the retaining wall has a vertical back and is smooth, Rankine's theory may be considered appropriate. One may use Rankine's theory even for an inclined back of the wall with a slight wall friction, provided the resultant thrust obtained by combining the thrust on an imaginary vertical plane through the heel of the wall and the weight of the additional wedge of soil standing on the back of the wall, does not have an obliquity greater than the wall friction angle.

If the backface of the wall is plane and the wall friction is not inconsiderable and the soil shows a tendency to slide along the back of the wall, the use of Coulomb's theory is appropriate.

Coulomb's wedge theory with plane rupture faces should not be used for the estimation of passive resistance especially in the case of structures such as sheet pile walls, wherein passive earth resistance plays a major role. Coulomb's theory with curved rupture surfaces, such as the logarithmic spiral, should be used.

For cantilever and counterfort walls, Rankine's theory is used; for gravity and semi-gravity walls, Coulomb's theory is preferred.

13.9 ILLUSTRATIVE EXAMPLES

Example 13.1: A retaining wall, 6 m high, retains dry sand with an angle of friction of 30° and unit weight of 1.62 g/cm³. Determine the earth pressure at rest. If the water table rises to the top of the wall, determine the increase in the thrust on the wall. Assume the submerged unit weight of sand as 1.00 g/cm³.

(a) Dry backfill:

$$\phi = 30° \qquad H = 6 \text{ m}$$

$$K_0 = 1 - \sin 30° = 0.5$$

(Also $K_0 = 0.5$ for medium dense sand)

$$\sigma_0 = K_c \gamma \cdot H$$

$$= \frac{0.5 \times 1.62 \times 600}{1000} \text{ kg/cm}^2$$

$$= 0.496 \text{ kg/cm}^2$$

Thrust per metre length of the wall = $0.496 \times 100 \times 100 \times \frac{1}{2} \times 6$ kg = **14,880 kg**

(b) Water level at the top of the wall

The total lateral thrust will be the sum of effective and neutral lateral thrusts.

Effective lateral earth thrust, $P_0 = \frac{1}{2} K_0 \gamma \cdot H^2$

$$= \frac{1}{2} \times 0.5 \times \frac{1}{1000} \times 600 \times 100 \times 100 \times 6 \text{ kg/m. run}$$

$$= 9,000 \text{ kg/m. run}$$

Neutral lateral pressure $P_w = \frac{1}{2}\gamma_w H^2$

$$= \frac{1}{2} \times \frac{1}{1000} \times 600 \times 100 \times 100 \times 6 \text{ kg/m. run}$$

$$= 18,000 \text{ kg/m. run}$$

Total lateral thrust = 27,000 kg/m. run
Increase in thrust = **12,120 kg/m. run.**
This represents an increase of about **81.5%** over that of dry fill.

Example 13.2: What are the limiting values of the lateral earth pressure at a depth of 3 metres in a uniform sand fill with a unit weight of 2000 kg/m³ and a friction angle of 35°? The ground surface is level.

(S.V.U.—B.E. (R.R.)—Feb., 1976)

If a retaining wall with a vertical back face is interposed, determine the total active thrust and the total passive resistance which will act on the wall.

Depth, H = 3 m
$\left.\begin{array}{l} \gamma = 2000 \text{ kg/m}^3 \\ \phi = 35° \end{array}\right\}$ for sand fill with level surface.

Limiting values of lateral earth pressure:

Active pressure $= K_a \cdot \gamma H = \dfrac{1 - \sin 35°}{1 + \sin 35°} \times 2000 \times 3$

$$= 0.271 \times 6000$$

$$= \textbf{1,626 kg/m}^2$$

Passive pressure $= K_p \cdot \gamma H = \dfrac{1 + \sin 35°}{1 - \sin 35°} \times 2000 \times 3$

$$= 3.690 \times 6000$$

$$= \textbf{22,140 kg/m}^2$$

Total active thrust per metre run of the wall

$$P_a = \frac{1}{2}\gamma H^2 K_a = 1626 \times \frac{1}{2} \times 3 = \textbf{2,439 kg}$$

Total passive resistance per metre run of the wall

$$P_p = \frac{1}{2}\gamma H^2 \cdot K_p = 22140 \times \frac{1}{2} \times 3 = \textbf{33,210 kg.}$$

Example 13.3: A gravity retaining wall retains 12 m of a backfill, $\gamma = 17.7$ kN/m^3 $\phi = 25°$ with a uniform horizontal surface. Assume the wall interface to be vertical, determine the magnitude and point of application of the total active pressure. If the water table is at a height of 6 m, how far do the magnitude and the point of application of active pressure changed?

(S.V.U.—Four-year B.Tech.—Oct., 1982)

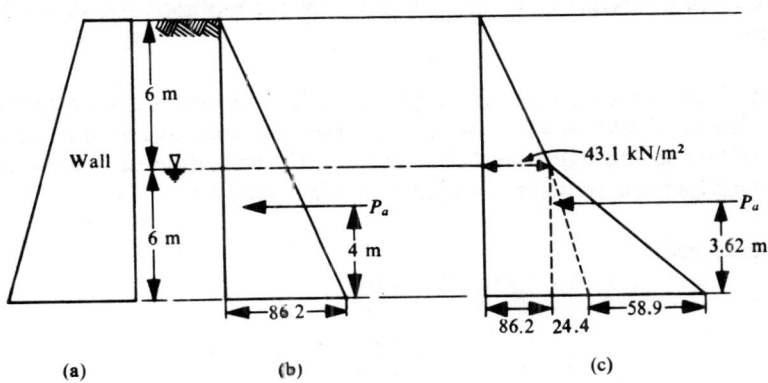

Fig. 13.53 Retaining wall and pressure distribution (Ex. 13.3)

(*a*) Dry cohesionless fill:

$$H = 12 \text{ m} \quad \phi = 25° \quad \gamma = 17.7 \text{ kN/m}^3$$

$$\therefore \quad K_a = \frac{1 - \sin 25°}{1 + \sin 25°} = 0.406$$

Active pressure at base of wall $= K_a \cdot \gamma H = \dfrac{1 - \sin 25°}{1 + \sin 25°} \times 17.7 \times 12$

$$= 86.2 \text{ kN/m}^2$$

The distribution of pressure is triangular as shown in Fig. 13.53(*b*).

Total active thrust per metre run of wall $= \frac{1}{2} \gamma H^2 K_a = \frac{1}{2} \times 12 \times 86.2 = \mathbf{517.2 \text{ kN}}$

This acts at $(1/3)H$ or 4 m above the base of the wall.

(*b*) Water table at 6 m from surface:

Active pressure at 6 m depth $= 0.406 \times 17.7 \times 6 = 43.1$ kN/m^2

Active pressure at the base of the wall $= K_a (\gamma \cdot 6 + \gamma' \cdot 6) + \gamma_w \cdot 6$

$$= 0.406 (17.7 \times 6 + 10 \times 6) + 9.81 \times 6 = 67.5 + 58.9 = 126.4 \text{ kN/m}^2$$

(This is obtained by assuming γ above the water table to be 17.7 kN/m^2 and the submerged unit weight, γ', in the bottom 6 m zone, to be 10 kN/m^2.
The pressure distribution is shown in Fig. 13.53(*c*).

Total active thrust per metre run = Area of the pressure distribution diagram

$$= \frac{1}{2} \times 6 \times 63.1 + 6 \times 43.1 + \frac{1}{2} \times 6 \times 24.4 + \frac{1}{2} \times 6 \times 58.9$$

$$= 129.3 + 258.6 + 73.2 + 176.7 = \mathbf{637.8\ kN}$$

The height of its point of application above the base is obtained by taking moments.

$$\bar{z} = \frac{(129.3 \times 8 + 258.6 \times 3 + 73.2 \times 2 + 176.7 \times 2)}{637.8} = \mathbf{3.62\ m}$$

Total thrust increases by **120.6 kN** and the point of application gets lowered by **0.38 m**.

Example 13.4: A wall, 5.4 m high, retains sand. In the loose state the sand has a void ratio of 0.63 and $\phi = 27°$, while in the dense state, the corresponding values of void ratio and ϕ are 0.36 and 45° respectively. Compare the ratio of active and passive earth pressure in the two cases, assuming $G = 2.64$.

(*a*) Loose State:

$$G = 2.64 \quad e = 0.63$$

$$\gamma_d = \frac{G \cdot \gamma_w}{(1+e)} = \frac{2.64 \times 1}{(1+0.63)} = 1.62\ t/m^3$$

$$\phi = 27°$$

$$K_a = \frac{1-\sin 27°}{1+\sin 27°} = 0.376; \quad K_p = \frac{1+\sin 27°}{1-\sin 27°} = 2.663$$

Active pressure at depth H m $= K_a \cdot \gamma \cdot H = 0.376 \times 1.62 H = 0.609 \cdot H\ t/m^2$

Passive pressure at depth H m $= K_p \cdot \gamma H = 2.663 \times 1.62 H = 4.314 H\ t/m^2$

(*b*) Dense State:

$$G = 2.64 \quad e = 0.36$$

$$\gamma_d = \frac{2.64 \times 1}{(1+0.36)} = 1.94\ t/m^3$$

$$\phi = 45°$$

$$K_a = \frac{1-\sin 45°}{1+\sin 45°} = 0.172; \quad K_p = \frac{1+\sin 45°}{1-\sin 45°} = 5.828$$

Active pressure at depth H m $= 0.172 \times 1.94 H = 0.334 H\ t/m^2$

Passive pressure at depth H m $= 5.828 \times 1.94 H = 11.306 H\ t/m^2$

Ratio of active pressure in the dense state to that in the loose state $= \dfrac{0.334}{0.609} = \mathbf{0.55}$

Ratio of passive resistance in the dense state to that in the loose state $= \dfrac{11.306}{4.314}$
$= \mathbf{2.62}$

Example 13.5: A smooth backed vertical wall is 6.3 m high and retains a soil with a bulk unit weight of 18 kN/m³ and $\phi = 18°$. The top of the soil is level with the top of the wall and is horizontal. If the soil surface carries a uniformly distributed load of 4.5 kN/m², determine the total active thrust on the wall per lineal metre of the wall and its point of application.

$$H = 6.3 \text{ m} \qquad \gamma = 18 \text{ kN/m}^3 \qquad \phi = 18°$$

$$q = 45 \text{ kN/m}^2$$

$$K_a = \frac{1 - \sin 18°}{1 + \sin 18°} = 0.528$$

Active pressure due to weight of soil at the base of wall = $K_a \gamma H$

$$= 0.528 \times 18 \times 6.3$$

$$= 59.9 \text{ kN/m}^2$$

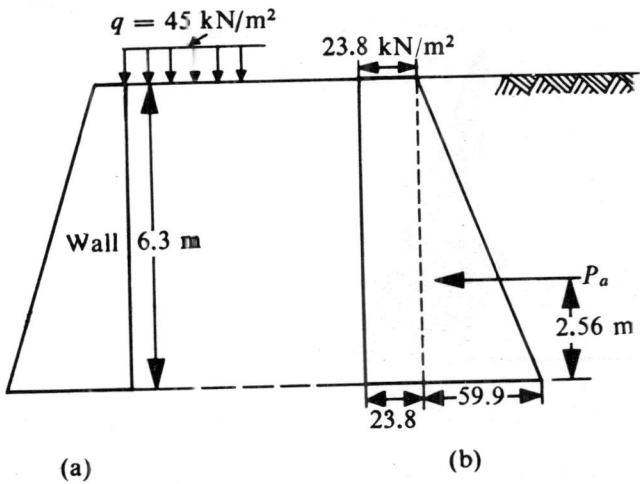

Fig. 13.54 Retaining wall and pressure distribution (Ex. 13.5)

Active pressure due to uniform surcharge = $K_a \cdot q$

$$= 0.528 \times 45$$

$$= 23.8 \text{ kN/m}^2$$

The former will have triangular distribution while the latter will have rectangular distribution with depth. The resultant pressure distribution diagram will be as shown in Fig. 13.54(*b*).

Total active thrust per lineal metre of wall,

P_a = Area of pressure distribution diagram = $\frac{1}{2}K_a\gamma H^2 + K_a qH$

$= \frac{1}{2} \times 59.9 \times 6.3 + 23.8 \times 6.3 = 188.7 + 149.9 = \mathbf{338.6\ kN}$

The height of its point of application above the base may be obtained by taking moments:

$$\bar{z} = \frac{\left(188.7 \times \frac{1}{3} \times 6.3 + 149.9 \times \frac{1}{2} \times 6.3\right)}{338.6} \text{ m} = \mathbf{2.56\ m}$$

Example 13.6: A vertical wall with a smooth face is 7.2 m high and retains soil with a uniform surcharge angle of 9°. If the angle of internal friction of the soil is 27°, compute the active earth pressure and passive earth resistance assuming $\gamma = 2$ g/cm³.

$H = 7.2$ m $\beta = 9°$

$\phi = 27°$ $\gamma = 2$ g/cm³ = 2 t/m³

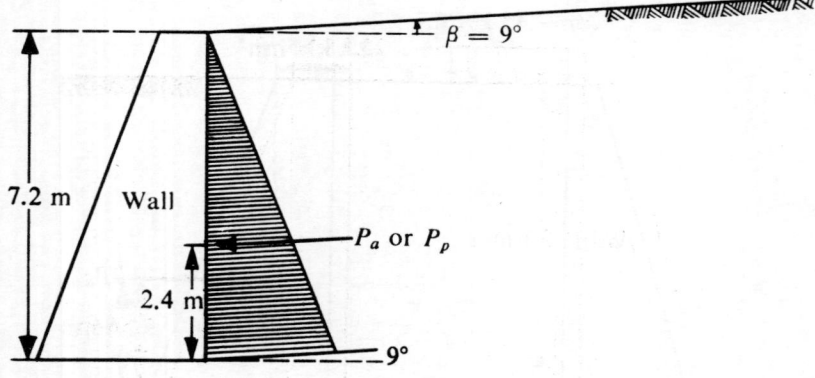

Fig. 13.55 Retaining wall with inclined surcharge and pressure distribution (Ex. 13.6)

According to Rankine's theory,

$$K_a = \cos\beta \left(\frac{\cos\beta - \sqrt{\cos^2\beta - \cos^2\phi}}{\cos\beta + \sqrt{\cos^2\beta - \cos^2\phi}}\right)$$

$$= \cos 9° \left(\frac{\cos 9° - \sqrt{\cos^2 9° - \cos^2 27°}}{\cos 9° + \sqrt{\cos^2 9° - \cos^2 27°}}\right)$$

$$= 0.988 \times 0.397 = 0.392$$

$$K_p = \cos\beta \left(\frac{\cos\beta + \sqrt{\cos^2\beta - \cos^2\phi}}{\cos\beta - \sqrt{\cos^2\beta - \cos^2\phi}}\right) = 0.988 \times \frac{1}{0.397} = 2.488$$

Total active thrust per metre run of the wall
$$P_a = \frac{1}{2}\gamma H^2 \cdot K_a = \frac{1}{2} \times 2 \times 7^2 \cdot 2 \times 0.392 = \mathbf{20.32 \text{ t}}$$
Total passive resistance per metre run of the wall
$$P_p = \frac{1}{2}\gamma H^2 \cdot K_p = \frac{1}{2} \times 2 \times 7^2 \cdot 2 \times 2.488 = \mathbf{128.98 \text{ t}}$$

The pressure is considered to act parallel to the surface of the backfill soil and the distribution is triangular for both cases. The resultant thrust thus acts at a height of $(1/3)H$ or 2.4 m above the base at 9° to horizontal, as shown in Fig. 13.55.

Example 13.7: The Rankine formula of active earth pressure for a vertical wall and a level fill is much better known than the general form and sometimes it is used even when it does not apply. Determine the percentage error introduced by assuming a level fill when the angle of surcharge actually equals 20°. Assume a friction angle of 35° and the wall vertical. Comment on the use of the erroneous result.

(S.V.U.—B.E. (R.R.)—Nov., 1974)

$\phi = 35°$
Active pressure coefficient of Rankine for inclined surcharge:
$$K_{ai} = \cos\beta \cdot \left(\frac{\cos\beta - \sqrt{\cos^2\beta - \cos^2\phi}}{\cos\beta + \sqrt{\cos^2\beta - \cos^2\phi}}\right)$$
when $\beta = 0°$ for horizontal surface of the backfill,
$$K_a = \frac{1 - \sin\phi}{1 + \sin\phi}$$
K_{ai} for $\beta = 20°$ and $\phi = 35°$ is given by
$$K_{ai} = \cos 20° \left(\frac{\cos 20° - \sqrt{\cos^2 20° - \cos^2 35°}}{\cos 20° + \sqrt{\cos^2 20° - \cos^2 35°}}\right) = 0.322$$
K_a for $\beta = 0°$ and $\phi = 35°$ is given by
$$K_a = \frac{1 - \sin 35°}{1 + \sin 35°} = 0.271$$
Percentage error in the computed active thrust by assuming a level fill when it is actually inclined at 20° to horizontal
$$= \left(\frac{0.322 - 0.271}{0.322}\right) \times 100 = \mathbf{15.84}$$
The thrust is underestimated by assuming a level fill, obviously.

Example 13.8: A retaining wall 9 m high retains a cohesionless soil, with an angle of internal friction 33°. The surface is level with the top of the wall. The unit weight of the top 3 m of the fill is 2.1 t/m³ and that of the rest is 2.7 t/m³. Find the magnitude and point of application of the resultant active thrust.

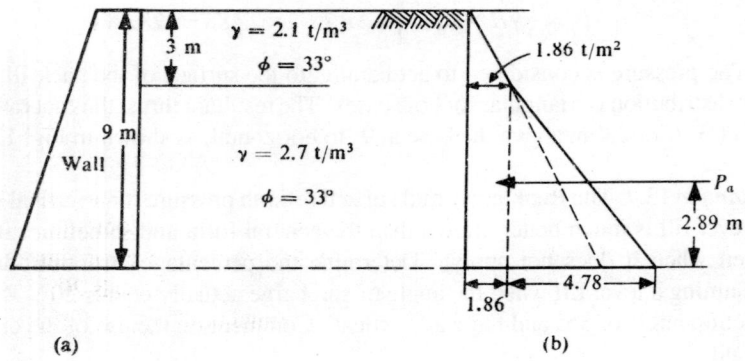

Fig. 13.56 Lateral pressure due to stratified backfill (Ex. 13.8)

It is assumed that $\phi = 33°$ for both the strata of the backfill.

∴ $K_a = \dfrac{1-\sin 33°}{1+\sin 33°} = 0.295$, for both the strata of the backfill.

Active pressure at 3 m depth

$$K_a \cdot \sigma_v = 0.295\,(2.1 \times 3) = 1.86 \text{ t/m}^2$$

Active pressure at the base of the wall

$$K_a \cdot \sigma_v = 0.295(2.1 \times 3 + 2.7 \times 6) = 6.64 \text{ t/m}^2$$

The variation of pressure is linear, with a break in the slope at 3 m depth, as shown in Fig. 13.56 (*b*). The total active thrust per metre run, P_a, is given by the area of the pressure distribution diagram.

∴ $P_a = \dfrac{1}{2} \times 3 \times 1.86 + 6 \times 1.86 + \dfrac{1}{2} \times 6 \times 4.78 = \mathbf{28.3\ t}$

The height, above the base, of the point of application of this thrust is obtained by taking moments about the base

$$\bar{z} = \dfrac{(2.79 \times 7 + 11.16 \times 3 + 14.34 \times 2)}{28.3} \text{ m} = \mathbf{2.89\ m}$$

Example 13.9: A retaining wall, 7.5 m high, retains a cohesionless backfill. The top 3 m of the fill has a unit weight of 18 kN/m³ and $\phi = 30°$ and the rest has a unit weight of 24 kN/m³ and $\phi = 20°$. Determine the pressure distribution on the wall.

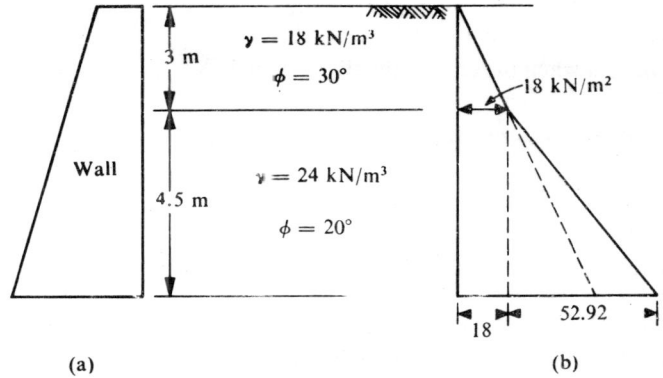

Fig. 13.57 Stratified backfill with different K_a-values for different layers (Ex. 13.9)

K_a for top layer $= \dfrac{1-\sin 30°}{1+\sin 30°} = 1/3$

K_a for bottom layer $= \dfrac{1-\sin 20°}{1+\sin 20°} = 0.49$

Active pressure at 3 m depth

$$K_{a_1} \cdot \frac{1}{3} \times 3 \times 18 = 18 \text{ kN/m}^2$$

Active pressure at the base of wall

$$K_{a_1} \cdot 3 \times 18 + K_{a_2} \cdot 4.5 \cdot 24 = \frac{1}{3} \times 3 \times 18 + 0.49 \times 4.5 \times 24$$

$$= 70.92 \text{ kN/m}^2$$

The pressure distribution with depth is shown in Fig. 13.57(b).
Total active thrust, P_a, per metre run of the wall
= Area of the pressure distribution diagram

$$= \frac{1}{2} \times 3 \times 18 + 4.5 \times 18 + \frac{1}{2} \times 4.5 \times 52.92$$

$$= 27 + 81 - 119 = 227 \text{ kN}$$

The height of the point of application of this thrust above the base of the wall is obtained by taking moments, as usual

$$\bar{z} = \frac{(27 \times 5.5 + 81 \times 2.25 + 119 \times 1.5)}{227} \text{ m} = \mathbf{2.24 \text{ m}}$$

Example 13.10: Excavation was being carried out for a foundation in a plastic clay with a unit weight of 2.25 g/cm³. Failure occurred when a depth of 8.10 m was reached. What is the value of cohesion if $\phi = 0°$?

600 *Geotechnical Engineering*

$$\phi = 0° \quad \gamma = 2.25 \text{ g/cm}^3$$

Failure occurs when the critical depth, H_c, which is $\dfrac{4c}{\gamma} \cdot \sqrt{N_\phi}$ is reached.

Since $\phi = 0$, $N_\phi = \tan^2(45° + \phi/2) = 1$

$$\frac{4c \times 1000}{2.25 \times 1} \times 1 = 8.10 \times 100$$

$\therefore \quad$ Cohesion, $c = \dfrac{2.25 \times 810}{4000}$ kg/cm^2 = **0.456 kg/cm²**

Example 13.11: A sandy loam backfill has a cohesion of 12 kN/m² and $\phi = 20°$. The unit weight is 17.0 kN/m³. What is the depth of the tension cracks?

Depth of tension cracks, z_c, is given by

$$z_c = \frac{2c}{\gamma} \cdot \sqrt{N_\phi} \quad \phi = 20°$$

$\therefore \quad \sqrt{N_\phi} = \tan(45° + \phi/2) = \tan 55° = 1.428$

$$c = 12 \text{ kN/m}^2$$

$$\gamma = 17.0 \text{ kN/m}^3$$

$\therefore \quad z_c = \dfrac{2 \times 12}{17.0} \times 1.428$ m

$$= 2.00 \text{ m}$$

Example 13.12: A retaining wall with a smooth vertical back retains a purely cohesive fill. Height of wall is 12 m. Unit weight of fill is 20 kN/m³. Cohesion is 1 N/cm². What is the total active Rankine thrust on the wall? At what depth is the intensity of pressure zero and where does the resultant thrust act?

$$H = 12 \text{ m} \quad \gamma = 20 \text{ kN/m}^3 \quad \phi = 0°$$

$$N_\phi = \tan^2(45° + \phi/2) = 1$$

$$c = 1 \text{ N/cm}^2 = 10 \text{ kN/m}^2$$

$$z_c = \frac{2c}{\gamma} = \frac{2 \times 10}{20} = 1 \text{ m}$$

Lateral Earth Pressure and Stability of Retaining Walls 601

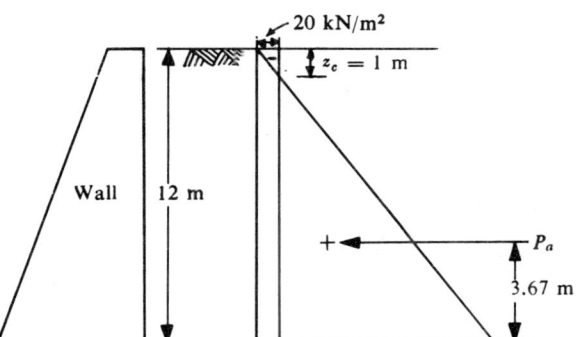

Fig. 13.58 Retaining wall with purely cohesive fill (Ex. 13.12)

∴ The intensity of pressure is zero at a depth of 1 m from the surface.

$$\frac{\gamma H}{N_\phi} = \frac{20 \times 12}{1} = 240 \text{ kN/m}^2$$

$$\frac{2c}{\sqrt{N_\phi}} = \frac{2 \times 10}{\sqrt{1}} = 20 \text{ kN/m}^2$$

The net pressure diagram is shown in Fig. 13.58(b).

The total active thrust may be found by ignoring the tensile stresses, as the area of the positive part of the pressure diagram.

$$P_a = \frac{1}{2} \times 220 \times 11 = 1{,}210 \text{ kN/metre run}$$

This acts at a height of 11/3 m or **3.67 m** from the base of the wall.

Example 13.13: A smooth vertical wall 5 m high retains a soil with $c = 2.5 \text{ N/cm}^2$, $\phi = 30°$, and $\gamma = 18 \text{ kN/m}^3$. Show the Rankine passive pressure distribution and determine the magnitude and point of application of the passive resistance.

$$H = 5 \text{ m} \qquad \phi = 30° \qquad c = 2.5 \text{ kN/cm}^2 = 25 \text{ kN/m}^2$$

$$\gamma = 18 \text{ kN/m}^3$$

$$N_\phi = \tan^2\left(45° + \frac{30°}{2}\right) = 3$$

Pressure at the base:

$$\gamma H \cdot N_\phi = 18 \times 5 \times 3 = 270 \text{ kN/m}^2$$

$$2c\sqrt{N_\phi} = 2 \times 25 \times \sqrt{3} = 86.6 \text{ kN/m}^2$$

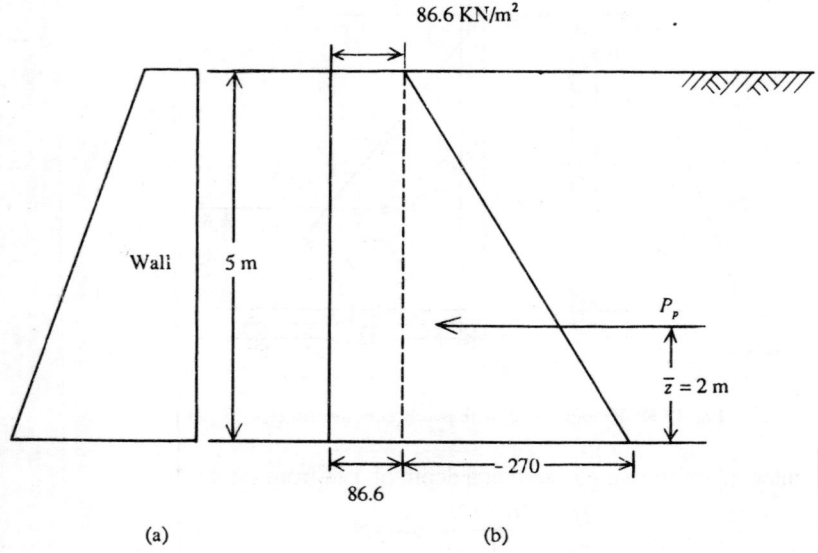

Fig. 13.59 Passive pressure of a $c - \phi$ soil (Ex. 13.13)

The distribution of the first component is triangular and that of the second component is rectangular with depth and the pressure distribution is as shown in Fig. 13.59(b).

The total passive resistance, P_p, on the wall per metre run is obtained as the area of the pressure distribution diagram.

$$\therefore \quad P_p = 5 \times 86.6 + \frac{1}{2} \times 270 \times 5 = 433.0 + 675.0 = b\mathbf{1,108 \text{ kN}}$$

The height of the point of application above the base is obtained by taking moments as usual.

$$\therefore \quad \bar{z} = \frac{(433 \times 5/2 + 675 \times 5/3)}{1108} \text{ m} = \mathbf{2.00 \text{ m}}$$

Example 13.14: A retaining wall 9 m high retains granular fill weighing 1.8 g/cm³ with level surface. The active thrust on the wall is 18 t per metre length of the wall. The height of the wall is to be increased and to keep the force on the wall within allowable limits, the backfill in the top-half of the depth is removed and replaced by cinders. If cinders are used as backfill even in the additional height, what additional height may be allowed if the thrust on the wall is to be limited to its initial value? The unit weight of the cinders is 0.9 g/cm³. Assume the friction angle for cinders the same as that for the soil.

$$H = 9 \text{ m}$$

$$\gamma = 1.8 \text{ g/cm}^3 = 1.8 \text{ t/m}^3$$

$$P_a = 18 \text{ t/m run}$$

Initially,

$$P_a = \frac{1}{2}\gamma H^2 \cdot K_a$$

$$\therefore \quad 18 = \frac{1}{2} \times 1.8 \times 9^2 \times K_a$$

$$K_a = \frac{2 \times 18}{1.8 \times 9^2} = 0.247$$

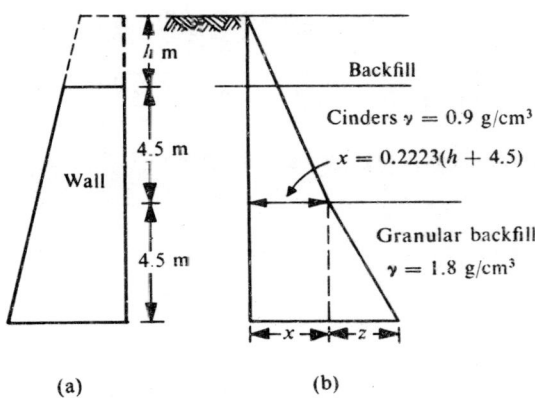

(a)   (b)

Fig. 13.60 Retaining wall with backfill partly of cinders (Ex. 13.14)

Let the increase in the height of wall be h m.
The depth of cinders backfill will be $(h + 4.5)$ m and the bottom 4.5 m is granular backfill with $K_a = 0.247$. Since the friction angle for cinders is taken to be the same as that for the granular soil, K_a for cinders is also 0.247, but γ for cinders is 0.9 t/m^3.

The intensity of pressure at $(h + 4.5)$m depth $= 0.247 \times 0.9 \, (h + 4.5)$ t/m^2

$$= 0.2223 \, (h + 4.5) \text{ t/m}^2$$

Intensity of pressure at the base $= 0.247 \, [0.9 \, (h + 4.5) + 1.8 \times 4.5]$ t/m^2

$$= 0.2223 \, (h + 4.5) + 2 \text{ t/m}^2$$

Total thrust $P_a' = 0.1112 \, (h + 4.5)^2 + 0.2223 \times 4.5 \, (h + 4.5) + \frac{1}{2} \times 4.5 \times 2$

Equating this to the initial value P_a, or 18 t, the following equation is obtained:

$$0.1112 h^2 + 2h - 6.75 = 0$$

Solving, $h = 2.90$ m

Thus, the height of the wall may be increased by **2.90 m** without increasing the thrust.

Example 13.15: A gravity retaining wall retains 12 m of a backfill. $\gamma = 18$ kN/m³, $\phi = 30°$ with a uniform horizontal backfill. Assuming the wall interface to be vertical, determine the magnitude of active and passive earth pressure. Assume the angle of wall friction to be 20°. Determine the point of action also.

(S.V.U.—Four-year B.Tech.—Dec., 1982)

Since wall friction is to be accounted for, Coulomb's theory is to be applied.

$$\gamma = 18 \text{ kN/m}^3 \text{ and } H = 12 \text{ m}$$

$$K_a = \frac{\sin^2(\alpha + \phi)}{\sin^2\alpha \cdot \sin(\alpha - \delta)\left[1 + \sqrt{\frac{\sin(\phi + \delta) \cdot \sin(\phi - \beta)}{\sin(\alpha - \delta) \cdot \sin(\alpha + \beta)}}\right]^2}$$

$\alpha = 90°$ and $\beta = 0°$ in this case. $\phi = 30°$ and $\delta = 20°$

$$\therefore \quad K_a = \frac{\cos^2\phi}{\cos\delta\left[1 + \sqrt{\frac{\sin(\phi + \delta) \cdot \sin\phi}{\cos\delta}}\right]^2}$$

$$= \frac{\cos^2 30°}{\cos 20°\left[1 + \sqrt{\frac{\sin 50° \cdot \sin 30°}{\cos 20°}}\right]^2} = 0.132$$

$$K_p = \frac{\sin^2(\alpha - \phi)}{\sin^2\alpha \cdot \sin(\alpha + \delta)\left[1 - \sqrt{\frac{\sin(\phi + \delta) \cdot \sin(\phi + \beta)}{\sin(\alpha + \delta) \cdot \sin(\alpha + \beta)}}\right]^2}$$

Putting $\alpha = 90°$ and $\beta = 0°$,

$$K_p = \frac{\cos^2\phi}{\cos\delta\left[1 - \sqrt{\frac{\sin(\phi + \delta) \cdot \sin\phi}{\cos^2\delta}}\right]^2}$$

$$= \frac{\cos^2 30°}{\cos 20°\left[1 - \sqrt{\frac{\sin 50° \cdot \sin 30°}{\cos 20°}}\right]^2} = 2.713$$

$$P_a = \tfrac{1}{2}\gamma H^2 \cdot K_a = \tfrac{1}{2} \times 18 \times 12^2 \times 0.132 = \mathbf{171 \text{ kN/m}}$$

$$P_p = \tfrac{1}{2}\gamma H^2 \cdot K_p = \tfrac{1}{2} \times 18 \times 12^2 \times 2.713 = \mathbf{3.516 \text{ kN/m.}}$$

Both P_a and P_p act at a height of $(1/3)H$ or **4 m** above the base of the wall and are inclined at 20° above and below the horizontal, respectively.

Example 13.16: A retaining wall is battered away from the fill from bottom to top at an angle of 15° with the vertical. Height of the wall is 6 m. The fill slopes upwards at an angle 15° away from the rest of the wall. The friction angle is 30° and wall friction angle is 15°. Using Coulomb's wedge theory, determined the total active and passive thrusts on the wall, per lineal metre assuming $\gamma = 20$ kN/m³.

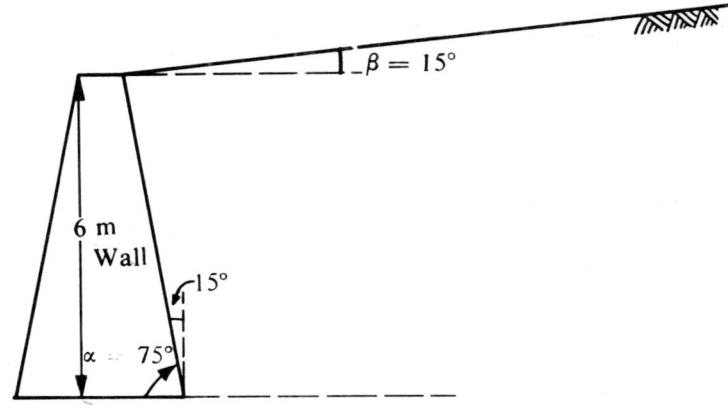

Fig. 13.61 Battered wall with inclined surcharge (Ex. 13.15)

$$H = 6 \text{ m}$$
$$\beta = 15°$$
$$\alpha = 75° \text{ from Fig. 13.61}$$
$$\phi = 30°$$
$$\delta = 15°$$
$$\gamma = 20 \text{ kN/m}^2$$

$$K_a = \frac{\sin^2(\alpha + \phi)}{\sin^2\alpha \cdot \sin(\alpha - \delta)\left[1 + \sqrt{\frac{\sin(\phi + \delta) \cdot \sin(\phi - \beta)}{\sin(\alpha - \delta) \cdot \sin(\alpha + \beta)}}\right]^2}$$

$$= \frac{\sin^2 105°}{\sin^2 75° \cdot \sin 60°\left[1 + \sqrt{\frac{\sin 45° \cdot \sin 15°}{\sin 60° \cdot \sin 90°}}\right]^2} = 0.542$$

$$K_p = \frac{\sin^2(\alpha - \phi)}{\sin^2\alpha \cdot \sin(\alpha + \delta)\left[1 - \sqrt{\frac{\sin(\phi + \delta) \cdot \sin(\phi + \beta)}{\sin(\alpha + \delta) \cdot \sin(\alpha + \beta)}}\right]^2}$$

$$= \frac{\sin^2 45°}{\sin^2 75° \cdot \sin 90°\left[1 - \sqrt{\frac{\sin 45° \cdot \sin 45°}{\sin 90° \cdot \sin 90°}}\right]^2} = 6.242$$

Total active thrust, P_a, per lineal metre of the wall

$$= \frac{1}{2}\gamma H^2 \cdot K_a = \frac{1}{2} \times 20 \times 6^2 \times 0.542 = \mathbf{195 \ kN}$$

Total passive resistance, P_p, per lineal metre of the wall

$$= \frac{1}{2}\gamma H^2 \cdot K_p = \frac{1}{2} \times 20 \times 6^2 \times 6.242 = \mathbf{2,247 kN}$$

Example 13.17: A vertical retaining wall 10 m high supports a cohesionless fill with $\gamma = 1.8$ g/cm^3. The upper surface of the fill rises from the crest of the wall at an angle of 20° with the horizontal. Assuming $\phi = 30°$ and $\delta = 20°$, determine the total active earth pressure using the analytical approach of Coulomb.

(S.V.U.—B.Tech. (Part-time)—Sep., 1982)

$\gamma = 1.8$ g/cm^3 = 1.8 t/m^3 $H = 10$ m. $\phi = 30°$ $\delta = 20°$ $\beta = 20°$ $\alpha = 90°$

$$\psi = \alpha - \delta = 90° - 20° = 70°$$

$$\phi + \delta = 30° + 20° = 50°$$

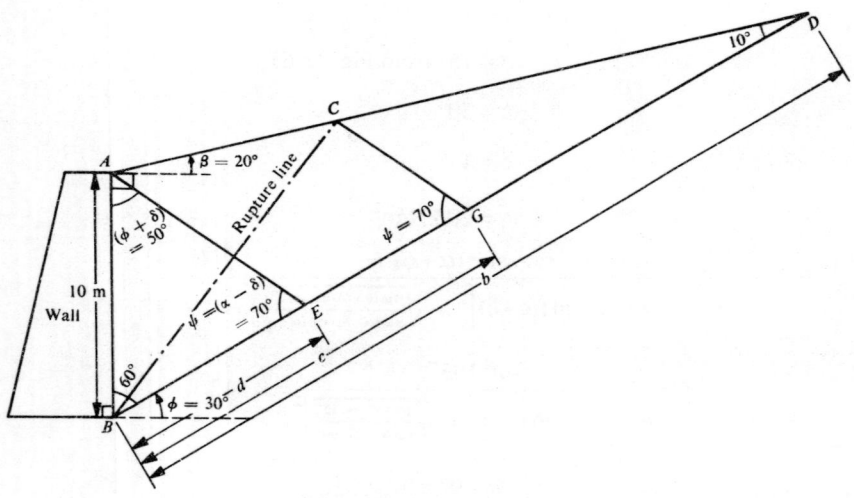

Fig. 13.62 Coulomb's analytical approach (Ex. 13.17)

$\overline{AB} = H = 10$ m $a = \dfrac{10}{\sin 70°} \cdot \sin 60° = 9.216$ m

$d = \dfrac{10}{\sin 70°} \cdot \sin 50°$ (from $\triangle ABE$) $= 8.152$ m

$$b = \frac{10}{\sin 10°} \cdot \sin 110° \text{ (from } \triangle ABD) = 54.115 \text{ m}$$

$$c = \sqrt{bd} = \sqrt{8.152 \times 54.115} \text{ m} = 21.003 \text{ m}$$

$$x = \frac{ab}{(b+c)} = \frac{9.216 \times 54.115}{75.118} \text{ m} = 6.639 \text{ m}$$

$$P_a = \frac{1}{2}\gamma x^2 \cdot \sin \psi = \frac{1}{2} \times 1.8 \times (6.639)^2 \sin 70° \text{ t/m run}$$

$$= 37.3 \text{ t/m}$$

Example 13.18: A retaining wall, 3.6 m high, supports a dry cohesionless backfill with a plane ground surface sloping upwards at a surcharge angle of 10° from the top of the wall. The back of the wall is inclined to the vertical at a positive batter angle of 9°. The unit weight of the backfill is 1.89 t/m³ and $\phi = 30°$. Assuming wall friction angle of 12°, determine the total active thrust by Rebhann's method.

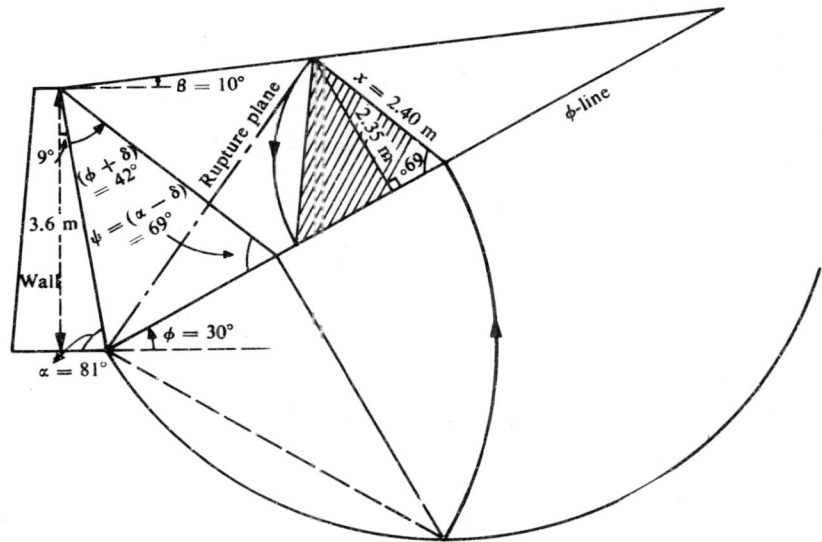

Fig. 13.63 Rebhann's construction for active thrust (Ex. 13.18)

$H = 3.6$ m $\quad \phi = 30°$ $\quad \delta = 12°$ $\quad \beta = 10°$ $\quad \alpha = 81°$ $\quad \gamma = 1.89$ t/m³

$\psi = \alpha - \delta = 69°$ $\quad P_a = \frac{1}{2}\gamma x^2 \sin \psi = \frac{1}{2} \times 1.89 \times 2^2 \cdot 40 \times \sin 69°$

$$= 5.1 \text{ t/m run}$$

608 Geotechnical Engineering

Example 13.19: A retaining wall with a vertical back 5 m high supports a cohesionless backfill of unit weight of 19 kN/m³. The upper surface of the backfill rises at an angle of 10° with the horizontal from the crest of the wall. The angle of internal friction for the soil is 30°, and the angle of wall friction is 20°. Determine the total active pressure per lineal metre of the wall and mark the direction and point of application of the thrust. Use Rebhann's graphical method.

(S.V.U.—B.E., (Part-time)—Apr., 1982)

$H = 5$ m $\phi = 30°$ $\delta = 20°$ $\beta = 10°$ $\alpha = 90°$ $\gamma = 19$ kN/m³
$\psi = \alpha - \delta = 70°$

$$P_a = \tfrac{1}{2}\gamma x^2 \sin\psi = \tfrac{1}{2} \times 19 \times 3^2 \times \sin 70°$$

$$= 80.34 \text{ kN/m}.$$

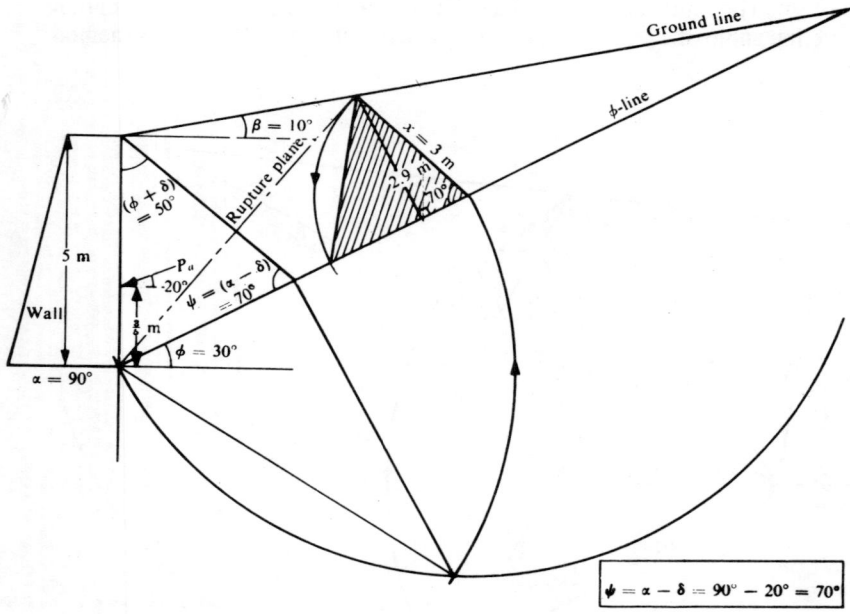

Fig. 13.64 Rebhann's construction for active thrust (Ex. 13.19)

Example 13.20: A retaining wall 8 m high is battered at a positive angle of 15° and retains a cohesionless backfill rising at 20° with the horizontal from the crest of the wall. $\phi = 30°$ and $\delta = 20°$. Bulk density of the fill is 17.1 kN/m³. Determine the active thrust, its direction and point of action on the wall by Rebhann's method.

(S.V.U.—B.E. (N.R.)—Sep., 1967)

$H = 8$ m $\alpha = 90° - 15° = 75°$ $\beta = 20°$ $\phi = 30°$ $\delta = 20°$
$\psi = \alpha - \delta = 55°$

Since the φ-line does not meet the ground surface within the drawing, the special case when β ≈ φ is applied, the construction being performed with an arbitrary location, D_1 for D. AG is drawn parallel to A_1G_1, G being on the φ-line. Triangle CGL, the weight of which gives P_a, is completed, with $\angle CGL = \psi$.

$$P_a = \frac{1}{2}\gamma x^2 \cdot \sin\psi = \frac{1}{2} \times 17.1 \times (7.0)^2 \sin 55° = \mathbf{343 \text{ kN/m run}}$$

The location of P_a is at $(1/3) H$ or **2.67 m** above the base and the direction is at δ or 20° with the horizontal.

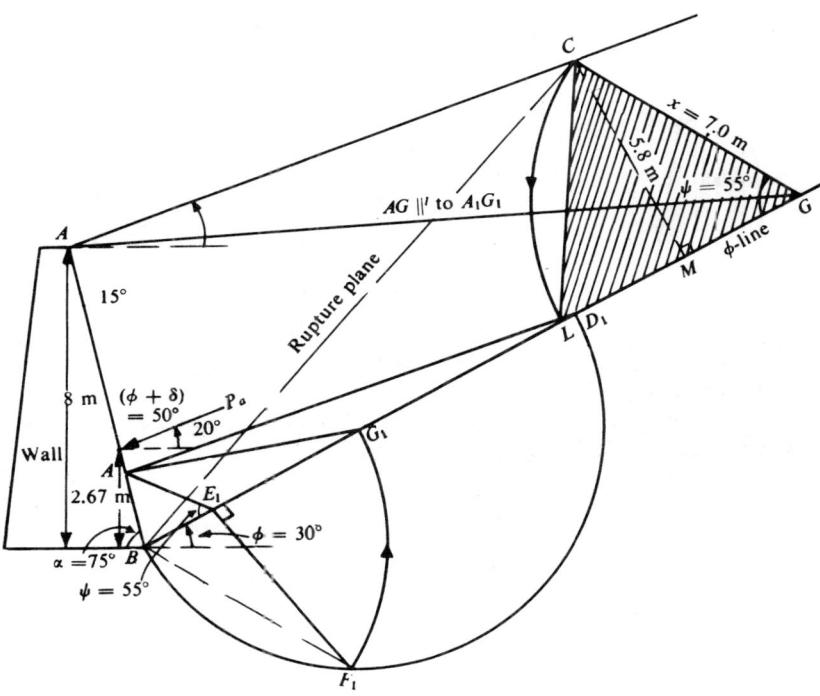

Fig. 13.65 Special case of Rebhann's construction when β ≈ φ (Ex. 13.20)

Example 13.21: A retaining wall 3.6 m high supports a dry cohesionless backfill with a plane ground surface sloping upwards at a surcharge angle of 20° from the top of the wall. The back of the wall is inclined to the vertical at a positive batter angle of 9°. The unit weight of the backfill is 1.89 t/m³ and φ = 20°. Assuming a wall friction angle of 12°, determine the total active thrust by Rebhann's method.

$H = 3.6$ m $\phi = 20°$ $\beta = 20°$ $\alpha = 81°$ $\delta = 12°$ $\psi = \alpha - \delta = 69°$

Since $\beta = \phi$, the special case of Rebhann's construction for this condition is applied. The triangle CGL is constructed from any arbitrary point C.

$P_a = \frac{1}{2}\gamma x^2 \cdot \sin\psi = \frac{1}{2} \times 1.89 \times (3.85)^2 \sin 69° = \mathbf{13.1\ t/m\ run}$

The rupture surface cannot be located in this case.

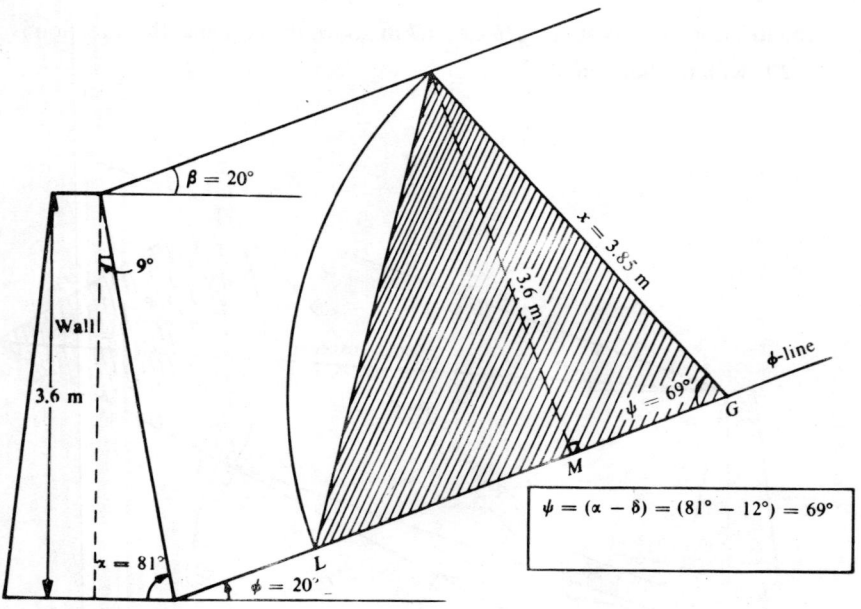

Fig. 13.66 Special case of Rebhann's construction when $\beta = \phi$ (Ex. 13.21)

Example 13.22: A masonry wall with vertical back has a backfill 5 m behind it. The ground level is horizontal at the top and the ground water table is at ground level. Calculate the horizontal pressure on the wall using Coulomb's earth pressure theory. Assume the unit weight of saturated soil is 15.3 kN/m³. Cohesion = 0. $\phi = 30°$. Friction between wall and earth = 20°.

(S.V.U.—B.E., (N.R.)—Sep., 1968)

$H = 5$ m $c = 0$ $\phi = 30°$ $\delta = 20°$ $\gamma_{sat} = 15.3$ kN/m³

$$K_a = \frac{\sin^2(\alpha+\phi)}{\sin^2\alpha \cdot \sin(\alpha-\delta)\left[1+\sqrt{\frac{\sin(\phi+\delta)\cdot\sin(\phi-\beta)}{\sin(\alpha-\delta)\cdot\sin(\alpha+\beta)}}\right]^2}$$

since $\alpha = 90°$ and $\beta = 0°$ in this case,

$$K_a = \frac{\cos^2\phi}{\cos\delta\left[1+\sqrt{\frac{\sin(\phi+\delta)\cdot\sin\phi}{\cos\delta}}\right]^2}$$

$$= \frac{\cos^2 30°}{\cos 20° \left[1 + \sqrt{\frac{\sin 50° \cdot \sin 30°}{\cos 20°}}\right]^2} = 0.132$$

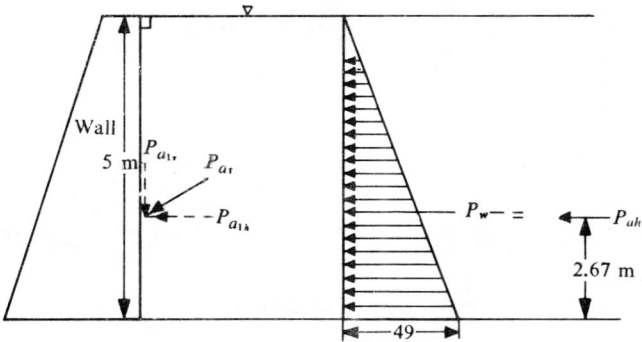

Fig. 13.67 Horizontal thrust on wall from submerged fill from Coulomb's theory (Ex. 13.22)

$P_{a_1} = \frac{1}{2}\gamma' H^2 \cdot K_a = \frac{1}{2} \times (15.30 - 9.8) \times 5^2 \times 0.132$ kN/m $= 9.06$ kN/m

(since $\gamma' = \gamma_{sat} - \gamma_w$)

This is inclined at 20° with the horizontal.

∴ Its horizontal component, $P_{a_{1h}} = 9.06 \times \cos 20° = 8.50$ kN/m

Water pressure, $P_w = \frac{1}{2}\gamma_w H^2 = \frac{1}{2} \times 9.81 \times 5^2$ kN/m $= 122.6$ kN/m

Total horizontal thrust, $P_{ah} = P_{a_{1h}} + P_w = 8.5 + 122.6 = \mathbf{131.1}$ **kN/m.**

This will act at $(1/3) H$ or **2.67 m** above the base of the wall, as shown in Fig. 13.67.

Example 13.23: A retaining wall 4.5 m high with a vertical back supports a horizontal fill weighing 18.60 kN/m³ and having $\phi = 32°$, $\delta = 20°$, and $c = 0$. Determine the total active thrust on the wall by Culmann's method.

(S.V.U.—B.E. (R.R.)—Sep., 1978)

$\gamma = 18.6$ kN/m³ $\phi = 32°$ $c = 0$ $\delta = 20°$ for the fill

Active thrust, $P_a = FF'$

≈ 51.5 kN/m. run

Check:

K_a from Coulomb's formula $= \dfrac{\cos^2 32°}{\cos 20°\left[1+\sqrt{\frac{\sin 52° \cdot \sin 32°}{\cos 20°}}\right]^2}$

$= 0.2755$

$P_a = \frac{1}{2}\gamma H^2 K_a = \frac{1}{2} \times 18.6 \times 4^2 \times 5 \times 0.2755 = 51.9$ kN/m

The Culmann value agrees excellently with this value.

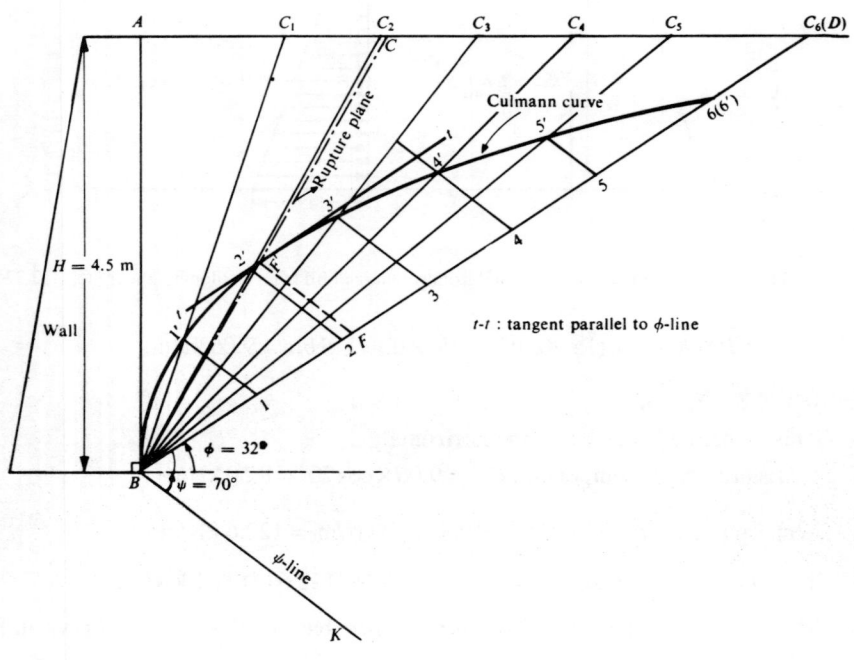

Fig. 13.68 Culmann's method (Ex. 13.23)

Example 13.24: A retaining wall with its face inclined at 75° with horizontal is 10 m high and retains soil inclined at a uniform surcharge angle of 10°. If the angle of internal friction of the soil is 36°, wall friction angle 18°, unit weight of soil 15 kN/m³, and a line load of intensity 90 kN per metre run of the wall acts at a horizontal distance of 5 m from the crest, determine the active thrust on the wall by Culmann's method.

t' t': Tangent to the modified Culmann line, parallel to ϕ-line.

α = 75°   ϕ = 36°   δ = 18°   β = 10°   γ = 15 kN/m³.

Active thrust, P_a, on the wall per metre run = Vector $F'G'$ = **360 kN**.

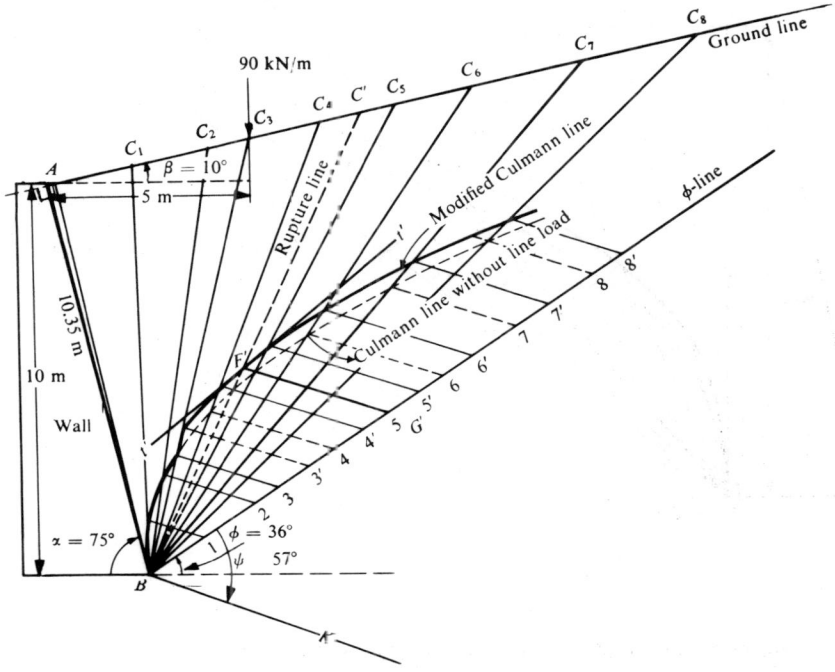

Fig. 13.69 Culmann's method for line load on backfill (Ex. 13.24)

Example 13.25: A retaining wall 4.5 m high with vertical back supports a backfill with horizontal surface. The unit weight of the fill is 18 kN/m³ and the angle of internal friction is 36°. The angle of wall friction may be taken as 18°. A footing running parallel to the retaining wall and carrying a load of 18 kN/m is to be constructed. Find the safe distance of the footing from the face of the wall so that there is no increase in lateral pressure on the wall due to the load of the footing.
tt: tangent to the Culmann-curve without line load, parallel to the φ-line

The safe distance beyond which the line load does not increase the lateral pressure is **3.5 m** in this case.

The Culmann-curve without the line load is drawn as usual. Now the modified Culmann-curve, with the line load included at C_1, C_2, C_3 ..., the borders of each of the wedges such as ABC_1, ABC_2, ABC_3, ..., is drawn. A tangent *tt* to the Culmann-curve without the line load is drawn parallel to φ-line to meet the modified curve with line load in F'. BF' is joined and produced to meet the surface in C, which gives the critical position of the line load, beyond which location, it does not affect the lateral pressure.

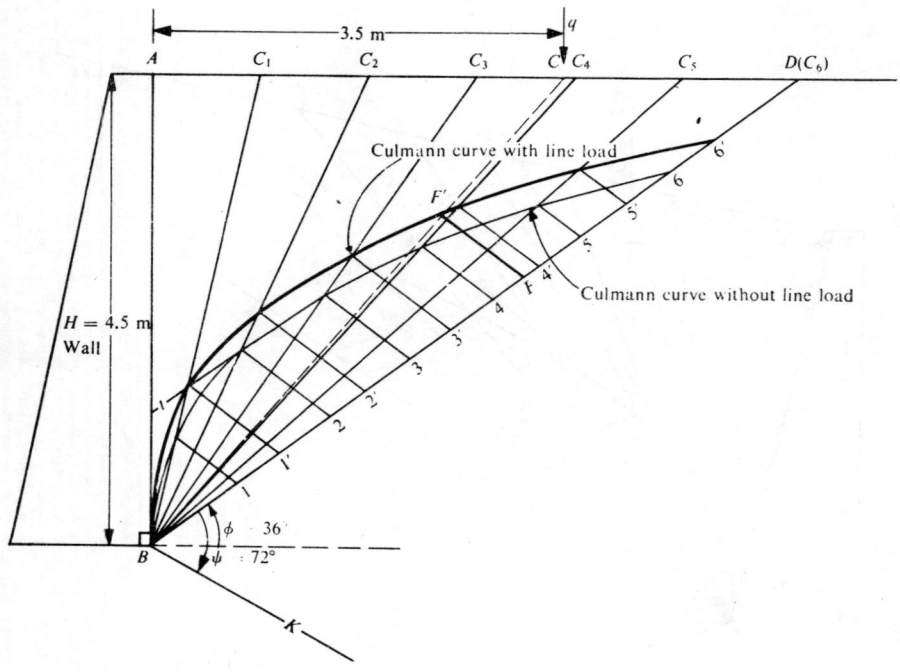

Fig. 13.70 Location of critical position of line load by Culmann's method (Ex. 13.25)

Example 13.26: A masonry retaining wall is 1.5 m wide at the top, 3.5 m wide at the base and 6 m high. It is trapezoidal in section and has a vertical face on the earth side. The backfill is level with top. The unit weight of the fill is 16 kN/m³ for the top 3 m and 23 kN/m³ for the rest of the depth. The unit weight of masonry is 23 kN/m³. Determine the total lateral pressure on the wall per metre run and the maximum and minimum pressure intensities of normal pressure at the base. Assume $\phi = 30°$ for both grades of soil.

(S.V.U.—Four-year B.Tech.—July, 1984)

$\phi = 30°$ $K_a = \dfrac{1 - \sin 30°}{1 + \sin 30°} = 1/3$

Horizontal pressure of soil at 3 m depth $= K_a \gamma_1 H_1 = \frac{1}{3} \times 16 \times 3 = 16$ kN/m²

Lateral earth pressure at 6 m depth $= K_a(\gamma_1 H_1 + \gamma_2 H_2)$

$$= \frac{1}{3}(16 \times 3 + 18 \times 3) = 34 \text{ kN/m}^2$$

Total active thrust per metre run of the wall,

$P_a = \frac{1}{2} \times 3 \times 16 + 3 \times 16 + \frac{1}{2} \times 3 \times 18 = 24 + 48 + 27 = 99$ kN

Lateral Earth Pressure and Stability of Retaining Walls 615

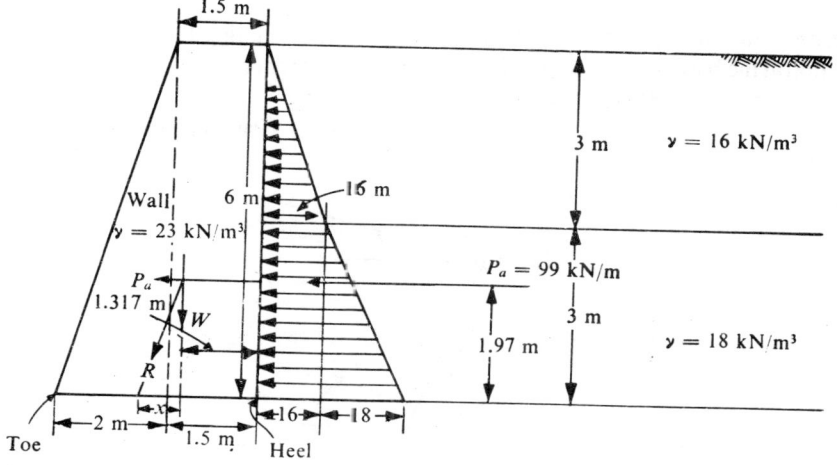

Fig. 13.71 Retaining wall (Ex. 13.26)

Let $\bar{z}$ metres be the height of the point of action above its base.

By taking moments about the base, $\bar{z} = \dfrac{(24 \times 4 + 48 \times 1.5 + 27 \times 1)}{99} = 1.97$ m

Weight of wall per metre run, $W = 6 \times 1.5 \times 23 + \dfrac{1}{2} \times 2 \times 6 \times 23$

$$= 207 + 138 = 345 \text{ kN}$$

Let the distance of its point of action from the vertical face be $\bar{x}$ m.

By moments, $\bar{x} = \dfrac{\left(207 \times 0.75 + 138 \times \tfrac{13}{6}\right)}{345} = 1.317$ m

Let x metres be the distance from the line of action of W to the point where the resultant strikes the base.

$\dfrac{x}{1.97} = \dfrac{P_a}{W}$ $\therefore$ $x = \dfrac{1.97 \times 99}{345} = 0.565$ m

Eccentricity, $e = (1.317 + 0.565 - 1.750) = 0.132$ m

Since this is less than $(1/6)\,b$ or $(1/6) \times 3.5$ m, no tension occurs at the base.

Vertical pressure intensity at the base, $\sigma = \dfrac{W}{b}\left(1 \pm \dfrac{6e}{b}\right) = \dfrac{345}{3.5}\left(1 \pm \dfrac{6 \times 0.132}{3.5}\right)$

or $\sigma_{max} = 120.88$ kN/m² at the toe

and $\sigma_{min} = 76.26$ kN/m, at the heel.

Example 13.27: A trapezoidal masonry retaining wall 1 m wide at top and 3 m wide at its bottom is 4 m high. The vertical face is retaining soil ($\phi = 30°$) at a surcharge angle of 20° with the horizontal. Determine the maximum and minimum intensities of pressure at the base of the retaining wall. Unit weights of soil and masonry are 20 kN/m³ and 24 kN/m³ respectively. Assuming the coefficient of friction at the base of the wall as 0.45, determine the factor of safety against sliding. Also determine the factor of safety against overturning.

(S.V.U.—B.E., (Part-time)—Dec., 1981)

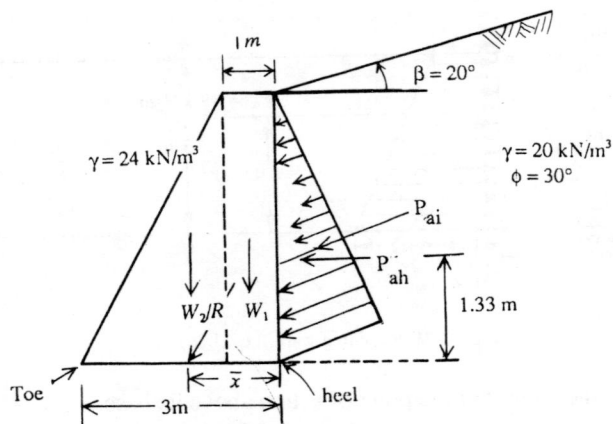

Fig. 13.72 Retaining wall (Ex. 13.27)

For backfill,

$$\gamma = 20 \text{ kN/m}^3 \quad \phi = 30° \quad \beta = 20°$$

$$K_{ai} = \cos\beta \cdot \frac{(\cos\beta - \sqrt{\cos^2\beta - \cos^2\phi})}{(\cos\beta + \sqrt{\cos^2\beta - \cos^2\phi})}$$

$$= \cos 20° \cdot \frac{(\cos 20° - \sqrt{\cos^2 20° - \cos^2 30°})}{(\cos 20° + \sqrt{\cos^2 20° - \cos^2 30°})}$$

$$= 0.414$$

$P_{ai} = \frac{1}{2}\gamma H^2 \cdot K_{ai} = \frac{1}{2} \times 20 \times 4^2 \times 0.414 = 66.24$ kN/m

This acts at 1.33 m above the base, at an angle of 20° with the horizontal.

$P_{ah} = P_{ai} \cdot \cos\beta = 66.24 \cos 20° = 62.25$ kN/m

$P_{av} = P_{ai} \cdot \sin\beta = 66.24 \sin 20° = 22.66$ kN/m

W_1, wt of the rectangular portion of the wall $= 1 \times 4 \times 24 = 96$ kN

W_2, wt of the triangular portion of the wall $= \frac{1}{2} \times 2 \times 4 \times 24 = 96$ kN

W_1 acts at 0.50 m and W_2 at 1.67 m from the vertical face.

$$\Sigma V = W_1 + W_2 + P_{av} = 96 + 96 + 22.66 = 214.66 \text{ kN}$$

The distance of the point where the resultant strikes the base from the heel,

$$\bar{x} = \frac{\Sigma M}{\Sigma V} = \frac{(96 \times 0.50 + 96 \times 1.67 + 62.25 \times 1.33)}{214.66} = 1.357 \text{ m}$$

$e = \frac{b}{2} - \bar{x} = 1.500 - 1.357 = 0.143$ m

σ_{max}, at the heel $= \dfrac{\Sigma V}{b}\left(1+\dfrac{6e}{b}\right) = \dfrac{214.66}{3}\left(1+\dfrac{6\times 0.143}{3}\right) = 92 \text{ kN/m}^2$

σ_{min}, at the toe $= \dfrac{\Sigma V}{b}\left(1-\dfrac{6e}{b}\right) = \dfrac{214.66}{3}\left(1-\dfrac{6\times 0.143}{3}\right) = 51 \text{ kN/m}^2$

These are intensities of normal pressures at the base.

Check for sliding:

Factor of safety against sliding, $\eta_s = \dfrac{\mu \cdot N}{T}$

$$= \dfrac{0.45 \times 214.66}{162.25} = 1.55$$

This is O.K.

Check for overturning:

Factor of safety against overturning, η_0

$$= \dfrac{\text{Restoring moment about the toe}}{\text{Overturning moment about the toe}}$$

$$= \dfrac{(96\times 2.5 + 96\times 1.33 + 22.66\times 3)}{62.25\times 1.33}$$

$$= 5.25$$

This is excellent.

SUMMARY OF MAIN POINTS

1. The property of soil by virtue of which it exerts lateral pressure influences the design of earth-retaining structures, the most common of them being a retaining wall.
2. The limiting values of lateral pressure occur when the wall yields away from the backfill (or moves toward the fill); these are knwon as the 'active' and the 'passive' states. The pressure exerted when there is no movement is called the 'at-rest' pressure, which is intermediate between the active and the passive values.

   Very little yield is adequate to cause active conditions, but relatively greater movement is necessary to mobilise passive resistance.
3. The classical earth pressure theories of Rankine (1857) and Coulomb (1776) stood the test of time. Rankine considered the plastic equilibrium of a soil

when there is stretching and compression of the mass, and applied the relationships between the principal stresses so derived for determining the pressure on the wall. Coulomb considered straightaway a wall and a backfill and the equilibrium of the sliding wedge for deriving the total thrust on the wall. The former neglected wall friction, while the latter considered it.

The distribution of pressure is considered to be triangular with depth; in the case of uniform surcharge, however, it will be rectangular.

4. Rankine assumes a conjugate relationship between stresses in the case of an inclined backfill surface.
5. There will exist a 'tensile' zone near the surface of a cohesive fill. The depth of this zone is given by $\frac{2c}{\gamma} \cdot \sqrt{N_\phi}$. and the 'critical' depth or the depth up to which the soil may stand unsupported is $\frac{4c}{\gamma} \cdot \sqrt{N_\phi}$. Tension cracks occur in the tension zone and these may cause some relief of pressure in the active case.
6. The Coulomb wedge theory which assumes a plane rupture surface introduces significant errors in the estimation of passive earth resistance, although the error is small in the estimation of active thrust. Thus, it is generally recommended that analysis based on curved rupture surfaces (for example, Terzaghi's logarithmic spiral method) be used for passive resistance.
7. The Poncelet construction based on Rebhann's condition and Poncelet rule, and the Culmann's graphical approach are versatile graphical solutions to Coulomb's wedge theory and are popularly used in view of the complexity of the analytical expressions derived by Coulomb. The charts and tables prepared by Caquot and Kerisel, and Jumikis are also relevant in this context.
8. The angle of wall friction will usually range between $\frac{1}{2}\phi$ and $\frac{3}{4}\phi$; Terzaghi recommends $\left(\frac{2}{3}\right)\phi$ in the absence of data.
9. The stability considerations for gravity retaining walls are:
   (a) The maximum pressure on the base should be less than the safe bearing power of the foundation soil;
   (b) no tension should develop anywhere in the wall;
   (c) the factor of safety against sliding must be adequate; and
   (d) the factor of safety against overturning must be adequate.

The nature of yield of the wall influences the wall design very much; for example, the yield at the bottom of a sheeting supporting a trench causes arching-active conditions, in which the distribution of pressure varies significantly from the active case, although the total thrust value remains the same.

REFERENCES

1. Alam Singh and B.C. Punmia: *Soil Mechanics and Foundations*, Standard Book House, Delhi-6, 1970.
2. I. Alpan: *The Empirical Evaluation of the Coefficients K_o and K_{or}*, Soils and Foundation, Vol. VIII, No. 1, 1967.
3. A.L. Bell: *Lateral Pressure and Resistance of Clay and the Supporting Power of Clay Foundations*, Minutes, Proceedings of the Institution of Civil Engineers, London, Vol. 199, 1915.
4. E.W. Brooker and H.O. Ireland: *Earth Pressure at Rest Related to Stress History*, Canadian Geotechnical Journal, Vol. II, No. 1, 1965.
5. A. Caquot and J. Kérisel: *Traite de mechanique des Sols*, Gauthier—Villars, Paris, 1949. *Traite de mechanics des règles de soils*, 3rd ed., Gauthier—Villars, Paris, 1956.
6. C.A. Coulomb: *Essai sur une application des règles des maximis et minimis à quelques problèmes de statique relatifs à l'architecture*, Mém. Acad. Roy. Pres. divers savants, Vol. 7, Paris, 1776.
7. K. Culmann: *Die graphische statik*, Mayer und Zeller, Zurich, 1866.
8. F. Engesser: *Geometrische Erddruck theorie*, Z. Bauwesen, Vol. 30, 1980.
9. A.M. Fraser: *The Influence of Stress Ratio on Compressibility and Pore Pressure Coefficients in Compacted Soils*, Ph. D. Thesis, London, 1957.
10. J. Jaky: *The Coefficient of Earth Pressure at Rest*, Jl. Soc. of Hungarian Architects & Engineers, 1944.
11. A.R. Jumikis: *Soil Mechanics*, D. Van Nostrand Co., Inc., New Jersey, USA, 1962.
12. T.C. Kenney: *Discussion on Proc. paper 1732*, Proc. ASCE, Vol. 85, No. SM-3, 1959.
13. A. Kezdi: *Erddruck Theorien*, Springer Verlag, Berlin, 1962.
14. T.W. Lambe and R.V. Whitman: *Soil Mechanics*, John Wiley & Sons, Inc., NY, USA, 1969.
15. V.N.S. Murthy: *Soil Mechanics and Foundation Engineering*, Dhanpat Rai & Sons, Delhi-6, 1974.
16. J.V. Poncelet: *Mém. sur la stabilité des revêtments et de leurs Foundations*, Mém. de l' officier du génié, Vol 13, 1840.
17. W.J.M. Rankine: *On the Stability of Loose Earth*, Philosophical Transactions, Royal Society, London, Vol. 147, 1857.
18. G. Rebhann: *Theorie des Erddruckes und der Füttermauern*, Vienna, 1871.
19. J. Resal: *La Poussée des terres*, Paris, 1910.
20. S.B. Sehgal: *A Text Book of Soil Mechanics*, Metropolitan Book Co., Pvt., Ltd., Delhi-6, 1967.
21. Shamsher Prakash, Gopal Ranjan, and Swami Saran: *Analysis Design of Foundations and Retaining Structures*, Sarita Prakashan, Meerut, 1979.
22. G.N. Smith: *Elements of Soil Mechanics for Civil and Mining Engineers*, Crosby Lockwood Staples, London, 1974.

23. V.V. Sokolovski: *Statics of Granular Media*, Translated from the Russian by J.K. Luscher, Pergamon Press, London, 1965.
24. D.W. Taylor: *Fundamentals of Soil Mechanics*, John Wiley & Sons, Inc., NY, USA, 1948.
25. K. Terzaghi: *A Fundamental Fallacy in Earth Pressure Computations'*, Journal of Boston Society of Civil Engineers, Apr., 1936.
26. K. Terzaghi: *Theoretical Soil Mechanics*, John Wiley & Sons, Inc., NY, USA, 1943.
27. G.P. Tschebotarioff: *Soil Mechanics, Foundations, and Earth Structures*, McGraw-Hill Book Co., New York, USA.

QUESTIONS AND PROBLEMS

13.1 Write notes on:
  (a) Rankine earth pressure theory.
  (b) Rebhann's construction.
      (S.V.U.—Four year B.Tech.—Sept., 1983, B.E., (R.R.)—Sep., 1978)
  (c) Culmann method.
      (S.V.U.—Four-year B.Tech.—Dec., 1982)
  (d) Coefficient of passive earth pressure.
      (S.V.U.—B.Tech., (Part-time)—June, 1982)
13.2 (a) Distinguish between 'active' and 'passive' earth pressure.
  (b) Explain clearly Rebhann's graphical construction method to evaluate the earth pressure on a retaining wall. What are the advantages or disadvantages of Culmann's graphical method as compared to Rebhann's graphical method? Illustrate your answer by working out an example, assuming suitable data.
      (S.V.U.—B.Tech., (Part-time)—Sept., 1983)
13.3 What are the design criteria to be satisfied for the stability of a gravity retaining wall? Indicate briefly how you will ensure the same.
      (S.V.U.—B.Tech., (Part-time)—Sept., 1983)
13.4 Differentiate critically between Rankine and Coulomb theories of earth pressure.
      (S.V.U.—Four-year B.Tech.—Apr., 1983, B.E., (R.R.)—Feb., 1976, Nov., 1973)
13.5 Explain (i) active, (ii) passive and (iii) at rest conditions in earth pressure against a retaining wall.
      (S.V.U.—Four-year B.Tech.—Dec., 1982, B.E., (Part-time)—Dec., 1981)
13.6 With the aid of Mohr's circle diagram explain what is meant by active and passive Rankine states in a cohesionless soil with a horizontal surface. Hence obtain an expression for the intensity of active earth pressure behind a vertical wall and explain why for this condition there is an implied assumption of smooth wall.
      (S.V.U.—B.E., (R.R.)—Sept., 1978)

13.7 Describe Culmann's graphical method of finding earth pressure and explain the classical theory of earth pressure on which this procedure is based. Explain how surcharge will affect earth pressure in active and passive states.
(S.V.U.—B.E., (R.R.)—Nov., 1975)

13.8 Describe the wedge theory for determining active earth pressure and evaluate the assumptions.
Discus the advantages.
(S.V.U.—B.E., (R.R.)—May., 1975, Nov., 1974, May, 1971, Nov., 1969)

13.9 Explain Rankine's theory of earth pressure. For what types of retaining walls and soils may this theory be used?
(S.V.U.—B.E., (R.R.)—May, 1970)

13.10 Indicate an analytical or graphical method to calculate the active earth pressure due to a cohesive soil ($c = \phi$ soil) against a rigid retaining well.
(S.V.U.—B.E., (R.R.)—May, 1969)

13.11 Derive a general expression for active earth pressure by the wedge theory behind a vertical wall due to a cohesionless soil with a level surface.
(S.V.U.—B.E., (N.R.)—May, 1969)

13.12 A wall with a smooth vertical back and 9 metres high retains a moist cohesionless soil with a horizontal surface. The soil weighs 1.50 g/cm^3 and has an angle of internal friction of $30°$. Determine the total earth pressure at rest and its location. If, subsequently, the water table rises to the ground surface, determine the increase in earth pressure at rest. Assume effective unit weight of soil as 0.9 g/cm^3.

13.13 Determine the active and passive earth pressure given the following data: Height of retaining wall = 10 m; $\phi = 25°$; $\gamma_d = 1.7$ t/m^3. Ground water table is at the top of the retaining wall.
(S.V.U.—Four-year B.Tech.—Dec., 1982)

13.14 A retaining wall 12 metres high is proposed to hold sand. The values of void ratio and ϕ in the loose state are 0.63 and $30°$ while they are 0.42 and $40°$ in the dense state. Assuming the sand to be dry and that its grain specific gravity is 2.67, compare the values of active and passive earth pressures in both the loose and dense states.

13.15 A retaining wall with a smooth vertical back, 4.5 m high, retains a dry cohesionless backfill level with the top of the wall. $\gamma = 18.6$ kN/m^3 and $\phi = 30°$. The backfill carries a uniformly distributed surcharge of 20.6 kN/m^2. Determine the magnitude and point of application of the total active thrust per lineal metre of the wall.

13.16 A 9 metre high retaining wall retains a soil with the following properties: $\phi = 36°$; $\gamma = 18$ kN/m^3. What will be the increase in horizontal thrust if the soil slopes up from the top of the wall at an angle of $36°$ to the horizontal than when it has a horizontal surface?

13.17 A wall with a smooth vertical back 9 m high supports a purely cohesive soil with $c = 1$ t/m^2 and $\gamma = 1.8$ g/cm^3. Determine the total Rankine active thrust per metre run, the position of zero pressure and the distance of the centre of pressure from the base.

13.18 A retaining wall, 4.5 m high, retains a soil with $c = 2$ N/cm², $\phi = 30°$ and $\gamma = 20$ kN/m³, with horizontal surface level with the top of the wall. The backfill carries a surcharge of 20 kN/m². Compute the total passive earth resistance on the wall and its point of application.

13.19 A R.C. retaining wall holds dry sand fill 6 m in height. The unit weight of sand is 1.93 g/cm³ and $\phi = 32°$. Assuming the back of the wall to be vertical, calculate the earth pressure on the wall by Rebhann's construction $\delta = 16°$.

(S.V.U.—B.E., (R.R.)—Dec., 1968)

13.20 A gravity retaining wall retains 12 m of a backfill. $\gamma = 1.8$ t/m³, $\phi = 25°$ and wall friction $= 0°$. Using Culmann's method, determine the active earth thrust. If the water-table is 6 m from the top, determine how the earth pressure gets affected.

(S.V.U.—Four-year B.Tech.—Apr., 1983)

13.21 A masonry retaining wall of trapezoidal section with the vertical face on the earth side is 1.5 m wide at the top and 3.5 m wide at the base and is 5.0 m high. It retains a sand fill sloping at 2 horizontal to 1 vertical. The unit weight of sand is 18 kN/m² and $\phi = 30°$. Find the maximum and minimum pressure at the base of the wall assuming the unit weight of masonry as 23 kN/m³.

(S.V.U.—B.E., (Part-time)—Apr., 1982)

14

BEARING CAPACITY

14.1 INTRODUCTION AND DEFINITIONS

The subject of bearing capacity is perhaps the most important of all the aspects of geotechnical engineering. Loads from buildings are transmitted to the foundation by columns, by load-bearing walls or by such other load-bearing components of the structures. Sometimes the material on which the foundation rests is ledge, very hard soil or bed-rock, which is known to be much stronger than is necessary to transmit the loads from the structure. Such a ledge, or rock, or other stiff material may not be available at reasonable depth and it becomes invariably necessary to allow the structure to bear directly on soil, which will furnish a satisfactory foundation, if the bearing members are properly designed. It is here that the subject of bearing capacity assumes significance. A scientific treatment of the subject of bearing capacity is necessary to enable one to understand the factors upon which it depends.

A number of definitions are relevant in this context:

Foundation: The lowest part of a structure which is in contact with soil and transmits loads to it.

Foundation soil or bed: The soil or bed to which loads are transmitted from the base of the structure.

Footing: The portion of the foundation of the structure, which transmits loads directly to the foundation soil.

Bearing capacity: The load-carrying capacity of foundation soil or rock which enables it to bear and transmit loads from a structure.

Ultimate bearing capacity: Maximum pressure which a foundation can withstand without the occurrence of shear failure of the foundation.

Gross bearing capacity: The bearing capacity inclusive of the pressure exerted by the weight of the soil standing on the foundation, or the 'surcharge' pressure, as it is sometimes called.

Net bearing capacity: Gross bearing capacity minus the original overburden pressure or surcharge pressure at the foundation level; obviously, this will be the same as the gross capacity when the depth of foundation is zero, i.e., the structure is founded at ground level.

Safe bearing capacity: Ultimate bearing capacity divided by the factor of safety. The factor of safety in foundation may range from 2 to 5, depending upon the importance of the structure, and the soil profile at the site. The factor of safety should be applied to the net ultimate bearing capacity and the surcharge pressure due to depth of the foundation should then be added to get the safe bearing capacity.

It is thus the maximum intensity of loading which can be transmitted to the soil without the risk of shear failure, irrespective of the settlement that may occur.

Allowable bearing pressure: The maximum allowable net loading intensity on the soil at which the soil neither fails in shear nor undergoes excessive or intolerable settlement, detrimental to the structure.

14.2 BEARING CAPACITY

The conventional design of a foundation is based on the concept of bearing capacity or allowable bearing pressure.

14.2.1 Criteria for the Determination of Bearing Capacity

The criteria for the determination of bearing capacity of a foundation are based on the requirements for the stability of the foundation. These are stated as follows:

(*i*) Shear failure of the foundation or bearing capacity failure, as it is sometimes called, shall not occur. (This is associated with plastic flow of the soil material underneath the foundation, and lateral expulsion of the soil from underneath the footing of the foundation); and,

(*ii*) The probable settlements, differential as well as total, of the foundation must be limited to safe, tolerable or acceptable magnitudes. In other words, the anticipated settlement under the applied pressure on the foundation should not be detrimental to the stability of the structure.

These two criteria are known as the shear strength criterion, and settlement criterion, respectively. These are independent criteria and hence require independent investigation. The design value of the safe bearing capacity, obviously, would be the smaller of the two values, obtained from these two criteria. This has already been defined as the allowable bearing pressure.

14.2.2 Factors Affecting Bearing Capacity

Bearing capacity is governed by a number of factors. The following are some of the more important ones which affect bearing capacity:

(*i*) Nature of soil and its physical and engineering properties;

(*ii*) Nature of the foundation and other details such as the size, shape, depth below the ground surface and rigidity of the structure;

(*iii*) Total and differential settlements that the structure can withstand without functional failure;

(*iv*) Location of the ground water table relative to the level of the foundation; and

(*v*) Initial stresses, if any.

In view of the wide variety of factors that affect bearing capacity, a systematic study of the factors involved in a logical sequence is necessary for proper understanding.

14.3 METHODS OF DETERMINING BEARING CAPACITY

The following methods are available for the determination of bearing capacity of a foundation:
  (I) Bearing capacity tables in various building codes
  (II) Analytical methods
  (III) Plate bearing tests
  (IV) Penetration tests
  (V) Model tests and prototype tests
  (VI) Laboratory tests

Bearing capacity tables have been evolved by certain agencies and incorporated in building codes. They are mostly based on past experience and some investigations.

A number of analytical approaches, based on the work of Rankine, Fellenius, Housel, Prandtl, Terzaghi, Meyerhof, Skempton, Hansen and Balla may be used. Some of these would be dealt with in later sections.

Plate bearing tests are load tests conducted in the field on a plate. These involve effort and expense. There are also certain limitations to their use.

Penetration tests are conducted with devices known as 'Penetrometers', which measure the resistance of soil to penetration. This is correlated to bearing capacity.

Model and prototype tests are very cumbersome and costly and are not usually practicable. Housel's approach is based on model tests.

Laboratory tests which are simple, may be useful in arriving at bearing capacity, especially of pure clays.

14.4 BEARING CAPACITY FROM BUILDING CODES

The traditional approach to the bearing capacity problems is illustrated by the building codes of many large cities, such as New York and Boston. Practically, all codes give lists of soil types and the respective safe or allowable bearing capacity. Some values may be subject to modifications under designated conditions. It is presumed that the soil can support the indicated pressure with safety against shear failure and without undue settlement. This is perhaps the basis for the widespread but incorrect notion that the bearing capacity depends mainly on the characteristics of the soil in question.

Actually the bearing capacity depends on a number of factors as stated in a previous section and this should never be lost sight of. Thus, the value stated for the bearing capacity is, at best, a rough estimate, based on past experience of construction in the area, rather than a sound basis for design. For the general use of buildings, perhaps the tabular values, which are valid under a definite set of simple and easily defined conditions, may be modified for known departures from the specified conditions.

The tabular values of bearing capacity are also known as "Presumptive bearing capacities" and are included in several Civil Engineering Handbooks. The ISI have specified these values in their code of practice "IS: 1904-1978 Code of practice for structural safety of Building Foundations—Second Revision". Excerpts of these recommendations are given below:

Table 14.1 Safe bearing capacity (IS: 1904-1978 revised)

S.No.	Type of rock or soil	Safe bearing capacity kN/m^2 (t/m^2)	Remarks
	I. ROCKS		
1.	Rocks without laminations and defects —e.g., granite, trap, diorite	3240 (330)	
2.	Laminated rocks, e.g., sandstone and limestone, in sound condition	1620 (165)	
3.	Residual deposits of shattered and broken bed rock and hard shale, cemented material	880 (90)	
4.	Soft Rock	440 (45)	
	II. COHESIONLESS SOILS		
5.	Gravel, sand & gravel, compact and offering high resistance to penetration when excavated by tools	440 (45)	See note 2
6.	Coarse sand, compact and dry	440 (45)	Dry means that the GWL is at a depth not less than width of the foundation below the base of the foundation.
7.	Medium sand, compact and dry	245 (25)	
8.	Fine sand, silt (dry lumps easily pulverised by fingers)	150 (15)	
9.	Loose gravel or sand-gravel mixture; loose coarse to medium sand, dry	245 (25)	See note 2
10.	Fine sand, loose and dry	100 (10)	
	III. COHESIVE SOILS		
11.	Soft shale, hard or stiff clay, dry	440 (45)	Susceptible to long-term consolidation settlement
12.	Medium clay, readily indented with a thumb nail	245 (25)	
13.	Moist clay and sand-clay mixture which can be indented with strong thumb pressure	150 (15)	
14.	Soft-clay indented with moderate thumb pressure	100 (10)	
15.	Very soft clay which can be penetrated easily with the thumb	50 (5)	
16.	Black cotton soil or other shrinkable or expansive clay in dry condition (50% saturation)	—	See note 3. To be determined after investigation
	IV. PEAT		
17.	Peat	—	See note 3 and note 4 To be determined after investigation
	V. MADE-UP GROUND		
18.	Fills or made-up ground	—	See note 2 and note 4 To be determined after investigation

Note 1. Values listed in the table are from shear consideration only.

Note 2. Values are very much rough for the following reasons:
   (a) Effect of characteristics of foundations (that is, effect of depth, width, shape, roughness, etc...) has not been considered.
   (b) Effect of range of soil properties (that is, angle of internal friction, cohesion, water table, density, etc.) has not been considered.
   (c) Effect of eccentricity and inclination of loads has not been considered.

Note 3. For non-cohesive soils, the values listed in the table shall be reduced by 50 per cent, if the water table is above or near the base of footing.

Note 4. Compactness or looseness of non-cohesive soils may be determined by driving the cone of 65 mm dia and 60° apex angle by a hammer of 65 kg falling from 75 cm. If corrected number of blows (N) for 30 cm penetration is less than 10, the soil is called loose, if N lies between 10 and 30, it is medium, if more than 30, the soil is called dense.

Limitations of Bearing Capacity values for building codes

The following are the limitations of the bearing capacity values specified in building codes:
   (i) By specifying a value or a range for bearing capacity, the concept is unduly oversimplified.
   (ii) The codes tacitly assume that the allowable bearing capacity is dependent only on the soil type.
   (iii) The effects of many soil characteristics which are likely to influence the bearing capacity are ignored.
   (iv) The codes do not indicate the method used to obtain the bearing capacity values.
   (v) The codes assume that the bearing capacity is independent of the size, shape and depth of foundation. All these factors are known to have significant bearing on the values.
   (vi) Building codes are usually not up-to-date.

However, the values given in codes are used in the preliminary design of foundations.

14.5 ANALYTICAL METHODS OF DETERMINING BEARING CAPACITY

The following analytical approaches are available:
1. The theory of elasticity—Schleicher's method.
2. The classical earth pressure theory—Rankine's method, Pauker's method and Bell's method.
3. The theory of plasticity—Fellenius' method, Prandtl's method, Terzaghi's method, Meyerhof's method, Skempton's method, Hansen's method and Balla's method.

Some of these methods will be discussed in the following subsections.

14.5.1 The Theory of Elasticity—Schleicher's Method

Based on the theory of elasticity and Boussinesq's stress distribution, Schleicher (1926) integrated the vertical stresses caused by a uniformly distributed surface load and obtained an expression for the elastic settlement, s, of soil directly underneath a perfectly elastic bearing slab as follows:

$$s = K \cdot q \cdot \sqrt{A} \frac{(1-v^2)}{E} \qquad \ldots(\text{Eq. 14.1})$$

where K = shape coefficient or influence value which depends upon the degree of stiffness of the slab, shape of bearing area, mode of distribution of the total load and the position of the point on the slab where the settlement is sought;
q = net pressure applied from the slab on to the soil;
A = area of the bearing slab;
E = moduls of elasticity of soil; and
v = Poisson's ratio for the soil.

It may be noted that settlements are not the same at all points under an elastic slab, while settlements are the same under all points of a rigid slab. If, in Eq. 14.1, $\dfrac{E}{(1-v^2)}$ is designated as a constant, C, Schleicher's equation reduces to:

$$s = K \cdot \frac{q \cdot \sqrt{A}}{C} \qquad \ldots(\text{Eq. 14.2})$$

The maximum settlement occurs at the centre of circular and rectangular bearing areas and the minimum value occurs at the periphery of the circle or at corners of the rectangle.

Schelicher's shape coefficients, K, are given in Table 14.2.

Table 14.2 Schleicher's shape coefficients or influence factors (Jumikis, 1962)

Shape of bearing area	Side ratio a/b	Center point M, $K_{max} = K_M$	Free corner point A, $K_{min} = K_A$	Mid-point of short side B, K_B	Mid-point of long side C, K_C	K (average)
Circle	—	1.13	0.72	0.72	0.72	0.96
Square	1.0	1.12	0.56	0.76	0.76	0.95
Rectangle	1.5	1.11	0.55	0.73	0.79	0.94
"	2	1.08	0.54	0.69	0.79	0.92
"	3	1.03	0.51	0.64	0.78	0.88
"	5	0.94	0.47	0.57	0.75	0.82
"	10	0.80	0.40	0.47	0.67	0.71
"	100	0.40	0.20	0.22	0.36	0.37
"	1000	0.173	0.087	0.093	0.159	0.163
"	10000	0.069	0.035	0.037	0.065	0.066

If, in the Schleicher's equation 14.2 above, the tolerable settlement, s, the shape coefficient K, the size A of the loading area and the soil properties included under C, are known, the bearing capacity q can be calculated as,

$$q = \frac{s \cdot C}{K \cdot \sqrt{A}} \qquad \ldots(\text{Eq. 14.3})$$

The elastic settlement equation also permits deriving the following rule:

$$\frac{s_1}{s_2} = \sqrt{\frac{A_1}{A_2}} \qquad \ldots(\text{Eq. 14.4})$$

where s_1 and s_2 are settlements brought about by two bearing areas of similar shape but of different sizes, A_1 and A_2 respectively, with equal contact pressures.

This rule, expressing a model law, is useful in calculating the settlement of a prototype foundation, if the settlement attained by a model with the same contact pressure has been measured.

14.5.2 The Classical Earth Pressure Theory—Rankine's, Pauker's, and Bell's Methods

The classical earth pressure theory assumes that on exceeding a certain stress condition, rupture surfaces are formed in the soil mass. The stress developed upon the formation of the rupture surfaces is treated as the ultimate bearing capacity of the soil.

The bearing capacity may be determined from the relation between the principal stresses at failure. The pertinent methods are those of Rankine, Pauker and Bell.

Rankine's method

This method, based on Rankine's earth pressure theory, is too approximate and conservative for practical use. However, it is given just as a matter of academic interest.

Rankine uses the relationship between principal stresses at limiting equilibrium conditions of soil elements, one located just beneath the footing and the other just outside it as shown in Fig. 14.1.

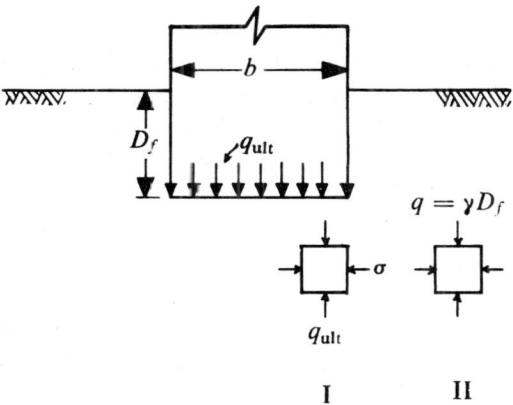

Fig. 14.1 Rankine's method for bearing capacity of a footing

In element *I*, just beneath the footing, at the base level of the foundation, the applied pressure q_{ult} is the major principal stress; under its influence, the soil adjacent to the element tends get pushed out, creating active conditions. The active pressure is σ on the vertical faces of the element. From the relationship between the principal stresses at limiting equilibrium relating to the active state, we have:

$$\sigma = q_{ult} \cdot K_A = q_{ult}\left(\frac{1-\sin\phi}{1+\sin\phi}\right) \qquad \text{...(Eq. 14.5)}$$

In element *II*, just outside the footing, at the base level of the foundation, the tendency of the soil adjacent to the element is to compress, creating passive conditions. The pressure σ on the vertical faces of the element will thus be the passive resistance. This will thus be the major principal stress and the corresponding minor principal stress is q (= γD_f), the vertical stress caused by the weight of a soil column on it, or the surcharge due to the depth of the foundation. From the relationship between the principal stresses at limiting equilibrium relating to the passive state, we have,

$$\sigma = q \cdot K_p = \gamma D_f \cdot K_p = \gamma D_f\left(\frac{1+\sin\phi}{1-\sin\phi}\right) \qquad \text{...(Eq. 14.6)}$$

The two values of σ may be equated from Eqs. 14.5 and 14.6 to get a relationship for q_{ult}:

$$q_{ult} = \gamma D_f\left(\frac{1+\sin\phi}{1-\sin\phi}\right)^2 \qquad \text{...(Eq. 14.7)}$$

This gives the bearing capacity of the footing. It does not appear to take into account the size of the footing. Further the bearing capacity reduces to zero for $D_f = 0$ or for a footing founded at the surface. This is contrary to facts.

Equation 14.7 is rewritten sometimes, to give D_f, which is termed the minimum depth required for a foundation:

$$D_f = \frac{q}{\gamma}\left(\frac{1-\sin\phi}{1+\sin\phi}\right)^2 \qquad \text{...(Eq. 14.8)}$$

An alternative approach based on Rankine's earth pressure theory which takes into account the size *b* of the footing is as follows (Fig. 14.2).

It is assumed that rupture in the soil takes place along *CBD* and *CFG* symmetrically. The failure zones are made of two wedges as shown. It is sufficient to consider the equilibrium of one half.

Wedge *I* is Rankine's active wedge, pushed downwards by q_{ult} on *CA*; consequently the vertical face *AB* will be pushed outward.

Wedge *II* is Rankine's passive wedge. The pressure *P* on face *AB* of wedge *I* will be the same as that which acts on face *AB* of wedge *II*; consequently, the soil wedge *II* is pushed up. The surcharge, $q = \gamma D_f$, due to the depth of footing resists this.

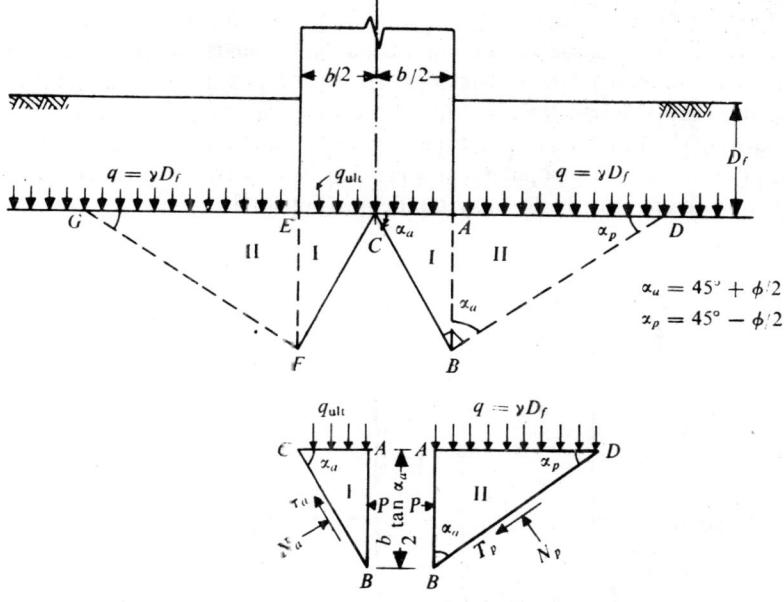

Fig. 14.2 Rankine's method taking into account the size of the footing

From wedge II,

$$\overline{AB} = \frac{b}{2}\tan \alpha_a = \frac{b}{2}\tan(45° + \phi/2) = \frac{b}{2}\sqrt{N_\phi}$$

$$P = \frac{1}{2} \cdot \gamma \frac{b^2}{4} \cdot N_\phi^2 + \gamma D_f \cdot \frac{b}{2} N_\phi^{3/2} \qquad \text{...(Eq. 14.9)}$$

from Rankine's theory for the case with surcharge. From Wedge I, similarly,

$$P = \frac{1}{2} \cdot \gamma \cdot \frac{b^2}{4} \cdot \frac{N_\phi}{N_\phi} + q_{ult} \cdot \frac{b}{2} \sqrt{N_\phi} \cdot \frac{1}{N_\phi} = \frac{1}{2} \gamma \cdot \frac{b^2}{4} + q_{ult} \cdot \frac{b}{2} \cdot \frac{1}{\sqrt{N_\phi}}$$

...(Eq. 14.10)

Equating the two values of P, we get

$$q_{ult} = \frac{1}{2} \cdot \gamma \cdot \frac{b}{2} \sqrt{N_\phi}(N_\phi^2 - 1) + \gamma D_f N_\phi^2 \qquad \text{...(Eq. 14.11)}$$

This is written as $q_{ult} = \frac{1}{2}\gamma \cdot bN_\gamma + \gamma D_f N_q$ ...(Eq. 14.12)

where

$$N_\gamma = \frac{1}{2}\sqrt{N_\phi}(N_\phi^2 - 1) \qquad \text{...(Eq. 14.13)}$$

and

$$N_q = N_\phi^2, \qquad \text{...(Eq. 14.14)}$$

Both are known as "bearing capacity factors".

Pauker's method

Colonel Pauker, a Russian military engineer, is credited to have derived one of the oldest formulae for the bearing capacity of a foundation in cohesionless soil and the minimum depth of foundation. He was supposed to have used his formula in the 1850's during the construction of fortifications and sea-batteries for the Czarist Naval base of Kronstadt (Pauker, 1889—reported by Jumikis, 1962). His theory was once very popular and was extensively used in Czarist Russia, before the revolution. The theory is set out below (Fig. 14.3):

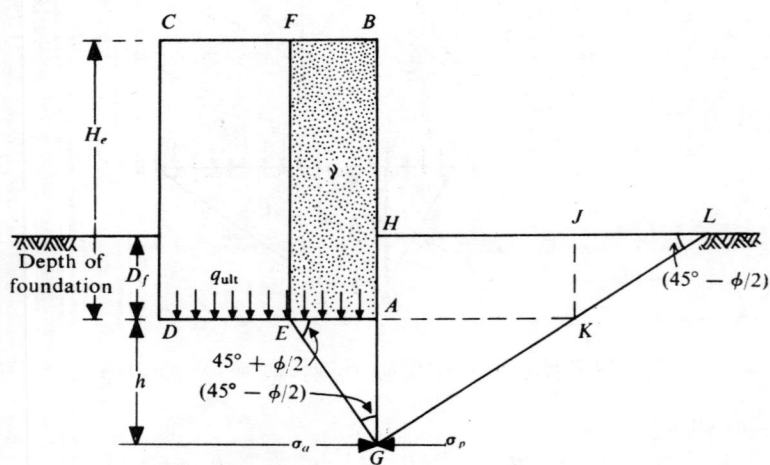

Fig. 14.3 Pauker's method of determination of bearing capacity

Pauker considered the equilibrium of a point, say, G, in the soil mass underneath the base of the footing, as shown, at a depth h below the base, the depth of foundation being D_f below the ground surface. The strip foundation is assumed to transmit a pressure of q_{ult} to the soil at its base.

The classical earth presure theory for an ideal soil is used under the following assumptions:

(*i*) The soil is cohesionless.

(*ii*) The contact pressure, q_{ult}, is replaced by an equilvalent height, H_e, of soil of unit weight, γ, the same as that of the foundation soil:

$$H_e = \frac{q_{ult}}{\gamma} \qquad \ldots(\text{Eq. 14.15})$$

(*iii*) At imminent failure, it is assumed that a part $AEFB$, obtained by drawing GE at $(45°-\phi/2)$ with respect to GA (G being chosen vertically below A), tears off from the rest of the soil mass.

Bearing Capacity

(*iv*) Under the influence of the weight of the equivalent layer of height H_e, the soil to the left of the vertical section GA tends to be pushed out, inducing active earth pressure on GA.

(*v*) The soil to the right of GA tends to get compressed, thus offering passive earth resistance against the active pressure.

(*vi*) The equilibrium condition at G is determined by that of soil prisms GEA and GHJK. The friction of the soil on the imaginary vertical section, GA, is ignored. In other words, the earth pressures act normal to GA, i.e., horizontally.

(*vii*) If sliding of soil from underneath the footing is to be avoided, the condition stated by Pauker is

$$\sigma_p \geq \sigma_a \qquad \ldots(\text{Eq. 14.16})$$

By Rankine's earth pressure theory,

$$\sigma_p = \gamma(D_f + h)\tan^2(45° + \phi/2) \qquad \ldots(\text{Eq. 14.17})$$

$$\sigma_a = \gamma(H_e + h)\tan^2(45° - \phi/2) \qquad \ldots(\text{Eq. 14.18})$$

where ϕ is the angle of internal friction of the soil.

Equation 14.16 now reduces to

$$\frac{(D_f + h)}{(H_e + h)} \geq \tan^4(45° - \phi/2) \qquad \ldots(\text{Eq. 14.19})$$

[by dividing by $(H_e + h)\tan^2(45° + \phi/2)$ and noting that

$$\frac{\tan^2(45° - \phi/2)}{\tan^2(45° + \phi/2)} = \tan^4(45° - \phi/2).]$$

The most dangerous point G is that for which $\dfrac{(D_f + h)}{(H_e + h)}$ is a minimum.

By inspection, one can see that this is minimum when $h = 0$; that is to say, the critical point is A itself.

Eq. 14.19 reduces to the form:

$$\frac{D_f}{H_e} \geq \tan^4(45° - \phi/2) \qquad \ldots(\text{Eq. 14.20})$$

This is known as Pauker's equation and is written as:

$$D_f = H_e \tan^4(45° - \phi/2) \qquad \ldots(\text{Eq. 14.21})$$

or, noting $H_e = \dfrac{q_{ult}}{\gamma}$, $D_f = \dfrac{q_{ult}}{\gamma} \cdot \tan^4(45° - \phi/2) \qquad \ldots(\text{Eq. 14.22})$

This may be written in the following form also:

$$q_{ult} = \gamma D_f \tan^4(45° + \phi/2) \qquad \ldots(\text{Eq. 14.23})$$

In the first form it may be used to determine the minimum depth of foundation and in the second, to determine the ultimate bearing capacity.

It is interesting to observe that Eqs. 14.22 and 14.23 are identical to Eqs. 14.8 and 14.7 respectively of Rankine, except for the difference in their trigonometric form.

Thus, the limitations and deficiencies of Rankine's approach in respect of Eqs. 14.7 and 14.8 apply equally to Pauker's equations.

Bell's method

Bell (1915) modified Pauker-Rankine formula to be applicable for cohesive soils; both friction and cohesion are considered in this equation. With reference to Fig. 14.1, from the stresses on element *I*,

$$\sigma = q_{ult} \tan^2(45° - \phi/2) - 2c \tan(45° - \phi/2) \qquad \text{...(Eq. 14.24)}$$

or
$$\sigma = \frac{q_{ult}}{N_\phi} - \frac{2c}{\sqrt{N_\phi}} \qquad \text{...(Eq. 14.25)}$$

with the usual notation, $N_\phi = \tan^2(45° + \phi/2)$.

This is from the relationship between the principal stresses in the active Rankine state of plastic equilibrium.

From the stresses on element *II*,

$$\sigma = \gamma D_f \tan^2(45° + \phi/2) + 2c \tan(45° + \phi/2) \qquad \text{...(Eq. 14.26)}$$

or
$$\sigma = \gamma D_f N_\phi + 2c \sqrt{N_\phi} \qquad \text{...(Eq. 14.27)}$$

Equating the two values of σ for equilibrium, we have:

$$q_{ult} = \gamma D_f \tan^4(45° + \phi/2) + 2c \tan(45° + \phi/2)[1 + \tan^2(45° + \phi/2)]$$

$$\text{...(Eq. 14.28)}$$

or
$$q_{ult} = \gamma D_f N_\phi^2 + 2c \sqrt{N_\phi}(1 + N_\phi) \qquad \text{...(Eq. 14.29)}$$

This is Bell's equation for the ultimate bearing capacity of a $c - \phi$ soil at a depth D_f.

If $c = 0$, this reduces to Eq. 14.23 or 14.7. For pure clay, with $\phi = 0$, Bell's equation reduces to

$$q_{ult} = \gamma D_f + 4c \qquad \text{...(Eq. 14.30)}$$

If D_f is also zero, $\qquad q_{ult} = 4c \qquad \text{...(Eq. 14.31)}$

This value is considered to be too conservative as will be shown later on.

The limitation of Bell's equation that the size of the foundation is not considered may be overcome as in the case of Rankine's equation by considering soil wedges instead of elements.

Figure 14.2 may be employed for this purpose. Proceeding on exctly similar lines as in the Rankine approach, one gets:

$$q_{ult} = \frac{1}{2}\gamma \cdot \frac{b}{2} \cdot \sqrt{N_\phi}(N_\phi^2 - 1) + \gamma D_f \cdot N_\phi^2 + 2c \sqrt{N_\phi}(N_\phi + 1) \qquad \text{...(Eq. 14.32)}$$

or
$$q_{ult} = \frac{1}{2}\gamma \cdot b \cdot N_\gamma + \gamma D_f \cdot N_q + c \cdot N_c \qquad ...(Eq.\ 14.33)$$

where
$$N_\gamma = \frac{1}{2}\sqrt{N_\phi}(N_\phi^2 - 1) \qquad ...(Eq.\ 14.34)$$

$$N_q = N_\phi^2 \qquad ...(Eq.\ 14.35)$$

and
$$N_c = 2\sqrt{N_\phi}(N_\phi + 1) \qquad ...(Eq.\ 14.36)$$

Equations 14.34 and 14.35 are identical to Eqs. 14.13 and 14.14, already given. N_γ, N_q and N_c are known as 'bearing capacity factors'.

If $c = 0$, Eq. 14.32 reduces to Eq. 14.11 of Rankine.

If $\phi = 0$, we have $q_{ult} = 4c + D_f$, the same as Eq. 14.30, for pure clay.

In other words, for pure clay, the size of foundation does not affect bearing capacity.

14.5.3 Fellenius' Method

The Fellenius method of circular failure surfaces (Fellenius, 1939) may be used to determine the ultimate bearing capacity of highly cohesive soils. The failure is assumed to take place by slip and the consequent heaving of a mass of soil is on one side only as shown in Fig. 14.4. Model tests and observation of failure surfaces confirm this; the possible reasons being lack of homogeneity of soil and a slight unintended eccentricity of loading.

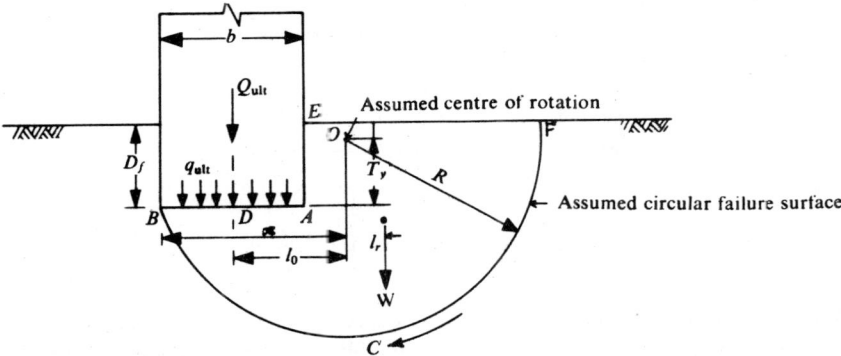

Fig. 14.4 Fellenius' method of determining bearing capacity

A trial cylindrical failure surface is chosen with centre O, the co-ordinates of which are x and y with respect to B, the outer edge of the base of the footing. The weight W of the soil mass within the slip surface for unit length of the footing and its line of action are determined. The total cohesive force C, resisting the slip surface is determined ($C = c \cdot \overrightarrow{BF}$). $Q_{ult}(= q_{ult} \cdot b \cdot l$, since unit length of the footing is considered). It will tend to cause slip, and W and C will tend to resist slip.

At imminent failure, their moments about the centre of rotation must balance:

$$Q_{ult} \cdot l_0 = W \cdot l_r + C \cdot R \qquad \ldots \text{(Eq. 14.37)}$$

or
$$Q_{ult} = W \cdot \frac{l_r}{l_0} + C \cdot \frac{R}{l_0} \qquad \ldots \text{(Eq. 14.38)}$$

$$\therefore \quad q_{ult} = \frac{W \cdot l_r}{bl_0} + \frac{C \cdot R}{bl_0} = \frac{(Wl_r + CR)}{bl_0} \qquad \ldots \text{(Eq. 14.39)}$$

This procedure is repeated for several possible slip surfaces and the minimum value of q_{ult} so obtained is the bearing capacity of the footing.

The method is considered most suitable and satisfactory for cohesive soils, although it can be extended to allow for friction. Wilson (1941) extended it by preparing a chart for locating the centre of the most critical circle, applicable only for cohesive soils and for footings founded below the ground surface. The co-ordinates of the centre of the most critical circle, x and y, with respect to the outer edge, B, of the footing may be obtained from Wilson's chart, shown in Fig. 14.5.

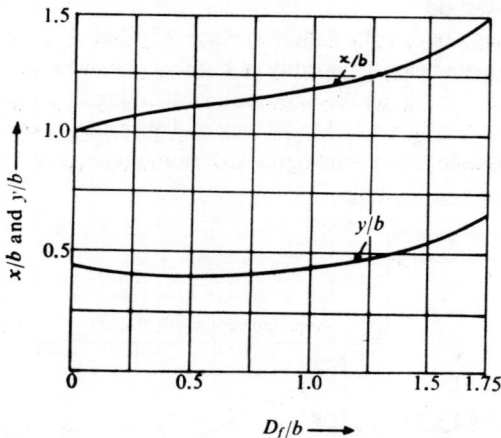

Fig. 14.5 Wilson's chart for location of centre of critical circular arc for use with Fellenius' method

Wilson found that the net ultimate bearing capacity by this method has an almost exactly linear variation with the depth to breadth ratio upto a value of 1.5 for this ratio. Wilson's results lead to the following equation for the net ultimate bearing capacity of long footings below the surface of highly cohesive soils:

$$q_{net\ ult} = 5.5c(1 + 0.38 D_f/b) \qquad \ldots \text{(Eq. 14.40)}$$

It can be demonstrated that the critical circle for a surface footing is as shown in Fig. 14.6 and that the ultimate bearing capacity is given by:

$$q_{ult} = 5.5c \qquad \ldots \text{(Eq. 14.41)}$$

The method is particularly useful when the properties of soil vary in the failure zone; in this case Wilson's critical circle may be tried first and other circles nearby may be analysed later to arrive at a reasonably quick solution.

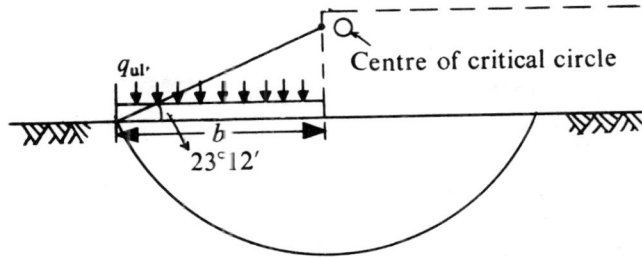

Fig. 14.6 Location of critical circle for surface footing in Fellenius' method

14.5.4 Prandtl's Method

Prandtl analysed the plastic failure in metals when punched by hard metal punchers (Prandtl, 1920). This analysis has been adapted to soil when loaded to shear failure by a relatively rigid foundation (Prandtl, 1921). The bearing capacity of a long strip footing on the ground surface may be determined by this theory, illustrated in Fig. 14.7.

The assumptions in Prandtl's theory are:

(i) The soil is homogeneous, isotropic and weightless.
(ii) The Mohr-Coulomb equation for failure envelope $\tau = c + \sigma \tan \phi$ is valid for the soil, as shown in Fig. 14.7(b).
(iii) Wedges I and III act as rigid bodies. The zones in Sectors II deform plastically. In the plastic zones all radius vectors or planes through A and B are failure planes and the curved boundary is a logarithmic spiral.
(iv) Wedge I is elastically pushed down, tending to push zones III upward and outward, which is resisted by the passive resistance of soil in these zones.
(v) The stress in the elastic zone I is transmitted hydrostatically in all directions.

It may be noted that the section is symmetrical up to the point of failure, with an equal chance of failure occurring to either side. (That is why, the section to one side, say to the left, is shown by dashed lines). The equilibrium of the plastic sector is considered by Prandtl.

Let BC be r_0. The equation to a logarithmic spiral is:

$r = r_0 e^{\theta \tan \phi}$, where θ is the spiral angle.

Then $BD = r_0 e^{(\pi/2) \tan \phi}$, since $\angle CBD = 90° = \pi/2$ rad.

From the Mohr's circle for $c - \phi$ soil, Fig. 14.7(b), the normal stress corresponding to the cohesion intercept is:

$$\sigma_i = c \cot \phi \qquad \ldots \text{(Eq. 14.42)}$$

This is termed the 'initial stress', which acts normally to BC in view of assumption (v); also q_{ult}, the applied pressure is assumed to be transferred normally on to BC.

Thus the force on BC is

$$(\sigma_i + q_{ult}) \overrightarrow{BC} \text{ or } r_0 (\sigma_i + q_{ult})$$

638 Geotechnical Engineering

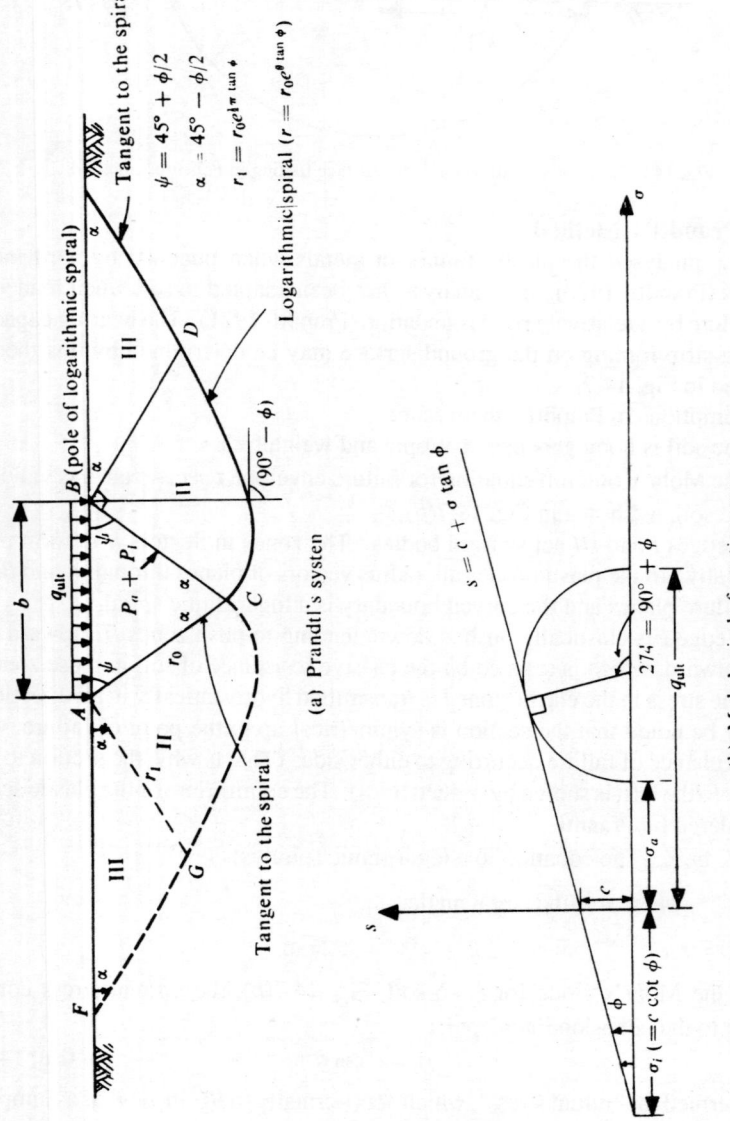

Fig. 14.7 Prandtl's method of determining bearing capacity of a $c - \phi$ soil

Moment, M_0, of this force about B is

$$r_0(\sigma_i + q_{ult}) \times \frac{r_0}{2}$$

Substituting for σ_i,

$$M_0 = \frac{r_0^2}{2}(c \cot\phi + q_{ult}), \text{ counterclockwise} \qquad \text{...(Eq. 14.43)}$$

The passive resistance P_p on the face BD is given by

$$P_p = \sigma_i \cdot N_\phi \cdot \overrightarrow{BD} \qquad \text{...(Eq. 14.44)}$$

where $N_\phi = \tan^2(45° + \phi/2) = \dfrac{1+\sin\phi}{1-\sin\phi}$

This is because σ_i, due to cohesion alone is transmitted by the wedge BDE. Its moment about B, M_r, is,

$$M_r = P_p \cdot \frac{\overrightarrow{BD}}{2} = \sigma_i N_\phi \cdot \frac{\overrightarrow{BD}^2}{2} = c \cot\phi \cdot N_\phi \cdot \frac{1}{2} \cdot r_0^2 e^{\pi\tan\phi}$$

...(Eq. 14.45)

For equilibrium of the plastic zone, equating M_0 and M_r, and rearranging,

$$q_{ult} = c \cot\phi (N_\phi \cdot e^{\pi\tan\phi} - 1) \qquad \text{...(Eq. 14.46)}$$

This is Prandtl's expression for ultimate bearing capacity of a $c - \phi$ soil.

Apparently this leads one to the conclusion that if $c = 0$, $q_{ult} = 0$. This is ridiculous since it is well known that even cohesionless soils have bearing capacity. This anomaly arises chiefly owing to the assumption that the soil is weightless. This was later rectified by Terzaghi and Taylor.

For purely cohesive soils, $\phi = 0$ and the logarithmic spiral becomes a circle and Prandtl's analysis for this special case leads to an indeterminate quantity. But, by applying L'Hospital's rule, for taking limit one finds that

$$q_{ult} = (\pi + 2) c = 5.14c \qquad \text{...(Eq. 14.47)}$$

Interestingly, this agrees reasonably with the Fellenius' solution for this case.

Terzaghi's correction

Terzaghi proposed a correction to the bearing capacity expression of Prandtl with a view to removing the anomaly that the bearing capacity is zero when cohesion is zero. He suggested that the weight of the soil involved be considered by adding a factor c' to the original quantity c in Prandtl's equation.

$$c' = \gamma H_1 \tan\phi \qquad \text{...(Eq. 14.48)}$$

where H_1 = equivalent height of soil material

$$= \frac{\text{Area of wedges and sector}}{\text{length } \widehat{CDE}}$$

and γ = unit weight of soil.

The area of wedges and sector obviously means one-half of the system; the idea is that a soil mass of equivalent height, H_1 moves during shear and offers frictional resistance.

Thus, the corrected expression for bearing capacity is

$$q_{ult} = (c + c') \cot \phi \, (N_\phi e^{\pi \tan \phi} - 1) \qquad \ldots \text{(Eq. 14.49)}$$

or

$$q_{ult} = (c \cot \phi + \gamma H_1)(N_\phi e^{\pi \tan \phi} - 1) \qquad \ldots \text{(Eq. 14.50)}$$

Taylor's correction

Taylor (1948) suggested a correction factor for $c \cot \phi$ as follows:

$$q_{ult} = \left(c \cot \phi + \frac{1}{2} \gamma b \sqrt{N_\phi}\right)(N_\phi e^{\pi \tan \phi} - 1) \qquad \ldots \text{(Eq. 14.51)}$$

Taylor's correction is simple and easy to apply, while Terzaghi's correction is more logical but more difficult to calculate. However, nothing was said as to how Taylor's correction factor was derived.

Taylor has also attempted to include the effect of overburden pressure in the case of a footing founded at a depth D_f below the ground surface, proceeding in an exactly similar way as is done in deriving Prandtl's equation (Eq. 14.46). The additional value, q'_{ult}, of the bearing capacity in this case is

$$q'_{ult} = \gamma D_f N_\phi \cdot e^{\pi \tan \phi} \qquad \ldots \text{(Eq. 14.52)}$$

The general equation for the bearing capacity of a footing founded at a depth D_f below the ground surface is then given by,

$$q_{ult} = \left(c \cot \phi + \frac{1}{2} \gamma b \sqrt{N_\phi}\right)(N_\phi \cdot e^{\pi \tan \phi} - 1) + \gamma D_f N_\phi e^{\pi \tan \phi}$$

$$\ldots \text{(Eq. 14.53)}$$

according to Taylor.

However, Jumikis (1962) prefers Terzaghi's correction in the final expression as follows:

$$q_{ult} = (c + c') \cot \phi \, (N_\phi e^{\pi \tan \phi} - 1) + \gamma z \, N_\phi e^{\pi \tan \phi} \qquad \ldots \text{(Eq. 14.54)}$$

where γz is considered to be the surcharge at the base level of the footing, either because of the depth of the footing below the ground or because of any externally applied surcharge load.

Discussion of Prandtl's Theory

(i) Prandtl's theory is based on an assumed compound rupture surface, consisting of an arc of a logarithmic spiral and tangents to the spiral.

(ii) It is developed for a smooth and long strip footing, resting on the ground surface.

(iii) Prandtl's compound rupture surface corresponds fairly well with the mode of failure along curvilinear rupture surfaces observed from experiments.

In fact, for $\phi = 0°$, Prandtl's rupture surface agrees very closely with Fellenius' rupture surface (Taylor, 1948).

(iv) Although the theory is developed for a $c - \phi$ soil, the original Prandtl expression for bearing capacity reduces to zero when $c = 0$, contradicting common observations in reality. This anomaly arises from the fact that the weight of the soil wedge directly beneath the base of the footing is ignored in Prandtl's analysis.

This anomaly is sought to be rectified by the Terzaghi/Taylor correction.

(v) For a purely cohesive soil, $\phi = 0$, and Prandtl's equation, at first glance, leads to an indeterminate quantity; however this difficulty is overcome by the mathematical technique of evaluating a limit under such circumstances.

Then, for $\phi = 0$, $q_{ult} = (2 + \pi) c = 5.14 c$

(vi) Prandtl's expression, as originally derived, does not include the size of the footing.

14.5.5 Terzaghi's Method

Terzaghi's method is, in fact, an extension and improved modification of Pandtl's (Terzaghi, 1943). Terzaghi considered the base of the footing to be rough, which is nearer facts, and that it is located at a depth D_f below the ground surface ($D_f \leq b$, where b is the width of the footing).

The analysis for a strip footing is based on Fig. 14.8.

The soil above the base of the footing is replaced by an equilvalent surcharge, $q(= \gamma D_f)$. This substitution simplifies the computations very considerably, the error being unimportant and on the safe side. This, in effect, means that the shearing resistance of the soil located above the base is neglected. (For deep foundations, where $D_f > b$, this aspect becomes important and cannot be ignored).

The zone of plastic equilibrium, $CDEFG$, can be subdivided into *I* a wedge-shaped zone located beneath the loaded strip, in which the major principal stresses are vertical, *II* two zones of radial shear, BCD and ACG, emanating from the outer edges of the loaded strip, with their boundaries making angles $(45° - \phi/2)$ and $(45° + \phi/2)$ with the horizontal. and *III* two passive Rankine zones, AGF and BDE, with their boundaries making angles $(45° - \phi/2)$ with the horizontal.

The soil located in zone *I* is in a state of elastic equilibrium and behaves as if it were a part of the sinking footing, since its tendency to spread laterally is resisted by the friction and adhesion between the soil and the base of the footing. This leads one to the logical conclusion that the tangent to the surface of sliding at C will be vertical. Also AC and DC are surfaces of sliding and hence, they must intersect at an angle of $(90° - \phi)$; therefore, the boundaries AC and BC must rise at an angle ϕ to the horizontal. The footing cannot sink into the ground until the pressure exerted onto the soil adjoining the inclined boundaries of zone *I* becomes

642 Geotechnical Engineering

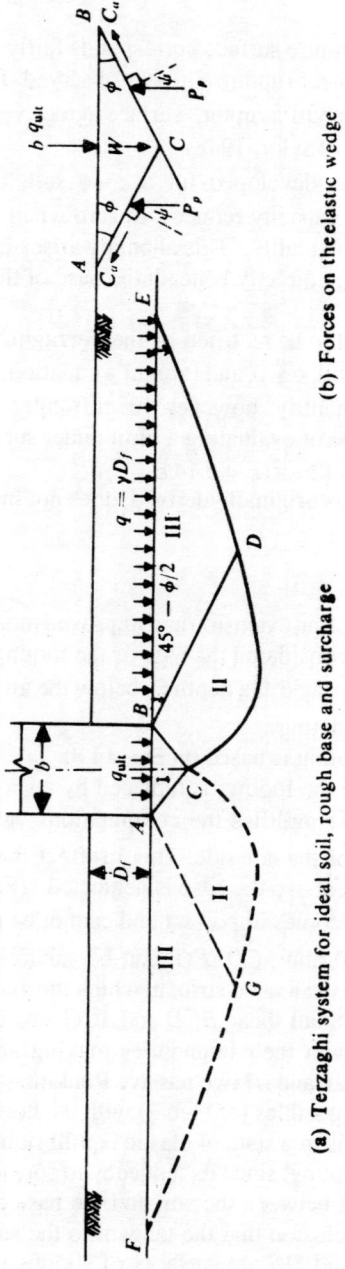

Fig. 14.8 Terzaghi's method for bearing capacity of strip footing

(a) Terzaghi system for ideal soil, rough base and surcharge

(b) Forces on the elastic wedge

equal to the passive earth pressure. This pressure P_p, acts at an angle ϕ ($\phi = \delta =$ *wall friction angle*) to the normal on the contact face; that is, vertically, in this case.

The adhesion force C_a on the faces AC and BC is given by

$$C_a = \frac{b}{2\cos\phi} \cdot c \qquad \text{...(Eq. 14.55)}$$

where c is a unit cohesion of the soil, with shearing resistance of the soil being defined by Coulomb's equation $s = c + \sigma \tan \phi$.

Considering a unit length of the footing and the equilibrium of the wedge ABC, the vertical components of all forces must sum up to zero.

The weight of the soil in the wedge is given by

$$W = \frac{\gamma b^2 \tan \phi}{4} \qquad \text{...(Eq. 14.56)}$$

Hence, for $\Sigma V = Q$,

$$b \cdot q_{ult} + \frac{\gamma b^2 \tan \phi}{4} - 2P_p - bc \tan \phi = 0 \qquad \text{...(Eq. 14.57)}$$

or

$$q_{ult} = \frac{2P_p}{b} + c \tan \phi - \frac{b \tan \phi}{4} \qquad \text{...(Eq. 14.58)}$$

This equation represents the solution to the problem if P_p is known.

For the simpler case of $D_f = 0$ and $c = 0$, $q = 0$—that is, if the base of the footing rests on the horizontal surface of a mass of cohesionless sand, we have

$$P_p = \frac{1}{2\cos\delta\sin\alpha} \cdot \gamma H^2 \cdot K_p$$

In this case, $H = \frac{b}{2}\tan\phi$, $\delta = \phi$, $K_p = K_{p\gamma}$, and $\alpha = 180 - \phi$.

$$\therefore \quad P_p = \frac{1}{2}\frac{\gamma b^2}{4} \cdot \frac{\tan\phi}{\cos^2\phi} \cdot K_{p\gamma} \qquad \text{...(Eq. 14.59)}$$

Here $K_{p\gamma}$ is the coefficient of passive earth pressure for $c = 0$, $\alpha = 180° - \phi$, and $\delta = \phi$; that is, it is the value purely due to the weight of the soil.

Substituting Eq. 14.59 into Eq. 14.56, and putting $c = 0$,

$$(q_{ult})_{c=0} = \frac{1}{4} \cdot \gamma b \tan \phi \left(\frac{K_{p\gamma}}{\cos^2\phi} - 1 \right) \qquad \text{...(Eq. 14.60)}$$

or

$$(q_{ult})_{c=0} = \frac{1}{2} \cdot \gamma b \cdot N_\gamma \qquad \text{...(Eq. 14.61)}$$

wherein

$$N_\gamma = \frac{1}{2} \tan \phi \left(\frac{K_{p\gamma}}{\cos^2\phi} - 1 \right) \qquad \text{...(Eq. 14.62)}$$

The value of $K_{p\gamma}$ is obtained by means of the spiral or the friction circle method. Since the angle of wall friction δ and the slope angle α of the contact face are equal to ϕ and to $(180° - \phi)$ respectively, the value of $K_{p\gamma}$ and hence of N_γ depend only on ϕ; thus, the values of N_γ for various values of ϕ may be established once for all.

N_γ is called the "bearing-capacity factor" expressing the effect of the weight of the soil wedge, ABC, of a cohesionless soil.

For the calculation of the bearing capacity of a cohesive soil, the computation P_p involves a considerable amount of labour. Terzaghi, therefore, advocated a simplified approach, which is based on the equation

$$P_{pn} = \frac{H}{\sin \alpha}(cK_{pc} + qK_{pq}) + \frac{1}{2}\gamma H^2 \cdot \frac{K_{p\gamma}}{\sin \alpha} \qquad \text{...(Eq. 14.63)}$$

where P_{pn} = normal component of the passive earth pressure on a plane contact face with a height H,

α = slope angle of the contact face, and

K_{pc}, K_{pq}, and $K_{p\gamma}$ = coefficients whose values are independent of H and γ. In the present case,

$$H = \frac{b}{2}\tan\phi, \ \alpha = 180° - \phi, \ \delta = \phi.$$

Also the total passive earth pressure P_p on the contact face is equal to $\dfrac{P_{pn}}{\cos \delta}$.

$$\therefore \quad P_p = \frac{P_{pn}}{\cos \delta} = \frac{P_{pn}}{\cos \phi} = \frac{b}{2\cos^2 \phi}(cK_{pc} + qK_{pq}) + \frac{1}{2}\gamma \cdot \frac{b^2}{4} \cdot \frac{\tan \phi}{\cos^2 \phi} \cdot K_{p\gamma}$$

$$\text{...(Eq. 14.64)}$$

where $(cK_{pc} + qK_{pq}) = p_{pn}$ is the normal component of the passive earth pressure comprehending the effect of cohesion and surcharge.

Combining this equation with Eq. 14.58, we have

$$q_{ult} = c\left(\frac{K_{pc}}{\cos^2 \phi} + \tan \phi\right) + q\frac{K_{pq}}{\cos^2 \phi} + \frac{\gamma b}{4}\tan\phi\left(\frac{K_{p\gamma}}{\cos^2 \phi} - 1\right) \qquad \text{...(Eq. 14.65)}$$

wherein K_{pc}, K_{pq}, and $K_{p\gamma}$ are pure numbers whose values are independent of b.

If the soil wedge, ABC, is assumed weightless ($\gamma = 0$) (Prandtl, 1920), Eq. (14.65) takes the form

$$q_{ult_c} + q_{ult_q} = c\left(\frac{K_{pc}}{\cos^2 \phi} + \tan \phi\right) + q \cdot \frac{K_{pq}}{\cos^2 \phi}$$

$$= cN_c + q \cdot N_q \qquad \text{...(Eq. 14.66)}$$

The factors N_c and N_q are pure numbers whose values depend only on the value ϕ in Coulomb's equation. The value q_{ult_c} represents the bearing capacity of the weightless soil, if the surcharge q were equal to zero ($\gamma = 0$ and $q = 0$), and q_{ult_q} is the bearing capacity exclusively due to the surcharge q ($\gamma = 0$ and $c = 0$).

On the other hand, if $c = 0$ and $q = 0$, γ being greater than zero, the bearing capacity is given by Eqs. 14.61 and 14.62:

$$q_{ult} = \frac{1}{2}\gamma b N_\gamma = \frac{1}{4}\gamma b \tan\phi \left(\frac{K_{p\gamma}}{\cos^2\phi} - 1\right)$$

(However, it should be noted that the failure surface for this condition is somewhat above that for $\gamma = 0$, and also the exact mathematical shape is not known).

If the values c, D_f, and γ are greater than zero,

$$q_{ult} = q_{ult_c} + q_{ult_q} + q_{ult_\gamma} = cN_c + \gamma D_f N_q + \frac{1}{2}\gamma b N_\gamma \quad \text{...(Eq. 14.67)}$$

This is called "Terzaghi's general bearing capacity formula". (The discrepancy arising out of the difference in failure surfaces for the two conditions—$\gamma = 0$ and $\gamma > 0$—is considered inconsequential).

The coefficients $N_c, N_q,$ and N_γ are called "bearing capacity factors" for shallow continuous footings. Since their values depend only on the angle of shearing resistance ϕ they can be computed once for all.

The problem of N_c and N_q has been rigorously solved by means of Airy's stress function (Prandtl 1920, Reissner, 1924), for the condition $\gamma = 0$:

$$N_c = \cot\phi \left[\frac{a_\theta^2}{2\cos^2(45° + \phi/2)} - 1\right] \quad \text{...(Eq. 14.68)}$$

and

$$N_q = \frac{a_\theta^2}{2\cos^2(45° + \phi/2)} \quad \text{...(Eq. 14.69)}$$

wherein $a_\theta = e^{(3\pi/4 - \phi/2)\tan\phi}$ ...(Eq. 14.70)

The values N_c and N_q depend only on the value of ϕ.

The critical load per unit length of the strip footing is given by

$$Q_{ult} = b \cdot q_{ult} \quad \text{...(Eq. 14.71)}$$

Also, $\quad N_c = \cot\phi (N_q - 1) \quad$ ...(Eq. 14.72)

For a purely cohesive soil, $\phi = 0$

$N_c = \frac{3}{2}\pi + 1 = 5.7$ (obtained by applying L' Hospital's rule, since

$$N_c = \infty \times 0 \text{ for } \phi = 0) \quad \text{...(Eq. 14.73)}$$

$$N_q = 1 \qquad \ldots(\text{Eq. 14.74})$$

and
$$N_\gamma = 0 \qquad \ldots(\text{Eq. 14.75})$$

Thus, the bearing capacity of a strip footing with a rough base on the ground surface is given by

$$q_{ult} = 5.7c \qquad \ldots(\text{Eq. 14.76})$$

This compares very well with the corresponding value from Prandtl's equation for a continuous footing with a smooth base.

For a strip footing at a depth D_f in a purely cohesive soil

$$q_{ult} = 5.7c + \gamma D_f \qquad \ldots(\text{Eq. 14.77})$$

Equation 14.67, along with the bearing capacity factors N_c, N_q and N_γ are valid for 'general shear failure'. An explanation of 'general shear failure' and 'local shear failure', as given by Terzaghi, is set out below:

Before the load is applied, the soil beneath the base of the footing is in a state of elastic equilibrium. *When the load is increased beyond a certain critical value, the soil gradually passes into a state of plastic equilibrium.* During this process of transition both the distribution of soil reactions over the base of the footing and the orientation of the principal stresses in the soil beneath the footing change. The transition starts at the outer edges of the base and spreads outwards. If the machanical properties of the soil are such that the strain which precedes the failure of the soil by plastic flow is very small, the footing does not sink into the ground until a state of plastic equilibrium indicated in Fig. 14.8(*a*) has been reached. The failure occurs by sliding in the two outward directions. The corresponding relation between load and settlement is shown by the solid curve C_1 in Fig. 14.9. This type of failure is called 'general shear failure'. This is applicable to dense and stiff soils.

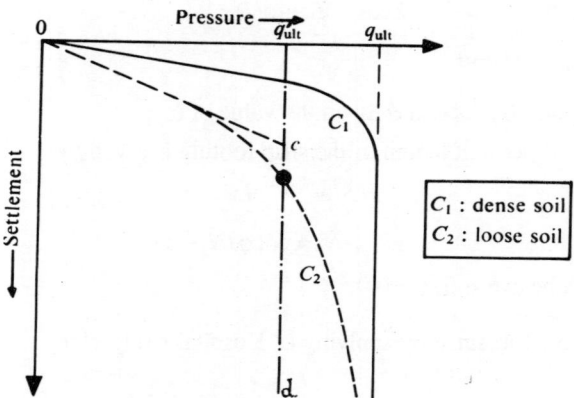

Fig. 14.9 Relation between pressure and settlement for dense (C_1) and loose (C_2) soil

On the other hand, if the mechanical properties of the soil are such that the plastic flow is preceded by a very important strain, the approach to general shear failure is associated with a rapidly increasing settlement and the relation between stress and settlement is approximately as indicated by dashed curve C_2 in Fig. 14.9. The criterion that the slope of the settlement curve should increase steeply for failure of soil is satisfied even before the failure spreads to the surface. Hence, this type of failure will be called 'local shear failure'. This is applicable for very loose or very compressible soils.

The curve C_2 may be idealised as Ocd, a broken line, which represents the stress-strain relation of an ideal plastic material whose shear parameters c' and ϕ' are smaller than values c and ϕ for curve C_1. Based on available data on stress-strain relations, Terzaghi suggests the following values for c' and ϕ'.

$$c' = (2/3)\, c \qquad \ldots \text{(Eq. 14.78)}$$

and
$$\tan \phi' = (2/3) \tan \phi \qquad \ldots \text{(Eq. 14.79)}$$

The corresponding values of the bearing capacity factors are designated N_c', N_q' and N_γ', which are less than the corresponding values for general shear failure. Also c' and ϕ' must be used wherever c and ϕ occur in the computation for bearing capacity.

Hence, for local shear failure,

$$q'_{ult} = (2/3)\, cN_c' + \gamma D_f N_q' + \frac{1}{2} \gamma b\, N_\gamma' \qquad \ldots \text{(Eq. 14.80)}$$

If the stress-strain relations are intermediate between C_1 and C_2 in Fig. 14.9, the bearing capacity is intermediate between q_{ult} and q'_{ult}.

Terzaghi's bearing capacity factors are plotted in Fig. 14.10 for general shear as well as for local shear. As a general guideline, if failure occurs at less than 5% strain and if density index is greater than 70%, general shear failure may be assumed, if the strain at failure is 10% to 20% and if the density index is less than 20%, local shear failure may be assumed, and, for intermediate situations, linear interpolation of the factors may be employed.

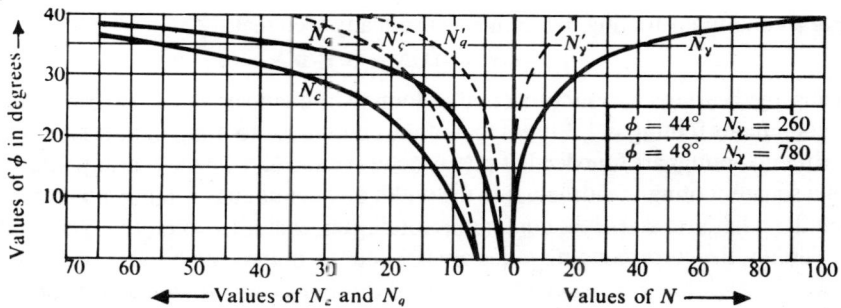

Fig. 14.10 Terzaghi's bearing capacity factors (Terzaghi, 1943)
(Full lines for general shear and dashed lines for local shear)

The bearing capacity factors of Terzaghi are tabulated in Table 14.3 for certain values of ϕ:

Table 14.3 Terzaghi's bearing capacity factors

Angle of shearing resistance $\phi°$	N_c	N_q	N_γ
0	5.7	1.0	0.0
5	7.3	1.6	1.5
10	9.6	2.7	1.2
15	12.9	4.4	2.5
20	17.7	7.4	5.0
25	25.1	12.7	9.7
30	37.2	22.5	19.7
35	57.8	41.4	42.4
40	95.7	81.3	100.4
45	172.3	173.3	297.5
50	347.5	415.1	1153.0

Bearing capacity of shallow circular and square footings

By repeating the reasoning which led to Eq. 14.67, the bearing capacity of circular footings has been proposed by Terzaghi as follows, from the analysis of experimental data available.

$$q_{ult_c} = 1.3\, cN_c + \gamma D_f N_q + 0.3\, \gamma d\, N_\gamma \qquad ...(Eq.\ 14.81)$$

where d = diameter of the circular footing.
The critical load for the footing is given by

$$Q_{ult_c} = \left(\frac{\pi d^2}{4}\right) \cdot q_{ult_c} \qquad ...(Eq.\ 14.82)$$

Similarly, the bearing capacity of a square footing of side b is:

$$q_{ult_s} = 1.3\, cN_c + \gamma D_f N_q + 0.4\, \gamma b\, N_\gamma \qquad ...(Eq.\ 14.83)$$

The critical load for the footing is given by

$$Q_{ult_s} = (b^2) \cdot q_{ult_s} \qquad ...(Eq.\ 14.84)$$

For a continuous footing of width b, it is already seen that,

$$q_{ult} = cN_c + \gamma D_f N_q + 0.5\, \gamma b\, N_\gamma$$

Thus, the bearing capacity of a circular footing of diameter equal to the width of a continuous footing is 1.3 times that of the continuous footing, or at least nearly so, if the footings are founded in a purely cohesive soil ($\phi = 0$); the bearing capacity of a square footing of side equal to the width of a continuous footing also bears a similar relation to that of the continuous footing under similar conditions just cited.

Further, the corresponding ratios are 0.6 and 0.8 in the case of circular footing and square footing, respectively, when the footings are founded in a purely cohesionless soil ($c = 0$).

The "benefit" of surcharge or depth of foundation, as it is called, is only marginal in the case of footings on purely cohesive soils, since N_q is just equal to 1; in fact, the increase in bearing capacity due to depth is just equal to the surcharge γD_f and it is only the difference between the gross and net values of bearing capacities. However, this benefit or increase in bearing capacity is significant in the case of cohesionless soils or $c - \phi$ soils ($\phi > 0$), especially when the angle of shearing resistance and hence N_q-value are very high as for dense sands.

While the bearing capacity of a footing in pure sand may be increased either by increasing the width or depth below ground at a given density index, the value may be increased by densification. However, these avenues are not useful in the case of footings in pure clays.

The differences in the bearing capacity values arising out of differences in the size of the footing and in the shape of the footing are termed 'size effects' and 'shape-effects', respectively.

14.5.6 Meyerhof's Method

The important difference between Terzaghi's and Meyerhof's approaches is that the latter considers the shearing resistance of the soil above the base of the foundation, while the former ignores it. Thus, Meyerhof allows the failure zones to extend up to the ground surface (Meyerhof, 1951). The typical failure surface assumed by Meyerhof is shown in Fig. 14.11.

The significant zones are: Zone I   ABC   ... elastic
Zone II BCD   ... radial shear
Zone III BDEF   ... mixed shear

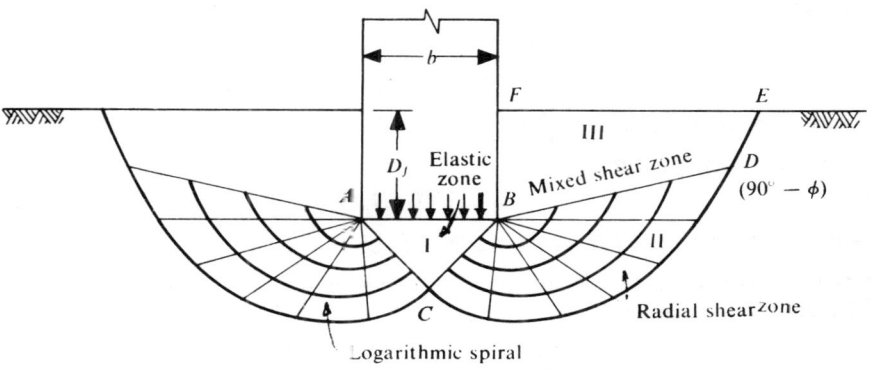

Fig. 14.11 Meyerhof's method for the bearing capacity of shallow foundation

Meyerhof's equation for the bearing capacity of a strip footing is of the same general form as that of Terzaghi:

$$q_{ult} = cN_c + \gamma D_f N_q + \frac{1}{2}\gamma b N_\gamma \qquad \text{...(Eq. 14.85)}$$

wherein N_c, N_q and N_γ are "Meyerhof's bearing capacity factors", which depend not only on ϕ, but also on the depth and shape of the foundation and roughness of the base.

Meyerhof's factors are more difficult to obtain that Terzaghi's, and have been presented in the form of charts by Meyerhof.

The nature of the failure surface assumed by Meyerhof implies the occurrence of a substantial downward movement of the footing before the full value of the shearing resistance is mobilised. This may be more probable in the case of purely cohesive soils as indicated by the available experimental evidence.

Hence, the values of N_c from Meyerhof's theory for cohesive soils are given as follows:

For strip footings: $N_c = 5.5\,(1 + 0.25\,D_f/b)$ ...(Eq. 14.86)

with a limiting value of 8.25 for N_c for $D_f/b > 2.5$.

For square or circular footings: $N_c = 6.2\,(1 + 0.32\,D_f/b)$ ...(Eq. 14.87)

with a limiting value of 9.0 for N_c for $D_f/b > 2.5$.

(b is the side of a square or diameter of circular footing).

14.5.7 Skempton's Method

Skempton proposed equations for bearing capacity of footings founded in purely cohesive soils based on extensive investigations (Skempton, 1951). He found that the factor N_c is a function of the depth of foundation and also of its shape. His equations may be summarised as follows:

The net ultimate bearing capacity is given by:

$$q_{\text{net ult}} = c \cdot N_c \qquad \text{...(Eq. 14.88)}$$

wherein N_c is given as follows:

Strip footings:

$$N_c = 5\,(1 + 0.2\,D_f/b) \qquad \text{...(Eq. 14.89)}$$

with a limiting value of N_c of 7.5 for $D_f/b > 2.5$.

square or circular footings:

$$N_c = 6\,(1 + 0.2 D_f/b) \qquad \text{...(Eq. 14.90)}$$

with a limiting value of N_c of 9.0 for $D_f/b > 2.5$.

(b is the side of square or diameter of circular footing).

Rectangular footings:

$$N_c = 5\left(1 + 0.2\frac{b}{L}\right)\left(1 + 0.2\frac{D_f}{b}\right) \qquad \text{...(Eq. 14.91)}$$

for $D_f/b \leq 2.5$, and

$$N_c = 7.5\,(1 + 0.2\,b/L) \qquad \text{...(Eq. 14.92)}$$

for $D_f/b > 2.5$,

wherein b = width of the rectangular footing, and
L = length of the rectangular footing.

In fact, the experimental relationships deduced by Skempton are not exactly linear with respect to D_f/b, but straight lines are fitted for the sake of simplicity.

For a surface footing of square or circular shape on purely cohesive soil

$$Q_{net\,ult} = 6c \qquad \text{(Eq. 14.93)}$$

as against $7.4c$ from Terzaghi's theory.

It must be noted that Terzaghi's theory is limited to shallow foundations wherein $D_f/b \leq 1$, but Skempton's equations do no suffer from such a limitation.

14.5.8 Brinch Hansen's Method

Brinch Hansen (1961) has proposed the following semi-empirical equation for the bearing capacity of a footing, as a generalisation of the Terzaghi equation:

$$q_{ult} = \frac{Q_{ult}}{A} = cN_c s_c d_c i_c + qN_q s_q d_q i_q + \frac{1}{2}\gamma b N_\gamma s_\gamma i_\gamma \qquad \text{...(Eq. 14.94)}$$

where

Q_{ult} = vertical component of the total load (= V),
A = effective area of the footing (this will arise for inclined and eccentric loads, when the area A is transformed to an estimated equivalent rectangle with sides b and L, such that the load is central to the area),
q = overburden pressure at the foundation level (= $\gamma \cdot D_f$),
N_c, N_q and N_γ = bearing capacity factors of Hansen, given as follows:

$$N_q = N_\phi \cdot e^{\pi \tan \phi} \qquad \text{...(Eq. 14.95)}$$

$$N_c = (N_q - 1) \cot \phi \qquad \text{...(Eq. 14.96)}$$

$$N_\gamma = 1.8 \ (N_q - 1) \tan \phi \qquad \text{...(Eq. 14.97)}$$

($N_\phi = \tan^2 (45° + \phi/2)$, with the usual notation.)
s'' = shape factors
d'' = depth factors, and

i'' = inclination factors.

The bearing capacity factors of Hansen, shape factors, depth factors, and inclination factors are given in Tables 14.4 to 14.7.

It has been found that Hansen's theory gives a better correlation for cohesive soils than the Terzaghi theory, although it may not give good results for cohesionless soils.

Table 14.4 Brinch Hansen's bearing capacity factors

Angle of shearing Resistance $\phi°$	Hansen's Bearing Capacity Factors		
	N_c	N_q	N_γ
0	5.14	1.00	0
5	6.49	1.57	0.09
10	8.34	2.47	0.47

(contd.)

15	10.98	3.94	1.42
20	14.83	6.40	3.54
25	20.72	10.66	8.11
30	30.14	18.40	18.08
35	46.13	33.29	40.69
40	95.41	75.32	64.18
45	133.89	134.85	240.85
50	266.89	318.96	681.84

Table 14.5 Brinch Hansen's shape factors

Type of footing	Hansen's shape factors		
	s_c	s_q	s_γ
Continuous (Width b)	1.0	1.0	1.0
Rectangular ($b \times L$)	$1 + 0.2 \dfrac{b}{L}$	$1 + 0.2 \dfrac{b}{L}$	$1 - 0.4 \dfrac{b}{L}$
Square (Size b)	1.3	1.2	0.8
Circular (Diameter b)	1.3	1.2	0.6

Table 14.6 Brinch Hansen's depth factors

d_c	d_q	d_γ
$1 + 0.35 D_f/b$	$1 + 0.35 D_f/b$	1.0

$$d_q = d_c \text{ for } \phi > 25°$$

$$d_q = 1.0 \text{ for } \phi = 0°$$

Table 14.7 Brinch Hansen's inclination factors

i_c	i_q	i_γ
$1 - \dfrac{H}{2c_a bL}$	$1 - 0.5 \dfrac{H}{V}$	$(i_q)^2$

Limitation: $H \leq V \tan \delta + c_a\, bL$.

where H and V = horizontal and vertical components of total load
δ = angle of friction between base of footing and soil
c_a = adhesion between footing and soil
L = length of footing parallel to H

Revised values of inclination factors:

$$i_c = i_q = \left(1 - \frac{H}{V + A \cdot c\, \cot\phi}\right)^2 \quad \text{....(Eq. 14.98)}$$

$$i_\gamma = i_q^2 \quad \text{...(Eq. 14.99)}$$

But, for $\phi = 0°$,

$$i_c = i_q = 0.5 + 0.5\sqrt{1 - \frac{H}{Ac}} \qquad \text{...(Eq. 14.100)}$$

14.5.9 Balla's Method

Balla has proposed a theory for the bearing capacity of continuous footings (Balla. 1962). The theory appears to give values which are in good agreement with field test results for footings founded in cohesionless soils.

The form of the bearing capacity equation is the same as that of Terzaghi:

$$q_{ult} = c \cdot N_c + \gamma D_f N_q + \frac{1}{2} \gamma b N_\gamma$$

But the equations for the bearing capacity factors are cumbersome to solve without the aid of a digital computer. Therefore, it is generally recommended that Balla's charts be used for the determination of these factors.

The charts are shown in Figs. 14.12 and 14.13:

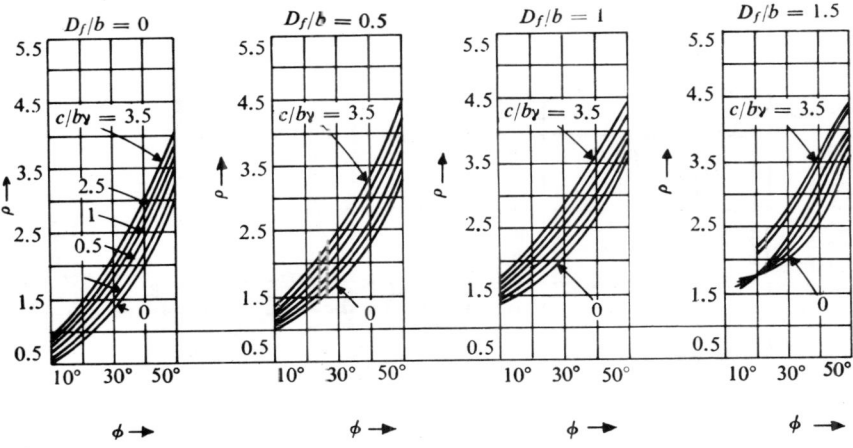

Fig. 14.12 Balla's charts for the parameter ρ

The steps involved are as follows:

(i) The ratios D_f/b and $\dfrac{c}{\gamma b}$ are determined.

(ii) The parameter ρ is found from Fig. 14.12, the appropriate chart for the particular value of D_f/b being picked, since the value of φ is known and $\dfrac{c}{\gamma b}$ is also known.

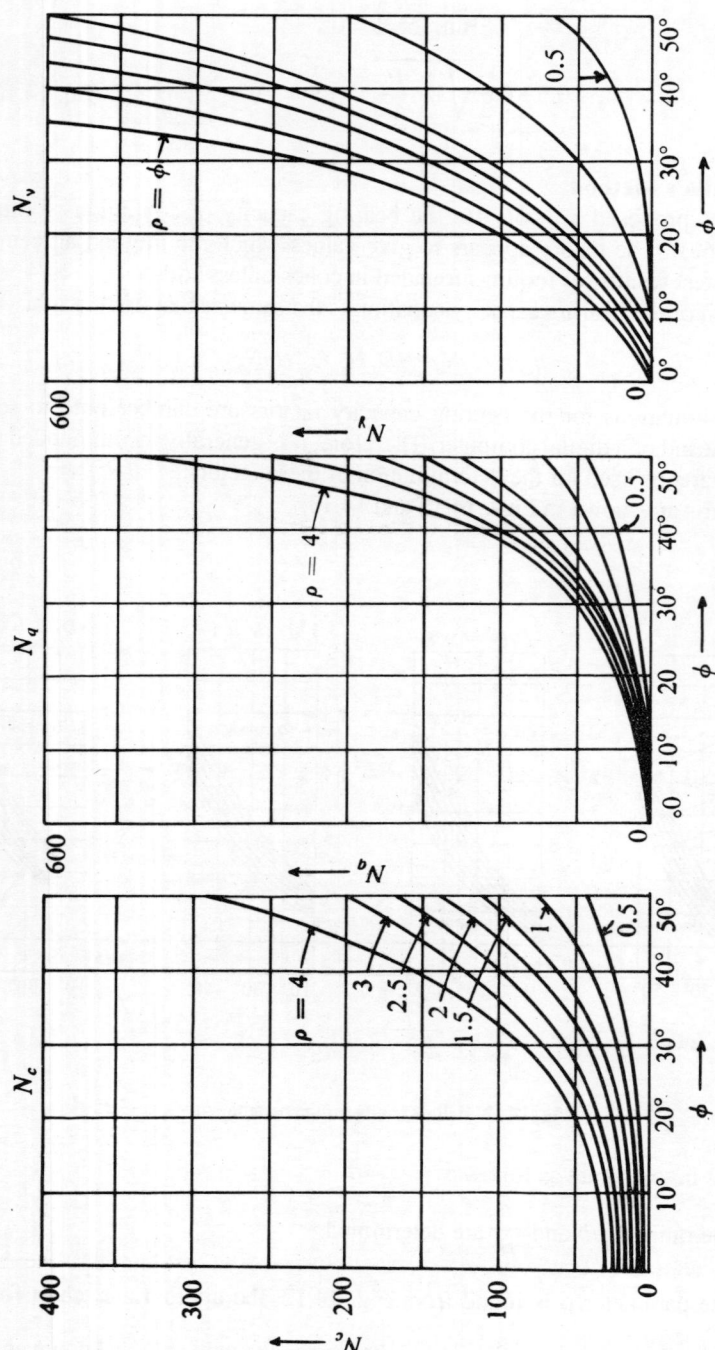

Fig. 14.13 Balla's charts for the bearing capacity factors

($\rho = \dfrac{R}{b}$, where R is the radius of the rupture surface).

(iii) The bearing capacity factors of Balla N_c, N_q, and N_γ are determined from the charts of Fig. 14.13, for the particular value of ρ determined in (ii) above and for the known value of ϕ.

(iv) The ultimate bearing capacity is determined by using these factors and other relevant quantities in Balla's formula.

The limitations are that it should be used when $\dfrac{D_f}{b} \le 1.5$ and that it is applicable to continuous footings only.

14.6 EFFECT OF WATER TABLE ON BEARING CAPACITY

The Terzaghi equation for bearing capacity,

$$q_{ult} = cN_c + \gamma D_f N_q + \frac{1}{2}\gamma b N_\gamma,$$

contains the unit weight, γ, and the cohesion, c, of the soil directly, and its angle of shearing resistance, ϕ, indirectly, since the bearing capacity factors, N_c, N_q, and N_γ depend upon the value of ϕ.

Water in soil is known to affect its unit weight and also the shear parameters c and ϕ. When the soil is submerged under water, the effective unit weight γ' is to be used in the computation of bearing capacity. Similarly, the effective stress parameters, c' and ϕ', obtained from an appropriate test in the laboratory, on saturated sample of the soil, are to be used.

However, the effect of water table on the shear parameters of the foundation soil is usually considered small and hence, ignored. But the effective unit weight γ' is roughly half the saturated unit weight; consequently there will be about 50% reduction in the value of the corresponding term in the bearing capacity formula.

It should be now obvious that the location of the ground water table and its seasonal fluctuations have a bearing on the capacity of a foundation. There will be no effect or reduction in the bearing capacity if the water table is located at a sufficient depth below the base of the footing. In fact, this minimum depth below the base of the footing is set at a value equal to the width of the footing since the maximum depth of the zone of shear failure below the base is not expected to exceed this value ordinarily. If the water table is above this level, there will be a reduction in the bearing capacity. If the water table is at the level of the base of the footing, γ' is to be used for γ in the third term, which indicates the contribution of the weight of the soil in the elastic wedge beneath the base of the footing, since the entire wedge is submerged; that is to say, a reduction factor of 0.5 is to be applied to the third term. For any location of the water table intermediate between the base of the footing and a depth equal to the width of the footing below its base, a suitable linear interpolation of the necessary reduction is suggested.

If the water table is above the base of the footing, the reduction factor for the third term is obviously limited to the maximum of 0.5. Further, in such a case, reduction will be indicated for the second term, which indicates the contribution to the bearing capacity of the surcharge due to the depth of the foundation. Proceeding with a similar logic, one comes to the conclusion that the maximum reduction of 0.5 is indicated for the second term when the water table is at the ground level itself (or above it), since γ' is to be used for γ in the second term. While no reduction in the second term is required when the water table is at or below the base of the footing, a proportionate reduction, with a suitable linear interpolation, is indicated when the water table is at a level intermediate between the ground level and the base of the footing. Thus, both the second and third terms will be modified in this case.

The first term, cN_c, does not get affected significantly by the location of the water table; except for the slight change due to the small reduction in the value of cohesion in the presence of water.

In the case of purely cohesive soils, since $\phi \approx 0°$, $N_q = 1$ and $N_\gamma = 0$, the net ultimate bearing capacity is given by $c \cdot N_c$, which is virtually unaffected by the water table, if it is below the base of the footing. Even if the water table is at the ground level, only the gross bearing capacity is reduced by 50% of the surcharge term γD_f ($N_q = 1$), while the net value is again only $c \cdot N_c$.

In the case of purely cohesionless soils, since $c = 0$, and $\phi > 0$, and N_q and N_γ are significantly high, there is a substantial reduction in both the gross and net values of the bearing capacity if the water table is at or near the base of the footing and more so if it is at or near the ground surface.

For locations of ground water table within a depth of the width of the foundation below the base and the ground level, the equation for the ultimate bearing capacity may be modified as follows:

$$q_{ult} = {}^*c'N_c + \gamma D_f N_q R_q + \tfrac{1}{2}{}^{**}\gamma b N_\gamma \cdot R_\gamma \qquad \ldots \text{(Eq. 14.101)}$$

where $c' =$ effective cohesion (may be taken as c itself, in the absence of sufficient data).

N_c, N_q, and $N_\gamma =$ bearing capacity factors based on the effective value of friction angle ϕ' and,

R_q and $R_\gamma =$ reduction factors for the terms involving N_q and N_γ owing to the effect of water table.

R_q and R_γ may be obtained as follows, from Fig. 14.14:

---

*appropriate multiplying factor should be used for isolated footings.
**Appropriate shape factor.

$$z_q = 0 \ldots R_q = 0.5 \qquad\qquad z_\gamma = 0 \ldots R_\gamma = 0.5$$

$$z_q = D_f \ldots R_q = 1.0 \qquad\qquad z_\gamma = b \ldots R_\gamma = 1.0$$

These conditions and the linear relationship of the chart are also expressed by the following equations:

$$R_q = 0.5\left(1 + \frac{z_q}{D_f}\right) \qquad \ldots \text{(Eq. 14.102)}$$

$$R_\gamma = 0.5\left(1 + \frac{z_\gamma}{b}\right) \qquad \ldots \text{(Eq. 14.103)}$$

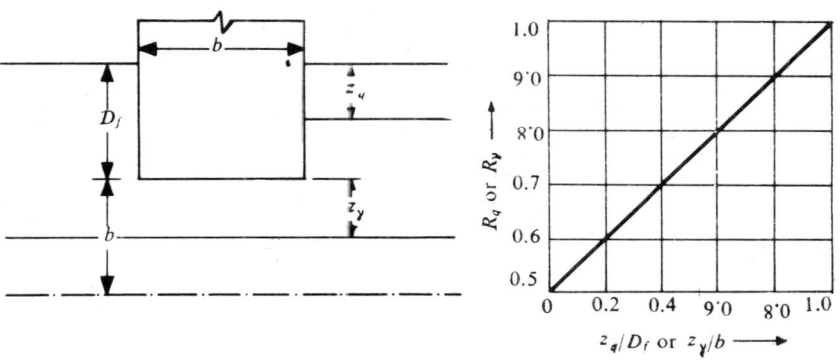

(a) Location of water table   (b) Reduction factors R_q and R_γ

Fig. 14.14 Effect of water table on bearing capacity

Note: For $z_q > D_f$ (the water table is below the base of the footing), R_q is limited to 1.0. For $0 \le z_q \le D_f$ (the water table is above the base of the footing), R_γ is limited to 0.5. For $z_q > (D_f + b)$ or $z_\gamma > b$, R_q as well as R_γ are limited to 1.0. For $z_q = 0$, R_q as well as R_γ are limited to 0.5.

14.7 SAFE BEARING CAPACITY

Safe bearing capacity, as already defined in Sec. 14.1, is the maximum pressure intensity that the soil will safely transmit without the risk of shear failure irrespective of settlement that may occur.

The value of the safe bearing capcacity is determined by applying a suitable factor of safety to the ultimate bearing capacity, determined by any one of the methods available. The ultimate value is composed of three terms—one due to the cohesion of the soil, another due to the weight of the soil in the elastic zone, and a third due to the depth of foundation or surcharge. The contribution due to this third term is a factor N_q times the surcharge or original overburden pressure, γD_f.

Since the soil has already been subjected to this original overburden pressure, there is no need to apply a factor of safety greater than unity to this component of the ultimate bearing capacity.

It has also been seen that the gross bearing capacity minus the original overburden pressure or surcharge pressure at the level of the base of the foundation is called the net bearing capacity (Sec. 14.1).

Thus, the following procedure may be specified for arriving at the safe bearing capacity:

(i) The surcharge pressure, γD_f, is deducted from the gross ultimate bearing capacity q_{ult}, to give the net ultimate bearing capacity, $q_{net\,ult}$.

(ii) The net ultimate bearing capacity is divided by the chosen factor of safety η, to give the net safe bearing capacity, q_{ns}.

(iii) Finally, the surcharge pressure is added to the net safe bearing capacity, to give the safe bearing capacity, q_s, against shear failure.

That is to say,

$$q_{net\,ult} = q_{ult} - D_f \qquad \ldots(\text{Eq. 14.104})$$

$$q_{ns} = \frac{q_{net\,ult}}{\eta} \qquad \ldots(\text{Eq. 14.105})$$

$$q_s = q_{ns} + \gamma D_f = \frac{q_{net\,ult}}{\eta} + \gamma D_f$$

or
$$q_s = \frac{(q_{ult} - \gamma D_f)}{\eta} + \gamma D_f \qquad \ldots(\text{Eq. 14.106})$$

or
$$q_s = \frac{q_{ult}}{\eta} + \frac{\gamma D_f(\eta - 1)}{\eta} \qquad \ldots(\text{Eq. 14.107})$$

An expression for $q_{net\,ult}$ is sometimes written by combining γD_f terms, somewhat as follows:

$$q_{net\,ult} = q_{ult} - \gamma D_f = cN_c + \gamma D_f(N_q - 1) + \frac{1}{2}\gamma b\, N_\gamma \qquad \ldots(\text{Eq. 14.108})$$

However, the procedure involved in Eqs. 14.104 to 14.106 is advocated to avoid possible confusion.

In the case of a surface footing founded on a purely cohesive soil, ($D_f = 0$, $\phi = 0°$, $N_q = 1$, and $N_\gamma = 0$), the safe bearing capacity is obviously $\dfrac{c \cdot N_c}{\eta} \left(= \dfrac{q_{ult}}{\eta} \right)$ itself; it is $\dfrac{q_{ult}}{\eta}$ for a surface footing founded on any soil, for that matter.

Bearing Capacity 659

The factor of safety generally chosen is 3. However, it may sometimes be as high as 5. Apparently this may appear to be high, but is not so high when one remembers the innumerable factors and imponderables involved in the determination of the bearing capacity. It can be easily demonstrated that, even with a small variation of ϕ, the value of q_{ult} may change considerably. This justifies the statement made above.

The allowable bearing pressure (Sec. 14.1) is the smaller of the two values—the safe bearing capacity to avoid the risk of shear failure and the maximum net allowable pressure that will produce settlements of the structure within tolerable limits.

More of this will be seen in Sec. 14.13 since the bearing capacity of sands is governed invariably by settlement rather than shear failure.

14.8 FOUNDATION SETTLEMENTS

Settlement—total settlement and differential settlement of foundations and consequently of the structures above the foundations—and its determination has been presented in detail in Chapter 11 on "Settlement Analysis".

Some of the important aspects related to the bearing capacity of footings will be summarised in the subsections to follow.

14.8.1. Source of Settlement

Foundation settlements may be caused due to some or a combination of the following factors:
  (*i*) Elastic compression of the foundation and the underlying soil, giving rise to what is known as 'immediate', 'contact', 'initial', or 'distortion' settlement,
  (*ii*) Plastic compression of the underlying soil, giving rise to consolidation, settlement of fine grained soils, both primary and secondary,
  (*iii*) Ground water lowering, especially repeated lowering and raising of ground water level in loose granular soils and drainage without adequate filter protection,
  (*iv*) Vibration due to pile driving, blasting and oscillating machinery in granular soils,
  (*v*) Seasonal swelling and shrinkage of expansive clays,
  (*vi*) Surface erosion, creep or landslides in earth slopes,
  (*vii*) Miscellaneous sources such as adjacent excavation, mining subsidence and underground erosion.

The settlements from the first two sources alone may be predicted with a fair degree of confidence.

14.8.2 Bearing Capacity Based on Tolerable Settlement

As discussed in Sec. 14.2, the bearing capacity of a foundation is based on two criteria—the pressure that might cause shear failure of the foundation soil and the maximum allowable pressure such that the settlements produced are not more than the tolerable values.

The first criterion has already been discussed in detail. For the second criterion, the tolerable values of the total and differential settlements which a particular structure, on a particular type of foundation in a given soil, can undergo without sustaining any harmful effects are to be decided upon. These values have already been specified, basing on experience and judgement (Chapter 11). Once the limiting values of settlement are fixed, the procedure involves determining that pressure which causes settlements just equal to the limiting values. This is the allowable bearing capacity on the basis of the settlement criterion. (It is to be noted that there is no need to apply a further factor of safety to this pressure, since it would have been applied even at the stage of fixing up tolerable settlement values).

The smaller pressure of the values obtained from the two criteria is termed the 'allowable bearing pressure', which is used for design of the foundation.

The bearing capacity based on the settlement criterion may be determined from the field load tests or plate load tests (dealt with in the next section), standard penetration tests or from the charts prepared by authorities like Terzaghi and Peck, based on extensive investigations. More of this will be seen in the section on bearing capacity of sands.

14.8.3 Construction Practices to Avoid Differential Settlement

A few construction practices are recommended, based on experience, to avoid or minimise detrimental differential settlements.

(i) Suitable design of the structure and foundation...desired degree of flexibility of the various component parts of a large structure may be introduced in the construction.

(ii) Choice of a suitable type of foundation for the structure and the foundation soil conditions...e.g., large, heavily loaded structures on relatively weak and non-uniform soils may be founded on 'mat' or 'raft' foundations. Sometimes, piles and pile foundations may be used to bypass weak strata.

(iii) Treatment of the foundation soil...to encourage the occurrence of settlement even before the construction of the structure, e.g., (a) Dewatering and drainage, (b) Sand drains and (c) Preloading.

(iv) Provision of plinth beams and lintel beams at plinth level and lintel level in the case of residential buildings to be founded on weak and compressible strata.

14.9 PLATE LOAD TESTS

Perhaps the most direct approach to obtain information on the bearing capacity and the settlement characteristics at a site is to conduct a load test. As tests on prototype foundation are not practicable in view of the large loading required, the time factor involved and the high cost of a full-scale test, a short-term model loading test, called the 'plate load test' or 'plate bearing test', is usually conducted. This is a semi-direct method since the differences in size between the test and the structure are to be properly accounted for in arriving at meaningful interpretation of the test results.

The test essentially consists in loading a rigid plate at the foundation level, increasing the load in arbitrary increments, and determining the settlements corresponding to each load after the settlement has nearly ceased each time a load increment is applied.

The nature of the load applied may be gravity loading or dead weights on an improvised platform or reaction loading by using a hydraulic jack. The reaction of the jack load is taken by a cross beam or a steel truss anchored suitably at both ends. The test set-up with a jack is shown in Fig. 14.15.

Test plates are usually square or circular, the size ranging from 300 to 750 mm (side or diameter); the minimum thickness recommended is 25 mm for providing sufficient rigidity. If the loading set-up is a platform with dead weights, the kentledge may be in the form of sand bags, scrap iron or ingots or any other convenient heavy material. Jack-loading is superior in terms of accuracy and uniformity of loading. Settlement of the test plate is measured by means of at least two or three dial gauges with a least count of 0.02 mm.

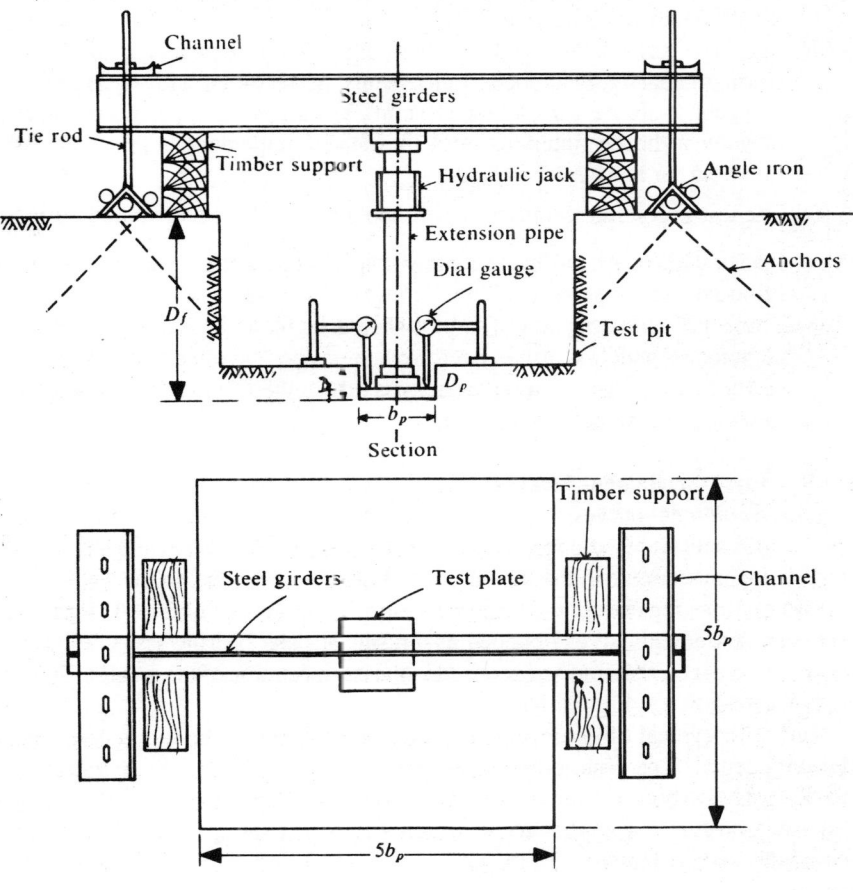

Fig. 14.15 Set-up for plate load test

The test pit should be at least five times as wide as the test plate and the bottom of the test plate should correspond to the proposed foundation level. At the centre of the pit, a small square hole is made the size being that of the test plate and the depth being such that,

$$\frac{D_p}{b_p} = \frac{D_f}{b} \qquad \ldots(Eq.\ 14.109)$$

where D_f and b are the depth and width of the proposed foundation.

Bigger size plates are preferred in cohesive soils. The test procedure is given in IS: 1888—1971 (Revised). The procedure, in brief, is as follows:

(i) After excavating the pit of required size and levelling the base, the test plate is seated over the ground. A little sand may be spread below the plate for even support. If ground water is encountered, it should be lowered slightly below the base by means of pumping.

(ii) A seating pressure of 7.0 kN/m² (70 g/cm²) is applied and released before actual loading is commenced.

(iii) The first increment of load, say about one-tenth of the anticipated ultimate bearing capacity, is applied. Settlements are recorded with the aid of the dial gauges after 1 min., 4 min., 10 min., 20 min., 40 min., and 60 min., and later on at hourly intervals until the rate of settlement is less than 0.02 mm/hour, or at least for 24 hours.

(iv) The test is continued until a load of about $1\frac{1}{2}$ times the anticipated ultimate load is applied. According to another school of thought, a settlement at which failure occurs or at least 2.5 cms should be reached.

(v) From the results of the test, a plot should be made between pressure and settlement, which is usually referred to as the "load-settlement curve", rather loosely. The bearing capacity is determined from this plot, which is dealt with in the next subsection.

14.9.1 Load-Settlement Curves

Load-Settlement curves or pressure-settlement curves to be more precise, are obtained as a result of loading tests either in the laboratory or in the field, oedometer tests being an example in the laboratory and plate bearing test, in the field.

Information regarding the bearing capacity may be obtained from the pressure-settlement curves obtained as a result of plate bearing test in the field; however, great care should be exercised in the interpretation of the results. Typical curves are shown in Fig. 14.16.

Curve I is typical of dense sand or gravel or stiff clay, wherein general shear failure occurs. The point corresponding to failure is obtained by extrapolating backwards (as shown in the figure), as a pronounced departure from the straight line relationship that applies to the initial stages of loading is observed. (This coincides approximately with the point up to which the range of proportionality extends).

Curve *II* is typical of loose sand or soft clay, wherein local shear failure occurs. Continuous steepening of the curve is observed and it is rather difficult to pinpoint failure; however, the point where the curve becomes suddenly steep is located and treated as that corresponding to failure.

Curve *III* is typical of many $c - \phi$ soils which exhibit characteristics intermediate between the above two. Here also the failure point is not easy to locate and the same criterion as in the case of Curve *II* is applied.

Thus, it is seen that, except in a few cases, arbitrary location of failure point becomes inevitable in the interpretation of load test results.

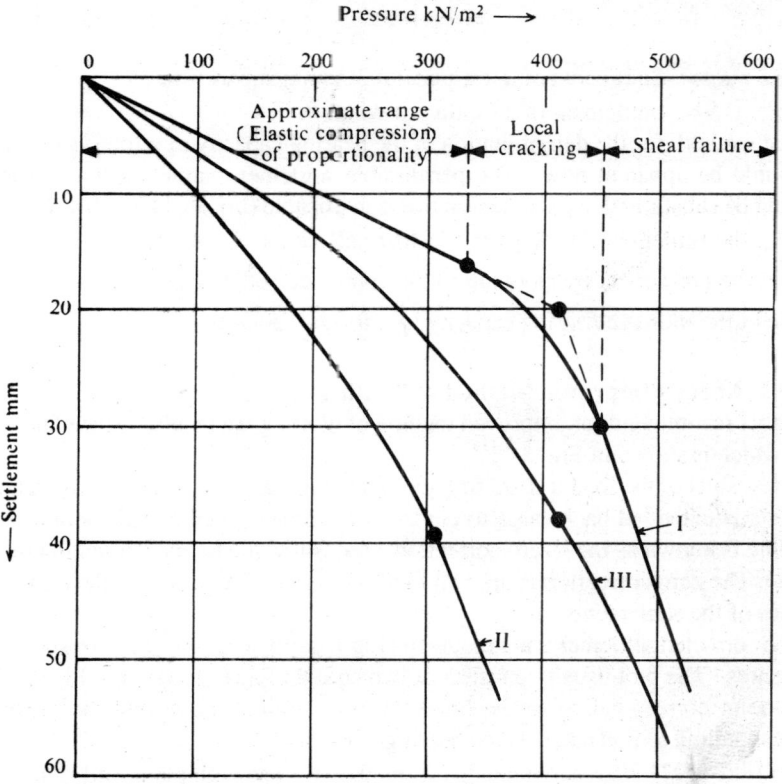

Fig. 14.16 Typical load-settlement curves from plate load tests

Determination of bearing capacity from plate load test

The size effect has been empirically evolved in the form of the following equation (Terzaghi and Peck, 1948):

$$\frac{S}{S_p} = \left[\frac{b(b_p + 0.3)}{b_p(b + 0.3)}\right]^2 \qquad \ldots(\text{Eq. } 14.110)$$

where S = settlement of the proposed foundation (mm),
S_p = settlement of the test plate (mm), $\Big]$ (same units)
b = size of the proposed foundation (m), and
b_p = size of the test plate (m).

This is applicable for sands.

However, the relationship is simpler for clays, since the modulus value E_s, for clays is reasonably constant:

$$\frac{S}{S_p} = \frac{b}{b_p} \qquad \ldots\text{(Eq. 14.111)}$$

Equation 14.110 may be put in a slightly simplified form as follows:

$$S = S_p \left[\frac{2b}{b+0.3}\right]^2 \qquad \ldots\text{(Eq. 14.112)}$$

where S_p = Settlement of a test plate of 300 mm square size,
and S = Settlement of a footing of width b.

The method for the determination of the bearing capacity of a footing of width b should be apparent now. The permissible settlement value, such as 25 mm, should be substituted in the equation that is applicable (Eq. 14.110 to 14.112); and the S_p, the settlement of the plate must be calculated. From the load-settlement curve, the pressure corresponding to the computed settlement S_p, is the required value of the ultimate bearing capacity, q_{ult}, for the footing.

14.9.2 Abbet's Improved Method of Plotting

Abbet recommends an improved method of plotting the results of the plate load test which is shown in Fig. 14.17.

The results of the load test are first plotted to natural scale and the early straight portion is extended backwards to cut the settlement axis; the settlement at zero loading is known as the "zero correction" (possibly due to uneven seating of the plate). The zero correction is applied to all settlement values to get the corrected values of the settlements.

The corrected settlements are plotted on log-log graph against the corresponding pressures. The plot usually consists of two straight lines as shown in Fig. 14.17. The point corresponding to the break gives the failure point and the pressure corresponding to it is taken as the bearing capacity.

IS: 1888–1971 also recommends this method for use with plate load tests.

14.9.3 Limitations of Plate Load Tests

Although the plate load test is considered to be an excellent approach to the problem of determining the bearing capacity by some engineers, it suffers from the following limitations:

(i) Size effects are very important. Since the size of the test plate and the size of the prototype foundation are very different, the results of a plate load test do not directly reflect the bearing capacity of the foundation.

Bearing Capacity 665

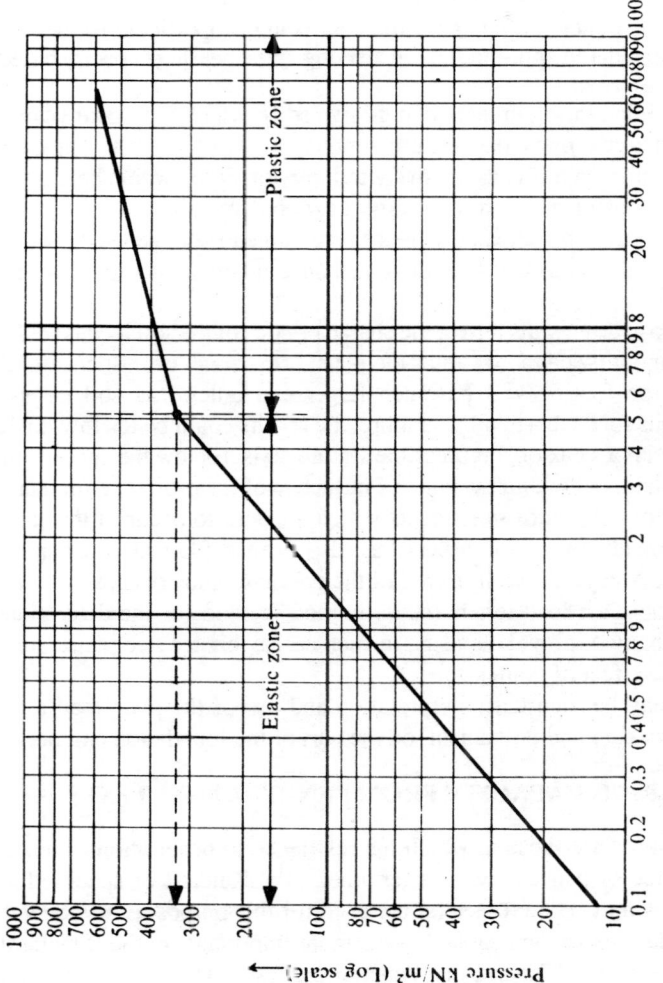

Fig. 14.17 Improved method of plotting of plate load test results (After Abbet)

666 Geotechnical Engineering

The bearing capacity of footings in sands varies with the size of footing; thus, the scale effect gives rather misleading results in this case. However, this effect is not pronounced in cohesive soils as the bearing capacity is essentially independent of the size of footing in such soils.

The settlement versus size relationship is rather complex in the case of cohesionless soils (Terzaghi and Peck, 1948); however, in the case of cohesive soils, this relation is rather simple, the settlement being proportional to the size. This should be considered appropriately in arriving at the bearing capacity based on the settlement criterion.

(*ii*) Consolidation settlements in cohesive soils, which may take years, cannot be predicted, as the plate load test is essentially a short-term test. Thus, load tests do not have much significance in the determination of allowable bearing pressure based on settlement criterion with respect to cohesive soils.

(*iii*) Results from plate load test are not recommended to be used for the design of strip footings, since the test is conducted on a square or circular plate and shape effects enter.

(*iv*) The load test results reflect the characteristics of the soil located only within a depth of about twice the width of the plate. This zone of influence in the case of a prototype footing will be much larger and unless the soil is essentially homogeneous for such a depth and more, the results could be terribly misleading. For example, if a weak or compressible stratum exists below the zone of influence of the test plate, but within the zone of influence of the prototype foundation, the plate test may not record settlements which are sure to occur in the case of the prototype foundation. This aspect has also been explained in Chapter 11 on "Settlement Analysis", with the aid of the pressure bulb concept.

Perhaps the plate load test is the only good method for the determination of bearing capacity of gravel deposits; in such cases, bigger size plates are used to minimise the effect of grain size.

Thus, it may be seen that interpretation and use of the plate load test results requires great care and judgement, on the part of the foundation engineer.

14.10 BEARING CAPACITY FROM PENETRATION TESTS

Penetration tests are those in which the resistance to penetration of a soil for a standard value of penetration is determined in a standard or specified manner. Devices known as 'penetrometers' are used for this purpose. A wide variety of these tests has become available, but the more important are the 'Standard Penetration Test' and the 'Dutch Cone Penetration Test'. These are more commonly employed for cohesionless soils. More detailed information on these devices and test will be provided in Chapter 18 on "Soil Exploration".

At this juncture, it is sufficient to know that the standard penetration test results are commonly in the form of 'Penetration Number', N, which indicates the number of blows required to cause 300 mm penetration of a split-spoon sampler into the soil under test by means of a 65 kg hammer falling through 750 mm.

This value has been correlated to Terzaghi's bearing capacity factors, density index and angle of shearing resistance, ϕ (Peck, Hansen, and Thornburn, 1953). Terzaghi and Peck have prepared charts for allowable bearing pressure, based on a standard allowable settlement for footings of known widths on sand, whose N-values are known. (Terzaghi and Peck, 1948). These correlations and charts will be presented in Sec. 14.13.

* 14.11 BEARING CAPACITY FROM MODEL TESTS—HOUSEL'S APPROACH

Housel (1929) has suggested, based on extensive experimental investigations, a practical method of determining the bearing capacity of a prototype foundation in a foundation soil which is reasonably homogeneous in depth by means of two or more small-scale model tests. It is assumed that the load-carrying capacity of a foundation for a predetermined allowable settlement consists of two distinct components—one which is carried by the soil column directly beneath the foundation, and the other which is carried by the soil around the perimeter of the foundation. The first component is a function of the area and the second, a function of the perimeter of the foundation.

This concept is expressed by the formula

$$W = q_s \cdot A = \sigma \cdot A + mP \qquad \text{...(Eq. 14.113)}$$

where $W =$ total ultimate load which the foundation can carry (kN),
 $q_s =$ bearing capacity of the foundation (kN/m^2) for a specified settlement,
 $\sigma =$ contact pressure developed under the bearing area of the foundation (or an experimental constant) (kN/m^2),
 $m =$ perimeter shear (or an empirical experimental constant) (kN/m)
 $A =$ bearing area of the foundation (m^2), and
 $P =$ perimeter of the foundation (m).

Equation 14.109 may be modified as

$$q_s = \sigma + m \cdot \frac{P}{A} \qquad \text{...(Eq. 14.114)}$$

or
$$q_s = mx + \sigma \qquad \text{...(Eq. 14.115)}$$

wherein x represents the perimeter-area ratio, P/A. Housel assumes that σ and m are constant for different loading tests on the same soil for a specific settlement, which would be tolerated by the prototype foundation. Hence, he suggested that σ and m be determined by conducting small-scale model tests by loading two or more test plates or model footings which have different areas and different perimeters and measuring the total load required to produce the specified allowable settlement in each case, at the proposed level of the foundation.

This gives two or more simultaneous equations from which σ and m may be determined. Then the bearing capacity of the proposed prototype foundation may be calculated from Eq. 14.112, by substituting for x, the perimeter-area ratio of the proposed foundation. Thus this procedure involves a kind of extrapolation from models to the prototype.

The method is commonly known as "Housel's Perimeter shear method" or "Housel's Perimeter-Area Ratio method".

14.12 BEARING CAPACITY FROM LABORATORY TESTS

The bearing capacity of a cohesive soil can also be evaluated from the unconfined compression strength. From the concept of shearing strength, the bearing capacity of a cohesive soil is the value of the major principal stress at failure in shear. This stress at failure is called the unconfined compression strength, q_u:

$$\sigma_1 = q_u = 2c \tan(45° + \phi/2) \qquad \ldots(\text{Eq. 14.116})$$

When $\phi = 0°$, for a purely cohesive soil,

$$q_u = 2c \qquad \ldots(\text{Eq. 14.117})$$

This applies at the ground surface, *i.e.*, when $D_f = 0$. The ultimate bearing capacity may be divided by a suitable factor of safety, say 3, to give the safe bearing capacity.

Casagrande and Fadum, (1944) suggest this procedure as an indirect check of the ultimate bearing capacity of cohesive soil, since the allowable soil pressures commonly specified in building codes are conservative from the standpoint of safety against rupture of the clay.

Further, bearing capacity problems may be studied experimentally from the shape of the rupture surface developed in the soil at failure. Experimental results may be translated to prototype structures by means of the theory of similitude, or modelling. (Jumikis, 1956 and 1961).

14.13 BEARING CAPACITY OF SANDS

The net ultimate bearing capacity of a footing in sand which is required in the proportioning of the size, is given by:

$$q_{\text{net-ult}} = \alpha \gamma b \cdot N_\gamma + \gamma D_f (N_q - 1) \qquad \ldots(\text{Eq. 14.118})$$

where $\alpha =$ Shape factor, which is given as
0.5 for continuous footing of width b,
0.4 for square footing of side b, and
0.3 for circular footing of diameter b.

Thus, the net ultimate bearing capacity depends upon

(*i*) the unit weight of soil, γ, and

(*ii*) the angle of shearing resistance, ϕ (since N_γ and N_q depend upon ϕ), besides the size, b, and depth, D_f, of the footing.

An appropriate value of γ must be used depending upon the condition of the soil with regard to water content and its location relative to ground water table.

The angle of shearing resistance of a cohesionless soil or sand is known to be dependent upon the density index the density index is correlated to the penetration resistance value, or the standard penetration number, N. Peck, Hanson, and Thornburn (1953) have provided a chart for evaluating ϕ, and the bearing capacity factors of Terzaghi, N_q and N_γ, from the standard penetration number, N. They have also given a simple chart relating N to ϕ. This is shown in Fig. 14.18 and the former in Fig. 14.19. If settlement is of no consequence, it is possible to think in terms of ultimate bearing capacity according to Terzaghi's formula by using these charts. But this procedure is not popular.

The necessary corrections for the observed value of N are to be applied before use in conjunction with these charts. These corrections, required for fine grained soils like silts below the water table and for the effect of overburden pressure, are dealt with in Chapter 18.

The charts of Figs. 14.18 and 14.19 do not apply to gravels or those soils containing a large percentage of gravels.

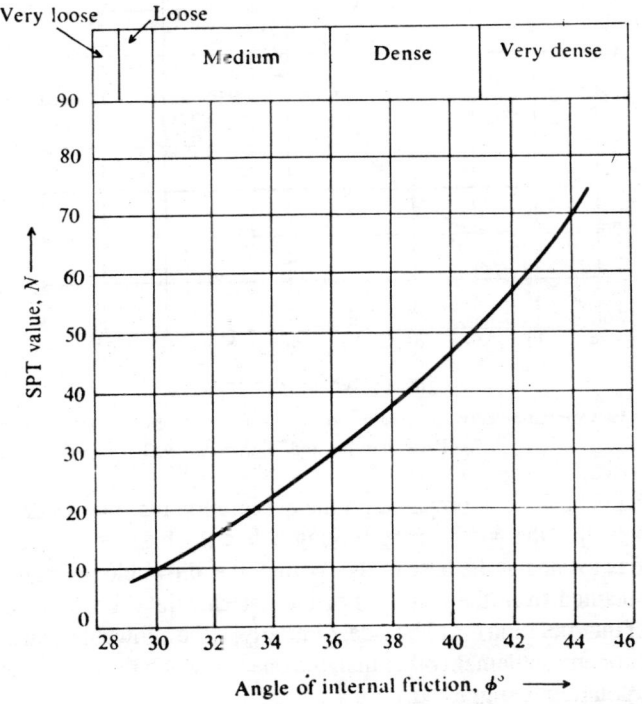

Fig. 14.18 Approximate correlation between N-value and ϕ for granular soils (After Peck, Hanson and Thornburn)

If settlement criterion governs the allowable bearing pressure, as it invariably does in the case of footings in sands, the design charts given by Terzaghi and Peck

(1948) or that given by Peck, Hanson and Thornburn (1953) may be used for the determination of allowable bearing pressure for a specific allowable settlement of 25 mm or 40 mm, as the case may be. These are shown in Figs. 14.20 to 14.22.

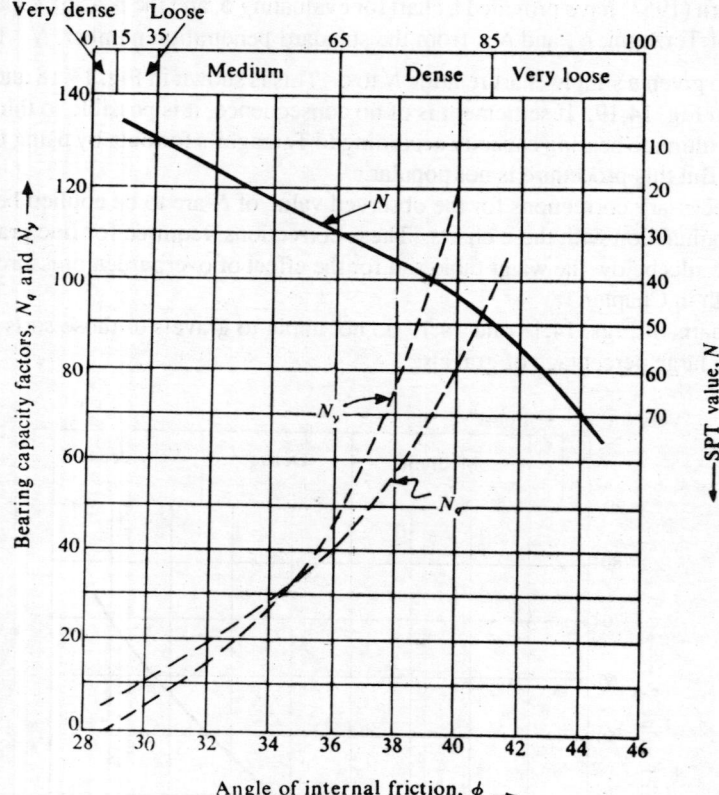

Fig. 14.19 Correlation between N-value, ϕ, and the bearing capacity factors (N_q and N_γ) (After Peck, Hanson, and Thornburn, 1953)

These charts have been prepared on the assumption that the water table is at a depth greater than the width of the footing below the base of the footing. If the water table is located at the base of the footing, the allowable pressure is taken as half that obtained from the charts. For any intermediate position of the water table, linear interpolation may be made. Similarly, if the allowable bearing pressure is required for any settlement other than the one for which the charts are prepared, linear interpolation is suggested.

Teng (1969) has proposed the following equation for the graphical relationship of Terzaghi and Peck (Fig. 14.20) for a settlement of 25 mm:

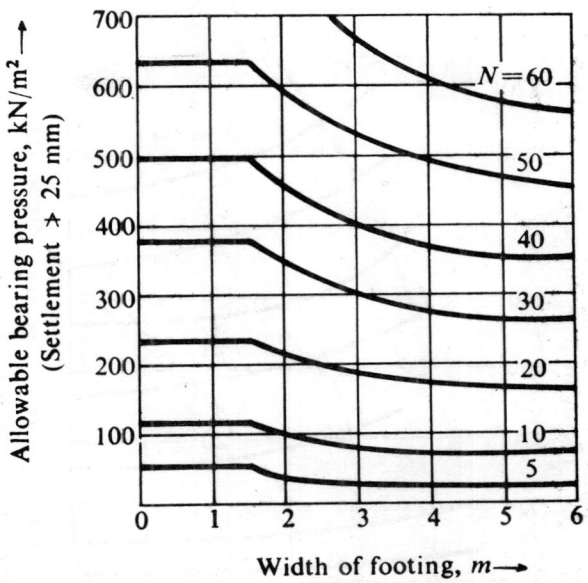

Fig. 14.20 Allowable bearing pressure for 25 mm settlement from *SPT* values (After Terzaghi and Peck, 1948)

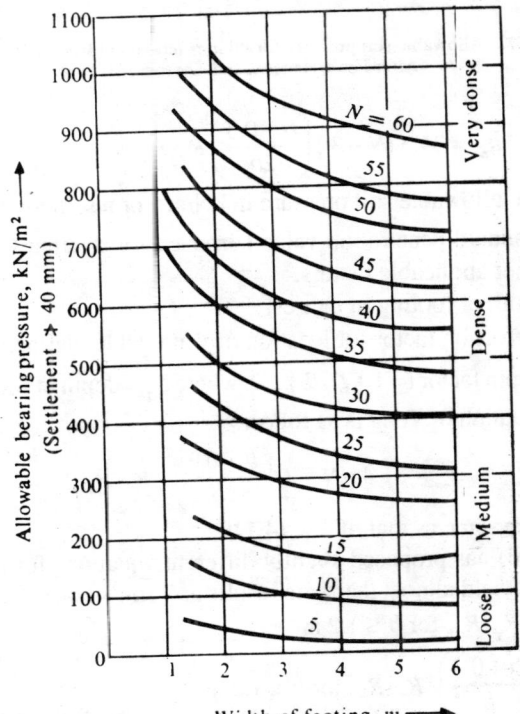

Fig. 14.21 Allowable bearing pressure for 40 mm settlement from *SPT* values (After Terzaghi and Peck, 1948)

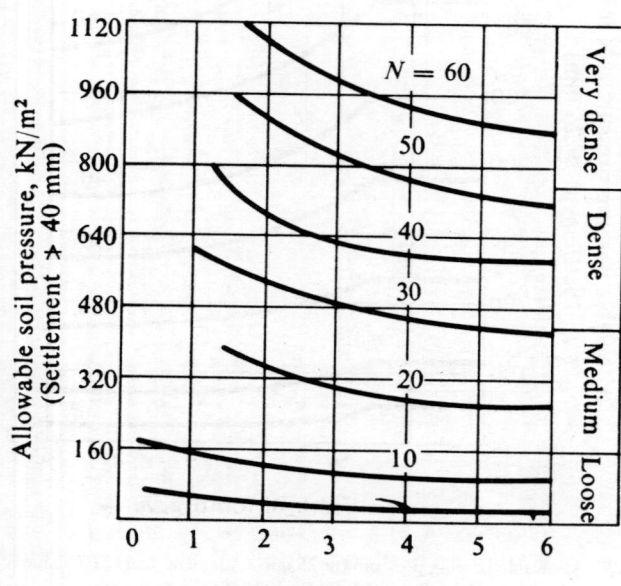

Fig. 14.22 Allowable soil pressure for 40 mm settlement from *SPT* values
(After Peck, Hanson, and Thornburn)

$$q_{na} = 34.3(N-3)\left(\frac{b+0.3}{2b}\right)^2 R_\gamma \cdot R_d \qquad \text{...(Eq. 14.119)}$$

where q_{na} = net allowable soil pressure in kN/m² for a settlement of 25 mm,
N = Standard penetration value corrected for overburden pressure and other applicable factors,
b = width of footing in metres,
R_γ = correction factor for location of water table, defined in Eq. 14.102,
and R_d = Depth factor (= $1 + D_f/b$) ≤ 2. where D_f = depth of footing in metres.

The modified equation of Teng is as follows:

$$q_{na} = 51.45(N-3)\left(\frac{b+0.3}{2b}\right)^2 \cdot R_\gamma \cdot R_d \qquad \text{...(Eq. 14.120)}$$

The notation is the same as that of Eq. 14.119.

Meyerhof (1956) has proposed slightly different equations for a settlement of 25 mm, but these yield almost the same results as Teng's equation:

$$q_{na} = 12.25 NR_\gamma \cdot R_d, \text{ for } b \leq 1.2 \text{ m} \qquad \text{...(Eq. 14.121a)}$$

$$q_{na} = 8.17N\left(\frac{b+0.3}{b}\right) \cdot R_\gamma \cdot R_d, \text{ for } b > 1.2 \text{ m}. \qquad \text{...(Eq. 14.121b)}$$

The notation is the same as those of Eqs. 14.119 and 14.120.
Modified equation of Meyerhof is as follows:

$$q_{na} = 18.36\, NR_\gamma \cdot R_d, \text{ for } b \leq 1.2 \text{ m} \qquad \text{...(Eq. 14.122a)}$$

$$q_{na} = 12.25\, N \left(\frac{b+0.3}{b}\right) R_\gamma \cdot R_d, \text{ for } b > 1.2 \text{ m} \qquad \text{...(Eq. 14.122b)}$$

The modified equations of Teng and Meyerhof are based on the recommendation of Bowles (1968).

The I.S. code of practice gives Eq. 14.122 for a settlement of 40 mm; but, it does not consider the depth effect.

Teng (1969) also gives the following equations for bearing capacity of sands based on the criterion of shear failure:

$$q_{net\,ult} = 1/6[3N^2\, b\, R_\gamma + 5(100 + N^2)\, D_f \cdot R_q] \qquad \text{...(Eq. 14.123)}$$

(for continuous footings)

$$q_{net\,ult} = 1/6[2N^2\, b\, R_\gamma + 6(100 - N^2)\, D_f \cdot R_q] \qquad \text{...(Eq. 14.124)}$$

(for square or circular footings)

Here again,

$q_{net\,ult}$ = net ultimate soil pressure in kN/m²,

N = Standard penetration value, after applying the necessary corrections,

b = width of continuous footing (side, if square, and diameter, if circular in metres),

D_f = depth of footing in metres, and

R_γ and R_q = correction factors for the position of the ground water table, defined in Eq. 14.102.

With a factor of safety of 3, the net safe bearing capacity q_{ns}, is given by

$$q_{ns} = \frac{1}{18}[3N^2\, b\, R_\gamma + 5(100 + N^2)\, D_f R_q] - \frac{2}{3}\gamma \cdot D_f \qquad \text{...(Eq. 14.125)}$$

(for continuous footings)

$$q_{ns} = \frac{1}{18}[2N^2\, b\, R_\gamma + 6(100 + N^2)\, D_f R_q] - \frac{2}{3}\gamma \cdot D_f \qquad \text{...(Eq. 14.126)}$$

(for square or circular footings)

If γ is not known with any degree of confidence, the second term may be ignored, although this may introduce some error in the value of q_{ns}. The smaller of the two values q_{na} and q_{ns} will be used for design.

14.14 BEARING CAPACITY OF CLAYS

For pure clays, $\phi = 0°$.

$$q_{ut} = cN_c + \gamma D_f = 5.7c + \gamma D_f$$

∴ $q_{net\,ult} = 5.7c$, for continuous footings.

$q_{net\,ult} = 1.3 \times 5.7c = 7.4\,c$ ...(Eq. 14.127)

(for square or circular footings, c being the cohesion.)

These are from Terzaghi's theory. Alternatively, Skempton's equations may be used. (Eqs. 14.91 and 14.92).

Skempton's equations are prferred for rectangular footings in pure clay (Eqs. 14.91 and 14.92)

Correlation of cohesion and consistency of clays with N-values is not reliable. Unconfined compression test is recommended for evaluating cohesion.

Overconsolidated or precompressed clays might show hair cracks and slickensides. Load tests are recommended in such cases.

Settlements of footings in clays may be calculated or predicted by the use of Terzaghi's one-dimensional consolidation. Long-term load tests also may be used but they are highly cumbersome and time-consuming.

The bearing capacity of footings in clays is practically unaffected by the size of the foundation.

14.15 RECOMMENDED PRACTICE (I.S)

The safe bearing capacity may be obtained from the relevant table given in IS: 1904-1978 (Second Revision). Where data for characteristics of a soil (cohesion, angle of internal friction, density, etc) are available, the safe bearing capacity may be calculated from consideration of shear failure (Terzaghi's theory). A factor of safety of three shall be adopted. Safe bearing pressure for sands may be obtained from the standard penetration resistance values (corrected for the presence of ground water and for overburden pressure), from both considerations of shear and settlement as given by Terzaghi and Peck, by Peck, Hanson and Thornburn, and by Teng.

Details are given in the Appendices of IS: 1904-1978 (Second Revision), already cited.

14.16 ILLUSTRATIVE EXAMPLES

Example 14.1: Calculate the elastic settlement of a rectangular foundation, 6 m $\times$ 12 m, on a uniform sand with $E = 20{,}000$ kN/m^2 and Poisson's ratio, $\nu = 0.2$. The contact pressure is 200 kN/m^2. The settlements are to be calculated at the centre, mid-point of long side, and mid-point of short side, and at the free corner.

Also compute the allowable bearing pressure, if the maximum settlement is restricted to 40 mm.

Side ratio of rectangle = 12/6 = 2
The elastic settlement is given by Schleicher as

k = shape factor = 1.08, 0.79, 0.69, and 0.54 for the centre, mid-point of long side, mid-point of short-side, and free corner respectively. (Table 14.2)

∴ Settlement at the centre

$$= 1.08 \times 200 \times \sqrt{12 \times 6} \frac{(1-0.2^2)}{20,000} \times 1000 \text{ mm} = \mathbf{88 \text{ mm}}.$$

Settlement at the mid-point of long-side

$$= 0.79 \times 200 \times \sqrt{72} \frac{(1-0.2^2)}{20,000} \times 1000 \text{ mm} = \mathbf{64.4 \text{ mm}}$$

Settlement at the mid-point of short-side

$$= 0.69 \times 200 \times \sqrt{72} \frac{(1-0.2^2)}{20,000} \times 1000 \text{ mm} = \mathbf{56.2 \text{ mm}}.$$

Settlement at the free corner

$$= 0.54 \times 200 \times \sqrt{72} \frac{(1-0.2^2)}{20,000} \times 1000 \text{ mm} = \mathbf{44 \text{ mm}}$$

If the maximum settlement is restricted to 40 mm, the centre settlement should not exceed this value.
Then the allowable bearing pressure:

$$q = \frac{sE}{\sqrt{A}(1-v^2)} = \frac{40 \times 20,000}{1.08 \times \sqrt{72}(1-0.2^2)} \approx \mathbf{90 \text{ kN/m}^2}$$

Example 14.2: What is the minimum depth required for a foundation to transmit a pressure 6 t/m² in a cohesionless soil with $\gamma = 1.8$ t/m³ and $\phi = 18°$? What will be the bearing capacity if a depth of 1.5 m is adopted according to Rankine's approach?

$$\gamma = 1.8 \text{ t/m}^3 \qquad \phi = 18° \qquad q = 6 \text{ t/m}^2$$

Minimum depth of foundation, according to Rankine,

$$D_f = \frac{q}{\gamma} \left(\frac{1-\sin\phi}{1+\sin\phi}\right)^2$$

$$= \frac{6}{18}\left(\frac{1-\sin 18°}{1+\sin 18°}\right)^2$$

$$= 0.93 \text{ m} \approx \mathbf{1 \text{ m}}$$

If $D_f = 1.5$ m,

$$q_{ult} = \gamma D_f \left(\frac{1+\sin\phi}{1-\sin\phi}\right)^2$$

$$= 1.8 \times 1.5 \left(\frac{1+\sin 18°}{1-\sin 18°}\right)^2$$

$$= \mathbf{9.68 \text{ t/m}^2}$$

Example 14.3: Calculate the ultimate bearing capacity of a strip footing, 1 m wide, in a soil for $\gamma = 18$ kN/m^3, $c = 20$ kN/m^2, and $\phi = 20°$, at a depth of 1 m. Use Rankine's and Bell's approaches.

$\phi = 20°$

In Rankine's approach, cohesion is not considered.

$$q_{ult} = \frac{1}{2}\gamma b N_\gamma + \gamma D_f N_q$$

where

$$N_\gamma = \frac{1}{2}\sqrt{N_\phi}(N_\phi^2 - 1) \text{ and } N_q = N_\phi^2$$

$$N_\phi = \tan^2(45° + \phi/2) = \tan^2 55° = 2.04$$

$$N_\gamma = \frac{1}{2} \times \tan 55°(\tan^4 55° - 1) = 2.256$$

$$N_q = \tan^4 55° = 4.16$$

$\therefore \quad q_{ult} = \frac{1}{2} \times 18 \times 1 \times 2.256 + 18 \times 1 \times 4.16 = \mathbf{95.2 \text{ kN/m}^2}$

In Bell's approach, cohesion is also considered.

$$q_{ult} = cN_c + \frac{1}{2}\gamma b N_\gamma + \gamma D_f N_q$$

where $N_c = 2\sqrt{N_\phi}(N_\phi + 1)$, $N_\gamma = \frac{1}{2}\sqrt{N_\phi}(N_\phi^2 - 1)$, and $N_q = N_\phi^2$

$\therefore \quad N_c = 2\tan 55°(\tan^2 55° + 1) = 8.682$

$$N_\gamma = \frac{1}{2}\tan 55°(\tan^4 55° - 1) = 2.256$$

$$N_q = \tan^4 55° = 4.16$$

$\therefore \quad q_{ult} = 20 \times 8.682 + \frac{1}{2} \times 18 \times 1 \times 2.256 + 18 \times 1 \times 4.16$

$$= \mathbf{268.8 \text{ kN/m}^2}$$

Example 14.4: A strip footing, 1.5 m wide, rests on the surface of a dry cohesionless soil having $\phi = 20°$ and $\gamma = 1.90$ t/m^3. If the water table rises temporarily to the surface due to flooding, calculate the percentage reduction in the ultimate bearing capacity of the soil. Assume $N_\gamma = 5.0$.

(S.V.U.—B.E., (Part-Time)—Apr., 1982)

$\phi = 20°$ $\quad N_\gamma = 5.0 \quad b = 1.5$ m $\quad D_f = 0$
Dry cohesionless soil, $\therefore c = 0$

$$q_{ult} = cN_c + \frac{1}{2}\gamma b N_\gamma + \gamma D_f N_q$$

$$= \frac{1}{2}\gamma b N_\gamma \text{ in this case.}$$

$$= \frac{1}{2} \times 1.90 \times 1.5 \times 5.0 = 7.13 \text{ t/m}^2$$

If the water table rises temporarily to the surface due to flooding, reduction factors R_w and R'_w shall be applied as the maximum values for the N_γ and N_q terms respectively.

In this case, $R_w = 0.5$ is applied for N_γ-term.

$$\therefore \quad q_{ult} = \frac{1}{2}\gamma b N_\gamma \cdot R_w = \frac{1}{2} \times 1.90 \times 1.5 \times 5.0 \times 0.5$$

$$= 3.56 \text{ t/m}^2$$

The percentage reduction in the ultimate bearing capacity is thus 50 due to flooding and consequent complete submergence.

(*Note:* γ_{sat} is assumed to be γ itself here, and $\gamma' \approx \frac{1}{2}\gamma_{sat}$)

Example 14.5: A continuous footing of width 2.5 m rests 1.5 m below the ground surface in clay. The unconfined compressive strength of the clay is 150 kN/m². Calculate the ultimate bearing capacity of the footing. Assume unit weight of soil is 16 kN/m³.

(S.V.U.—B.E., (R.R.)—May, 1969)

Continuous footing $b = 2.5$ m $D_f = 1.5$ m
Pure clay.
$\phi = 0°$ $q_u = 150$ kN/m² $\gamma = 16$ kN/m³

$$c = \frac{q_u}{2} = 75 \text{ kN/m}^2$$

For $\phi = 0°$, Terzaghi's factors are: $N_\gamma = 0$, $N_q = 1$, and $N_c = 5.7$.

$$q_{ult} = cN_c + \frac{1}{2}\gamma b N_\gamma + \gamma D_f N_q$$

$$= cN_c + \gamma D_f N_q, \text{ in this case.}$$

$$\therefore \quad q_{ult} = 5.7 \times 75 + 16 \times 1.5 \times 1$$

$$= 451.5 \text{ kN/m}^2 \approx \mathbf{450 \text{ kN/m}^2}$$

Geotechnical Engineering

Example 14.6: Compute the safe bearing capacity of a continuous footing 1.8 m wide, and located at a depth of 1.2 m below ground level in a soil with unit weight $\gamma = 20$ kN/m³, $c = 20$ kN/m², and $\phi = 20°$. Assume a factor of safety of 2.5. Terzaghi's bearing capacity factors for $\phi = 20°$ are $N_c = 17.7$, $N_q = 7.4$, and $N_\gamma = 5.0$, what is the permissible load per metre run of the footing?

$b = 1.8$ m   continuous footing $D_f = 1.2$ m

$\gamma = 20$ kN/m³ $c = 20$ kN/m²

$\phi = 20°$ $N_c = 17.7$

$N_q = 7.4$ $N_\gamma = 5.0$ $\eta = 2.5$

$$q_{ult} = cN_c + \frac{1}{2}\gamma b N_\gamma + \gamma D_f N_q$$

$$= 20 \times 17.7 + \frac{1}{2} \times 20 \times 1.8 \times 5.0 + 20 \times 1.2 \times 7.4$$

$$= 621.6 \text{ kN/m}^2$$

$$q_{net\ ult} = q_{ult} - \gamma D_f$$

$$= 621.6 - 20 \times 1.2$$

$$= 597.6 \text{ kN/m}^2$$

$$q_{net\ safe} = \frac{q_{net\ ult}}{\eta} = \frac{597.6}{2.5} = 239 \text{ kN/m}^2$$

$$q_{safe} = q_{net\ safe} + \gamma D_f$$

$$= 239 + 20 \times 1.2$$

$$= 263 \text{ kN/m}^2$$

Permissible load per metre run of the wall = 263×1.8 kN

$$= 473.5 \text{ kN}$$

Example 14.7: What is the ultimate bearing capacity of a square footing resting on the surface of a saturated clay of unconfined compressive strength of 1 kg/cm² ?
(S.V.U.—Four-year B.Tech.—Apr., 1983)

Square footing.
  Saturated clay,
    $\phi = 0°$ $D_f = 0$.
Terzaghi's factors for $\phi = 0°$ are: $N_c = 5.7$, $N_q = 1$, and $N_\gamma = 0$.

$$q_u = 1 \text{ kg/cm}^2$$

$$\therefore \quad c = \frac{1}{2} q_u = 0.5 \text{ kg/cm}^2$$

$$q_{ult} = 1.3 \, cN_c = 1.3 \times 0.5 \times 5.7 = 3.7 \text{ kg/ cm}^2$$

$$\therefore \quad q_{ult} = 37 \text{ kg/cm}^2$$

Example 14.8: Determine the ultimate bearing capacity of a square footing of 1.5 m size, at a depth of 1.5 m, in a pure clay with an unconfined strength of 1.5 kg/cm². $\phi = 0°$ and $\gamma = 1.7$ g/cm³.

(S.V.U.—Four-year B.Tech.,—Sept., 1983)

Square footing $b = 1.5$ m = 150 cm $\quad D_f = 1.5$ m = 150 cm
Pure clay $\phi = 0°$ $\quad q_u = 1.5$ kg/cm², $\gamma = 1.7$ g/cm³

$$c = \frac{q_u}{2} = 0.75 \text{ kg/cm}^2$$

Terzaghi's factors for $\phi = 0°$ are $N_c = 5.7$, $N_q = 1$, and $N_\gamma = 0$.

$$\therefore \quad q_{ult} = 1.3 \, c \, N_c + \gamma D_f N_q + 0.4 \gamma b \, N_\gamma$$

$$= 1.3 \, cN_c + \gamma D_f N_q, \text{ in this case}$$

$$q_{ult} = 1.3 \times 0.75 \times 5.7 + \frac{1.7}{1000} \times 150 \times 1$$

$$= 5.8 \text{ kg/cm}^2$$

$$\therefore \quad q_{ult} = 58 \text{ t/m}^2$$

Example 14.9: A square footing, 1.8 m × 1.8 m, is placed over loose sand of density 1.6 g/cm³ and at a depth of 0.8 m. The angle of shearing resistance is 30°. $N_c = 30.14$, $N_q = 18.4$, and $N_\gamma = 15.1$. Determine the total load that can be carried by the footing.

(S.V.U.—Four-year B.Tech.,—Apr., 1983)

Square footing $b = 1.8$ m
$$\gamma = 1.6 \text{ g/cm}^3, \quad c = 0, \quad \phi = 30°, \quad D_f = 0.8 \text{ m}$$
$$N_c = 30.14, \quad N_q = 18.4, \quad N_\gamma = 15.1$$
$$q_{ult} = 1.3 \, cN_c + 0.4 \gamma b \, N_\gamma + \gamma D_f N_q$$
$$= 0.4 \, \gamma b \, N_\gamma + \gamma D_f N_q, \text{ in this ca}$$

∴ $q_{ult} = 0.4 \times 1.6 \times 1.8 \times 15.1 + 1.6 \times 0.8 \times 18.4$

$= 17.4 + 19.9$

$= 37.3 \text{ t/m}^2$

The ultimate load that can be carried by the footing

$= q_{ult} \times \text{Area}$

$= 37.3 \times 1.8 \times 1.8 \text{ t}$

$= \mathbf{120.85 \text{ t}}$

Example 14.10: Compute the safe bearing capacity of a square footing 1.5 m × 1.5 m, located at a depth of 1 m below the ground level in a soil of average density 20 kN/m³. $\phi = 20°$, $N_c = 17.7$, $N_q = 7.4$, and $N_\gamma = 5.0$. Assume a suitable factor of safety and that the water table is very deep. Also compute the reduction in safe bearing capacity of the footing if the water table rises to the ground level.

(S.V.U.—B.Tech., (Part-time)—Sept., 1983)

$b = 1.5$ m Square footing $D_f = 1$ m

$\gamma = 20$ kN/m³ $\phi = 20°$ $N_c = 17.7$, $N_q = 7.4$, and $N_\gamma = 5.0$

Assume $c = 0$ and $\eta = 3$

$q_{ult} = 1.3 \, c \, N_c + 0.4 \, \gamma \, b \, N_\gamma + \gamma D_f N_q$

$= 0.4 \, \gamma \, b \, N_\gamma + \gamma D_f N_q$, in this case.

$= 0.4 \times 20 \times 1.5 \times 5.0 + 20 \times 1 \times 7.4$

$= 60 + 148 = 208 \text{ kN/m}^2$

$q_{\text{net ult}} = q_{ult} - \gamma D_f = 208 - 20 \times 1 = 188 \text{ kN/m}^2$

$q_{\text{safe}} = \dfrac{q_{\text{net ult}}}{\eta} + \gamma D_f = \dfrac{188}{3} + 20 \times 1 = 83 \text{ kN/m}^2$

If the water table rises to the ground level,

$R_w = 0.5 = R'_w$

∴ $q_{ult} = 0.4 \, \gamma \, b \, N_\gamma \cdot R_w + \gamma D_f N_q \cdot R'_w$

$= 0.4 \times 20 \times 1.5 \times 5.0 \times 0.5 + 20 \times 1 \times 7.4 \times 0.5$

$= 30 + 74 = 104 \text{ kN/m}^2$

$q_{\text{net ult}} = q_{ult} - \gamma' D_f = 104 - 10 \times 1 = 94 \text{ kN/m}^2$

$q_{\text{safe}} = \dfrac{q_{\text{net ult}}}{\eta} + \gamma' D_f = \dfrac{94}{3} + 10 \times 1 = 41 \text{ kN/m}^2$

Percentage reduction in safe bearing capacity

$$= \frac{42}{83} \times 100 \approx 50$$

Example 14.11: A foundation, 2.0 m square is installed 1.2 m below the surface of a uniform sandy gravel having a density of 19.2 kN/m³, above the water table and a submerged density of 10.1 kN/m³. The strength parameters with respect to effective stress are $c' = 0$ and $\phi' = 30°$. Find the gross ultimate bearing capacity for the following conditions:
  (*i*) Water table is well below the base of the foundation (*i.e.*, the whole of the rupture zone is above the water table);
  (*ii*) Water table rises to the level of the base of the foundation; and
  (*iii*) the water table rises to ground level.
    (For $\phi = 30°$, Terzaghi gives $N_q = 22$ and $N_\gamma = 20$)
    (S.V.U.—B.Tech., (Part-time)—Sept., 1982)

Square $b = 2$ m $D_f = 1.2$ m $c' = 0$ $\phi' = 30°$
$\gamma = 19.2$ kN/m³ $\gamma' = 10.1$ kN/m³ $N_q = 22$ $N_\gamma = 20$
  (*i*) Water table is well below the base of the foundation:
    $q_{ult} = 1.3 c N_c + 0.4 \gamma b N_\gamma + \gamma D_f N_q = 0.4 \gamma b N_\gamma + \gamma D_f N_q$, in this case.

  or $q_{ult} = 0.4 \times 19.2 \times 2 \times 20 + 19.2 \times 1.2 \times 22 = $ **814 kN/m²**
  (*ii*) Water table rises to the level of the base of the foundation:
    $q_{ult} = 0.4 \gamma' b N_\gamma + \gamma D_f N_q$

    $= 0.4 \times 10.1 \times 2 \times 20 + 19.2 \times 1.2 \times 22 = $ **668 kN/m²**
  (*iii*) Water table rises to the ground level:
    $q_{ult} = 0.4 \gamma' b N_\gamma + \gamma' D_f N_q$

    $= 0.4 \times 10.1 \times 2 \times 20 + 10.1 \times 1.2 \times 22 = $ **428 kN/m²**
Thus, as the water table rises, there is about 20% to 50% decrease in the ultimate bearing capacity.

Example 14.12: The footing of a column is 2.25 m square and is founded at a depth of 1 m on a cohesive soil of unit weight 17.5 kN/m³. What is the safe load for this footing if cohesion = 30 kN/m²; angle of internal friction is zero and factor of safety is 3. Terzaghi's factors for $\phi = 0°$ are $N_c = 5.7$, $N_q = 1$, and $N_\gamma = 0$.
    (S.V.U.—B.E., (R.R.)—Feb., 1976)

Square $b = 2.25$ m $D_f = 1$ m $\gamma = 17.5$ kN/m³
$q_{ult} = 1.3 c N_c + 0.4 \gamma b N_\gamma + \gamma D_f N_q$

$= 1.3 c N_c + \gamma D_f N_q$, in this case since $N_\gamma = 0$

$$\therefore \quad q_{ult} = 1.3 \times 30 \times 5.7 + 17.5 \times 1 \times 1$$

$$= 239.8 \text{ kN/m}^2$$

$$q_{net\,ult} = q_{ult} - \gamma D_f$$

$$= 239.8 - 17.5 = 222.3 \text{ kN/m}^2$$

$$q_{safe} = \frac{q_{net\,ult}}{\eta} + \gamma D_f$$

$$= \frac{222.3}{3} + 17.5 = \mathbf{91.6 \text{ kN/m}^2}$$

Safe load on the footing:

$$= q_{safe} \times \text{Area}$$

$$= 91.6 \times 2.25 \times 2.25 \approx \mathbf{463 \text{ kN}}$$

Example 14.13: What is the ultimate bearing capacity of a circular footing of 1 m diameter resting on the surface of a saturated clay of unconfined compression strength of 100 kN/m² ? What is the safe value if the factor of safety is 3?

Diameter, $D = 1 \quad D_f = 0$

$$q_u = 1.3\,c\,N_c + 0.3\,\gamma D\,N_\gamma + \gamma D_f N_q$$

For saturated clay,

$$\phi = 0°$$

$$\therefore \quad N_c = 5.7 \quad N_q = 1 \quad N_\gamma = 0$$

Also

$$c = \frac{1}{2} q_u = 50 \text{ kN/m}^2$$

$$\therefore \quad q_{ult} = 1.3 \times 50 \times 5.7$$

$$= 370.5 \text{ kN/m}^2$$

$$q_{safe} = \frac{q_{ult}}{\eta}, \text{ in this case,}$$

since

$$N_q = 1$$

$$\therefore \quad q_{safe} = \frac{370.5}{3} = \mathbf{123.5 \text{ kN/m}^2}$$

Example 14.14: A circular footing is resting on a stiff saturated clay with $q_u = 250$ kN/m². The depth of foundation is 2 m. Determine the diameter of the footing if the column load is 600 kN. Assume a factor of safety as 2.5. The bulk unit weight of the soil is 20 kN/m³.

(S.V.U.—Four-year B.Tech.—Dec., 1982)

Circular footing: $\phi = 0°$, $N_c = 5.7$, $N_q = 1$, $N_\gamma = 0$

$q_u = 250$ kN/m² $\quad c = \frac{1}{2} q_u = 125$ kN/m² $\quad D_f = 2$ m

Column load = 600 kN $\quad \eta = 2.5 \quad \gamma = 20$ kN/m³

$$q_{ult} = 1.3\, c\, N_c + 0.3\, \gamma D\, N_\gamma + \gamma D_f N_q$$

$$= 1.3\, c\, N_c + \gamma D_f N_q, \text{ in this case.}$$

$\therefore \quad q_{ult} = 1.3 \times 125 \times 5.7 + 20 \times 2 \times 1 = 966$ kN/m²

$q_{net\ ult} = q_{ult} - \gamma D_f = 966 - 20 \times 2$

$\qquad = 926$ kN/m²

$$q_{safe} = \frac{q_{net\ ult}}{\eta} + \gamma D_f$$

$$= \frac{926}{2.5} + 20 \times 2 = 410 \text{ kN/m}^2$$

Safe load on the column

$$= q_{safe} \times \text{Area} = 600 \text{ kN}$$

$\therefore \quad 600 = \dfrac{410 \times \pi d^2}{4}$

$\therefore \quad d = \sqrt{\dfrac{4 \times 600}{410}}$ m $= 1.365$ m

A diameter of **1.5 m** may be adopted in this case.

Example 14.15: A column carries a load of 1000 kN. The soil is a dry sand weighing 19 kN/m³ and having an angle of internal friction of 40°. A minimum factor of safety of 2.5 is required and Terzaghi factors are required to be used. ($N_\gamma = 42$ and $N_q = 21$).

(*i*) Find the size of a square footing, if placed at the ground surface; and,
(*ii*) find the size of a square footing required if it is placed at 1 m below ground surface with water table at ground surface. Assume $\gamma_{sat} = 21$ kN/m³.

(i) At ground surface:
$$\phi = 40° \text{ Dry sand}, N = 42 \ N_q = 21$$
Let the size of the footing be b m.
$$q_{ult} = 0.4\, b \times 19 \times 42$$
Since $D_f = 0$, $q_{net\,ult} = q_{ult}$
$$\therefore \quad q_{safe} = q_{net\,ult} = \frac{q_{ult}}{\eta} = \frac{0.4\, b \times 19 \times 42}{2.5} = 128\, b \text{ kN/m}^2$$
$$Q_{safe} = 128\, b \times b^2 = 1000$$
$$\therefore \quad b = \sqrt[3]{\frac{1000}{128}} \text{ m} = \mathbf{2 \text{ m}}$$

(ii) At 1 m below ground surface with water table at ground surface:
$$\gamma' = \gamma_{sat} - \gamma_w = (21 - 10) = 11 \text{ kN/m}^3$$
$$N = 42,\ N_\gamma = 21$$
$$q_{ult} = 0.4 \times \gamma b\, N_\gamma + \gamma D_f N_q$$
$$q_{ult} = 0.4\, b \times 11 \times 42 + 11 \times 1 \times 21 = (185\, b + 231) \text{ kN/m}^2$$
$$q_{net\,ult} = q_{ult} - \gamma D_f = 185\, b + 231 - 11 \times 1 = (185\, b + 220) \text{ kN/m}^2$$
$$q_{safe} = \frac{q_{net\,ult}}{\eta} + \gamma D_f = \left[\frac{(185\, b + 200)}{2.5} + 11\right] \text{ kN/m}^2$$
$$\therefore \quad Q_{safe} = \left[\frac{185\, b + 220}{2.5} + 11\right] b^2 = 1000$$

Solving by trial and error, $b = 2$ m.
$$b = \mathbf{2 \text{ m}.}$$

Example 14.16: What is the ultimate bearing capacity of a rectangular footing, 1 m × 2 m, on the surface of a saturated clay of unconfined compression strength of 100 kN/m² ?

Rectangular footing:
$b = 1$ m $L = 2$ m $D_f = 0$ $q_u = 100$ kN/m²
$$\therefore \quad c = \frac{1}{2} q_u = 50 \text{ kN/m}^2$$

Skempton's equation:
$$q_{net\,ult} = c \cdot N_c, \text{ where } N_c = 5\left(1 + 0.2\frac{b}{L}\right)(1 + 0.2\, D_f/b) \text{ for } D_f/b \leq 2.5)$$

$\therefore \quad N_c = 5\left(1 + 0.2 \times \dfrac{1}{2}\right) = 5.5$, since $D_f = 0$.

$\therefore \quad q_{\text{net ult}} = 5.5 \times 50 = 275 \text{ kN/m}^2$

Since $D_f = 0$, $q_{\text{ult}} = q_{\text{net ult}} = 275 \text{ kN/m}^2$

Example 14.17: What is the safe bearing capacity of a rectangular footing, 1 m × 2 m, placed at a depth of 2 m in a saturated clay having unit weight of 20 kN/m³ and unconfined compression strength of 100 kN/m²? Assume a factor of safety of 2.5.

Rectangular footing:

$b = 1 \text{ m} \qquad L = 2 \text{ m} \qquad D_f = 2 \text{ m} \qquad q_u = 100 \text{ kN/m}^2 \qquad \gamma = 20 \text{ kN/m}^3$

$D_f/b = \dfrac{2}{1} = 2 \qquad b/L = \dfrac{1}{2} \qquad c = \dfrac{1}{2} \qquad q_u = 50 \text{ kN/m}^3$

Skempton's equation:

$q_{\text{net ult}} = c \cdot N_c$, where $N_c = 5\left(1 + 0.2\dfrac{b}{L}\right)\left(1 + 0.2\dfrac{D_f}{b}\right)$ for $D_f/b \le 2.5$

Since $D_f/b = 2 < 2.5$,

$N_c = 5\left(1 + 0.2 \times \dfrac{1}{2}\right)(1 + 0.2 \times 2/1) = 7.7$

$\therefore \quad q_{\text{net ult}} = 7.7 \times 50 = 385 \text{ kN/m}^2$

$q_{\text{net safe}} = \dfrac{q_{\text{net ult}}}{\eta} = \dfrac{385}{2.5} = 154 \text{ kN/m}^2$

$q_{\text{safe}} = q_{\text{net safe}} + \gamma D_f = 154 + 20 \times 2 = 194 \text{ kN/m}^2$

Example 14.18: A steam turbine with base 6 m × 3.6 m weighs 10,000 kN. It is to be placed on a clay soil with $c = 135$ kN/m². Find the size of the foundation required if the factor of safety is to be 3. The foundation is to be 60 cm below ground surface.

Skempton's equation:

$q_{\text{net ult}} = 5c\left(1 + 0.2\dfrac{b}{L}\right)\left(1 + 0.2\dfrac{D_f}{b}\right)$ for $D_f/b \le 2.5$.

$D_f = 0.6 \text{ m}$

For $\phi = 0°$, $N_\gamma = 0$ and $N_q = 1$. Assume $\gamma = 18 \text{ kN/m}^3$.

Adopt $b/L = 0.6$, same as that for the turbine base.

$D_f/b = 0.6/b$

$$\text{Area, } A = bL = \frac{b^2}{0.6} = \left(\frac{5b^2}{3}\right) m^2$$

$$\therefore \quad q_{net\,ult} = 5 \times 135\,(1 + 0.2 \times 0.6)\left(1 + \frac{0.2 \times 0.6}{b}\right)$$

$$= 756\left(1 + \frac{0.12}{b}\right) kN/m^2$$

$$q_{safe} = \frac{q_{net\,ult}}{\eta} + \gamma D_f = \left[\frac{756\left(1 + \frac{0.12}{b}\right)}{3} + 18 \times 0.6\right] kN/m^2$$

$$Q_{safe} = q_{safe} \times A = \frac{5b^2}{3}\left[756\frac{\left(1 + \frac{0.12}{b}\right)}{3} + 10.8\right] kN$$

Equating Q_{safe} to 10,000, we have

$$420\,b^2\left(1 + \frac{0.12}{b}\right) + 18\,b^2 = 10,000$$

Solving for b,

$b = 4.72$ m, say 4.80 m.

($D_f/b < 2.5$ is satisfied)

$L = 4.8/0.6 = 8.0$ m

Hence, the size of the foundation required is **4.8 m × 8.0 m.**

Example 14.19: Calculate the ultimate bearing capacity, according to the Brinch Hansen's method, of a rectangular footing 2 m × 3 m, at a depth of 1 m in a soil for which $\gamma = 18$ kN/m³, $c = 20$ kN/m², and $\phi = 20°$. The ground water table is lower than 3 m from the surface. The total vertical load is 1350 kN and the total horizontal load is 75 kN at the base of the footing. Hansen's factors for $\phi = 20°$ are $N_c = 14.83$, $N_q = 6.40$, and $N_\gamma = 3.54$. Determine also the factor of safety.

Rectangular footing:

$\phi = 20°$, $N_c = 14.83$, $N_q = 6.40$, $N = 3.54$, $D_f = 1$ m

(Hansen's factors)

$c = 20$ kN/m² $q = \gamma D_f = 18 \times 1 = 18$ kN/m² $\gamma = 18$ kN/m³

$b = 2$ m $L = 3$ m $A = 6$ m² $H = 75$ kN $V = 1350$ kN

Hansen's formula:
$$q_{ult} = cN_c s_c d_c i_c + qN_q s_q d_q i_q + \frac{1}{2}\gamma b N_\gamma s_\gamma d_\gamma i_\gamma$$

Shape factors:
$$s_c = 1 + 0.2\, b/L = 1 + 0.2 \times \frac{2}{3} = 1.133$$

$$s_q = 1 + 0.2\, b/L = 1.133$$

$$s_\gamma = 1 - 0.4\, b/L = 1 - 0.4 \times \frac{2}{3} = 0.733$$

Depth factors:
$$d_c = 1 + 0.35\, D_f/b = 1 + 0.35 \times \frac{1}{2} = 1.175$$

$$d_q = 1 + 0.35\, D_f/b = 1.175$$

$$d_\gamma = 1.0$$

Inclination factors (Revised):
$$i_c = i_q = \left(1 - \frac{H}{V + A\, c\, \cot\phi}\right)^2$$

$$= \left(1 - \frac{75}{1350 + 6 \times 20 \times \cot 20°}\right)^2 = 0.913$$

$$i_\gamma = (i_q)^2 = 0.913^2 = 0.833$$

$$\therefore \quad q_{ult} = 20 \times 14.83 \times 1.133 \times 1.175 \times 0.913 + 18 \times 6.40 \times 1.175 \times 0.913$$

$$+ \frac{1}{2} \times 18 \times 2 \times 3.54 \times 0.733 \times 0.83$$

$$= 360.5 + 140.0 + 38.9 = \mathbf{539.5\ kN/m^2}$$

Vertical load that can be borne,
$$Q_{ult} = q_{ult} \times \text{Area} = 539.4 \times 6$$

$$= 3236\ \text{kN}$$

Factor of safety $= 3236/1350 \approx \mathbf{2.4.}$

Example 14.20: What is the ultimate bearing capacity of a footing resting on a uniform sand of porosity 40% and specific gravity 2.6, if $\phi = 30°$ (Hansen's factors: $N_c = 30.4, N_q = 18.4, N_\gamma = 18.08$ at a depth of 1.5 m under the following conditions:

(*i*) Size 2 m × 3 m, G.W.L. at 8 m below natural ground level; and
(*ii*) Size 2 m × 3 m, G.W.L. at 1.5 m below natural ground level.

(S.V.U.—Four-year B.Tech.—Dec., 1982)

Hansen's formula:

$$\phi = 30°; \quad N_c = 30.14; \quad N_q = 18.4; \quad N_\gamma = 18.08$$

$$q_{ult} = cN_c s_c d_c i_c + qN_q s_q d_q i_q + \frac{1}{2}\gamma b N_\gamma s_\gamma i_\gamma$$

Since the soil is sand, $c = 0$, and the first term vanishes.
$n = 40\%, \quad G = 2.60 \quad D_f = 1.5\ m$

$$e = \frac{n}{(1-n)} = \frac{0.4}{0.6} = 0.67$$

$$\gamma_d = \frac{G\gamma_w}{(1+e)} = \frac{2.60 \times 9.81}{1.67} = 15.30\ kN/m^3$$

$$\gamma' = \frac{(G-1)\gamma_w}{(1+e)} = \frac{1.60 \times 9.81}{1.67} = 9.42\ kN/m^3$$

Since $q = \gamma D_f$,

$$q_{ult} = \gamma D_f N_q s_q d_q i_q + \frac{1}{2}\gamma b N_\gamma s_\gamma i_\gamma$$

Shape factors:

$$s_q = 1 + 0.2\ b/L = 1 + 0.2 \times 2/3 = 1.33$$

$$s_\gamma = 1 - 0.4\ b/L = 1 - 0.4 \times 2/3 = 0.733$$

Depth factors:

$$d_q = 1 + 0.35\ D_f/b = 1 + 0.35 \times \frac{1.5}{2} = 1.263$$

$$d_\gamma = 1.0$$

Inclination factors:

$$i_q = i_\gamma = 1\ \text{for purely vertical loading.}$$

(*i*) G.W.L. at 8 m below natural ground level:

$$\gamma = \gamma_d\ \text{in both terms.}$$

$$\therefore \quad q_{ult} = 15.3 \times 1.5 \times 18.14 \times 1.133 \times 1.263 \times 1$$

$$+ \frac{1}{2} \times 15.3 \times 2 \times 18.08 \times 0.733 \times 1 \times 1$$

$$= 595.74 + 202.76 = \mathbf{798.5\ kN/m^2}$$

(*ii*) G.W.L. at 1.5 m below natural ground level:
$\gamma = \gamma_d$ in the first term and $\gamma = \gamma'$ in the second term.

$$\therefore \quad q_{ult} = 15.3 \times 1.5 \times 18.14 \times 1.133 \times 1.263 \times 1$$

$$+ \frac{1}{2} \times 9.42 \times 2 \times 18.08 \times 0.733 \times 1 \times 1$$

$$= 595.74 + 124.84 = \mathbf{720.6 \ kN/m^2}$$

Thus, there is about 10% decrease in bearing capacity as the water table rises to the level of the base of the footing.

Example 14.21: A foundation 2 m × 3 m is resting at a depth of 1 m below the ground surface. The soil has a unit cohesion of 10 kN/m², angle of shearing resistance of 30° and unit weight of 20 kN/m³. Find the ultimate bearing capacity using Balla's method.

Balla's equation:

$$q_{ult} = cN_c + qN_q + \frac{1}{2}\gamma b N_\gamma$$

$$D_f/b = \frac{1}{2} = 0.5$$

$$\frac{c}{\gamma b} = \frac{10}{20 \times 2} = 0.25$$

For $\phi = 30°$ and $\frac{c}{\gamma b} = 0.25$

$$\rho = 1.9 \text{ from the charts for } D_f/b = 0.5$$

For $\phi = 30°$ and $\rho = 1.9$, from the relevant charts, Balla's factors are:

$$N_c = 37$$
$$N_q = 25$$
$$N_\gamma = 64$$

Hence,

$$q_{ult} = 10 \times 37 + 20 \times 1 \times 25 + \frac{1}{2} \times 20 \times 2 \times 64$$

$$= 370 + 500 + 1280$$

$$= \mathbf{2{,}750 \ kN/m^2}$$

(**Note:** Strictly speaking, Balla's method is applicable only for continuous footings).

Example 14.22: A plate load test was conducted on a uniform deposit of sand and the following data were obtained:

Pressure kN/m²	50	100	200	300	400	500	600
Settlement mm	1.5	2.0	4.0	7.5	12.5	20.0	40.0

The size of the plate was 7.50 mm × 750 mm and that of the pit 3.75 m × 3.75 m × 1.5 m.

(i) Plot the pressure-settlement curve and determine the failure stress.

(ii) A square footing, 2 m × 2 m, is to be founded at 1.5 m depth in this soil. Assuming the factor of safety against shear failure as 3 and the maximum permissible settlement as 40 mm, determine the allowable bearing pressure.

(iii) Design of footing for a load of 2,000 kN, if the water table is at a great depth.

(i) The pressure-settlement curve is shown in Fig. 14.23

The failure point is obtained as the point corresponding to the intersection of the initial and final tangents. In this case, the failure stress is 500 kN/m².

$$\therefore \quad q_{ult} = 500 \text{ kN/m}^2$$

(ii) The value of q_{ult} here is given by $\frac{1}{2}\gamma b_p N_\gamma$

b_p, the size of test plate = 0.75 m

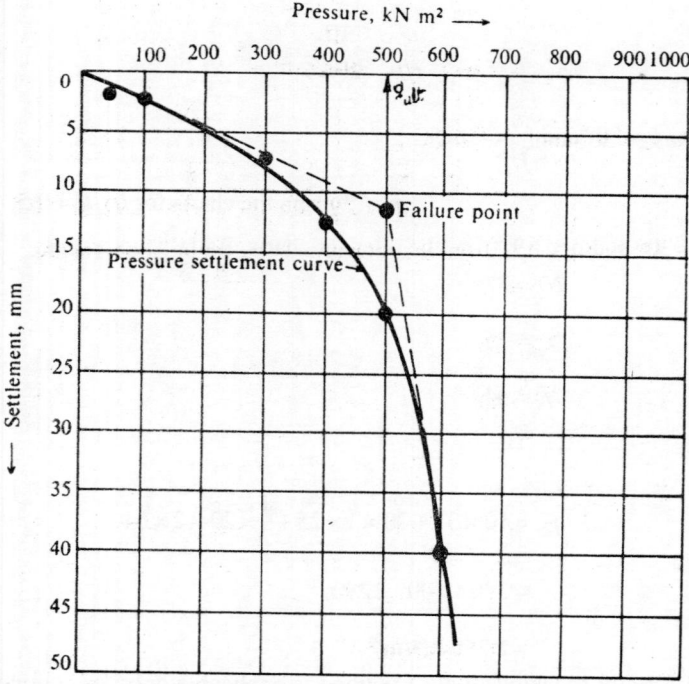

Fig. 14.23 Pressure-settlement curve (Ex. 14.22)

Assuming $\gamma = 20$ kN/m³,

$$500 = \frac{1}{2} \times 20 \times 0.75 \, N_\gamma$$

$$\therefore \quad N_\gamma = 500/7.5 \approx 6.7$$

$$\phi = 38°$$

$\therefore \quad N_q \approx 50$ from Terzaghi's charts.

For square footing of size 2 m and $D_f = 1.5$ m,

$$q_{\text{net ult}} = 0.4 \, \gamma b \, N_\gamma + \gamma D_f (N_q - 1)$$

$$= 0.4 \times 20 \times 2 \times 67 + 20 \times 1.5 \times 49$$

$$= 2,542 \text{ kN/m}^2$$

$$q_{\text{safe}} = \frac{2,542}{3} \approx 847 \text{ kN/m}^2 \text{ (for failure against shear)}$$

From settlement consideration,

$$\frac{S_p}{S} = \left[b_p \frac{(b + 0.30)}{b(b_p + 0.30)} \right]^2$$

$$S_p = S \left[\frac{b_p (b + 0.30)}{b(b_p + 0.30)} \right]^2$$

$$= 40 \left[\frac{0.75 \, (2 + 0.30)}{2.00 \, (0.75 + 0.30)} \right]^2$$

$$= 40 \left(\frac{0.75 \times 2.3}{2 \times 1.05} \right)^2 \text{ mm}$$

$$= 27 \text{ mm}$$

Pressure for a settlement of 27 mm for the plate (from Fig. 14.23) = 550 kN/m².
Allowable bearing pressure is the smaller of the values from the two criteria = **550 kN/m²**.

(*iii*) Design load = 2,000 kN
From Part (*ii*), it is known that a 2 m square footing can carry a load of $2 \times 2 \times 550 = 2,200$ kN.
Therefore, a **2 m** square footing placed at a depth of **1.5 m** is adequate for the design load.

Example 14.23: A loading test was conducted with a 300 mm square plate at depth of 1 m below the ground surface in a pure clay deposit. The water table is located at a depth of 4 m below the ground level. Failure occurred at a load of 45 kN.

692 Geotechnical Engineering

What is the safe bearing capacity of a 1.5 m wide strip footing at 1.5 m depth in the same soil? Assume $\gamma = 18$ kN/m³ above the water table and a factor of safety of 2.5.

The water table does not affect the bearing capacity in both cases.
Test plate:

$$q_{ult} = \frac{\text{Failure load}}{\text{Area of plate}} = \frac{45}{0.3 \times 0.3} = 500 \text{ kN/m}^2$$

$$\gamma = 18 \text{ kN/m}^3$$

For $\phi = 0°$, Terzaghi's factors are $N_c = 5.7$, $N_q = 1$, and $N_\gamma = 0$

$$\therefore \quad q_{ult} = 1.3 \, c \, N_c + \gamma D_f N_q$$

$$= 1.3 \times 5.7 c + 18 \times 1.0$$

$$= 7.4 \, c + 18 \text{ (kN/m}^2\text{, if } c \text{ is expressed in kN/m}^2\text{)}$$

$\therefore \quad 7.4 c + 18 = 500$

$$\therefore \quad c = \frac{482}{7.4} \approx 65 \text{ kN/m}^2$$

Strip footing:

$$q_{ult} = c \, N_c + \gamma D_f N_q$$

$$= 5.7c + \gamma D_f \text{ in this case.}$$

$$q_{net\ ult} = q_{ult} - \gamma D_f = 5.7c$$

$$\therefore \quad q_{net\ ult} = 5.7 \times 65 = 370.5$$

$$q_{net\ safe} = \frac{370.5}{2.5} \approx 148 \text{ kN/m}^2$$

$$q_{safe} = q_{net\ safe} + \gamma D_f = 148 + 18 \times 1.5$$

$$= 175 \text{ kN/m}^2$$

However, the net safe bearing capacity is used in the design of footings in clay.

Example 14.24: A continuous wall footing, 1 m wide, rests on sand, the water table lying at a great depth. The corrected N-value for the sand is 10. Determine the load which the footing can support if the factor of safety against bearing capacity failure is 3 and the settlement is not to exceed 40 mm.

Continuous footing:

$$b = 1 \text{ m}$$

Let us assume a minimum depth, D_f, of 0.5 m.

Assume $\gamma = 18$ kN/m^3
For $N = 10$, $\phi = 30°$ (Fig. 14.18 or 14.19, after Peck, Hanson, and Thornburn)
Assuming general shear failure,
for $\phi = 30°$, Terzaghi's factors are: $N_q = 22.5$ and $N_\gamma = 19.7$

$$\therefore \quad q_{net\,ult} = \frac{1}{2}\gamma b N_\gamma + \gamma D_f (N_q - 1)$$

$$= \frac{1}{2} \times 18 \times 1 \times 19.7 + 18 \times 0.5 \times 21.5 \approx 370 \text{ kN/m}^2$$

$$q_{net\,safe} = \frac{q_{net\,ult}}{\eta} = \frac{370}{3} \approx 123 \text{ kN/m}^2$$

For $N = 10$ and $b = 1$m, and for a permissible settlement of 40 mm, the allowable soil pressure = 120 kN/m^2 (From Fig. 14.22, after Peck, Hanson, and Thornburn)
∴ Allowable bearing pressure ≈ 120 kN/m^2
(i.e., shear failure governs the design).
Load which the footing can support = 120×1 kN/m = **120 kN/metre run.**

Example 14.25: What load can a 2 m square column carry in a dense sand ($\gamma = 20$ kN/m^3 and $\phi = 36°$) at a depth of 1 m, if the settlement is not to exceed 30 mm? Assume a factor of safety of 3 against shear and that the ground water table is at a great depth.

$\gamma = 20$ kN/m^3 $D_f = 1$ m $b = 2$ m (Square footing) $\phi = 36°$
For $\phi = 36°$, $N = 30$ (from Fig. 14.18, after Peck, Hanson and Thornburn)
For $N = 30$, $b = 2$ m permissible settlement = 40 mm,
Allowable soil pressure = 500 kN/m^2. (from Fig. 14.21 after Terzaghi and Peck)
For a permissible settlement of 30 mm,
Allowable pressure $= \dfrac{500 \times 30}{40} = 375$ kN/m^2

For $\phi = 36°$, $N_q = 43$ and $N_\gamma = 45$

$$\therefore \quad q_{net\,ult} = 0.4 \gamma b N_\gamma + \gamma D_f (N_q - 1)$$

$$= 0.4 \times 20 \times 2 \times 46 + 20 \times 1 = 42$$

$$= 736 + 840 = 1576 \text{ kN/m}^2$$

$$q_{safe} = \frac{1576}{3} = 525 \text{ kN/m}^2$$

∴ Settlement governs the design, and

$$q_{allowable} = 375 \text{ kN/m}^2$$

694   *Geotechnical Engineering*

Safe load for the column $= 375 \times 2 \times 2$

$$= 1500 \text{ kN.}$$

Example 14.26: A footing, 2 m square, is founded at a depth of 1.5 m in a sand deposit, for which the corrected value of N is 27. The water table is at a depth of 2 m from the surface. Determine the net allowable bearing pressure, if the permissible settlement is 40 mm and a factor of safety of 3 is desired against shear failure.

Settlement criterion:

$$N = 27 \quad b = 2 \text{ m} \quad D_f = 1.5 \text{ m}$$

Using Teng's equation for the graphical relationship of Terzaghi and Peck (Fig. 14.20) for a settlement of 25 mm,

$$q_{na} = 34.3 \, (N - 3) \left(\frac{b + 0.3}{2b}\right)^2 R_\gamma \cdot R_d$$

q_{na} is in kN/m² b = Width in m
R_γ is the correction factor for the location of water table.
R_d is the depth factor.

$$z_\gamma = (2 - 1.5) = 0.5 \text{ m} \quad \therefore \quad z_q > D_f$$

$$\therefore \quad \frac{z_\gamma}{b} = 0.5/2 = 0.25$$

$$R_q = 1.0 \text{ (limiting value)}$$

$$R = 0.5\left(1 + \frac{z_\gamma}{b}\right) = 0.5(1 + 0.25) = 0.625$$

$$R_d = 1 + \frac{D_f}{b} = 1 + \frac{1.5}{2} = 1.75$$

$$\therefore \quad q_{na} = 34.3 \times 24 \times \left(\frac{2.3}{4}\right)^2 \times 0.625 \times 1.75 \text{ kN/m}^2 \approx 298 \text{ kN/m}^2.$$

Since this is for a settlement of 25 mm,

q_{na}, for settlement of 40 mm $= 298 \times \dfrac{40}{25} \approx 476 \text{ kN/m}^2.$

Shear failure criterion:
   For a factor of safety of 3, Teng's equation for Terzaghi's bearing capacity equation is:

$$q_{ns} = \frac{1}{18}[2N^2 b \, R_\gamma + 6 \, (100 + N^2) D_f R_q],$$

neglecting $(2/3)\gamma D_f$. (for square footing)

$$= \frac{1}{18}[2\times 27^2 \times 2 \times 0.625 + 6(100+27^2)1.5 \times 1.0] \approx 516 \text{ kN/m}^2$$

Hence settlement governs the design and the allowable bearing pressure is **476 kN/m²**. (**Note:** This is conservative and if Bowles' recommendation is considered, it can be enhanced by 50%; in this case shear failure governs the design, and q_{safe} will be 516 kN/m²).

Example 14.27: Two load tests were conducted at a site—one with a 0.5 m square test plate and the other with a 1.0 m square test plate. For a settlement of 25 mm, the loads were found to be 60 kN and 180 kN, respectively in the two tests. Determine the allowable bearing pressure of the sand and the load which a square footing, 2 m × 2 m, can carry with the settlement not exceeding 25 mm.

$$q_{ult} = mx + \sigma$$

where x = Perimeter-area ratio, P/A

First test

$$x_1 = \frac{P_1}{A_1} = \frac{4 \times 0.5}{0.5 \times 0.5} = 8 \text{ m}^{-1}$$

$$q_1 = \frac{60}{0.5 \times 0.5} = 240 \text{ kN/m}^2$$

Second Test

$$x_2 = \frac{P_2}{A_2} = \frac{4 \times 1}{1 \times 1} = 4 \text{ m}^{-1}$$

$$q_2 = \frac{180}{1 \times 1} = 180 \text{ kN/m}^2$$

$\therefore \quad 240 = 8m + \sigma$ ...(1) $\quad 180 = 4m + \sigma$ ...(2)

Solving Eqs (1) and (2) simultaneously,

$$m = 40 \text{ (kN/m)} \quad \text{and} \quad \sigma = 20 \text{ (kN/m}^2)$$

Prototype footing:

$$x = P/A = \frac{4 \times 2}{2 \times 2} = 2 \text{ m}^{-1}$$

$$\therefore \quad q = mx + \sigma$$

$$= 40 \times 2 + 20$$

$$= 100 \text{ kN/m}^2$$

This is the allowable bearing pressure for a settlement of 25 mm. Load which the footing can carry,

$$Q_s = q_s \times \text{Area} = 100 \times 2 \times 2 = \textbf{400 kN}.$$

SUMMARY OF MAIN POINTS

1. The load-carrying capacity of a foundation to transmit loads from the structure to the foundation soil is termed its 'bearing capacity'. The criteria for the determination of the bearing capacity are avoidance of the risk of shear failure of the soil and of detrimental settlements of the foundation.

    Safe bearing capacity is the ultimate value divided by a suitable factor of safety; the allowable bearing pressure is the smaller safe capacity from the two criteria of shear failure and settlement.
2. The factors on which the bearing capacity depends are the size, shape and depth of the foundation and soil characteristics, including the location of the GWT relative to the foundation.
3. The methods of determination of bearing capacity are selection from building codes, analytical methods, plate load tests, penetration tests, model tests, and laboratory tests.
4. Of the analytical methods, Schleicher's is based on the theory of elasticity, Rankine's and Bell's are based on Rankine's classical theory of earth pressure, Fellenius', Prandtl's, Terzaghi's, Meyerhof's, Skempton's and Brinch Hansen's methods are based on the theory of plasticity.
5. Fellenius' method is based on circular slip surfaces and plastic equilibrium of the soil mass within the slip surface. Terzaghi's theory is based on composite rupture surface (logarithmic spiral and plane) and is the most popular.
6. As the footing is loaded to failure, the soil first reaches 'local shear' and then 'general shear'. Local shear occurs when the soil in a zone becomes plastic. General shear occurs when all the soil along a slip surface is at failure. In loose sand, local shear occurs at a much lower stress than does general shear. In dense sand, local shear occurs at a stress only slightly less than that which causes general shear.
7. Shape effect causes the bearing capacity of isolated square, circular and rectangular footings to be somewhat different from that for continuous footings; in general, the capacity of these will be about 20 to 30% more.

    Skempton's theory relates to the bearing capacity of rectangular footings in pure clay. Brinch Hansen's general bearing capacity equation takes into account the size and shape effects, depth effect and the effect of inclined loads in any kind of soil.
8. Plate load tests and penetration tests are semiempirical approaches, which reflect field experience; as such, theoretical methods should be used in conjunction with these empirical approaches, wherever feasible.

    The 'Standard Penetration Number' has been correlated to ϕ, bearing capacity factors and allowable bearing pressure for specified settlements; this approach is more suited to cohesionless soils.
9. Bearing capacity and settlement of a footing on sand are related both to footing size and depth of embedment and to soil properties. The capacity increases significantly with increase in size of footing and depth of embedment. Settlement increases somewhat with size.

Bearing capacity of a footing on clay is practically independent of size of footing. Even the depth of embedment causes the capacity to increase just by the difference between gross pressure and net pressure. As such, the benefit of the depth of embedment is considered marginal, and only the net allowable capacity itself is used for design purposes.

REFERENCES

1. Abbet: *American Civil Engineering Practice.* Vol. I.
2. Alam Singh and B.C. Punmia: *Soil Mechanics and Foundations,* Standard Book House, Delhi-6, 1970.
3. A.L. Balla: *Bearing Capacity of Foundations,* Proceedings, American Society of Civil Engineers, Vol. 88, 1962.
4. A.L. Bell: *The Lateral Pressure and Resistance of Clay, and the Supporting Power of Clay Foundation.* Proc. Institution of Civil Engineers, London, 1915.
5. Bharat Singh and Shamsher Prakash: *Soil Mechanics and Foundation Engineering,* Nem Chand & Bros., Roorkee, India, 1963.
6. Joseph E. Bowles: *Foundation Analysis and Design,* McGraw Hill Book Co., N.Y., USA, 1968.
7. A. Casagrande and R.E. Fadum: *Application of Soil Mechanics in Designing Building Foundations,* Transactions, ASCE, Vol. 109, 1944.
8. W. Fellenius: *Erdstatische Berechnungen mit Reibung und Kohäsion, Adhäsion, und uncer Annahme Kreiszylindrischer Gleitflächen,* Rev. Ed., Ernst, Berlin, 1939.
9. J. Brinch Hansen: *A General Formula for Bearing Capacity,* Danish Geotechnical Institute Bulletin No. 11, 1961.
10. W.S. Housel: *A Practical Method for the Selection of Foundations Based on Fundamental Research in Soil Mechanics,* University of Michigan Engineering Research Bulletin, No. 13, October, 1929.
11. IS: 1904-1978: Indian Standard Code of Practice for *Structural Safety of Buildings, Shallow Foundations,* Second Revision, 1978.
12. IS: 1888-1971 (Revised): Indian Standard Code of Practice for *Load Test on Soils,* 1971.
13. A.R. Jumikis: *Rupture Surfaces in Dry Sand Under Oblique Loads,* Proceedings, ASCE, Jan., 1956.
14. A.R. Jumikis: *Soil Mechanics,* D. Van Nostrand Co., Princeton NJ, USA, 1962.
15. T.W. Lambe and R.V. Whitman: *Soil Mechanics,* John Wiley & Sons, Inc., NY, USA, 1969.
16. G.G. Meyerhof: *Ultimate Bearing Capacity of Foundations,* Geotechnique, Vol. 2, No. 4, London, 1951.
17. G.G. Meyerhof: *Penetration Tests and Bearing Capacity of Cohesionless Soils,* Proceedings, ASCE, Vol. 82, 1956.

18. V.N.S. Murthy: *Soil Mechanics and Foundation Engineering*, Dhanpat Rai & Sons, Delhi-6, 1974.
19. H.P. Oza: *Soil Mechanics and Foundation Engineering*, Charotar Book Stall, Anand, India, 1969.
20. H.E. Pauker: *An Explanatory Report on the Project of a Sea-battery*, (in Russian), Journal of the Ministry of Ways and Communications, St. Petersburg, Sept., 1889.
21. R.B. Peck, W.E. Hanson, T.H. Thornburn: *Foundation Engineering*, John Wiley & Sons, Inc., NY, USA, 1953 (2nd ed., 1973).
22. L. Prandtl: *Über die Härte Plastischer Körper*, Nachrichten von der Königlichen Gesellschaft der Wissenschaft zu Göttingen, Berlin, 1920.
23. W.J.M. Rankine: *A Manual of Applied Mechanics*, Charles Griffin and Co., London, 1885.
24. Reissner: *Zum Erddruk Problem*, Proceedings, First International Congress of Applied Mechanics, Delft, Holland, 1924.
25. F. Schliecher: *Zur Theorie des Baugrundes*, Der Bauingenieur, nos. 48 and 49 Nov., and Dec., 1926.
26. S.B. Sehgal: *A Text Book of Soil Mechanics*, Metropolitan Book Co., Pvt., Ltd., Delhi-6, 1967.
27. Shamsher Prakash and Gopal Ranjan: *Problems in Soil Engineering*, Sarita Prakashan, Meerut, India, 1976.
28. Shamsher Prakash, Gopal Ranjan and Swami Saran: *Analysis and Design of Foundations and Retaining Structures*, Sarita Prakashan, Meerut, India, 1979.
29. A.W. Skempton: *Bearing Capacity of Clays*, Building Research Congress, Div. 1, 1951.
30. G.N. Smith: *Elements of Soil Mechanics for Civil and Mining Engineers*, 3rd SI ed., Crosby Lockwood Staples, London, 1974.
31. M.G. Spangler: *Soil Engineering*, International Text Book Co., Scranton, USA, 1951.
32. D.W. Taylor: *Fundamentals of Soil Mechanics*, John Wiley & Sons, Inc., NY, USA, 1948.
33. W.C. Teng: *Foundation Design*, Prentice Hall of India Pvt., Ltd., New Delhi, 1969.
34. K. Terzaghi: *Theoretical Soil Mechanics*, John Wiley & Sons, Inc., , USA, 1943.
35. K. Terzaghi and R.B. Peck: *Soil Mechanics in Engineering Practice*, John Wiley & Sons, Inc., NY, USA, 1948.
36. G. Wilson: *The Calculation of the Bearing Capacity of Footings on Clay*, Journal of the Institution of civil Engineers, London, Nov., 1941.

QUESTIONS AND PROBLEMS

14.1 To obtain a higher bearing capacity, either width of the footing could be increased or the depth of foundation can be increased. Discuss critically the relative merits and demerits.

(S.V.U.—Four year B.Tech.,—Sept., 1983)

14.2 Discuss the various factors that affect the bearing capacity of a shallow footing. Write brief critical notes on settlement of foundations. How do you ascertain whether a foundation soil is likely to fail in local shear or in general shear?

(S.V.U.—B.Tech., (Part-time)—Sept., 1983)

14.3 Discuss the various types of foundations and their selection with respect to different situations.

(S.V.U.—Four year B.Tech.,—Dec., 1982)

Discuss the effect of shape on the bearing capacity. Differentiate between safe bearing capacity and allowable soil pressure. Write critical notes on (i) foundations on black cotton soils and (ii) Penetration tests.

(S.V.U.—Four year B.Tech.,—Apr., 1983)

14.4 Bring out clearly the effect of ground water table on the safe bearing capacity.

(S.V.U.—Four year B.Tech.,—Dec., 1982, Oct., 1982)

Describe the procedure of determining the safe bearing capacity based on the standard penetration test.

Write brief critical notes on (i) floating foundation and (ii) factors affecting depth of foundation.

(S.V.U.—Four year B.Tech.,—Sept., 1983)

14.5 What are the criteria for deciding the depth of foundations? Write brief critical notes on tolerable settlements for buildings.

(S.V.U.—Four year B.Tech.,—Dec., 1982)

14.6 Explain 'general shear failure' and 'Local shear failure'. Differentiate between (i) Shallow foundation and deep foundation, (ii) Gross and net bearing capacity, (iii) Safe bearing capacity and allowable soil pressure.

(S.V.U.—Four year B.Tech.,—Oct., 1982)

14.7 What are the assumptions made in Terzaghi's analysis of bearing capacity of a continuous footing?

(S.V.U.—B.E., (Part-time)—Apr., 1982)

14.8 (a) Explain the recommended construction practices to avoid detrimental differential settlement in large structures.

(b) What is meant by bearing capacity of soil? How will you determine it in the field? Describe the procedure bringing out its limitations.

(c) Write brief critical notes on:
  (i) Standard Penetration test
  (ii) General shear failure and local shear failure of shallow foundations.

(S.V.U.—B.E., (Part-time)—Dec., 1981)

14.9 (a) Describe Terzaghi's theory of bearing capacity of shallow strip foundations. Define the three bearing capacity factors and give their values for '$\phi = 0$' case.

(b) Explain how this procedure is modified for square and rectangular footings. How is local shear failure accounted for?

(S.V.U.—B.E., (R.R.)—Nov., 1975)

14.10 (a) Give the algebraic equations showing the variation of safe bearing capacity of soil (for clay and sand to be given separately) in shallow foundation with:
  (i) depth of foundation;
  (ii) width of foundation; and
  (iii) position of water table.
(b) Give the approximate formula you will use for the design of:
  (i) square footing;
  (ii) circular footing; and
  (iii) rectangular footing.

(S.V.U.—B.E., (R.R)—Dec., 1968)

14.11 A strip footing, 1 m wide, rests on the surface of a dry cohesionless soil having $\phi = 25°$ and $\gamma = 1.8$ t/m^3. What is the ultimate bearing capacity? What is the value, if there is complete flooding? Assume $N_\gamma = 10$.

14.12 Compute the safe bearing capacity of a 1.2 m wide strip footing resting on a homogeneous clay deposit at a depth of 1.2 m below ground level. The soil parameters are $c = 4$ t/m^2, $\phi = 0°$, and average unit weight of soil = 2 t/m^3. Factor of safety = 3.

(S.V.U.—B.Tech., (Part-time)—Sept., 1983)

14.13 Compute the safe bearing capacity of a continuous footing 1.5 m wide, at a depth of 1.5 m, in a soil with $\gamma = 18$ kN/m^3, $c = 18$ kN/m^2, and $\phi = 25°$. Terzaghi's factors for $\phi = 25°$ are $N_c = 25$, $N_q = 12.5$, and $N_\gamma = 10$. What is the safe load per metre run if the factor of safety is 3?

14.14 What is the safe bearing capacity of a square footing resting on the surface of a saturated clay of unconfined compression strength of 90 kN/m^2? Factor of safety is 3.

14.15 What is the ultimate bearing capacity of a square footing, 1 m × 1 m, resting on a saturated clay of unconfined compression strength of 1.8 kg/cm^2 and a bulk density of 1.8 g/cm^3, at a depth of 1.5 m?

(S.V.U.—Four year B.Tech—Oct., 1982)

14.16 Compute the allowable bearing capacity of a square footing of 2 m size resting on dense sand of unit weight 20 kN/m^3. The depth of foundation is 1 m and the site is subject to flooding. The bearing capacity factors are: $N_c = 55$, $N_q = 38$, and $N_\gamma = 45$.

(S.V.U.—B.E., (Part-time)—Dec., 1981)

14.17 What is the safe bearing capacity of a circular footing of 1.5 m diameter resting on the surface of a saturated clay of unconfined compression strength of 120 kN/m^2, if the factor of safety is 3?

14.18 A three-storey building is to be constructed on a sand beach. Ground water rises to a maximum of 3 m below ground level. The beach sand has the following properties: $\gamma_d = 17.5$ kN/m^3, $\phi = 32°$ ($N_c = 40$, $N_q = 25$, $N_\gamma = 30$).

The maximum column load will be 700 kN. Determine the sizes of footing for depths of 1 m and 2 m using a factor of safety of 3. Settlements are not to be considered. Evaluate the two alternatives from practical considerations (difficulties of construction and cost).

(S.V.U.—B.Tech., (Part-time)—Sep., 1982)

14.19 A circular footing rests on a pure clay with $q_u = 270$ kN/m², at a depth of 1.8 m. Determine the diameter of the footing if it has to transmit a load of 720 kN. Assume the bulk unit weight of soil as 18 kN/m³ and the factor of safety as 3.

14.20 Determine the size of a square footing at the ground level to transmit a load of 900 kN in sand weighing 18 kN/m³ and having an angle of shearing resistance of 36° ($N_\gamma = 46$, $N_q = 43$). Factor of safety is 3. What will be the modification in the result, if the footing may be placed at a depth of 1 m below ground surface? Assume, in this case, the water table may rise to the ground surface. $\gamma' = 9$ kN/m³.

14.21 Determine the net ultimate bearing capacity of a rectangular footing, 1.2 m × 3.0 m, placed at 1.8 m below the ground in a saturated clay with a unit weight of 90 kN/m². Use Skempton's approach.

14.22 A machine with base 6 m × 3 m weighs 9,000 kN. It is to be placed on a clay with cohesion of 120 kN/m², at a depth of 1 m. $\gamma = 20$ kN/m³. Assuming a factor of safety of 3, design a rectangular footing foundation to support this machine.

14.23 What is the ultimate bearing capacity of a rectangular footing, 1.75 × 3.50 m, at a depth of 1.5 m in a soil for which $c = 30$ kN/m², $\phi = 15°$, and $\gamma = 18$ kN/m³. Brinch Hansen's factors are $N_c = 10.98$, $N_q = 3.94$, and $N_\gamma = 1.42$. The water table is deep. The vertical load is 1500 kN and the horizontal load is 150 kN at the base of the footing. Determine also the factor of safety.

14.24 A plate bearing test was conducted in a pure cohesive soil with 30 cm square plate at a depth of 1.5 m below the ground level. The water table was found to be at 6 m depth. Failure occurred at a load of 50 kN. Find the factor of safety if a 1.2 m wide wall footing carries 140 kN/m run and the foundation is at a depth of 2 m below ground level.
(S.V.U.—B.Tech., (Part-time)—Sept., 1982)

14.25 What is the allowable load for 1.8 m square column in a dense sand ($\gamma = 20$ kN/m³ and $\phi = 40°$) at a depth of 1.2 m, if the settlement is not to exceed 30 mm? Factor of safety against shear failure is 3. Water table is at a great depth.

14.26 A 1.8 m square column is founded at a depth of 1.8 m in sand, for which the corrected N-value is 24. The water table is at a depth of 2.7 m. Determine the net allowable bearing pressure for a permissible settlement of 40 mm and a factor of safety of 3 against shear failure.

14.27 Two load tests were performed at a site-one with a 50 cm square plate and the other with a 75 cm square plate. For a settlement of 15 mm, the loads were recorded as 50 kN and 90 kN, respectively in the two tests. Determine the allowable bearing pressure of the sand and the load which a square footing, 1.5 m size, can carry with the settlement not exceeding 25 mm.

15
SHALLOW FOUNDATIONS

15.1 INTRODUCTORY CONCEPTS ON FOUNDATIONS

The ultimate support for any structure is provided by the underlying earth or soil material and, therefore, the stability of the structure depends on it. Since soil is usually much weaker than other common materials of construction, such as steel and concrete, a greater area or volume of soil is necessarily involved in order to satisfactorily carry a given loading. Thus, in order to impart the loads carried by structural members of steel or concrete to soil, a load transfer device is necessary. The structural foundation serves the purpose of such a device. A foundation is supposed to transmit the structural loading to the supporting soil in such a way that the soil is not overstressed and that serious settlements of the structure are not caused (Chapter 14). The type of foundation utilised is closely related to the properties of the supporting soil, since the performance of the foundation is based on that of the soil, in addition to its own. Thus, it is important to recognise that it is the soil-foundation system that provides support for the structure; the components of this system should not be viewed separately. The foundation is an element that is built and installed, while the soil is the natural earth material which exists at the site.

Since the stability of structure is dependent upon the soil-foundation system, all forces that may act on the structure during its lifetime should be considered. In fact, it is the worst combination of these that must be considered for design. Typically, foundation design always includes the effect of dead loads plus the live loads on the structures. Other miscellaneous forces that may have to be considered result from the action of wind, water, heat, ice, frost, earthquake and explosive blasts.

15.2 GENERAL TYPES OF FOUNDATIONS

The various types of structural foundations may be grouped into two broad categories—shallow foundations and deep foundations. The classification indicates the depth of the foundation relative to its size and the depth of the soil providing most of the support. According to Terzaghi, a foundation is shallow if its depth is equal to or less than its width and deep when it exceeds the width.

Shallow Foundations 703

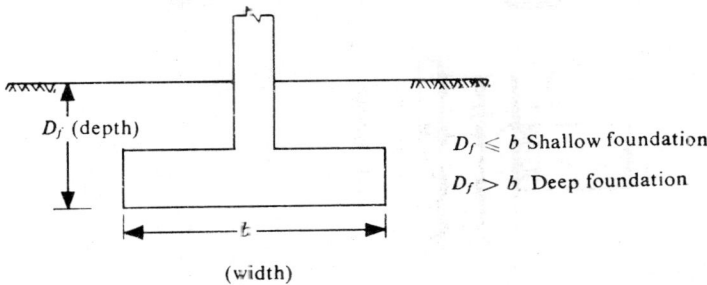

Fig. 15.1 Foundation-shallow or deep (Terzaghi)

Further classification of shallow foundations and deep foundations is as follows:

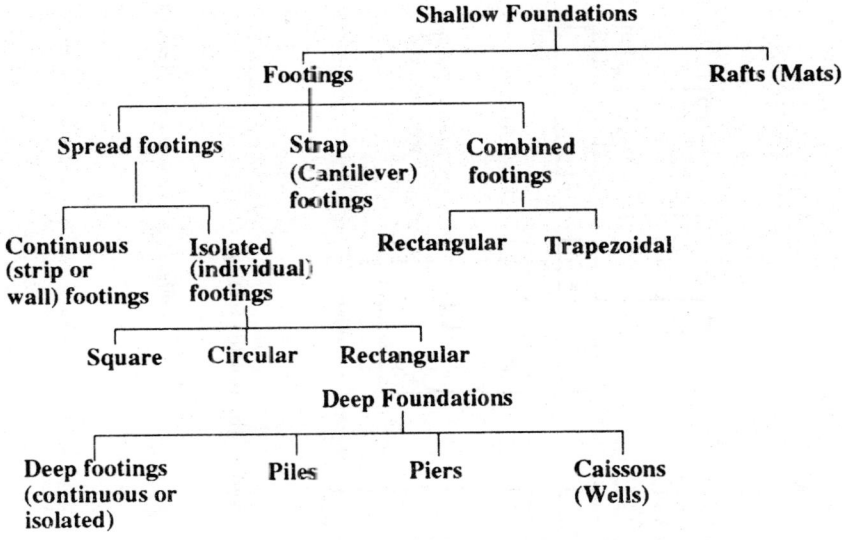

The 'floating foundation', a special category, is not actually a different type, but it represents a special application of a soil mechanics principle to a combination of raft-caisson foundation, explained later.

A short description of these with pictorial representation will now be given.

Spread footings

Spread footing foundation is basically a pad used to "spread out" loads from walls or columns over a sufficiently large area of foundation soil. These are constructed as close to the ground surface as possible consistent with the design

704 *Geotechnical Engineering*

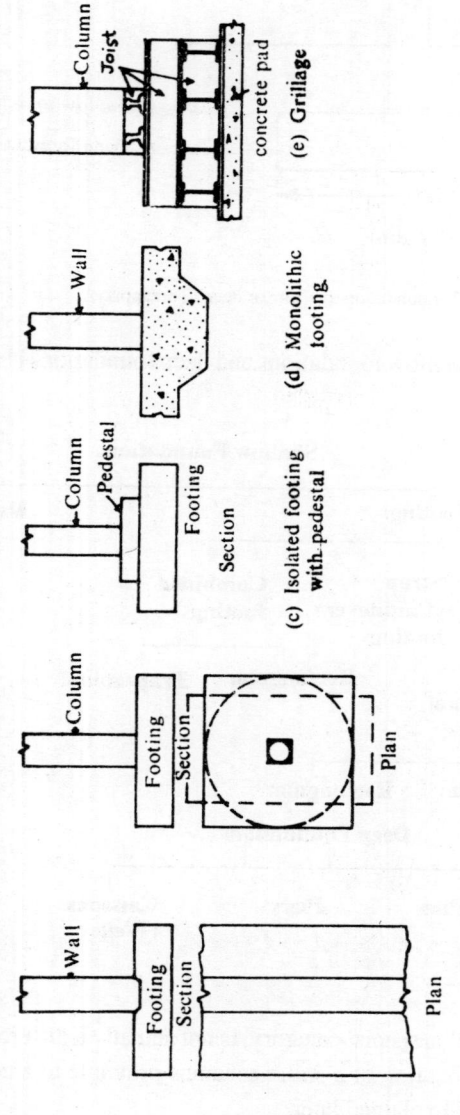

Fig. 15.2 Common types of spread footings

Shallow Foundations 705

requirements, and with factors such as frost penetration depth and possibility of soil erosion. Footings for permanent structures are rarely located directly on the ground surface. A spread footing need not necessarily be at small depths; it may be located deep in the ground if the soil conditions or design criteria require.

Spread footing required to support a wall is known as a continuous, wall, or strip footing, while that required to support a column is known as an individual or an isolated footing. An isolated footing may be square, circular, or rectangular in shape in plan, depending upon factors such as the plan shape of the column and constraints of space.

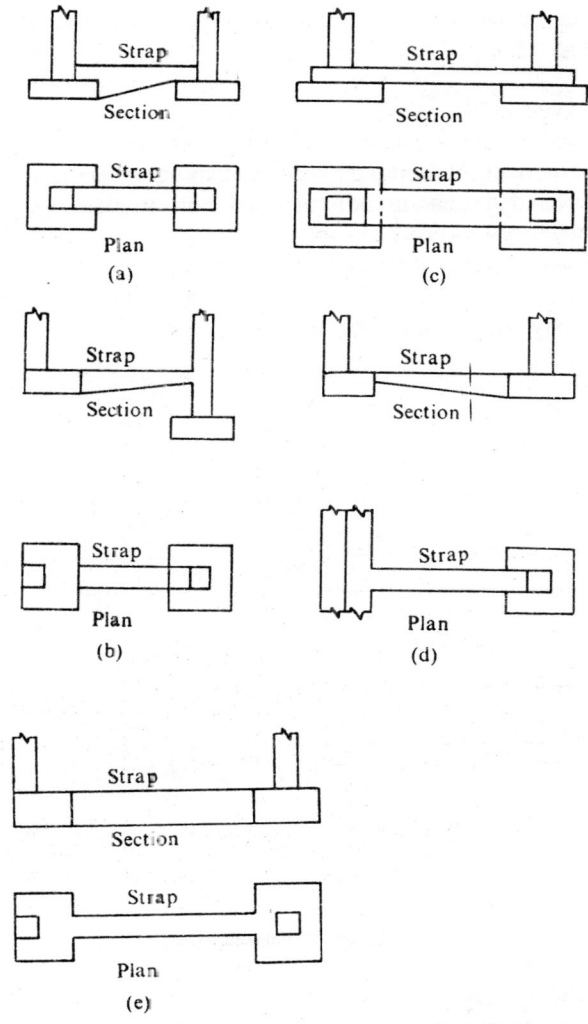

Fig. 15.3 Common arrangements of strap beams in strap footings

706 *Geotechnical Engineering*

If the footing supports more than one column or wall, it will be a strap footing, combined footing or a raft foundation.

The common types of spread footings referred to above are shown in Fig. 15.2. Two miscellaneous types—the monolithic footing, used for watertight basement (also for resisting uplift), and the grillage foundation, used for heavy loads are also shown.

Strap footings

A 'strap footing' comprises two or more footings connected by a beam called 'strap'. This is also called a 'cantilever footing' or 'pump-handle foundation'. This may be required when the footing of an exterior column cannot extend into an adjoining private property. Common types of strap beam arrangements are shown in Fig. 15.3.

Combined footings

A combined footing supports two or more columns in a row when the areas required for individual footings are such that they come very near each other. They are also preferred in situations of limited space on one side owing to the existence of the boundary line of private property.

The plan shape of the footing may be rectangular or trapezoidal; the footing will then be called 'rectangular combined footing' or 'trapezoidal combined footing', as the case may be. These are shown in Fig. 15.4.

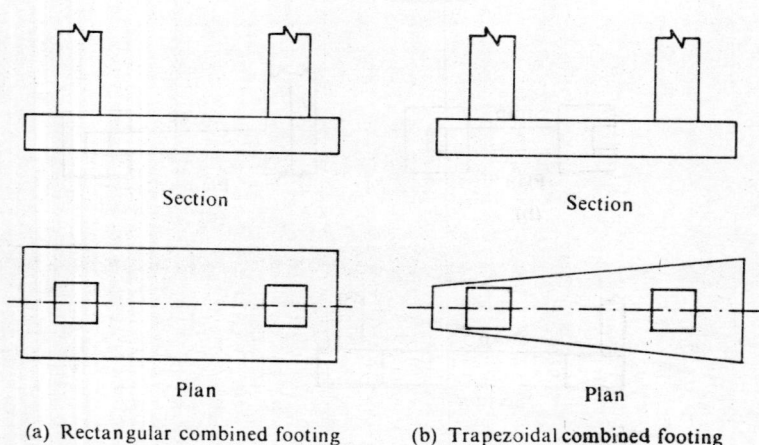

(a) Rectangular combined footing   (b) Trapezoidal combined footing

Fig. 15.4 Combined footings

Raft foundations (Mats)

A raft or mat foundation is a large footing, usually supporting walls as well as several columns in two or more rows. This is adopted when individual column footings would tend to be too close or tend to overlap; further, this is considered suitable when differential settlements arising out of footings on weak soils are to be minimised. A typical mat or raft is shown in Fig. 15.5.

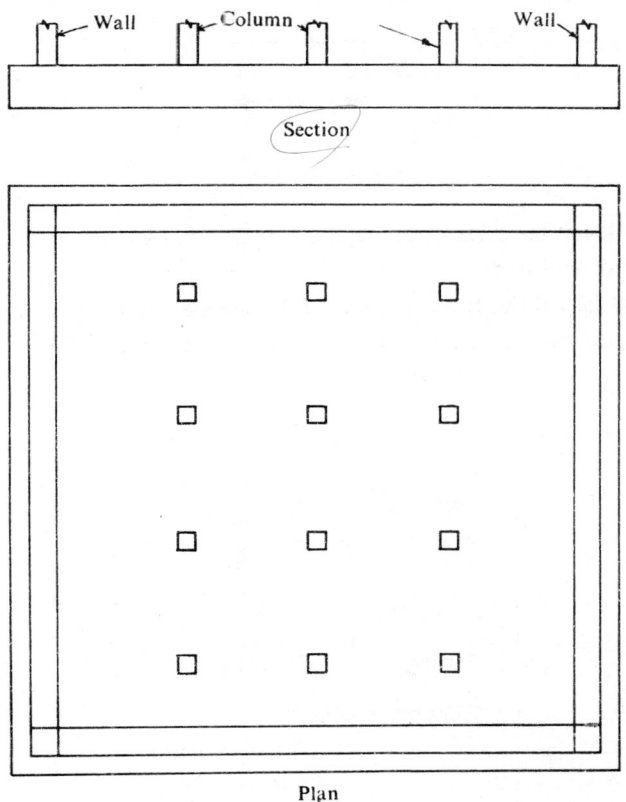

Fig. 15.5 Raft (Mat) foundation (Flat slab type)

Deep footings

According to Terzaghi, if the depth of a footing is less than or equal to the width, it may be considered a shallow foundation. Theories of bearing capacity have been considered for these in Chapter 14. However, if the depth is more, the footings are considered as deep footings (Fig. 15.6); Meyerhof (1951) developed the theory of bearing capacity for such footings.

Pile foundations

Pile foundations are intended to transmit structural loads through zones of poor soil to a depth where the soil has the desired capacity to transmit the loads. They are somewhat similar to columns in that loads developed at one level are transmitted

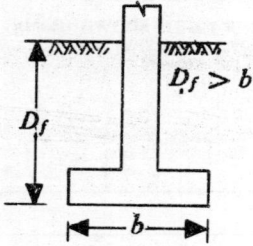

Fig. 15.6 Deep footing

to a lower level; but piles obtain lateral support from the soil in which they are embedded so that there is no concern with regard to buckling and, it is in this respect that they differ from columns. Piles are slender foundation units which are usually driven into place. They may also be cast-in-place (Fig. 15.7).

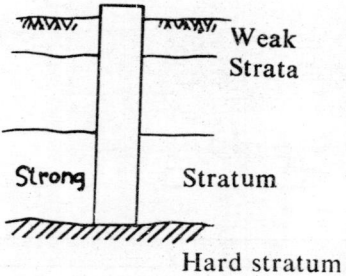

Fig. 15.7 Pile foundation

A pile foundation usually consists of a number of piles, which together support a structure. The piles may be driven or placed vertically or with a batter. More detailed treatment of this type of foundation is given in Chapter 16.

Pier foundations

Pier foundations are somewhat similar to pile foundations but are typically larger in area than piles. An opening is drilled to the desired depth and concrete is poured to make a pier foundation (Fig. 15.8). Much distinction is now being lost between the pile foundation and pier foundation, adjectives such as 'driven', 'bored', or 'drilled', and 'precast' and 'cast-in-situ', being used to indicate the method of installation and construction. Usually, pier foundations are used for bridges.

Caissons (Wells)

A caisson is a structural box or chamber that is sunk into place or built in place by systematic excavation below the bottom. Caissons are classified as 'open' caissons, 'pneumatic' caissons, and 'box' or 'floating' caissons. Open caissons may be box-type or pile-type.

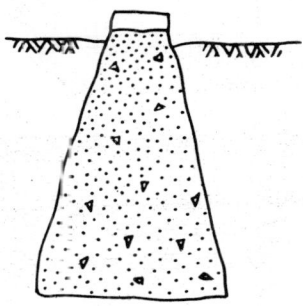

Fig. 15.8 Pier foundation

The top and bottom are open during installation for open caissons. The bottom may be finally sealed with concrete or may be anchored into rock.

Pneumatic caisson is one in which compressed air is used to keep water from entering the working chamber; the top of the caisson is closed. Excavation and concreting is facilitated to be carried out in the dry. The caisson is sunk deeper as the excavation proceeds and on reaching the final position, the working chamber is filled with concrete.

Box or floating caisson is one in which the bottom is closed. It is cast on land and towed to the site and launched in water, after the concrete has got cured. It is sunk into position by filling the inside with sand, gravel, concrete or water. False bottoms or temporary bases of timber are sometimes used for floating the caisson to the site. The various types of caissons are shown in Fig. 15.9.

Floating foundation

The floating foundation is a special type of foundation construction useful in locations where deep deposits of compressible cohesive soils exist and the use of piles is impractical. The concept of a floating foundation requires that the substructure be assembled as a combination of a raft and caisson to create a rigid box as shown in Fig. 15.10.

710 *Geotechnical Engineering*

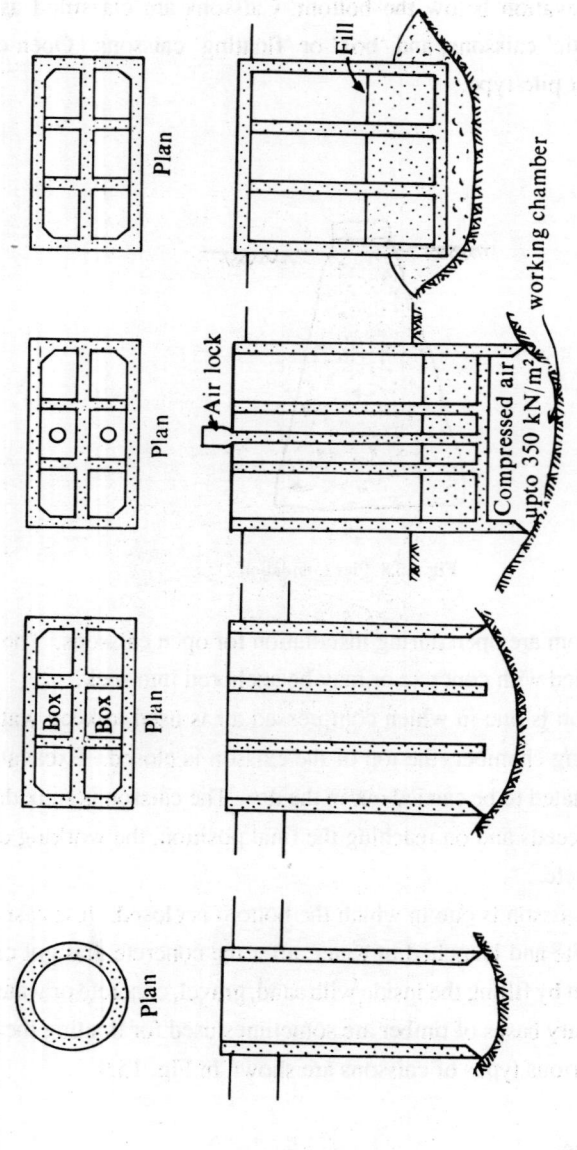

Fig. 15.9 Types of caissons (After Teng, 1976)

(a) Pile-type open caisson (b) Box-type open caisson (c) Pneumatic caisson (d) Box (floating) caisson

Shallow Foundations 711

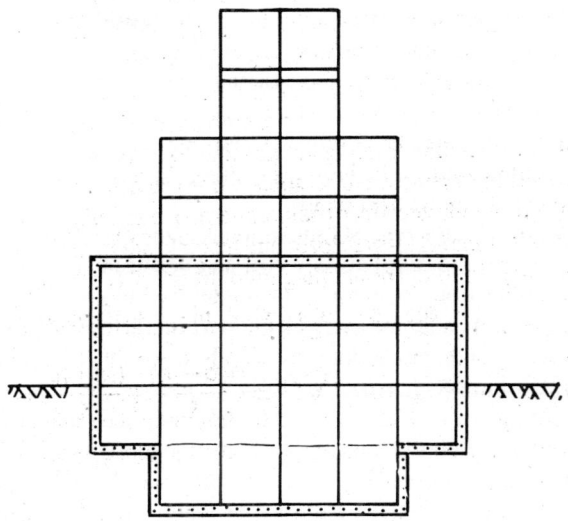

Fig. 15.10 Rigid box caisson foundation using floating foundation concept
(McCarthy, 1977)

This foundation is installed at such a depth that the total weight of the soil excavated for the rigid box equals the total weight of the planned structure. Theoretically speaking, therefore, the soil below the structure is not subjected to any increase in stress; consequently, no settlement is to be expected. However, some settlement does occur usually because the soil at the bottom of the excavation expands after excavation and gets recompressed during and after construction.

15.3 CHOICE OF FOUNDATION TYPE AND PRELIMINARY SELECTION

The type of foundation most appropriate for a given structure depends upon several factors: (*i*) the function of the structure and the loads it must carry, (*ii*) the subsurface conditions, (*iii*) the cost of the foundation in comparison with the cost of the superstructure. These are the principal factors, although several other considerations may also enter into the picture.

There are usually more than one acceptable solution to every foundation problem in view of the interplay of several factors. Judgement also plays an important part. Foundation design is enriched by scientific and engineering developments; however, a strictly scientific procedure may not be possible for practising the art of foundation design and construction.

The following are the essential steps involved in the final choice of the type of foundation:

1. Information regarding the nature of the superstructure and the probable loading is required, at least in a general way.
2. The approximate subsurface conditions or soil profile is to be ascertained.
3. Each of the customary types of foundation is considered briefly to judge whether it is suitable under the existing conditions from the point of view of the criteria for stability—bearing capacity and settlement. The obviously unsuitable types may be eliminated, thus narrowing down the choice.
4. More detailed studies, including tentative designs, of the more promising types are made in the next phase.
5. Final selection of the type of foundation is made based on the cost—the most acceptable compromise between cost and performance.

The design engineer may sometimes be guided by the successful foundations in the neighbourhood. Besides the two well known criteria for stability of foundations —bearing capacity and settlement—the depth at which the foundation is to be placed, is another important aspect.

For small loading on good soils, spread footings could be selected. For columns, individual footings are chosen unless they come too close to one another, in which case, combined footings are used.

For a series of closely spaced columns or walls, continuous footings are the obvious choice. When the footings for rows of columns come too close to one another, a raft foundation will be the obvious choice. In fact, when the area of all the footings appears to be more than 50 per cent of the area of the structure in plan, a raft should be considered. The total load it can take will be substantially greater than footings for the same permissible differential settlement.

In case a shallow foundation does not answer the problem on hand, in spite of choosing a reasonable depth for the foundation, some type of deep foundation may be required. A pier foundation is justified in the case of very heavy loading as in bridges. Piles, in effect, are slender piers, which are used to bypass weak strata and transmit loading to hard strata below. As an alternative to raft foundation, the economics of bored piles is considered.

After the preliminary selection of the type of the foundation is made, the next step is to evaluate the distribution of pressure, settlement, and bearing capacity.

Certain guidelines are given in Table 15.1 with regard to the selection of the type of foundation based on soil conditions at a site. For the design comments it is assumed that a multistorey commercial structure, such as an office building, is to be constructed.

Table 15.1 Appropriate foundation types for certain soil conditions

S.No.	Soil conditions	Appropriate type of foundation and location	Design comments
1.	Compact sand deposit extending to great depth.	D_f > frost depth and depth of erosion	Spread footings most appropriate for conventional needs. Piles may be required only if unusual forces such as uplift are expected.

Table 15.1 (Contd.)

#	Soil Profile	Foundation Sketch	Recommendation
2.	Firm clay or silty clay extending to great depth	D_f > frost depth and zone of swelling and shrinkage	Spread footings most appropriate for conventional needs. Piles may be used only if unusual forces such as uplift are expected.
3.	3 m Firm clay / Soft clay extending to great depth	D_f > frost depth and zone of swelling and shrinkage	Spread footing appropriate for low or medium loading, if not too close to soft clay. Deep foundations may be required for heavy loading.
4.	Loose sand extending to great depth	D_f > frost depth and depth of erosion	Spread footings may settle excessively. Raft foundation may be appropriate. Spread footings may be used if the sand is compacted by vibro-floatation. Driven piles or augered cast-in-place piles may also be used.
5.	7.5 m (Soft) / 7.5 m (Medium) / (firmer) / Soft clay but firmness increasing with depth extending to great depth	or	Friction piles or piers would be satisfactory if some settlement could be tolerated. Long piles would reduce settlement. Raft foundation or floating foundation may also be considered.
6.	20 m Soft clay / Rock		Deep foundation-piles, piers, caissons—bearing directly on/in the rock.
7.	2.5 m Compact sand / 3.5 m Medium clay / Hard clay extending to great depth		Spread footings in upper sand layer would probably experience large settlement because of underlying soft clay layer. Drilled piers with a bell formed in the hard clay layer, or other pile foundation may be considered.
8.	6 m Soft clay / Medium dense sand extending to great depth	Auger pile or bulb type pile	Best solution is deep foundation. Cast-in-place type piles such as auger-piles or bulb piles into sand layer may be appropriate.

714 *Geotechnical Engineering*

Table 15.1 (Contd.)

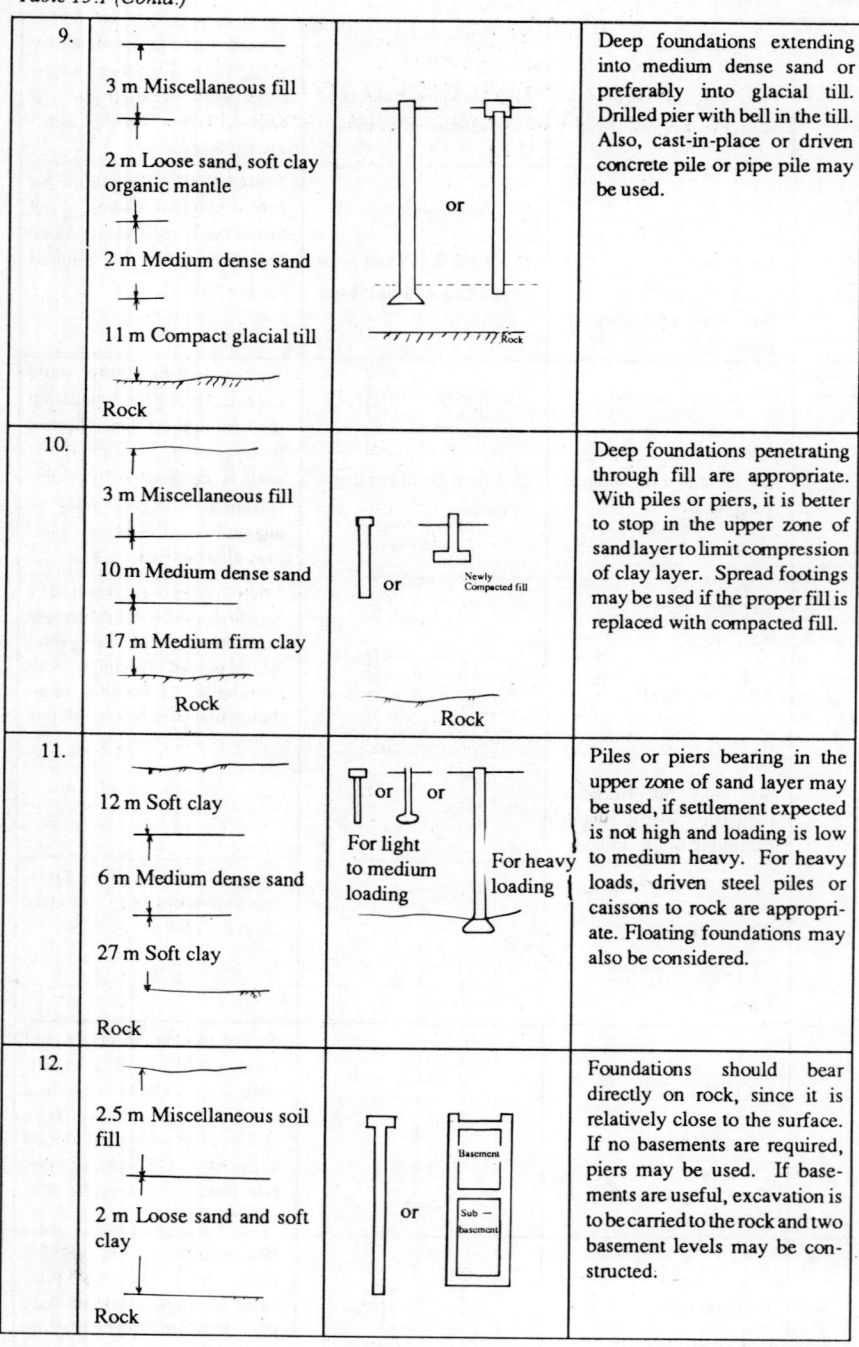

9.	3 m Miscellaneous fill 2 m Loose sand, soft clay organic mantle 2 m Medium dense sand 11 m Compact glacial till Rock	Deep foundations extending into medium dense sand or preferably into glacial till. Drilled pier with bell in the till. Also, cast-in-place or driven concrete pile or pipe pile may be used.
10.	3 m Miscellaneous fill 10 m Medium dense sand 17 m Medium firm clay Rock	Deep foundations penetrating through fill are appropriate. With piles or piers, it is better to stop in the upper zone of sand layer to limit compression of clay layer. Spread footings may be used if the proper fill is replaced with compacted fill.
11.	12 m Soft clay 6 m Medium dense sand 27 m Soft clay Rock	Piles or piers bearing in the upper zone of sand layer may be used, if settlement expected is not high and loading is low to medium heavy. For heavy loads, driven steel piles or caissons to rock are appropriate. Floating foundations may also be considered.
12.	2.5 m Miscellaneous soil fill 2 m Loose sand and soft clay Rock	Foundations should bear directly on rock, since it is relatively close to the surface. If no basements are required, piers may be used. If basements are useful, excavation is to be carried to the rock and two basement levels may be constructed.

(After McCarthy, 1977)

15.4 SPREAD FOOTINGS

Spread footings are the most widely used type among all foundations because they are usually more economical than others. Least amount of equipment and skill are required for the construction of spread footings. Further, the conditions of the footings and the supporting soil can be readily examined.

Other types of foundations are more favourable when the soil has a very low bearing capacity or when excessive settlements are expected to result due to the presence of compressible strata within the active zone.

15.4.1 Common Types of Spread Footings

A spread footing is a type of shallow foundation used to support a wall or a column. In the former case, it is called a continuous or wall footing and in the latter, it is called an isolated or individual footing. The commonly used variations of individual footings are illustrated in Fig. 15.11.

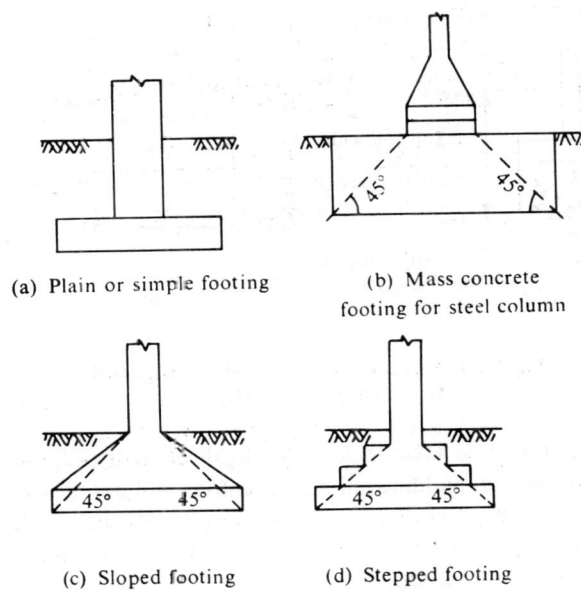

(a) Plain or simple footing

(b) Mass concrete footing for steel column

(c) Sloped footing

(d) Stepped footing

Fig. 15.11 Common variations of individual footings

The base area of the footing is governed by the bearing capacity of the soil. The plain footing is usually of reinforced concrete and is used to support a reinforced concrete column. The mass concrete footing is used to support a steel column. Usually the sloped footing will be of the same material as that for the column; alternatively, it can be of reinforced concrete. The stepped footing is used either

716 *Geotechnical Engineering*

for a column or for a wall. All the steps may be of concrete or the bottom most step alone may be of concrete, the others being of the same material as for the column.

15.4.2 Depth of Footings

The important criteria for deciding upon the depth at which footings have to be installed may be set out as follows:

1. Footings should be taken below the top (organic) soil, miscellaneous fill, debris or muck.

If the thickness of the top soil is large, two alternatives are available:

(a) Removing the top soil under the footing and replacing it with lean concrete; and, (b) removing the top soil in an area larger than the footing and replacing it with compacted sand and gravel; the area of this compacted fill should be sufficiently large to distribute the loads from the footing on to a larger area.

The choice between these two alternatives, which are shown in Fig. 15.12(a) and (b) will depend upon the time available and relative economy.

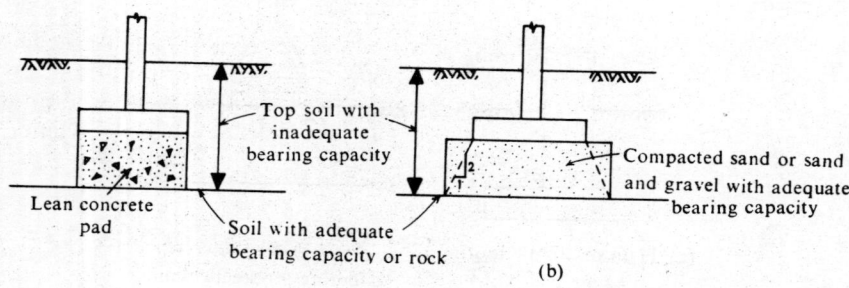

Fig. 15.12 Alternatives when top soil is of large thickness
(After Teng, 1976)

2. Footings should be taken below the depth of frost penetration. Interior footings in heated buildings in cold countries will not be affected by frost. The minimum depths of footings from this criterion are usually specified in the local building codes of large cities in countries in which frost is a significant factor in foundation design.

The damage due to frost action is caused by the volume change of water in the soil at freezing temperatures. Gravel and coarse sand above water level, containing less than 3% fines, cannot hold water and consequently are not subjected to frost action. Other soils are subjected to frost heave within the depth of frost penetration.

In tropical countries like India, frost is not a problem except in very few areas like the Himalayan region.

3. Footings should be taken below the possible depth of erosion due to natural causes like surface water run off. The minimum depth of footings on this count is usually taken as 30 cm for single and two-storey constructions, while it is taken as 60 cm for heavier construction.
4. Footings on sloping should be constructed with a sufficient edge distance (minimum 60 cm to 90 cm) for protecting against erosion (Fig. 15.13).

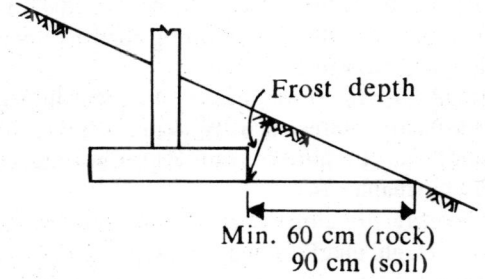

Fig. 15.13 Edge distance for floating on sloping ground

5. The difference in elevation between footings should not be so great as to introduce undesirable overlapping of stresses in soil. The guideline used for this is that the maximum difference in elevation should be maintained equal to the clear distance between two footings in the case of rock and equal to half the clear distance between two footings in the case of soil (Fig. 15.14). This is also necessary to prevent disturbance of soil under the higher footing due to the excavation for the lower footing.

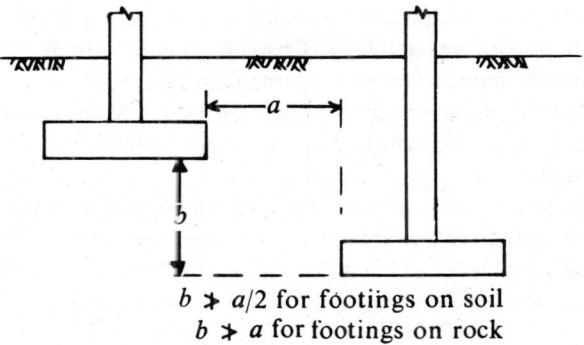

Fig. 15.14 Footings at different elevations—restrictions

15.4.3 Bearing Capacity of Soils Under Footings

Granular Soils

Bearing capacity of granular soils depends upon the unit weight γ and the angle of internal friction ϕ of the soil, both of which vary primarily with the density index of the soil. Dense soils have large values of γ and ϕ and consequently high bearing capacity. Loose soils, on the other hand, have small γ and ϕ values and low bearing power.

The density index of granular soils *in situ* is generally determined by standard penetration tests. The relationship between N-values and ϕ-values established empirically by Peck, Hanson, and Thornburn may be used, and later the relevant Terzaghi equations may be applied to get the bearing capacity.

In conventional design, the allowable bearing capacity should be taken as the smaller of the following two values:

(i) *Bearing capacity based on shear failure:* This is the ultimate bearing capacity divided by a suitable factor of safety; usually a value of 3 is used for normal loading and 2 for maximum load. Empirical equations for bearing capacity in terms of N-value may be used (Chapter 14).

(ii) *Allowable bearing pressure based on tolerable settlement:* Empirical equation given by Terzaghi and Peck may be used in terms of N-values—for the net allowable bearing pressure.

The value may be modified by using the linear relationship with permissible settlement, if it is desired for a different value of the permissible settlement.

If $D_f/b > 1$, the value is obtained by multiplying by the factor $(1 + D_f/b)$.

The allowable bearing pressure is taken as the smaller of (i) and (ii) finally.

Cohesive soils

The ultimate bearing capacity of cohesive soils depends primarily on their shear strength (or consistency). This may be determined by any one of the following:

(i) *Standard penetration tests:* For conservative design of small jobs, the correlation between standard penetration value, consistency and allowable bearing capacity given by Terzaghi and Peck (Chapter 14) may be used.

(ii) *Unconfined compression tests:* For medium jobs, the shear strength obtained from unconfined compression tests should be used. Skempton's equation for bearing capacity is used in which cohesion is taken as half the unconfined compression strength.

(iii) *Triaxial tests:* For large jobs, the shear strength may be determined from triaxial tests on undisturbed samples. The shear parameters are obtained by plotting the data from triaxial tests. Drainage conditions in the field are to be simulated in the laboratory and careful interpretation of the results is required.

Silts

Silt is often a poor foundation soil and should be avoided for supporting footings. Apparent cohesion, exhibited by moist silt disappears on immersion. Plate load tests at about the ground water level are advocated in this case.

Compacted fills

Bearing capacity for compacted fills must be determined both before and after compaction.

Shallow Foundations

Organic Soils

Organic soils are not suitable for supporting footings. Highly organic soils settle unduly even under their own weight, both by consolidation and by decay or decomposition of the organic matter.

Rocks

Generally speaking, rocks can withstand pressures greater than concrete can do. Rocks with fissures, folds, faults and bedding planes are exceptions to this. Shales may become clay or silt on soaking. Weathered rocks are treacherous and lose strength on wetting.

15.4.4 Settlement of Footings

If the allowable bearing pressure is determined based on the smaller value from the two criteria—shear strength and permissible settlement—footings on granular soils do not suffer detrimental settlement.

Footings on clay will experience settlement which consists of three components (Skempton and Bjerrum, 1957):

$$S = S_i + S_c + S_s \qquad \ldots\text{(Eq. 15.1)}$$

where,

S = Total settlement,
S_i = Immediate elastic settlement,
S_c = Consolidation settlement due to primary compression, and
S_s = Settlement due to secondary compression of the clay.

These and other details of settlement analysis have already been dealt with exhaustively (Chapter 11).

15.4.5 Proportioning Sizes of Footings and Choice of Column Loads

A structure is usually supported on a number of columns. These columns usually carry different loads depending upon their location with respect to the structure. Differential settlements are minimised by proportioning the footings for the various columns so as to equalise the average bearing pressure under all columns.

But each column load consists of dead load plus live load. The full live load does not act all the time; further live loads such as those due to heavy wind do not produce significant settlement since they act only for short durations; this is especially true in the case of cohesive soils. Hence, dead load plus full live load is not a realistic criterion for producing equal settlement.

What is known as the 'service load' is a better criterion. This is the actual load expected to act on the foundation during the normal service of the structure, i.e., for most of the time. In ordinary buildings, this is taken as the dead load plus one half the live load; a larger fraction of the live load should be used for warehouses and other industrial structures.

The following procedure is given by Teng (1976) based on the recommendations of Peck, Hanson and Thornburn (1974):

(i) Dead load, inclusive of self-weight of column and estimated value for footing, is noted for each column footing.
(ii) The live load for each column is calculated (appropriate values are chosen from the relevant I.S. Codes of Practice).
(iii) The ratio of live load to dead load is calculated for each column footing; the maximum value of this ratio is noted.
(iv) The allowable bearing pressure of the soil is determined by the procedures given in Chapter 14.
(v) For the footing with the largest live load to dead load ratio, the area of footing required is calculated by dividing the total load (dead load plus maximum live load) by the allowable bearing pressure of the soil.
(vi) The service load for the column with the maximum live load to dead load ratio is computed by adding the appropriate fraction of the live load to the dead load.
(vii) The allowable bearing pressure to be used for all the other column footings is obtained by dividing the service load for the column with maximum live load to dead load ratio by the area of the footing for this column (This pressure will be obviously somewhat less than the computed allowable bearing pressure of step (iv).
(viii) Service loads for all other columns are computed.
(ix) The area of the footing for each of the other columns is obtained by dividing the corresponding service load by the reduced allowable bearing pressure of step (vii).

The advantage in this procedure is that the allowable bearing pressure of the soil is never exceeded under any circumstances and the reduced or service loads, which are effective during most of the time are expected to result in equal settlements.

The procedure, as standardised by the ISI, is set out in "IS: 1080-1980 Code of Practice for Design and Construction of simple spread foundations (First Revision)".

15.4.6 Footings Subjected to Moments—Eccentric Loading

Footings supporting axially loaded columns and which are symmetrically placed with respect to the columns will be subjected to uniform soils pressures. However, footings may often have to resist not only axial loads but also moment about one or both axes. The moment may exist at the bottom of an axially loaded column, whence it is transmitted to the footing; alternatively, it may be produced by an axial vertical load located eccentrically from the centroid of the base of the footing, positioned unsymmetrically with respect to the column. If the moment in the first case is equal to the product of the axial load and eccentricity in the second, the soil pressure distribution will be just identical. Thus, the substitution of an equivalent eccentric load for a real moment is considered a convenient method which simplifies computations in some cases.

Foundations for retaining walls may have to resist moments due to the active earth pressure and those for bridge piers may have to resist moments produced primarily by wind and traction on the superstructure. These foundations also have to be treated in a somewhat similar manner as footings subjected to moments.

Once the soil reactions are determined, the design data such as critical moments and shears may be obtained as a prerequisite for the structural design. Fundamental to all these computations are the laws of statics. The distribution of vertical soil pressure at the base must satisfy the requirements of statics that (i) the total upward soil reaction must be equal to the sum of the downward loads on the base, and (ii) the moment of the resultant vertical load about any point must equal the moment of the total soil reaction about the same point. In addition, an adequate horizontal soil reaction must be available, by virtue of frictional resistance at the base, to oppose the resultant horizontal load.

Ordinary footings are commonly assumed to act as rigid structures. This assumption leads to the conclusion that the vertical settlement of the soil beneath the base must have a planar distribution since a rigid foundation remains plane when it settles. Another assumption is that the ratio of pressure to settlement is constant, which also leads to the conclusion regarding the planar distribution of

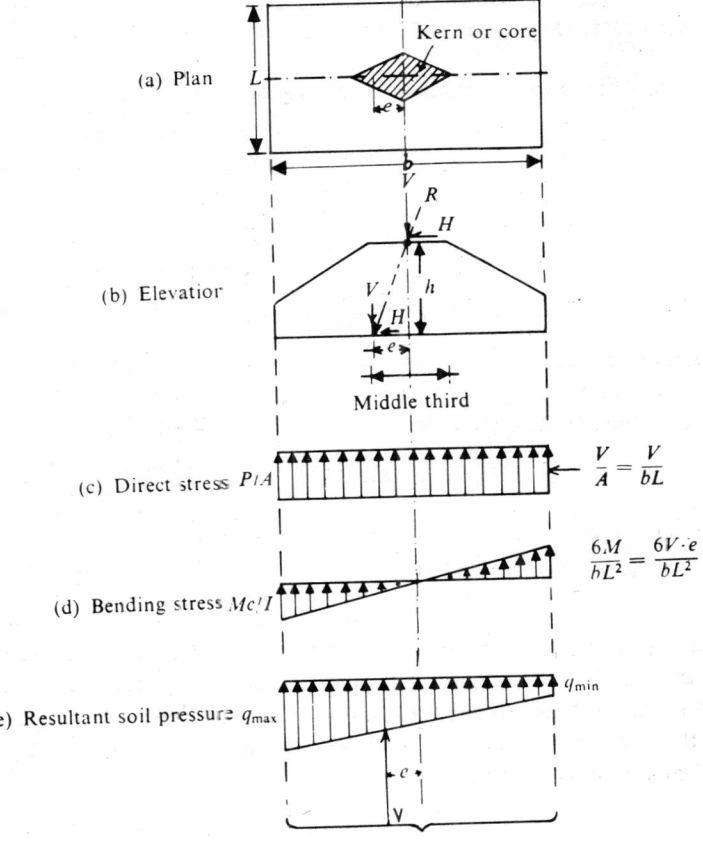

Fig. 15.15 Footing subjected to moment—Resultant force within the middle-third of base

soil pressure. Although, neither of these assumptions is strictly valid, each is considered to be sufficiently accurate for ordinary purposes of design.
Two distinct cases arise:
(1) Resultant force within the middle third of the base; and,
(2) Resultant force outside the middle third of the base.

Resultant force within the middle third of the base
A footing subjected to moment is shown in Fig. 15.15.
The forces acting on the footing including self-weight are resolved into V and H. The moment M may be expressed as $M = H \cdot h = V \cdot e$.
We have, $e = M/V$ ...(Eq. 15.2)
This equation enables one to determine the eccentricity of the resultant of all forces acting on the base regardless of how complicated the conditions of loading may be.

If the vertical load V acts alone, it produces uniform soil pressure due to the direct stress as shown in Fig. 15.14(c). If the horizontal load H acts alone, it produces a shear that must be resisted by the soil at the base and also a moment which produces soil pressure distribution shown in Fig. 15.14(d), due to bending.

The resultant soil pressure will be the combined effect of V and M is shown in Fig. 15.14(e).

The maximum and minimum soil pressure are obtained as:

$$q = \frac{V}{bL}\left(1 \pm \frac{6e}{b}\right) \qquad \text{...(Eq. 15.3)}$$

$$q_{max} = \frac{V}{bL}\left(1 + \frac{6e}{b}\right) \qquad \text{...(Eq. 15.4)}$$

$$q_{min} = \frac{V}{bL}\left(1 - \frac{6e}{b}\right) \qquad \text{...(Eq. 15.5)}$$

Equation 15.3 is merely a special form of the basic formula for the resultant stress on a section subjected to a direct load P and a moment M, expressed in strength of materials, in the form:

$$f = \frac{P}{A} \pm \frac{Mc}{I}$$

The maximum eccentricity for no tension to occur in the base is obtained by equating q_{min} to zero, and solving for e:

$$e_{max} = b/6 \qquad \text{...(Eq. 15.6)}$$

Since the eccentricity can occur to either side of the middle depending upon the direction of H, the resultant force should fall within the middle-third of the base in order that no tensile stresses occur anywhere in the base.

If the eccentricity occurs with respect to the axis which bisects the other dimension L of the footing.

$$q = \frac{V}{bL}\left(1 \pm \frac{6e}{L}\right) \qquad \text{...(Eq. 15.7)}$$

$$e_{max} = L/6 \qquad \text{...(Eq. 15.8)}$$

Shallow Foundations

This leads to the concept of 'kern' or 'core' of a section, which is the zone within which the resultant should fall for the entire base to be subjected to compression.

For a rectangular section, the kern is a centrally located rhombus with the diagonals equal to one-third of the breadth and length; for a circular section it is a concentric circle with diameter one-fourth of that of the circle.

Most footings are designed so that the resultant of the loads falls within the kern and the soil reaction everywhere is compressive. However, in certain cases such as the design of the base slab of a cantilever retaining wall, the resultant may fall outside the kern, and the distribution of pressure shown in Fig. 15.15 must be used for the structural design of the footing.

Resultant force outside the middle-third of the base

If the horizontal component of the total load increases beyond a certain limit in relation to the vertical component, the resultant force falls outside the middle-third of the base, the eccentricity being more than the limiting value of one-sixth the size of the base. It must be remembered that soil cannot provide tensile reaction; it just loses contact with the footing in the zone of tension. This situation is shown in Fig. 15.16.

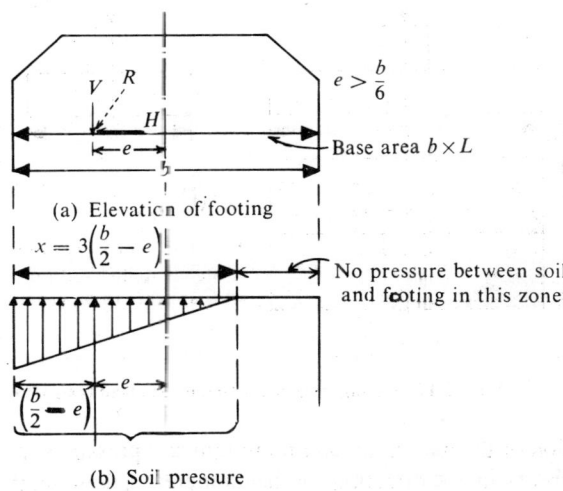

Fig. 15.16 Footing subjected to moment-resultant force outside the middle-third of the base

From the laws of statics, the total upward force must be equal to V and also collinear with V. That is to say:

$$V = \frac{q_{max} \times L}{2} \qquad \text{...(Eq. 15.9)}$$

and

$$\frac{x}{3} = \left(\frac{b}{2} - e\right) \qquad \text{...(Eq. 15.10)}$$

Also,
$$q_{max} = \frac{V}{A}\left[\frac{4b}{3b-6e}\right] \qquad \ldots(\text{Eq. 15.11})$$

where $A = bL$.

Equation 15.9 reveals that the maximum soil pressure is merely twice the average pressure produced by V acting on the area xL.

Moment about both axes

When moments act simultaneously about both axes, for example when a vertical load acts at an eccentricity with respect to both the axes, as shown in Fig. 15.17, the soil pressure is given by the following equation:

$$q = \frac{V}{A} \pm \frac{M_b c_b}{I_b} \pm \frac{M_L \cdot c_L}{I_L} \qquad \ldots(\text{Eq. 15.12})$$

This is under the assumption that the entire base is under compression.

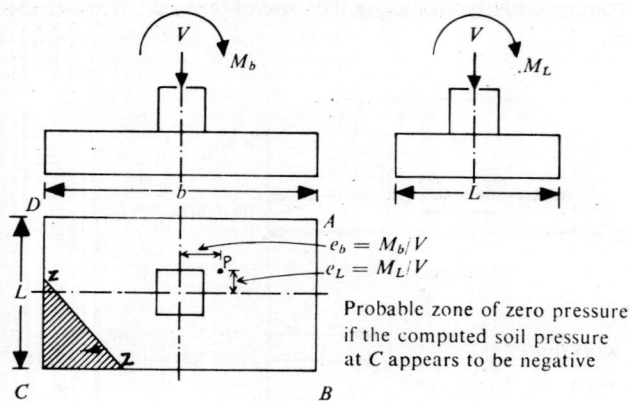

Fig. 15.17 Footing subjected to moments about both axes

The location of the maximum and minimum soil pressures may be determined readily by observing the directions of the moments. Likewise, the proper signs in Eq. 15.12 may be determined by inspection for any other point on the base of the footing.

If the minimum soil pressure computed appears to be negative, there exists a zone like CZZ in which the footing loses contact with the soil and hence, there will be no pressure in the zone. Equation 15.12 will not be applicable to this case. For the determinations of soil pressures for this situation, the reader is referred to Peck, Hanson, and Thornburn (1974), who give an excellent trial and error procedure.

Useful width concept

For the determination of the bearing capacity of an eccentrically loaded footing, the concept of 'useful width' has been introduced. By this concept, the portion of the footing which is symmetrical about the load is considered useful and the other portion is simply assumed superfluous for the convenience of computation (Teng, 1976). This is illustrated in Fig. 15.18.

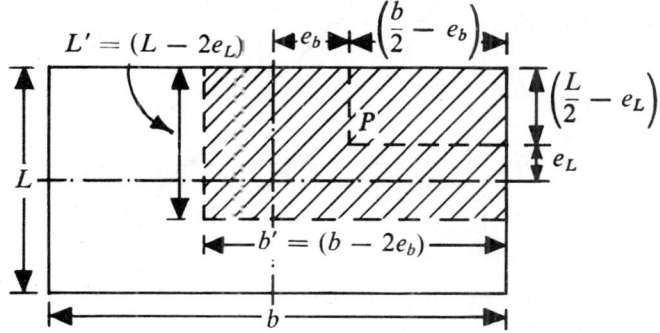

Fig. 15.18 Useful width concept for eccentrically loaded footings

If the eccentricities are e_b and e_L, as shown, the useful widths b' and L' are:

$$b' = b - 2e_b \qquad L' = L - 2e_L \qquad \text{...(Eq. 15.13)}$$

The equivalent area A' is considered to be subjected to a central load for the determination of bearing capacity:

$$A' = b'L' = (b - 2e_b)(L - 2e_L) \qquad \text{...(Eq. 15.14)}$$

The procedure may be used even if the eccentricity is with respect to one of the axes only.

This concept simply means that the bearing capacity of a footing decreases linearly with the eccentricity of load. This is almost true in the case of cohesive soils; however, the relationship is parabolic rather than linear in the case of granular soils (Meyerhof, 1953).

Therefore, it is considered better to use a reduction factor R_e for getting the reduced bearing capacity due to eccentricity of loading:

$$q'_{ult} = q_{ult} \cdot R_e \qquad \text{...(Eq. 15.15)}$$

where

q'_{ult} = bearing capacity of an eccentrically loaded footing size $b \times L$,

q_{ult} = bearing capacity of a centrally loaded footing of size $b \times L$, and

R_e = reduction factor for eccentricity.

If there is eccentricity about both axes, the product of the two factors must be used.

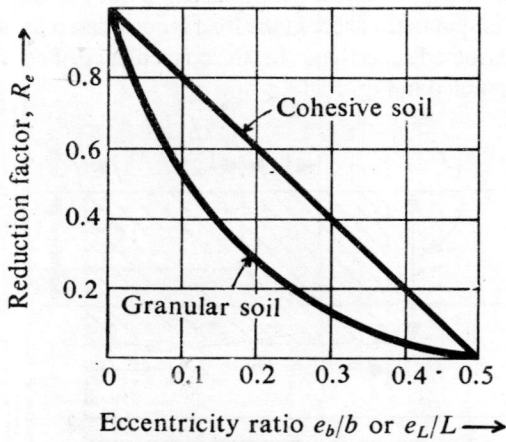

Fig. 15.19 Reduction factor for eccentrically loaded footings

Footings with unsymmetrical shapes

The assumption till now has been that at least one axis of symmetry exists for the footing in plan. If an unsymmetrical section is involved under eccentric loading, computation of soil pressures becomes a problem, since Eq. 15.12 is not applicable even though the entire base may be in compression. However, the errors involved in using Eq. 15.12 may not be intolerable for design, unless the footing is greatly unsymmetrical.

15.4.7 Inclined Loading

The conventional procedure of analysing the stability of footings subjected to inclined loading consists in resolving the load into a vertical component V and a horizontal component H, and dealing with the effect of each separately. The soil pressure due to the vertical load is considered to be uniform and the stability against ultimate failure is analysed in the usual way.

The stability against the horizontal load is analysed by ensuring a minimum factor of safety against sliding at the base, which is defined as the ratio between the total resistance to sliding and the applied horizontal force. The total horizontal resistance usually consists of passive resistance of the soil and a frictional resistance F at the base, which is dependent upon the coefficient of friction between the base of the footing and the soil beneath it. This is illustrated in Fig. 15.20.

Shallow Foundations 727

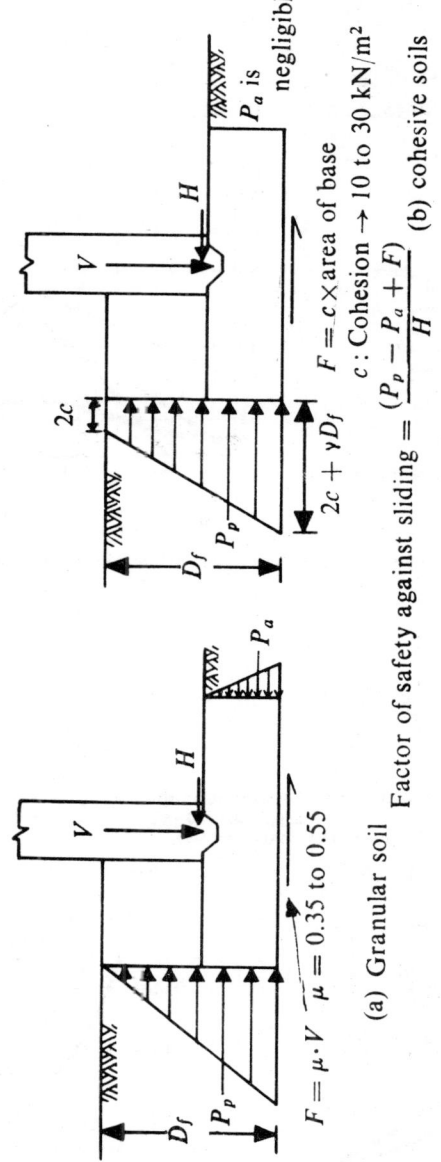

Fig. 15.20 Conventional method of analysis of footings subjected to inclined loads

Janbu (1957) proposed an analysis which is a direct extension of the Terzaghi theory with an additional factor N_h, in addition to Terzaghi's factors N_c, N_γ and N_q:

$$\frac{(R+N_h \cdot H)}{A} = cN_c + \gamma D_f N_q + \frac{1}{2}\gamma bN_\gamma \qquad \ldots(\text{Eq. 15.16})$$

where A = area of base of footing.

The notation and values are shown in Fig. 15.21.

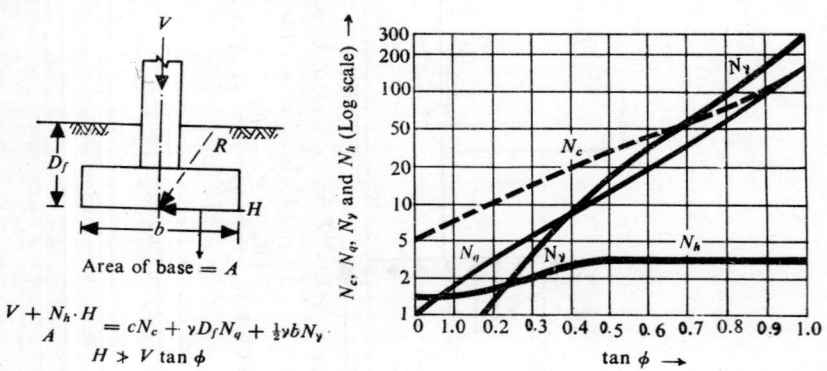

Fig. 15.21 Continuous footing subjected to inclined load (After Janbu, 1957)

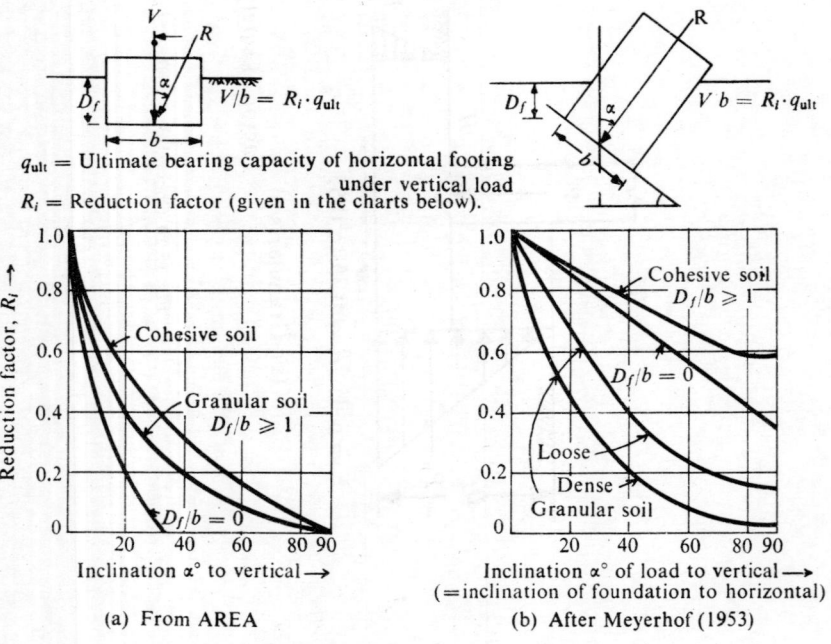

(a) From AREA

(b) After Meyerhof (1953)

Fig. 15.22 Footings subjected to inclined load (a) Horizontal foundation (AREA) (b) Inclined foundation (Meyerhof)

Shallow Foundations 729

Meyerhof (1953) proposed an analysis of footings subjected to inclined loads and constructed convenient charts, shown in Fig. 15.22. The load is considered to act vertically and the bearing capacity is obtained by the normal procedure. It is then corrected by multiplying by the factor R_i.

15.4.8 Footings on Slopes

Meyerhof (1957) again proposed an equation for the bearing capacity of footings on sloping ground as follows:

$$q_{ul} = cN_{cq} + \frac{1}{2}\gamma bN_{\gamma q} \qquad \text{...(Eq. 15.17)}$$

The values of the bearing capacity factors N_{cq} and $N_{\gamma q}$ for continuous footings are given in Fig. 15.23. These factors vary with the slope of the ground, the relative position of the footing and the angle of internal friction of the soil.

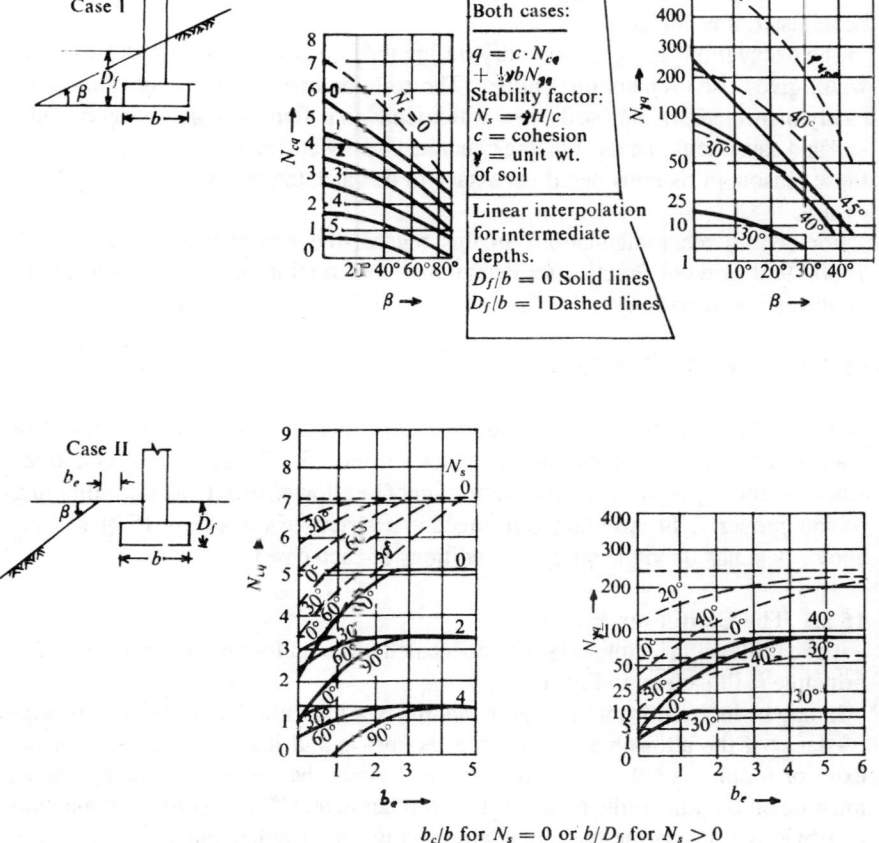

b_e/b for $N_s = 0$ or b/D_f for $N_s > 0$

Fig. 15.23 Bearing capacity of continuous footings on slopes
(After G.G. Meyerhof)

Footings must be constructed only on slopes which are stable. The stability of the slope itself may be endangered by the construction of footings.

15.4.9 Construction of Spread Footings

Footings are relatively simple to construct. The inspection of subsoil conditions, the relative depth of footings and dewatering of excavation when necessary require special attention. Depending on the nature of soil, form work may be required for the sides.

The average soil condition based on the soil boring results must be ascertained. As the foundation is constructed, the actual soil conditions encountered must be checked with respect to the boring analysis.

Adjacent footings should be constructed such that their difference of levels, if any, does not introduce undue additional stress at the lower footings and also that the lower footing does not affect the stability of the upper one. This difficulty is generally avoided by keeping the difference in the elevations of footings not greater than one-half the clear distance between the footings. It is always a good practice to construct the lower footings first, so that the elevation of the upper footing may be adjusted if necessary.

The excavation should be kept dry during the construction period because free water gives rise to many difficulties. The soil conditions under water cannot be easily inspected. In clay soils, free water tends to soften the upper portion of the soil and cause settlements. Placing concrete under water also poses problems. For these reasons, it is considered necessary to dewater the excavations where necessary.

For certain recommendations in this regard, the reader is referred to "IS: 1080-1980 Code of Practice for design and construction of simple spread foundations (First Revision)".

15.5 STRAP FOOTINGS

A relatively common type of combined construction is the 'strap footing' or 'cantilever footing', as has already been seen in Sec. 15.2. This is usually employed when the footing of an exterior column cannot be allowed to extend into adjoining private property. Straps may be arranged in a variety of ways (Fig. 15.2), and the choice depends upon the specific conditions of each case.

15.5.1 The Cantilever Principle

The cantilever principle is largely concealed in actual footings of this type. This principle is illustrated in Fig. 15.24.

It may be inferred from this figure that the two individual footings is a problem of statics if the allowable soil pressure is known and if the dimension b of the exterior footing is either fixed or assumed. Also, the centroid of the two areas must lie on the line of the action of the resultant load. This requirement may not be obvious because the two areas are usually found rather independently from reactions determined from the principles of statics.

Shallow Foundations 731

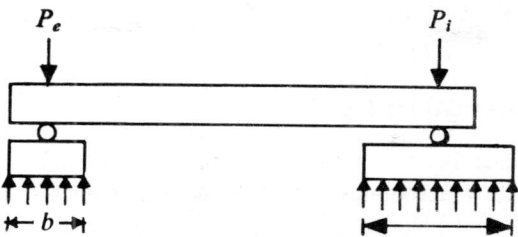

Fig. 15.24 Cantilever principle of strap footing

15.5.2 Basis for Design of Strap Footings
Strap footings are designed based on the following assumptions:
(i) The strap footing is considered to be infinitely stiff. It serves to transfer the column loads onto the soil with equal and uniform soil pressure under both the footings.
(ii) The strap is a pure flexure member and does not directly take soil reaction. The soil below the strap will be loosened up in order that the strap does not rest on the soil and exert pressure.

With these assumptions, the procedure of design is simple. With reference to Fig. 15.25, it may be given as follows:

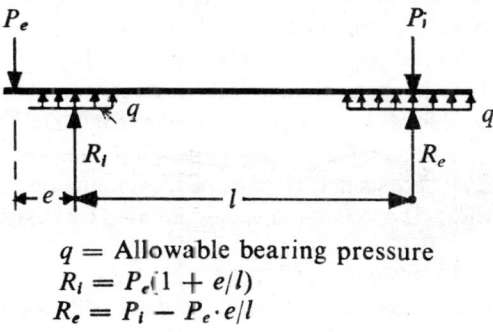

q = Allowable bearing pressure
$R_i = P_e(1 + e/l)$
$R_e = P_i - P_e \cdot e/l$

Fig. 15.25 Design of strap footing

Assume a trial value of e and compute the reactions R_i and R_e from statics. The tentative areas of the footing are equal to the reactions R_i and R_e divided by the allowable bearing pressure q. The value of e is computed with tentative sizes. These steps are repeated until the trial value of e is identical with the final value.

The shearing force and bending moment in the strap are determined, the strap being designed to withstand the maximum values of these.

732 Geotechnical Engineering

Each of these footings is assumed to be subjected to uniform soil pressure and designed as simple spread footings. Under the assumptions given above, the resultant of the column loads P_e and P_i would coincide with the centre of gravity of the areas of the two footings.

15.6 COMBINED FOOTINGS

The use of combined footings is appropriate either when two columns are spaced so closely that individual footings are not practicable or when a wall column is so close to the property line that it is impossible to center an individual footing under the column.

A combined footing is so proportioned that the centroid of the area in contact with the soil lies on the line of action of the resultant of the loads applied to the footing; consequently, the distribution of soil pressure is reasonably uniform. In addition, the dimensions of the footing are chosen such that the allowable soil pressure is not exceeded. When these criteria are satisfied, the footing should neither settle nor rotate excessively.

A combined footing may be of rectangular shape or of trapezoidal shape in plan. These are usually constructed using reinforced concrete.

15.6.1 Rectangular Combined Footing

A combined footing is usually given a rectangular shape if the rectangle can extend beyond each column the necessary distance to make the centroid of the rectangle coincide with the point at which the resultant of the column loads intersects the base.

If the footing is to support an exterior column at the property line where the projection has to be limited, provided the interior column carries the greater load, the length of the combined footing is established by adjusting the projection of the footing beyond the interior column. The width is then obtained by dividing the sum of the vertical loads by the product of the length and the allowable soil pressure. A rectangular combined footing is shown in Fig. 15.26.

The B.M. and S.F. diagrams may be sketched, assuming that the column loads are concentrated loads. The maximum values are used for design.

15.6.2 Trapezoidal Combined Footing

When the two column loads are unequal, the exterior column carrying higher load and when the property line is quite close to the exterior column, a trapezoidal combined footing is used. It may be used even when the interior column carries higher load; but the width of trapezoid will be higher in the inner side. The location of the resultant of the column loads establishes the position of the centroid of the trapezoid. The length is usually limited by the property line at one end and adjacent construction, if any, at the other.

The width at either end of the trapezoid can be determined from the solution of two simultaneous equations—one expressing the location of the centroid of the trapezoid and the other equating the sum of the column loads to the product of the allowable soil pressure and the area of the footing.

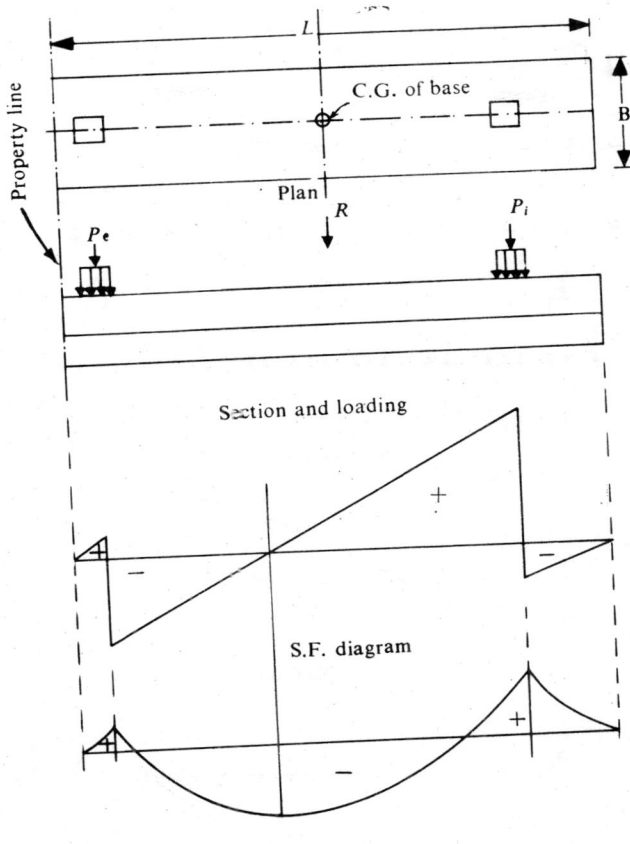

Fig. 15.26 Rectangular combined footing

The resulting pressure distribution is linear or uniformly varying (and not uniform) as shown in Fig. 15.27.

In order to determine B_1 and B_2, the following is the procedure:

$$B_1 + B_2 = \frac{2}{L}\left(\frac{P_e + P_i}{q_a}\right) \qquad \ldots\text{(Eq. 15.18)}$$

where q_a = allowable soil pressure.

By taking moments about the property line or left edge, and on simplifying,

$$\frac{2B_1 + B_2}{B_1 + B_2} = \frac{3}{L}\left[e_1 + \frac{2P_i L'}{(P_e + P_i)}\right] \qquad \ldots\text{(Eq. 15.19)}$$

L' and e_1 are as indicated in Fig. 15.27.

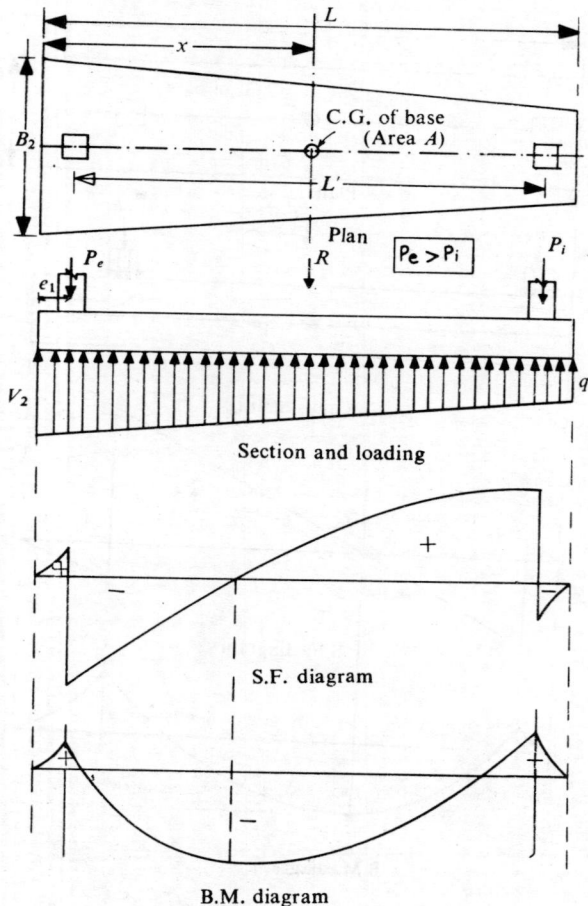

Fig. 15.27 Trapezoidal combined footing

B_1 and B_2 may be solved from Eqs. 15.18 and 15.19 since the quantities on the right-hand sides are known.

The solution leads to:

$$B_1 = \frac{2A}{L}\left(\frac{3\bar{x}}{L} - 1\right) \qquad \text{...(Eq. 15.20)}$$

and

$$B_2 = \frac{2A}{L} - B_1 \qquad \text{...(Eq. 15.21)}$$

The pressure intensities, q_1 and q_2 are calculated, once B_1 and B_2 are obtained:

Shallow Foundations 735

$$q_1 = B_1 \cdot q_a \qquad q_2 = B_2 \cdot q_a \qquad \ldots\text{(Eq. 15.22)}$$

The bending moment and shear force diagrams can be easily sketched now, as shown in Fig. 15.27. The maximum values are used for the purposes of design. From Eq. 15.20,

$$B_1 = 0 \text{ when } \bar{x} = L/3$$

For a rectangular shape,

$$\bar{x} = L/2$$

Thus, a trapezoidal combined footing solution exists when $\bar{x}$ is such that:

$$\frac{L}{3} < \bar{x} < \frac{L}{2}$$

In tentative designs, whenever the distance $\bar{x}$ approaches $L/3$, or is less than $L/3$, the length L should be increased by increasing the projection beyond the inner column.

15.7 RAFT FOUNDATIONS

A 'raft' or a 'mat' foundation is a combined footing which covers the entire area beneath of a structure and supports all the walls and columns. This type of foundation is most appropriate and suitable when the allowable soil pressure is low, or the loading heavy, and spread footings would cover more than one half the plan area. Also, when the soil contains lenses of compressible strata which are likely to cause considerable differential settlement, a raft foundation is well-suited, since it would tend to bridge over the erratic spots, by virtue of its rigidity. On occasions, the principle of floating foundation may be applied best in the case of raft foundations, in order to minimise settlements.

15.7.1 Common Types of Raft Foundations

Common types of raft foundations in use are illustrated in Fig. 15.28.

Figure 15.28(*a*) represents a true raft which is a flat concrete slab of uniform thickness throughout the entire area; this is suitable for closely spaced columns, carrying small loads. (*b*) represents a raft with a portion of the slab under the thickened column; this provides sufficient strength for relatively large column loads. (*c*) is a raft with thickened bands provided along column lines in both directions; this provides sufficient strength, when the column spacing is large and column loads unequal. (*d*) represents a raft in which pedestals are provided under each column; this alternative serves the same purpose as (*b*).(*e*) represents a two-way grid structure made of cellular construction and of intersecting structural steel construction (Teng, 1949). (*f*) represents a raft wherein basement walls have been used as ribs or deep beams.

A raft foundation usually rests directly on soil or rock; however, it may rest on piles as well, if hard stratum is not available at a reasonably small depth.

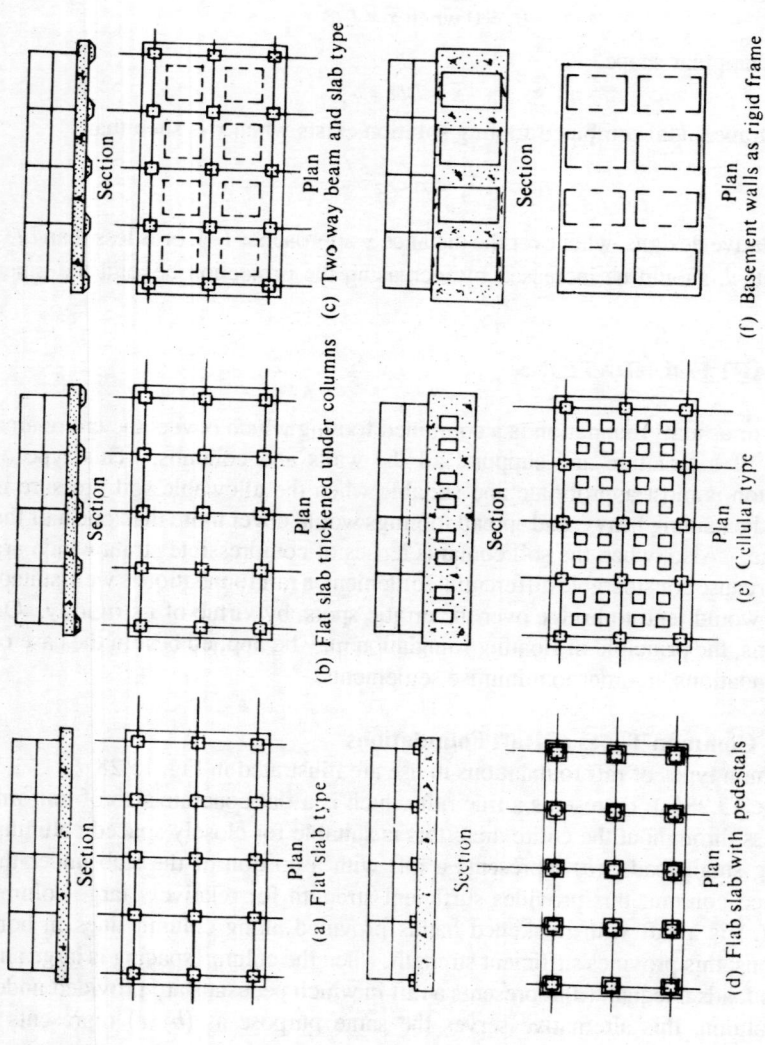

Fig. 15.28 Common types of raft foundations (Teng, 1976)

15.7.2 Bearing Capacity of Rafts on Sands

Since the bearing capacity of sand increases with the size of the foundation and since rafts are usually of large dimensions, a bearing capacity failure of raft on sand is practically ruled out. As a raft bridges over loose pockets and eliminates their influence, the differential settlements are much smaller than those of a footing under the same pressure. Hence, higher allowable soil pressures may be used for design of rafts on sands.

Terzaghi and Peck (1948), as also Peck, Hanson, and Thornborn (1974), recommend an increase of 100% over the value allowed for spread footings. The design charts developed for the bearing capacity from N-values for footings on sands may be used for this purpose. The effect of the location of water table is treated as in the case of footings.

15.7.3 Bearing Capacity of Rafts on Clays

The net ultimate bearing capacity is divided by the factor of safety to obtain the net allowable soil pressure for a footing. The same principle is applicable to rafts on clay. Accordingly, the factor of safety, η, in terms of net soil pressure, is given by

$$\eta = \frac{cN_c}{(q - \gamma D_f)} \qquad \text{...(Eq. 15.23)}$$

where, c = unit cohesion
N_c = bearing capacity factor for cohesion,
q = gross soil pressure or contact pressure,
γ = unit weight of soil,
D_f = depth of raft below ground surface.

It is obvious that the factor of safety is very large for rafts established at such depths that γD_f is nearly equal to q. In fact, the theoretical value of η is infinite, when γD_f equals q; in such a case, the raft is said to be a 'fully compensated foundation' (Peck, Hanson and Thornburn, 1974).

15.7.4 Coefficient of Subgrade Reaction

The 'coefficient of subgrade reaction' or 'subgrade modulus' is defined as the ratio between the pressure and the settlement at a given point:

$$k = \frac{q}{S'} \qquad \text{...(Eq. 15.24)}$$

where, k = coefficient of subgrade reaction in N/mm^3,
q = pressure against the footing or raft at a given point in N/mm^2,
and S' = settlement at the particular point in mm.

In other words, the coefficient of subgrade reaction is the pressure required to produce unit settlement. This is difficult to determine for clayey soils in view of the long time required for the consolidation settlement to occur. Equation 15.24 is based on the following assumptions:

(i) k is independent of pressure.
(ii) k is the same at every point of the footing or mat.

Actually, a number of factors affect the value of coefficient of subgrade reaction (Terzaghi, 1955):

Effect of size
The value of k decreases with increasing width of footing.

$$k = k_1 \left(\frac{b+0.3}{2b}\right)^2 \quad \ldots \text{granular soils} \quad \ldots \text{(Eq. 15.25)}$$

$$k = \frac{k_1}{b} \quad \ldots \text{cohesive soils} \quad \ldots \text{(Eq. 15.26)}$$

where k = coefficient of subgrade reaction for a very long footing of width b m, and
k_1 = coefficient of subgrade reaction for a very long footing of width 1 m.

Equation 15.25 is established from experiments and Eq. 15.26 from the pressure bulb concept.

Effect of shape
For footings with the same width b under the same pressure q and supported on the same soil, k decreases with increasing length L of the footing.

$$k = \frac{k_s(1+b/L)}{1.5} \quad \ldots \text{(Eq. 15.27)}$$

where k = coefficient of subgrade reaction for a rectangular footing, size $b \times L$,
and k_s = coefficient of subgrade reaction for square footing, $b \times b$.

This indicates that k value for an infinitely long footing is equal to two-thirds of that for a square footing.

Effect of depth
The elastic modulus, E, of sand increases with depth and it may be expressed by:

$$E = c \cdot \gamma \cdot z \quad \ldots \text{(Eq. 15.28)}$$

where c = constant, depending on the properties of sand,
γ = unit weight of sand, and
z = depth.

$$E = \frac{\text{Average stress}}{\text{Average strain}} = \frac{\tfrac{1}{2}q}{S/\text{Depth of pressure bulb}} = c\gamma(D_f + b/2)$$

...(Eq. 15.29)

$$\therefore \quad k' = q/S = c\gamma(1 + 2D_f/b) \quad \ldots \text{(Eq. 15.30)}$$

where k' = coefficient of subgrade reaction at depth D_f.
If $\quad D_f = 0, \quad k' = c$

$$\therefore \quad k' = (1 + 2D_f/b) \quad (k' < 2k) \quad \ldots \text{(Eq. 15.31)}$$

This indicates that the settlement of a footing is reduced to one-half, if it is lowered from the ground surface to a depth equal to one-half of the width of the footing.

A general equation may now be written to include the effect of size and depth for square footings:

On granular soils

$$k = k_1 \left(\frac{b+0.3}{b}\right)^2 (1+2D_f/b) \qquad \ldots(\text{Eq. 15.32})$$

with $k \not> 2k_1 \left(\dfrac{b+0.3}{b}\right)^2$

On cohesive soils

The modulus of elasticity for a purely cohesive soil is practically constant throughout the depth. Therefore, the depth has no effect on the value of modulus of subgrade reaction.

On $c - \phi$ soils:

$$k = k_a \left(\frac{b+0.3}{b}\right)^2 (1+2D_f/b) + k_b/b \qquad \ldots(\text{Eq. 15.33})$$

k_a and k_b must be evaluated by at least two tests using two different sizes, say 300 mm square and 600 mm square.

15.7.5 General Considerations in the Design of Rafts

The conditions under which a raft foundation is suitable have already been discussed. In its simplest form a raft consists of a reinforced-concrete slab that supports the columns and walls of a structure and that distributes the load therefrom to the underlying soil. Such a slab is usually designed as a continuous flat-slab floor supported without upward deflection at the columns and walls. The soil pressure acting against the slab is commonly assumed to be uniformly distributed and equal to the total of all column loads, divided by the area of the raft. The moments and shears in the slab are determined by the use of appropriate coefficients listed in codes for the design of flat-slab floors.

On account of erratic variations in compressibility of almost every soil deposit, there are likely to be correspondingly erratic deviations of the soil pressure from the average value. Since the moments and shears are determined on the basis of the average pressure, it is considered good practice to provide the slab more reinforcement than the theoretical requirement and to use the same percentage of steel at top and bottom (Peck, Hanson, and Thornburn, 1974).

The flat slab analogy is valid only if the differential settlement between columns is small and furthermore, if the pattern of the differential settlement is erratic rather than systematic. Also, even if deep-seated or systematic settlements are negligible, the flat-slab analogy is likely to lead to uneconomical design unless the columns are more or less equally spaced and equally loaded. Otherwise, differential settlements may lead to substantial redistribution of moments in the slab.

Under such circumstances, rafts are sometimes designed on the basis of the concept of the modulus of subgrade reaction, which implies that soil is considered to be analogous to a bed of closely and equally spaced elastic springs of equal stiffness in its stress-strain behaviour. Evaluation of the modulus of subgrade reaction, k, for design is not a simple problem since k is known to vary in a complex manner on the shape and size of the loaded area, as well as on the magnitude and position of near-by loaded areas. [For IS procedure, refer "IS: 2950 (Part-I–1973 Code of Practice for Design and Construction of Raft Foundation—Part-I Design"].

If a raft covers a fairly large area and significantly increases the stresses in an underlying deposit of compressible clay, it is likely to experience large systematic differential settlements. For these to be avoided, strength of the slab alone is not sufficient, but stiffness is also required. However, a stiff raft is likely to be subjected to bending moments far in excess of those corresponding to the flat-slab or subgrade modulus analyses (Peck, Hanson, and Thornburn, 1974). These moments may require deep beams or trusses. Thus, the raft in such instances may be considered to consist of two almost independent elements: the base slab, which may still be designed by the flat-slab analogy; and the stiffening members, which have the function of preventing most of the differential settlement of the points of support for the base slab.

It has been known that contact pressure distributions in sand and clay are different from the uniform distribution commonly assumed in conventional raft design. In the case of sand, maximum pressure occurs at the middle and minimum, if any, occurs at the edges; in the case of clay, minimum pressure occurs in the middle and maximum (in fact, sometimes, very high) pressure occurs at the edges. It is also interesting to note that the pressure under a raft on clay may vary with time (Teng, 1949), and the worst conditions expected are to be considered for design.

It is unlikely that the edge pressure will exceed twice the average pressures.

As an alternative to the relatively high cost of a stiff raft of large-size above a compressible deposit, substantial economy can be realised by designing a flexible raft and superstructure that can deform without damage into the shape corresponding to the compression of the subsoil. It may often prove preferable to accept the deformations if the cost of a stiff foundation can be avoided. The design of a flexible raft foundation cannot be readily based on the calculation of stresses in the slab. Instead, it is necessary to estimate the maximum curvature to which the raft may be flexed, and to select the thickness of the slab and the quantum of reinforcement such that the slab will not develop cracks large enough to permit a serious leakage of ground water. As an approximate guideline, 1% of steel may be provided in each of two directions at right-angles to each other, equally divided between the top and bottom of the slab. The thickness of the slab should not be generally greater than 1% of the radius of curvature, though local increases of thickness near columns and walls may be required to prevent shear failures.

15.7.6 Construction of Raft Foundations

Raft foundations are invariably constructed of reinforced concrete. They are poured in small areas such as 10 m × 10 m to avoid excessive shrinkage cracks. Construction joints are carefully located at places of low shear stress—such as the centre lines between columns. Reinforcements should be continuous across points. If a bar is spliced, adequate lap is provided. Shear keys may be provided along joints so that the shear stress across the joint is safely transmitted. If necessary, the raft may be thickened to provide sufficient strength at the joints.

* 15.8 FOUNDATIONS ON NON-UNIFORM SOILS

It is generally assumed that the subsoil is relatively uniform either to a very great depth or else to a limited depth where a firm base is encountered. In reality, such situations are so uncommon as to be considered rare exceptions. The procedures of foundation design are not often directly applicable to practical problems; but these may be modified to give reliable indications of the probable behaviour of foundations on non-uniform deposits.

Most subsoils consist either of definite strata or more or less lenticular elements. On the basis of preliminary information, such as that from exploratory borings together with standard penetration tests and simple laboratory tests, it is possible to identify deposits which are sufficiently strong and incompressible. This would enable one to concentrate on the weaker or more compressible strata, so as to ascertain their influence on the behaviour of the proposed foundation. The load-carrying capacity of the doubtful materials is ascertained and based on failure or permissible settlement. Usually this information is adequate for a selection of the proper type of foundation. Sometimes, more elaborate exploratory procedures and soil tests may be required to provide the basis for a sound decision.

Stresses may be computed using Newmark's chart or by some simplified procedure. Although the chart is based on the assumption that the material is homogeneous, the errors due to stratification or other irregularities are not likely to be significant enough to invalidate the predictions of the probable behaviour of the soil.

In the following subsections, the more important kinds of non-uniform soil deposits will be discussed.

15.8.1 Soft or Loose Strata Overlying Firm Strata

This situation is relatively simple to deal with since an unsatisfactory character of the materials is likely to be apparent and rarely overlooked. The important decision is whether or not a footing foundation may be used. This may be determined by computing the safe load on the basis of the soft deposit assuming that it extends to a great depth. If the computed safe load is too small, or the computed settlement too great, footings should be eliminated from consideration. Provision of piles or piers or using a fully compensated or floating raft foundation are the two possible alternatives.

15.8.2 Dense or Stiff Layer Overlying Soft Deposit

The implications of the presence of a soft deposit at some depth below firm strata are not very obvious, as when the soft materials are at shallow depth. If the firm stratum is not sufficiently thick, footings or rafts may exert sufficient pressure to break into the underlying soft soil. Even if the overlying firm layer is of sufficient thickness to prevent such a failure, the settlement of the structure due to consolidation of the soft deposits may be excessive.

If the loading does not exceed the safe capacity of the underlying soft deposit, failure by breaking through the overlying stiff crust will be highly improbable. If the footings are widely spaced and the firm layer fairly thin with respect to the width of the footings, the stress at the top of the soft layer can be considerably decreased by increasing the size of the footings. On the other hand, if the footings are spaced rather closely and the firm layer is comparatively thick, the distribution of pressure at the top of the soft layer cannot be altered radically by changing the contact pressure (Peck, Hanson, and Thornburn, 1974).

Even if the safe load on the soft soil beneath the firm layer is not exceeded, the settlement of a footing or raft may be excessive. The settlement may be computed according to the procedures given earlier, and if it is excessive, one of the other types of foundations must be adopted.

If the computed settlement is not excessive and if the firm layer is thick enough to prevent a bearing capacity failure the footing can be designed as if the soft deposit were not present.

15.8.3 Alternating Soft and Stiff Layers

If a deposit contains a number of weak layers, bearing capacity and settlement computations may be made for each. If the structure cannot be supported on footings, piles or piers may be used to transmit the loads to one of the firm strata at sufficient depth to provide a satisfactory foundation. This depth may be determined from computations. The choice between piles and piers, or of the type of pile to be used, is likely to depend on the difficulty that may be experienced in driving through the firm strata involved. Conclusions with regard to this aspect have to be based on the results of driving test piles.

Excavation to compensate for part or all of the weight of the structure may permit the use of raft. This alternative should be considered along with others.

15.8.4 Irregular Deposits

If the subsoil consists of lenticular or wedge-shaped masses, it is rarely possible to make an accurate estimate of bearing capacity or settlement. In such cases, it is better to determine the general character of the deposit by means of numerous subsurface soundings supplemented by a few borings and soil tests. The purpose is to form an idea regarding the size and distribution of the softer elements and to judge the most unfavourable combination of elements that can be reasonably expected. The estimate of settlement should be based on the assumption that the most unfavourable conditions may occur in the most highly stressed portion of the soil. (Peck, Hanson, and Thornburn, 1974).

15.9 ILLUSTRATIVE EXAMPLES

Example 15.1: A building is supported on nine columns as shown in Fig. 15.29 and column loads are indicated. Determine the required areas of the column footings:

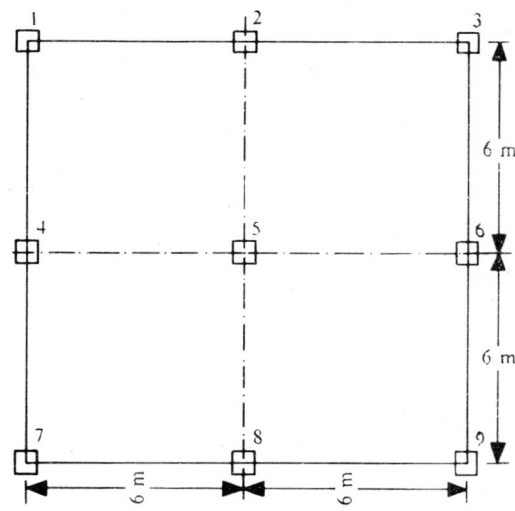

Fig. 15.29 Building founded on columns (Ex. 15.1)

Column No.	1	2	3	4	5	6	7	8	9
Dead Load (kN)	180	360	240	300	600	360	180	360	210
Max. Live Load (kN)	180	400	210	300	720	360	120	300	180

At the selected depth of 1.5 m the allowable bearing capacity is 270 kN/m². $\gamma = 20$ kN/m³.

Dead load plus maximum live load, maximum live load to dead load ratio, reduced live load and dead load plus reduced live load are all determined and tabulated for all the columns (A reduction factor of 50% is used for LL).

Column No.	1	2	3	4	5	6	7	8	9
Dead Load (kN)	180	360	240	300	600	360	180	360	210
Max. LL (kN)	180	400	210	300	720	360	120	300	180
DL + Max. LL (kN)	360	760	450	600	1320	720	300	660	390
Max. LL/DL	1.00	1.11	0.88	1.00	1.20	1.00	0.67	0.83	0.86
Reduced LL (kN)	90	200	105	150	360	180	60	150	90
DL + Reduced LL (kN)	270	560	345	450	960	540	240	510	300

744 Geotechnical Engineering

Column No. 5 has the maximum LL to DL ratio of 1.20 and hence it governs the design.

Assuming the thickness of the footing as 1 m,

allowable soil pressure corrected for the weight of the footing = $(270 - 1 \times 20) = 250$ kN/m^2

$$\therefore \text{ Area of footing for column No. 5} = \frac{1320}{250} = 5.28 \text{ m}^2$$

Reduced Load for this column = 960 kN

$$\text{Reduced allowable pressure} = \frac{\text{Reduced load}}{\text{Area}} + \text{Weight of footing}$$

$$= \frac{960}{5.28} + 20 = 182 + 20 \approx \mathbf{200 \text{ kN/m}^2}$$

The footing sizes will be obtained by dividing the reduced loads, for each column by the corrected reduced allowable pressure of $\frac{960}{5.28}$ or 182 kN/m^2.

The results are tabulated below:

Column No.	1	2	3	4	5	6	7	8	9
Reduced Load (kN)	270	560	345	450	960	540	240	510	300
Corrected reduced soil pressure (kN/m^2)	182	182	182	182	182	182	182	182	182
Required area (m^2)	1.49	3.07	1.90	2.48	5.28	2.97	1.32	2.80	1.65
Size of footing (m^2)	1.25	1.75	1.40	1.60	2.30	1.75	1.20	1.70	1.30

The thickness of the footing may be varied somewhat with loading. This will somewhat alter the reduced allowable pressures for different footings. The areas of the footings will get increased slightly. However, this refinement is ignored in tabulating the sizes of the square footings.

The structural design of the footings may now be made.

Example 15.2: Compute the ultimate load that an eccentrically loaded square footing of width 2.1 m with an eccentricity of 0.35 m can take at a depth of 0.5 m in a soil with $\gamma = 18$ kN/m^3, $c = 9$ kN/m^2 and $\phi = 36°$, $N_c = 52$; $N_q = 35$; and $N_\gamma = 42$.

Conventional approach (Peck, Hanson, and Thornburn, 1974):

For $\phi = 36°$, $N_c = 52$ $N_q = 35$ $N_\gamma = 42$

q_{ult} for axial loading $= 1.3cN_c + \gamma D_f N_q + 0.4\gamma b N_\gamma$

$= 1.3 \times 9 \times 52 + 18 \times 0.5 \times 35 + 0.4 \times 18 \times 2.1 \times 42$

$= 608.4 + 315 + 635.04 \approx 1558$ kN/m^2

Eccentricity ratio, $e/b = 0.35/2.10 = 1/6$.
If the ultimate load is Q_{ult},

$$\text{maximum soil pressure} = 2 \cdot q_{av} = \frac{2 \times Q_{ult}}{\text{Area}}$$

$$= \frac{2 \times Q_{ult}}{2.1 \times 2.1}$$

Equating q_{ult} to this value, $1558 = \dfrac{2Q_{ult}}{4.41}$

$$\therefore \quad Q_{ult} = 1558 \times 4.41 \approx \mathbf{3435 \text{ kN}}$$

Useful width concept:
$$b' = b - 2e = 2.10 - 2 \times 0.35 = 1.40 \text{ m}$$

Since the eccentricity is about only one axis,

$$\text{effective area} = 1.40 \times 2.10 = 2.94 \text{ m}^2$$

$$\therefore \quad q_{ult} = 1.3 \times 9 \times 52 + 18 \times 0.5 \times 35 + 0.4 \times 18 \times 1.4 \times 42$$

$$= 608.4 + 315 + 423.36 \approx 1347 \text{ kN/m}^2$$

$$\therefore \quad Q_{ult} = q_{ult} \times \text{effective area}$$

$$= 1347 \times 2.94 \approx \mathbf{3960 \text{ kN.}}$$

There appears to be significant difference between the results obtained by the two methods. The conventional approach is more conservative.

Example 15.3: Proportion a strap footing for the following data:
Allowable pressures:
  150 kN/m² for DL + reduced LL
  225 kN/m² for DL + LL

Column loads

	Column A	Column B
DL	540 kN	690 kN
LL	400 kN	810 kN

Proportion the footing for uniform pressure under DL + reduced LL. Distance c/c of columns = 5.4 m
Projection beyond column A not to exceed 0.5 m.

DL + reduced LL:
  for column A ... 740 kN
  for column B ... 1095 kN

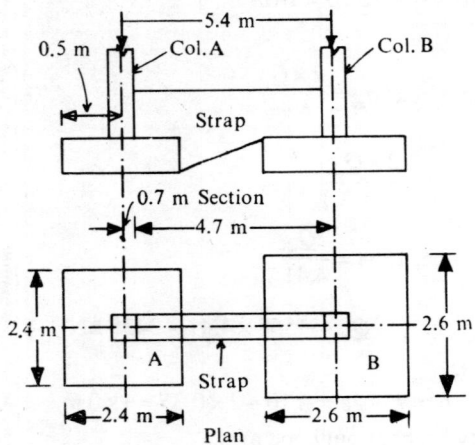

Fig. 15.30 Strap footing (Ex. 15.3)

Footing A
Assume a width of 2.4 m. Eccentricity of column load with respect to the footing
= (1.2 − 0.5) = 0.7 m
c/c of footings (assuming footing B to be centrally placed with respect to column B) = 5.4 − 0.7 = 4.7 m

Enhanced load $= 740 \times \dfrac{5.4}{4.7}$ kN = 850 kN

Area required = 850/150 = 5.67 m²
Width: 5.67/2.40 = 2.36 m.
Use **2.4 m × 2.4 m** footing (actual area 5.76 m²).

Footing B
Load on Column B = 1095 kN

Net load $= 1095 - 740 \times \dfrac{0.7}{4.7} = 985$ kN.

Area required = 950/150 = 6.57 m²
Use **2.6 m × 2.6 m** footing (actual area 6.76 m²)
Soil pressure under DL + LL:

Footing A: Load $= 940 \times \dfrac{54}{47} = 1080$ kN   Pressure $= \dfrac{1080}{5.76} = 187.5$ kN/m²

Footing B: Load $= 1500$ kN $- 940 \times \dfrac{0.7}{47} = 1360$ kN   Pressure $= \dfrac{1360}{6.76} \approx 201$ kN/m²

These are less than 225 kN/m². Hence O.K.

Example 15.4: Proportion a rectangular combined footing for uniform pressure under dead load plus reduced live load, with the following data:

Allowable soil pressures:
150 kN/m² for DL + reduced LL
225 kN/m² for DL + LL

Column loads:

	Column A	Column B
DL	540 kN	690 kN
LL	400 kN	810 kN

Distance c/c of columns = 5.4 m
Projection of footing beyond Column A = 0.5 m

Total Column Loads	Column A	Column B	Total
DL + reduced LL	740 kN	1095 kN	1835 kN
DL + LL	940 kN	1500 kN	2440 kN

For uniform pressure under DL + reduced LL:
Let the distance of a resultant of column loads from Column A be $\bar{x}$ m.

$$\bar{x} = \frac{1095 \times 5.4}{1835} \text{ m} = 3.22 \text{ m}$$

Length $L = 2(3.22 + 0.50) = 7.44$ m   Use **7.50 m**

Width $B = \dfrac{1835}{150 \times 7.5}$ m = 1.63 m say **1.65 m**

Soil pressure under DL + LL:
Let the distance of resultant of column loads from Column A be $\bar{x}_1$ m

$$\bar{x}_1 = \frac{1500 \times 5.4}{24.40} \text{ m} = 3.32 \text{ m}$$

3.25 m from Column A to c.g. of footing $\therefore e = 0.07$ m

$$q_{max} = \frac{2440}{7.5 \times 1.65} \cdot 1 + \frac{6 \times 0.07}{7.5} = 208 \text{ kN/m}^2 < 225 \text{ kN/m}^2 \text{ O.K.}$$

$$q_{min} = \frac{2440}{7.5 \times 1.65} \cdot 1 - \frac{6 \times 0.07}{7.5} = 186 \text{ kN/m}^2$$

Structural design of the footing will have to follow.

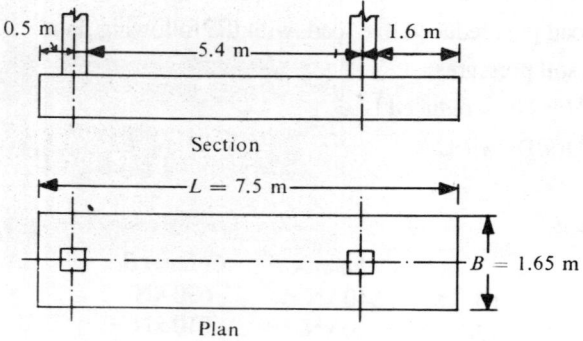

Fig. 15.31 Rectangular combined footing (Ex. 15.4)

Example 15.5: Proportion a trapezoidal combined footing for uniform pressure under dead load plus reduced live load, with the following data:
Allowable soil pressures:
150 kN/m² for DL + reduced LL
225 kN/m² for DL + LL

Column loads

	Column A	Column B
DL	540 kN	690 kN
LL	400 kN	810 kN

Distance c/c of columns = 5.4 m
Projection of footing beyond column not to exceed 0.5 m.

Resultant Column Loads	Column A	Column B	Total
DL + reduced LL	740 kN	1095 kN	1835 kN
DL + LL	940 kN	1500 kN	2440 kN

For uniform pressure under DL + reduced LL:
Let us use equal projections beyond columns A & B.

$$L' = 5.4 \text{ m}$$

$$L = L' + 2e_1 = 5.4 + 2 \times 0.5 = \textbf{6.4 m.}$$

Total area required, $A = 1835/150 = 12.23$ m²
$\bar{x}$, distance of resultant column load from the left edge = 3.22 + 0.5 = 3.72

Shallow Foundations 749

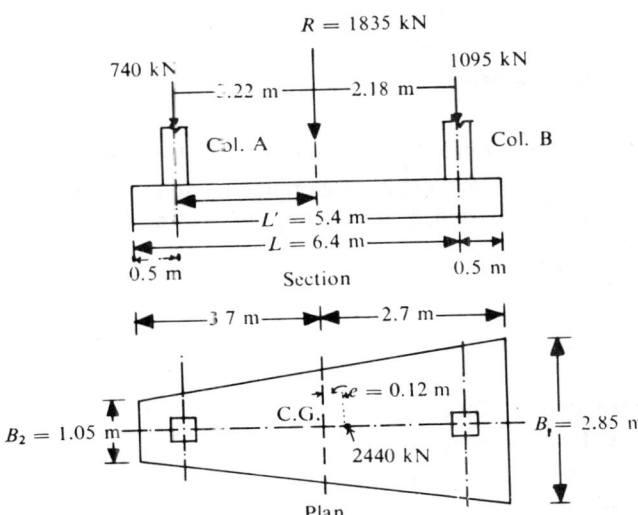

Fig. 15.32 Trapezoidal combined footing (Ex. 15.5)

$$\therefore \quad B_1 = \frac{2A}{L}\left(\frac{3\bar{x}}{L} - 1\right) = \frac{2 \times 12.23}{6.4}\left(\frac{3 \times 3.72}{6.4} - 1\right) = \mathbf{2.85 \text{ m}}$$

$$B_2 = \frac{2A}{L} - B_1 = \frac{2 \times 12.23}{6.4} - 2.85 = \mathbf{1.05 \text{ m}}$$

Total area provided $= \dfrac{2.85 + 1.05}{2} \times 6.4 = 12.48 \text{ m}^2$ (O.K.)

Total DL + LL = 2440 kN

Location of c.g. from $B_1 = (6.4/3)\dfrac{2 \times 1.05 + 2.05}{3.90} = 2.70 \text{ m}$

Location of resultant DL + LL $= \dfrac{940 \times 5.4}{2440} = 2.08$ m from Col. B

or   2.58 m from B

$e = 0.12$ m

Moment of inertia of the section about longer edge

$$= \frac{1}{3} \times 1.05 \times 6.4^3 + \frac{1}{12} \times 1.80 \times 6.4^3 = 131.09 \text{ m}^4$$

M.I. about an axis through c.g. = $131.09 - 12.48 \times 2.70 = 40.1 \text{ m}^4$

$$q_{max} = \frac{2440}{12.48} + \frac{2440 \times 0.12 \times 2.7}{40.1} = 215 \text{ kN/m}^2 < 225 \text{ kN/m}^2 \quad \text{(O.K.)}$$

$$q_{min} = \frac{2440}{12.48} - \frac{2440 \times 0.12 \times 2.7}{40.1} = 176 \text{ kN/m}^2$$

The structural design of the footing has now to follow.

Example 15.6: A raft, 9 m × 27 m, is founded at a depth of 3 m in sand with a value of $N = 25$ up to great depth. Determine the total load which the raft can support. If the raft is designed as a floating foundation, what will be the load it can support?
Assume $\gamma = 18$ kN/m^3.

Allowable soil pressure for a footing for $N = 25$ is 330 kN/m^2
(from Terzaghi and Peck's charts for 40 mm settlement)
Allowable soil pressure for a raft = 2 × 330 = 660 kN/m^2
(According to Peck, Hanson, and Thornburn)
Total load which the raft can support = 660 × 9 × 27 = **160,380 kN**
If the raft is designed as a floating foundation,
the soil pressure = relief of stress due to excavation, = 3 × 18 = 54 kN/m^2
Total load which the raft can support in that case, = 54 × 9 × 27 ≈ **13,120 kN**.

SUMMARY OF MAIN POINTS

1. Foundations are categorised as shallow foundations ($D_f/b \leq 1$) and deep foundations. Shallow foundations are either footings or rafts. Footings may be spread footings which may be continuous for walls, or isolated for columns; the shape of the latter being square, circular or rectangular. Strap footings and combined footings are used to support more than one column; the latter may be rectangular or trapezoidal in shape.
   A floating foundation is not a type but a concept whereby the relief of stress due to excavation is made nearly equal to the structure load.
2. The choice of type of foundation for a given situation may sometimes involve difficult judgement; relative economy of possible types must be studied before a final choice is made. No rigid rules can be made but only approximate guidelines stated.
3. Bearing capacity of footings on sands is invariably governed by settlement criterion, while that on clays by shear failure.
4. The settlement of footings may be considered to consist of contributions due to immediate or elastic compression, consolidation and secondary compression.
5. Proportioning of several footings supporting a structure is done such that settlement is nearly equal for all footings under service loads, which are a judicious combination of dead and live loads.

6. Eccentrically loaded footings or footings subjected to moments are usually designed based on the *useful width concept;* according to this the area symmetrical to the applied load is considered to be the effective or useful area.
7. Combined footings may be rectangular or trapezoidal in plan shape; the latter are used when space restrictions due to the proximity of the property line exist, and when the column loads are very unequal.
8. Raft foundations are preferred on poor soils where spread footings are not practicable; they are designed either by the conventional rigid approach, assuming uniform contact pressure or by the concept of modulus of subgrade reaction approach.
9. Foundation design in non-uniform soil deposits is rather complex; especially so when a dense or stiff layer overlies a soft deposit; great care is required in coming to conclusions in such cases.

REFERENCES

1. Alam Singh and B.C. Punmia: *Soil Mechanics and Foundations,* Standard Book House, Delhi-6, 1970.
2. Bharat Singh and Shamsher Prakash: *Soil Mechanics and Foundation Engineering,* Nem Chand & Bros., Roorkee, U.P., India, 1976.
3. IS: 1080-1980: *Code of Practice for Design and Construction of Simple Spread Foundations* (First Revision), New Delhi, 1980.
4. IS: 2950 (Part-I) 1973: *Code of Practice for Design and Construction of Raft Foundations—Part I - Design,* ISI, New Delhi, 1973.
5. G.A. Leonards: *Foundation Engineering,* ed., McGraw-Hill Book Co., NY, USA, 1962.
6. D.F. McCarthy: *Essentials of Soil Mechanics and Foundations,* Reston Book Co., Va., USA, 1977.
7. G.G. Meyerhof: *Ultimate Bearing Capacity of Foundations,* Geotechnique, Vol. 2., 1951.
8. G.G. Meyerhof: *The Bearing Capacity of Footings Under Eccentric and Inclined Loads,* Proceedings—Third International Conference on Soil Mechanics and Foundation Engineering, Zurich, 1953.
9. G.G. Meyerhof: *Ultimate Bearing Capacity of Footings on Slopes,* Proceedings, Fourth International Conference on Soil Mechanics and Foundation Engineering, London.
10. V.N.S. Murthy: *Soil Mechanics and Foundation Engineering,* Dhanpat Rai & Sons, Delhi-6, 2nd edn., 1977.
11. H.P. Oza: *Soil Mechanics and Foundation Engineering,* Charotar Book Stall, Anand, India, 1969.
12. R.B. Peck, W.E. Hanson, and T.H. Thornburn: *Foundation Engineering,* John Wiley & Sons Inc., NY, USA, 2nd ed., 1974.
13. S.B. Sehgal: *A Textbook of Soil Mechanics,* Metropolitan Book House Pvt. Ltd., Delhi-6, 1967.

752 *Geotechnical Engineering*

14. Shamsher Prakash and Gopal Ranjan: *Problems in Soil Engineering*, Sarita Prakashan, Meerut, India, 1976.
15. Shamsher Prakash, Gopal Ranjan and Swami Saran: *Analysis and design of Foundations and Earth-Retaining Structures*, Sarita Prakashan, Meerut, India, 1979.
16. A.W. Skempton and L. Bjerrum: *A contribution to Settlement Analysis of Foundations in Clay*, Geotechnique, London, 1957.
17. W.C. Teng: *A Study of Contact Pressure Against a Large Raft Foundation*, Geotechnique, London, 1949.
18. W.C. Teng: *Foundation Design*, Prentice Hall of India Pvt., Ltd., New Delhi, 1976.
19. K. Terzaghi and R.B. Peck: *Soil Mechanics in Engineering Practice*, John Wiley & Sons Inc., NY, USA, 1948.
20. K. Terzaghi: *Evaluation of Coefficient of Subgrade Reaction*, Geotechnique, London, 1955.

QUESTIONS

15.1 (a) What is the function of a 'foundation'?
 (b) Write an explanatory note on the general types of foundations, with suitable sketches.
15.2 (a) What are the general considerations in the choice of the foundation type?
 (b) How is the depth of the foundation determined?
15.3 (a) How is the settlement of footings estimated?
 (b) Write a note on the methods of proportioning of footings for equal settlement.
15.4 (a) How are eccentrically loaded footings designed?
 (b) Write a note on the 'useful width concept'.
15.5 (a) Explain the circumstances under which a strap footing is used.
 (b) What is the basis for design of strap footings?
15.6 (a) What are the conditions under which combined footings are used?
 (b) When is a trapezoidal combined footing preferred to as rectangular one? Explain how it is proportioned.
15.7 (a) What is a 'raft foundation'? When is it preferred?
 (b) Explain the concept of floating foundation applied to a raft.
15.8 (a) Explain the conventional rigid approach to the design of a raft foundation.
 (b) What is the coefficient of subgrade reaction? On what factors does it depend?
15.9 (a) Explain how a foundation may be designed when a dense stratum overlies a loose one.
 (b) How is a foundation designed when soft and stiff layers alternate at a site?
15.10 (a) A building is supported symmetrically on nine columns, spaced at 4.5 m c/c.

At the chosen depth of 2 m, the allowable bearing capacity is 300 kN/m^2; $\gamma = 18$ kN/m^3.

(b) The column loads are as given below:

Column No.	1	2	3	4	5	6	7	8	9
DL (kN)	200	350	250	270	500	350	150	350	200
LL (kN)	200	400	200	270	650	350	120	300	180

Proportion the footings for uniform pressure and for equal settlement.

15.11 What is the ultimate load which an eccentrically loaded square footing of 2 m size with an eccentricity of 0.40 m can take at a depth of 0.6 m in a soil with $\gamma = 20$ kN/m^3, $c = 12$ kN/m^2, and $\phi = 30°$, $N_c = 30$, $N_q = 18$, and $N_\gamma = 15$.

15.12 Proportion a strap footing for the following data:
Allowable soil pressures:
for DL + reduced LL : 180 kN/cm^2
for DL + LL : 270 kN/m^2

	Column A	Column B
DL	500 kN	660 kN
LL	400 kN	840 kN

Distance c/c of columns: 5 m
Projection beyond column A not to exceed 0.5 m.

15.13 Proportion a rectangular combined footing for the data of Problem 15.12.

15.14 Proportion a trapezoidal combined footing for the data of Problem 15.12, if the projection beyond both columns cannot exceed 0.5 m.

15.15 A raft, 8 m × 24 m, is founded at a depth of 4 m in sand with a value of $N = 20$ up to great depth. What is the total load which the raft can support? What will be the total capacity if it is to act as a floating foundation at this depth?
Assume $\gamma = 20$ kN/m^3.

16
PILE FOUNDATIONS

16.1 INTRODUCTION

Deep foundations are employed when the soil strata immediately beneath the structure are not capable of supporting the load with tolerable settlement or adequate safety against shear failure. Merely extending the level of support to the first hard stratum is not sufficient, although this is a common decision that is reached. Instead, the deep foundation must be engineered in the same way as the shallow foundation so that the soil strata below remain safe and free of deleterious settlement.

Two general forms of deep foundation are recognised:
1. Pile foundation
2. Pier, caisson or well foundation.

Piles are relatively long, slender members that are driven into the ground or cast-in-situ. Piers, caissons or wells are larger, constructed by excavation and are sunk to the required depth; these usually permit visual examination of the soil or rock on which they rest. In effect they are deep spread footings or mats. They are normally used to carry very heavy loads such as those from bridge piers or multi-storeyed buildings. A sharp distinction between piles and piers is impossible because some foundations combine features of both.

Piles have been used since prehistoric times. The Neolithic inhabitants of Switzerland, 12,000 years ago, drove wooden poles in the soft bottoms of shallow lakes and on them erected their homes, high above marauding animals and warring neighbours. Pile foundations were used by Romans; Vitruvius (59 A.D.) records the use of such foundations.

Today, pile foundations are much more common than any other type of deep foundation, where the soil conditions are unfavourable.

16.2 CLASSIFICATION OF PILES

Piles may be classified in a number of ways based on different criteria:
  (*a*) Function or action
  (*b*) Composition and material
  (*c*) Installation

16.2.1 Classification Based on Function or Action
Piles may be classified as follows based on the function or action:

End-bearing piles

Used to transfer load through the pile tip to a suitable bearing stratum, passing soft soil or water.

Friction piles

Used to transfer loads to a depth in a frictional material by means of skin friction along the surface area of the pile.

Tension or uplift piles

Used to anchor structures subjected to uplift due to hydrostatic pressure or to overturning moment due to horizontal forces.

Compaction piles

Used to compact loose granular soils in order to increase the bearing capacity. Since they are not required to carry any load, the material may not be required to be strong; in fact, sand may be used to form the pile. The pile tube, driven to compact the soil, is gradually taken out and sand is filled in its place thus forming a 'sand pile'.

Anchor piles

Used to provide anchorage against horizontal pull from sheetpiling or water.

Fender piles

Used to protect water-front structures against impact from ships or other floating objects.

Sheet piles

Commonly used as bulkheads, or cut-offs to reduce seepage and uplift in hydraulic structures.

Batter piles

Used to resist horizontal and inclined forces, especially in water front structures.

Laterally-loaded piles

Used to support retaining walls, bridges, dams, and wharves and as fenders for harbour construction.

16.3.2. Classification Based on Material and Composition
Piles may be classified as follows based on material and composition:

Timber piles

These are made of timber of sound quality. Length may be up to about 8 m; splicing is adopted for greater lengths. Diameter may be from 30 to 40 cm. Timber piles perform well either in fully dry condition or submerged condition. Alternate wet and dry conditions reduce the life of a timber pile; to overcome this, creosoting is adopted. Maximum design load is about 25 tonnes.

Steel piles

These are usually H-piles (rolled H-shape), pipe piles, or sheet piles (rolled sections of regular shapes). They may carry loads up to 100 tonnes or more.

Concrete piles

These may be 'precast' or 'cast-in-situ'. Precast piles are reinforced to withstand handling stresses. They require space for casting and storage, more time to cure and heavy equipment for handling and driving.

Cast-in-situ piles are installed by pre-excavation, thus eliminating vibration due to driving and handling. The common types are Raymond pile, MacArthur pile and Franki pile.

Composite piles

These may be made of either concrete and timber or concrete and steel. These are considered suitable when the upper part of the pile is to project above the water table. Lower portion may be of untreated timber and the upper portion of concrete. Otherwise, the lower portion may be of steel and the upper one of concrete.

16.2.3 Classification Based on Method of Installation

Piles may also be classified as follows based on the method of installation:

Driven piles

Timber, steel, or precast concrete piles may be driven into position either vertically or at an inclination. If inclined they are termed 'batter' or 'raking' piles. Pile hammers and pile-driving equipment are used for driving piles.

Cast-in-situ piles

Only concrete piles can be cast-*in-situ*. Holes are drilled and these are filled with concrete. These may be straight-bored piles or may be 'under-reamed' with one or more bulbs at intervals. Reinforcements may be used according to the requirements.

Driven and cast-in-situ piles

This is a combination of both types. Casing or shell may be used. The Franki pile falls in this category.

16.3 USE OF PILES

The important ways in which piles are used are as follows:
  (*i*) To carry vertical compressive loads,
  (*ii*) To resist uplift or tensile forces, and
  (*iii*) To resist horizontal or inclined loads.

Bearing piles are used to support vertical loads from the foundations of buildings and bridges. The load is carried either by transferring to the incompressible soil or rock below through soft strata or by spreading the load through soft strata that are incapable of supporting concentrated loads from shallow footings. The former type are called point-bearing piles, while the latter are known as friction-piles.

Tension piles are used to resist upward forces in structures subjected to uplift, such as buildings with basements below the ground water level, aprons of dams or buried tanks. They are also used to resist overturning of walls and dams and for anchors of towers, guywires and bulkheads.

Laterally loaded piles support horizontal or inclined forces such as the foundations of retaining walls, bridges, dams, and wharves and as fenders in harbour construction.

In case the lateral loads are of large magnitude they may be more effectively resisted by batter piles, driven at an inclination. Closely spaced piles or thin sheet piles are used as cofferdams, seepage cut-offs and retaining walls. Piles may be used to compact loose granular soils and also to safeguard foundations against scouring. These are illustrated in Fig. 16.1.

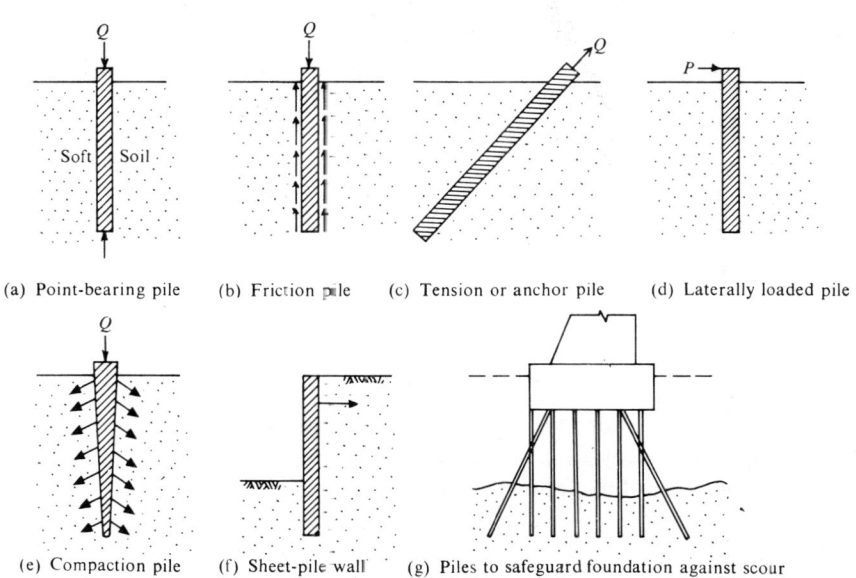

Fig. 16.1 Uses of piles

16.4 PILE DRIVING

The operation of forcing a pile into the ground is known as 'pile driving'. The oldest method and the most widely used even today is by means of a hammer. The equipment used to lift the hammer and allow it to fall on to the head of the pile is known as the 'pile driver'. The Romans used a stone block hoisted by an A-frame derrick with slave or horse power. While such a simple pile-driving rig is still in use today with mechanical power, the more common equipment consists of essentially a crawler-mounted crane, shown schematically in Fig. 16.2. Attached to the boom are the 'leads', which are just two parallel steel channels fastened together by U-shaped spacers and stiffened by trussing. The leads are braced against the crane with a stay, which is usually adjustable to permit driving of batter piles. A steam generator or air compressor is required for steam hammers.

The most important feature of the driving rig, from an engineering point of view, is its ability to guide the pile accurately. It must be rugged and rigid enough to keep the pile and hammer in alignment and plumb inspite of wind, underground obstructions and the movement of the pile hammer.

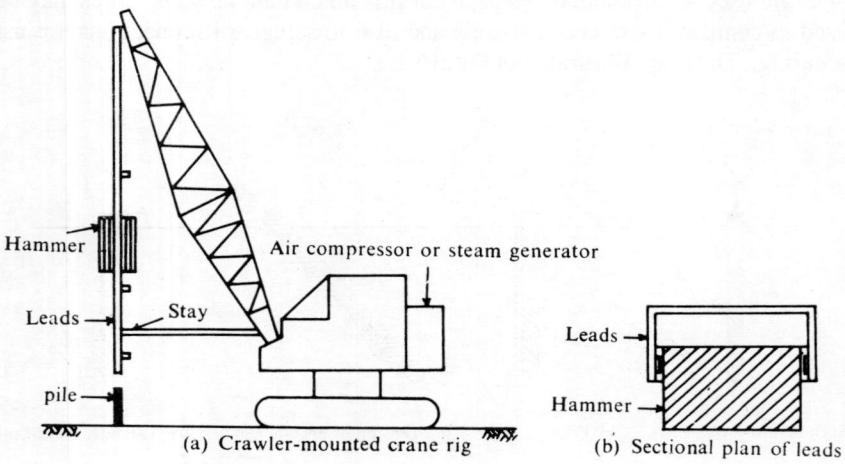

Fig. 16.2 Pile driver with crawler-mounted crane rig

Pneumatic-tyred motor crane rigs are used for highway work; rail-mounted rigs are available for railway work; and barge-mounted rigs are used for marine work.

Pile hammers are of the following types:
 (i) Drop hammer
 (ii) Single-acting hammer (steam or pneumatic)
 (iii) Double-acting hammer (steam or pneumatic)
 (iv) Diesel hammer (internal combustion)
 (v) Vibratory hammer

Drop hammer

This is the simplest type. The hammer, ram or monkey is raised by pulley and winch and allowed to fall on the top of the pile.

The drop hammer is simple but very slow and is used for small jobs only.

Pile Foundations

Single-acting hammer

In this type, the hammer is raised by steam or compressed air and is allowed to drop on to the pile head. The hammer is usually heavy and rugged, weighing 1000 to 10000 kg. The height of fall may be about 60 to 90 cm. The blows may be delivered much more rapidly than in the case of drop hammer.

Double-acting hammer

In this type, steam or air pressure is employed to lift the ram and then accelerate it downward. The blows are more rapid; from 90 to 240 blows per minute, thus reducing the time required to drive the pile, and making the driving easier. The weight of the ram may be 1000 to 2500 kg.

This type of hammer loses its effectiveness with wear and poor valve adjustment. The energy delivered in each blow varies greatly with the steam or air pressure. If the number of blows per minute is approximately the rated value, the pressure is probably correct.

Steam operation is more efficient, particularly with circulating steam generators. If the hammer is to be operated under water, which is possible with enclosed double-acting types, compressed air operation is necessary.

The advantage of power hammers is that the blows follow in rapid succession, keeping the pile in continuous motion and reducing the effect of impact, thus minimising the damage to the pile head.

Diesel hammer

This works on the internal combustion of diesel oil. Energy is provided both for raising the hammer and for downward stroke. This type is self-contained, economical, and simple. The energy delivered per blow is relatively high, considering the weight of the hammer, as it is developed by a high-velocity blow. The disadvantage is that the energy per hammer blow varies with the resistance offered by the pile and is difficult to evaluate. Thus, the diesel hammer is best adopted to conditions where controlled energy is not critical.

A double-acting hammer activated by hydraulic pressure is somewhat faster and lighter compared to equivalent steam hammers because the operating pressure is much higher. The compact hydraulic pump system is easier to move than the bulky air compressor or steam generator.

A heavy single-acting hammer may be more effective sometimes than a light double-acting hammer, since the hammer may bounce back due to high velocity, in soils of high resistance to penetration.

Vibratory hammer

The driving unit vibrates at high frequency and thus, the driving is quick and quiet. A variable speed oscillator is used for the purpose of creating resonance conditions. This allows easy penetration of the pile with a relatively small driving effort. This method is popular in the U.S.S.R.

Most pile hammers require the use of driving heads, helmets, or 'pile caps' that distribute the force of the blow more evenly over the pile head. A 'cushion', consisting of a pad of resilient material such as wood, fibre, or plastic, is interposed between the pile head and pile cap, as the top portion of the pile cap.

Piles are ordinarily driven to a resistance measured by the number of blows required for the last 1 cm. of penetration depending upon the material and weight of the pile.

16.5 PILE CAPACITY

The ultimate bearing capacity of a pile is the maximum load which it can carry without failure or excessive settlement of the ground. The allowable load on a pile is the load which can be imposed upon it with an adequate margin of safety; it may be the ultimate load divided by a suitable factor of safety, or the load at which the settlement reaches the allowable value.

The bearing capacity of a pile depends primarily on the type of soil through which and/or on which it rests, and on the method of installation. It also depends upon the cross-section and length of the pile.

The pile shaft is a structural column that is fixed at the point and usually restrained at the top. The elastic stability of piles, or their resistance against buckling, has been investigated both theoretically and by load tests (Bjerrum, 1957). Both theory and experience demonstrate that buckling rarely occurs because of the effective lateral support of the soil; it may occur only in extremely slender piles in very soft clays or in piles that extend through open air or water. Therefore, the ordinary pile in sand or clay may be designed as though it were a short column.

The pile transfers the load into the soil in two ways. Firstly, through the tip-in compression, termed 'end-bearing' or 'point-bearing'; and, secondly, by shear along the surface, termed 'skin friction'. If the strata through which the pile is driven are weak, the tip resting on a hard stratum transfers most part of the load by end-bearing; the pile is then said to be an end-bearing pile. Piles in homogeneous soils transfer the greater part of their load by skin friction, and are then called friction piles; however, nearly all piles develop both end-bearing and skin friction.

The following is the classification of the methods of determining pile capacity:
 (*i*) Static analysis
 (*ii*) Dynamic analysis
 (*iii*) Load tests on pile
 (*iv*) Penetration tests

The first two are theoretical approaches and the last two are field or practical approaches.

16.5.1 Static Analysis

The ultimate bearing load of a pile is considered to be the sum of the end-bearing resistance and the resistance due to skin friction:

$$Q_{up} = Q_{eb} + Q_{sf} \qquad \ldots \text{(Eq. 16.1)}$$

where Q_{up} = ultimate bearing load of the pile,
 Q_{eb} = end-bearing resistance of the pile, and
 Q_{sf} = skin-friction resistance of the pile.

However, at low values of load Q_{eb} will be zero, and the whole load will be carried by skin friction of soil around the pile. Q_{eb} and Q_{sf} may be analysed separately; both are based upon the state of stress around the pile and on the shear patterns that develop at failure. Meyerhof (1959) and Vesic (1967) proposed certain failure surfaces for deep foundations. According to Vesic, only punching shear failure occurs in deep foundations irrespective of the density index of the soil, so long as the depth to width ratio is greater than 4 (This is invariably so for pile foundations).

$$Q_{eb} = q_b \cdot A_b \qquad \ldots\text{(Eq. 16.2)}$$

$$Q_{sf} = f_s A_s \qquad \ldots\text{(Eq. 16.3)}$$

Here, q_b = bearing capacity in point-bearing for the pile,
f_s = unit skin friction for the pile-soil system,
A_b = bearing area of the base of the pile, and
A_s = surface area of the pile in contact with the soil.

The general form of the equation for q_b presented by various investigators is:

$$q_b = cN_c + \frac{1}{2}\gamma b N_\gamma + q \cdot N_q \qquad \ldots\text{(Eq. 16.4)}$$

which is the same form as the bearing capacity of shallow foundations.

For piles in sands:

$$q_b = \frac{1}{2}\gamma b N_\gamma + q \cdot N_q \qquad \ldots\text{(Eq. 16.5)}$$

(for square or rectangular piles)

$$q_b = 0.3\gamma D N_\gamma + q \cdot N_q \qquad \ldots\text{(Eq. 16.6)}$$

(for circular piles with diameter D)

With driven piles the term involving the size of the pile is invariably negligible compared with the surcharge term $q \cdot N_q$. Thus, for all practical purposes,

$$q_b = q \cdot N_q \qquad \ldots\text{(Eq. 16.7)}$$

The surcharge pressure q is given by

$q = \gamma_z$ if $Z < Z_c$, and

$$q = \gamma \cdot Z_c \text{ if } Z > Z_c \qquad \ldots\text{(Eq. 16.8)}$$

Z being the embedded length of pile and Z_c the critical depth.

This indicates that the vertical stress at the tip of a long pile tends to reach a constant value and the depth beyond which the stress does not increase linearly with depth is called the critical depth. This is due to the mechanics of transfer of load from a driven pile to the surrounding soil. Large-scale tests by Vesic (1967) in the U.S.A. and Kerisel (1967) in France indicate that the critical depth Z_c is a function of density index. For $I_D < 30\%$, $Z_c = 10\,D$; for $I_D > 70\%$, $Z_c = 30\,D$; and, for intermediate values, it is nearly proportional to density index (D is the dimension of the pile cross-section).

The bearing capacity factor N_q is related to the angle of internal friction of the sand in the vicinity of the pile tip (several pile diameters above and below the pile tip), and the ratio of the pile depth to pile width. Values of N_q presented by different investigators show a wide range of variation because of the assumptions made in defining the shear zones near the pile tip; for example, while Meyerhof assumes the shear zones to extend back to the pile shaft, Vesic assumes punching shear in which the shear zones do not extend to the pile shaft. Values of N_q attributed to Berezantzev et al (1961), which take into account the effect of z/b ratio, are believed to be the most applicable for the most commonly encountered field conditions. The angle of internal friction for the soil in the vicinity of the pile tip is determined from the standard penetration test. If Dutch cone resistance data are available, these values are correlated directly to the end-bearing resistance of the pile, q_b.

Values of N_q given by different investigators are shown in Fig. 16.3.

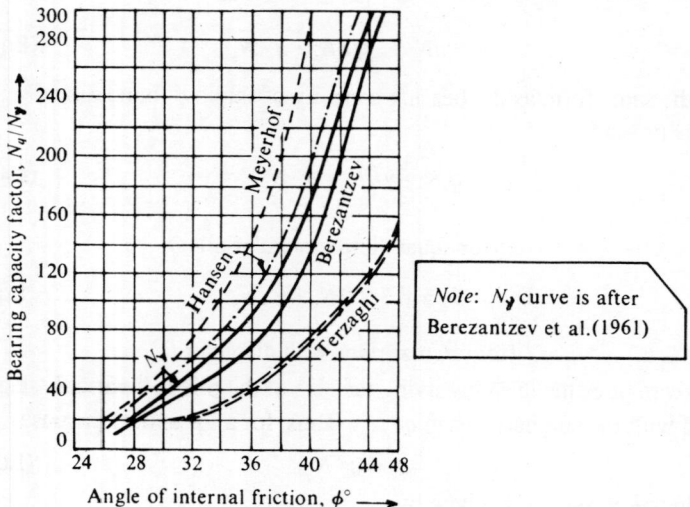

Fig. 16.3 Bearing capacity factor N_q for piles in sand (Nordlund, 1963)

These values of N_q are based on the assumption that the soil above the pile tip is comparable to the soil below the pile tip. If the pile penetrates the compact layer only slightly, and loose material exists above the compact soil, an N_q value for a shallow foundation will be more appropriate than a value from Fig. 16.3.

If Eq. 16.5 or 16.6 is to be used, the value of N_γ for a deep foundation can be conservatively taken as twice the N_γ value used for shallow foundation; otherwise, it may be taken from Fig. 16.3, the values given by Berezantzev et al.

According to Nordlund and Tomlinson (1969), Berezantzev's values of N_q increase rapidly for high values of ϕ. Further, the decrease in N_q with increase in z/b also will be significant for high values of ϕ.

Vesic's equation for q_b is:

$$q_b = 3q \cdot N_q \qquad \text{...(Eq. 16.9)}$$

where
$$N_q = e^{3.8\phi \tan \phi} \cdot N_\phi \qquad \text{...(Eq. 16.10)}$$

with the usual notation for $N_\phi [= \tan^2(45° + \phi/2)]$.
Vesic's values of N_q are given in Table 16.1:

Table 16.1 Vesic's values of N_q for deep foundation

$\phi°$	0	5	10	15	20	25	30	35	40	45	50
N_q	1.0	1.2	1.6	2.2	3.3	5.3	9.5	18.7	42.5	115.4	4.22

For *piles in clays*, q_b is given by:

$$q_b = cN_c + q \qquad \text{...(Eq. 16.11)}$$

since $\qquad N_q = 1$ and $N_\gamma = 0$ for $\phi = 0°$

N_c ranges from 6 to 10 depending upon the stiffness of the clay; a value of 9 is taken for N_c conventionally.

It is also considered that q is not significant compared to cN_c. Hence, for all practical purposes

$$q_b = 9c \qquad \text{...(Eq. 16.12)}$$

(for piles in clay)

The general form for the unit skin friction resistance, f_s, is given by

$$f_s = c_a + \sigma_h \tan \delta \qquad \text{...(Eq. 16.13)}$$

where c_a = adhesion, which is independent of the normal pressure on the contact area. Cohesion c is used if the shearing is between soil and soil;

σ_h = average lateral pressure of soil against the pile surface; and

δ = angle of wall friction, which depends upon the material of the pile and the condition of its surface.

σ_h is given by

$$\sigma_h = K_s \cdot q \qquad \text{...(Eq. 16.14)}$$

where K_s = coefficient of earth pressure.
For loose sand ($I_D < 30\%$), $K_s = 1$ to 3, and
for dense sand ($I_D > 70\%$), $K_s = 2$ to 5
For piles in sands:
$f_s = \sigma_h \tan \delta$, c_a being zero.

The values of $\tan \delta$ may be determined by direct shear tests in which one half of the shear box is replaced by the same material as the pile surface.

Representative values of the coefficient of friction between sand and various pile materials are shown in Table 16.2.

Table 16.2 Coefficient of friction between sand and pile materials (McCarthy, 1977)

S. No.	Material	Coefficient of friction tan δ
1.	Wood	0.4
2.	Concrete	0.45
3.	Steel, smooth	0.2
4.	Steel, rusted	0.4
5.	Steel, corrugated	tan φ

Values of ratio δ/ϕ as determined by Potyondy from shear box tests are shown in Table 16.3.

Table 16.3 Values of δ/ϕ for pile materials in contact with dense and dry sand (After Potyondy, 1961)

S. No.	Material and surface condition	δ/ϕ
1.	Wood, parallel to grain	0.76
2.	Wood, perpendicular to grain	0.88
3.	Rough concrete, cast against soil	0.98
4.	Smooth concrete, poured in form work	0.80
5.	Steel, smooth	0.54
6.	Steel, rusted	0.76

For piles in clays:

$$f_s = c_a, \text{ since tan } \delta \text{ is zero.}$$

The adhesion c_a may be expressed as

$$c_a = \alpha \cdot c \qquad \ldots \text{(Eq. 16.15)}$$

where α is called the 'adhesion factor', which varies with the consistency of the clay.

When a pile is driven in soft clay, the soil around gets remoulded and loses some of its strength. However, it regains almost its full strength within a few weeks of driving, through consolidation. Since piles will be usually loaded a few months after driving, this reduction in strength soon after driving does not pose any problem. However, if piles are to be loaded soon after driving, the remoulded shear strength is to be considered.

When piles are driven into stiff clays, the soil close to the pile may get remoulded and this may also create a slight gap between the pile and the soil; consequently the adhesion is always smaller than cohesion and α will be less than unity.

The adhesion factors for different pile materials and consistency of the clay are shown in Table 16.4.

Table 16.4 Adhesion factors for piles in clay (Tomlinson, 1969)

S. No.	Material of pile	Consistency of clay	Cohesion (kN/m²)	Adhesion factor
1.	Wood and concrete	Soft	0 – 35	0.90 to 1.00
		Medium	35 – 70	0.60 to 0.90
		Stiff	70 – 140	0.45 to 0.60
2.	steel	Soft	0 – 35	0.45 to 1.00
		Medium	35 – 70	0.10 to 0.50
		Stiff	70 – 140	0.50

Under-reamed Piles

An 'under-reamed' pile is one with an enlarged base or a bulb; the bulb is called 'under-ream'. There could be one or more under-reams in a pile; in the former case, it is called a single under-reamed pile and in the latter, it is said to be a multi-under-reamed pile (Fig. 16.4).

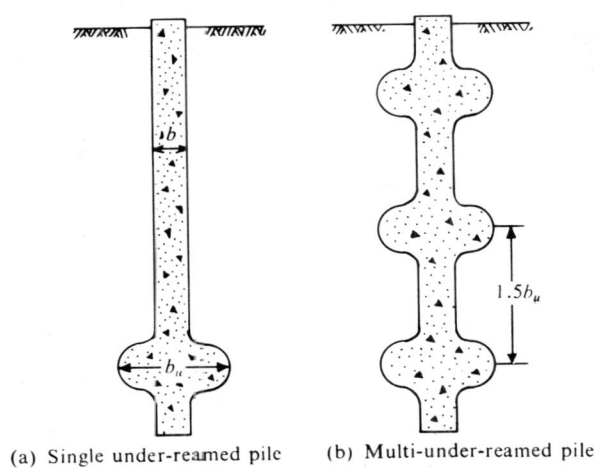

(a) Single under-reamed pile (b) Multi-under-reamed pile

Fig. 16.4 Under-reamed piles

Under-reamed piles are cast-in-situ piles, which may be installed both in sandy and in clayey soils. The sides may be stabilised, if necessary, by the use of bentonite slurry, sometimes called 'drilling mud'. The under-reams are formed by a special under-reaming equipment. The ratio of bulb size to the pile shaft size may be 2 to 3; usually a value of 2.5 is used. The bearing capacity of the pile increases because of the increased base area; the more the number of under-reams the more the capacity. Field tests indicate that an under-reamed pile is more economical than a straight bored pile for a given load.

The load capacity of an under-reamed pile may be found in much the same way as for driven piles [Murthy, 1977 and IS: 2911 (Part I)-1964]:

$$Q_{up} = Q_{eb} + Q_{sf} = q_b \cdot A_b + f_s \cdot A_s \qquad \ldots(\text{Eq. 16.16})$$

where q_b = unit point-bearing capacity of a bulb
f_s = unit skin-friction
A_b = area of section of the bulb
A_s = surface area of the embedded pile shaft.

This is for a single under-reamed pile.

For a multi-under-reamed pile,

$$Q_{up} = q_b \cdot A_b + f_s A_s + \bar{f_s}\bar{A_s} \qquad \ldots(\text{Eq. 16.17})$$

where q_b = unit point bearing resistance,
f_s = unit skin-friction,

$\bar{f}_s$ = unit frictional resistance between soil and soil,

A_b = area of section of the lowest bulb,

A_s = surface area of the embedded portion of pile above the top bulb, and

$\overline{A}_s$ = surface area of a cylinder of diameter b_u and height equal to the distance between the centres of the extreme bulbs.

This is based on the assumption that the soil between the bulbs might move together with the bulbs at ultimate load. Many factors are involved in the type of failure that may occur and this is only an intelligent guess.

16.5.2 Dynamic Analysis

Dynamic analysis aims at establishing a relationship between pile capacity and the resistance offered to driving with a hammer. This is appropriate for piles penetrating soils such as sands and hard clays that will not develop pore water pressures during installation. In saturated fine-grained soils, high pore pressures develop due to vibration caused by driving; in such cases, the predicted capacities from dynamic analysis will be different from the value attained after the dissipation of excess pore pressures.

The loading and failure produced by driving with a hammer occurs in a fraction of a second, whereas in the structure the load is applied over a fairly long period. A fixed relation between dynamic and long-term capacity can exist only in a soil for which shear strength is independent of the rate of loading. This is nearly true in the case of dry sand and also in medium dense wet sand with coarse grains. In clays and in loose fine-grained saturated soils, the strength depends upon the rate of shear; in such soils, dynamic analysis of pile capacity cannot be valid.

Wave equation method

As the top of the pile is struck with the pile hammer, the impact energy of the blow causes a stress wave to be transmitted through the length of the pile. Some of the force is absorbed by the surrounding soil, while some is imparted to the soil at the pile tip. What has come to be known as the 'wave equation method' involves the application of the wave transmission theory to determine the load-carrying capacity developed by the pile and also to determine the maximum stresses that occur within the pile during driving.

The dynamic process of pile driving is analogous to the impact of a concentrated mass upon an elastic rod. The rod is restrained partially by skin friction and by end bearing at the tip. The system can be approximated by a lumped mass elastic model (Ramot, 1967), as shown in Fig. 16.5. The distributed mass of the pile is represented by a series of small concentrated masses, linked by springs that simulate the longitudinal resistance of the pile. The skin friction can be represented by a rheological model of damping or surface restraint that includes friction and elastic distortion also.

When the hammer strikes the pile cap, a force is generated that accelerates the cap (W_c) and compresses it. This transfers a certain force to the top segment of the pile (W_1), and causes it to accelerate, slightly after the acceleration of the pile cap. The compressive force induced in the top segment accelerates the next segment

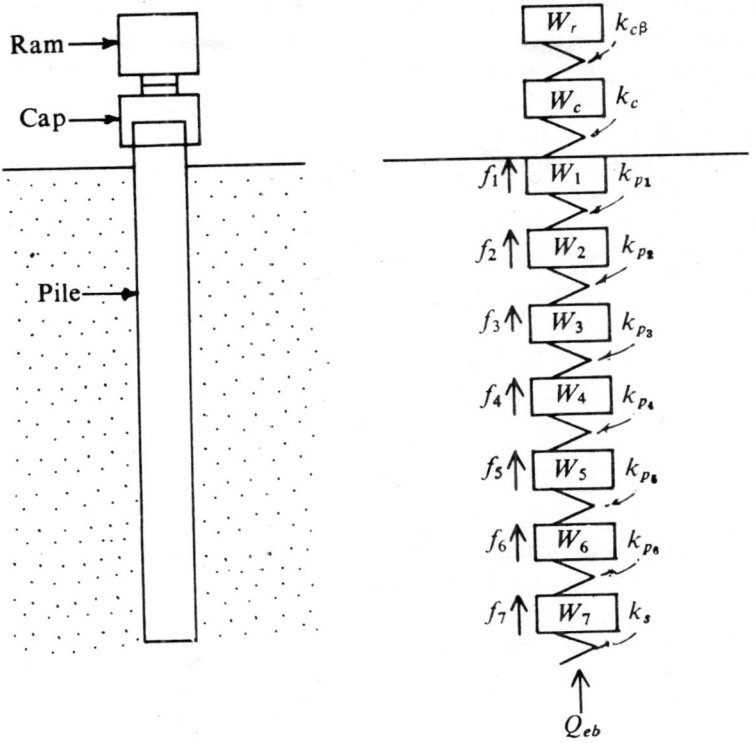

Fig. 16.5 Wave equation method for pile capacity

of the pile (W_2). Thus, a wave of compression moves down the pile. The vertical force at any instant is equal to the compression of the spring. The force wave is partially dissipated in overcoming skin friction on the way down; at the bottom, the remaining force overcomes end-bearing. In order that the pile penetrate deeper, the force of the wave must exceed the summation of ultimate values of skin friction and end-bearing. If this does not happen, the pile is said to meet 'refusal'. The shape of the wave depends upon the rigidity of the pile. If the maximum stress produced exceeds the strength of the pile, the pile will get damaged due to 'over driving'.

To obtain a solution for the wave equation, it is necessary to know the approximate values of length, cross-section, elastic properties and weight of the pile and the pile hammer characteristics, and to assign suitable values for spring constants and soil damping. By analysing changing conditions for successive small increments of time, the effects of the travelling force wave are simulated. This requires numerical integration, which is conveniently handled by a computer. Generally, the analysis is used for diagnosing causes of unusual driving behaviour or as a guide for more efficient choice of equipment or pile.

Approximate methods — dynamic pile driving formulae

Approximate methods of dynamic analysis or the socalled 'dynamic pile-driving formulae', have been used for more than a century to predict pile capacity. These are developed from work-energy theory and are simpler to apply than the wave equation. Hence, these formulae are still useful in predicting pile capacity from simple observations of driving resistance.

The basic assumption underlying the pile formulae is that the kinetic energy delivered by the pile hammer is transferred to the pile and the soil and, accomplishes useful work by forcing the pile into the soil against its dynamic resistance.

Thus, the basic work-energy equation is as follows:

$$W_h \times H \times \eta = Q_{up} \times s + \text{Energy losses} \qquad \ldots(\text{Eq. 16.18})$$

where W_h = weight of pile hammer,

H = height of fall of hammer,

η = efficiency of hammer,

Q_{up} = ultimate capacity of the pile, and

s = penetration (set) of the pile per blow.

This relationship is solved for Q_{up}, assumed to be equal to the capacity of the pile under sustained loading.

Energy losses include those due to elastic compression of the pile, soil, pile cap and cushion and by heat generation. Variations in the several pile formulae result from the different methods to account for the energy losses, which is also the major uncertainty in this approach.

The assumption of work-energy theory does not properly consider the effect of impact on a long member such as a pile; however, some of the formulae which consider the losses in an empirical manner, have shown reliability for predicting the pile capacity in cohesionless soil.

Engineering News formula, Danish formula and Hiley's formula are the more commonly used pile formulae.

Engineering News formula

The 'Engineering News' formula (Wellington, 1886) was derived from observations of driving of timber piles in sand with a drop hammer. The general form of this equation is as follows:

$$Q_{up} = \frac{W_h H}{(s + C)} \qquad \ldots(\text{Eq. 16.19})$$

where s = final penetration (set) per blow. It is taken as average penetration per blow for the last 5 blows or 20 blows depending on whether the hammer is a drop hammer or steam hammer;

C = empirical constant (representing the temporary elastic compression of the helmet, pile and soil)

This is a dimensionally correct equation—H, s and C should be in the same units. Then the units of Q_{up} will be those of W_h.

A factor of safety of 6 was introduced to make up for any inaccuracies arising from the use of arbitrary values for the constant, while arriving at the allowable load on the pile.

$$Q_{ap} = \frac{W_h \cdot H}{6(s+C)} \qquad ...(Eq.\ 16.20)$$

The value of C (in cm) is taken as 2.5 for drop hammer, and 0.25 for steam hammer.

For convenience in practical use, Eq. 16.20 may be transformed into mixed units as follows:

$$Q_{ap} = \frac{500\ W_h \cdot H}{3(s+25)} \text{ for drop hammer} \qquad ...(Eq.\ 16.21a)$$

$$Q_{ap} = \frac{500\ W_h \cdot H}{3(s+2.5)} \text{ for steam hammer} \qquad ...(Eq.\ 16.21b)$$

where H and s are respectively in metres and millimetres, and Q_{ap} in the same units as W_h.

For double acting steam hammers:

$$Q_{ap} = \frac{(W_h + ap) \cdot H}{6(s+2.5)} \qquad ...(Eq.\ 16.22)$$

where W_h = weight of hammer (newtons),
  a = effective area of piston (mm^2),
  p = mean effective steam pressure (N/mm^2),
  H = height of fall of hammer (metres)
  s = final penetration of pile per blow (mm), and
  Q_{ap} = allowable load on the pile (kN).

(Note: This equation has mixed units).

Numerous load tests on piles indicate that the real safety factor of this formula averages 2 instead of the assumed value of 6; it can be as low as 2/3 and as high as 20.

For wood piles driven with drop hammers and for lightly loaded short piles driven with steam hammers, the Engineering News formula gives a crude indication of pile-capacity. For other conditions, it can be very misleading.

Hiley's formula

This is considered to be the most complete one, as described by Chellis (1961). Hiley (1930) takes into account the energy losses in Eq. 16.18 as follows:
1. Elastic compression of pile, pile cap and soil:

$$\frac{1}{2} Q_{up}(c_1 + c_2 + c_3) \text{ or } Q_{up} \cdot C$$

where $C = \frac{1}{2}(c_1 + c_2 + c_3)$

  c_1 = elastic compression of pile

c_2 = elastic compression of pile cap
c_3 = elastic compression of soil.

This is on the assumption of gradual application of the load.

2. The loss of energy during the impact of the pile and hammer:

This depends upon the coefficient of restitution, C_r, for the system which may vary between 0.25 and 0.90, depending upon the materials involved. The available energy after impact is given by multiplying the energy of the hammer by

$$\left(\frac{W_h + C_r^2 W_p}{W_h + W_p} \right)$$

The loss of energy is therefore given by

$$W_h \cdot H \eta \left[1 - \frac{W_h + C_r^2 W_p}{W_h + W_p} \right],$$

or

$$\frac{W_p W_h \cdot H \cdot \eta (1 - C_r^2)}{(W_h + W_p)}$$

Substituting these losses into the energy Eq. 16.18, and simplifying, one obtains

$$Q_{up} = \frac{W_h \cdot H \cdot \eta}{(s + C)} \times \frac{(1 + C_r^2 R)}{(1 + R)} \qquad \ldots \text{(Eq. 16.23)}$$

where $R = W_p/W_h$.

This is referred to as the Hiley formula.

The allowable load Q_{ap} may be obtained by dividing Q_{up} by a suitable factor of safety, which may be 2 to 2.5. (The formula is dimensionally homogeneous).

For long and rigid piles, the Hiley formula is conservative since a fraction of the total weight of the pile is accelerated at one time, as demonstrated by wave analysis. The weight W_p in the formula is then taken as the weight of pile cap plus the weight of the top portion of the pile; Chellis suggests that $\left(\frac{1}{2} W_p \right)$ be taken for W_p.

The Hiley formula is considered to be reasonably accurate for piles driven in cohesionless soil.

The various quantities used in Eq. 16.23 are obtained as follows:

(*i*) Elastic compression of pile (c_1)

This is computed from the equation

$$C_1 = \frac{Q_{up} L_e}{AE} \qquad \ldots \text{(Eq. 16.24)}$$

where L_e = embedded length of pile,
A = cross-sectional area of pile, and
E = modulus of elasticity of the material of the pile.

(*ii*) Elastic compression of pile cap (c_2)

Table 16.5 c_2 values

Material of pile	Driving stress (kN/m²)	Value of c_2
Precast concrete pile with packing at head	3000 to 15000	0.12 to 0.50
Steel H pile	3000 to 15000	0.04 to 0.16
Timber pile without cap	3000 to 15000	0.05 to 0.20

(iii) Elastic compression of soil (c_3)

The average value is taken as 0.1 cm (it may range from 0 for hard soil to 0.2 for soft soil).

(iv) Efficiency of hammer

Table 16.6 Hammer efficiencies

Type of hammer	Efficiency of hammer
Drop hammer	1.00
Single-acting steam hammer	0.75 to 0.85
Double-acting steam hammer	0.85
Diesel hammer	1.00

(v) Coefficient of restitution (C_r)

Table 16.7 C_r-values

Material	Value of C_r
Wood pile	0.25
Wood cushion on steel pile	0.32
C.I. Hammer on concrete pile without cap	0.40
C.I. Hammer on steel pipe without cushion	0.55

In I.S: 2911 (Part I)-1964 the Hiley-formula is given in a slightly different form:

$$Q_{up} = \frac{W_h H \eta \eta_b}{(s + C/2)} \quad \ldots(\text{Eq. 16.25})$$

where, η_b = efficiency of hammer blow (or ratio of energy after impact to the energy of the hammer before impact)

η_b is given by

$$\eta_b = \left(\frac{W_h + C_r^2 W_p}{W_h + W_p}\right), \text{ when } W > C_r W_p \quad \ldots(\text{Eq. 16.26a})$$

$$= \left(\frac{W_h + C_r^2 W_p}{W_h + W_p}\right) - \left(\frac{W_h - C_r W_p}{W_h + W_p}\right)^2, \text{ when } W < C_r W_p$$

$$\ldots(\text{Eq. 16.26b})$$

Here, $C = C_1 + C_2 + C_3$

C_1 = temporary compression of dolly and packing

$= 1.77 \dfrac{Q_{up}}{A}$, where the driving is without dolly or helmet and cushion about 2.5 cm thick,

$C_1 = 9.05 \dfrac{Q_{up}}{A}$, where the driving is with short dolly up to 60 cm long helmet and cushion up to 7.5 cm thick.

C_2 = temporary compression of pile

$= 0.0657 \dfrac{Q_{up} L}{A}$

C_3 = temporary compression of ground

$= 3.55 \dfrac{Q_{up}}{A}$

L = length of pile in metres
A = area of cross-section of pile in cm^2.

This is applicable for friction piles.

For point-bearing piles, a value $\left(\frac{1}{2} W_p\right)$ is substituted for W_p. The value $\eta . H$ is also referred to as the effective fall of hammer.

Danish Formula

The Danish formula is

$$Q_{up} = \dfrac{W_h \cdot H \cdot \eta}{\left(s + \frac{1}{2} s_o\right)} \qquad \text{...(Eq. 16.27)}$$

where s_o = elastic compression of the pile

$$s_o = \dfrac{2 W_h H L}{A E} \qquad \text{...(Eq. 16.27a)}$$

Here L, A and E refer to the length, area of cross-section, and modulus of elasticity of the pile.

A factor of safety of 3 is recommended for use in conjunction with this formula.

Comments on the use of dynamic pile-driving formulae:

1. In general, dynamic pile driving formulae appear to be more applicable to piles driven into cohesionless soils. However, Vesic (1967) suggests that the value of C in Engineering News formula should be taken 1 cm for steel pipe piles and 1.5 cm for precast concrete piles, since the results would otherwise be too conservative.

2. According to Vesic (1967), Hiley's formula does not give consistent results for piles in cohesionless soils.

3. A basic objection to the use of these formulae is that dynamic resistance of a soil is very much different from its static resistance. However, formulae such as the Engineering News formula and the Danish formula have gained popularity owing to their simplicity.

4. If the pile is driven into saturated loose sand and silt, liquefaction might result, reducing the pile capacity.
5. Dynamic pile driving formulae are not considered to be applicable to piles driven into cohesive soils. This is because there may be apparent increase in driving resistance due to the development of pore pressures, while there may be a tendency for it to decrease later on due to dissipation of pore pressure.
6. Pile driving in cohesive soils disturbs the soil structure and consequently the resistance tends to decrease due to remoulding in sensitive clays although there might be some regaining of strength with passage of time, due to thixotropy.

Thus, it is evident that some degree of caution and judgement are called for in the use of these formulae.

16.5.3 Load Test on Pile

Load test on a pile is one of the best methods of determining the load-carrying capacity of a pile. It may be conducted on a driven pile or cast-in-situ pile, on a working pile or a test pile, and on a single pile or a group of piles. A working pile is one which forms part of the foundation, while a test pile is one which is used primarily to check estimated capacities (as predetermined by other methods).

The aim of a pile load test is invariably to determine the vertical load capacity; however, in certain special cases the test may be used to obtain the uplift capacity or lateral load capacity. Load test on a pile group is expensive and may be conducted only in the case of important projects.

Both cohesive and cohesionless soils will have their properties altered by pile driving. In clays, the disturbance causes remoulding and consequent loss of strength. With passage of time, much of the original strength will be regained. The effect of pile driving in sand is to create a temporary condition wherein extra resistance is developed, which is lost later by stress relaxation. Hence, the test should be conducted only after a lapse of a few weeks in clays and at least a few days in sands, in order that the results obtained be more meaningful for design.

Load may be applied by using a hydraulic jack against a supported platform (Fig. 16.6a), or against a reaction girder secured to anchor piles (Fig. 16.6b). Sometimes a proving ring is preferred for better accuracy in obtaining the load. Instead of reaction loading, gravity loading may also be used; but the former is given better uniformity in loading. Measurement for pile settlement is related to a fixed reference mark. The support for the reference mark has to be located outside the zone that could be affected by pile movements.

The most common procedure is the test in which the load is maintained slowly. About five to eight equal increments are used until the load reaches about double the design value. Time-settlement data are recorded for each load increment. Each increment is maintained until the rate of settlement becomes a value less than 0.25 mm per hour. The final load is maintained for 24 hours.

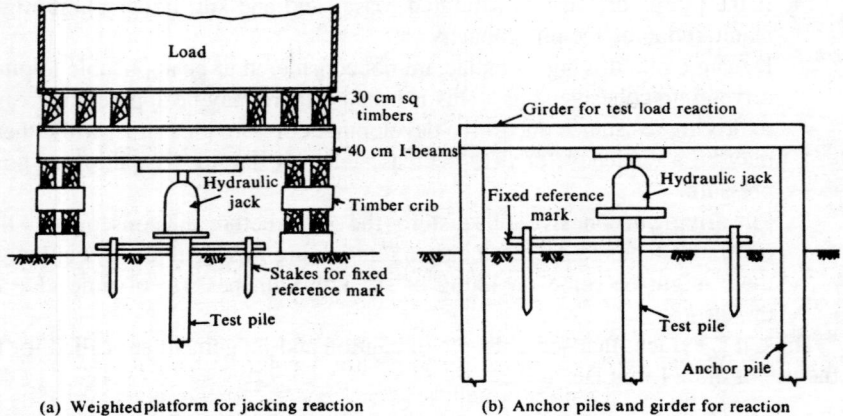

Fig. 16.6 Typical pile load test arrangements

Another procedure is the constant-strain rate method. In this method, the load is increased such that the settlement occurs at a predetermined rate such as 0.5 mm per minute. This test is considerably faster than the other approach.

Other procedures include cyclic loading, where each load increment is repeatedly applied and removed. Settlements are recorded at every increment or decrement of load. These help in separating elastic and plastic settlements, and also point-bearing and frictional resistances.

The load-settlement curve is obtained from the data. Often the definition of 'failure-load' is arbitrary. It may be taken when a predetermined amount of settlement has occurred or where the load-settlement plot is no longer a straight line. If the ultimate load could be found, a suitable factor of safety—2 to 3—may be used to determine the allowable load.

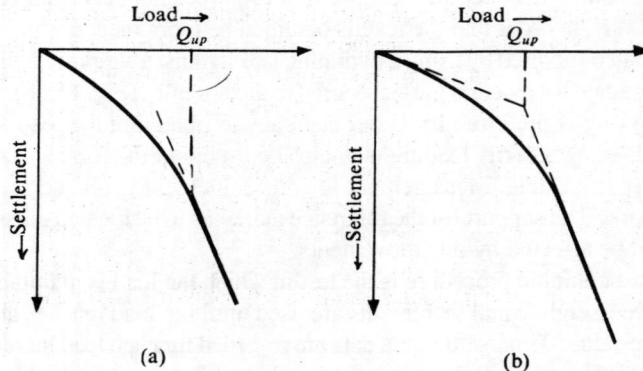

Fig. 16.7 Determination of ultimate load from load-settlement curve for a pile

The ultimate load may be determined as the abscissa of the point where the curved part of the load-settlement curve changes to a steep straight line (Fig. 16.7a). Alternatively, the ultimate load is the abscissa of the point of intersection of initial and final tangents of the load-settlement curve (Fig. 16.7b).

Another method in use for the slow test is to plot both load and settlement values on logarithmic scale. The results typically plot as two straight lines (Fig. 16.8). The intersection of the straight lines is taken as failure load for design purposes although this may not be the actual load at which failure occurs.

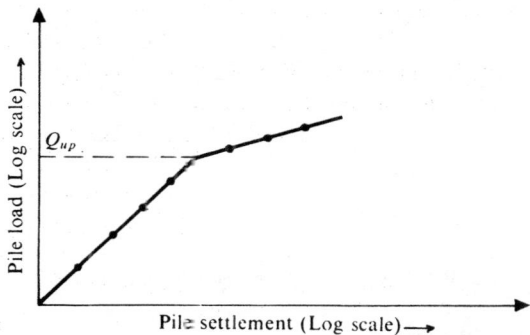

Fig. 16.8 Log-Log plot of pile load-settlement curve and determination of failure

The allowable load on a single pile may be obtained as one of the following [I.S: 2911 (Part I)-1964]:
1. 50% of the ultimate load at which the total settlement is equal to one-tenth the diameter of the pile.
2. Two-thirds of the load which causes a total settlement of 12 mm.
3. Two-thirds of the load which causes a net (plastic) settlement of 6 mm (total settlement minus elastic settlement).

In the case of the cyclic load test, the load is raised up to a particular level, released to zero and again raised to a higher value and released to zero. Settlements are recorded at each increment or decrement of load. A typical plot of a cyclic load test data will look as shown in Fig. 16.9.

The total settlement S of the pile head at any load may be written as

$$S = S_e + S_p \qquad \ldots\text{(Eq. 16.28)}$$

and

$$S = \Delta L + S_{es} + S_p \qquad \ldots\text{(Eq. 16.29)}$$

where S_e = elastic compression of the pile and soil at the base,
S_p = plastic compression of the soil at the base
S_{es} = elastic compression of the soil at the base, and
ΔL = elastic compression of the pile
(This is based on the assumption that plastic compression of the pile is negligible).

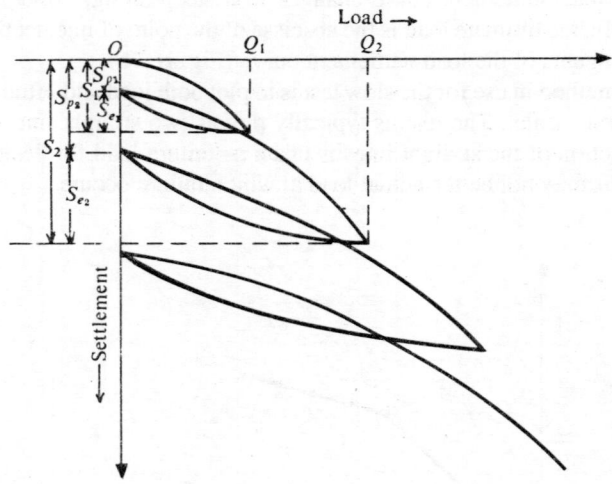

Fig. 16.9 Cyclic load test on a pile-load vs. settlement

ΔL may be obtained from the equation

$$\Delta L = \frac{(Q - Q_f/2)L}{AE} \qquad \ldots\text{(Eq. 16.30)}$$

where Q = load on the pile,
Q_f = frictional resistance component, and
L, A, E = length, area of section and modulus of elasticity of the pile.

The total settlement may easily be separated into the elastic and plastic components by removal of the load and observation of the net settlement. The elastic settlement is got by deducting the net settlement after removal of load from the total value of the settlement under the load (Eq. 16.28).

The elastic settlement of the soil at the base is obtained by subtracting the elastic settlement of the pile from the total elastic component of the settlement (Eq. 16.29).

The separation of the applied load at any load level into the point-bearing and frictional components is based on an experimental finding of Van Wheele (1957). Until the load reaches a certain value, the point-bearing component will be zero. With increase in load, both the friction and point-bearing components go on increasing. The frictional component attains a maximum value at a particular load level and thereafter the point-load component goes on increasing with further increase in load. Van Wheele found that the point-load component increases linearly with the elastic compression of the soil at the base and that the straight line showing this linear relationship is parallel to the straight line portion of the curve between the load on the pile and elastic compression of the soil.

Since the right-hand side of Eq. 16.30 contains Q_f which is not known to start with, a sort of a 'trial and error' procedure is employed to determine Q_f and Q_b (frictional and point-bearing components) corresponding to any pile load Q. The procedure is illustrated in Fig. 16.10.

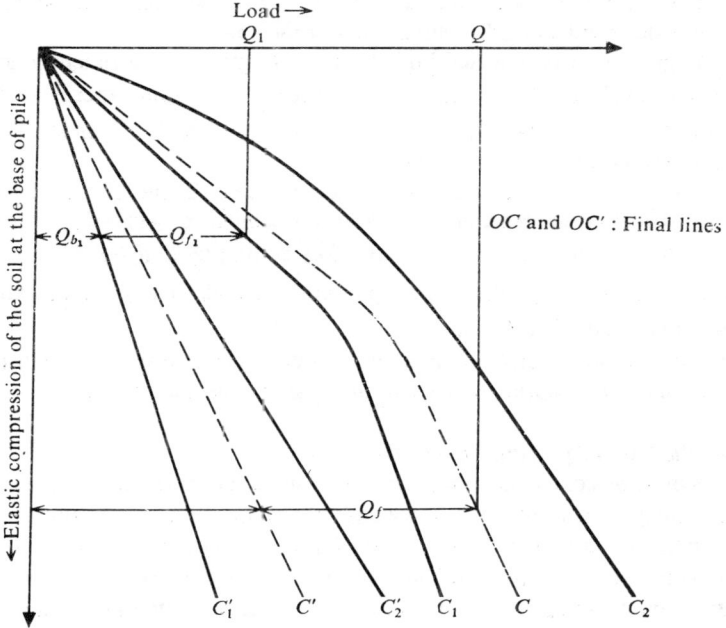

Fig. 16.10 Separation of point-bearing and skin-friction resistances from cyclic load test data

1. Since Q_f is not known to start with, ΔL is assumed to be zero. Then the elastic compression of the soil at the base is equal to the total elastic recovery of pile head. A curve OC_1 is drawn between the pile load and the elastic compression of the soil, calculated in this manner.
2. A line OC_1' is drawn from the origin O parallel to the straight portion of the curve OC_1. This line is supposed to divide the pile load Q into the components Q_b and Q_f.
3. For different loads $Q_1, Q_2, ...$, components Q_{f_1}, Q_{f_2} are determined, as shown in Fig. 16.10.
4. The values of ΔL corresponding to different values of Q_f are computed from Eq. 16.30.
5. The elastic compressions of the soil are obtained by deducting these values of ΔL from the corresponding values of elastic recovery of the pile head.

778 *Geotechnical Engineering*

6. A modified curve OC_2 is now drawn between pile load and elastic compression of soil.
7. Through the origin, a line OC_2' is drawn parallel to the straight line portion of OC_2.
8. Steps 3 through 7 are repeated to get the next modified curve between the pile load and the elastic compression of the soil.
9. The procedure is repeated until a curve which gives sufficiently accurate values of Q_b and Q_f is obtained. It has been found from experience that the third curve gives the desired degree of accuracy. The I.S. Code in this regard also recommends this procedure.

I.S: 2911 (Part I)-1964 recommends factors of safety 2 and 2.5 on the ultimate values of skin friction resistance and point resistance, respectively. Hence, the allowable load on the pile may be obtained by adding $Q_f/2$ and $Q_b/2.5$, where Q_f and Q_b correspond to the values corresponding to a load causing a total settlement of one-tenth of the pile diameter.

It should be obvious that the settlement required to cause ultimate point resistance is greater than that required to cause ultimate skin resistance.

16.5.4 Pile Capacity from Penetration Tests

Results of static cone penetration test and standard penetration test are also used to determine pile load capacity. In the case of a static cone penetration test, a 60° cone with a base area of 100 mm² attached to one end of a rod housed in a pipe and the pipe itself are pushed down alternately at a slow constant rate and the resistance encountered by each is recorded by means of pressure gauges. The pressure offered by the cone is recorded as penetration resistance q_c and that oferred by the pipe as skin friction resistance f_c.

The values of q_c may also be obtained indirectly from the Standard Penetration Number N, through correlations between N-value and the static cone penetration resistance q_c-value.

Franki Pile Company has suggested the following:

Table 16.8 Correlation between q_c and N (Franki Pile Co.)

Type of soil	q_c/N
Silty clay	3
Sandy clay	4
Silty sand	5
Clayey sand	6

Schmertmann (1970) suggests the following:

Table 16.9 Correlation between q_c and N (Schmertmann, 1970)

Type of soil	q_c/N
Sandy silts, silts	2.0
Silty sands, fine to medium sands	3.5
Coarse sands	5.5
Sandy gravel, gravel	9.0

These values are applicable for N-values equal to or greater than 75 q_c will be obtained in kg/cm².

Piles in granular soils

Meyerhof (1959) has shown that the ultimate point-bearing capacity of a pile, q_b, ranges from 2/3 to $1\frac{1}{2}$ times the static cone penetration resistance, q_c.

On the average, therefore,

$$q_b = q_c \qquad \text{...(Eq. 16.31)}$$

Meyerhof proposed the following equation:

$$q_b = q_c = 4N \qquad \text{...(Eq. 16.32)}$$

(q_b and q_c will be obtained in kg/cm²)

$$q_b = q_c = 400\,N \qquad \text{...(Eq. 16.33)}$$

(q_b and q_c will be in kN/m²)

Similarly, it has been found that the observed values of skin friction, f_s, varies from $1\frac{1}{4}$ to 3 times the static skin friction, f_c.

On the average, therefore,

$$f_s = 2f_c \qquad \text{...(Eq. 16.34)}$$

However, for small displacement piles such as H-piles,

$$f_s = f_c \qquad \text{...(Eq. 16.35)}$$

Meyerhof suggests the following:

$$f_s = q_c/200 \qquad \text{...(Eq. 16.36)}$$

$$f_s = N/50 \text{ or } f_s = 2N \qquad \text{...(Eq. 16.37)}$$

(kg/cm²)   (kN/m²)

For H-piles,

$$f_s = N \qquad \text{...(Eq. 16.38)}$$

(kN/m²)

According to De Beer, point resistance of bored piles in granular soils is less than that of driven piles, almost about one-third. This may be because of lack of compaction at the base and the disturbance of soil at the base during boring.

Piles in cohesive soils

Relationship between unit cohesion, c and the static cone penetration resistance, q_c, is as follows:

$$\frac{q_c}{18} < c < \frac{q_c}{15} \text{ for } q_c < 20 \text{ kg/cm}^2 \text{ (2000 kN/m}^2\text{)}$$

(for normally consolidated clays)

$$q_c/22 < c < q_c/26 \text{ for } q_c > 25 \text{ kg/cm}^2 \text{ (2500 kN/m}^2\text{)}$$

(for over consolidated clays)

If q_c is not known directly, the N-value may be obtained from the correlations between q_c and N, and it may be used for determining c.

Once the c-value is known, Eqs. 16.12 and 16.15 may be used to obtain q_b and f_s through c and c_a.

16.5.5 Negative Skin Friction

'Negative skin friction' or 'down drag' is a phenomenon which occurs when a soil layer surrounding a portion of the pile shaft settles more than the pile. This condition can develop where a soft or loose soil stratum located anywhere above the pile tip is subjected to new compressive loading. If a soft or loose layer settles after the pile has been installed, the skin-friction-adhesion developing in this zone is in the direction of the soil movement, pulling the pile downward, as shown in Fig. 16.11. Extra loading is thus imposed on the pile.

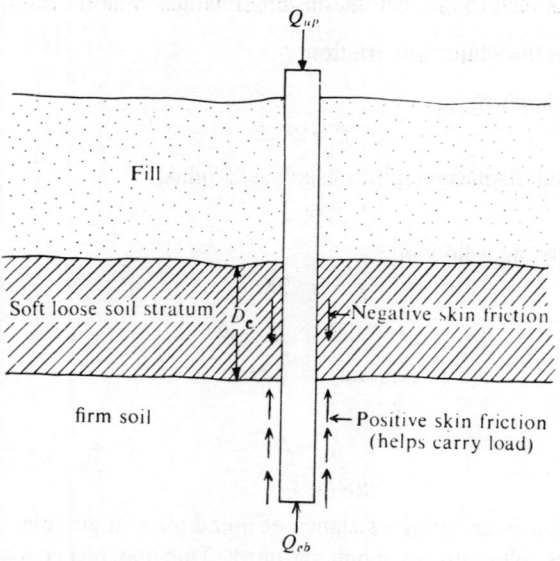

Fig. 16.11 Negative skin friction on a pile

Negative skin friction may also occur by the lowering of ground water which increases the effective stress inducing consolidation and consequent settlement of the soil surrounding the pile.

It is necessary to subtract negative skin friction force from the total load that the pile can support. In such a case the factor of safety will be modified as follows:

$$\text{Factor of safety} = \frac{\text{Ultimate pile load capacity}}{\text{Working load} + \text{Negative skin friction force}}$$

Sometimes this may also be written as

$$\text{Factor of safety} = \frac{\text{Ultimate pile load capacity} - \text{Negative skin friction force}}{\text{Working load}}$$

Values of negative skin force are computed in just the same way as positive skin friction.

For cohesive soils:

$$Q_{nf} = P \cdot D_n \cdot c \qquad \ldots(\text{Eq. 16.39})$$

where Q_{nf} = negative skin friction force on the pile,
 P = perimeter of the pile section,
 D_n = depth of compressible layer settling in relation to the pile
and c = unit cohesion of soil layer which is settling.

For cohesionless soils:

$$Q_{nf} = \frac{1}{2} P D_c^2 \cdot \gamma K \tan \delta \qquad \ldots(\text{Eq. 16.40})$$

where γ = unit weight of soil in the compressible zone,
 K = earth pressure coefficient ($K_a < K < K_p$), and
 δ = angle of wall friction ($\phi/2 < \delta < \phi$).

Sometimes negative skin friction may develop even in the zone of the fill, if the fill itself is settling under its self-weight.

When a large magnitude of negative skin friction force is anticipated, a protective sleeve or coating may be provided for the section that is embedded in the settling soil. Skin friction is thus eliminated for this section of the pile and a down drag is prevented. Negative skin force may be computed even for pile groups.

16.5.6 Factor of Safety

Where load tests are not performed, it is usual practice to apply a factor of safety of two to determine the design load.

Piles subject to uplift develop resistance to pull-out only by skin friction. Point-bearing resistance does not apply, but the weight of the pile may be included in the uplift capacity. Generally, a larger factor of safety is employed for uplift than for conventional downward loading. The strength of pile to pile-cap connection becomes critical in the case of uplift forces, since tensile force at this location negates any pull-out resistance of the pile.

$$Q_{ap} = \frac{(Q_{sf} + W_p)}{\eta} \qquad \ldots(\text{Eq. 16.41})$$

where W_p is the weight of the pile and factor of safety η is to be not less than 2.

782 Geotechnical Engineering

16.6 PILE GROUPS

A structure is never founded on a single pile. Piles are ordinarily closely spaced beneath structures; consequently, the action of the entire pile group must be considered. This is particularly important when purely friction piles are used.

The bearing capacity of a pile group is not necessarily the capacity of the individual pile multiplied by the number of piles in the group; the phenomenon by virtue of which this discrepancy occurs is known as 'Group action of piles'.

16.6.1 Number of Piles and Spacing

Usually driven piles are not used singly beneath a column or a wall because of the tendency of the pile to wander laterally during driving and consequent uncertainty with regard to centering the pile beneath the foundation. In cases where unplanned eccentricities result, failure may occur either at the connection between the pile and column or within the pile itself. Hence, piles for walls are commonly installed in a staggered arrangement to both sides of the centre line of the wall. For a column, at least three piles are used in a triangular pattern, even for small loads. When more than three piles are required in order to obtain adequate capacity, the arrangement of piles is symmetrical about the point or area of load application.

Representative pile group patterns for wall and column loads are indicated in Fig. 16.12.

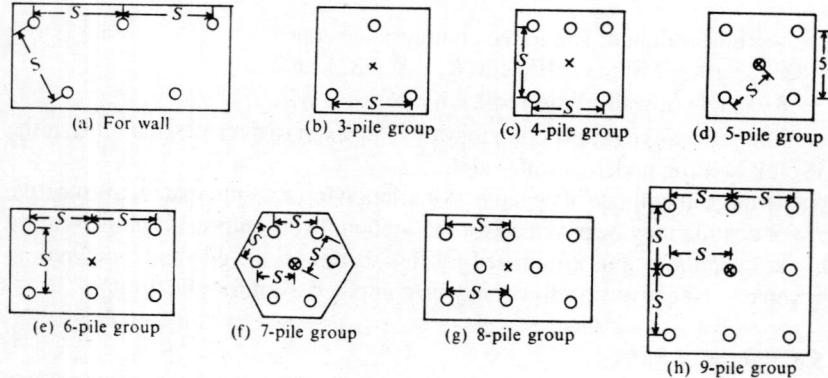

Fig. 16.12 Representative pile group patterns

Column and wall loads are usually transferred to the pile group through a pile cap, which is typically a reinforced concrete slab structurally connected to the pile heads to help the group act as a unit (Fig. 16.13).

The requirement for group arrangement of driven piles does not apply to bored piles. Drilled shafts can be installed quite accurately. A single large-diameter drilled shaft pile is commonly used to support columns in residential buildings. This may be used when the three-pile configuration yields unnecessarily extra load carrying capacity in the case of driven piles.

The spacing of piles in a group depends upon a number of factors such as the overlapping of stresses of adjacent piles, cost of foundation and the desired efficiency of the pile group.

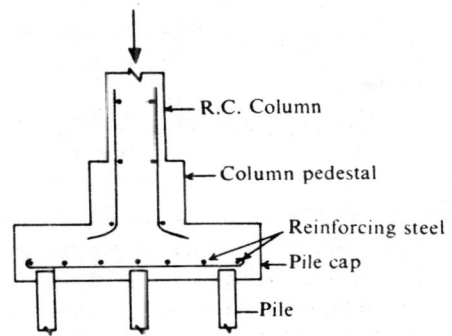

Fig. 16.13 Pile cap for a R.C. column

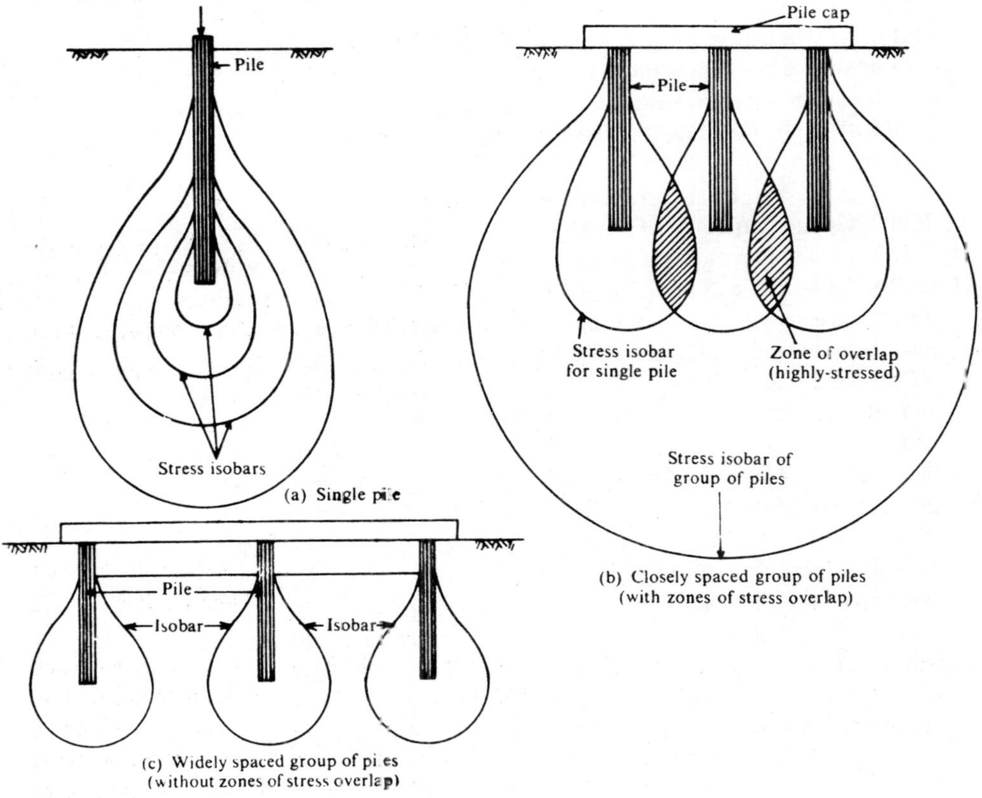

Fig. 16.14 Stress isobars of single pile and groups of piles

(a) Single pile

(b) Closely spaced group of piles (with zones of stress overlap)

(c) Widely spaced group of piles (without zones of stress overlap)

The stress isobars of a single pile carrying a concentrated load will be somewhat as shown in Fig. 16.14(a). When piles are driven in a group, there is a possibility of stress isobars of adjacent piles overlapping each other as shown in Fig. 16.14(b). Since, the overlapping might cause failure either in shear or by excessive settlement, this possibility may be averted by increasing the spacing as shown in Fig. 16.14(c). Large spacings are not advantageous since a bigger size of pile cap would increase the overall cost of the foundation.

In the case of driven piles there will be greater overlap of stresses due to the displacement of soil. If piles are driven in loose sands, compaction takes place and hence, the spacing may be small. However, if piles are driven in saturated silt or clay, compaction does not take place but the piles may experience uplift. To avoid this, greater spacing may be adopted. Smaller spacings may be used for cast-in-situ piles in view of less disturbance.

Point-bearing piles may be more closely spaced than friction piles. The minimum spacing of piles is usually specified in building codes. The spacing may vary from $2d$ to $6d$ for straight uniform cylindrical piles, d being the diameter of the pile. For friction piles, the recommended minimum spacing is $3d$. For point-bearing piles passing through relatively compressible strata, the minimum spacing is $2.5d$ when the piles rest in compact sand or gravel; this should be $3.5d$ when the piles rest in stiff clay. The minimum spacing may be $2d$ for compaction piles.

Piles should be, in general, driven proceeding outward from the centre, except in soft clay or very soft soil; in the latter case, the pile driving proceeds from the periphery of the foundation to the centre to prevent the lateral flow of soil during driving.

16.6.2 Group Capacity of Piles

The capacity of a pile group is not necessarily the capacity of the individual pile multiplied by the number of individual piles in the group. Disturbance of soil during the installation of the pile and overlap of stresses between the adjacent piles, may cause the group capacity to be less than the sum of the individual capacities.

Conversely, the soil between individual piles may become "locked in" due to densification from driving and the group may tend to behave as a unit or an equivalent single large pile. Densification and improvement of the soil surrounding the group can also occur. These factors tend to provide the group a capacity greater than the sum of the capacities of individual piles. The capacity of the equivalent large pile is analysed by determining the skin friction resistance around the embedded perimeter of the group and calculating the end-bearing resistance by assuming a tip area formed by this block, as in Fig. 16.15.

To determine the capacity of a pile group, the sum of the capacities of the individual piles is compared with the capacity of the single large equivalent (block) pile; the smaller of the two values is taken. Applying an appropriate factor of safety to this chosen value, the design load of the pile group is obtained.

Pile Foundations 785

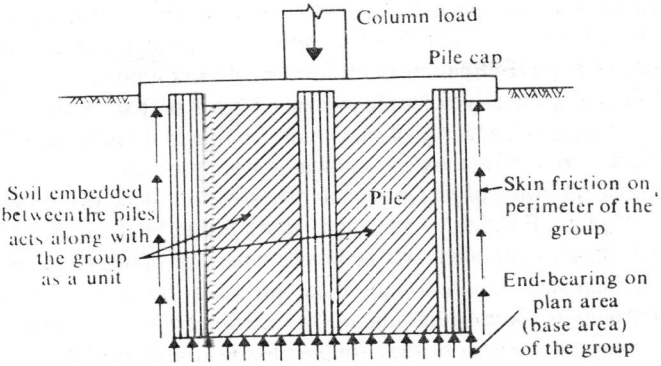

Fig. 16.15 Single equivalent large pile concept for a group (block failure)

The skin friction resistance of the single large equivalent pile (block) is obtained by multiplying the surface area of the group by the shear strength of the soil around the group. The end-bearing resistance is computed by using the general bearing capacity equation of Terzaghi. The bearing capacity factors for deep foundations are used when the length of the pile is at least ten times the width of the group; otherwise, the factors for shallow foundations are used.

Obviously, the capacity of the equivalent large pile is affected by soil type and properties, besides spacing of piles. Generally speaking, there is a greater tendency for the group to act as a block or large single unit when the piles are close and the soil is firm or compact.

Pile groups in cohesionless soils

For driven piles embedded in cohesionless soils, the capacity of the large equivalent pile (block) will be almost always greater than the sum of the capacities of individual piles, in view of the densification that occurs during driving. Consequently, for design, the group capacity is taken as the sum of the individual pile capacities or the product of the number of piles in the group and the capacity of the individual pile.

This procedure is not applicable, if the pile tips rest on compressible soils such as clays; in such cases, the pile group capacity is governed by the shear strength and compressibility of clay soil, rather than on the characteristics of the cohesionless soil.

Bored piles or cast-in-situ concrete piles are constructed by boring a hole of required diameter and depth and pouring in of concrete. Boring is accompanied invariably by some degree of loosening of the soil. In view of this, the group capacity of such piles will be somewhat less than the sum of individual pile capacities typically — about two-thirds of it. It may also be taken as the sum of individual pile capacities approximately.

Pile groups in cohesive soils

When piles are driven into clay soils, there will be considerable remoulding especially when the soil is soft and sensitive. The soil between the piles may also heave since compaction cannot be easily achieved in soils of such low permeability. Bored piles are generally preferred to driven piles in such soils. However, if driven piles are to be used, spacing of piles must be relatively large and the driving so adjusted as to minimise the development of pore pressure.

The mode of failure of pile groups in cohesive soils depends primarily upon the spacing of piles. For smaller spacings, 'block failure' may occur; in other words, the group capacity as a block will be less than the sum of individual pile capacities. For larger spacings, failure of individual piles may occur; or, it is to say that the group capacity is given by the sum of the individual pile capacities, which will be smaller than the strength of the group acting as a unit or a block. The limiting value of the spacing for which the group capacities obtained from the two criteria —block failure and individual pile failure—are equal is usually considered to be about 3 pile-diameters.

Negative skin friction Q_{ng} may be computed for a group in cohesive soils as follows:

Individual pile action:

$$Q_{ng} = nQ_{nf} = nP.D_n.c \qquad \ldots\text{(Eq. 16.42)}$$

Notations are the same as for Eq. 16.39.

Block action:

$$Q_{ng} = c.D_n.P_g + \gamma.D_n.A_g \qquad \ldots\text{(Eq. 16.43)}$$

Here, P_g and A_g are the perimeter and area of the pile block.

$[P_g = 4B$ and $A_g = B^2$, where B is the overall width of the block].

The larger of the two values for Q_{ng} is chosen as the negative skin friction.

16.6.3 Pile Group Efficiency

The 'efficiency', η_g, of a pile group is defined as the ratio of the group capacity, Q_g, to the sum of the capacities of the number of piles, n, in the group:

$$\eta_g = \frac{Q_g}{(n \cdot Q_p) \text{ or } (\Sigma Q_p)} \qquad \ldots\text{(Eq. 16.44)}$$

where Q_p = capacity of individual pile.

Obviously, the group efficiency depends upon parameters such as the types of soil in which the piles are embedded and on which they rest, method of installation, and spacing of piles.

Vesic (1967) has shown that end-bearing resistance is virtually unaffected by group action. However, skin friction resistance increases with increase in spacing for pile groups in sands. For pile groups in clay, the skin friction component of the resistance decreases for certain pile spacings. Thus, in general, efficiencies of pile groups in clay tend to be less than unity. Interestingly, Vesic's experimental investigations on pile groups in sands indicate group efficiencies greater than unity.

Sowers *et al* (1961) have shown that the optimum spacing at which the group efficiency is unity for long friction piles in clay is given by

$$S_o = 1.1 + 0.4n^{0.4} \qquad \text{...(Eq. 16.45)}$$

where S_o = optimum spacing in terms of pile diameters; S_o is 2 to 3 pile diameters centre to centre,

and n = number of piles in the group.

The actual efficiency, η_g, at the theoretical optimum spacing is

$$\eta_g = 0.5 + \frac{0.4}{(n-0.9)^{0.1}} \qquad \text{...(Eq. 16.46)}$$

This has been found to be 0.85 to 0.90, rather than unity. Since a factor of safety is used for design, the error in assuming the real efficiency to be 1 at optimum spacing is inconsequential.

A number of empirical equations for pile group efficiency are available. There is no acceptable formula and these should be used with caution as they may be no better than a good guess. These formulae yield efficiency values less than unity, and as such, will not be applicable to closely spaced friction piles in cohesionless soils and to piles through soft material resting on a firm stratum.

The Converse-Labarre formula, the Feld's rule and the Seiler-Keeny formula are given here:

Converse-Labarre Formula

$$\eta_g = 1 - \frac{\phi}{90} \left[\frac{m(n-1) + n(m-1)}{mn} \right] \qquad \text{...(Eq. 16.47)}$$

where η_g = efficiency of pile group,

$\phi = \tan^{-1} \dfrac{d}{s}$ in degrees, d and s being the diameter and spacing of piles,

m = number of rows of piles, and
n = number of piles in a row } (interchangeable)

Feld's rule

According to "Feld's rule", the value of each pile is reduced by one-sixteenth owing to the effect of the nearest pile in each diagonal or straight row of which the particular pile is a member. This is illustrated in Fig. 16.16.

Seiler-Keeney Formula

The efficiency of a pile group, η_g, is given by

$$\eta_g = \left[1 - 0.479 \left(\frac{s}{s^2 - 0.093} \right) \left(\frac{m+n-2}{m+n-1} \right) \right] + \frac{0.3}{(m+n)} \qquad \text{...(Eq. 16.48)}$$

Here m, n, and s stand for the number of rows of piles, number of piles in a row and pile spacing, respectively.

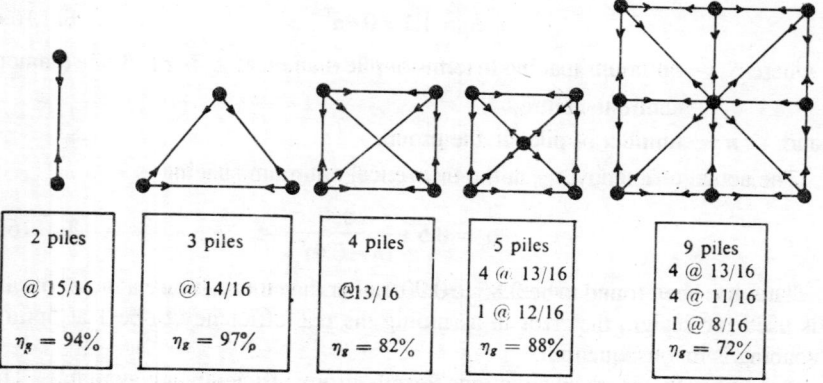

Fig. 16.16 Efficiencies of pile groups using Feld's rule

16.7 SETTLEMENT OF PILES AND PILE GROUPS

Settlement of piles and pile groups cannot be evaluated with any degree of confidence.

16.7.1 Settlement of Single Pile

Vesic has proposed an equation for computing the settlement of a single pile in cohesionless soil, based on experiments on test piles of different sizes embedded in sands with different density index values. These tests were conducted on driven piles, bored piles and jacked piles (jacked piles are those that are pushed into the soil by means of the static pressure of a jack).

The equation for the settlement is

$$S = S_p + S_f \qquad \text{...(Eq. 16.49)}$$

where S = total settlement,

S_p = settlement of pile tip, and

S_f = settlement due to deformation of pile shaft.

Further,

$$S_p = \frac{C_w Q_p}{(1+I_D)^2 q_b} \qquad \text{...(Eq. 16.50)}$$

where Q_p = point load,

I_D = density index of sand,

q_b = unit resistance in point-bearing, and

C_w = settlement coefficient ...   0.04 for driven piles,
    0.18 for bored piles, and
    0.05 for jacked piles.

$$S_r = (Q_p + \alpha Q_f)\frac{L}{AE} \qquad \text{...(Eq. 16.51)}$$

where, Q_f = Friction load,
L = Length of pile,
A = Area of cross-section of pile,
E = Modulus of elasticity of pile material, and
α = Coefficient which depends on the distribution of skin friction along the shaft and is usually taken as 0.6.

Settlement of a single pile in clay is more difficult to evaluate. An assumption is required to be made regarding the level to which the load is transferred. If the pile penetrates homogeneous clay, the load may be taken to be transmitted to a depth of $\frac{2}{3}$ of the embedded length of the pile from the surface. Concepts of stress distribution in soil (Boussinesq's, for example), coupled with the settlement due to consolidation will be used. If the pile penetrates a weak stratum and is embedded into a firm stratum, the load may be taken to be transmitted to a level at a depth of $\frac{2}{3}$ of the depth of embedment into the firm stratum from the top of the firm stratum.

If the pile penetrates a weak stratum and rests on the top of a firm stratum, the load may be taken to be transferred to the top of the firm stratum. The computation of the settlement will involve that due to consolidation of the compressible strata present below the level to which the load is assumed to be transmitted.

16.7.2 Settlement of Groups of Piles

The computation of settlement of groups of piles is more complex than that for a single pile.

Settlement of pile groups in sands

Settlement of pile groups are found to be many times that of a single pile. The ratio, F_g, of the settlement of a pile group to that of a single pile is known as the group settlement ratio.

$$F_g = S_g/S \qquad \text{...(Eq. 16.52)}$$

where F_g = group settlement ratio,
S_g = settlement of pile group, and
S = total settlement of individual pile.

Vesic has obtained the relation between F_g and $\frac{B}{d}$, where B is the width of the pile group (centre to centre of outermost piles), and d is the diameter of the pile (only pile groups, square in plan, are considered). This is shown in Fig. 16.17.

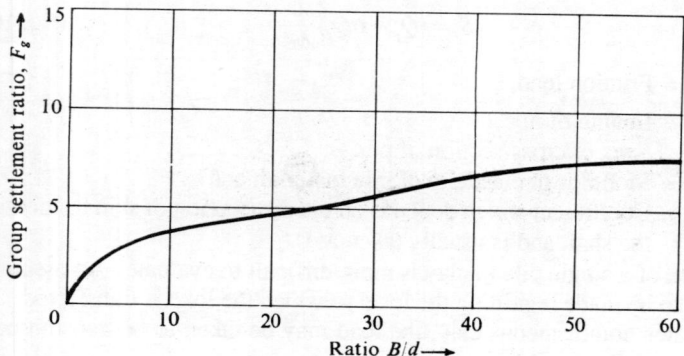

Fig. 16.17 Group settlement ratio vs. B/d
(After Vesic, 1967)

These results have been obtained for medium dense sand. For sands with other density indices the results could be different.

Settlement of pile groups in clay

The equation for consolidation settlement may be used treating the pile group as a block or unit. The increase in stress is to be evaluated appropriately under the influence of the load on the pile group.

When the piles are embedded in a uniform soil (friction and end-bearing piles), the total load is assumed to act at a depth equal to two-thirds the pile length. Conventional settlement analysis procedures assuming the Boussinesq or Westergaard stress distribution are then applied to compute the consolidation settlement of the soil beneath the pile tip.

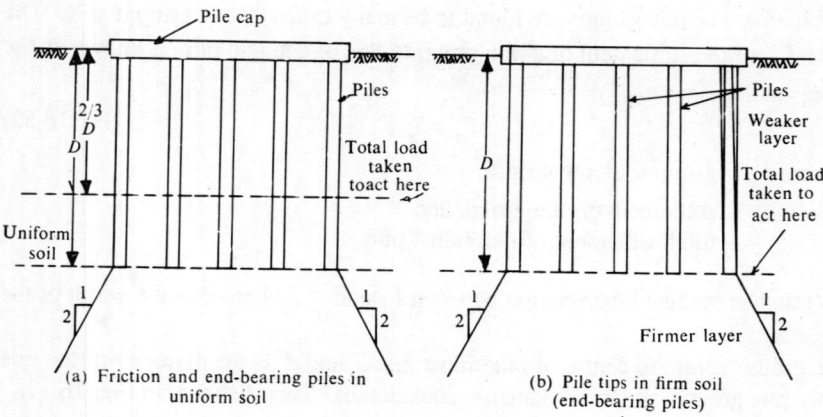

Fig. 16.18 Conditions assumed for settlement of pile groups in clay

When the piles are resting on a firmer stratum than the overlying soil (end-bearing piles), the total load is assumed to act at the pile tip itself. If the piles are embedded into the firmer layer in this case, the load is assumed to be transmitted to a depth equal to two-thirds of the embedment from the top of the firmer layer. The rest of the settlement analysis procedure is applicable. These are illustrated in Fig. 16.18.

The total pressure may be assumed to be distributed on a slope of 2 vertical to 1 horizontal, for the purpose of computation of increment of stress, in an approximate manner.

16.8 LATERALLY LOADED PILES

Piles and pile groups may be subjected to vertical loads, lateral loads or a combination of both. If the lateral loads act at an elevation considerably higher than the base of the foundation, there will be significant moments acting on it.

Vertical piles may be relied upon to resist large magnitudes of lateral loads. The lateral load capacity of a vertical pile depends upon the nature of the soil, the size of the pile, and the conditions at the pile head. If the pile head is fixed rigidly in a pile cap, its lateral load capacity will be more than when it is free.

Extensive theoretical and experimental studies have beeen made on laterally loaded piles by Reese and Matlock (1960), Palmer and Brown (1954), and Murthy (1964). Most of these are based on the concept of coefficient of sub grade reaction, which is the pressure required to cause unit deflection.

Winkler's hypothesis

Most of the theoretical solutions for laterally loaded piles involve the concept of 'coefficient of subgrade reaction', or 'soil modulus' as it is sometimes called, based on Winkler's (1867) hypothesis that a soil medium may be approximated by a series of infinitely closely spaced independent elastic springs, which is only an approximation of a beam on an elastic foundation (Fig. 16.19).

Terzaghi recommends that the coefficient of sub grade reaction be taken to be constant with respect to the depth for preconsolidated clays. For normally loaded clays and silts, Reese and Matlock, and Vesic assume that it varies linearly with depth. The constant of proportionality is termed the coefficient of soil modulus variation.

$$k = k' \cdot z \qquad \text{...(Eq. 16.53)}$$

where k = coefficient of subgrade reaction at depth z, and
k' = coefficient of soil modulus variation
Units for k are $N/cm^2/cm$ or N/cm^3
Units for k' are $N/cm^3/m$.

These will be dependent upon the unconfined compression strength in the case of clays and upon the density index in the case of sands. The values of k and k' are best determined from a full-sized pile load test.

For long flexible piles, the lateral deflection is very nearly zero for most of the length of the pile and hence the length of the pile is not important.

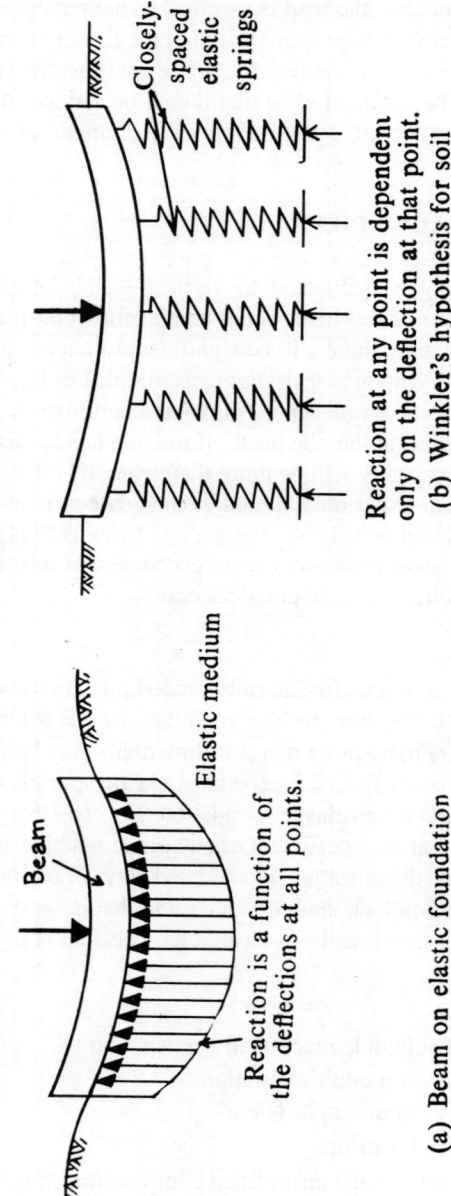

Fig. 16.19 Winkler's hypothesis for a soil medium

For short piles, the flexural rigidity of the pile loses its significance, the pile tends to rotate as a unit, acting somewhat as a rigid member.

Reese and Matlock use in their analysis the relative stiffness factor, T, which is the ratio of the stiffnesses of the pile and of the soil:

$$T = \left(\frac{EI}{k'}\right)^{1/5} \qquad \ldots(\text{Eq. 16.54})$$

They give the deflection, bending moment and the soil pressure for the pile in terms of non-dimensional coefficients in the form of tables and graphs.

Broms (1964) gives an analysis for laterally loaded piles in cohesive soils. Detailed treatment of these analyses is outside the scope of this book.

16.9 BATTER PILES

Batter piles combined with vertical piles are most effective for resisting large horizontal thrusts. Such combinations have been commonly used to support retaining walls, bridge piers and abutments, tall structures subjected to wind loads and as anchors for wharves, bulkheads and other waterfront structures. The batter may be up to 30° with the vertical. Depending on the direction of the lateral force relative to the direction of inclination with respect to the vertical, the batter may be termed 'positive' or 'negative'. If the tendency of the force is to 'right' the pile (bring it nearer vertical), the batter is considered positive; otherwise, it is considered negative (Fig. 16.20).

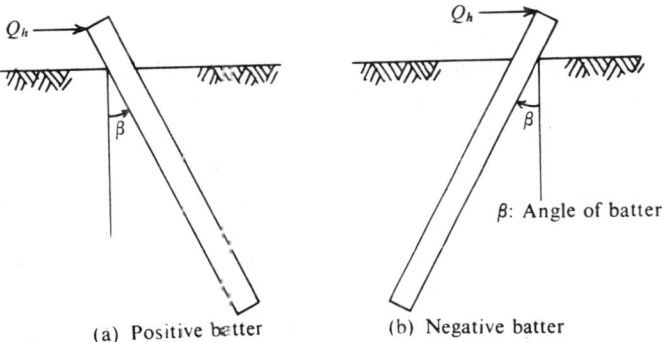

Fig. 16.20 Batter piles with positive batter and negative batter

A rational analysis of the action of batter piles is difficult because the problem is statically indeterminate to a high degree. One approximate method assumes the piles to be hinged at their tips and at their butts. A batter and vertical pile combination that is usually employed in sheet-pile bulkhead construction is shown in Fig. 16.21.

794 Geotechnical Engineering

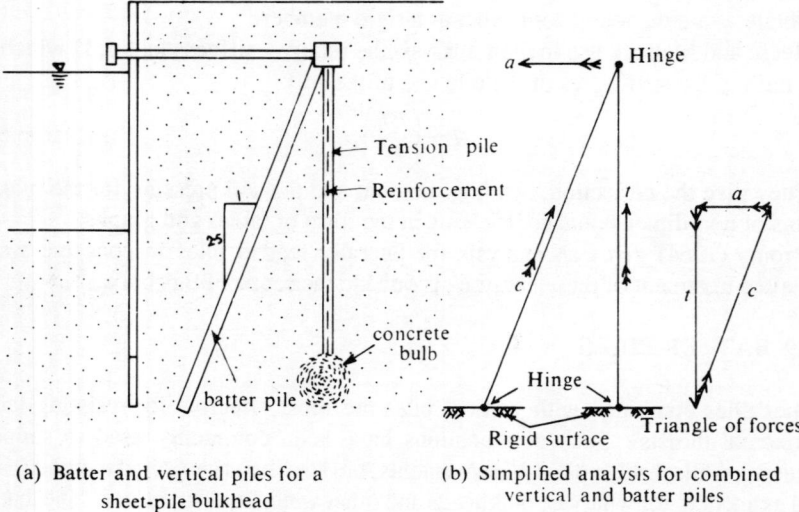

(a) Batter and vertical piles for a sheet-pile bulkhead

(b) Simplified analysis for combined vertical and batter piles

Fig. 16.21 Batter and vertical pile combination

Hrennikoff (1949) and Vesic (1970) have advanced theoretical analyses for batter piles. The former assumes the coefficient of a sub grade reaction as constant with the depth for all soils, while the latter assumes it as constant for clays and as varying linearly with depth for sands. 'Pile constants' and 'foundation constants' have been assumed by them in their analyses. If these constants could be evaluated to a reasonable degree of accuracy, the theories may be expected to yield satisfactory results. Murthy (1964) tried to evaluate some of the constants for certain angles of batter, both positive and negative.

16.10 DESIGN OF PILE FOUNDATIONS

The design of a pile foundation consists of assuming a design, then checking the proposed design for safety and revising it until it is satisfactory. The final design is selected on the basis of cost and time available for construction.

Sometimes piles may be valueless in some locations and may be even harmful under certain circumstances. For example, a layer of reasonably firm soil over a deep stratum of soft soil might act as a natural mat to distribute the load of a shallow footing. The driving of piles into the firm layer might break it up or remould it. The result could be a concentration of stress in the soft stratum, leading to excessive settlement.

16.10.1 Selection of Length of Piles

Selection of the approximate length of the pile is made from a study of the soil profile and the strength and compressibility of the soil strata. End-bearing piles must reach a stratum that is capable of supporting the entire foundation load without failure or undue settlement and friction piles must be long enough to distribute the stresses through the soil mass so as to minimise settlement and obtain adequate safety for the piles.

16.10.2 Selection of Type of Pile and Material of Pile

The points to be considered in the selection of type of a pile and material of pile are: (*i*) the loads, (*ii*) time available for completion of the job, (*iii*) the characteristics of the soil strata involved, (*iv*) the ground water conditions, (*v*) the availability of equipment, and (*vi*) the statutory requirements of building codes.

If the structure is a bridge abutment or a water-front structure, the characteristics of flow of water and scour the must be considered.

16.10.3 Pile Capacity

The pile capacity both for an individual pile and for groups of piles shall be determined in accordance with the procedures outlined earlier. An appropriate factor of safety shall be applied to determine the allowable load.

16.10.4 Pile Spacing

The piles are placed so that the capacity of the pile group acting as a unit is equal to the sum of the capacities of the individual piles.

It is impossible to construct piles in exactly the required location or angle because they tend to drift out of line when hard or soft spots are encountered. The tolerance at the top could be from 5 cm to 15 cm. A pile may be permitted to be out of plumb by 1 to 2% of the length.

16.10.5 Inspection and Records

Competent engineering inspection and keeping complete records of the driving of every pile is an essential part of any important job.

All details such as those relating to the hammer, pile, number of blows, penetration, length driven, heaving and shrinkage of adjacent ground, details of pile cap, shall be recorded.

16.11 CONSTRUCTION OF PILE FOUNDATIONS

The construction of a pile foundation involves two steps, namely the installation of piles and the making of pile caps. The second step is relatively simple and is similar to the construction of footings.

Installation of piles would depend upon whether they are driven or cast-in-place. Some details regarding the equipment required to install piles by driving them into soil have already been given. Water jetting is used to assist penetration of the piles.

Cast-in-place piles are mostly concrete piles of standard types such as the Raymond pile and the Franki pile, so called after the piling firms which standardised their construction.

Damage due to improper driving may be avoided if driving is stopped when the penetration reaches the desired resistance.

796 Geotechnical Engineering

Some degree of tolerance in alignment has to be permitted since piles can never be driven absolutely vertical and true to position.

A pile may be considered defective if it is damaged by driving or is driven out of position, is bent or bowed along its length. A defective pile must be withdrawn and replaced by another pile. It may be left in place and another pile may be driven adjacent to it.

Pile driving may induce subsidence, heave, compaction, and disturbance of the surrounding soil. These effects are to be carefully studied so as to understand thier bearing on the capacity of the pile.

16.12 ILLUSTRATIVE EXAMPLES

Example 16.1: A timber pile was driven by a drop hammer weighing 3 tonnes with a free fall of 1.2 m. The average penetration of the last few blows was 5 mm. What is the capacity of the pile according to Engineering News Formula?

Allowable load on the pile

$$Q_{ap} = \frac{500\, W_h \cdot H}{3(s+25)} \text{ for drop hammer,}$$

H being in metres and s being in mm.

$$W_h = 3\,t \qquad H = 1.2\,m \qquad s = 3\,mm$$

$$\therefore \quad Q_{ap} = \frac{500 \times 3 \times 1.2}{3 \times (5+25)}\, t = \mathbf{20\ t}$$

Example 16.2: A pile is driven with a single acting steam hammer of weight 15 kN with a free fall of 900 mm. The final set, the average of the last three blows, is 27.5 mm. Find the safe load using the Engineering News Formula.
Allowable load on the pile

$$Q_{ap} = \frac{500\, W_h \cdot H}{3(s+2.5)} \text{ for steam hammer,}$$

H being in metres and s in mm.

$$W_h = 15\,kN \qquad H = 900\,mm = 0.9\,m$$

$$\therefore \quad Q_{ap} = \frac{500 \times 15 \times 0.9}{3(27.5+2.5)}\,kN = \mathbf{75\ kN}$$

Example 16.3: A pile is driven in a uniform clay of large depth. The clay has an unconfined compression strength of 0.9 kg/cm². The pile is 30 cm diameter and 6 m long. Determine the safe frictional resistance of the pile, assuming a factor of safety of 3. Assume the adhesion factor $\alpha = 0.7$.

Cohesion of clay $= \frac{1}{2} \times 0.9 = 0.45$ kg/cm^2

Frictional resistance $= \alpha \cdot c \, A_s$

$\qquad = 0.7 \times 0.45 \times \pi \times 30 \times 600$ kg $= 17,813$ kg

∴ Safe frictional resistance $= \dfrac{17,813}{3} \approx 5930$ kg

Example 16.4: A group of 16 piles of 50 cm diameter is arranged with a centre to centre spacing of 1.0 m. The piles are 9 m long and are embedded in soft clay with cohesion 30 kN/m^2. Bearing resistance may be neglected for the piles—Adhesion factor is 0.6. Determine the ultimate load capacity of the pile group.

$\qquad n = 16 \qquad d = 50$ cm $\qquad L = 9$ m $\qquad s = 1.5$ m

Width of group, $B = (1 \times 3 + 0.50) = 3.5$ m

For block failure:

$\qquad Q_g = c \times$ Perimeter $\times$ Length

$\qquad\quad = c \times 4\,B \times 2$

$\qquad\quad = 30 \times 4 \times 3.5 \times 9$

$\qquad\quad = 3780$ kN

For piles acting individually:

$\qquad Q_g = n(\alpha \cdot c\, A_s)$

$\qquad\quad = 16 \times 0.6 \times 30 \times \pi \times 0.5 \times 9$

$\qquad\quad = 4,070$ kN

Hence the foundation is governed by block failure and the ultimate load capacity is **3,780 kN**.

Example 16.5: A square group of 9 piles was driven into soft clay extending to a large depth. The diameter and length of the piles were 30 cm and 9 m respectively. If the unconfined compression strength of the clay is 9 t/m^2, and the pile spacing is 90 cm centre to centre, what is the capacity of the group? Assume a factor of safety of 2.5 and adhesion factor of 0.75.

Block failure:

Since, it is a square group, 3 rows of 3 piles each will be used.

$$Q_g = c' \cdot N_c \cdot A_g + c \cdot P_g \cdot L$$

Here, cohesion $c = c' = \dfrac{9}{2}$ t/m^2 $= 4.5$ t/m^2

N_c is taken as 9.

$L = 9$ m $\qquad B = 2s + d = 2 \times 0.9 + 0.3 = 2.1$ m

$P_g = 4B = 8.4$ m

$A_g = B^2 = 2.1^2$ m$^2 = 4.41$ m^2

Substituting,
$$Q_g = 4.5 \times 9 \times 4.41 + 4.5 \times 8.4 \times 9$$
$$= 178.6 + 340.0 = 518.6 \text{ t}$$

Individual pile failure:
$$Q_g = n[q_{eb} \cdot A_b + f_s \cdot A_s]$$
$$= n[cN_c A_b + \alpha \cdot c A_s]$$
$$= 9\left[4.5 \times 9 \times \frac{\pi}{4} \times 0.3^2 + 0.75 \times 4.5 \times \pi \times 0.3 \times 9\right]$$
$$= 9[2.863 + 28.63] = 283.5 \text{ t}$$

In this case, individual pile failure governs the design. Allowable load on the pile group
$$= \frac{283.5}{2.5} \text{ t}$$
$$\approx 113 \text{ t}$$

Example 16.6: Design a square pile group to carry 400 kN in clay with an unconfined compression strength of 60 kN/m². The piles are 30 cm diameter and 6 m long. Adhesion factor may be taken as 0.6.

$$\text{Cohesion, } c = \frac{60}{2} = 30 \text{ kN/m}^2$$

Frictional resistance of one pile
$$= \alpha \cdot c \cdot \pi d \cdot L$$
$$= 0.6 \times 30 \times \pi \times 0.3 \times 6 = 101.8 \text{ kN}$$

Safe load per pile $= \dfrac{101.8}{3} \approx 34$ kN (with a factor of safety of 3)

Approximate number of piles required $= \dfrac{400}{34} \approx 12$

Let us try a square 16-pile group with centre-to-centre spacing of 60 cm.

$$\text{Efficiency } \eta_g = 1 - \frac{\phi}{90°}\left[\frac{m(n-1)+n(m-1)}{mn}\right]$$

$$\phi = \tan^{-1}(d/s) = \tan^{-1}(d/2d) = 26.565$$

$$= 1 - \frac{26.565}{90}\left[\frac{4 \times 3 + 4 \times 3}{4 \times 4}\right] = 0.56$$

Safe load on the pile group = $0.56 \times 16 \times 34 \approx$ **305 kN** with a factor of safety of 3.

It will be 400 kN with $\eta = \dfrac{305 \times 3}{400} = 2.3$

Check for block failure:
Frictional resistance of the block
$$= 2.1 \times 4 \times 6 \times 30$$
$$= 1512 \text{ kN}$$

Safe load with $\eta = 2.5$ is $\dfrac{1512}{2.5} \approx 605$ kN

Hence, the safe load may be taken as **400 kN** for the group, although the factor of safety falls short of 2.5 slightly.

Example 16.7: A 16-pile group has to be arranged in the form of a square in soft clay with uniform spacing. Neglecting end-bearing, determine the optimum value of the spacing of the piles in terms of the pile diameter, assuming a shear mobilisation factor of 0.6.

At the optimum spacing, efficiency of the pile group is unity.
Let d and s be the diameter and spacing of the piles. Let L be their length.
Width of the block for a 16-pile square group,
$$B = 3s + d$$
Group capacity for block failure
$$= 4L(3s + d) \times c$$
where c is the unit cohesion of the soil.
Group capacity based on individual pile failure
$$= n[(0.6c)(\pi d L)]$$
$$= 16 \times 0.6 \, \pi d \, Lc$$

Equating these two,
$$4Lc(3s + d) = 16 \times 0.6 \, \pi d \, Lc$$
$$12s + 4d = 9.6 \, \pi d$$
$$s = \dfrac{(9.6\pi - 4)}{12} d = 2.18 \, d$$

∴ The optimum spacing is about **2.2 d**.

Example 16.8: A square pile group of 9 piles passes through a recently filled up material of 4.5 m depth. The diameter of the pile is 30 cm and pile spacing is 90 cm centre to centre. If the unconfined compression strength of the cohesive material is 60 kN/m² and unit weight is 15 kN/m³, compute the negative skin friction of the pile group.

800 Geotechnical Engineering

Individual piles:

$$\text{Cohesion} = \frac{1}{2} \times 60 = 30 \text{ kN/m}^2$$

Negative skin friction

$$Q_{ng} = n \times PD_n c$$

$$= 9 \times \pi \times 0.3 \times 4.5 \times 30 = 1145 \text{ kN}$$

Block:

$$B = 2s + d = 2 \times 0.9 + 0.3 = 2.1 \text{ m}$$

$$P_g = 4B = 8.4 \text{ m}$$

$$A_g = B^2 = 4.41 \text{ m}^2$$

Negative skin friction

$$Q_{ng} = CD_n P_g + \gamma \cdot D_n A_g$$

$$= 30 \times 4.5 \times 8.4 + 15 \times 4.5 \times 4.41 = (1134 + 298)$$

$$= 1432 \text{ kN}$$

∴ Negative skin friction (the larger of the two values) = **1,432 kN**.

Example 16.9: A reinforced cement concrete pile weighing 30 kN (including helmet and dolly) is driven by a drop hammer weighing 30 kN with an effective fall of 0.9 m. The average penetration per blow is 15 mm. The total temporary elastic compression of the pile, pile cap and soil may be taken as 18 mm. Coefficient of restitution 0.36. What is the allowable load on the pile with a factor of safety of 2? Use Hiley's formula.

$$W_p = 30 \text{ kN} \qquad W_h = 30 \text{ kN} \qquad R = \frac{W_p}{W_h} = 1$$

Effective fall = ηH = 0.9 m = 900 mm

$s = 15$ mm $C = \frac{1}{2}$ (total elastic compression of pile, pile cap and soil)

$$C_r = 0.36 = \frac{1}{2} \times 18 = 9 \text{ mm}$$

$$Q_{ap} = \frac{Q_{up}}{2} = \frac{1}{2}\left[\frac{W_h \cdot H \cdot \eta}{(s+C)}\left(\frac{1+C_r^2 \cdot R}{1+R}\right)\right] \qquad \text{by Hiley's formula}$$

$$= \frac{1}{2}\left[\frac{30 \times 900 \times 1}{(15+9)} \times \frac{1+(0.36)^2 \times 1}{(1+1)}\right]$$

$$= \frac{1}{2} \cdot \frac{30}{24} \cdot 900 \times \frac{1.1296}{2}$$

$$= 317.7 \text{ kN}$$

Approximate safe load may be taken as **315 kN**.

SUMMARY OF MAIN POINTS

1. Two general forms of deep foundation are recognised: Pile foundation and pier foundation. Piles are long, slender members used to bypass soft strata and transmit loads to firmer strata situated below.
2. Piles, other than sheet piles which are commonly used for reducing seepage, derive their capacity from end-bearing of the tip and skin friction of the surrounding soil against them.

    Piles may be of timber, steel, concrete or composite. Concrete piles may be precast or cast-in-place; the former are driven by pile hammers—drop, steam, pneumatic, diesel or vibratory.
3. Pile capacity may be obtained by static analysis—bearing capacity theories such as those of Meyerhof and Vesic for deep foundations—or by dynamic analysis.

    The Engineering News Formula and Hiley's formula are the most commonly used dynamic pile formulae; the former is simple, while the latter is more complete. Load test on a pile is one of the best approaches for determining pile capacity.
4. Negative skin friction tends to develop when a soil layer surrounding a pile settles more than the pile. This decreases the factor of safety.
5. Pile groups need not have a capacity equal to the number of piles multiplied by individual piles. Usually group capacity is smaller than this and the ratio is termed group efficiency; the group efficiency is less than unity for piles in clays, especially where skin friction is involved. This may be occasionally greater than unity for piles in sands. Pile spacing is an important factor in this regard.

    Block failure must be checked for pile groups in clay.
6. Settlement of a pile group is many times that of an individual pile.
7. Laterally loaded piles are analysed based on Winkler's hypothesis and the concept of the coefficient of sub grade reaction.
8. Batter piles are used to carry large lateral loads.
9. Pile-driving records must be carefully kept and studied for a proper evaluation of a pile or pile group.

REFERENCES

1. Alam Singh & B.C. Punmia: *Soil Mechanics and Foundations*, Standard Book House, Delhi-6, 1970.
2. V.G. Berezantzev, K.S. Khristoforov and V.N. Golubkov: *Load Bearing Capacity and Deformation of Piled Foundations*, Proc. Fifth International Conference on Soil Mechanics and Foundation Engineering, Paris, 1961.
3. Bharat Singh and Shamsher Prakash: *Soil Mechanics and Foundation Engineering*, Nem Chand & Brothers, Roorkee, 1976.
4. L. Bjerrum: *Norwegian Experiences with Steel Piles to Rock*, Geotechnique, June 1957.

5. B.B. Broms: *Lateral Resistance of Piles in Cohesive Soils*, Jl. of Soil Mechanics and Foundations Division, ASCE., Vol. 90, No. SM2, Mar., 1964.
6. P.L. Capper, W.F. Cassie and J.D. Geddes: *Problems in Engineering Soils*, E & F.N. Spon Ltd., 1971.
7. R.D. Chellis: *Pile Foundations*, McGraw Hill Book Co., Inc., NY., USA, 1961.
8. A. Hiley: *Pile Driving Calculations with Notes on Driving Forces and Ground Resistance*, The Structural Engineer, Vol.8, July-Aug. 1930.
9. A. Hrennikoff: *Analysis of Pile Foundations with Batter Piles*, Trans. ASCE, Vol. 115, 1950.
10. I.S: 2911 (Part I)—1964: *Code of Practice for Design and Construction of Pile Foundations Part I Load-bearing Concrete Piles*, 1964.
11. J. Kerisel: *Deep Foundations—Basic Experimental Facts*, Proc. of the Conference on Deep Foundations, Mexico City, 1964; also *J.L. Kerisel: Vertical and Horizontal Bearing Capacity of Deep Foundations in Clay*, symposium on bearing capacity and settlement of Foundations, Duke Uny. Duham N.C., USA., 1967.
12. G.A. Leonards: *Foundation Engineering*, McGraw-Hill Book Co., Inc., NY., USA., 1962.
13. D.F. McCarthy: *Essentials of Soil Mechanics & Foundations*, Reston Publishing Inc., Reston, Va., USA, 1977.
14. G.G. Meyerhof: *Compaction of Sands and Bearing Capacity of Piles*, Jl. of Soil Mechanics and Foundations Division, Proc. ASCE, Dec. 1959.
15. V.N.S. Murthy: *Behaviour of Batter Piles Subjected to Lateral Loads*, Ph.D. Thesis, I.I.T., Kharagpur, 1964.
16. V.N.S. Murthy: *Soil Mechanics and Foundation Engineering*, Dhanpat Rai & Sons, Delhi-6, 1977.
17. R.L. Nordlund: *Bearing Capacity of Piles in Cohesionless Soils*, Jl. of Soil Mechanics and Foundations Dn., Proc., ASCE., May, 1963.
18. H.P. Oza: *Soil Mechanics & Foundation Engineering*, Charotar Book Stall, Anand, 1969.
19. L.A. Palmer and P.P. Brown: *Deflection, Moment, and Shear by the Method of Difference Equation*, ASTM symposium on Lateral Load Tests on Piles, 1954.
20. R.B. Peck, W.E. Hanson, T.H. Thornburn: *Foundation Engineering*, John Wiley & Sons, Inc., NY., USA., 1974
21. J.G. Potyondy: *Skin Friction Between Cohesive Granular Soils and Construction Material*, Geotechnique Vol.XI., No.4, Dec., 1961.
22. T. Ramot: *Analysis of Pile Driving by the Wave Equation*, Foundation Facts, Raymond Concrete Pile Co., NY., 1967.
23. L.C. Reese and H. Matlock: *Generalised Solutions for Laterally Loaded Piles*, Soil Mechanics and Foundations Dn., Proc., ASCE, 1960.
24. J.H. Schmertmann: *Static Cone to Compute Settlement Over Sand*, Jl. of Soil Mechanics and Foundations Dn., Proc., ASCE, 1970.
25. Shamsher Prakash and Gopal Ranjan: *Problems in Soil Engineering*, Sarita Prakashan, Meerut, 1976.

26. Shamsher Prakash, Gopal Ranjan and Swami Saran: *Analysis and Design of Foundations and Retaining Structures*, Sarita Prakashan, Meerut, 1979.
27. G.N. Smith: *Elements of Soil Mechanics for Civil & Mining Engineers*, Crosby Lockwood Staples, London, 1974.
28. G.I. Sowers, L. Wilson, B. Martin and M. Fausold: *Model Tests of Friction Pile Groups in Homogeneous Clay*, Proc. Fifth International Conference on Soil Mechanics and Foundation Engineering, Paris, 1961.
29. G.B. Sowers and G.F. Sowers: *Soil Mechanics and Foundations*, 3rd ed., Macmillan Company, 1970.
30. D.W. Taylor: *Fundamentals of Soil Mechanics*, John Wiley & Sons, Inc., NY., USA, 1948.
31. W.E. Teng: *Foundation Design*, Prentice Hall of India Pvt. Ltd., New Delhi, 1976.
32. B.P. Verma: *Problems in Soil Mechanics*, Khanna Publishers, Delhi-6, 1972.
33. A.S. Vesic: *Ultimate Loads and Settlement of Deep Foundations in Sand*, Proceedings, Symposium on Bearing Capacity and Settlement, Duke Uny, Durham, N.C., USA., 1967.
34. A.S. Vesic: *Load Transfer in Pile Soil System*, Proc. Design and Installation of Pile Foundation and Cellular Structures, Lehigh Valley, Pa., Lehigh Uny., Envo Publishing Co., USA., 1970.
35. A.N. Wellington: *Engineering News Formula for Pile Capacity*, Engineering News Record, 1988.
36. E. Winkler: *Die Lehre von der Elastizität und Festigkeit*, Prague, 1867.

QUESTIONS AND PROBLEMS

16.1 Write brief critical notes on the Engineering News Formula.
(S.V.U.—B.E., (N.R.)—Sep., 1967)

16.2 Write brief critical notes on the bearing capacity of piles.
(S.V.U.—B.E., (R.R.)—Nov. 1974 and May, 1975)

16.3 Explain the function of pile foundation and show how the bearing capacity of the foundation can be estimated.
(S.V.U.—B.E., (R.R.)—Dec. 1970)

16.4 Explain the basic difference in the bearing capacity computation of shallow and deep foundations. How are skin friction and point resistance of a pile computed?
(S.V.U.—B.E., (R.R.)—Nov. 1973)

16.5 Outline the procedure to determine the bearing capacity of a single driven pile and that of a group of piles in a thick layer of soft clay.
(S.V.U.—B.E., (R.R.)—May, 1971)

16.6 Distinguish between driven and bored piles. Explain why the settlement of a pile foundation (pile group) will be many times that of a single pile even though the load per pile on both cases is maintained the same.
(S.V.U.—B.Tech., (Part-time)—Sep., 1982)

804   *Geotechnical Engineering*

16.7 Give a method to determine the bearing capacity of a pile in clay soil. What is group effect and how will you estimate the capacity of a pile group in clay?

(S.V.U.—B.E., (R.R.)—May, 1970)

16.8 What are the various methods used for determining the capacity of (*i*) a driven pile and (*ii*) a cast-in-situ pile?

16.9 What is the basis on which the dynamic formulae are derived? Mention two well known dynamic formulae and explain the symbols involved

(S.V.U.—B.E., (N.R.)—Mar., 1966)

16.10 A wood pile of 10 m length is driven by a 1500 kg drop hammer falling through 3 m to a final set equal to 1.25 cm per blow. Calculate the safe load on the pile using the Engineering News Formula.

(S.V.U.—B.E., (N.R.)—Mar. 1966)

16.11 A precast concrete pile is driven with a 30 kN drop hammer with a free fall of 1.5 m. The average penetration recorded in the last few blows is 5 mm per blow. Estimate the allowable load on the pile using the Engineering News Formula.

16.12 What will be the penetration per blow of a pile which must be obtained in driving with a 3 t steam hammer falling through 1 m allowable load is 25 tonnes?

16.13 A 30 cm diameter pile penetrates a deposit of soft clay 9 m deep and rests on sand. Compute the skin friction resistance. The clay has a unit cohesion of 0.6 kg/cm^2. Assume an adhesion factor of 0.6 for the clay.

16.14 A square pile 25 cm size penetrates a soft clay with unit cohesion of 75 kN/m^2 for a depth of 18 m and rests on stiff soil. Determine the capacity of the pile by skin friction. Assume an adhesion factor of 0.75.

16.15 A 30 cm square pile, 15 m long, is driven in a deposit of medium dense sand ($\phi = 36°$, $N_\gamma = 40$ and $N_q = 42$). The unit wt. of sand is 15 kN/m^3. What is the allowable load with a factor of safety of 3? Assume lateral earth pressure coefficient = 0.6.

16.16 A square pile group of 9 piles of 25 cm diameter is arranged with a pile spacing of 1 m. The length of the piles is 9 m. Unit cohesion of the clay is 75 kN/m^2. Neglecting bearing at the tip of the piles determine the group capacity. Assume adhesion factor of 0.75.

16.17 Determine the group efficiency of a rectangular group of piles with 4 rows, 3 piles per row, the uniform pile spacing being 3 times the pile diameter. If the individual pile capacity is 100 kN, what is the group capacity according to this concept?

16.18 A square pile group of 16 piles passes through a filled up soil of 3 m depth. The pile diameter is 25 cm and pile spacing is 75 cm. If the unit cohesion of the material is 18 kN/m^2 and unit weight is 15 kN/m^3, compute the negative skin friction on the group.

17
SOIL STABILISATION

17.1 INTRODUCTION

'Soil Stabilisation', in the broadest sense, refers to the procedures employed with a view to altering one or more properties of a soil so as to improve its engineering performance.

Soil Stabilisation is only one of several techniques available to the geotechnical engineer and its choice for any situation should be made only after a comparison with other techniques indicates it to be the best solution to the problem.

It is a well known fact that, every structure must rest upon soil or be made of soil. It would be ideal to find a soil at a particular site to be satisfactory for the intended use as it exists in nature, but unfortunately, such a thing is of rare occurrence.

The alternatives available to a geotechnical engineer, when an unsatisfactory soil is met with, are (i) to bypass the bad soil (e.g., use of piles), (ii) to remove bad soil and replace with good one (e.g., removal of peat at a site and replacement with selected material), (iii) redesign the structure (e.g., floating foundation on a compressible layer), and (iv) to treat the soil to improve its properties.

The last alternative is termed soil stabilisation. Although certain techniques of stabilisation are of a relatively recent origin, the art itself is very old. The original objective of soil stabilisation, was, as the name implies, to increase the strength or stability of soil. However, techniques have now been developed to alter almost every engineering property of soil. The primary aim may be to alter the strength and/or to reduce its sensitivity to moisture changes.

The most common application of soil stabilisation is the strengthening of the soil components of highway and airfield pavements.

17.2 CLASSIFICATION OF THE METHODS OF STABILISATION

A completely consistent classification of soil stabilisation techniques is difficult. Classifications may be based on the treatment given to soil, on additives used, or on the process involved.

Broadly speaking, soil stabilisation procedures may be brought under the following two heads:
   I. Stabilisation without additives
   II. Stabilisation with additives

Stabilisation without additives may be 'mechanical'—rearrangement of particles through compaction or addition or removal of soil particles. It may be by 'drainage' —drainage may be achieved by the addition of external load, by pumping, by electro-osmosis, or by application of a thermal gradient—heating or cooling.

Stabilisation with additives may be cement stabilisation (that is, soil cement), bitumen stabilisation, or chemical stabilisation (with fly ash, lime, calcium or sodium chloride, sodium silicate, dispersants, physico-chemical alteration involving ion-exchange in clay-minerals or injection stabilisation by grouting with soil, cement or chemicals).

The appropriate method for a given situation must be chosen by the geotechnical engineer based on his experience and knowledge. Comparative laboratory tests followed by limited field tests, should be used to select the most economical method that will serve the particular problem on hand. Field-performance data may help in solving similar problems which arise in future.

It must be remembered, however, that soil stabilisation is not always the best solution to a problem.

17.3 STABILISATION OF SOIL WITHOUT ADDITIVES

Some kind of treatment is given to the soil in this approach; no additives are used. The treatment may involve a mechanical process like compaction and a change of gradation by addition or removal of soil particles or processes for drainage of soil.

17.3.1 Mechanical Stabilisation

'Mechanical stabilisation' means improving the soil properties by rearrangement of particles and densification by compaction, or by changing the gradation through addition or removal of soil particles.

Rearrangement of particles—Compaction

The process of densification of a soil or 'compaction', as it is called, is the oldest and most important method. In addition to being used alone, compaction constitutes an essential part of a number of other methods of soil stabilisation.

The important variables involved in compaction are the moisture content, compactive effort or energy and the type of compaction. The most desirable combination of the placement variables depends upon the nature of the soil and the desired properties. Fine-grained soils are more sensitive to placement conditions than coarse-grained soils.

Compaction has been shown to affect soil structure, permeability, compressibility characteristics and strength of soil and stress-strain characteristics (Leonards, 1962). Soil compaction has already been studied in some detail in Chapter 12.

Change of gradation—addition or removal of soil particles

The engineering behaviour of a soil depends upon (among other things) the grain-size distribution and the composition of the particles. The properties may be significantly altered by adding soil of some selected grain-sizes and, or by removing some selected fraction of the soil. In other words, this approach consists in manipulating the soil fractions to obtain a suitable grading, which involves mixing coarse material or gravel (called 'aggregate'), sand, silt, and clay in proper proportions so that the mixture when compacted attains maximum density and strength. It may involve blending of two or more naturally available soils in suitable proportions to achieve the desired engineering properties for the mixture after necessary compaction.

Soil materials can be divided into two fractions, the granular fraction or the 'aggregate', retained on a 75-micron I.S. Sieve, and the fine soil fraction or the 'binder', passing this sieve. The aggregate provides strength by internal friction and hardness or incompressibility, while the binder provides cohesion or binding property, water-retention capacity or imperviousness and also acts as a filler for the voids of the aggregate.

The relative amounts of aggregate and binder determine the physical properties of the compacted stabilised soil. The optimum amount of binder is reached when the compacted binder fills the voids without destroying all the grain-to-grain contacts of coarse particles. Increase in the binder beyond this limit results in a reduction of internal friction, a slight increase in cohesion and greater compressibility. Determination of the optimum amount of binder is an important component of the design of the mechanically stabilised mixture.

Mechanical stabilisation of this type has been largely used in the construction of low-cost roads. Guide specifications have been developed based on past experience, separately for base courses and surface courses.

The grading obtained by a simple rule given by Fuller has been found to be satisfactory:

$$\text{Percent passing a particular sieve} = 100 \sqrt{\frac{d}{D}}$$

...(Eq. 17.1)

where d = aperture size of the sieve, and

D = size of the largest particle.

Suggested gradings for mechanically stabilised base and surface courses for roads are given in Table 17.1.

If the primary aim is to reduce the permeability of a soil, sodium montmorillonite—a clay mineral, called "bentonite"—may be added. For example, the permeability of a silty sand could be reduced from 10^{-4} cm/s to 10^{-9} cm/s by the addition of 10% of bentonite. However, it must be remembered that bentonite is costly and its effectiveness may be reduced by flowing water, and wetting and drying. Naturally available local clay can be blended with pervious soils to result in a more nearly permanent blanket; this may be a much cheaper and superior approach, if such a material is available in the proximity of the site.

Geotechnical Engineering

Table 17.1 Suggested gradings for mechanically stabilised base and surface courses (adapted from HMSO, 1952)

I.S. Sieve size	Per cent passing				Base or surface course	
	Base course Max. size		Surface course Max. size		Max. size	
	80 mm	40 mm	20 mm	20 mm	10 mm	5 mm
80 mm	100	—	—	—	—	—
40 mm	80–100	100	—	—	—	—
20 mm	60–80	80–100	100	100	—	—
10 mm	45–65	55–80	80–100	80–100	100	—
5 mm	30–50	40–60	50–75	60–85	80–100	100
2.36 mm	—	30–50	35–60	45–70	50–80	80–100
1.18 mm	—	—	—	35–60	40–65	50–80
600 micron	10–30	15–30	15–35	—	—	30–60
300 micron	—	—	—	20–40	20–40	20–45
75 micron	5–15	5–15	5–15	10–25	10–25	10–25

Note: 1. Not less than 10% should be retained between each pair of successive sieves specified, excepting the largest pair.

2. Material passing I.S. Sieve No. 36 shall have the following properties:

For base courses:
   Liquid limit $\not> 25\%$
   Plasticity Index $\not> 6\%$

For surface courses:
   Liquid limit $\not> 35\%$
   Plasticity Index: between 4 and 9.

Gravel is used for base courses of pavements and for filter courses. The presence of fines to an extent more than the optimum might make the gravel unsatisfactory. The limit for the fines may be 3 to 7%, depending upon its intended use. The upper limit is for filter purposes.

An obvious treatment, for a gravel with larger amount of fines than desired, is to wash out excess fines. This may sound very easy, but it is not that simple in practice. When supplies of dirty gravel along with a satisfactory source of water are available locally, the procedure of removal of fines from gravel by washing is employed.

Mehra's method of stabilisation

The procedure advocated by S.R. Mehra (IRC, 1976) for mechanical stabilisation has been widely used in the construction of base and surface courses for low-cost roads. The method involves the use of just three sieves for the mechanical analysis (instead of ten as advocated by ASTM and HMSO) and only the plasticity index. The three sieves are equivalents of I.S. Sieves of sizes 1.18 mm, 300–μ and 75–μ.

Mehra recommends a compacted thickness of 76 mm for the base course with a minimum of 50% sand (fraction between 300–μ and 75–μ), and a plasticity index of 5 to 7. The surface course also should be 76 mm thick with a minimum of one-third portion of sand in the soil mixture to which brick aggregate is added at

50% of the soil mixture. The plasticity index for the mix is recommended to be 9.5 to 12.5 for roads without surface treatments, and 8 to 10 for roads with surface treatments.

Although the grading is not conducive to producing the densest mix, it is supposed to yield a satisfactory road surface under mixed traffic conditions prevailing in India.

Mehra's method has been very popular, especially in the northern parts of the country.

17.3.2 Stabilisation by Drainage

Generally speaking, the strength of a soil generally decreases with an increase in pore water and in the pore water pressure. Addition of water to a clay causes a reduction of cohesion by increasing the electric repulsion between particles. The strength of a saturated soil depends directly on the effective or intergranular stress. For a given total stress, an increase in pore water pressure results in a decrease of effective stress and consequent decrease in strength.

Thus, drainage of a soil is likely to result in an increase in strength which is one of the primary objectives of soil stabilisation.

The methods used for drainage for this purpose are:
1. application of external load to the soil mass,
2. drainage of pore water by gravity and/or pumping, using well-points, sand-drains, etc.,
3. application of an electrical gradient or electro-osmosis; and,
4. application of a thermal gradient.

Application of external load to the soil mass

The aim is to squeeze out pore water. The common load is by way of adding an earth surcharge. Other miscellaneous techniques are also used.

Drainage of pore water by gravity and/or pumping

Well-points are used to drain pore water either by gravity and/or pumping (Barron, 1948).

Vertical sand drains or sand piles (Rutledge and Johnson, 1958) are used to expedite drainage of a soil stratum. The diameter of the sand-drains may be 40 to 50 cm and the spacing may be 2 to 3 m. A drainage blanket is placed on top and a surcharge fill is placed on top of this blanket.

A proper design of sand-drain installation involves the determination of diameter and spacing of sand drains, the thickness of the drainage blanket, and, amount and duration of surcharge fill loading (Terzaghi, 1943).

Application of electrical gradient or electro-osmosis

When a direct electric current is passed through a saturated soil, water moves towards the cathode. If this is removed the soil undergoes consolidation. This phenomenon is called "electro-osmosis".

In addition to electro-osmotic consolidation, passage of electric current can cause ion exchange, alteration of arrangement of the particles, and electro-chemical decomposition of the electrodes. The combination of these changes brought about in the soil is called 'electrical stabilisation'. This procedure has been successfully employed to increase skin friction of piles (Casagrande, 1952 and 1953).

Application of thermal gradient

Heating or cooling a soil can cause significant changes in its properties. The main drawback of thermal stabilisation is the cost involved, which makes it seldom cost-competitive with other techniques.

Even a slight increase in temperature can reduce the electric repulsion between clay particles and can cause slight increase in strength. Temperature, in excess of 100°C, drives off the adsorbed moisture on clay particles, thereby increasing its strength. Temperatures of the order of 400 to 600°C, cause irreversible changes in the clay minerals, making them less water-sensitive. Heat has been known to change an expansive clay to an essentially non-expansive type.

Soviet engineers have used thermal stabilisation for deep deposits of partially saturated loess soil up to about 12 m (Lambe, 1962). The method consisted in burning a mixture of liquid fuel and air injected through a network of pipes. Cylinders of solidified soil about 2.7 m in diameter were formed which served as pile foundations. Rumanian engineers have also used this technique for cohesive soils (Beles and Stanculescu, 1958). The heat was provided by burning liquid or gas fuel in a unit lowered into a boring.

Freezing pore water in a wet soil increases its strength. Clay soil requires temperatures much lower than 0°C for this purpose. Ice piles, ice coffer dams and underpinning buildings by this approach are examples.

17.4 STABILISATION OF SOIL WITH ADDITIVES

Stabilisation of soil with some kind of additive is very common. The mode and degree of alteration necessary depend on the nature of the soil and its deficiencies. If additional strength is required in the case of cohesionless soil, a cementing or a binding agent may be added and if the soil is cohesive, the strength can be increased by making it moisture-resistant, altering the absorbed water films, increasing cohesion with a cement agent and adding internal friction. Compressibility of a clay soil can be reduced by cementing the grains with a rigid material or by altering the forces of the adsorbed water films on the clay minerals. Swelling and shrinkage may also be reduced by cementing, altering the water adsorbing capacity of the clay mineral and by making it moisture-resistant. Permeability of a cohesionless soil may be reduced by filling the voids with an impervious material or by preventing flocculation by altering the structure of the adsorbed water on the clay mineral; it may be increased by removing the fines or modifying the structure to an aggregated one.

A satisfactory additive for soil stabilisation must provide the desired qualities and, in addition, must meet the following requirements: Compactibility with the soil material, permanency, easy handling and processing, and low cost.

Many additives have been employed but with varying degrees of success. No material has been found to meet all the requirements, and most of the materials are expensive.

17.4.1 Types of Additives Used

The various additives used fall under the following categories:

(i) *Cementing materials*: Increase in strength of the soil is achieved by the cementing action of the additive. Portland cement, lime, fly-ash and sodium silicate are examples of such additives.

(ii) *Water-proofers*: Bituminous materials prevent absorption of moisture. These may be used if the natural moisture content of the soil is adequate for providing the necessary strength. Some resins also fall in this category, but are very expensive.

(iii) *Water-retainers*: Calcium chloride and sodium chloride are examples of this category.

(iv) *Water-repellents or retarders*: Certain organic compounds such as stearates and silicones tend to get absorbed by the clay particles in preference to water. Thus, they tend to keep off water from the soil.

(v) *Modifiers and other miscellaneous agents*: Certain additives tend to decrease the plasticity index and modify the plasticity characteristics. Lignin and lignin-derivatives are used as dispersing agents for clays.

17.4.2 Cement Stabilisation

Portland cement is one of the most widely used additives for soil stabilisation. A mixture of soil and cement is called "soil-cement". If a small percentage of cement is added primarily to reduce the plasticity of fat soils, the mixture is said to be a "cement-modified soil". If the soil-cement has enough water which facilitates pouring it as mortar, it is said to be a 'plastic soil cement". It is used in canal linings.

The chemical reactions of cement with the silicious soil in the presence of water are believed to be responsible for the cementing action. Many of the grains of the coarse fraction get cemented together, but the proportion of clay particles cemented is small.

Almost any inorganic soil can be successfully stabilised with cement; organic matter may interfere with the cement hydration.

Soil-cement has been widely used for low-cost pavements for highways and airfields, and as bases for heavy traffic. Generally, it is not recommended as a wearing coarse in view of its low resistance to abrasion.

Factors affecting soil-cement

The important factors which affect the properties of soil-cement are the nature of the soil, cement content, compaction, and the method of mixing.

Nature of the soil

Almost all soils, devoid of organic matter and capable of being pulverised, can be stabilised with the addition of cement. The requirement of cement will increase with the increase in specific surface of the soil; in other words, it increases with the fines content. Expansive clays are difficult to deal with. Well-graded soils with less than 50% of particles finer than 75-µ and a plasticity index less than 20 are most suitable for this method of stabilisation. Approximate limits of gradation of soil for economic stabilisation with cement are obtained by research (HRB, 1943).

Soils containing more than 2% of organic matter are generally considered to be unsuitable, since the strength of soil-cement is reduced by the organic matter interfering with the hydration of cement. The presence of sulphates also renders a soil unsuitable for stabilisation with cement.

The nature of the exchangeable ions on the soil grains is an important factor. Calcium is the most desirable ion for the case of cement stabilisation. Addition of less than 1% of lime or calcium chloride may render a soil more suitable for stabilisation with cement in spite of the presence of organic matter.

Cement content

The normal range of cement content used is 5 to 15% by weight of the dry soil, finer soils requiring greater quantity of cement. The more the cement content, the greater is the strength of the resulting soil-cement. A compressive strength of 2000 to 3000 kN/m^2 as obtained from a test on a cylinder of soil-cement after 7 days of curing must be satisfactory. High early strength cement yields better results than ordinary cement.

Compaction

Adequate compaction is essential. Optimum moisture content is to be used in the process as there is no problem of stability as for concrete.

Mixing

Uniform mixing will lead to strong soil-cement. The efficiency of mixing depends upon the type of plant used. Mixing should not be done after hydration has begun.

Admixtures

Addition of about 0.5 to 1.0% of certain chemicals such as lime or calcium chloride to soil-cement has been found to accelerate the set and to improve the properties of the final products (Lambe, Michaels and Moh, 1959).

Designing and testing soil-cement

Soil-cement mix design consists of selecting the amount of cement, the amount of water and the compaction density to be achieved in the field. The thickness of the stabilised soil and also the placement conditions are to be decided upon. The thickness of a soil-cement base is to be taken equal to that required for a granular base for a good subgrade, and as equal to 75% of that required for a granular base for a poor subgrade. Usually 15 to 20 cm thickness is adequate for a soil-cement base.

The procedures adopted for design are:
  (i) Complete method, using moisture-density, freeze-thaw, and wet-dry tests;
  (ii) Short-cut method for sandy soils, using moisture-density and strength tests in conjunction with the charts prepared by the Portland Cement Association; and
  (iii) Rapid method, using moisture-density tests and visual inspection.

Field construction of soil-cement

The steps in the construction of soil-cement are:
  (i) Pulverization of soil
  (ii) Adding water and cement
  (iii) Mixing
  (iv) Spreading and compaction
  (v) Finishing and curing

In advanced countries most of the operations are mechanised. The method may be mixed-in-place, travelling plant, or stationary plant type.

Applications

Soil-cement is used as base course for pavements for light traffic. Pressed soil-cement blocks can be used in place of bricks. Rammed earth walls with just 2% cement are used for low-cost housing. Soil-cement blocks may be used in place of concrete blocks for pitching of banks of canals or canal linings.

17.4.3 Bitumen Stabilisation

Bituminous materials such as asphalts and tars have been used for soil stabilisation. This method is better suited to granular soils and dry climates.

'Bitumens' are nonaqueous systems of hydrocarbons which are completely soluble in 'Carbon disulphide'.

'Asphalts' are natural materials or refined petroleum products, which are bitumens.

'Tars' are bituminous condensates produced by the destructive distillation of organic materials such as coal, oil, lignite and wood. (Lambe, 1962).

Most bitumen stabilisation has been with asphalt. Asphalt is usually too viscous to be incorporated directly with soil. Hence, it is either heated or emulsified or cut back with a solvent like gasoline, to make it adequately fluid.

Tars are not emulsified but are heated or cut back prior to application.

Soil-asphalt is used mostly for base courses of roads with light traffic.

Bitumen stabilises soil by one or both of two mechanisms: (i) binding soil particles together, and (ii) making the soil water-proof and thus protecting it from the deleterious effects of water.

Obviously, the first mechanism occurs in cohesionless soils, and the second in cohesive soils, which are sensitive to water. Asphalt coats the surfaces of soil particles and protects them from water. It also plugs the voids in the soil, inhibiting a flow of pore water.

Bitumen stabilisation may produce one of the following:
  (i) Sand-bitumen
  (ii) Soil-bitumen
  (iii) Water-proof mechanical stabilisation
  (iv) Oiled earth

Sand-bitumen

Cohesionless soils like sand stabilised by bitumen are called sand-bitumen. The primary function of bitumen is to bind the sand grains. The quantity of asphalt may range from 5 to 12%.

Soil bitumen

Cohesive soil stabilised by bitumen is referred to as soil-bitumen; the primary objective is to make it water-proof and preserve its cohesive strength.

For best results the soil must conform to the following requirements:
Max. size : less than one-third the compacted thickness
Passing 4.75 mm sieve : greater than 50%
Passing 425-μ Sieve : 35 to 100%
Passing 75-μ sieve : 10 to 30%
Liquid limit : less than 40%
Plasticity Index : less than 18

Water-proof mechanical stabilisation

Small quantities of bitumen—1 to 3%—may be added to mechanically stabilised soils to make them water-proof.

Oiled earth

Slow-curing and medium-curing road oils are sprayed to make the earth water-resistant and resistant to abrasion. The oils penetrate a short depth into the soil without any mechanical mixing.

Admixtures

The addition of small quantities of phosphorus pentoxide or certain amines was found to improve the effectiveness of asphalt as a soil stabiliser.

The construction of soil-asphalt is very much similar to that of soil-cement; usual thickness ranges from 15 to 20 cm as with soil-cement.

17.4.4 Chemical Stabilisation

Chemical stabilisation refers to that in which the primary additive is a chemical. The use of chemicals as secondary additives to increase the effectiveness of cement and of asphalt has been mentioned earlier.

Lime and salt have found wide use in the field. Some chemicals are used for stabilising the moisture in the soil and some for cementation of particles. Certain aggregates and dispersants have also been used.

Lime stabilisation

Lime is produced from natural limestone. The hydrated limes, called 'slaked limes', are the commonly used form for stabilisation.

In addition to being used alone, lime is also used in the following admixtures, for soil stabilisation:

(*i*) Lime-fly ash (4 to 8% of hydrated lime and 8 to 20% of fly-ash)
(*ii*) Lime-portland cement
(*iii*) Lime-bitumen

The use of lime as a soil stabiliser dates back to Romans, who used it in the construction of the 'Appian way' in Rome. This road has given excellent service and is maintained as a traffic artery even today.

There are two types of chemical reactions that occur when lime is added to wet soil. The first is the alteration of the nature of the adsorbed layer through ion exchange of calcium for the ion naturally carried by the soil, or a change in the double layer on the soil colloids. The second is the cementing action or pozzolanic action which requires a much longer time. This is considered to be a reaction between the calcium with the available reactive alumina or silica from the soil.

Lime has the following effects on soil properties:

Lime generally increases the plasticity index of low-plasticity soils and decreases that of highly plastic soils; in the latter case, lime tends to make the soil friable and more easily handled in the field.

It increases the optimum moisture content and decreases the maximum compacted density; however, there will be an increase in strength. About 2 to 8% of lime may be required for coarse-grained soils, and 5 to 10% for cohesive soils.

Certain sodium compounds (e.g., sodium hydroxide and sodium sulphate), as secondary additives, improve the strength of soil stabilised with lime.

Lime may be applied in the dry or as a slurry. Better penetration is obtained when it is used as a slurry. The construction of lime-stablised soils is very much similar to that of soil-cement. The important difference is that, in this case, no time limitation may be placed on the operations, since the lime-soil reactions are slow. Care should be taken, however, to prevent the carbonation of lime. Lime stabilisation has been used for bases of pavements.

Salt stabilisation

Calcium chloride and sodium chloride have been used for soil stabilisation. Calcium chloride is hygroscopic and deliquiscent. It absorbs moisture from the atmosphere and retains it. It also acts as a soil flocculant. The action of sodium chloride is similar.

The effect of salt on soil arises from colloidal reactions and the alteration of the characteristics of soil water. Salt lowers the vapour pressure of pore water and also the freezing point; the frost heave will be reduced because of the latter phenomenon.

The main disadvantage is that the beneficial effects of salt are lost, if the soil gets leached.

Lignin and chrome-lignin stabilisation

Lignin is one of the major constituents of wood and is obtained as a by-product during the manufacture of paper from wood. Lignin, both in powder form and in the form of sulphite liquor, has been used as an additive to soil for many years. A concentrated solution, partly neutralised with calcium bases, known as Lignosol, has also been used.

The stabilising effects of lignin are not permanent since it is soluble in water; hence periodic applications may be required. In an attempt to improve the action of lignin, the 'Chrome-lignin process'' was developed (Smith, 1952). The addition of sodium bichromate or potassium bichromate to the sulphite waste results in the formation of an insoluble gel.

If the lignin is not neutralised, it is acid and acts as a soil aggregant; when neutralised as with Lignisol, it acts as a dispersant. Chrome lignin imparts considerable strength to soils as a cementing agent (Lambe, 1962).

Stabilisers with water-proofers

It is well known that cohesive soils possess considerable strength when they are dry. When they have access to water, they imbibe it and lose strength. Waterproofers, i.e., chemicals which prevent the deleterious effects of water on soils, are useful in such cases. Siliconates, amines and quaternary ammonium salts fall in this category.

Water-proofers do not increase the strength, but help the soil retain its strength even in the presence of water.

Stabilisation with natural and synthetic resins

Certain natural as well as synthetic resins, which are obtained by polymerisation of organic monomers, have also been used for soil stabilisation. They act primarily as water-proofers. Vinol resin and Rosin, both of which are obtained from pinetrees, are the commonly used natural resins. Aniline-furfural, polyvinyl alcohol (PVA), and calcium acrylate are commonly used synthetic resins. Asphalt and lignin, which are also resinous materials, have already been discussed separately.

Aggregants and dispersants

Aggregants and dispersants are chemicals which bring about modest changes in the properties of soils containing fine grains. These materials function by altering the electrical forces between the soil particles of colloidal size, but provide no cementing action. They affect the plasticity, permeability and strength of the soil treated. Low treatment levels are adequate for the purpose.

Aggregants increase the net electrical attraction between adjacent fine-grained soil particles and tend to flocculate the soil mass. Inorganic salts such as calcium chloride and ferric chloride, and polymers such as Krilium are important examples. Changes in adsorbed water layers, ion-exchange phenomena and increase in ion concentration are the possible mechanisms by which the aggregants work.

Dispersants are chemicals which increase the electrical repulsion between adjacent fine-grained soil particles, reduce the cohesion between them, and tend to cause them to disperse. Phosphates, sulphonates and versanates are the most common dispersants, which tend to decrease the permeability. Ion exchange and anion adsorption are the possible mechanisms by which the dispersants work.

Miscellaneous chemical stabilisers

Sodium silicate can be used as a primary stabiliser as well as a secondary additive to conventional stabilisers such as cement. Injection is the usual process by which this is used. Phosphoric acid also has been used to some extent.

Molasses, tung oil, sodium carbonate, paraffin and hydrofluoric acid are some miscellaneous chemicals which have been considered but have not received any extensive application.

17.4.5 Injection Stabilisation

Injection of the stabilising agent into the soil is called 'Grouting'. This process makes it possible to improve the properties of natural soil and rock formations, without excavation, processing, and recompaction. Grouting may have one of the two objectives—to improve strength properties or to reduce permeability. This is achieved by filling cracks, fissures and cavities in the rock and the voids in soil with a stabiliser that is initially in a liquid state or in suspension and which subsequently solidifies or precipitates.

Injection is a very common technique in the oil industry; petroleum engineers frequently use this method for sealing or operating wells. Injection techniques, unfortunately, are rather complex. The selection of proper grout material and appropriate technique can normally be made best only after field exploration and testing. The results of the injection process are rather difficult to assess. *Grouting must be called an art rather than a science.*

Grouting materials

The following have been used for grouting:
 (*i*) Soil
 (*ii*) Portland cement
 (*iii*) Bitumen
 (*iv*) Chemicals

Bitumen is not used as much as other materials for this purpose.

The properties of the grout must fit the soil or rock formation being injected. The dimensions of the pores or fissures determine the size of grout particle so that these can penetrate.

The following are the guide lines recommended:

$$D_{85} \text{ (Grout)} < \frac{1}{15} D_{15} \text{ (Soil)} \qquad \ldots\text{(Eq. 17.1)}$$

$$D_{85} \text{ (Grout)} < \frac{1}{3} B \text{ (Fissure)} \qquad \ldots\text{(Eq. 17.2)}$$

818 Geotechnical Engineering

$$R_g \, (= D_{85} \text{ Grout} / D_{15} \text{ Soil}) > 15 \quad \ldots \text{(Eq. 17.3)}$$

R_g : Groutability ratio.

(Johnson, 1958; Kennedy, 1958)

Viscosity and rate of hardening are important characteristics of the grout material. Low viscosity and slow hardening permits penetration to thin fissures and small voids whereas high viscosity and rapid hardening restrict flow to large voids.

The grout must not be unduly diluted or washed away by groundwater. Insoluble or rapid setting grouts are used in situations where there is ground water flow.

'Mudjacking' is a form of soil injection used to raise highway pavements, railway tracks and even storage tanks. This technique consists of injecting a mixture of soil, Portland cement and water to shallow depths at relatively low pressures.

Cement grouting has been used to stabilise rock formations, as also alluvial sands and gravels.

A number of chemicals have been used for grouting; among them, sodium silicate in water, known as 'water glass' is the most common. This solution contains both free sodium hydroxide and colloidal silicic acid. The addition of certain salts such as calcium chloride, magnesium chloride, ferric chloride and magnesium sulphate, or of certain acids such as hydrochloric acid and sulphuric acid, results in the formation of an insoluble silica gel.

On ageing, the gel shrinks and cracks. Hence, the effectiveness of silicate injection in the presence of ground water remains doubtful.

The grouting plant includes the material handling system, mixers, pumps and delivery hoses. The mixing of components is done by a proportioning valve or pump at the point of injection. A perforated pipe is driven into soil to the level of grouting; if it is rock, grouting holes are drilled.

The injection pattern depends on the purpose of grouting. Generally a grid pattern is used. The spacing may be 6 to 15 m. Sufficient pressure is used to force the grout into voids and fissures.

17.5 CALIFORNIA BEARING RATIO

The strength of the subgrade is an important factor in the determination of the thickness required for a flexible pavement. It is expressed in terms of its 'California Bearing Ratio', usually abbreviated as 'C.B.R.'.

The CBR value is determined by an empirical penetration test devised by the California State Highway Department (U.S.A.), and derives its name thereof. The results obtained by these tests are used in conjunction with empirical curves, based on experience, for the design of flexible pavements.

The test is arbitrary and the results give an empirical number, expressed usually in per cent, which may not be directly related to fundamental properties governing the shear strength of soils, such as cohesion and angle of internal friction. However, attempts have been made recently to correlate CBR value with the bearing capacity and plasticity index of the soil.

The California bearing ratio (CBR) is defined as the rate of the force per unit area required to penetrate a soil mass with a standard circular plunger of 50 mm diameter at the rate of 1.25 mm/min to that required for the corresponding penetration of a standard material.

The standard material is crushed stone and the load which has been obtained from a test on it is the standard load, this material being considered to have a CBR of 100%.

The CBR value is usually determined for penetrations of 2.5 mm and 5 mm. Where the ratio at 5 mm is consistently higher than that at 2.5 mm, the value at 5 mm is used. Otherwise, the value at 2.5 mm is used, which is more common.

The CBR test is usually carried out in the laboratory either on undisturbed samples or on remoulded samples, depending upon the condition in which the subgrade soil is likely to be used. Efforts shall be put in to simulate in the laboratory the pressure and the moisture conditions to which the subgrade is expected to be subjected in the field.

17.5.1 Determination of CBR Value

The C.B.R. Test as standardised by ISI [IS: 2720 (Part-XVI)-1979—Laboratory Determination of CBR] is as follows:

The apparatus consists of a cylindrical mould of 150 mm inside diameter and 175 mm in height. It is provided with a detachable metal extension collar 50 mm in height and a detachable perforated base plate 10 mm thick. A circular metal spacer disc 148 mm in diameter and 47.7 mm in height is also provided. A handle for screwing into the disc to facilitate its removal is also available. A standard metal rammer (IS: 9198-1979) is used for compaction for preparing remoulded specimens. The apparatus is shown in Fig. 17.1

One annular metal weight and several slotted weights weighing 24.5 N (2.5 kg) each (147 mm in diameter with a central hole 53 mm in diameter) are used for providing the necessary surcharge pressure.

A metal penetration plunger, 50 mm in diameter and not less than 100 mm long, is used for penetrating the specimen in the mould. If it is necessary to use a plunger of greater length, a suitable extension rod may be used. Dial gauges reading to 0.01 mm are used to record the penetration.

I.S. Sieves (20 mm and 4.75 mm), and other general apparatus such as a mixing bowl, straight edge, scales, soaking tank or pan, drying oven, filter paper, dishes and a calibrated measuring jar are also required.

A loading machine of capacity 50 kN (5,000 kg approx.) in which the rate of displacement of 1.25 mm/min can be maintained is necessary.

Preparation of test specimen

The test may be performed on undisturbed specimens or on remoulded specimens which may be compacted either statically or dynamically.

820 Geotechnical Engineering

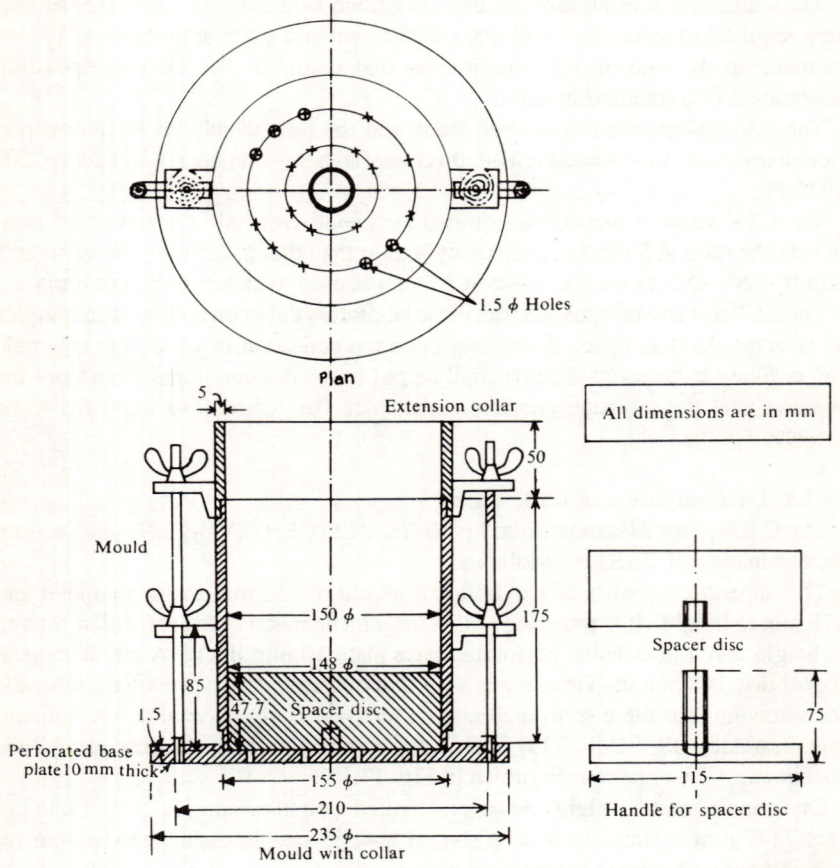

Fig. 17.1 CBR apparatus [IS: 2720 (Part-XVI)-1979]

Undisturbed specimens shall be obtained by fitting to the mould, the steel cutting edge of 150 mm internal diameter and pushing the mould as gently as possible into the ground. When the mould is sufficiently full of soil, it shall be removed by underdigging. The top and bottom undersurfaces are then trimmed to give the desired length to the specimen.

If the specimen is loose in the mould, the annular cavity shall be filled with paraffin wax thus ensuring that the soil receives proper support from the sides of the mould during the penetration test. The density of the soil and the water content of the soil must be determined by one of the available standard methods.

Remoulded specimens must be prepared in such a way that the dry density and water content correspond to those values at which the CBR value is desired. The material shall pass a 20-mm IS sieve. Allowance for larger material shall be made by replacing it by an equal amount of material which passes a 20 mm IS sieve but is retained on 4.75 mm IS Sieve.

Statically compacted specimens may be obtained by placing the calculated mass of soil in the mould and pressing in the displacer disc, a filter paper being placed between the disc and the soil. The pressing may be stopped when the top of the displacer disc is flush with the rim of the mould.

Dynamically compacted specimens may be obtained by using the standard metal rammer in accordance with "IS: 2720 (Part VII)-1974—Determination of water content—dry density relation using light compaction" or "IS: 2720 (Part VIII)-1974—Determination of water content—dry density relation using heavy compaction". The mould with the extension collar attached shall be clamped to the base plate. The spacer disc shall be inserted over the base plate and a disc of coarse filter paper placed on the top of the spacer disc. After compacting the soil into the mould, the extension collar shall be removed and the top of the sample struck off level with the rim of the mould by means of a straight edge. The perforated base plate and spacer disc shall be removed for recording the mass of the mould and the compacted soil. A disc of coarse filter paper shall be placed on the perforated base plate, the mould and the compacted soil shall be inverted, and the perforated base plate clamped to the mould with the compacted soil in contact with the filter paper.

In both cases of compaction, if soaking of the sample is required, representative samples of the material shall be taken both before compaction and after compaction for determination of water content.

If the sample is not to be soaked, representative sample of the material after the penetration shall be taken for the determination of the water content.

Test procedure

The mould containing the specimen, with the base plate in position, shall be placed on the lower plate of the loading machine. Surcharge weights, sufficient to produce a pressure equal to the weight of the base material and the pavement, shall be placed on the specimen. If the specimen has been soaked previously, the surcharge shall be equal to that used during the soaking period. The annular weight above which the slotted weights are placed prevents the upheaval of the soil into the slots of the weights. The plunger shall be seated under a load of 39.2 N (4 kg) so that, full contact is established between surface of the specimen and plunger. The dial gauges of the proving ring and those for penetration are set to zero. The seating load for the plunger is ignored for the purpose of showing the load penetration relation. Load shall be applied such that the rate of penetration is approximately 1.25 mm/min. Load readings shall be recorded at penetrations of 0, 0.5, 1.0, 1.5, 2.0, 2.5, 4.0, 5.0, 7.5, 10.0 and 12.5 mm. The maximum load and penetration shall be recorded if it occurs for a penetration of less than 12.5 mm. The plunger shall be raised and detached from the loading machine. About 0.5 N (50 g) of soil shall be collected from the top 30 mm layer of the specimen and the water content determined as per IS: 2720 (Part-II)-1973. The presence of any oversize particles shall be verified which may affect the results if they happen to be located directly below the penetration plunger.

The penetration test may be repeated for the reverse end of the sample as a check. The set-up is shown schematically in Fig. 17.2.

822 Geotechnical Engineering

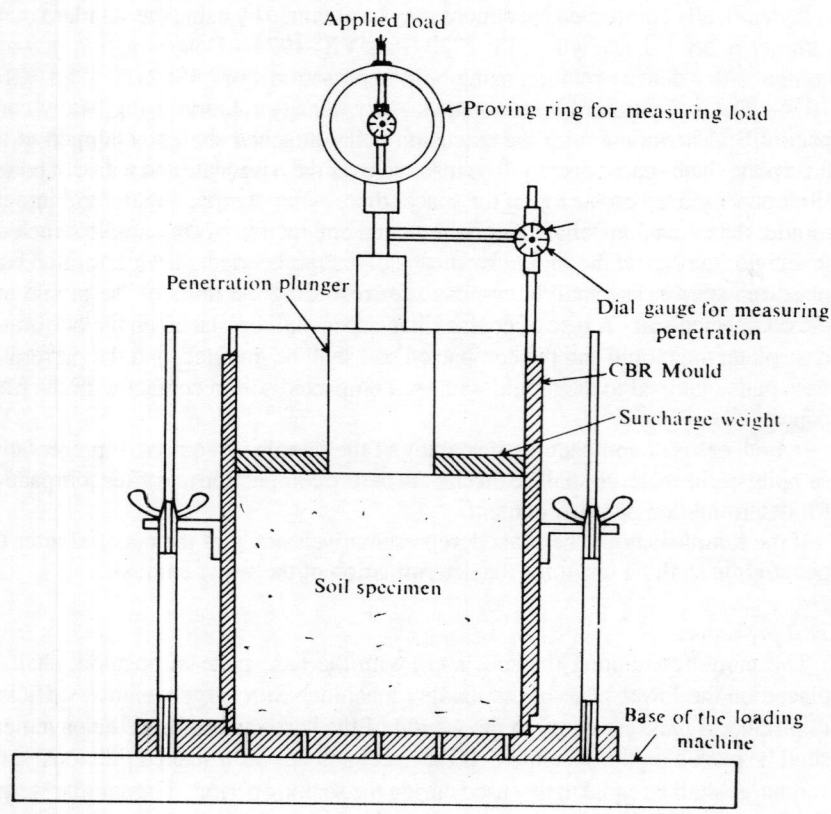

Fig. 17.2 Schematic of the set-up for CBR test

Load-penetration curve

Load vs penetration curve is plotted. This curve will be mainly convex upwards although the initial portion of the curve may be concave upwards due to surface irregularities. A correction shall then be applied by drawing a tangent to the upper curve at the point of contraflexure. The corrected curve shall be taken to be this tangent plus the convex portion of the original curve with the origin of penetrations shifted to the point where the tangent cuts the horizontal penetration axis as illustrated in Fig. 17.3.

CBR value

Corresponding to the penetration value at which the CBR is desired, corrected load shall be taken from the load penetration curve and the CBR calculated as follows:

$$\text{CBR} = \frac{P_T}{P_S} \times 100 \qquad \ldots(\text{Eq. 17.4})$$

where P_T = Corrected unit (or total) test load corresponding to the chosen penetration from the load-penetration curve, and

P_S = standard unit (or total) load for the same depth of penetration as for P_T taken from Table 17.2.

The CBR values are usually calculated for penetrations of 2.5 mm and 5 mm. Generally, the CBR value at 2.5 mm penetration will be greater than that at 5 mm penetration and in such a case the former shall be taken as the CBR value for design purposes. If the CBR value corresponding to a penetration of 5 mm exceeds that for 2.5 mm the test shall be repeated. If identical results follow, the CBR corresponding to 5 mm penetration shall be taken for design.

The CBR value shall be reported correct to the first decimal place.

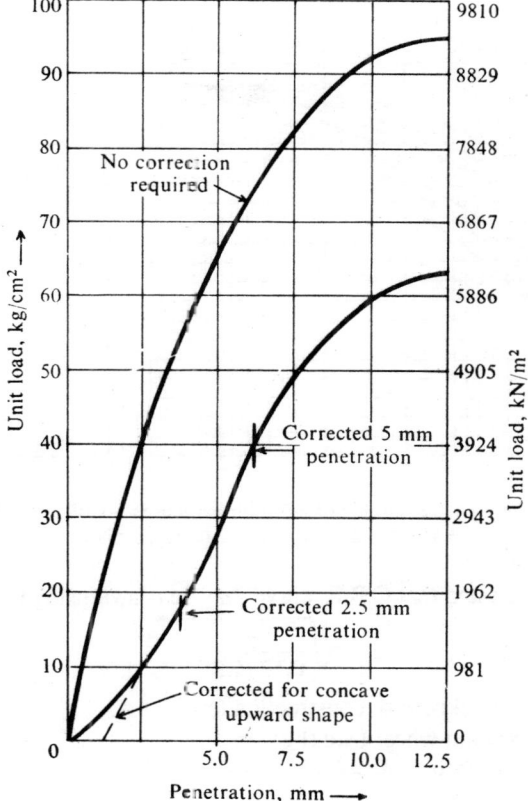

Fig. 17.3 Load vs penetration curves for CBR test

Table 17.2 Standard load

Depth of Penetration (mm)	Unit standard load		Total standard load	
	kg/cm^2	kN/m^2	kg	kN
2.5	70	6,867	1370	13.44
5.0	105	10,300	2055	20.16
7.5	134	13,145	2630	25.80
10.0	162	15,892	3180	31.20
12.5	183	17,952	3600	35.32

17.5.2 Use of CBR

Design curves have been developed by different authorities for determining the appropriate thickness of construction above subgrade materials of known CBR for different wheel loads and traffic conditions. This approach is one of the popular ones for the design of flexible pavements.

Typical design charts developed by the Road Research Laboratory, London, which are also used in India, are shown in Fig. 17.4 and 17.5.

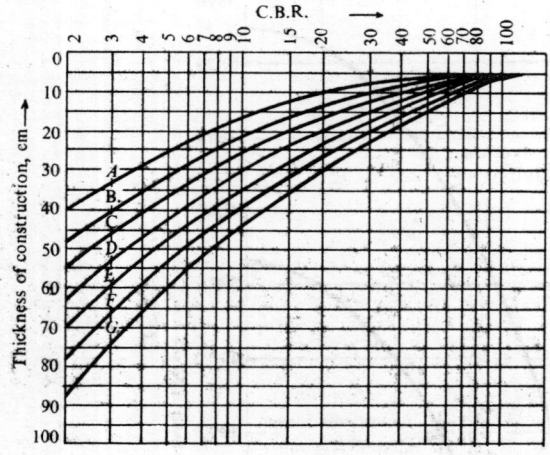

Curve	No. of vehicles/day ($> 3t$)
A	0–15
B	15–45
C	45–150
D	150–450
E	450–1500
F	1500–4500
G	Over 4500

Fig. 17.4 Design charts for flexible pavements—CBR method
(After R.R.L., London & Alam Singh, 1967)

It has been suggested that C.B.R. curve may be drawn using the equation:

$$d = \sqrt{\frac{W}{0.57\,(\text{CBR})} - \frac{A}{x}} \qquad \ldots\text{(Eq. 17.5)}$$

where, d = total thickness of construction (cm),
W = maximum wheel load (kg), and
A = area of tyre contact (cm^2)

(Hansen, 1959)

17.6 ILLUSTRATIVE PROBLEMS

Example 17.1: Following are the results obtained in a CBR test. Determine the CBR value:

Soil Stabilisation 825

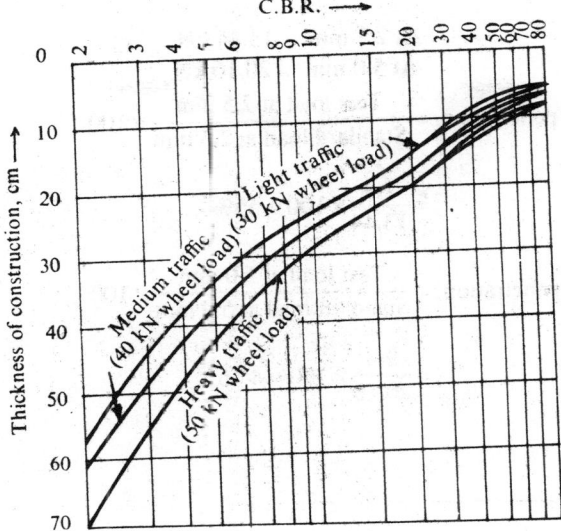

Fig. 17.5 Use of CBR in the design of flexible pavements

Penetration (mm)	0.60	1.20	1.80	2.40	3.60	4.80	7.20
Load (kN)	3.2	5.0	6.4	7.3	8.5	9.4	10.6

(S.V.U.—Four-year B.Tech.—Apr., 1983)

Load versus penetration curve is drawn as shown in Fig. 17.6.

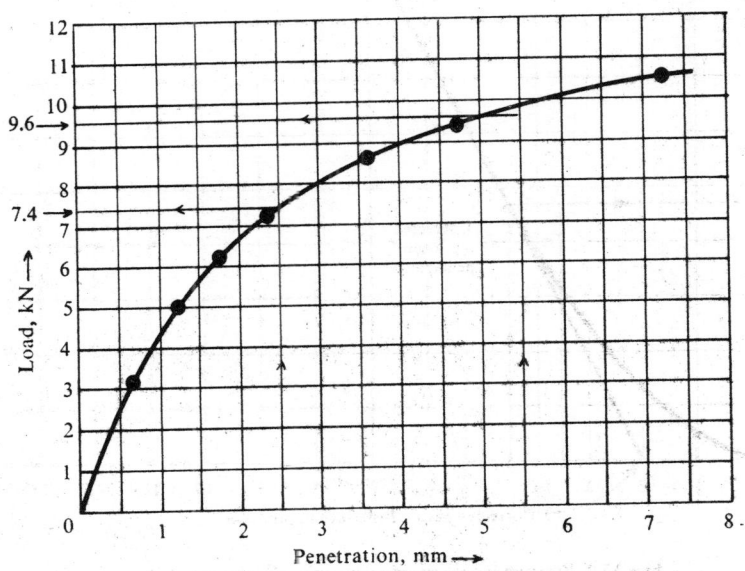

Fig. 17.6 Load versus penetration curve (Example 17.1)

826   *Geotechnical Engineering*

Standard Load:

$$\text{At } 2.5 \text{ mm} \ldots 13.44 \text{ kN}$$
$$\text{At } 5.0 \text{ mm} \ldots 20.16 \text{ kN}$$

$$\text{CBR at 2.5 mm penetration} = \frac{\text{Test load at 2.5 mm}}{\text{Standard load at 2.5 mm}} \times 100$$

$$= \frac{7.4}{13.44} \times 100 = 55\%$$

$$\text{CBR at 5.0 mm penetration} = \frac{\text{Test load at 5.0 mm}}{\text{Standard load at 5.0 mm}} \times 100$$

$$= \frac{9.6}{20.16} \times 100 = 47.6$$

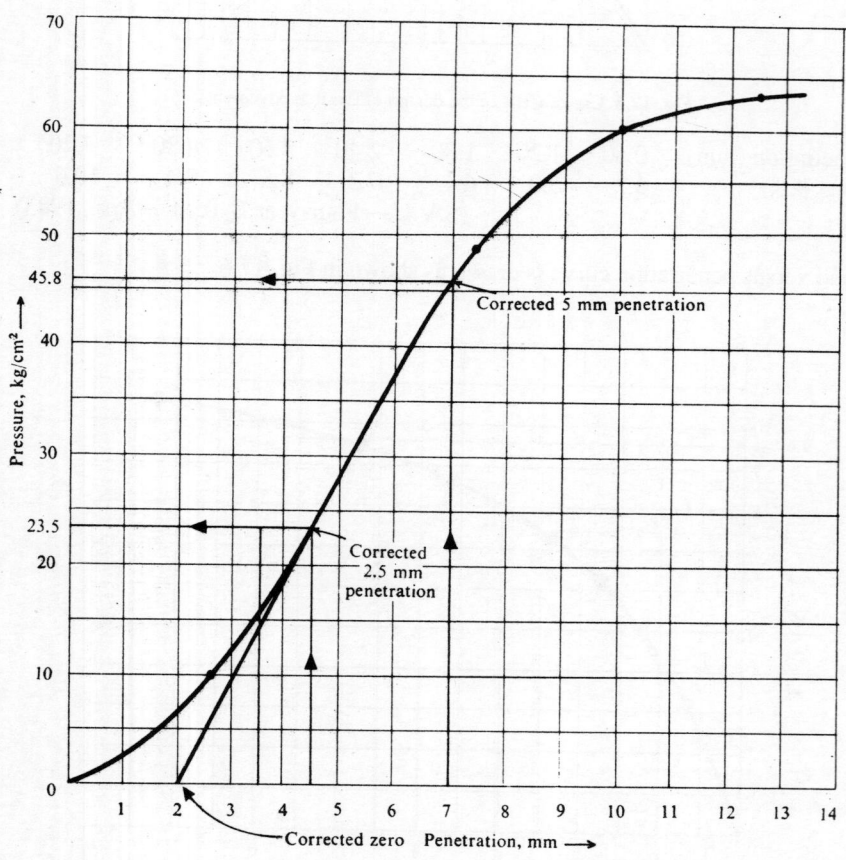

Fig. 17.7 Pressure versus penetration curve (Example. 17.2)

Since, the CBR is greater at 2.5 mm penetration, it is to be taken.
$$\therefore CBR = 55\%.$$

Example 17.2: The following observations are made in a standard CBR test. Plot the values and evaluate the CBR value of the soil.

Penetration (mm)	2.5	5.0	7.5	10.0	12.5
Load on piston (kg/cm^2)	10	28	49	60	63

(S.V.U.—B.E., (Part-time)—Dec., 1981)

Standard Load:
$$\text{At } 2.5 \text{ mm penetration} \ldots 70 \text{ kg/cm}^2$$
$$\text{At } 5.0 \text{ mm penetration} \ldots 105 \text{ kg/cm}^2$$

The observations are plotted as shown in Fig. 17.7.

Note. Since the initial portion of the curve is concave upwards, correlation is applied to the penetration; the corrected zero being 2 mm.
The test pressures at corrected values of penetration of 2.5 and 5.0 mm are obtained as 23.5 and 45.8 kg/cm^2, respectively:
C.B.R. at 2.5 mm penetration = 23.5/70 = 33.6%
C.B.R. at 5.0 mm penetration = 45.8/105 = **43.6%**

Since the value at 5 mm is greater, it is to be reported as the CBR assuming that it will be consistently higher even when the test is repeated.

Example 17.3: A road expected to carry medium traffic is to be constructed in an area where the CBR value of the subgrade is 10%. The base material chosen has a CBR value of 50%. No sub-base is to be provided. Determine the thickness required for the base course and for the surface course.

CBR value of subgrade = 10%; CBR value of base material = 50%
Using Fig. 17.5, for medium traffic, thickness of construction required for CBR 10% = 25 cm, and thickness required for CBR 50% = 10 cm, nearly.
$\therefore$ Thickness of the surface course (over the base) = **10 cm.**
Thickness of the base course (over the subgrade)
= Total thickness over the subgrade – Thickness of surface course = (25 – 10) cm
= **15 cm.**

SUMMARY OF MAIN POINTS

1. 'Soil stabilisation' means treatment of a soil with the objective of improving its engineering properties. This may be done without any additives or with one or more additives.
2. Mechanical stabilisation involves rearrangement of particles and densification by compaction or by changing the gradation by addition or removal of some of the fractions.

It may also include stabilisation by effecting drainage of the soil including the application of thermal or electrical gradient.

3. Cement stabilisation is one of the most widely used among the methods in which additives are used. The soil-cement, thus obtained, is used primarily as a base for pavements. Cohesive soils are most suited for this treatment, although finer grains need more cement.

   Bitumen stabilisation is also commonly used especially for granular soils. It may also be used for waterproofing of cohesive soils.

4. Chemical stabilisation involves a chemical as the primary additive. Lime and salt have been very commonly used. Lime modifies the plasticity characteristics of clays, and is also used as a secondary additive along with bitumen or cement.

   Natural and synthetic resins, lignin, chrome-lignin and certain aggregants and dispersants are also used as additives to soil.

5. Stabilisation by grouting or injection is also an important technique. Soil, cement, bitumen, or some chemicals may be used for grouting either soil or rock.

6. California Bearing Ratio (CBR) is an empirical concept which is used to indicate the strength of a subgrade soil. It is defined as the ratio of the load or pressure required to cause an arbitrary penetration under standard test conditions to that required by a standard crushed rock material. It may be determined in the laboratory or *in-situ*.

   It is used for the design of flexible pavements with the aid of design charts developed specifically for the purpose.

REFERENCES

1. Alam Singh: *Soil Engineering in Theory and Practice*, Asia Publishing House, Bombay, 1967.
2. R.A. Barron: *Consolidation of Fine-grained Soils by Drain Wells*, Transactions, ASCE, 1948.
3. A.A. Beles and I.I. Stanenlescu: *Thermal Treatment as a Means of Improving the Stability of Earth Masses*, Geotechnique, Dec., 1958.
4. Bharat Singh and Shamsher Prakash: *Soil Mechanics and Foundation Engineering*, Nem Chand Brothers, Roorkee, 4th ed., 1976.
5. L. Casagrande: *Electro-osmotic Stabilisation of Soils*, Journal of Boston Society of Civil Engineers, 1952.
6. L. Casagrande: *Review of Past and Current Work on Electro-osmotic Stabilisation of Soils*, Harvard Soil Mechanics Series, Dec., 1953.
7. J.B. Hansen: *Developing a Set of C.B.R. Curves*, U.S. Army Corps of Engineers, Instruction Report No. 4, 1959.
8. HMSO: *Soil Mechanics for Road Engineers*, Her Majesty's Stationery Office, London, U.K., 1952.
9. HRB: *Use of Soil-Cement Mixtures for Base Courses*, HRB war-time problems, No. 7, Washington, D.C., 1943.

10. IRC: *Tentative Guide Lines for the Use of Low Grade Aggregates and Soil Aggregate Mixture in Road Pavement Construction*, IRC:63-1976, Indian Roads Congress, New Delhi.
11. IS: 2720 (Part XVI)-1979: *Laboratory Determination of CBR*, New Delhi, 1979.
12. S.J. Johnson: *Cement and Clay Grouting of Foundations: Grouting with Clay-Cement Grouts*, Jl. of SMFE, Division, ASCE, Feb., 1958.
13. T.B. Kennedy: *Pressure Grouting Fine Fissures*, Jl. of SMFE Division, ASCE, Aug., 1958.
14. T.W. Lambe: *Soil Stabilisation*, Chapter 4; *Foundation Engineering*, Ed. G.A. Leonards, McGraw-Hill Book Co., NY, USA, 1962.
15. T.W. Lambe, A.S. Michaels, Z.C. Moh: *Improvement of Soil-cement with Alkaline and Metal Compounds*, Highway Research Board (HRB), Washington, D.C., 1959.
16. G.A. Leonards: *Foundation Engineering*, McGraw-Hill Book Co., NY, USA, 1962.
17. H.P. Oza: *Soil Mechanics and Foundation Engineering*, Charotar Book Hall, Anand, India, 1969.
18. P.C. Rutledge and S.J. Johnson: *Review of Uses of Vertical Sand Drains*, HRB Bulletin, 173, 1958.
19. J.C. Smith: *The Chrome-lignin process and Ion-exchange studies*, Proceedings, M.I.T. Soil Stabilisation Conference, 1952.
20. G.B. Sowers and G.F. Sowers: *Introductory Soil Mechanics and Foundations*, 3rd ed., Collier—Macmillan Ltd., London, 1970.
21. K. Terzaghi: *Theoretical Soil Mechanics*, John Wiley & Sons Inc., NY, USA, 1943.

QUESTIONS AND PROBLEMS

17.1 Discuss the different methods of improving the bearing capacity of weak soils.
(S.V.U.—B.E., (Part-time)—Apr., 1982)

17.2 Describe the different steps involved in the process of soil stabilisation using cement as the additive.
(S.V.U.—B.E., (R.R.)—Sep., 1978)

17.3 Outline the basic principles of soil stabilisation and briefly outline the various methods of stabilisation.
(S.V.U.—B.E., (R.R.)—May, 1975)

17.4 (*a*) Describe the basic principles involved in successful stabilisation of (*i*) sand, (*ii*) inert clay and (*iii*) expansive soil.
(*b*) Describe a method suitable to stabilise a highway fill foundation in hilly terrain with high rainfall.
(S.V.U.—B.E., (R.R.)—Nov., 1974)

17.5 Write brief critical notes on the principles of soil stabilisation.
(S.V.U.—B.E., (R.R.)—Nov., 1973, Nov & June, 1972)

17.6 What are the methods available for soil stabilisation? Describe its use for roads.

(S.V.U.—B.E., (R.R.)—May, 1970)

17.7 Write brief critical notes on methods suitable for stabilising black cotton clay.

(S.V.U.—B.E., (N.R.)—Sep., 1967)

17.8 'CBR Test is an arbitrary test', justify. Describe the procedure for conducting the CBR test in the field. Surcharge loads are used in the CBR test; explain their significance.

(S.V.U.—Four-year B.Tech.—Sep., 1983)

17.9 Write brief critical notes on the CBR test and its use.

(S.V.U.—B.Tech., (Part-time)—Sep., 1983 B.E., (R.R.)—Dec., 1970)

17.10 (a) Under what circumstances is CBR test conducted under saturated condition?

(b) Under what circumstances is CBR test conducted on:

(i) Compacted samples and (ii) Undisturbed samples.

(S.V.U.—Four-year B.Tech. —Apr., 1983)

17.11 What is CBR test? What are its uses? Describe in detail the CBR test with neat sketches.

(S.V.U.—Four-year B.Tech., —Oct., 1982)

17.12 Give a critical review of the methods of evaluating the CBR of subgrade soils.

(S.V.U.—B.E., (Part-time)—Dec., 1981)

17.13 Briefly describing the test procedure, explain how CBR is used to design flexible pavements.

(S.V.U.—B.E., (R.R.)—Sep., 1978, Nov., 1975, June, 1972, Dec., 1981)

17.14 Describe the standard procedures for California Bearing Ratio test in the laboratory and in the field.

(S.V.U.—B.E., (R.R.)—May, 1975)

17.15 Describe the "CBR" test. What is the standard for comparison in this test?

(S.V.U.—B.E., (R.R.)—May, 1970)

17.16 Describe the CBR test procedure and explain the usefulness of the test in pavement design.

(S.V.U.—B.E., (R.R.)—Dec., 1971 and Dec., 1970)

17.17 The following results were obtained for a CBR test on compacted soil:

Penetration (mm)	1.25	2.50	3.75	5.00	7.50	10.00	12.50
Load (kN)	0.32	1.32	2.40	3.32	4.32	4.96	5.35

Assuming the standard loads as 14.72 kN and 22.07 kN for penetrations of 2.5 mm and 5.0 mm respectively, determine CBR for the soil.

(S.V.U.—B.E., (R.R.)—May, 1969)

17.18 The CBR values of a proposed subgrade, base and sub-base are 6%, 15%, and 50% respectively. Determine the thicknesses required for the base, sub-base and surface courses for light traffic conditions.

18
SOIL EXPLORATION

18.1 INTRODUCTION

A fairly accurate assessment of the characteristics and engineering properties of the soils at a site is essential for proper design and successful construction of any structure at the site. The field and laboratory investigations required to obtain the necessary data for the soils for this purpose are collectively called *soil exploration*.

The choice of the foundation and its depth, the bearing capacity, settlement analysis and such other important aspects depend very much upon the various engineering properties of the foundation soils involved.

Soil exploration may be needed not only for the design and construction of new structures, but also for deciding upon remedial measures if a structure shows signs of distress after construction. The design and construction of highway and airport pavements will also depend upon the characteristics of the soil strata upon which they are to be aligned.

The primary objectives of soil exploration are:
 (*i*) Determination of the nature of the deposits of soil,
 (*ii*) determination of the depth and thickness of the various soil strata and their extent in the horizontal direction,
 (*iii*) the location of groundwater and fluctuations in GWT,
 (*iv*) obtaining soil and rock samples from the various strata,
 (*v*) the determination of the engineering properties of the soil and rock strata that affect the performance of the structure, and
 (*vi*) determination of the *in-situ* properties by performing field tests.

18.2 SITE INVESTIGATION

'Site investigation' refers to the procedure of determining surface and subsurface conditions in the area of proposed construction. Thus, this term has a broader connotation than 'soil exploration', and includes the latter.

Information regarding surface conditions is necessary for planning construction technique. Accessibility of a site to men, materials and equipment is affected by the surface topography. The nature and extent of vegetative cover will determine costs relating to site clearance. Availability of water and electrical power, proximity to major transporation routes, environmental protection regulations of various agencies and availability of sufficient area for post-construction use may be the other factors which could affect construction procedures.

Information on subsurface conditions existing at a site is also an important requirement. The possible need for dewatering will be revealed by the subsurface investigation. Necessity for bracing of excavations for foundations can be established.

The importance of adequate site investigation and soil exploration cannot be overemphasised because the lack of it could lead to increased costs due to unforeseen difficulties and the consequent modifications in the design and execution of the project. Usually, the cost of a thorough investigation and exploration programme will be less than 1% of the total cost of the entire project.

Site investigations may involve one or more of the following preliminary steps:
1. Reconnaissance
2. Study of maps
3. Aerial photography

18.2.1 Reconnaissance

Reconnaissance involves an inspection of the site and study of the topographical features. This will yield useful information about the soil and ground-water conditions and also help the engineer plan the programme of exploration. The topography, drainge pattern, vegetation and land use provide valuable information. Ground-water conditions are often reflected in the presence of springs and the type of vegetation. The water levels in wells and ponds may indicate ground-water, but these can be influenced by intensive use or by irrigation in the proximity of the area.

Valuable information about the presence of fills and knowledge of any difficulties encountered during the building of other nearby structures may be obtained by inquiry. Aerial reconnaissance is also undertaken if the area is large and the project is a major one.

Reconnaissance investigation gives a preliminary idea of the soils and other conditions involved at the site and its value should not be underestimated. Further study may be avoided if reconnaissance reveals the inadequacy or unsuitability of the site for the proposed work for any glaring reasons.

18.2.2 Study of Maps

Information on surface and subsurface conditions in an area is frequently available in the form of maps. Such sources in India are the Survey of India and Geological Survey of India, which provide topographical maps, often called 'toposheets'. Soil conservation maps may also be available.

A geological study is essential. The primary purpose of such a study is to establish the nature of the deposits underlying the site. The types of soil and rock likely to be encountered can be determined, and the method of exploration most suited to the situation may be selected. Faults, folds, cracks, fissures, dikes, sills and caves, and such other defects in rock and soil strata may be indicated. Data on the availability of natural resources such as oil, gas and minerals will have to be considered carefully during the evaluation of a site. Legal and engineering problems may arise where structures settle because of mine workings in their proximity.

Seismic potential or potential seismic activity is a major factor in structural design in many regions of the world, especially in the construction of major structures such as dams and nuclear power plants. Maps are now available showing the earthquake zones of different degrees of vulnerability. A lot of work has been done in this regard by the ' Centre for Research and Training in Earthquake Engineering'' located at Roorkee in India.

18.2.3 Aerial Photography

Aerial photography is now a fairly well-developed method by which site investigation may be conducted for any major project. Air photo interpretation is the estimation of underground conditions by relating landform development and plant growth to geology as reflected in aerial photographs.

Photographs are obtained in sequence by flying in more or less straight lines across a site with a two-thirds overlap in the direction of flight and one-quarter overlap between successive flight lines. For general mapping, a scale of 1 : 20,000 may be adequate but for more detailed work larger scales obtained by low altitude photography are necessary.

Analysis consists of an identification of all the natural and man-made features and their grouping by geological association. Finally the probable geological formation of soil or rock is determined from the total pattern of associations.

The features include topography, stream patterns, erosion details, colour and tone, vegetation, man-made features, natural and man-made foundations and micro details in topography such as sink holes, rock outcrops and accumulations of boulders. Each of these features is used to associate with a particular type of rock or soil stratum. For example, vegetation differences frequently reflect both drainage and soil character.

Land form represents the total effect of environment and geological history on the underlying rock and soil foundations. The study of evolutionary processes that produce a given land form is termed 'geomorphology'. Once the land form is identified, the geological associations are defined.

Air photo interpretation requires a thorough grounding in geology, geomorphology, agriculture and hydrology. The technique, though highly specialised, is a valuable preview and supplement to site reconnaissance.

18.3 SOIL EXPLORATION

The subsoil exploration should enable the engineer to draw the soil profile indicating the sequence of the strata and the properties of the soils involved.

In general, the methods available for soil exploration may be classified as follows:
1. Direct methods ... Test pits, trial pits or trenches
2. Semi-direct methods ... Borings
3. Indirect methods ... Soundings or penetration tests and geophysical methods

In an exploratory programme, one or more of these methods may be used to yield the desired information.

18.3.1 Test Pits

Test pits or trenches are open type or accessible exploratory methods. Soils can be inspected in their natural condition. The necessary soils samples may be obtained by sampling techniques and used for ascertaining strength and other engineering properties by appropriate laboratory tests.

Test pits will also be useful for conducting field tests such as the plate-bearing test.

Test pits are considered suitable only for small depths—up to 3m; the cost of these increases rapidly with depth. For greater depths, especially in pervious soils, lateral supports or bracing of the excavations will be necessary. Ground water table may also be encountered and may have to be lowered.

Hence, test pits are usually made only for supplementing other methods or for minor structures.

18.3.2 Boring

Making or drilling bore holes into the ground with a view to obtaining soil or rock samples from specified or known depths is called 'boring'.

The common methods of advancing bore holes are:
1. Auger boring
2. Auger and shell boring
3. Wash boring
4. Percussion drilling ⎫ more commonly employed for sampling in rock
5. Rotary drilling ⎬ strata.

Auger Boring

'Soil auger' is a device that is useful for advancing a bore hole into the ground. Augers may be hand-operated or power-driven; the former are used for relatively small depths (less than 3 to 5 m), while the latter are used for greater depths. The soil auger is advanced by rotating it while pressing it into the soil at the same time. It is used primarily in soils in which the bore hole can be kept dry and unsupported. As soon as the auger gets filled with soil, it is taken out and the soil sample collected.

Two common types of augers, the post hole auger and the helical auger, are shown in Fig. 18.1.

Auger and Shell Boring

If the sides of the hole cannot remain unsupported, the soil is prevented from falling in by means of a pipe known as 'shell' or 'casing'. The casing is to be driven first and then auger; whenever the casing is to be extended, the auger has to be withdrawn, this being an impediment to quick progress of the work.

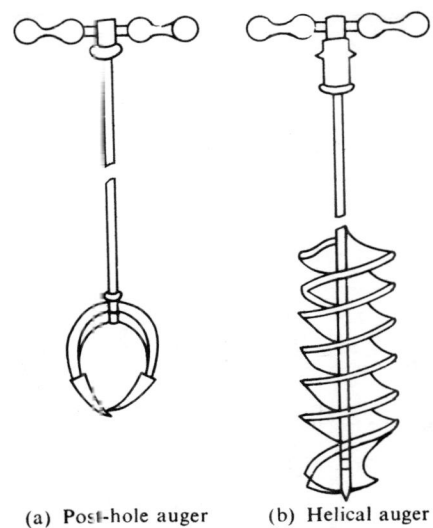

(a) Post-hole auger   (b) Helical auger

Fig. 18.1 Soil augers

An equipment called a 'boring rig' is employed for power-driven augers, which may be used up to 50 m depth. (A hand rig may be sufficient for borings up to 25 m in depth). Casings may be used for sands or stiff clays. Soft rock or gravel can be broken by chisel bits attached to drill rods. Sand pumps are used in the case of sandy soils.

Wash Boring

Wash boring is commonly used for exploration below ground water table for which the auger method is unsuitable. This method may be used in all kinds of soils except those mixed with gravel and boulders. The set-up for wash boring is shown in Fig. 18.2.

Initially, the hole is advanced for a short depth by using an auger. A casing pipe is pushed in and driven with a drop weight. The driving may be with the aid of power. A hollow drill bit is screwed to a hollow drill rod connected to a rope passing over a pulley and supported by a tripod. Water jet under pressure is forced through the rod and the bit into the hole. This loosens the soil at the lower end and forces the soil-water suspension upwards along the annular surface between the rod and the side of the hole. This suspension is led to a settling tank where the soil particles settle while the water overflows into a sump. The water collected in the sump is used for circulation again.

The soil particles collected represent a very disturbed sample and is not very useful for the evaluation of the engineering properties. Wash borings are primarily

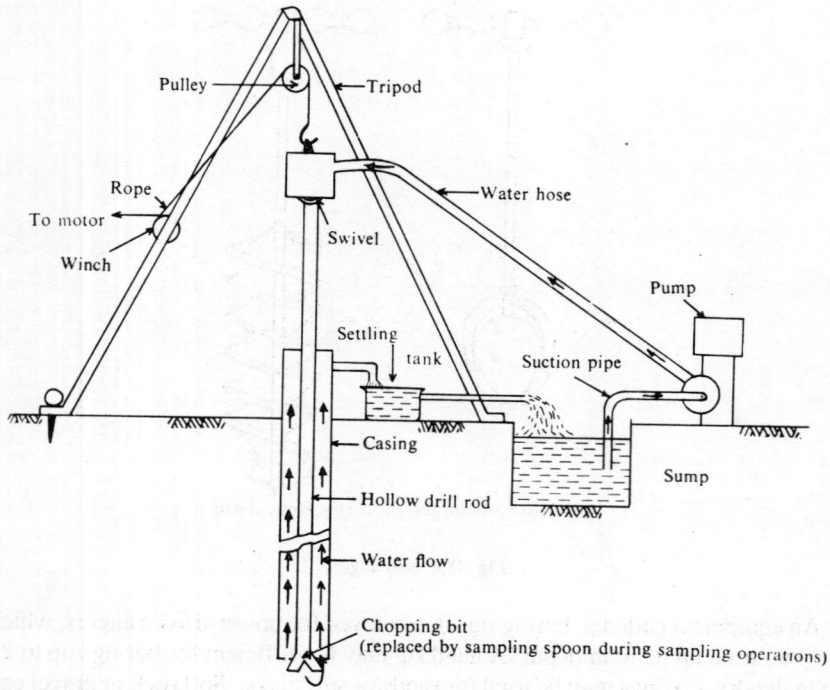

Fig. 18.2 Set-up for Wash Boring

used for advancing bore holes; whenever a soil sample is required, the chopping bit is to be replaced by a sampler.

The change of the rate of progress and change of colour of wash water indicate changes in soil strata.

Percussion Drilling

A heavy drill bit called 'churn bit' is suspended from a drill rod or a cable and is driven by repeated blows. Water is added to facilitate the breaking of stiff soil or rock. The slurry of the pulverised material is bailed out at intervals. The method cannot be used in loose sand and is slow in plastic clay.

The formation gets badly disturbed by impact.

Rotary Drilling

This method is fast in rock formations. A drill bit, fixed to the lower end of a drill rod, is rotated by power while being kept in firm contact with the hole. Drilling fluid or bentonite slurry is forced under pressure through the drill rod and it comes up bringing the cuttings to the surface. Even rock cores may be obtained by using suitable diamond drill bits. This method is not used in porous deposits as the consumption of drilling fluid would be prohibitively high.

18.3.3 Planning an Exploration Programme

The planning of an exploration programme depends upon the type and importance of the structure and the nature of the soil strata. The primary purpose vis-á-vis the cost involved should be borne in mind while planning a programme. The depth, thickness, extent, and composition of each of the strata, the depth of the rock, and the depth to the ground water table are important items sought to be determined by an exploration programme. Further, approximate idea of the strength and compressibility of the strata is necessary to make preliminary estimates of the safety and expected settlement of the structure.

The planning should include a site plan of the area, a layout plan of proposed structures with column locations and expected loads and the location of bore holes and other field tests. A carefully planned programme of boring and sampling is the crux of any exploration job. Resourceful and intelligent personnel trained in the principles of geology and geotechnical engineering are necessary.

The two important aspects of a boring programme are 'spacing of borings' and 'depth of borings'.

18.3.4 Spacing of Borings

The spacing of borings, or the number of borings for a project, is related to the type, size, and weight of the proposed structure, to the extent of variation in soil conditions that permit safe interpolation between borings, to the funds available, and possibly to the stipulations of a local building code.

It is impossible to determine the spacing of borings before an investigation begins, since it depends on the uniformity of the soil deposit. Ordinarily a preliminary estimate of the spacing is made. Spacing is decreased if additional data are necessary and is increased if the thickness and depth of the different strata appear about the same in all the borings.

The following spacings are recommended in planning an exploration programme:

Table 18.1 Spacing of borings (Sowers and Sowers, 1970)

S.No.	Nature of the project	Spacing of borings (metres)
1.	Highway (subgrade survey)	300 to 600
2.	Earth dam	30 to 60
3.	Borrow pits	30 to 120
4.	Multistorey buildings	15 to 30
5.	Single story factories	30 to 90

Note: For uniform soil conditions, the above spacings are doubled; for irregular conditions, these are halved.

"IS: 1892-1979—Code of Practice for Subsurface Investigation for Foundations" has made the following recommendations:

For a compact building site covering an area of about 0.4 hectare, one bore hole or trial pit in each corner and one in the centre should be adequate. For smaller and less important buildings even one bore hole or trial pit in the centre will suffice. For very large areas covering industrial and residential colonies, the geological nature of the terrain will help in deciding the number of bore holes or trial pits. Cone penetration tests may be performed at every 50 m by dividing the area in a grid pattern and number of bore holes or trial pits decided by examining the variation in penetration curves. The cone penetration tests may not be possible at sites having gravelly or boulderous strata. In such cases geophysical methods may be suitable.

18.3.5 Depth of Borings

In order to furnish adequate information for settlement predictions, the borings should penetrate all strata that could consolidate significantly under the load of the structure. This necessarily means that, for important and heavy structures such as bridges and tall buildings, the borings should extend to rock. For smaller structures, however, the depth of boring may be estimated from the results of previous investigations in the vicinity of the site, and from geologic evidence.

Experience indicates that damaging settlement is unlikely to occur when the additional stress imposed on the soil due to the weight of the structure is less than 10% of the initial stress in the soil due to self-weight. E.De Beer of Belgium adopted this rule for determining the so-called 'critical depth of boring' (Hvorslev, 1949). Based on this, recommended depths of borings for buildings are about 3.5 m and 6.5 m for single- and two-storey buildings. For dams and embankments, the depth ranges between half the height to twice the height depending upon the foundation soil.

According to IS: 1892-1979: "The depth of exploration required depends upon the type of the proposed structure, its total weight, the size, shape and disposition of the loaded areas, soil profile and the physical properties of the soil that constitutes each individual stratum. Normally, it should be one and half times the width of the footing below foundation level. If a number of loaded areas are in close proximity, the effect of each is additive. In such cases, the whole area may be considered as loaded and exploration should be carried out up to one and half times the lower dimension. In any case, the depth to which seasonal variations affect the soil should be regarded as the minimum depth for the exploration of the sites. But, where industrial processes affect the soil characteristics, this depth may be more. The presence of fast-growing and water-seeking trees also contributes to the weathering processes . . ."

The depth of exploration at the start of the work may be decided as given in Table 18.2, which may be modified as exploration proceeds, if required.

18.3.6 Boring Log

Information on subsurface conditions obtained from the boring operation is typically presented in the form of a boring record, commonly known as "boring log". A continuous record of the various strata identified at various depths of the

Table 18.2 Depth of Exploration (IS: 1892-1979)

S.No.	Type of foundation	Depth of exploration
1.	Isolated spread footings or raft or adjacent footings with clear spacing equal or greater than four times the width	One and half times the width
2.	Adjacent footings with clear spacing less than twice the width	One and half times the length
3.	Adjacent rows of footings	
	(i) With clear spacing between rows less than twice the width	Four and half times the width
	(ii) With clear spacing between rows greater than twice the width	Three times the width
	(iii) With clear spacing between rows greater than or equal to four times the width	One and half times the width
4.	Pile and Well foundations	One and half times the width of structure from bearing level (toe of pile or bottom of well)
5.	Road cuts	Equal to the bottom width of the cut
6.	Fill	Two metres below the ground level or equal to the height of the fill whichever is greater

boring is presented. Description or classification of the various soil and rock types encountered, and data regarding ground water level have to be necessarily given in a pictorial manner on the log. A "field" log will consist of this minimum information, while a "lab" log might include test data presented alongside the boring sample actually tested.

RECORD OF BORING [IS : 1892-1979]

Name of boring organization:

Bored for .... Location-site ....
Ground level .... Boring No. ....
Type of boring .... Soil sampler used ....
Diameter of boring .... Date started ....
Inclination: Vertical .... Date completed ....
Bring: .... Recorded ....

Description of strata	Soil classification	Thickness of stratum	Depth from GL	R.L. of lower contact	Samples			GWL	Remarks
					Type	No	Depth and thickness of sample		
Fine to medium sand with practically no binder	SP		—1 m —2 m 2.7 m		Undisturbed	1	—1 m 1.4 m 1.7 m —2 m		
Silty clays of medium plasticity no coarse or medium sands	CI		—3 m —4 m —5 m		Undisturbed	2	—3 m 4 m 4.3 m —5 m	Not struck upto 6 m depth	

Fig. 18.3 Sample Boring Log

Sometimes a subsurface profile indicating the conditions and strata in all borings in series is made. This provides valuable information regarding the nature of variation or degree of uniformity of strata at the site. This helps in delineating between "good" and "poor" areas.

The standard practice of interpolating between borings to determine conditions surely involves some degree of uncertainty.

A sample record sheet for a boring is shown in Fig. 18.3. A site plan showing the disposition of the borings should be attached to the records.

18.4 SOIL SAMPLING

'Soil Sampling' is the process of obtaining samples of soil from the desired depth at the desired location in a natural soil deposit, with a view to assessing the engineering properties of the soil for ensuring a proper design of the foundation. The ultimate aim of the exploration methods described earlier, it must be remembered, is to obtain soil samples besides obtaining all relevant information regarding the strata. The devices used for the purpose of sampling are known as 'soil samplers'.

Determination of ground water level is also considered part of the process of soil sampling.

18.4.1 Types of Samples

Broadly speaking, samples of soil taken out of natural deposits for testing may be classified as:

Disturbed samples, and undisturbed samples, depending upon the degree of disturbance caused during sampling operations.

A disturbed sample is that in which the natural structure of the soil gets modified partly or fully during sampling, while an undisturbed sample is that in which the natural structure and other physical properties remain preserved. 'Undisturbed', in this context, is a purely relative term, since a truly undisturbed sample can perhaps be never obtained as some little degree of disturbance is absolutely inevitable even in the best method of sampling devised till date.

Disturbed samples may be further subdivided as: (*i*) Non-representative samples, and (*ii*) Representative samples. Non-representative samples consist of mixture of materials from various soil or rock strata or are samples from which some mineral constituents have been lost or got mixed up.

Soil samples obtained from auger borings and wash borings are non-representative samples. These are suitable only for providing qualitative information such as major changes in subsurface strata.

Representative samples contain all the mineral constituents of the soil, but the structure of the soil may be significantly disturbed. The water content may also have changed. They are suitable for identification and for the determination of certain physical properties such as Atterberg limits and grain specific gravity.

Undisturbed samples may be defined as those in which the material has been subjected to minimum disturbance so that the samples are suitable for strength tests and consolidation tests. Tube samples and chunk samples are considered to fall in this category.

Besides using a suitable tube sampler for the purpose, undisturbed samples may be obtained as 'chunks' from the bottom of test pits, provided the soil possesses at least some cohesion.

The soil at the bottom of the pit is trimmed as a chunk to the required shape and size approximately. A cylindrical container open at both ends is placed carefully over this chunk after covering the top with paraffin wax. The bottom is scooped with a steel spatula and trimmed after reversing the box along with the sample. Paraffin wax is again used to seal the face and any gaps in the sides, before transporting it to the laboratory. The procedure will become obvious from Fig. 18.4.

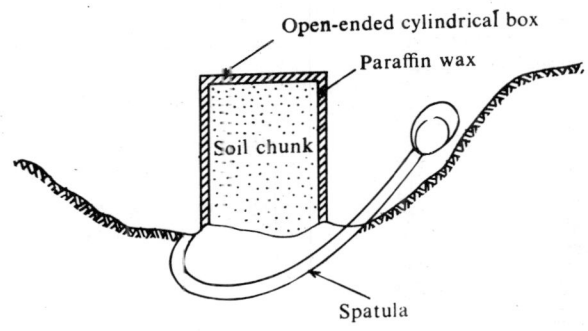

Fig. 18.4 Obtaining a chunk sample

18.4.2 Types of Samples

Soil samplers are classified as 'thick wall' samplers and 'thin wall' samplers. Split spoon sampler (or split tube sampler) is of the thick-wall type, and 'shelby' tubes are of the thin-wall type.

Depending upon the mode of operation, samplers may be classified as the open drive sampler, stationary piston sampler and rotary sampler.

Open drive sampler can be of the thick wall type as well as of the thin wall type. The head of the sampler is provided with valves to permit water and air to escape during driving. The check valve helps to retain the sample when the sampler is lifted. The tube may be seamless or may be split in two parts; in the latter case it is known as the split tube or split spoon sampler.

Stationary piston sampler consists of a sampler with a piston attached to a long piston rod extending up to the ground surface through drill rods. The lower end of the sampler is kept closed with the piston while the sampler is lowered through the bore hole. When the desired elevation is reached, the piston rod is clamped, thereby keeping the piston stationary, and the sampler tube is advanced further into the soil. The sampler is then lifted and the piston rod clamped in position. The piston prevents the entry of water and soil into the tube when it is being lowered, and also helps to retain the sample during the process of lifting the tube. The sampler is, therefore, very much suited for sampling in soft soils and saturated sands.

Rotary samplers are of the core barrel type (USBR, 1960) with an outer tube provided with cutting teeth and a removable thin liner inside. It is used for sampling in stiff cohesive soils.

18.4.3 Sample Disturbance

The design features of a sampler, governing the degree of disturbance of a soil sample are the dimensions of the cutting edge and those of the sampling tube, the characteristics of the non-return valve and the wall friction. In addition, the method of sampling also affects the sample disturbance. The lower end of a sampler with the cutting edge is shown in Fig. 18.5.

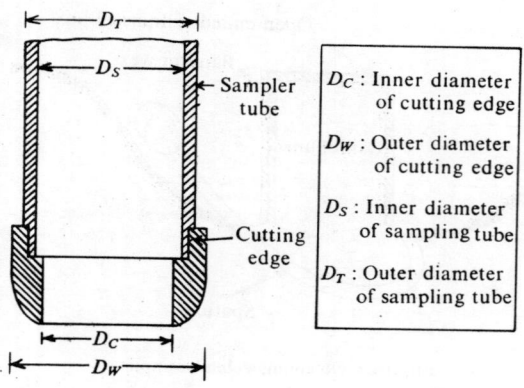

D_C : Inner diameter of cutting edge

D_W : Outer diameter of cutting edge

D_S : Inner diameter of sampling tube

D_T : Outer diameter of sampling tube

Fig. 18.5 Sampling tube with cutting edge

The following are defined with respect to the diameters marked in Fig. 18.5:

Area Ratio, $$A_r = \frac{(D_w^2 - D_c^2)}{D_c^2} \times 100\%$$ ...(Eq. 18.1)

Inside clearance, $$C_I = \frac{(D_s - D_c)}{D_c} \times 100\%$$ ...(Eq. 18.2)

Outside clearance, $$C_0 = \frac{(D_w - D_T)}{D_T} \times 100\%$$ ...(Eq. 18.3)

The walls of the sampler should be kept smooth and properly oiled to reduce wall friction in order that sample disturbance be minimised. The non-return valve should have a large orifice to allow the air and water to escape quickly and easily when driving the sampler.

Area ratio is the most critical factor which affects sample disturbance; it indicates the ratio of displaced volume of soil to that of the soil sample collected. If A_r is less than 10%, the sample disturbance is supposed to be small. A_r may be as high as 30% for a thick wall sampler like split spoon and may be as low as 6 to 9% for thin wall samplers like shelby tubes. The inside clearance, C_I, should not be more than 1 to 3%, the outside clearance C_O should also not be much greater than C_I. Inside clearance allows for elastic expansion of the soil as it enters the tube, reduces frictional drag on the sample from the wall of the tube, and helps to retain the core. Outside clearance facilitates the withdrawal of the sample from the ground.

The recovery ratio $R_r = L/H$ ...(Eq. 18.4)

where, L = length of the sample within the tube, and
H = depth of penetration of the sampling tube.

This value should be 96 to 98% for a satisfactory undisturbed sample. This concept is more commonly used in the case of rock cores.

18.4.4 Split-spoon Sampler

The split spoon sampler is basically a thick-walled steel tube, split length wise. The sampler as standardised by the I.S.I. (IS: 2131-1972— Standard Penetration Test for soils) is shown in Fig. 18.6.

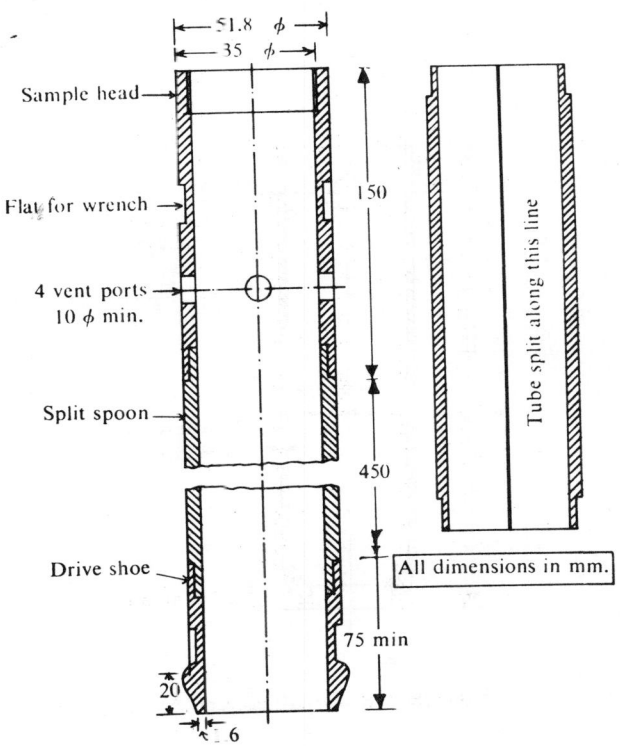

Fig. 18.6 Split spoon sampler (I.S.)

A drive shoe attached to the lower end serves as the cutting edge. A sample head may be screwed at the upper end of split spoon. The standard size of the spoon sampler is of 35 mm internal and 50.8 mm external diameter. The sampler is lowered to the bottom of the bore hole by attaching it to the drill rod. The sampler is then driven by forcing it into the soil by blows from a hammer. The assembly of the sampler is then extracted from the hole and the cutting edge and coupling at the top are unscrewed. The two halves of the barrel are separated and the sample is thus exposed. The sample may be placed in a glass jar and sealed, after visual examination.

If samples need not be examined in the field, a liner is inserted inside the split spoon. After separating the two halves, the liner with the sample is sealed with wax.

18.4.5 Thin-walled Sampler

Thin-walled sampler, as standardised by the ISI (I.S.: 2132-1972 Code of Practice for Thin-walled Tube Sampling of Soils), is shown in Fig. 18.7.

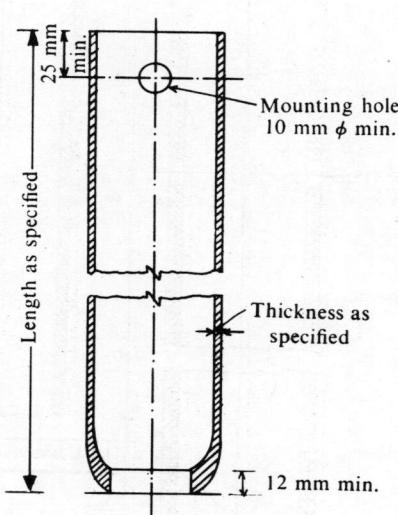

Fig. 18.7 Thin-walled sampler (I.S.)

The sampling tube shall be made of steel, brass, or aluminium. The lower end is levelled to form a cutting edge and is tapered to reduce wall friction. The salient dimensions of three of the sampling tubes are given in Table 18.3.

Table 18.3 Requirements of sampling tubes (I.S: 2132-1972)

Inside diameter, mm	38	70	100	Area ratio in this case is
Outside diameter, mm	40	74	106	$\dfrac{(D_e^2 - D_i^2)}{D_i^2}$
Minimum effective length available for soil sample, mm	300	450	450	where D_e = External dia.
Area Ratio, A, %	10.9	11.8	12.4	D_i = Internal dia.

Note: Sampling tubes of intermediate or larger diameters may also be used.

After having extracted the sample in the same manner as in the case of split spoon type, the tube is sealed with wax on both ends and transported to the laboratory.

18.4.6 Ground Water Level

Determination of the location of ground water is an essential part of every exploratory programme. Ordinarily, it is measured in the exploratory borings; however, it may sometimes become necessary to make borings purely for this purpose, when artesian or perched ground water is expected, or the use of drilling mud obscures ground water.

A correct indication of the general ground water level is found by allowing the water in the boring to reach an equilibrium level. In sandy soils, the level gets stabilised very quickly—within a few hours at the most. In clayey soils it will take many days for this purpose. Hence, standpipes or piezometers are used in clays and silt. A piezometer is an open-ended tube (may be about 50 mm in diameter) perforated at its end. The tube is packed around with gravel and sealed in position with puddle clay. Observations must be taken for several weeks until the water level gets stabilised. The arrangement is shown in Fig. 18.8.

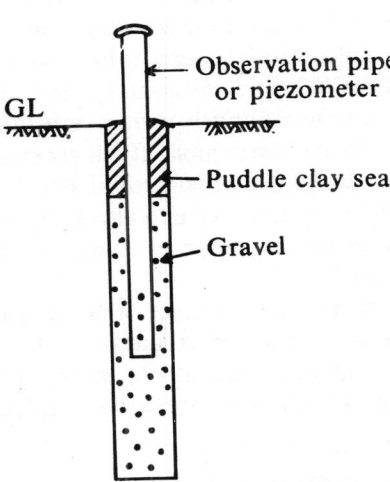

Fig. 18.8 Piezometer for observation of GWL in a bore hole

In the case of impermeable clays, pressure measuring devices are used.

The elevation of ground water table affects the design of the foundation, since the bearing capacity and a few other engineering properties of the soil strata depend upon it.

18.5 SOUNDING AND PENETRATION TESTS

Subsurface soundings are used for exploring soil strata of an erratic nature. They are useful to determine the presence of any soft pockets between drill holes and also to determine the density index of cohesionless soils and the consistency of cohesive soils at various desired depths from the ground surface.

Methods of sounding normally consist of driving or pushing a standard sampling tube or a cone. The devices involved are also termed 'penetrometers', since they are made to penetrate the subsoil with a view to measuring the resistance to penetration of the soil strata, and thereby try to identify the soil and some of its engineering characteristics. The necessary field tests are also called 'penetration tests'.

If a sampling tube is used to penetrate the soil, the test is referred to as the Standard Penetration Test (SPT, for brevity). If a cone is used to penetrate the soil, the test is called a 'Cone penetration test'. Static and dynamic cone penetration tests are used depending upon the mode of penetration—static or dynamic.

A field test called 'Vane Test' is used to determine the shearing strength of the soil located at a depth below the ground.

18.5.1 Standard Penetration Test (SPT)

The Standard Penetration Test (SPT) is widely used to determine the parameters of the soil *in-situ*. The test is especially suited for cohesionless soils as a correlation has been established between the SPT value and the angle of internal friction of the soil.

The test consists of driving a split-spoon sampler (Fig. 18.6) into the soil through a bore hole 55 to 150 mm in diameter at the desired depth. A hammer of 640 N (65 kg) weight with a free fall of 750 mm is used to drive the sampler. The number of blows for a penetration of 300 mm is designated as the "Standard Penetration Value" or "Number" N. The test is usually performed in three stages. The blow count is found for every 150 mm penetration. If full penetration is obtained, the blows for the first 150 mm are ignored as those required for the seating drive. The number of blows required for the next 300 mm of penetration is recorded as the SPT value. The test procedure is standardised by ISI and set out in "IS: 2131-1972 — Standard Penetration Test".

Usually SPT is conducted at every 2 m depth or at the change of stratum. If refusal is noticed at any stage, it should be recorded.

In the case of fine sand or silt below water-table, apparently high values may be noted for N. In such cases, the following correction is recommended (Terzaghi and Peck, 1948):

$$N = 15 + \frac{1}{2}(N' - 15) \qquad \text{....(Eq. 18.4)}$$

where N' = observed SPT value,
and N = corrected SPT value.

For SPT made at shallow levels, the values are usually too low. At a greater depth, the same soil, at the same density index, would give higher penetration resistance.

The effect of the overburden pressure on SPT value may be approximated by the equation:

$$N = N' \cdot \frac{350}{(\sigma + 70)} \quad \text{...(Eq. 18.5)}$$

where N' = observed SPT value,
N = corrected SPT value, and
σ = effective overburden pressure in KN/m², not exceeding 280 KN/m².

This implies that no correction is required if the effective overburden pressure is 280 kN/m².

Terzaghi and Peck also give the following correlation between SPT value, D_r, and ϕ:

Table 18.4 Correlation between N, D_r and ϕ

S.No.	Condition	N	D_r	ϕ
1.	Very loose	0 – 4	0 – 15%	Less than 28°
2.	Loose	4 – 10	15 – 35%	28° – 30°
3.	Medium	10 – 30	35 – 65%	30° – 36°
4.	Dense	30 – 50	65 – 85%	36° – 42°
5.	Very dense	Greater than 50	Greater than 85%	Greater than 42°

For clays the following data are given:

Table 18.5 Correlation between N and q_u

S. No.	Consistency	N	q_u (kN/m²)
1.	Very soft	0 – 2	Less than 25
2.	Soft	2 – 4	25 – 50
3.	Medium	4 – 8	50 – 100
4.	Stiff	8 – 15	100 – 200
5.	Very stiff	15 – 30	200 – 400
6.	Hard	Greater than 30	Greater than 400

The correlation for clays is rather unreliable. Hence, vane shear test is recommended for more reliable information.

18.5.2 Static Cone Penetration Test (Dutch Cone Test)

The static cone penetration test, which is also known as Dutch Cone test, has been standardised by the ISI and given in "IS: 4968 (Part-III)-1976—Method for subsurface sounding for soils—Part III Static cone penetration test".

Among the field sounding tests the static cone tests in a valuable method of recording variation in the *in-situ* penetration resistance of soils, in cases where the *in-situ* density is disturbed by boring operations, thus making the standard penetration test unreliable especially under water. The results of the test are also useful in determining the bearing capacity of the soil at various depths below the ground level. In addition to bearing capacity values, it is also possible to determine by this test the skin friction values used for the determination of the required lengths of piles in a given situation. The static cone test is most successful in soft or loose soils like silty sands, loose sands, layered deposits of sands, silts and clays as well as in clayey deposits. In areas where some information regarding the foundation strata is already available, the use of test piles and loading tests thereof can be avoided by conducting static cone penetration tests.

Experience indicates that a complete static cone penetration test up to depths of 15 to 20 m can be completed in a day with manual operations of the equipment, making it one of the inexpensive and fast methods of sounding available for investigation.

The equipment consists of a steel cone, a friction jacket, sounding rod, mantle tube, a driving mechanism and measuring equipment.

The steel cone shall be of steel with tip hardened. It shall have an apex angle of $60° \pm 15'$ and overall base diameter of 35.7 mm giving a cross-sectional area of 10 cm^2. The friction jacket shall be of high carbon steel. These are shown in Fig. 18.9.

The sounding rod is a steel rod of 15 mm diameter which can be extended with additional rods of 1 m each in length. The mantle tube is a steel tube meant for guiding the sounding rod which goes through it. It should be of one metre in length with flush coupling.

The driving mechanism should have a capacity of 20 to 30 kN for manually operated equipment and 100 kN for the mechanically operated equipment. The mechanism essentially consists of a rack and pinion arrangement operated by a winch. The reaction for the thrust may be obtained by suitable devices capable of taking loads greater than the capacity of the equipment.

The hand operated winch may be provided with handles on both sides of the frame to facilitate driving by four persons for loads greater than 20 kN. For the engine driven equipment the rate of travel should be such that the penetration obtained in the soil during the test is 10 to 15 mm/s.

Hydraulic pressure gauges should be used for indicating the pressure developed. Alternatively, a proving ring may also be used to record the cone resistance. Suitable capacities should be fixed for the gauges.

Basically, the test procedure for determining the static cone and frictional resistances consists of pushing the cone alone through the soil strata to be tested, then the cone and the friction jacket, and finally the entire assembly in sequence and noting the respective resistance in the first two cases. The process is repeated at predetermined intervals. After reaching the deepest point of investigation the entire assembly should be extracted out of the soil.

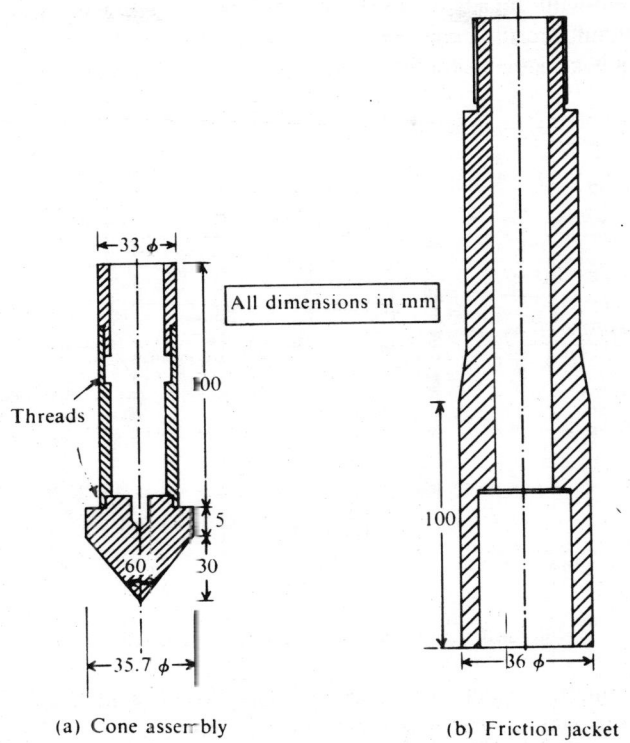

Fig. 18.9 Cone assembly and friction jacket for static cone penetration test (IS)

The results of the test shall be presented graphically, in two graphs, one showing the cone resistance in kN/m² with depth in metres and the other showing the friction resistance in kN/m² with depth in metres, together with a bore hole log.

The cone resistance shall be corrected for the dead weight of the cone and sounding rods in use. The combined cone and friction resistance shall be corrected for the dead weight of the cone, friction jacket and sounding rods. These values shall also be corrected for the ratio of the ram area to the base area of the cone.

The test is unsuitable for gravelly soils and for soils with standard penetration value N greater than 50. Also, in dense sands the anchorage becomes too cumbersome and expensive and for such cases dynamic cone penetration tests may be carried out.

18.5.3 Dynamic Cone Penetration Test

The dynamic cone penetration test is standardised by the ISI and given in "IS: 4968 (Part I)-1976— Method for Subsurface Sounding for Soils—Part I Dynamic method using 50 mm cone without bentonite slurry".

The equipment consists of a cone, driving rods, driving head, hoisting equipment and a hammer.

The cone with threads (recoverable) shall be of steel with tip hardened. The cone without threads (expendable) may be of mild steel. For the cone without threads, a cone adopter shall be provided. These are shown in Fig. 18.10.

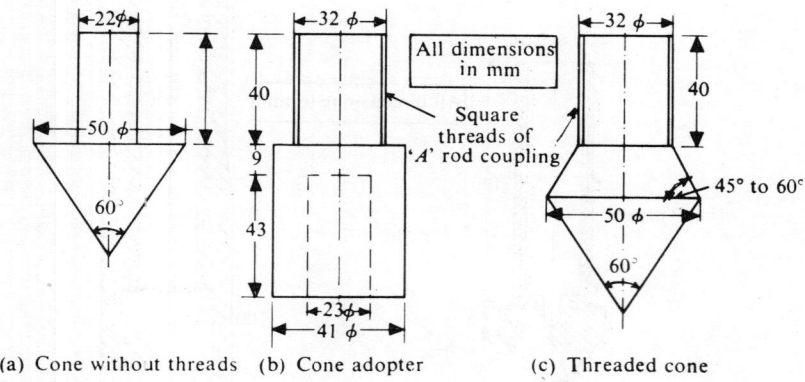

(a) Cone without threads (b) Cone adopter (c) Threaded cone

Fig. 18.10 Cone details for dynamic cone penetration test (IS)

The driving rods should be A rods of suitable length with threads for joining A rod coupling at either end. The rods should be marked at every 100 mm.

The driving head shall be of mild steel with threads at either end for a rod coupling. It shall have a diameter of 100 mm and a length of 100 to 150 mm.

Any suitable hoisting equipment such as a tripod may be used. A typical set-up using a tripod is shown in Fig. 18.11.

The hammer used for driving the cone shall be of mild steel or cast-iron with a base of mild steel. It shall be 250 mm high and of suitable diameter. The weight of the hammer shall be 640 N (65 kg).

The cone shall be driven into the soil by allowing the hammer to fall freely through 750 mm each time. The number of blows for every 100 mm penetration of the cone shall be recorded. The process shall be repeated till the cone is driven to the required depth. To save the equipment from damage, driving may be stopped when the number of blows exceeds 35 for 100 mm penetration.

When the depth of investigation is more than 6 m, bentonite slurry may be used for eliminating the friction on the driving rods. The cone used in this case is of 62.5 mm size and the details of the dynamic method using bentonite slurry, as standardised by the ISI, are available in "IS: 4968 (Part-II)-1976— Method for subsurface Sounding for Soils—Part II Dynamic method using 62.5 mm cone and bentonite slurry".

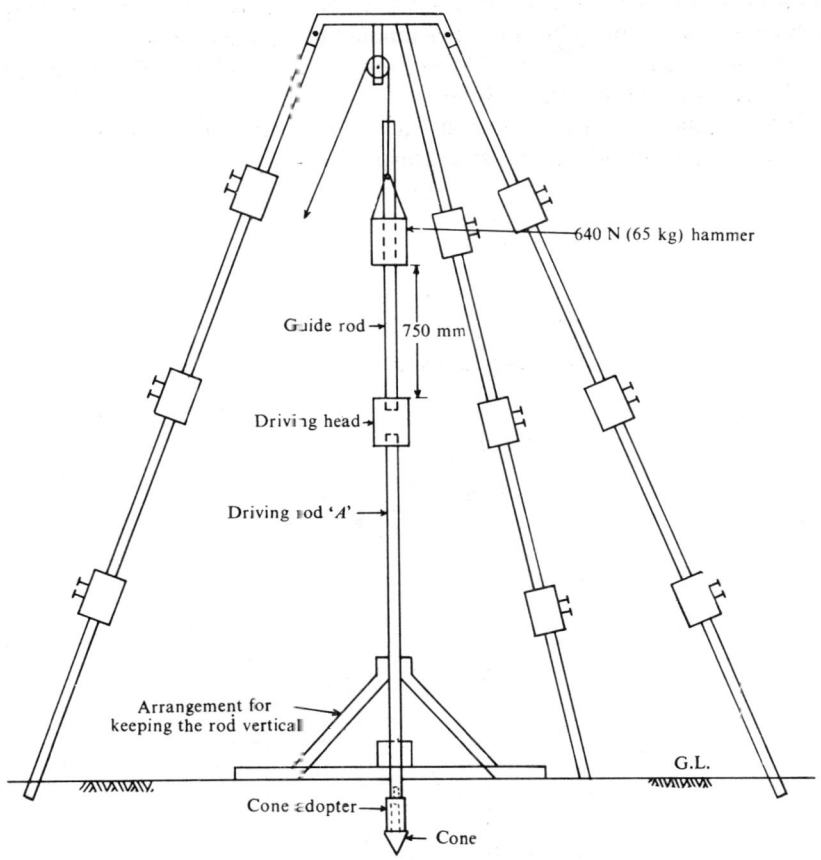

Fig. 18.11 Typical set-up for dynamic cone penetration test (IS)

Dynamic Cone Penetration test is a simple device for probing the soil strata and it has an advantage over the standard penetration test in that making of a bore hole is avoided. Moreover, the data obatained by cone test provides a continuous record of soil resistance.

Efforts are being made to correlate the cone resistance with the SPT value for different zones.

18.5.4 In-situ Vane Shear Test

In-situ vane shear test is best suited for the determination of shear strength of saturated cohesive soils, especially of sensitive clays, susceptible for sampling disturbances. The vane shear test consists of pushing a four-bladed vane in the soil and rotating it till a cylindrical surface in the soil fails by shear. The torque required to cause this failure is measured and this torque is converted to a unit shearing resistance of the cylindrical surface. The test may be conducted from the bottom of a bore hole or by direct penetration from ground surface.

The test, as standardised by the ISI, is given in "IS: 4434-1967—Code of Practice for *In-situ* Vane Shear Test for Soils".

The equipment consists of a shear vane, torque applicator, rods with guides, drilling equipment and jacking arrangement.

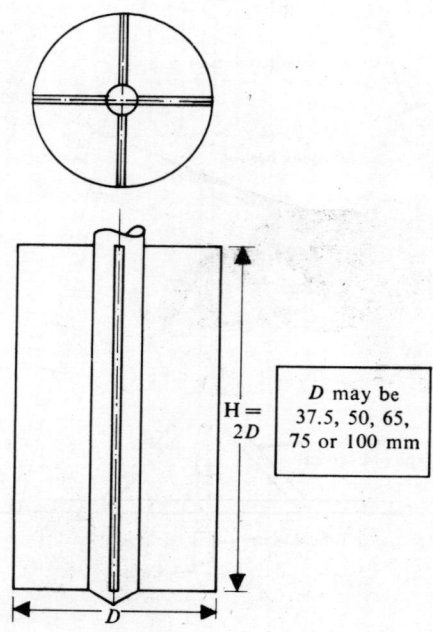

Fig. 18.12 Field Vane (IS)

The area ratio of the vane shall be kept as low as possible, and shall not exceed 15%, calculated as follows:

$$A_r = \frac{8t(D-d) + \pi d^2}{\pi D^2} \qquad \ldots(Eq.\ 18.6)$$

where, A_r = area ratio,

t = thickness of the blades of the vane,
D = overall diameter of the vane, and
d = diameter of central vane rod including any enlargement due to welding.

A diagrammatic vane test arrangement is shown in Fig. 18.13.

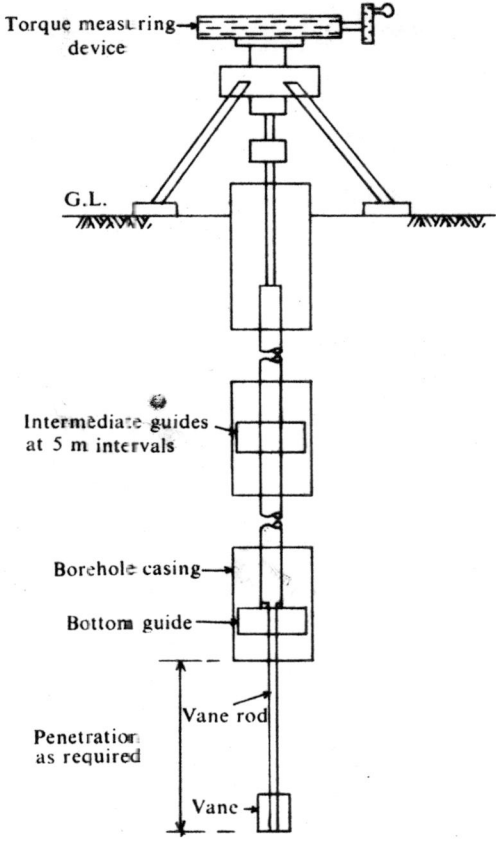

Fig. 18.13 Arrangement for Vane Test, from bottom of bore hole (IS)

The vane is pushed with a moderately steady force up to a depth of four-times the diameter of the bore hole or 50 cm, whichever is more, below the bottom. No torque shall be applied during the thrust. The torque applicator is tightened to the frame properly. After about 5 minutes, the gear handle is turned so that the vane is rotated at the rate of 0.1°/s. The maximum torque reading is noted when the reading drops appreciably from the maximum.

For a rectangular vane the shear strength of the soil is computed from the following formula:

$$\tau = \frac{T}{\pi D^2[(H/2)+(D/6)]} \qquad \text{...(Eq. 18.7)}$$

where, τ = shear strength (N/mm^2)
T = torque in mm-N,

D = overall diameter of the vane in mm, and
H = height of the vane in mm.

The assumptions involved are:
(i) shearing strengths in the horizontal and vertical directions are the same;
(ii) at the peak value, shear strength is equally mobilised at the end surface as well as at the centre;
(iii) the shear surface is cylindrical and has a diameter equal to the diameter of the vane; and
(iv) the shear stress distribution on the vane is as shown in Fig. 18.14.

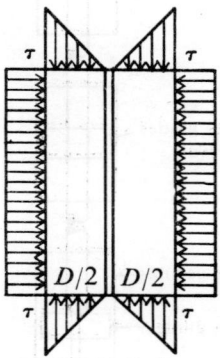

Fig. 18.14 Assumed stress distribution on blades of vane

For equilibrium, the applied torque, T = moment of resistance of the blades of the vane.

∴ T = surface area × surface stress × lever arm + end areas × average stress × lever arm.

$$= \pi D H \times \tau \times \frac{D}{2} + 2\left[\frac{\pi D^2}{4} \times \frac{\tau}{2} \times \frac{2}{3}D\right]$$

$$= \tau \left[\frac{\pi D^2 H}{2} + \frac{\pi D^3}{6}\right]$$

This leads to Eq. 18.7 for τ.

If $H = 2D$, Eq. 18.7 reduces to

$$\tau = \frac{6T}{7\pi D^3} \approx \frac{3T}{11 \cdot D^3} \qquad \text{...(Eq. 18.8)}$$

A shoe is used for protecting the vane if it is to penetrate direct from the ground surface.

A 100-mm vane is recommended for very soft soils. For moderately firm saturated soils, a 75-mm vane is recommended. The 50-mm vane is used infrequently but is intended for firm saturated soils.

The vane test may have a laboratory version also, the vane being relatively much smaller than the field vane.

18.6 INDIRECT METHODS—GEOPHYSICAL METHODS

The determination of the nature of the subsurface materials through the use of borings and test pits can be time-consuming and expensive. Considerable interpolation between checked locations is normally required to arrive at an area-wide indication of the conditions. Geophysical methods involve the technique of determining subsurface materials by measuring some physical property of the materials, and through correlations, using the values obtained for identifications. Most geophysical methods determine conditions over large distances and can be used to obtain rapid results. Thus, these are suitable for investigating large areas quickly, as in preliminary investigations.

A number of methods have been devised, but are mostly useful in the study of geologic structure and exploration for mineral wealth. However, two methods have been found to be useful for site investigation in the geotechnical engineering profession. They are the seismic refraction and the electrical resistivity methods. Although these have proven to be reliable, there are certain limitations as to the data that may be got; hence, spot checking with borings and pits has to be necessarily undertaken to complement the data obtained by the geophysical methods.

18.6.1 Seismic Refraction

When a shock or impact is made at a point on or in the earth, the resulting seismic (shock or sound) waves travel through the surrounding soil at speeds related to their elastic characteristics. The velocity is given by:

$$v = C\sqrt{\frac{Eg}{\gamma}} \qquad \text{...(Eq. 18.9)}$$

where, v = velocity of the shock wave,
 E = modulus of elasticity of the soil,
 g = acceleration due to gravity,
 γ = density of the soil, and
 C = a dimensionless constant involving Poissons's ratio.

The magnitude of the velocity is determined and is utilised to identify the material.

A shock may be created with a sledge hammer hitting a strike plate placed on the ground or by detonating a small explosive charge at or below the ground surface. The radiating shock waves are picked up by detectors, called 'geophones', placed in a line at increasing distances, d_1, d_2,..., from the origin of the shock (The geophone is actually a transducer, an electromechanical device that detects vibrations and converts them into measurable electric signals). The time required for the elastic wave to reach each geophone is automatically recorded by a 'seismograph'.

Some of the waves, known as *direct* or *primary* waves, travel directly from the source along the ground surface or through the upper stratum and are picked up first by the geophone. If the sub soil consists of two or more distinct layers, some of the primary waves travel downwards to the lower layer and get refracted as the surface. If the underlying layer is denser, the *refracted* waves travel much faster. As the distance from the source and the geophone increases, the refracted waves reach the geophone earlier than the direct waves. Figure 18.15 shows the diagrammatic representation of the travel of the primary and the refracted waves.

The distance of the point at which the primary and refracted waves reach the geophone simultaneously is called the 'critical distance' which is a function of the depth and the velocity ratio of the strata.

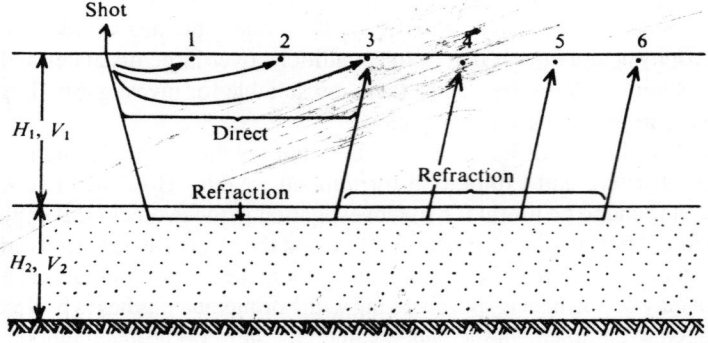

Fig. 18.15 Travel of primary and refracted waves

The results are plotted as a distance of travel versus time graph, known as the 'time-travel graph'. A simple interpretation is possible if each stratum is of uniform thickness and each successively deeper stratum has a higher velocity of transmission.

The reciprocal of the slope of the travel-time graph gives the velocity of the wave. The travel-time graph in the range beyond the critical distance is flatter than that in the range within that distance. The velocity in this range also can be computed in a similar manner. The break in the curve represents the point of simultaneous arrival of primary and refracted waves, or the critical distance. The travel-time graph appears somewhat as shown in Fig. 18.16.

In terms of the critical distance, d_c, and the velocities V_1 and V_2 in the upper soft layer of thickness H_1 and the lower hard layer respectively, the thickness of the upper layer may be written as follows:

$$H_1 = \frac{d_c}{2}\sqrt{\frac{(V_2-V_1)}{(V_2+V_1)}} \qquad \text{...(Eq. 18.10)}$$

The method can be extended to any situation with greater number of strata, provided each is successively harder than the one above. Typical wave velocities are given in Table 18.6.

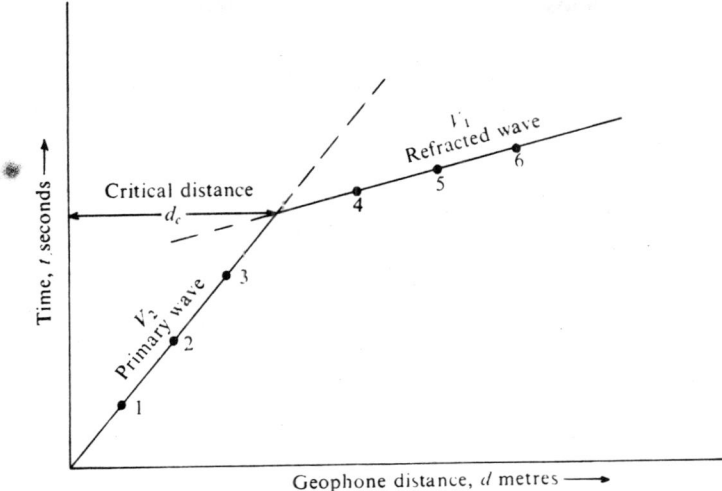

Fig. 18.16 Typical travel time graph for soft layer overlying hard layer

Table 18.6 Typical wave velocities for different materials (IS: 1892-1979 Appendix B)

Material	Velocity (m/s)	Material	Velocity (m/s)
Sand and top soil	180 to 365	Water in loose materials	1400 to 1830
Sandy clay	365 to 580	Shale	790 to 3350
Gravel	490 to 790	Sandstone	915 to 2740
Glacial till	550 to 2135	Granite	3050 to 6100
Rock talus	400 to 760	Limestone	1830 to 6100

There are certain significant limitations to the use of the seismic refraction method for determining the subsurface conditions. These are:

1. The method cannot be used where a hard-layer overlies a soft layer, because there will be no measurable refraction from a deeper soft layer. Test data from such an area would tend to give a single-slope line on the travel-time graph, indicating a deep layer of uniform material.
2. The method cannot be used in an area covered by concrete or asphalt pavement, since these materials represent a condition of hard surface over a softer stratum.
3. A frozen surface layer also may give results similar to the situation of a hard layer over a soft layer.
4. Discontinuities such as rock faults or earth cuts, dipping or irregular underground rock surface and the existence of thin layers of varying materials may also cause misinterpretation of test data.

18.6.2 Electrical Resistivity

Resistivity is a property possessed by all materials. The electrical resistivity method is based on the fact that in soil and rock materials the resistivity values differ sufficiently to permit that property to be used for purposes of identification.

Resistivity is usually defined as the resistance between opposite faces of a unit cube of the material. Each soil has its own resistivity depending upon the water content, compaction and composition; for example, the resistivity is high for loose dry gravel or solid rock and is low for saturated silt.

To determine the resistivity at a site, electrical currents are induced into the ground through the use of electrodes. Soil resistivity can then be measured by determining the change in electrical potential between known horizontal distances within the electric field created by the current electrodes.

The Wenner configuration with four equally spaced electrodes is simple and is popularly used. The four electrodes are placed in a straight line at equal distances as shown in Fig. 18.17.

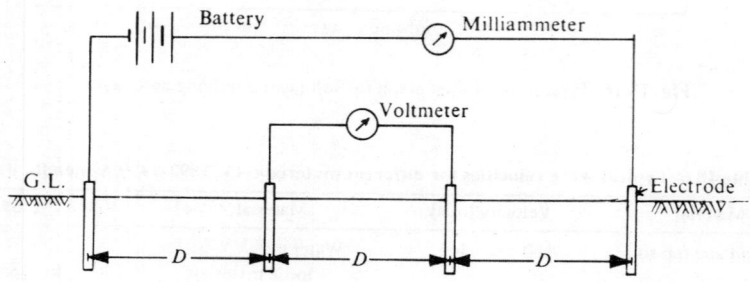

Fig. 18.17 Wenner configuration for electrical resistivity

A direct voltage, causing a current of 50 to 100 milliamperes typically, is applied between the outer electrodes and the potential drop is measured between the two inner electrodes by a null-point circuit that requires no flow of current at the instant of measurement.

In a semi-infinite homogeneous isotropic material the electrical resistivity, ρ, is given by:

$$\rho = 2\pi D \cdot \frac{E}{I} \qquad \ldots\text{(Eq. 18.11)}$$

where, D = distance between electrodes (m),
E = potential drop between the inner electrodes (Volts),
I = current flowing between the outer electrodes (Amperes), and
ρ = mean resistivity (ohm/m).

The calculated value is the apparent resistivity, which is a weighted average of all material within the zone created by the electrical field of the electrodes. The depth of material included in the measurement (depth of penetration) is approximately the same as the spacing between the electrodes.

Soil Exploration 859

It is necessary to make a preliminary trial on known formations, in order to be in a position to interpret the resistivity data for knowing the nature and distribution of soil formations. Average values of resistivity ρ for various rocks, minerals and soils are given in Table 18.7.

Table 18.7 Typical values of electrical resistivity of soils and rocks
(1 to 8 from IS : 1892-1979 Appendix B)

S. No.	Material	ρ (Ohm/m)	S. No.	Material	ρ (Ohm/m)
1.	Limestone (Marble)	10^2	7.	Limestones	120 – 400
2.	Quartz	10^{10}	8.	Clays	1 – 120
3.	Rock-salt	$10^6 - 10^7$	9.	Saturated inorganic clay or silt	10 – 50
4.	Granite	$5000 - 10^6$	10.	Saturated organic clay or salt	5 – 20
5.	Sandstone	35 – 4000	11.	Dry clays, silts	100 – 500
6.	Moraines	8 – 4000	12.	Dry sands, gravels	200 – 1000

Two different field procedures are used to obtain information on subsurface conditions. One method, known as "electrical profiling", is well-suited for establishing boundaries between different underground materials and has practical application in prospecting for sand and gravel deposits or ore deposits. The second method, called "electrical sounding", can provide information on the variation of subsurface conditions with depth and has application in site investigation for major civil engineering construction. It can also provide information on depth of water-table.

In electrical profiling, an electrode spacing is selected, and this same spacing is used in running different profile lines across an area, as in Fig. 18.18(a).

In electrical sounding, a centre location for the electrodes is selected and a series of resistivity readings is obtained by systematically increasing the electrode spacing, as shown in Fig. 18.18(b). Thus, information on layering of materials is obtained as the depth of information recovered is directly related to electrode sparing. This method is capable of indicating subsurface conditions where a hard-layer underlies a soft layer and also the situation of a soft layer underlying a hard layer.

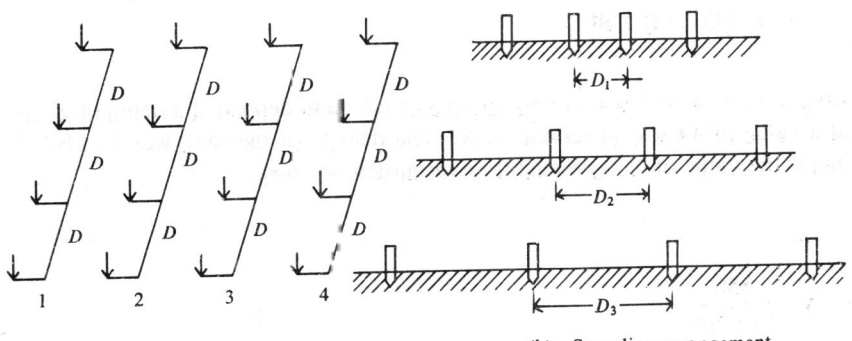

(a) Profiling arrangement (b) Sounding arrangement

Fig. 18.18 Electrode arrangement for electrical profiling and electrical sounding

18.7 ILLUSTRATIVE EXAMPLES

Example 18.1: One sampler has an area ratio of 8% while another has 16%; which of these samplers do you prefer and why?

(S.V.U.—B.Tech., (Part-time)—Sept., 1983)

The sampler with area ratio of 8% is to be preferred since the sample disturbance is inversely proportional to it. It is considered desirable that the area ratio be less than 10% for undisturbed sampling.

Example 18.2: Compute the area ratio of a thin walled tube sampler having an external diameter of 6 cm and a wall thickness of 2.25 mm. Do you recommend the sampler for obtaining undisturbed soil samples? Why?

(S.V.U.—Four-year B.Tech.—Dec., 1984)

External diameter, D_e = 6 cm = 60 mm

Wall thickness = 2.25 mm

∴ Internal diameter, D_i = (60 – 2 × 2.25) mm = 55.5 mm

Area ratio, $$A_r = \frac{[60^2 - (55.5)^2]}{(55.5)^2} = 16.88\%$$

Since the area ratio is more than 10%, the sampler is not recommended for obtaining undisturbed samples. The sample disturbance will not be insignificant.

Example 18.3: A SPT is conducted in fine sand below water table and a value of 25 is obtained for N. What is the corrected value of N?

Corrected $N = 15 + \frac{1}{2}(N' - 15)$

Here $N' = 25$

∴ $N = 15 + \frac{1}{2}(25 - 15) = \mathbf{20}$

Example 18.4: A SPT was conducted in a dense sand deposit at a depth of 22 m, and a value of 48 was observed for N. The density of the sand was 15 kN/m². What is the value of N, corrected for overburden pressure?

$N' = 48$

$$N = N' \cdot \frac{350}{(\sigma + 70)}$$

where σ = overburden pressure in kN/m².
Here, σ = 22 × 15 = 330 kN/m³

$$\therefore \text{ Corrected } N' = 48 \times \frac{350}{(330+70)} = \frac{48 \times 350}{400} = 42$$

Example 18.5: A vane, used to test a deposit of soft alluvial clay, required a torque 72 metre-newtons. The vane dimensions are $D = 100$ mm, and $H = 200$ mm. Determine a value for the undrained shear strength of the clay.

$D = 100$ mm $H = 200$ mm $= 2D$

Torque, $\qquad T = 72$ m N $= 72,000$ mm N.

$$\tau = \frac{T}{\pi D^2 \left(\frac{H}{2} + \frac{D}{6}\right)}$$

$$= \frac{6T}{7\pi D^3}, \text{ for } H = 2D$$

$$\therefore \quad \tau = \frac{6 \times 72000}{7 \times \pi \times 100 \times 100 \times 100} \text{ N/mm}^2$$

$$= 19.64 \times 10^{-3} \text{ N/mm}^2$$

$$= \frac{19.64 \times 10^{-3} \times 1000 \times 1000}{1000} \text{ kN/m}^2$$

$$= 19.64 \text{ kN/m}^2$$

Example 18.6: A seismic refraction study of an area has given the following data:

Distance from impact point to geophone (m)	15	30	60	90	120
Time to receive wave (s)	0.025	0.05	0.10	0.11	0.12

(a) Plot the time travel data and determine the seismic velocity for the surface layer and underlying layer.
(b) Determine the thickness of the upper layer.
(c) Using the seismic velocity information, give the probable earth materials in the two layers.
(a) The time-travel graph is shown in Fig. 18.19, Critical distance $d_c = 60$ m.

$$\text{Velocity in the upper layer, } V_1 = \frac{(60-15)}{(0.10-0.025)} = 600 \text{ m/s}$$

$$\text{Velocity in the lower layer, } V_2 = \frac{120-60}{(0.12-0.10)} = 3000 \text{ m/s}$$

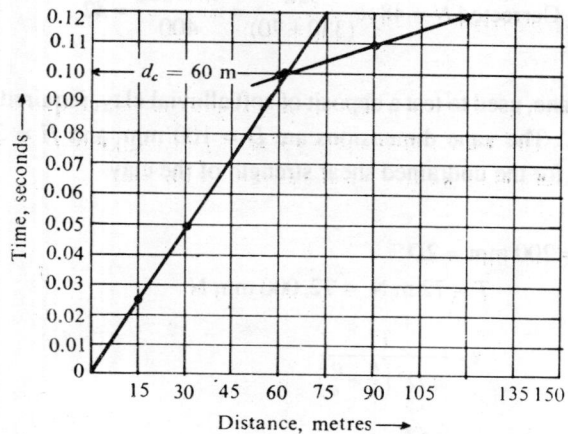

Fig. 18.19 Time-Travel graph (Example 18.6)

(b) Thickness of upper layer,

$$H_1 = \frac{d_c}{2}\sqrt{\frac{V_2 - V_1}{V_2 + V_1}} = (60/2)\sqrt{\frac{(3000 - 600)}{(3000 + 600)}} = \frac{30\sqrt{2}}{\sqrt{3}} = 10\sqrt{6} \text{ m} = \mathbf{24.5 \text{ m}}$$

(c) From the seismic velocity values, the probable materials are hard clay overlying *sound rock*.

SUMMARY OF MAIN POINTS

1. Site investigation and soil exploration involve field and laboratory investigations required to obtain necessary data for the soil strata existing at a site where an engineering construction is proposed. Reconnaissance, study of maps and aerial photography are the important steps in site investigation.
2. Test pits, trial pits or trenches are direct methods, borings are semi-direct methods, and soundings or penetration tests and geophysical methods are indirect methods.
3. Planning an exploratory programme involves the fixation of spacing and depth of bore holes. Record of boring data is usually given in the form of a boring log.
4. Taking out soil samples from soil strata for laboratory testing is known as 'soil sampling'. A sample may be disturbed or undisturbed (relatively speaking), the latter being necessary for the evaluation of certain engineering properties like strength and compressibility. Sample disturbance is dependent on a parameter called *area ratio* of the sampler. Thin-walled samplers are preferred for minimising sample disturbance.

5. Penetration tests commonly used are the standard penetration test and the cone test—static or dynamic. The standard penetration number is correlated to the density index and friction angle for granular soils. The *in-situ* vane shear test is used to determine the *in-situ* shearing strength of clayey soils.
6. Seismic refraction and electrical resistivity are the two most popular geophysical methods of soil exploration. Seismic refraction method utilises the variation of the velocity of propagation of shock waves through various earth materials; its significant limitation is that it fails when hard strata overly soft strata.

Electrical resistivity method utilises the variation of electrical resistivity with the composition of and presence of water in various earth materials. Wenner configuration is commonly used. Electrical profiling for areal coverage up to a certain depth and electrical sounding for evaluation of strata depthwise are used.

REFERENCES

1. M.J. Hvorslev: *Subsurface Exploration and Sampling of Soils for Civil Engineering Purposes*, U.S. Waterways Experiment station, Vicksburg, Miss., USA, 1949.
2. IS: 1892-1979: *Code of Practice for Subsurface Investigation for Foundations*.
3. IS: 2132-1972: *Standard Penetration Test for Soils*.
4. IS: 2132-1972: *Code of Practice for Thin-walled Tube Sampling of Soils*.
5. IS: 4434-1967: *Code of Practice for in-situ Vane Shear Test for Soils*.
6. IS: 4968 (Part I)-1976: *Method for Subsurface Sounding for Soils—Dynamic Method Using 50 mm Cone Without Bentonite Slurry*.
7. IS: 4968 (Part II)-1976: *Method for Subsurface Sounding for Soils—Dynamic Method Using 62.5 mm Cone and Benotonite Slurry*.
8. IS: 4968 (Part III)-1976: *Method for Subsurface Sounding for Soils*—Part III *Static Cone Penetration Test*.
9. D.F. McCarthy: *Essentials of Soil Mechanics and Foundations*, Reston Publishing Company, Reston, Virginia, USA, 1977.
10. G.B. Sowers and G.F. Sowers: *Soil Mechanics and Foundations*, 3rd ed., Collier Macmillan Company, Toronto, Canada, 1970.
11. K. Terzaghi and R.B. Peck: *Soil Mechanics in Engineering Practice*, John Wiley & Sons, Inc., NY, USA, 1948.
12. USBR: *Earth Manual*, U.S. Bureau of Reclamation, 1960.

QUESTIONS AND PROBLEMS

18.1 (a) Describe with a neat sketch how will you carry out the wash boring method of soil exploration. What are its merits and demerits?

(b) Explain the terms 'inside clearance' and 'outside clearance' as applied to a sampler. Why are they provided?

(S.V.U.—Four-year B.Tech.—Dec., 1984)

18.2 Write a short note on Geophysical exploration using electrical resistivity.

(S.V.U.—Four-year B.Tech.—Dec., 1984)

18.3 Under what circumstances are geophysical methods used in exploration? Discuss the usefulness of a dynamic cone penetration test and its limitations. Write a brief note on wash borings.

(S.V.U.—Four-year B.Tech.—Sep., 1983)

18.4 (a) Discuss with neat sketches any two boring methods used in soil exploration.

(b) Sketch a split-spoon sampler and explain its parts.

(S.V.U.—B.Tech. (Part-time)—Sep., 1983)

18.5 What are the various steps considered in the planning of sub-surface exploration programme?

Describe the standard penetration test. In what way is it useful in foundation design?

(S.V.U.—Four-year B.Tech.—Apr., 1983)

18.6 Write short notes on:

(a) Geophysical methods, (b) Penetration Tests.

(S.V.U.—Four-year B.Tech.—Apr., 1983)

18.7 Why are undisturbed samples required? Describe any one procedure of obtaining undisturbed samples for a multi-storeyed building project.

For what purpose are geophysical methods used? Describe any one method.

(S.V.U.—Four-year B.Tech.—Dec., 1982)

18.8 What are the advantages and disadvantages of accessible exploration? Discuss. Explain (a) Wash boring, (b) Split spoon sampler. Write a brief note on the precautions to be taken in transporting undisturbed samples.

(S.V.U.—Four-year B.Tech.—Oct., 1982)

18.9 (a) Explain and discuss the various factors that help decide the number and depth of bore holes required for subsoil exploration.

(b) What is 'N-value' of Standard Penetration Test? How do you find the relative density from 'N-value'? Explain the various corrections to be applied to the observed value of N.

(S.V.U.—B.Tech. (Part-time)—Sept., 1982)

18.10 (a) Enumerate the various methods of soil exploration and mention the circumstances under which each is best suited. What do you mean by undisturbed sample?

(b) Explain with a neat sketch the construction and use of a split spoon sampler.

(S.V.U.—B.E., (R.R.)—Feb., 1976)

18.11 Describe the "Standard Penetration Test" used in soil exploration. List the information that can be obtained by the test when made in (a) clay, (b) sand.

Comment on the correction factors to be used for N-values
(a) for sands for depth below ground level
(b) for fine sand under water table.
(S.V.U.—B.E., (R.R.)—May, 1969)

18.12 Write a brief critical note on vane shear test.
(S.V.U.—B.E., (N.R.)—Sep., 1967)

18.13 Two samplers have area ratios of 10.9% and 21%.
Which do you recommend for better soil sampling and why?

18.14 Compute the area ratio of a sampler with inside diameter 70 mm and thickness 2 mm. Comment.

18.15 A N-value of 35 was obtained for a fine sand below water-table. What is the corrected value of N?

18.16 A SPT was performed at a depth of 20 m in a dense sand deposit with a unit weight of 17.5 kN/m². If the observed N-value is 48, what is the N-value corrected for overburden?

18.17 A vane, 75 mm overall diameter and 150 mm high, was used in a clay deposit and failure occurred at a torque of 90 metre-newtons. What is the undrained shear strength of clay?

18.18 The inner diameters of a sampling tube and that of a cutting edge are 70 mm and 68 mm respectively, their outer diameters are 72 and 74 mm respectively. Determine the inside clearance, outside clearance and area ratio of the sampler.

ANSWERS TO NUMERICAL PROBLEMS

CHAPTER 2

(2.3). 86.96% (2.5). 1.41; 83.61% (2.6). 1.19; 2.97 (2.7). 0.53; 20%; 20.38 kN/m^3 (2.8). 70.04%; 2.22; 85.3% (2.9). The inconsistent value should be either e or γ since the determination of these values involves finding the volume of the soil sample, which may not be accurate. Thus, either $e = 7.83$ or $\gamma = 11.08$ kN/m^3 (1.13 g/cm^3). (2.10). 81.43%; 0.63; 16.24 kN/m^3; 10.23 kN/m^3; 20.04 kN/m^3 (2.11). 19.72 kN/m^3 (2.01 g/cm^3); 14.50 kN/m^3 (1.48 g/cm^3); 0.76 (2.12). 64.44%; 2.24; 77.67%

CHAPTER 3

(3.10). 2.70 (3.11). 2.54; 2.66; 4.5% (3.12). 13.72% (3.13). 57.2% (3.14). 67.55% (3.15). 19.47 kN/m^3 (1.985 g/cm^3); 17.70 kN/m^3 (1.804 g/cm^3) (3.16). 17.1 kN/m^3 (1.74 g/cm^3); 35.3%; 14.3%, 2.43 kN/m^3 (3.17). 8.25 s (3.18). 0.00671 mm; 40% (3.19). 1.0315 (3.20). 0.00586 mm; 31.63% (3.21). 42; 23.33% (3.22). 0.60; 0.41 (3.23). 51%; 2.81; 20% (3.24). 12.3%; 2.48; 0.79; 17.84 kN/m^3 (1.82 g/cm^3); 13.53 kN/m^3; 0.305; 18.64 kN/m^3 (1.9 g/cm^3) (3.25). 2.70; 1.56; 56% (3.26). 82 cm^3 (3.27). 16.86%; 1.86; 22.3% (3.28). 16%; 2.78 (3.29). 1.00; normal. (3.30). 10; extra-sensitive; flocculent.

CHAPTER 4

(4.1). CL; CI; CI–CH; MH or OH; CH–MH (4.2). SW–SM (4.3). 0.002 mm; 20. (well-graded)

CHAPTER 5

(5.3). At 2 m $-\sigma' = 33.24$ kN/m^2; $u = 0$; $\sigma = 33.24$ kN/m^2 At 6 m $-\sigma' = 71.28$ kN/m^2; $u = 39.24$ kN/m^2; $\sigma = 110.52$ kN/m^2 (5.9). 0.324 mm/s (5.13). 2.15×10^{-2} mm/s (5.14). 2.5 mm; No-3m 29 s. (5.16). 5.9×10^{-2} mm/s, the soil may be fine sand. (5.17). 0.48 mm/s (5.18). 0.11 mm/s; 11 m 21 s (5.19). 45 m 12 s (5.20). $k_h = 1.6 \times 10^{-1}$ mm/s $k_v = 625 \times 10^{-2}$ mm/s. (5.21). 21.7×10^{-2} mm/s $\left(\text{From } k \propto \dfrac{e^3}{1+e}\right)$ or 17.64×10^{-2} mm/s (From $k \propto e^2$). (5.23). 4.5×10^{-1} mm/s. (5.24). 9.3 mm/s. (5.26). 14.71 kN/m^2. (5.27). 3m (nearly) (5.28). 3.35 m (nearly) (5.29). 80 mm (5.30). At 3 m $-$ 29.43 kN/m^2; At 10 m $-$ 29.43 kN/m^2

Answers to Numerical Problems 867

CHAPTER 6

(6.10). Quick sand condition does not develop since the hydraulic gradient is less than the critical value. (6.11). 0.255 (6.12). 6.21; 0.33; 0.99 (6.13). 6 m; 2 m (6.14). 709.5 N/m^2; 354.75 N/m^2 (6.15). 300 ml/s/metre length (6.16). 45 l/s (6.17). 233.3 m^3/day (6.18). 101 m^3/day.

CHAPTER 7

(7.16). 4.5 cm (7.17). 0.565; 4.2 kg/cm^2; Overconsolidated. (7.18). 1.246 (7.19). 1.46 (7.20). 1.34 (7.21). Since the pore pressure is greater than the neutral pressure, consolidation is not complete. (7.22). 70% (7.23). 20 cm (7.24). 63.75 mm (7.25). $333\frac{1}{3}$ days. (7.26). 18.265 years (7.27). 1026 days (7.28). 5 cm (7.29). 5.87 cm; $441\frac{2}{3}$ days. (7.30). 20.27 cm; 10.14 cm; 62,500 days. (7.31). 0.63 (7.32). 780.8 days (7.33). 15.13 cm.

CHAPTER 8

(8.10). 30°; 300 kN/m^2 (normal), 173.2 kN/m^2 (shear); 200 kN/m^2 (max. shear) (8.11). 42°; 37° (8.12). 133.22 kN/m^2 (8.13). 38.22 kN/m^2; 6.77 kN/m^2; 57° with horizontal (major principal plane). (8.14). $c = 0.87$ kg/cm^2; $\phi = 36°$ (8.15). 435 kN/m^2 (8.16). 488 kN/m^2; 126 kN/m^2; 63° with horizontal (major principal stress). (8.17). $c = 120$ kN/m^2; $\phi = 16°$; 591 N; 330 kN/m^2. (8.18). 1.28 kg/cm^2 (8.19). $c = 65.4$ kN/m^2; $\phi = 18°$ (8.20). $c = 215$ kN/m^2; $\phi = 15°30'$ (8.21). $c = 0.70$ kg/cm^2; $\phi = 30°$ (8.22). 513.3 kN/m^2 (8.23). 38°30'; 64°15'. (8.24). $c = 91.3$ kN/m^2; $\phi = 10°$; 117.8 kN/m^2 (8.25). $c = 4$ t/m^2; $\phi = 19°15'$; $c' = 3$ t/m^2; $\phi' = 24°00'$. (8.26). $c = 80$ kN/m^2; $\phi = 7°30'$ (8.27). $\phi_{CIU} = 18°$; 2.8 kg/cm^2

CHAPTER 9

(9.7). $c = 16.96$ kN/m^2 (9.8). 1.9 (with respect to cohesion) (9.9). 12.5 m (9.10). 1.3 (9.11). $F_c = 1.5$; $F = 1.25$ (for equal mobilisation of friction and cohesion). (9.12). $\beta = 24°$ (nearly).

CHAPTER 10

(10.10). 0.06 t/m^2 (10.11). 0.033 t/m^2 (10.12). $\sigma_z = 2.39$ t/m^2 (Single point load), 2.21 t/m^2 (Four point loads), 2.16 t/m^2 (rigorous) (10.13). 38.2 kN/m^2; 18.3 kN/m^2 (10.14). 122.75 kN/m^2; 105.1 kN/m^2; 47.75 kN/m^2 (10.15). 58.2 kN/m^2 (10.16). Centre: 3.183 t/m^2 (Single point load), 2.565 t/m^2 (Four point loads), 2.448 t/m^2 (rigorous); Corner: 1.473 t/m^2 (Single point load), 1.662 t/m^2 (Four point loads), 1.451 t/m^2 (rigorous) (10.17). 24.52 kN/m^2 (rigorous) (10.18). 25 kN/m^2 (rigorous)

CHAPTER 11

(11.7). 90 mm   (11.8). 510 mm   (11.9). 25 cm   (11.10). $S_i = 72$ mm, $S_c = 362.3$ mm; Total settlement = 434.3 mm   (11.11). 137.44 mm   (11.12). 465.84 mm; 41.66 mm; 83.33 mm

CHAPTER 12

(12.13). 86.4%   (12.14). 14.3%; 2.16 g/cm^3 (12.15). 1940 m^3 (12.16). 109.5 m^3; (9.18). kl   (12.17). 6.5%; 2.066 g/cm^3 (12.18). 12%; 18.94 kN/m^3; 19%.

CHAPTER 13

(13.12). 30.38 t/m at 3 m above the base; 58.73 t/m   (13.13). 71.52 t/m; 180.6 t/m   (13.14). Active: Pressure in the dense state/pressure in the loose state = 0.747; Passive: Pressure in the dense state/pressure in the loose state = 1.760.   (13.15) 92.8 kN/m. run at 1.74 m above the base (13.16). 152%.   (13.17). 56 t/m (ignoring tensile stresses); 1.11 m; 2.63 m.   (13.18). 1189.4 kN/m at 1.87 m above the base. (13.19). 9.8 t/m. run (13.20). 60 t/m. run; 71.3 t/m. run.   (13.21). $\sigma_{max} = 116$ kN/m^2; (heel)

$\sigma_{min} = 79.2$ kN/m^2.
(toe)

CHAPTER 14

(14.11). 9 t/m^2; 4.5 t/m^2 (14.12). 10 t/m^2 (14.13). 488 kN/m. run   (14.14). 85 kN/m^2 (14.15). 54 t/m^2 (14.16). 290 kN/m^2 (14.17). 148 kN/m^2 (14.18). 1.60 m; 1.35 m   (14.19). 1.6 m   (14.20). 2m; 2m (since water table rises to the ground surface).   (14.21). 315 kN/m^2 (14.22). 4 m × 8 m   (14.23). 553.7 kN/m^2; 2.26. (14.24). 3.5   (14.25). 1944 kN   (14.26). 492 kN/m^2 (Shear failure governs the design)   (14.27). 200 kN/m^2; 450 kN.

CHAPTER 15

(15.10). Size ranges from 1 m to 2 m   (15.11). 1848 kN; 1987 kN (Useful width approach)   (15.12). 2.1 m sq. and 2.4 m sq.   (15.13). 1.5 m × 7.0 m   (15.14). 1 m and 2.5 m wide and 6 m long.   (15.15). 103,680 kN; 15,360 kN

CHAPTER 16

(16.10). 20 t   (16.11). 250 kN   (16.12). 17.5 mm   (16.13). 30.5 t   (16.14). 1012.5 kN   (16.15). 580 kN   (16.16). 3580 kN   (16.17). 852 kN   (16.18). 822 kN

CHAPTER 17

(17.17). 17% (17.18). 10 cm each.

CHAPTER 18

(18.13). First sampler ($A_r = 10.9\%$) (18.14). 11.8% (18.15). $N = 25$ (18.16). $N = 40$ (18.17). 58.2 kN/m² (18.18). $C_I = 2.94\%$; $C_O = 2.78\%$; $A_r = 18.43\%$.

Appendix A

A NOTE ON S.I. UNITS

S.I. is the abbreviation by which the Systéme Internationale d' Unités (International System of Units) is known. S.I is a 'coherent' system—the product or quotient of two or more of its units is the unit of the resultant quantity.
The base units of SI System are as follows:

Length

The metre is the length equal to 1,650,763.73 wavelengths in vacuum of the radiation corresponding to the transition between the levels $2p^{10}$ and $5d^5$ of the Krypton-86 atom.

Mass

The Kilogram is the unit of mass; it is equal to the mass of the international prototype of the kilogram.

Time

The second is the duration of 9,192,631,770 periods of the radiation corresponding to the transition between the two hyperfine levels of the ground state of the Caesium-133 atom.

Intensity of Electric Current

The ampere is that constant current which, if maintained in two straight parallel conductors of infinite length, of negligible circular cross-section and placed one metre apart in vacuum, would produce between these conductors a force equal to 2×10^{-7} newton per metre of length.

Thermodynamic Temperature

The Kelvin, unit of thermodynamic temperature, is the fraction 1/273.16 of the thermodynamic temperature of the triple point of water.

Luminous Intensity

The candela is the luminous intensity, in the perpendicular direction of a surface of 1/600,000 square metres of a black body at the temperature of freezing platinum under a pressure of 101,325 newtons per square metre.

Amount of Substance

The mole is the amount of substance of a system which contains as many elementary units as there are carbon atoms in 0.012 kilogramme of Carbon - 12. The elementary unit must be specified and may be an atom, a molecule, an ion, an electron, a photon, or a specified group of such entities.

The base and other units of the S.I. System are tabulated below:

Quantity	Unit	Symbol	Remarks
I Base Units :			
Length	metre	m	
Mass	kilogram	kg	
Time	second	s	
Electric Current	ampere	A	
Thermodynamic temperature	kelvin	K	Celsius scale may be used for practical purposes.
Luminous intensity	candela	cd	
Amount of substance	mole	mol	
II Supplementary Units			
Plane angle	radian	rad	
Solid angle	storadian	sr	Solid angle subtended at the cente of a sphere by a surface area equal to the square of the radius.
III Derived Units			
Force	newton	N	Force which produces an acceleration of 1 m/s^2 acting on a mass of 1 kg. Also, 1 kg = g N
Energy	joule	J	Work done by a force of 1 N over a displacement of 1 m. or 1 J = 1 Nm
Power	watt	W	Rate of work of 1 J/s
Flux	weber	Wb	
Flux density	tesla	T	
Frequency	hertz	Hz	Hz = 1 c/s (s^{-1})
Electric conductance	siemens	S	
Pressure, stress	pascal	Pa	1 Pa = 1 N/m^2

The SI Units specifically used in the field of Geotechnical Engineering are listed below, along with the MKS Units and the conversion factors:

S. No.	Quantity	MKS Unit	SI Unit	Abbreviation	Conversion factor
1.	Mass Density	kg (m) per cu m g (m) per cu cm	Megagram per cu m gm per millilitre	Mg/m³ g/ml	1 kg/m³ = 1 × 10⁻³ Mg/m³ 1 g/cm³ = 1 g/ml
2.	Unit Weight	kg (f) or t(f) per cu m g (f) per cu cm	kilonewton per cu m	kN/m³	1 t/m³ = 9.8 kN/m³ 1 g/m³ = 9.8 kN/m³
3.	Force	kg (f) t (f)	newton	N	1 kg = 9.8 N 1 t = 9.8 kN
4.	Pressure, Stress	kg (f) per sq cm t (f) per sq m	kilonewton per sq m	kN/m²	1 kg/cm² = 98 kN/m² 1 t/m² = 9.8 kN/m²
5.	Velocity, coefficient of permeability	cm per second m per second m per year	millimetre per second metre per second metre per year	mm/s m/s m/a	1 cm/s = 1 × 10 mm/s = 1 × 10⁻² m/s
6.	Rate of flow	cu m per second litre per second	cubic metre per second litre per second	m³/s l/s	1 cm³/s = 1 × 10⁻⁶ m³/s
7.	Coefficient of compressibility	Square metre per kg f	Sq m per-kilonewton	m²/kN	1 m²/kg = 102.041 m²/kN
8.	Coefficient of consolidation	Square metre per second	Sq m per second Sq. millimetre per second	m²/s mm²/s	1 m²/s = 1 × 10⁶ mm²/s
9.	Moment of force	kilogram (force) metre tonne (force) metre	kilonewton metre newton millimetre	kN m N mm	1 kg m = 9.8 × 10⁻³ kNm 1 t m = 9.8 kN m 1 kg m = 9.8 × 10³ N mm

General Rules and Conventions

Symbols
1. Full names of units, even when they are named after a person, are not written with a capital initial letter.
2. The symbol for a unit, named after a person, is written with a capital initial letter.
3. Symbols for other units are not written with capital letters.
4. Symbols for units are always used in the singular.
5. No punctuation marks should be used within or at the ends of symbols for units.
6. A space is left between the numeral and the symbol.
7. A space is left between the symbols for compound units—e.g., kW h; N m. (m N would denote metre newton while mN would denote millinewton)
8. The denominators of compound units are expressed preferably in the base units and not in their multiples or submultiples—e.g., MN/m^2 is preferred to N/mm^2.
9. The symbol of multiples and submultiples is to be used only conjointly with that of the unit and not with the full name of the unit—e.g., MW and not M watt.

Prefixes

Multiples or submultiples involving the factor 1000 (or $10^{\pm 3n}$ n being an integer) are preferentially used for the purpose of international consistency.
Common prefixes used are:

kilo (k)	10^3
mega (M)	10^6
giga (G)	10^9
milli (m)	10^{-3}
micro (μ)	10^{-6}
nano (n)	10^{-9}

Normally, double prefixes are avoided.

REFERENCES

1. Ramaswamy, G.S., and Rao. V.V.L., *SI Units: A Source Book*, Tata McGraw-Hill Co. Ltd., 1971.
2. IS: 3616: *Recommendations of the International System (SI) of Units*, ISI, New Delhi.
3. IS: SP5-1969: *Guide to SI Units*, ISI, New Delhi.

Appendix B

NOTATION

ENGLISH

A	:	Activity of clay (Ch. 3)
		Area of cross-section (Ch. 3)
		Pore pressure parameter (Ch. 8)
		Area of tyre contact (Ch. 17)
$A°$	:	Angstrom unit
A_o	:	Water adsorption capacity (Ch. 12)
$\overline{A}$	:	Product of parameters A & B (Ch. 8)
A_b	:	Bearing area of base of pile (Ch. 16)
A_g	:	Area of cross-section of pile block (Ch. 16)
A_r	:	Area ratio (Ch. 18)
A_s	:	Surface area of pile in contact with soil (Ch. 16)
$\overline{A}_s$	:	Surface area as proposed for underreamed piles (Ch. 16)
A_v		Area of cross-section of void space (Ch. 5)
a	:	Area of cross-section of stand pipe (Ch. 5)
		a particular distance (Ch. 6)
		Lever arm (Ch. 9)
		Radius of circular loaded area (Ch. 10)
		Effective area of piston (Ch. 16)
		Dimension (Ch. 15)
a_c	:	Air content (Ch. 2)
a_v	:	Coefficient of compressibility (Ch. 7)
B	:	Skempton's pore pressure parameter (Ch. 8)
		Width of loading (Ch. 10)
		Width of foundation (Ch. 18)
$\overline{B}$		Parameter (Ch. 9)
B_1, B_2	:	Widths of footing (Ch. 15)
b	:	Size of elemental square (Ch. 6)
		Breadth of slice (Ch. 9)
		Size (breadth) of footing (Ch. 14)
		Dimension (Ch. 15)
b'	:	Effective width (Ch. 13, Ch. 15)

C	:	Correction to hydrometer reading (Ch. 3)
		Shape factor (Ch. 5)
		Total cohesive force (Ch. 9)
		Correction factor (Ch. 11)
		Constant (Ch. 14)
		Empirical constant in pile-driving formulae (Ch. 16)
C_I	:	Inside clearance (Ch. 18)
C_O	:	Outside clearance (Ch. 18)
C_a	:	Total adhesion force (Ch. 13)
C_c	:	Coefficient of curvature (Ch. 3)
		Compression index (Ch. 7)
		Volume compressibility of soil (Ch. 8)
C_d	:	Deflocculating agent correction (Ch. 3)
C_e	:	Expansion index (Ch. 7)
C_m	:	Meniscus correction (Ch. 3)
C_r	:	Static cone resistance (Ch. 11)
		Coefficient of restitution (Ch. 16)
C_s	:	Constant (Ch. 6)
		Compressibility constant (Ch. 11)
C_t	:	Temperature correction (Ch. 3)
C_v	:	Coefficient of consolidation (Ch. 7)
		Volume compressibility of water (Ch. 8)
C_w	:	Settlement coefficient (Ch. 16)
C_u	:	Uniformity coefficient (Ch. 4)
CU	:	Consolidated undrained Test (Ch. 8)
CD	:	Consolidated drained Test (Ch. 8)
c	:	Percent clay-size particles (Ch. 3)
		cohesion (Ch. 8)
c_a	:	Unit adhesion (Ch. 16)
c_e	:	Effective cohesion (Hvorslev) (Ch. 8)
c_b, c_l	:	Extreme fibre distances (Ch. 15)
D	:	Diameter of soil particle (Ch. 3)
		Diameter of pipe pore (Ch. 5)
		Size of largest particle (Ch. 17)
		Electrode spacing (Ch. 18)
		Overall diameter of Shear Vane (Chapts. 8 and 18)
D_f	:	Depth of foundation (Ch. 14)
D_c, D_w	:	Diameters of cutting edge (Ch. 18)
D_s, D_t	:	Diameters of sampling tube (Ch. 18)

D_n	:	Depth of pressure bulb (Ch. 11)
		Depth of compressible layer (Ch. 16)
D_s	:	Effective particle size (Ch. 5)
D_{10}	:	10% finer size or effective size (Ch. 3)
D_{30}	:	30% finer size (Ch.3)
D_{60}	:	60% finer size (Ch. 3)
d	:	distance (Ch. 6)
		Aperature size of sieve (Ch. 17)
		Total thickness of construction (Ch. 17)
		Diameter of central vane rod (Ch. 18)
d_c	:	diameter of capillary (Ch. 5)
		critical distance (Ch. 18)
		depth factor (Ch. 14)
d_γ & d_q	:	depth factors (Ch. 14)
E	:	Modulus of elasticity of soil (Ch. 14)
		Potential drop (Ch. 18)
E_s	:	Modulus of elasticity of soil (Ch. 11)
e	:	Void ratio
		Lever arm for weight of slice (Ch. 9)
		Eccentricity of reaction (Ch. 13)
e_b, e_L	:	Eccentricities of load (Ch. 15)
e_o	:	Natural void ratio (Ch. 3)
		Initial void ratio (Ch. 11)
e_{max}		maximum void ratio (loosest state) (Ch. 3)
e_{min}	:	minimum void ratio (densest state) (Ch. 3)
F	:	Frictional resistance (Ch. 8)
		Factor of safety (Ch. 9)
F_g	:	Group settlement ratio (Ch. 16)
f_s	:	Unit skin friction (Ch. 16)
f_s	:	Unit frictional resistance between soil and soil (Ch. 16)
G	:	Specific gravity of soil solids or grain specific gravity (Ch. 2)
G_k	:	Specific gravity of Kerosene (Ch. 2)
G_m	:	Mass specific gravity (Ch. 2)
G_{ss}	:	Specific gravity of soil suspension (Ch. 2)
G_w	:	Specific gravity of water (Ch. 2)
G_{w_T}	:	Specific gravity of water at the temperature of the Test (Ch. 2)

$G_{w_{T_1}}; G_{w_{T_2}}$: Specific gravity of water at temperatures T_1^0 and T_2^0 respectively (Ch. 2)

G_{T_1}, G_{T_2} : Specific gravity of soil solids at temperatures T_1^0 and T_2^0 respectively (Ch. 2)

g : Acceleration due to gravity (Ch. 5)

H : Thickness of confined aquifer (Ch. 5)
Height of sample (Ch. 7)
Drainage path (Ch. 7)
Height of vane (Ch. 8)
Height of slope (Ch. 9)
Height of wall (Ch. 13)

H_c : Critical depth (Ch. 13)

H_e : Equivalent height of soil (Ch. 14)

H_s : Inclined height of wall face (Ch. 13)

h : Length of hydrometer bulb (Ch. 3)
Hydraulic head causing flow (Ch. 5)

h_c : capillary rise (Ch. 5)
depth of tension crack (Ch. 9)

h_t : height of water (Ch. 6)

h_p, h_e, h_v: Pressure head, elevation head, and velocity head (Ch. 5)

I : Current (Ch. 8)

I_b, I_L : Moments of inertia of footing (Ch. 15)

I_c : Consistency index (Ch. 3)

I_D : Density index

I_f : Flow index (Ch. 3)

I_L : Liquidity index (Ch. 3)

I_p : Plasticity index (Ch. 3)

I_s : Shrinkage index (Ch. 3)
Influence value (Ch. 11)

I_t : Influence value (Ch. 11)

I_T : Toughness index (Ch. 3)

I_σ : Influence value (Ch. 10)

i : hydraulic gradient (Ch. 5)

i_c, i_γ, i_q : Inclination factors (Ch. 14)

j : Seepage force per unit volume

K : Constant as specified (Ch. 3)
Specific or absolute permeability (Ch. 5)
Cohesion factor (Ch. 8)

K	:	Conjugate ratio (Ch. 13)
		Shape coefficient (Ch. 14)
K_A, K_a	:	Active earth pressure coefficient (Ch. 13)
K_B	:	Boussinesq's influence factor (Ch. 10)
K_I		Influence coefficient for line load (Ch. 10)
K_o	:	Coefficient of earth pressure at rest (Ch. 13)
K_p	:	Passive pressure coefficient (Ch. 13)
K_s	:	Coefficient of earth pressure (Ch. 16)
K_W	:	Westergaard's influence factor (Ch. 10)
k	:	Darcy's coefficient of permeability (Ch.5)
		Coefficient of subgrade reaction (Ch. 15)
k_o	:	Constant (Ch. 5)
k'	:	Coefficient of soil modulus variation (Ch. 16)
k_e	:	Effective permeability
k_p	:	Coefficient of percolation (Ch. 5)
L	:	Length of sample (Ch. 5)
		Size of elemental square (Ch. 6)
		Length of rectangular load (Ch. 10)
		Length of footing (Ch. 15)
L'	:	Effective length (Ch. 15)
L_e	:	Embedded length of pile (Ch. 16)
L_s	:	Lineal shrinkage (Ch. 3)
l	:	Length dimension (Ch. 8)
		Arc length (Ch. 9)
		Distance (Ch. 13)
l_c	:	Chord length (Ch.9)
l_0, l_r	:	Lever arm (Ch. 14)
l_s	:	Length of slices (Ch. 9)
M_B	:	Moment of force (Ch. 13)
M_o	:	Moment of force (Ch. 14)
M_r	:	Moment of resistance (Ch. 14)
M_b, M_L	:	Moments (Ch. 15)
m	:	Factor of safety with respect to total stresses (Ch. 9)
		factor for rectangular loading (Ch. 10)
		Perimeter shear factor (Ch. 14)
		number of rows of piles (Ch. 16)
m_v	:	modulus of volume change (Ch. 7)
N	:	Overall percentage of particles finer than D (Ch. 3)

N	:	Number of blows (determination of LL) (Ch. 3)
		Standard penetration number (Chs. 14 and 18)
		Normal force (Ch. 9)
		Taylor's stability number (Ch. 9)
		Normal component of Soil reaction (Ch. 13)
N'	:	Observed SPT value (Ch. 18)
N_c, N_q, N_γ	:	Bearing capacity factors (Ch. 14)
N_f	:	Percentage of particles finer than size D in a sample (Ch. 3)
N_ϕ	:	Flow value (Ch. 13)
n	:	Porosity
		Coefficient (Ch. 9)
		factor (Ch. 10)
		distance (Ch. 13)
		number of piles (Ch. 16)
n_a	:	Percent air voids (Ch. 2)
n_d	:	number of equipotential drops (Ch. 6)
n_f	:	number of flow channels (Ch. 6)
OCR	:	Over Consolidation Ratio (Ch. 7)
P	:	Force (Ch. 8)
		Perimeter of pile section (Ch. 16)
		Earth thrust (Ch. 13)
P_1	:	Earth thrust (Ch. 14)
P_a	:	Total active earth thrust
P_{ah}, P_{av}	:	Horizontal and vertical components of active earth thrust (Ch. 13)
P_e, P_i	:	Column loads (Ch. 15)
P_o	:	Total earth thrust at rest
P_g	:	Perimeter of pile group (Ch. 16)
P_p		Total passive earth resistance (Ch. 13)
P_s	:	Standard load (Ch. 17)
P_T	:	Corrected Test load (Ch. 17)
Pt	:	Peat (Ch. 4)
p	:	mean effective pressure (Ch. 16)
Q	:	Quantity of water collected (Ch. 5)
		Concentrated load (Ch. 10)
Q_h	:	horizontal force on pile (Ch. 16)
Q_{ult}	:	Total ultimate bearing capacity (Ch. 14)
Q_{up}	:	Ultimate bearing load on pile (Ch. 16)
Q_{eb}	:	End-bearing resistance of pile (Ch. 16)

Q_{sf}	:	Skin-friction resistance of pile (Ch. 16)
Q_{ap}	:	Allowable load on pile (Ch. 16)
Q_{nf}		Negative skin friction force on the pile (Ch. 16)
Q_{ng}	:	Net group capacity of piles (Ch. 16)
q	:	Discharge (Ch. 5 and Ch. 6)
		Rate of pumping (Ch. 5)
		Load intensity (Ch. 10)
		Surcharge stress (Ch. 13)
q'	:	Line load intensity (Ch. 10)
q_u	:	unconfined compression strength
q_{ult}	:	ultimate bearing capacity (Ch. 14)
q'_{ult}		ultimate bearing capacity under local shear failure (Ch. 14)
$q_{net\,ult}$	:	Net ultimate bearing capacity (Ch. 15)
q_b	:	bearing capacity of pile in point-bearing (Ch. 16)
q_{na}	:	Net allowable bearing pressure (Ch. 14)
q_s	:	Safe bearing capacity (Ch. 14)
		Bearing capacity of foundation for a specified settlement (Ch. 14)
q_{ns}	:	net safe bearing capacity (Ch. 14)
R	:	Shrinkage ratio (Ch. 3)
		Reynold's number (Ch. 5)
		Radius of influence (Ch. 5)
		Frictional resistance (Ch. 9)
		Length of polar ray (Ch. 10)
		Base reaction (Ch. 13)
		Ratio of weight of pile to weight of hammer (Ch. 16)
R_γ, R_d	:	Correction factors for water table (Ch. 14)
R_g	:	Groutability ratio (Ch. 17)
R_i, R_e	:	Base reactions (Ch. 14)
R_h	:	Corrected hydrometer reading (Ch. 3)
R_h	:	Observed hydrometer reading (Ch. 3)
r	:	radial distance (Ch. 5)
		radius of failure surface (Ch. 9)
		radius of friction circle (Ch. 9)
r_0	:	radius of central well (Ch. 5)
r_u	:	pore pressure ratio (Ch. 9)
S	:	Degree of saturation (Ch. 2)
		Specific surface area (Ch. 5)
		A particular distance (Ch. 6)
		Shearing resistance of base of slice (Ch. 9)

S	:	Shear component of soil reaction (Ch. 13)
		Settlement (Ch. 14 and Ch. 16)
S_c	:	Consolidation settlement (Ch. 7)
S_{cc}	:	Corrected consolidation settlement (Ch. 11)
S_e	:	Elastic compression of pile (Ch. 16)
S_{es}	:	Elastic compression of soil at base (Ch. 16)
S_f	:	Settlement of foundation (Ch. 11)
		Settlement due to deformation (Ch. 16)
S_i	:	Immediate settlement (Ch. 11)
S_n	:	Stability number (Ch. 9)
S_o	:	Optimum spacing of piles (Ch. 16)
S_p	:	Settlement of plate (Ch. 11)
		Settlement of pile tip (Ch. 16)
		Plastic compression of soil (Ch. 16)
S_r	:	Degree of shrinkage
S_s	:	Settlement due to secondary compression (Ch. 15)
S_t	:	Sensitivity of clay (Ch. 3)
s	:	shear strength (Ch. 8)
		Elastic settlement (Ch. 14)
		Penetration (set) of pile (Ch. 16)
		Pile spacing (Ch. 16)
S_o		Elastic compression of pile (Ch. 16)
s_c, s_γ, s_q	:	shape factors (Ch. 14)
T	:	Coefficient of transmissibility (Ch. 5)
		Time factor (Ch. 7)
		Tangential force (Ch. 9)
		Sliding resistance (Ch. 13)
		Relative stiffness factor (Ch. 16)
T_s	:	Surface Tension (Ch. 5)
t	:	Elapsed time (Ch. 3)
		Thickness of vane (Ch. 8 and Ch. 18)
U	:	Uniformity Coefficient (Ch. 3)
		Average consolidation ratio (Ch. 7)
		Total neutral force (Ch. 9)
U_z	:	Consolidation ratio at depth z (Ch. 7)
UU	:	Unconsolidated undrained Test (Ch. 8)
u	:	Pore water pressure
		Neutral stress (Ch.5)
		Excess pore pressure (Ch. 7)

V : Total volume of soil sample (Ch. 2)
Volume of suspension (Ch. 3)
V_a : Volume of air (Ch. 2)
V_d, V_m : Volume of soil at Shrinkage limit (Ch. 3)
V_h : Volume of hydrometer (Ch. 3)
V_i : Initial volume of soil sample (Ch. 3)
V_l : Volume of soil at liquid limit (Ch. 3)
V_p : Volume of soil at plastic limit (Ch. 3)
Volume of pipette sample (Ch. 3)
V_s : Volume of solids (Ch. 2)
Volumetric shrinkage (Ch. 3)
V_v : Volume of voids (Ch. 2)
V_w : Volume of water in the voids (Ch. 2)
V_1, V_2 : Velocity of shock waves (Ch. 18)
v : Terminal velocity (Ch. 3)
Velocity of flow (Ch. 5 and Ch. 6)
Velocity of shock waves (Ch. 18)
v_c : lower critical velocity (Ch. 5)
v_s : seepage velocity (Ch. 5)
W : Total weight of soil mass (Ch. 2)
Weight of soil slice (Ch. 9)
Weight of soil wedge (Ch. 13)
Maximum wheel load (Ch. 17)
W_a : Weight of air (negligible) (Ch. 2)
W_d : Weight of dry soil (Ch. 2)
W_D : Weight of soil finer than size D (Ch. 3)
W_f : Weight of fine soil fraction out of a total soil sample taken for combined sieve and sedimentation analysis (Ch. 3)
W_h : Weight of pile hammer (Ch. 16)
W_i : Initial weight of soil sample (Ch. 3)
W_m : Weight of soil sample at shrinkage limit (Ch. 3)
W_p : Weight of solids in pipette sample (Ch. 3)
Weight of pile (Ch. 16)
W_s : Weight of solids (Ch. 2)
$(W_s)_{sub}$: submerged weight of soil solids (Ch. 2)
W_v : Weight of material occupying void space (Ch. 2)
W_w : Weight of water (Ch. 2)

Appendix 883

w	:	Water content (Ch. 2)
		Weight of dispersing Agent (Ch. 3)
w_i	:	initial water content (Ch. 3)
w_L	:	Liquid limit (LL)
w_o	:	Optimum moisture content (Ch. 12)
w_p	:	Plastic limit (PL)
w_r	:	water content obtained by rapid moisture meter (Ch. 3)
w_s	:	shrinkage limit (SL)
w_α	:	water content corresponding to a penetration α (Ch. 3)
x	:	x-coordinate (Ch. 13)
		distance
$\bar{x}$	:	lever arm of reaction (Ch. 13)
y	:	y-coordinate
z	:	depth under consideration
z_c	:	critical depth (Ch. 8)

GREEK

α	:	Angle (Ch. 6)
		Angle of inclination of wall face with horizontal (Ch. 13)
		Shape factor (Ch. 14)
		Adhesion factor (Ch. 16)
		Depth of penetration (cone penetrometer method)
β	:	Angle of obliquity (Ch. 8)
		Angle of slope (Ch. 9)
γ	:	Bulk unit weight
γ'	:	Submerged or buoyant unit weight
γ_d	:	Dry unit weight
γ_i	:	Initial unit weight of suspension (Ch. 3)
γ_l	:	Equivalent unit weight (Ch. 13)
γ_o	:	Unit weight of water at 4°C (Ch. 2)
		Unit weight of soil in the natural state (Ch. 3)
γ_s		Unit weight of solids (Ch. 2)
γ_{sat}	:	Saturated unit weight (Ch. 2)
γ_w	:	Unit weight of water (Ch. 2)
γ_z	:	Unit weight of suspension at depth z at time t (Ch. 3)
γ_{max}	:	Maximum dry density (densest state) (Ch. 3)
γ_{min}	:	Minimum dry density (loosest state) (Ch. 3)
Δ	:	small increment to any quantity, e.g., Δa (Ch. 6)
ΔL	:	Elastic compression of pile (Ch. 16)
$\Delta \bar{\sigma}$	:	Increment of effective stress (Ch. 11)

δ	:	Angle (Ch. 9)
		Angle of wall friction (Ch. 13)
δ'	:	Angle of base friction (Ch. 13)
ε	:	Strain
ε_a	:	Axial strain (Ch. 8)
η	:	Factor of safety (Ch. 15)
		Hammer efficiency (Ch. 16)
η_g	:	Pile group efficiency (Ch. 16)
η_o	:	Factor of safety against overturning (Ch. 13)
η_s	:	Factor of safety against sliding (Ch. 13)
θ	:	Angle denoting orientation of plane (Ch. 8)
		Central angle of failure surface (Ch. 9)
μ	:	Micron (Ch. 8)
		Viscosity (Ch. 5)
		Coefficient of friction (Ch. 8)
μ_l	:	Viscosity of liquid (Ch. 3)
μ_w	:	Viscosity of water (Ch. 3)
υ	:	Kinematic viscosity (Ch. 5)
		Poisson's ratio (Ch. 10)
π	:	Pi - ratio of the circumference and the diameter of a circle
ρ		Mass density (Ch. 5)
		Balla's parameter (Ch. 14)
σ	:	Normal stress (Ch. 8)
		Effective overburden pressure (Ch. 18)
σ_1, σ_3	:	Principal stresses (Ch. 8)
σ_h	:	Horizontal stress or lateral pressure (Ch. 13)
$\bar{\sigma}$	:	Effective stress (Ch. 5, 7 and 11)
$\bar{\sigma}_0$	:	Initial effective overburden pressure (Ch. 1)
σ_l	:	conjugate lateral stresses (Ch. 5)
σ_v	:	vertical stress (Ch. 5)
$\bar{\sigma}_c$	:	Intergranular pressure in the capillary zone (Ch. 5)
τ	:	shear stress (Ch. 8)
ϕ	:	Phi—Velocity potential (Ch. 6)
		Angle of internal friction
ϕ_e	:	Hvorslev's parameter (Ch. 8)
ψ	:	Psi—Stream function (Ch. 6)
		Angle (Ch. 8)

SUBSCRIPTS

a	:	axial
		air
		allowable
		active
c	:	capillary
		cell
		consolidation
		critical
cr	:	critical
d	:	dry
		drained
e	:	effective
f	:	failure
h	:	horizontal
i	:	initial
m	:	mobilised
max	:	maximum
min	:	minimum
p	:	passive
s	:	solids
		shear
		static
T	:	Transformed
t	:	total
u	:	undrained
ult	:	ultimate
v	:	vertical
		void
w	:	water
		wall
x, y	:	directions
z	:	vertical direction
0	:	initial

SPECIAL

°	:	Degrees
$\int$	:	Integral
/	:	Per
∇	:	Water level

AUTHOR INDEX

Abbet 664
Alam Singh 500, 619, 824
Allen Hazen 65, 148
Alpan 525, 619
Atterberg 2, 66

Balla 653
Barron 809
Beles 810
Berezantzev 762
Bishop 138, 310, 315, 328, 388, 390
Bjerrum 310, 465
Boussinesq 2, 411
Bozozuk 469
Brinch Hansen 651
Broms 793
Brooker 525, 619
Brown 791
Burmister 101

Capper 449, 487
Caquot 554, 577, 619
Casagrande, A. 3, 71, 210, 223, 239, 252, 276
Casagrande, L. 215
Cassie 449, 487
Cauchy 205
Chellis 769
Converse 787
Coulomb 2, 305, 619
Culmann 2, 543, 565, 619

Darcy 2, 134
De Beer 457, 779, 838
Dupiut 143

Engessor 543 619

Fadum 276, 432
Feld 787
Fellenius 2, 384, 635
For chheimer 223
Fox 461, 487
Fraser 524, 619

Geddes 449, 487
Gilboy 239, 415
Goodier 410, 414
Gopal Ranjan 619
Gould 468, 487
Grim 5

Hanson 670, 719, 737, 740
Harr 223

Henkel 328
Hiley 769, 802
Housel 667
Hrennik of 794
Hvorslev 308, 838

Ireland 525, 619
Iterson 218

Jaky 524, 619
Janber 728
Johnson 809, 818
Jumikis 384, 404, 471, 473, 475, 570, 577
Jurgenson 429, 430

Keeney 787
Kennedy 818
Kenney 524, 619
Kerisel 701
Kezdi 525, 619
Kozeny 148, 210, 223

Labarre 787
Lambe 10, 138, 320, 408, 414, 523, 650, 810, 813
Lin 150
Loudon 148

MacDonald 469, 487
Martens 457
Matlock 791
McCarthy 47, 161, 408, 711, 714, 764
Mehra 808
Mayerhof 487, 649, 672, 707, 725, 728, 729, 761, 779
Meigh 457
Michaels 150
Mohr 2, 299, 304
Morgenstern 390, 408
Murthy 768, 791, 794
Muskat 135, 150

Newmark 431, 435
Nixon 457
Nordlund 762

Palmer 791
Peck 255, 256, 458, 469, 670, 719, 737, 740, 847
Poiseuille 149
Pokar 469
Polshin 469

888 Index

Poncelet 543, 557, 619
Potyondy 764
Prandtl 2, 637, 645
Proctor 3, 491, 515
Punmia 619

Ramot 766
Rankine 2, 526, 619
Rebann 2, 543, 555, 619
Reese 791
Reissner 645
Resal 526, 619
Reynolds 133
Riemonn 205
Rutledge 471, 809

Schaffernak 218, 233
Scheidegger 135
Schleicher 627
Schmertmann 255, 779
Sehgal 153
Seiler 787
Shamsher Prakash 619
Skempton 255, 327, 465, 469, 480 487
Smith 344
Sokolovski 554, 620
Sowers 469, 787, 837

Spangler 6, 429
Stanculescu 810
Steinbrenner 433, 460
Stokes 2, 54
Swamisaran 619

Taylor 164, 219, 223, 260, 396, 464, 568, 640
Teng 670, 710, 716, 719, 736
Terzaghi 1, 2, 255, 256, 262, 424, 458, 459, 469, 472, 487, 523, 584, 589, 641, 703, 737, 738, 809, 847.
Thiem 143
Thornburn 458, 670, 719, 737, 740
Timoshenko 410, 414
Tomlinson 762, 764
Tschebotarioff 555, 619

Van Wheel 776
Vesic 761, 763, 772, 794

Wellington 768
Westergaard 411, 419, 434
Whitman 32, 320 408, 523, 583, 650
Whitney 125
Wilson 636
Winkler 791

SUBJECT INDEX

Abbots' compaction test 499
Acid test 108
Active earth pressure 520, 521
  coefficient of 528
Active zone 473, 475
Activity of days 81
Adhesion 335
Absorbed water 129
Aerial photogrammetry 833
Air content 17
Allowable obliquity 296
  repose 330
  shearing resistance 306
Apparent cohesion 306, 334
Aquifer 142
  confined 142
  un-confined 142
Area correction 316
  ratio 842

Battery pile 793
Bearing capacity 623, 624
  ultimate 623
  gross 623
  net 623
  safe 623
  factors 631, 635, 645
  tables 625
Bentomite 109
Bishop's method 379, 388
Boring 834
  auger 834
  wash 834, 835
  percussion 834, 836
  rotary 834, 836
Boulders 109
Boussinesq's influence factor (coefficient) 414
Break in the backfill 568

Caissons (Wells) 709
Caliche 109
California bearing ratio (CBR) 120, 818
Capillarity 129, 156
Capillary rise 160
  water 128
Classical earth pressure theories 525
Clay 106
Coefficient of active earth pressure 528
  compressibility 245
  consolidation 265
  curvature 66

earth pressure at rest 524
passive earth resistance 529
percolation 137
subgrade reaction 737
transmissibility 142
Uniformity 65
volume compressibility 256
Cohesion 2, 295
  factor 308
Combined footings 706
  rectangular 732, 733
  trapezaidal 732, 734
Compaction 490
Compactive effort 492
Compressibility 237
Compression 237
  initial or elastic 456
  primary 456,
  secondary 456
Compression index 248
Cone of depression 143
Cone penetrometer 74
Conjugate harmonic functions 205
  ratio 536
Consistency index 70
  limits 2, 35, 66
Consolidated undrained test 309
Consolidation 237, 239
  ratio 258
Consolidometer 239
  fixed ring type 239, 240
  floating ring types 239, 240
Constant head premeameter 139
Contact pressure 472
Converse-Labarre formula 787
Core-cutter method 47, 50
Coulomb's law 306
  Wedge theory 525, 543
Critical depth 375, 539, 541
  hydraulic gradient 37, 225
  void ratio 332
Cyclic load test 776

Danish formula 772
Darcy's law 134
Deep footings 707
  Foundations 702
Degree of compaction 504
  consolidation 258
  saturation 16
  shrinkage 80
Density bottle 37

890 Index

index 35, 43
Dietert's test 500
Dilatancy test 113
Direct shear test 310
Dispertion test 107
Drained test 309
Drawdown curve 143
Dynamic cone penetration test 849
Dynamic pile driving formula 768

Earth pressure at rest 522, 523
Earth slope 370
  infinite 370
  finite 370, 378
Eccentric loading on footings 720
Effective size 65
Effective stress 132
  parameters 307
Electrical resistivity 858
Electro-osmosis 809
Engineering news formula 768
Equation of continuity 204
Equipotential line 193
Equivalent point load method 438
Exit gratient 196
Expansion index 248

Factor of safety against overturning 586
  sliding 586
Falling head (variable head) permeameter 140
Field's rule 787
Field (in-situ) compaction 500
Field consolidation line 255
Film moisture 129
Floating foundation 703
Flow channel 192
  index 73
  line 192
  value 538
Flownet 192
Footing 623
  on slopes 729
Foundation 2, 3, 623
  soil 623
Free water 128
Friction 295
  coefficient of 296
  angle of 296
Friction circle method 379, 391

General shear failure 646
Geophysical methods 855
Geostatic stresses 130
Grain size (particle-size) distribution curve 53, 54
Grain Specific gravity 19, 35
Gravel 106

Gravitational water 128
Ground water 128, 129
Group action of piles 782
Grouting 817

Harvard miniature compaction test 499
Held water 128, 129
Hiley's formula 769
Horizontal capillarity test 164
Hydrodynamic lag 253
Hydrometer analysis 59
Hygroscopic moisture 40, 129

Illite 36
Inclined Loading on footings 726
Indian standard compaction test 496
Influence diagram 416
Inside clearance 842
Insitu vane, shear test 851
Interlocking 297
Internal friction 2
Isochrones 266

Jodhpur mini-compactor test 500
Jodhpur permeameter 142

Kaolinite 36
Laplace's equation 204
Lateral earth pressure 518, 520
Laterally loaded piles 791
Laterite 4, 109
Leaching 4
Lineal shrinkage 81
Line load 421
Liquefaction 330
Liquid limit 66, 68, 71
Liquidity index 70
Load-settlement curves 662, 663
Load-test on pile 773
Loam 109
Local shear failure 647
Loess 109
Logarithm of time-fitting method 274, 276

Marl 109
Mass specific gravity 19
Mechanical stabilisation 806
Mehra's method 808
Modified proctor (modified ASHO) test 495
Modules (coefficient) of volume change 256
Mohr-coulmb theory 304
Mohr's circle 297
Mohr's strength theory 304
Montmorillonite 36
Moorum 109

Negative skin friction 780

Index 891

Neutral stress 132
Newmarks' influence chart 435, 437
Normally consolidated soil 251

Oedometer 239
One-dimensional consolidation 262
Optimum moisture content 492
Outside clearance 842
Overconsolidated soil 251
Overconsolidation ratio 252

Particle-size distribution 51
Passive earth resistance 520, 521
Peat 109
Penetration test 310, 625, 666
Percent air voids 17
Permiability 133
Phase 14
  diagram 14
Pier foundation 708
Piezometric head 137
Pile 754
  capacity 760
  driving 757
  foundation 707, 754
  group 782
  group efficiency 786
Pipette analysis 57
Piping 225
Plastic lag 253
  limit 66, 69, 75
Plasticity index 69
Plate bearing test 625
Plate load test 660
Point load 441
Poissons ratio 414
Poncelet rule 557
Poorly graded soil 64
Pore pressure parameters 327
  ratio 374
Porosity 15
Preconsolidation pressure 250
Pressure bulb 418
Pressure-void ratio relationship 258
Primary compression (consolidation) 276
Principal planes 297
Principal stress 297
  major 298
  minor 298
  intermediate 298
Proctor needle 504
Pycnometer 37, 41

Quartz 36
Quicksand 224

Radial flownet 219

Radius of influence 144
Raft (mat) foundation 706, 735, 736
Rammers 503
Rankine's theory 525, 526
Rapid drawdown 372, 387
Rapid moisture tester 40, 42
Rebhann's condition 557
Rebound curve 247
Recompression curve 249
Retaining walls 3, 518, 578
  gravity 518, 519, 579
  semi-gravity 579
  cantilever 518, 519, 580
  counterfort 518, 519, 581
  buttress 518, 519, 581
  grib 581, 582
Ring shear test 310
Rock flour 109
Rocks 4
  igneous 4
  sedimentary 4
  metamorphic 4
Rollers 501
  grid 501, 503
  pneumatic-tyred 501, 502
  sheepfoot 501, 502
  smooth wheeled 501
Rolling test 107

Sample disturbance 842
Sand-replacement method 47
Schleicher's method 627
Secondary consolidation 275, 279
Sedimentation (Wet) analysis 51, 54
Seepage 192
  velocity 136
Seismic refraction 855
Sensitivity of clays 82, 343
Settlement 35, 237, 453
  contact or distortion 457
  differential 453
  permissible 469
  time rate of 453
  total 453
Shaking test 107
Shallow foundation 702
Shape factor 194
Shear Strength of soil 295 ✓
Sheat pile walls 518
Shine test 108
Shrinkage test 35
  index 70
  limit 66, 69, 76
  ratio 79
Sieve analysis 51
Silt 36
Site investigation 831

892 Index

Skin friction 760
Slide 370
Soil 1
  aeoline 7, 110
  alluvial 7, 110
  black cotton 8, 109
  coarse-grained 36
  fine-grained 37
  glacial 7, 110
  lacustrine 7, 110
  marine 7, 110
  partially saturated 17
  residual 1, 6
  saturated 15
  transported 1, 6
Soil - cement 811
Soil exploration 831
Soil horizons 5
Soil moisture 128
Soil profile 5, 454
Soil sample 35, 840
  disturbed 35
  remoulded 35
  undisturbed 35
Soil stabilisation 805
Specific gravity 19
  of soil solids 19, 35
  of water 20
Split-spoon sampler 843
Spread footings 703, 704, 715
Square-root of time fitting method 274
Stability number 376, 396
Standard penetration test 457, 846
Standard proctor (AASHO) Test 494
Static cone penetration (Dutch cone) test 847
Stoke's law 54
Strap footing 705, 706, 730
Stream function 204
Strength (failure) envelope 306
Stress isobar 418
Strip load 423
Structural water 129
Structure 6, 8, 35
  single-grained 8
  honey-comb 9
  flocculent 9
Superficial velocity 136
Swedish method of slices 379, 380
Swelling 35, 238

Taylor's method 379, 396

Tension crack 380
Test pits 834
Texture 6, 11, 35
Tin-walled samples 844
Thixotropy 83
Time-compression curve 243
Time-factor 268
Top flow line 208
Torsion test 310
Total stress parameters 308
Trial wedge method 549
Triaxial compression test 310, 313
True shear parameters 308
Two is to one method 439

Unconfined compression strength 82
  test 310, 322
Unconsolidated undrained test 309
Underconsolidated soil 252
Underreamed piles 765
Uniform surcharge 533, 569
Unit weight 17
  bulk(mass) 17
  dry 19
  in-situ 47
  saturated 18
  submerged (buoyant) 18
Unit weight of solids 18
  of water 18
Useful width concept 725

Vane shear test 310, 325
Velocity potential 204
Vertical capillarity test 167
Vibrators 501, 503
Virgin compression curve 247
Viscous friction 336
Void ratio 16
Volumetric shrinkage 80

Wall friction 550
  angle of 550
Water (moisture) content 17, 35, 40
Wave equation method 766
Well-graded soil 64
Wenner configuration 858
Westergaords' influence coefficient 420
Winkler's hypothesis 791, 792

Zero-air voids line 493